Springer Series in Materials Science

Volume 253

The Springer Series in Materials Science covers the complete spectrum of materials physics, including fundamental principles, physical properties, materials theory and design. Recognizing the increasing importance of materials science in future device technologies, the book titles in this series reflect the state-of-the-art in understanding and controlling the structure and properties of all important classes of materials.

More information about this series at http://www.springer.com/series/856

Pavel G. Baranov · Hans Jürgen von Bardeleben
Fedor Jelezko · Jörg Wrachtrup

Magnetic Resonance of Semiconductors and Their Nanostructures

Basic and Advanced Applications

Springer

Pavel G. Baranov
Laboratory of Microwave Spectroscopy of Crystals
Ioffe Institute
St. Petersburg
Russia

Hans Jürgen von Bardeleben
Institut des Nanosciences de Paris-INSP
Université Pierre et Marie Curie and UMR 7588 au CNRS
Paris
France

Fedor Jelezko
Institut für Quantenoptik
Universität Ulm
Ulm, Baden-Württemberg
Germany

Jörg Wrachtrup
Physikalisches Institut
Universität Stuttgart
Stuttgart
Germany

ISSN 0933-033X ISSN 2196-2812 (electronic)
Springer Series in Materials Science
ISBN 978-3-7091-1156-7 ISBN 978-3-7091-1157-4 (eBook)
DOI 10.1007/978-3-7091-1157-4

Library of Congress Control Number: 2017932426

This Springer imprint is published by the registered company Springer-Verlag GmbH, AT part of Springer Nature
The registered company address is: Prinz-Eugen-Str. 8-10, 1040 Wien, Austria

Preface

Spectroscopy is a branch of physics that studies the interaction of an electromagnetic (EM) radiation with matter. The essence of spectroscopy is in the observation of the way in which atoms and molecules exchange energy with the outside world. The EM radiation is composed of two oscillating fields: electric field **E** and magnetic field **B**. The electric field **E** and the magnetic field **B** interact with the electric dipole moment and magnetic moment, respectively, and thereby cause transitions between different energy levels. Since the wavelengths of the electromagnetic radiation used in spectroscopy commonly exceed the size of atoms or molecules under study, one can ignore the fact that the magnitudes of the electric or the magnetic fields are not constant. The main task of spectroscopy is to determine the structure and to clarify the physical nature of the energy levels by spectral analysis of the response of matter to the impacts of electromagnetic radiation. Modern atomic physics and quantum mechanical approaches, including description of spin effects, have emerged from discoveries in the field of spectroscopy.

Magnetic resonance spectroscopy or radio spectroscopy examines the interaction of the magnetic component of electromagnetic radiation with magnetic moments existing in a material. The interaction with nuclear magnetic moments is the subject of nuclear magnetic resonance (NMR) spectroscopy, while that with magnetic moments of electrons is the subject of electron paramagnetic resonance (EPR) spectroscopy. Radio spectroscopy covers the EM radiation frequency range from zero to hundreds of gigahertz, i.e., from infinitely long to millimeter electromagnetic waves. The EM wave is characterized by a photon energy, oscillation frequency, or wavelength. The relation between the energy and frequency is named the Planck relation (German physicist Max Planck, 1900):

$$E = h\nu,$$

where h is Planck's constant, and ν is the frequency of electromagnetic oscillations. The Planck's constant is a physical constant that has the meaning of a quantum of action in quantum mechanics and is the proportionality constant between the energy (E) of a photon and the frequency (ν) of the associated electromagnetic wave. Since the frequency ν, wavelength λ, and speed of light c are related by $\lambda\nu = c$, each of these variables can be used as a characteristic of the electromagnetic radiation.

The electron paramagnetic resonance (EPR) was discovered in 1944 by the Russian physicist E. Zavoisky [1] in Kazan and has developed since into a major scientific technique. The first observation of an electron paramagnetic resonance was made in radiofrequency range, and Zavoisky's results were interpreted by Frenkel [2] as showing paramagnetic resonance absorption. Later experiments at higher frequencies showed the advantage of the use of high frequencies and high magnetic fields.

Excellent general review books on the EPR are available [3–33]. The electron paramagnetic resonance is observed in various systems with unpaired electrons carrying magnetic moments when an oscillating magnetic field causes transitions between electron levels. Systems of this kind are named paramagnetic, hence the name "electron paramagnetic resonance". As a rule, the level splitting is caused by an external magnetic field that interacts with the electron magnetic moments; however, levels may be split in some systems due to interactions within the system in a zero magnetic field.

The aim of the development of modern EPR spectroscopy is to increase the sensitivity and information content, i.e., resolving capabilities. These problems are solved simultaneously in several directions.

In the last decade, there has been a great interest in EPR at high frequencies and high-frequency EPR spectroscopy has seen a remarkable development (see e.g., [29]). Whereas 9 GHz (X-band) has remained the main frequency of operation for more then 50 years, it is now possible to perform EPR studies at frequencies as high as 95 GHz (W-band) and even higher. The main reason to go to high frequencies is the high absolute sensitivity and the high spectral resolution that can be obtained. The first aspect is important when small amounts of material are available such as in the case of thin layers, nanostructures, or biophysical and biochemical problems, and the second is of special significance to disentangle spectra that normally overlap at the conventional EPR frequency of 9 GHz.

EPR is a tool to manipulate electron spins in solids. Because of the limited sensitivity of conventional EPR, typically optically detected and electrically detected EPR is favored to detect small numbers of spins [32]. In both approaches, the spin state is transferred to a photon or charge state, respectively.

In spin-dependent optical emission or photoconductivity, the spin-to-photon or spin-to-charge transfer, respectively, is typically achieved via a spin-dependent process of recombination involving paramagnetic states of recombining partners. In optically detected magnetic resonance (ODMR), a microwave-induced repopulation of Zeeman sublevels is detected optically, i.e., there is a giant gain in sensitivity since an energy of optical quantum is by several orders of magnitude higher compared with microwave one, it becomes possible to detect a very small number of spins down to single spin! [34, 35] ODMR is a "trigger detection" in that the absorption of a resonance microwave photon triggers a change in emission (absorption) of an optical photon due to the selective feeding of the magnetic sublevels.

Until recently, the practical applications of semiconductors involved the use of charge- and spin-carrier ensembles. The capability to efficiently control spin states is the key question of semiconductor spintronics. The unique quantum properties of nitrogen-vacancy (NV) color centers in diamond [36] have opened a new era in spintronics: It has become possible to manipulate the spin states of a single atomic-sized center at room temperature using optically detected magnetic resonance. The optical detection of magnetic resonance in a single spin has become possible because of the existence of a unique cycle of optical alignment and, as a result, the creation of an inverse population of spin sublevels in the NV center ground state. Until recently, an NV center was the only known solid-state system in which such spin manipulations were possible.

The search for structures that exhibit unique quantum properties similar to those of NV centers in diamond and at the same time have broader functional capabilities is a highly hopeful task. The most promising material that may compete with diamond from the standpoint of spectroscopy of quantum systems is silicon carbide (SiC), which can be regarded as an artificial superlattice. A special feature of SiC is the existence of its different polytypes, and for each of the polytypes, the properties of spin color centers are unique; furthermore, even in one polytype, the center may be located in different nonequivalent positions in the lattice. This allows choosing the center with parameters (for instance, optical and microwave ranges) suited to a specific problem.

The great potential of the EPR spectroscopy cannot be fully realized with only conventional continuous-wave (CW) EPR. Continuous-wave EPR and pulsed magnetic resonance (EPR) are complementary, and the application of both gives a total picture of the spin phenomena under investigation. In the CW EPR, the magnitude of the magnetic field B_0 (static magnetic field) is swept, while the amplitude of the microwave field B_1 is constant with time. In the pulsed EPR experiments (time-resolved experiments), a time-dependent microwave pulse B_1 is applied in addition to a static magnetic field B_0. In the pulsed EPR spectroscopy, relaxation times can be directly measured by monitoring the magnetization on the same timescale in which relaxation occurs. The advantage of pulsed operation in addition to the recording of the relaxation times is that it is also possible to study photoexcited paramagnetic species including paramagnetic excited states (e.g., excitons) in combination with pulsed lasers. Moreover electron nuclear double resonance (ENDOR) experiments will become feasible in a much wider temperature range than for CW operation. ENDOR technique developed by Feher and Mims [6, 23] is very useful in the systems with not resolved hyperfine structure and makes possible the detection of nuclear magnetic resonance (NMR) through its effect on the electron paramagnetic resonance signal, thus using a high EPR sensitivity as compared with the NMR. The hyperfine coupling constants could be measured with much higher precision as compared with the EPR.

We are much obliged to our colleagues and coworkers who contributed many ideas and performed numerous experiments, and we are indebted to them for collaboration. One of us (P.G.B) thanks for collaboration Nikolai G. Romanov, Evgenii N. Mokhov, Andrei G. Badalyan, Ivan V. Ilyin, Vladimir A. Khramtsov,

Marina V. Muzafarova, Alexandra A. Soltamova, Victor A. Soltamov, Roman A. Babunts, Danil O. Tolmachev, Jan Schmidt, Celso de Mello Donegá, Ankie van Duijn-Arnold, Martinus T. Bennebroek, Oleg G. Poluektov, Sergei B. Orlinskii, Vladimir Dyakonov, Georgii V. Astakhov, Bruno Meyer, Albrecht Hofstaetter, Detlev M. Hofmann, Philipp Lavallard, Huib Blok, Edgar J.J Groenen, and Andries Meijerink and gratefully acknowledges a support of Russian Science Foundation under Agreement No. 14-12-00859.

We thank Anna P. Bundakova for help in writing the book in English and Yulia A. Uspenskaya for helping in the organization of the book in line with editorial rules. In Chap. 1 and partly in Chap. 2, we present the lectures delivered by P.G. Baranov to students of the Peter the Great St. Petersburg Polytechnic University and Ph.D. students in Ioffe Institute.

We are indebted to many authors around the world for kind permission to reproduce figures from their works.

Saint Petersburg, Russia — Pavel G. Baranov
Paris, France — Hans Jürgen von Bardeleben
Ulm, Germany — Fedor Jelezko
Stuttgart, Germany — Jörg Wrachtrup

References

1. Zavoisky, E.K.: Relaxation of liquid solutions for perpendicular fields. J. Phys. (USSR) **9**, 211–216 (1945)
2. Frenkel, J.: On the theory of relaxation losses, connected with magnetic resonance in solid bodies. J. Phys. (USSR) **9**, 299–304 (1945)
3. Low, W.: Paramagnetic Resonance in Solids. Academic, New York (1960)
4. Altshuler, S.A., Kozirev, B.M.: Electron Paramagnetic Resonance. Academic, New York (1964)
5. Orton, J.W.: Electron Paramagnetic Resonance. Iliffe Books, London (1968)
6. Pool, C.P., Jr.: Electron Spin Resonance. Comprehensive Treatise on Experimental Techniques. Wiley, New York, London, Sydney (1967); Poole, Ch.P.: Electron Spin Resonance, A Comprehensive Treatise on Experimental Techniques. Wiley, New York (1983)
7. Abragam, A., Bleaney, B.: Electron Paramagnetic Resonance of Transition Ions. Clarendon Press, Oxford (1970)
8. Wertz, J.E., Bolton, J.R.: Electron Spin Resonance: Elementary Theory and Practical Applications. McGraw-Hill, New York (1972). Wertz, J.E., Bolton, J.R.: Electron Spin Resonance: Elementary Theory and Practical Applications. Chapman and Hall, London (1986)
9. Geschwind S., Editor: Electron Paramagnetic Resonance. Plenum Press, New York (1972)
10. Sorin, L.: Electron Spin Resonance of Paramagnetic Crystals. Springer Verlag, (1973)
11. Pake, G.E.: The physical principles of electron paramagnetic resonance, 2nd Edition. W.A. Benjamin (1973)
12. Atherton, N.M.: Electron Spin Resonance—Theory and Applications. Wiley, New York (1973)
13. Molin, Y.N., Salikhov, K.M., Zamaraev, K.I.: Spin Exchange. Springer, Berlin (1980)

14. Wertz, J.E.: Electron Spin Resonance: Elementary Theory and Practical Applications. Springer Verlag (1986)
15. Poole, C.P. Jr., Farach, H.A.: The Theory of Magnetic Resonance. 2nd Edition. Wiley-Interscience, New York (1987), 1st Edition (1972)
16. Hoff, A.J.: Advanced EPR—Applications in Biology and Biochemistry. Elsevier (1989)
17. Pilbrow, J.R.: EPR of Transition Metal Ions. Oxford University Press (1991)
18. Dikanov, S.A., Tsvetkov, Y.: Electron Spin Echo Envelope Modulation (ESEEM) Spectroscopy. CRC Press, Oxford (1992)
19. Atherton, N.M.: Principles of Electron Spin Resonance. Horwood, E., Kemp, T.J.,(eds.), Ellis Horwood and Prentice Hall, London (1993)
20. Weil, J.A., Bolton, J.R., Wertz, J.E.: Electron Paramagnetic Resonance: Elementary Theory and Practical Applications. Wiley Interscience (1994)
21. Poole, C.P.: Electron Spin Resonance: A Comprehensive Treatise on Experimental Techniques. 2nd Edition, Dover Publications, New York (1997)
22. Eaton, G.R., Eaton, S.S., Salikhov, K.M.: Foundations of Modern EPR. World Scientific (1998).
23. Smith, G.M., Riedi, P.C.: Progress in High Field EPR. 17, Cambridge, UK: RSC 2000
24. Schweiger A., Jeschke G.: Principles of Pulse Electron Paramagnetic Resonance. Oxford University Press (2001)
25. Sajfutdinov, R.G., Larina, L.I., Vakulskaya, T.I., Voronkov, M.G.: Electron Paramagnetic Resonance in Biochemistry and Medicine. Springer Verlag (2001)
26. Weil J.A., Bolton J.R.: Electron Paramagnetic Resonance: Elementary Theory and Practical Applications. 2nd Edition, Wiley (2007)
27. Lund, A., Shiotani, M.: Principles and Applications of Electron Spin Resonance. Springer Verlag (2008)
28. Brustolon, M.R., Giamello, E. (eds.): Electron Paramagnetic Resonance Spectroscopy: A Practitioner's Toolkit. Wiley (2009)
29. Moebius, K., Savitsky, A.: High-field EPR Spectroscopy on Proteins and Their Model Systems. RSC Publishing (2009)
30. Misra, S.K. (ed.): Multifrequency Electron Paramagnetic Resonance: Theory and Applications. Wiley (2011)
31. Spaeth J.-M., Niklas J.R., Bartram R.H.: Structural Analysis of Point Defects in Solids: An Introduction to Multiple Magnetic Resonance Spectroscopy. Springer-Verlag (1992)
32. Carrington, A., McLachlan, A.D.: Introduction to Magnetic Resonance with Applications to Chemistry and Chemical Physics. Harper & Row, Publishers (1967)
33. Spaeth J.-M. Overhof H.: Point Defects in Semiconductors and Insulators: Determination of Atomic and Electronic Structure from Paramagnetic Hyperfine Interactions. Springer-Verlag Berlin Heidelberg (2003)
34. Köhler, J., Dosselhorst, J.A.J.M., Donckers, M.C.J.M., Groenen, E.J.J., Schmidt, J., Moerner, W.E.: Magnetic resonance of a single molecular spin. Nature **363**, 242–244 (1993)
35. Wrachtrup, J., von Borczyskowski, C., Bernard, J., Orrit, M., Brown, R.: Optical detection of magnetic resonance in a single molecule. Nature **363**, 244–245 (1993)
36. Gruber, A., Drabenstedt, A., Tietz, C., Fleury, L., Wrachtrup, J., von Borczyskowski, C.: Scanning confocal optical microscopy and magnetic resonance on single defect centres. Science **276**, 2012–2014 (1997)

Contents

Chapter 1
Basic Concepts of Electron Paramagnetic Resonance

1.1 Magnetic Dipole

The electron paramagnetic resonance is observable in substances that contain electronic magnetic dipoles. We list examples of systems in which there are unpaired electrons:

1. Transition and rare-earth elements, which have unfilled *d* and *f* shells, as well as the atoms and ions of elements in the periodic table, having unpaired *s* and *p* electrons.
2. Localized donor and acceptor states in semiconductor materials, nanostructures.
3. Conduction electron and holes in solids.
4. Point defects produced in a material by electromagnetic or particle irradiation, color centers.
5. Stable free radicals.
6. The excited state of defects that are not paramagnetic in the ground state, excitations (excitons, electron-hole pairs).
7. Systems for solar power engineering (photovoltaic).
8. Biological objects in which free radicals are involved in the metabolism, red ox processes occurring in single-electron states.
9. Products of the photosynthesis in which a primary oxidant and a reductant are formed in the initial primary photochemical act.
10. Metal proteins containing transition elements Fe, Mn, Cu, Ni, Co, etc.
11. Spin labels in biology.
12. Molecular oxygen having the paramagnetic ground state.

etc.

P.G. Baranov et al., *Magnetic Resonance of Semiconductors and Their Nanostructures*, Springer Series in Materials Science 253,
DOI 10.1007/978-3-7091-1157-4_1

1.1.1 Magnetic Dipole Moment

The elementary source of the magnetic field is the magnetic dipole, in contrast to the elementary source of the electric field, which can be created by positive or negative point charges. The magnetic monopole, a magnetic analogue of an electric charge, has never been observed.

The magnetic dipole can be represented as a planar closed loop that carries an electric current I (Fig. 1.1a). Its magnetic dipole moment, vector $\vec{\mu}$, is defined as a vector that points out of the plane of the current loop and has a magnitude equal to the product of the current and the loop area:

$$\vec{\mu} = IS\vec{n}\,(\text{SI system of units})\ \vec{\mu} = \frac{1}{c}IS\vec{n}\,(\text{CGS system of units}). \tag{1.1}$$

Here, I is the current in the loop, S is the loop area, and $\vec{n}$ is the normal to the loop in accordance with right-hand grip rule in relation to the current direction.

In SI units for the current-loop definition, the magnetic moment is measured in ampere–square meters (A m^2). In the equation for torque on a moment in the magnetic field, the moment is measured in joules per tesla (J T^{-1}), and these two representations are equivalent.

1.1.2 Magnetic Field Produced by a Magnetic Dipole Moment

A magnetic dipole produces a dipolar magnetic field $\vec{B}$ in the space surrounding the system (Fig. 1.1b), similarly to an electric dipole, which is the source of an electric field (using the dipole field as an approximation for points far from the loop),

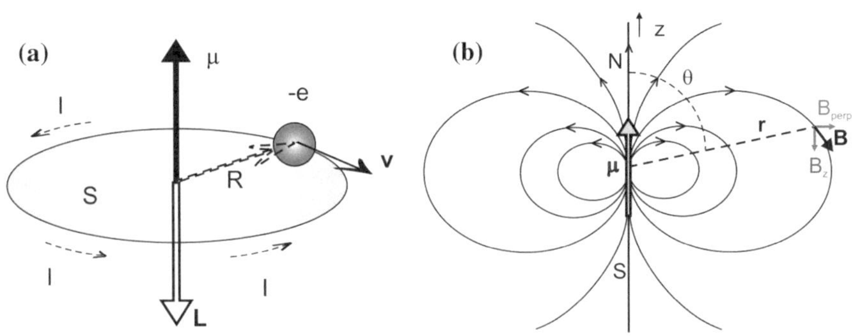

Fig. 1.1 **a** Magnetic moment $\vec{\mu}$ of a planar current having magnitude I and enclosing an area S. **b** External magnetic field produced by a magnetic dipole moment. $B_{\parallel} = (\mu/r^3)(3\cos^2\theta - 1)$, $B_{\perp} = (3\mu/r^3)\cos\theta\sin\theta$, $B = (B_{\parallel}^2 + B_{\perp}^2)^{1/2}$

$$\vec{B} = \frac{\mu_0}{4\pi}\frac{1}{r^3}\left[\frac{3(\vec{\mu}\cdot\vec{r})\cdot\vec{r}}{r^2} - \vec{\mu}\right], \text{ (SI system of units)}$$
$$\vec{B} = \frac{1}{r^3}\left[\frac{3(\vec{\mu}\cdot\vec{r})\cdot\vec{r}}{r^2} - \vec{\mu}\right] \text{ (CGS system of units).} \tag{1.2}$$

The magnetic field is symmetric about the direction of its magnetic dipole moment, and decreases as the inverse cube of the distance from the magnetic dipole. Without loss of generality by the problem, we direct the magnetic dipole along an arbitrary axis z, i.e., $\mu_z = \mu$, $\mu_x = \mu_y = 0$. As a result, we have the following expression for the magnetic field components $B_z = B_{\|} = (\mu/r^3)(3\cos^2\theta - 1)$, $B_x = B_y = B_{\perp} = (3\mu/r^3)\cos\theta\sin\theta$, $B = (B_{\|}^2 + B_{\perp}^2)^{1/2}$. We have used: $z = r\cos\theta$ and x $(y) = r\sin\theta$.

1.1.3 Magnetogyric Ratio

The magnetic moment is always associated with the angular momentum $\boldsymbol{G}$ (there is no magnetic moment without angular momentum) given by

$$\vec{G} = \vec{r} \times \vec{p}, \tag{1.3}$$

where $\vec{r}$ is the radius vector, and $\vec{p}$ is the momentum.

The ratio of the magnetic moment to angular momentum is called the magnetogyric ratio (or gyromagnetic ratio), which is commonly denoted by the symbol γ and is given by

$$\gamma = \frac{\vec{\mu}}{\vec{G}}. \tag{1.4}$$

In the special case in which the magnetic moment is parallel or antiparallel to the angular momentum (depending on the sign of a moving charge), the magnetogyric ratio can be represented as a scalar (with positive or negative sign). Then to turn with the current shown in Fig. 1.1a, we can consider only the absolute values of vectors. As a result, (1.1) simplifies to:

$$\mu = \frac{1}{c}IS, \; I = \frac{q}{2\pi R}v,$$

where q is the charge, $\frac{q}{2\pi R}$ is the charge density in the loop with the current, R is the loop radius, v is the velocity of the charge, and $S = \pi R^2$. Thus, the absolute value of the magnetic moment is

$$\mu = \frac{qvR}{2c}.$$

The angular momentum in a planar closed loop with current has, in accordance with (1.3), the form $G = m_q vR$, where m_q is the mass of the charge that moves at a speed v in a circle of radius R.

The magnitude of the magnetogyric ratio is given, in accordance with the (1.4), by

$$\gamma = \frac{\mu}{G} = \left(\frac{qvR}{2c}\right)/(m_q vR) = \frac{q}{2m_q\, c}.$$

This result is not of particular interest because the charge and the mass are unknown. However, we can represent the charge as a product of the electron charge e (the current is assumed to be due to motion of electrons) by the number of electrons N, i.e., $q = Ne$, and the mass of the same charge, as the product of the electron mass m_e by the same number of electrons $m_q = m_e N$. Then, we have

$$\gamma = \frac{e}{2m_e\, c}. \tag{1.5}$$

So, it is found that the magnetogyric ratio (1.5) depends only on universal constants. The amazing thing is that this classically defined relation can also be used in quantum mechanics! In fact, this relationship is true for the motion of an electron in an orbit within the atom! An atom with an electron in an orbit with an angular momentum is a small current loop, which suggests that it is also a magnetic dipole.

The motion of an electron in an orbit can be described by the laws of quantum mechanics. The quantum mechanics of the angular momentum is very different from its classical mechanics. The main difference is in that the angular momentum is quantized in the microscopic world, as evidenced by numerous experimental data. Therefore, the classical equations for the angular momentum are replaced in quantum mechanics by the equation for the eigenvalues. According to quantum-mechanical relations, the orbital angular momentum of an electron can be represented as

$$\hat{\vec{G}}_L = \hbar\hat{\vec{L}},$$

where $\hat{\vec{L}}$ is the operator of the orbital angular momentum (quantum number of orbital momentum), and $\hbar = \frac{h}{2\pi}$, where h is Planck's constant; thus, $\hbar$ is the fundamental unit of angular momentum. The 'hat' ^ on a symbol, like that on $\boldsymbol{G}$ and $\boldsymbol{L}$, will be used to indicate an operator. In obvious cases, we are going to use bold letters instead of the operator symbol. The angular momentum, as well as the total energy and momentum, are conserved in an isolated system (so-called "constants of motion"). Let us consider the operator for the total energy, $\hat{H}$, which is the sum of the kinetic and potential energy, the so-called Hamiltonian. Schrödinger's equation for stationary states has the form

$$\hat{H}\psi_n(x,y,z) = E_n\psi_n(x,y,z). \tag{1.6a}$$

$\hat{H}$ operates on the eigenfunctions $\psi_n(x,y,z)$ and E_n are the eigenvalues of the Hamiltonian which are the exact total energy of the nth state of the system, which are observed in experiments. The eigenfunctions $\psi_n(x,y,z)$ can be chosen in such a way that they are also eigenfunctions of the operator for the total angular momentum (more exactly, squared total angular momentum $\hat{\vec{L}}^2$) and the eigenfunctions of any one of the components of $\hat{\vec{L}}$, e.g., $\hat{L}_z$ and hence

$$\hat{\vec{L}}^2\psi_{n,L,M_L}(x,y,z) = \lambda_L\psi_{n,L,M_L}(x,y,z) \tag{1.6b}$$

and

$$\hat{L}_z\psi_{n,L,M_L}(x,y,z) = \lambda_{M_L}\psi_{n,L,M_L}(x,y,z), \tag{1.6c}$$

where $\lambda_L = L(L+1)$ and $\lambda_{ML} = M_L$ are the eigenvalues of $\hat{\vec{L}}^2$ and $\hat{L}_z$, respectively, which are determined only by the symmetry properties of the eigenfunctions. According to quantum mechanics, M_L can take the values L, $L-1$, $L-2$, … to $-L$. L is an integer and can not be negative, although its projection M_L may be negative. Expressions (1.6b) and (1.6c) can be rewritten using Dirac's notation (British physicist Paul Dirac, 1928) by a ket vector: $|L, M_L\rangle$, because the eigenstates associated with the angular momentum depend only on the values in parentheses

$$\begin{aligned} \hat{\vec{L}}^2|L,M_L\rangle &= L(L+1)|L,M_L\rangle, \\ \hat{L}_Z|L,M_L\rangle &= M_L|L,M_L\rangle. \end{aligned} \tag{1.6d}$$

Note, similarly expression (1.6a–1.6d) can be rewritten using Dirac's notation as

$$\hat{H}|n\rangle = E_n|n\rangle,$$

the angular bracket contains a label of the eigenvalue to which it corresponds.

We can write the expression for the operator of the orbital magnetic moment, which is proportional to the orbital angular momentum

$$\hat{\vec{\mu}}_L = \gamma\hat{\vec{G}}_L = \gamma\hbar\hat{\vec{L}} = \frac{e\hbar}{2m_e\,c}\hat{\vec{L}}.$$

The relationship

$$\frac{|e|\hbar}{2m_e\, c} = \mu_B$$

yields the quantity named the Bohr magneton. Then, the orbital magnetic moment is given by

$$\hat{\vec{\mu}}_L = -\mu_B \hat{\vec{L}}.$$

The negative sign appears because of the negative sign of the electron charge; this means that the orbital magnetic moment $\vec{\mu}_L$ and the orbital angular momentum $\vec{L}$ have opposite directions (Fig. 1.1a).

1.1.4 Electronic g-Factor of the Orbital and Spin Magnetic Moments

In radio spectroscopy a dimensionless quantity named the 'g-factor' is introduced instead of the magnetogyric ratio into the expression for the orbital magnetic moment. It can be written as

$$\hat{\vec{\mu}}_L = -g_L \mu_B \hat{\vec{L}} \tag{1.7}$$

and $g_L = 1$. In fact, the g-factor is the dimensionless magnetogyric ratio expressed in Bohr magnetons.

The magnetogyric ratio is inversely proportional to the mass of a moving charge. Because the electron has the minimum mass, it seems that the magnetogyric ratio for the orbital motion of an electron must be at a maximum. For example, the proton magnetogyric ratio is about three orders of magnitude smaller because the mass of the proton substantially exceeds that of the electron. However, the electron has, along with the orbital angular momentum, its own intrinsic angular momentum, named spin angular momentum, which is written as $\vec{S}$, or just spin for brevity (Dutch physicists, G. Uhlenbeck and S. Goudsmit, 1925). The magnitude of the spin is determined by a spin quantum number, s (for one electron), whose value is restricted to be ½ only. The spin angular momentum cannot be described in terms of the spatial wave functions $\psi_n(x, y, z)$, it has no classical counterpart and can be adapted in the quantum mechanics via the definition of a generalized angular momentum. For example, optical spectra of atoms show that the orbital angular momentum and spin angular momentum are additive, i.e., the physical quantities are of the same nature. In order to account for the spin, the spin coordinates S and M_S should be included in the wave function. In quantum mechanics, only the magnitude of the total angular momentum and one of its components can be simultaneously defined (in classical mechanics, the magnitude and direction of an angular momentum are defined simultaneously).

Thus, the electron is like a small bar magnet which would possess a magnetic dipole moment. Experiments show that the spin has the intrinsic magnetic moment $\hat{\vec{\mu}}_S$, and can be written in a form analogous to (1.7)

$$\hat{\vec{\mu}}_S = -g_S \mu_B \hat{\vec{S}} \tag{1.8}$$

where g factor is not unity: $g_S = 2.0023\ldots$. For condensed-matter experiments, it is sufficient to take $g_S = 2$.

The electron has no size and it is just a point. An electron can not be regarded as a current loop arising from one charged object orbiting another, like we did for the orbital motion in an atom. Thus, this intrinsic angular momentum (spin) is purely a quantum mechanical effect. The spin of the electron is ½, because only two states are observed in experiments (e.g., in Stern-Gerlach experiment, 1922).

$\vec{S}$ is the angular momentum of the electron spin. The quantities $\hat{\vec{L}}$ and $\hat{\vec{S}}$ are operators in quantum mechanics. The magnetic moment is directly proportional to the angular momentum and (1.7, 1.8) describe the magnetic moment operators for the orbital magnetic moment and the spin magnetic moment, respectively.

The electron spin appears only in the relativistic quantum mechanics, the so-called 'Dirac equation' in which, in contrast to Schrödinger's equation, the kinetic energy in the Hamiltonian is written in a relativistic way. The relativistic formulation of quantum mechanics shows the need to invoke an internal degree of freedom for an electron having the properties of an angular momentum. Because our consideration is restricted to classical quantum mechanics, we introduce the electron spin and, hence, the spin magnetic moment phenomenologically by assigning to an electron the intrinsic angular momentum $\hat{\vec{S}}$ and the intrinsic magnetic moment $\hat{\vec{\mu}}_S$ and placing it at the same point in space in which the electron charge resides.

Similar expressions can be written for the nuclear angular and magnetic moments of the hydrogen atom, that is, for the proton. In this case, the Bohr magneton is replaced by the nuclear magneton μ_N

$$\mu_N = \frac{e\hbar}{2m_p c}. \tag{1.9}$$

Here, account is taken that the proton charge is positive. The nuclear magneton is about 1840 times smaller than the Bohr magneton in accordance with the ratio between the proton and electron masses. Then the nuclear magnetic moment of the hydrogen atom can be written as

$$\hat{\vec{\mu}}_p = g_p \mu_N \hat{\vec{I}}, \tag{1.10}$$

where $\hat{\vec{I}}$ is the operator of the nuclear angular momentum of the proton, and g_p is the nuclear g factor of the proton. Thus, the magnetic moment of nuclei is about three orders of magnitude smaller than that of electrons. In general, the nuclear magnetic moment for various elements of the Periodic table and their isotopes can be written as

$$\hat{\vec{\mu}}_I = g_I \mu_N \hat{\vec{I}}. \tag{1.11}$$

The nuclear angular momentum I for different isotopes in the Periodic table can be set to be multiples of $I = 1/2$ and more, up to $I = 9/2$ and 5 in increments of 1/2. Nuclear g factors have been measured for various elements and their isotopes for the entire Periodic table and tabulated (see, e.g., Bruker tables, www.bruker-biospin.com). It should be noted that many elements have separate isotopes with zero angular momentum $I = 0$ and, consequently, zero magnetic moment.

1.2 Magnetic Moment of the Electron Shell in a Free Atom or Ion

1.2.1 Electronic g-Factor of the Orbital and Spin Magnetic Moments

Consider the electron states of a free atom (or a free ion). According to the quantum-mechanical description of the free atom, Schrödinger's equation (Austrian physicist Erwin Schrödinger, 1926), the electron is characterized by three quantum numbers: principal quantum number n, orbital (azimuthal) quantum number l, and the projection of the orbital angular momentum in the preferred direction (the magnetic quantum number), m_l. In addition, there are the electron spin s and the spin moment projection m_s, which should be added to n, l, m_l. The relationship between the integers n, l, m_l can be summarized by $n > 0$; $n - 1 \geq l \geq 0$; $+l \geq m_l \geq -l$. Finally, an electron in an atom should be associated with five quantum numbers: n, l, m_l, s, m_s. The magnetic properties of atoms (ions) are determined by the unpaired electrons in the outer shells.

Suppose an atom (ion) has n electrons. Each electron can be given its own orbital ($\boldsymbol{l}_1, \boldsymbol{l}_2, \ldots \boldsymbol{l}_n$) and spin ($\boldsymbol{s}_1, \boldsymbol{s}_2, \ldots \boldsymbol{s}_n$) angular momentum vectors, and these vectors are combined to form a resultant. In order to describe the states of composite systems that are combinations of subsystems with angular momenta $\boldsymbol{l}_1, \boldsymbol{l}_2, \ldots$, it is necessary to understand how the individual momenta combine to give a total angular momentum $\boldsymbol{L}$. For individual electrons we use small letters to denote the orbital and spin moments. In light atoms (with atomic number less than about 70), the Russell-Saunder coupling (*LS*-coupling) occurs. This form of coupling arises when all the orbital momenta combine to form the resultant angular momentum (addition rule of angular momenta) given by

$$\boldsymbol{L} = \boldsymbol{l}_1 + \boldsymbol{l}_2 + \cdots + \boldsymbol{l}_n.$$

In the same way, all the spin momenta combine to form their own resultant

$$\boldsymbol{S} = \boldsymbol{s}_1 + \boldsymbol{s}_2 + \cdots + \boldsymbol{s}_{\mathrm{n}}.$$

L and **S** can be found using (i) the addition rule for angular moments and (ii) the Pauli exclusion principle (Wolfgang Pauli, German physicist, 1925). The Pauli exclusion principle states that no two electrons in the same atom can have the same set of quantum numbers.

To find the lowest energy level in an atom (ion), the so-called Hund's rules are used. The first Hund's rules state that, for a given electron configuration: (i) the term with the greatest spin (multiplicity) has the lowest energy; (ii) if the spins of two terms are the same, then that with the greatest orbital angular momentum has the lowest energy.

Suppose there are two unpaired electrons. Each electron can be given its own orbital ($\mathbf{l}_1$, $\mathbf{l}_2$) and spin ($\mathbf{s}_1$, $\mathbf{s}_2$) angular momentum vectors, and these vectors combine to form a resultant. In accordance with the addition rule for angular moments, the total orbital and total spin angular momenta can only be:

$$\begin{aligned} L &= l_1 + l_2, l_1 + l_2 - 1, \ldots |l_1 - l_2|; \\ S &= s_1 + s_2, s_1 + s_2 - 1, \ldots |s_1 - s_2|. \end{aligned} \tag{1.12}$$

As an example, consider the case of two electrons in the states n_1p and n_2p, with $n_1 \neq n_2$, such as electrons $2p$ and $3p$, i.e., those with different principal quantum numbers. Here, $l_1 = 1$, $l_2 = 1$; $s_1 = 1/2$, $s_2 = 1/2$ and, in accordance with (1.12), the total orbital angular momentum can take the values $L = 2, 1, 0$ and the total spin moment can take the values $S = 1, 0$. When the angular momentum L is 0, 1, 2, 3, 4, 5, 6, 7 …, the state is labeled by a capital letter *S, P, D, F, G, H, I, K, …* state; here, of course, S has nothing to do with the quantity S used to denote the spin (similar notations are used to label orbitals according to their values of l). To indicate the status of a free atom (ion), the concept of "term" is introduced, written as ^{2S+1}L, where the corresponding letter of the alphabet is written instead of L, and the multiplicity of a state is defined by $2S + 1$, S being the symbol for the total spin rather than the symbol of state! A term symbol conveys information about the state and its multiplicity. Thus, in accordance with the addition rule for two p electrons n_1p and n_2p, with $n_1 \neq n_2$, we have the possible terms: ^3D, ^1D, ^3P, ^1P, ^3S, ^1S. In this case, the Pauli exclusion principle is not violated, since these electrons are distinguished by the principal quantum number n. In the case of $n_1 = n_2$, a state with two p electrons, np^2 (e.g., $2p^2$) the Pauli exclusion principle is to be taken into consideration and, as a result, not all of these terms are possible. Only terms ^1D, ^3P, ^1P, ^3S, ^1S can exist.

To choose a term with the lowest energy, which is of interest in most applications of EPR spectroscopy, it is necessary to use the first Hund's rule, under which the term with the maximum spin angular momentum has the lowest energy, and for the terms with equal spin moments, the term with the largest orbital angular momentum has the lower energy. Thus the minimum energy term is ^3P.

1.2.2 Spin-Orbit Interaction

Between the orbital angular momentum and the spin angular momentum there is the so-called spin-orbit coupling, which can be given in its simplest form by the term

$$\hat{H}_{SO} = \lambda \hat{\vec{L}} \cdot \hat{\vec{S}}, \tag{1.13}$$

where $\hat{H}_{SO}$ is the energy operator of the spin-orbit coupling, and λ is the constant of the spin-orbit coupling. The spin-orbit coupling is a purely relativistic effect caused by the interaction of the spin magnetic moment of an electron with the magnetic field generated at the location of the electron as a result of the relative motion of the positively charged atomic nucleus. The spin-orbit coupling constant λ grows with the mass of the atom as the velocity of the electron in its orbit increases, which enhances the relativistic effect.

Owing to the spin-orbit interaction, the term is split into subterms that differ in the total angular momentum of a free atom

$$\hat{\vec{J}} = \hat{\vec{L}} + \hat{\vec{S}}. \tag{1.14}$$

In accordance with the addition rule of angular momenta, J can take the values

$$J = L + S,\ L + S - 1, \ldots |L - S|.$$

It is possible to extend the term notation further by showing the value of J. Then we use a symbol $^{2S+1}L_J$ to define a "subterm" (level). The second Hund's rule states: if both the spin and the orbital angular momentum are the same and the electron shell is less than half-filled, the lowest energy is given by the lowest value of J. The converse is true if the shell is more than half-filled.

Consider a few examples of electronic configurations.

ns^1 : $l = 0$, $s = 1/2$; → $L = 0$, $S = 1/2$ → term ^2S (read 'doublet-S');

np^1 : $l = 1$, $s = 1/2$; → $L = 1$, $S = 1/2$ → term ^2P, the possible total angular momentum $J = 1/2$ and $J = 3/2$, under the influence of the spin-orbit interaction term ^2P splits into subterms $^2P_{1/2}$ and $^2P_{3/2}$; in line with the second Hund's rule, subterm $^2P_{1/2}$ has a lower energy, that is the lowest energy sublevel (Fig. 1.2a).

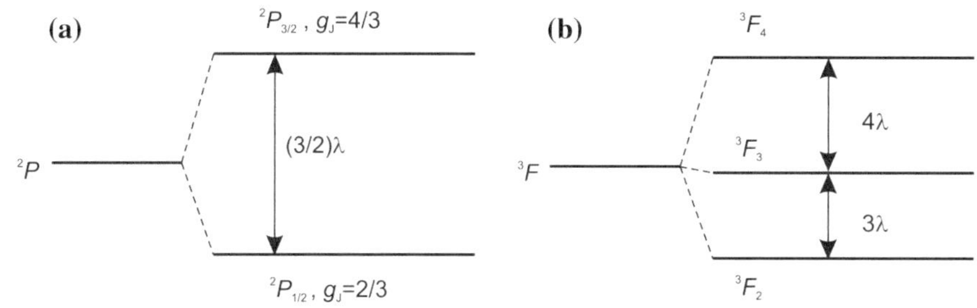

Fig. 1.2 **a** Illustration of the energy levels of a system with one unpaired p-electron and spin-orbit coupling. Term ^2P and two subterms $^2P_{1/2}$ and $^2P_{3/2}$ are shown. The spin-orbit splitting magnitude and Landé g-factors for subterms $^2\mathrm{P}_{1/2}$ and $^2\mathrm{P}_{3/2}$ are indicated. **b** The energy levels of an nd^2 electronic configuration with spin-orbit coupling. The spin-orbit splitting magnitudes are shown

We now find the magnitude of the spin-orbit interaction; in other words, we find the eigenvalues of the energy operator of the spin-orbit interaction (1.14). For this purpose, we recall a number of relations of quantum mechanics. When calculating properties of a free atom (free ion), which depend on the angular momentum, one uses orbitals that are eigenfunctions (eigenstates) of the squared orbital angular momentum and its z-component. The eigenstates for the angular momentum can be described in Dirac's notation by a ket vector: $|L, M_L\rangle$, $|S, M_S\rangle$, $|J, M_J\rangle$. Since the eigenstates associated with the angular momentum depend only on the values in parentheses, we consider only the properties associated with the angular momentum. We can summarise the results for the orbital, spin and total angular momentum operator equations as follows [see also 1.6a)]:

$$\begin{aligned}
\hat{\vec{L}}^2|L,M_L\rangle &= L(L+1)|L,M_L\rangle,\\
\hat{L}_Z|L,M_L\rangle &= M_L|L,M_L\rangle,\\
\\
\hat{\vec{S}}^2|S,M_S\rangle &= S(S+1)|S,M_S\rangle,\\
\hat{S}_Z|S,M_S\rangle &= M_S|S,M_S\rangle,\\
\\
\hat{\vec{J}}^2|J,M_J\rangle &= J(J+1)|J,M_J\rangle,\\
\hat{J}_Z|J,M_J\rangle &= M_J|J,M_J\rangle
\end{aligned} \tag{1.15}$$

Here $L(L + 1)$, $S(S + 1)$ and $J(J + 1)$ are the eigenvalues of the operators $\hat{\vec{L}}^2$, $\hat{\vec{S}}^2$ and $\hat{\vec{J}}^2$; and M_L, M_S and M_J are the eigenvalues of $\hat{L}_z$, $\hat{S}_z$ and $\hat{J}_z$, respectively. A complete set of commuting observables for states carrying an angular momentum (e.g., J) is provided by the operators $\hat{\vec{J}}^2$, $\hat{J}_z$. The eigenstates of these operators are labeled $|J,M_J\rangle$ where $J \geq 0$ is the angular momentum and M_J is the magnetic quantum number which gives the projection of $\hat{\vec{J}}$ on the quantization axis taken to be z. The possible values of these quantities are $J \geq 0$, integer or half-integer values only being allowed and $-J \leq M_J \leq J$, with successive values of M_J separated by unity. In what follows, the cap notation of operators will be omitted in the obvious cases. According to quantum mechanics, the angular momentum of a particle and its projection on a direction selected in the space (for example, the magnetic field) can only take discrete values. The orbital angular momentum L can take only integer values, while S and J can be either an integer or a half-integer. The eigenvalues of $\hat{L}_z$, $\hat{S}_z$ and $\hat{J}_z$, M_L, M_S and M_J are restricted to run from $+L$ to $-L$, $+S$ to $-S$ and $+J$ to $-J$, respectively, in integer steps. Thus, there are $(2L + 1)$ different values of M_L for a given L; $(2S + 1)$ different values of M_S for a given S and $(2J + 1)$ different values of M_J for a given J. Along with the length of the angular momentum (e.g., for J the length of the vector is equal to $[J(J + 1)]^{1/2}$) can be measured by only one projection, and similarly for L and S. The angular momentum can never be aligned exactly along the z-axis: the maximum value of its projection on the axis is J which is less than the length of the vector itself. For example, for $S = 1/2$, spin has, along with projections $M_S = 1/2$ and $-1/2$, a value that can be measured along with the projection. The length is 1.73 times the projection. This means that the spin axis is oriented at an angle of about 54° to z–axis. A surprising result is that the spin resists an attempt to align it with the z-axis. Being strange for a classical description of the angular momentum, this is true!

The magnetic moment is proportional to the angular momentum, hence the magnetic moment also takes a number of discrete values. Relations for the orbital angular momentum (1.15) were derived by considering the orbital motion of electrons in an orbit in the coordinate representation by solving Schrödinger's equation. The ratio for the spin can not be obtained in the same way, but the experiment shows that the orbital and spin angular momenta obey the same commutation rules and are additive, i.e., they have the same physical dimension (one can not add different values, e.g., mass to length). Thus, it does not matter whether one is dealing with the orbital angular momenta of electrons in atoms or with the spins of particles (electrons, nuclei). The quantum theory of angular momentum applies to all of these and allows understanding of a vast range of phenomena, and one cannot but be impressed by the efficiency and elegance of this theory in describing the phenomenon of the magnetic resonance.

1.2.3 Landé Interval Rule and Landé g-Factor

Raising the left- and right-hand parts of (1.12) to the square: $(\boldsymbol{J})^2 = (\boldsymbol{L} + \boldsymbol{S})^2$ and using relations (1.15), we have

$$J(J+1) = L(L+1) + S(S+1) + 2L \cdot S.$$

Rearranging,

$$L \cdot S = (1/2)[J(J+1) - L(L+1) - S(S+1)],$$

and substituting this expression into formula (1.14) for the spin-orbit interaction, we obtain the expression for the eigenvalues of the spin-orbit interaction

$$E_{so} = (1/2)\lambda[J(J+1) - L(L+1) - S(S+1)], \tag{1.16}$$

which leads to splitting of the neighboring sublevels corresponding to different subterms with J and $(J - 1)$ in the form

$$E_J - E_{J-1} = \lambda J. \tag{1.16a}$$

This expression is known as the Landé interval rule, a rule dealing with the relation between the electronic spin and orbit moments (Alfred Landé, German physicist, 1923): the spacing between two levels with multiplet numbers $J - 1$ and J is proportional to J. Each multiplet level is $2J + 1$ fold, the degeneracy being removed by the magnetic field.

If we apply (1.16a) to the case considered above for an np^1 configuration, we obtain the splitting between the subterms $^2P_{1/2}$ and $^2P_{3/2}$, equal to $E_{J=3/2} - E_{J=1/2} = (3/2)\lambda$ (Fig. 1.2a). It is amazing how quantum mechanics works: simple relations give accurate results that are easily verified by experiments!

We extend the examples.

(a) Configuration nd^1: $l = 2$, $s = 1/2$; $\rightarrow L = 2$, $S = 1/2 \rightarrow$ term ^2D; the possible total angular momenta $J = 3/2$ and $J = 5/2$ split under the influence of the spin-orbit interaction into subterms $^2D_{3/2}$ and $^2D_{5/2}$. In line with Hund's second rule, subterm $^2D_{3/2}$ is the lowest energy sublevel, the splitting between the subterms being $(5/2)\lambda$.
(b) Configuration nd^2: $l_1 = 2$, $s_1 = 1/2$; $l_2 = 2$, $s_2 = 1/2$. First, taking into account the Pauli exclusion principle and Hund's rule, we find the term corresponding to the minimum energy. We write down the corresponding quantum numbers for the maximum spin angular momentum, given that this should be the maximum orbital angular momentum:

for the first electron: $l_1 = 2$, $m_{l1} = 2$; $s_1 = 1/2$; $m_{s1} = 1/2$;
for the second electron: $l_2 = 2$, $m_{l2} = 1$; $s_2 = 1/2$; $m_{s2} = 1/2$.

The total orbital angular momentum L is equal to the maximum projection of the total orbital angular momentum, as $m_{l1} + m_{l2} = 3$, $L = 3$, for the spin moment $m_{s1} + m_{s2} = 1$, $S = 1$. Thus, the lower term has the form ^3F. As a result of the spin-orbit interaction (in line with the rules of addition of angular momenta) the term is split into three subterms $^3F_2$, $^3F_3$ and $^3F_4$ with the splittings between the levels equal to 3λ and 4λ (Fig. 1.2b).

According to (1.12), the total angular momentum of a free atom (ion) $\boldsymbol{J} = \boldsymbol{L} + \boldsymbol{S}$. This corresponds to a magnetic moment $\boldsymbol{\mu}_J$, which is the sum of the orbital magnetic moment $\boldsymbol{\mu}_L = -g_L\mu_B\boldsymbol{L}$, (1.7), and the spin magnetic moment $\boldsymbol{\mu}_S = -g_S\mu_B\boldsymbol{S}$, (1.8):

$$\boldsymbol{\mu}_J = \boldsymbol{\mu}_L + \boldsymbol{\mu}_S. \tag{1.17}$$

In quantum mechanics, the corresponding magnetic moments are the operators. We write

$$\boldsymbol{\mu}_J = -g_J\mu_B\boldsymbol{J}, \tag{1.18}$$

where g_J is the g-factor of the complete shell of a free atom for a particular subterm to be found. g_J can be expressed through J, L and S to give this subterm and is named the Landé g-factor (A. Landé, 1921) or the spectroscopic splitting factor.

Rewriting (1.17) in the form of $-g_J\mu_B\boldsymbol{J} = (-g_L\mu_B\boldsymbol{L}) + (-g_S\mu_B\boldsymbol{S})$, we have, upon reduction and substitution of $g_L = 1$ and $g_S = 2$, $g_J\boldsymbol{J} = \boldsymbol{L} + 2\boldsymbol{S}$. Multiplying the left- and right-hand parts of this expression by $\boldsymbol{J}$, we obtain the expression

$$g_J J(J+1) = \boldsymbol{L}\cdot\boldsymbol{J} + 2\boldsymbol{S}\cdot\boldsymbol{J}. \tag{1.19}$$

Next, we proceed to reveal the inner products in the same way as it was done when finding the splitting of the levels under the influence of spin-orbit interaction. From (1.12), we have two relations $\boldsymbol{L} = \boldsymbol{J} - \boldsymbol{S}$, and $\boldsymbol{S} = \boldsymbol{J} - \boldsymbol{L}$. For each of these relations, square the left and right side, the result is

$$\boldsymbol{L}\cdot\boldsymbol{J} = (1/2)[J(J+1) + L(L+1) - S(S+1)],$$
$$\boldsymbol{S}\cdot\boldsymbol{J} = (1/2)[J(J+1) + S(S+1) - L(L+1)].$$

Substitute this scalar product of vectors into (1.19) and obtain the Landé g-factor expressed in terms of the eigenvalues of the angular momenta

$$g_J = 1 + \frac{J(J+1) + S(S+1) - L(L+1)}{2J(J+1)}. \tag{1.20}$$

The Landé g-factor is only valid for free atoms (or ions) with a central potential. This formula can be tested for two limiting cases with (1) $L = 0$, i.e., $J = S$ and $g_J = g_S = 2.0$ and (2) $S = 0$, i.e., $J = L$ and $g_J = g_L = 1$. By way of example, one can calculate the various subterms g_J discussed above.

Again, we observe the unique features of quantum mechanics, which allows use of the simplest methods to calculate the fundamental characteristics of the atom. The g-factors enable an accurate calculation of the behavior of the energy levels of different subterms in external magnetic fields that can easily be verified by magnetic resonance experiments. This knowledge can find numerous practical applications.

We now pass to the section in which the interaction of the magnetic dipole with a magnetic field is considered and an attempt is made to solve the key problem of

EPR and NMR spectroscopy, to find the energy levels of magnetic dipoles in a magnetic field.

The essential results of this chapter can be summarized as follows. Systems with angular momenta $\boldsymbol{L}$, $\boldsymbol{S}$, $\boldsymbol{J}$, $\boldsymbol{I}$, posses a corresponding magnetic dipole moment which is fundamentally related to the angular momentum: orbital magnetic moment $\boldsymbol{\mu}_L = -g_L\mu_B\boldsymbol{L}$, (1.7), spin magnetic moment $\boldsymbol{\mu}_S = -g_S\mu_B\boldsymbol{S}$, (1.8), free atom (ion) magnetic moment $\boldsymbol{\mu}_J = -g_J\mu_B\boldsymbol{J}$, (1.18) and nuclear magnetic moment $\boldsymbol{\mu}_I = g_I\mu_N\boldsymbol{I}$, (1.11). Here g_L, g_S, g_I are the spectroscopic splitting factors of orbital, spin and nuclear magnetic dipole moments, and g_J is the spectroscopic splitting factor, Landé g-factor, (1.20), of the complete electronic shell of a free atom (ion) for a particular subterm $^{2S+1}L_J$. The Landé g-factor is only valid for free atoms (or ions) with a central potential.

1.3 Magnetic Dipole in a Magnetic Field

1.3.1 Electron Zeeman Interaction

In the absence of any magnetic field, the magnetic moment associated with the angular momentum J is randomly oriented, the energy of a particle with a magnetic moment is independent of the spatial orientation and all energy levels ($2J + 1$ levels) have the same energy, i.e., the energy levels are degenerate. When the magnetic moment starts to interact with the external magnetic field $\boldsymbol{B}$, the classical expression for the energy of the magnetic moment in a magnetic field is given by the scalar product of two vectors. Classically, the energy of a magnetic moment $\boldsymbol{\mu}$ in an external magnetic field $\boldsymbol{B}$ is given by

$$E = -\boldsymbol{\mu} \cdot \boldsymbol{B}. \tag{1.21}$$

The negative sign shows that, being parallel, $\boldsymbol{\mu}$ and $\boldsymbol{B}$ give the energetically most favourable arrangement, and the antiparallel arrangement is the least favourable.

The application of an external magnetic field $\boldsymbol{B}$ results in a splitting of the energy levels. The quantization of the energy levels is due to the quantum-mechanical nature of the electron angular momentum. The energies of quantum states with different projections of the magnetic moment in a magnetic field, which correspond to different discrete values of M_J, are not the same, i.e., the degeneracy is lifted. For a quantum mechanical system, the energy is replaced with the appropriate energy operator or Hamiltonian, which has the same form as (1.21), i.e., we have in terms of operators

$$\hat{H}_{Zeem} = -\hat{\vec{\mu}} \cdot \vec{B}, \tag{1.22}$$

where $\boldsymbol{\mu}$ is the operator of the magnetic moment (electronic or nuclear). The splitting between the energy states is named the electron Zeeman interaction (after the Dutch physicist Pieter Zeeman, who first discovered this splitting in 1896).

To discuss the principles of EPR, we start with the simple model of a two-level system for a paramagnetic centre with the smallest possible angular momentum $J = 1/2$. For definiteness, we consider the Hamiltonian for a free electron with the spin moment $S = 1/2$ in a magnetic field. Application of an external magnetic field $\boldsymbol{B}$ results in a splitting of the two energy levels because the electron spin S can only be oriented parallel or anti-parallel to the magnetic field vector. The potential energy of this system is derived from the classical expression for the energy of a magnetic dipole in a magnetic field (1.22) and is described by the spin Hamilton operator

$$\hat{H}_{Zeem} = g_S\mu_B\hat{\vec{S}}\cdot\vec{B}, \tag{1.23}$$

where the operator of the spin magnetic moment

$$\hat{\vec{\mu}}_S = -g_S\mu_B\hat{\vec{S}}$$

is used, and the dimensionless g-value of the free electron is $g_S = 2.00$.

Since the only preferred direction in this system is the direction of the external magnetic field, we can, without loss of generality, direct the z-axis along the magnetic field, i.e., $B \parallel z$: $B_z = B$, $B_x = B_y = 0$. The scalar product simplifies and the Hamiltonian for the Zeeman energy becomes

$$\hat{H}_{Zeem} = g_S\mu_B\hat{S}_Z B, \tag{1.23a}$$

Our goal is to find the energy levels of the system under consideration. The energies for the two spin states are then given by the eigenvalues of the Hamiltonian in (1.23a) and characterized by the spin quantum numbers $M_S = \pm 1/2$. It is necessary to find the eigenvalues of the Hamiltonian operator

$$\hat{H}_{Zeem}|S, M_S\rangle = E_{Zeem}|S, M_S\rangle, \tag{1.24}$$

where $|S,M_S\rangle$ are the eigenvectors (eigenfunctions). The only operator in (1.23a) is S_Z and it is clear that the basic equation to be considered is the following [see (1.15)]

$$\hat{S}_Z|S, M_S\rangle = M_S|S, M_S\rangle.$$

For the case of $S = 1/2$, M_S can take only two values $M_S = 1/2$ and $M_S = -1/2$. Therefore, it is convenient to distinguish the wave functions only by M_S, with their designations being $|M_S\rangle = |+1/2\rangle$ and $|-1/2\rangle$. The frequently used notation (which we use, too) is $|+1/2\rangle = |\alpha\rangle$ and $|-1/2\rangle = |\beta\rangle$ and, as a result, (1.24) can be represented as

$$g_S\mu_B\hat{S}_Z B|\alpha\rangle = \frac{1}{2}g_S\mu_B B|\alpha\rangle \text{ and } g_S\mu_B\hat{S}_Z B|\beta\rangle = -\frac{1}{2}g_S\mu_B B|\beta\rangle,$$

where 1/2 and −1/2 are the eigenvalues of the equations

$$\hat{S}_Z|\alpha\rangle = \frac{1}{2}|\alpha\rangle \text{ and } \hat{S}_Z|\beta\rangle = -\frac{1}{2}|\beta\rangle.$$

The eigenvalues of the Hamiltonian $\hat{H}_{Zeem}$ will give the energy levels

$$E_\alpha = 1/2g_S\mu_B B \text{ and } E_\beta = -1/2g_S\mu_B B. \tag{1.25}$$

As already mentioned, the electron system is doubly degenerate in the absence of a magnetic field and introduction of a magnetic field lifts this degeneracy. Thus, there are two states that are degenerate in zero field, and their separation grows linearly with increasing magnetic field. The magnetic field perturbs the electron energy. This effect can be accounted for in a quite general way by using the perturbation theory. To calculate the energy levels in the general case, the perturbation theory for degenerate states should be applied. It is necessary to write the matrix elements of the type

$$\langle M_S|\hat{H}_{Zeem}|M'_S\rangle$$

and construct a determinant from these. In the simplest case of the two-level system with $S = 1/2$, we can write the energy matrix as follows:

$\langle M_S\|\hat{H}_{Zeem}\|M'_S\rangle$	$\|\alpha\rangle$	$\|\beta\rangle$
$\langle\alpha\|$	$1/2g_S\mu_B B$	0
$\langle\beta\|$	0	$-1/2g_S\mu_B B$

In the calculation, we used the orthogonality of the wave functions:

$$\langle M_S|M'_S\rangle = 1 \text{ for } M_S = M'_S \text{ and } \langle M_S|M'_S\rangle = 0 \text{ for } M_S \neq M'_S,$$

i.e., $\langle\alpha|\alpha\rangle = \langle\beta|\beta\rangle = 1$ and $\langle\alpha|\beta\rangle = \langle\beta|\alpha\rangle = 0$. The matrix is diagonal because we have chosen the wave functions that are eigenfunctions of the Hamiltonian $\hat{H}_{Zeem}$; otherwise, the matrix would be non-diagonal and its diagonalization would be necessary for finding the energy levels and eigenfunctions of the Hamiltonian.

In the standard procedure of the perturbation theory for degenerate states, the determinant of the matrix elements is made in accordance with the above table (by subtracting a variable E from each diagonal element and setting to zero the resulting secular determinant). The two roots of the secular determinant will be the energy levels. The secular equation is found to be

$$\begin{vmatrix} 1/2 g_S \mu_B B - E & 0 \\ 0 & -1/2 g_S \mu_B B - E \end{vmatrix} = 0. \tag{1.26}$$

On expanding the determinant, we find the energy levels. In the above simple case of a diagonal matrix, the energy levels in the form $E_1 = E_\alpha$ and $E_2 = E_\beta$ can be immediately obtained. The same levels were obtained above by using (1.25). The splitting between the two energy states (electron Zeeman splitting) is proportional to the magnitude of $\boldsymbol{B}$, as illustrated by Fig. 1.3a.

The energy difference between the two Zeeman states is given by

$$\Delta E = g_S \mu_B B. \tag{1.27}$$

According to Planck's equation, electromagnetic energy can be absorbed or emitted if

$$\Delta E = h\nu, \tag{1.28}$$

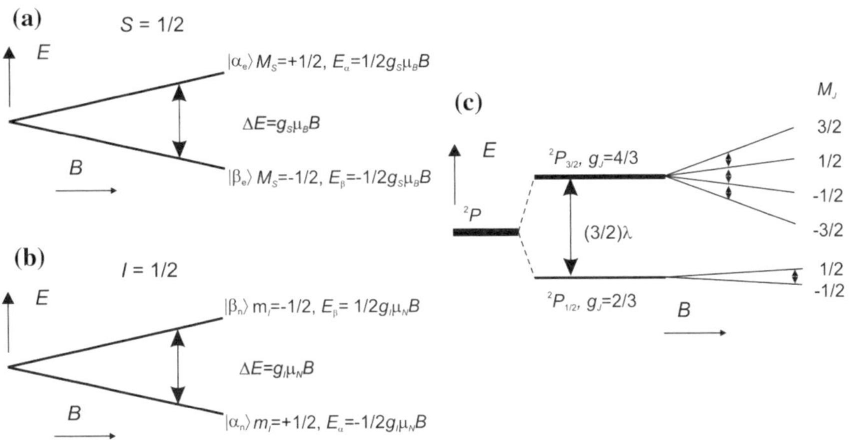

Fig. 1.3 **a** Illustration of the electron Zeeman splitting for an S = 1/2 system with one unpaired electron in an external magnetic field B. For a given irradiation frequency ν, a magnetic dipole transition between the Zeeman levels (indicated by the *arrow*) occurs if the resonance condition $\nu = g_S\mu_B B/h$ ($\omega = g_S\mu_B B/\hbar$) is satisfied. **b** Illustration of the nuclear Zeeman splitting for proton (I = 1/2) in an external magnetic field $\boldsymbol{B}$. For a given irradiation frequency ν, a magnetic dipole transition between the Zeeman levels (indicated by the *arrow*) occurs if the resonance condition $\nu = g_p\mu_N B/h$ ($\omega = g_p\mu_N B/\hbar$) is satisfied. **c** Illustration of the energy levels of a system with one unpaired *p*-electron and spin-orbit coupling without and with applied magnetic field. The magnetic dipole transitions for each subterm, indicated by the red arrow, occur if the resonance condition $\nu = g_J\mu_B B/h$ ($\omega = g_J\mu_B B/\hbar$) is satisfied

where ν is the frequency of the electromagnetic radiation.

In the common EPR experiment, an electromagnetic field of frequency ν and a variable magnetic field are applied to this system (for experimental reasons in a EPR experiment, the microwave frequency is usually held constant and the magnetic field is swept linearly). If the energy of the irradiation field matches the energy gap ΔE, transitions between the two spin states can be induced (shown by the arrow in Fig. 1.3a), i.e., the spin can be flipped from one orientation to the other. In this case, the resonance condition is satisfied:

$$\nu = g_S \mu_B B/\mathrm{h}. \tag{1.27a}$$

In applications in which the frequency is expressed in terms of radians per second ("angular frequency"), instead of cycles per second, the reduced Planck constant is used. It is equal to the Planck constant divided by 2π, and is denoted $\hbar$ ("h-bar"): $\hbar = \frac{h}{2\pi}$. The energy of a photon with an angular frequency ω, where ω = 2πν, is given by

$$\Delta E = \hbar\omega, \tag{1.28a}$$

As a result, (1.27a) can also be expressed as

$$\omega = g_S \mu_B B/\hbar \tag{1.27b}$$

The frequency ω corresponds to the frequency of the electron paramagnetic resonance. Expression (1.27b) can be represented in another form if we replace $g_S \mu_B$ with $\gamma\hbar$, where γ is the magnetogyric ratio:

$$\omega_L = \gamma B. \tag{1.29}$$

Here, the frequency ω_L is the so-called Larmor frequency. As shown below, this frequency corresponds to the precession of the magnetic moment around the magnetic field in the classical description of the system.

The lower energy level corresponds to the direction of the magnetic moment along the external magnetic field. Figure 1.3a shows that the state with the lower energy corresponds to the projection of the angular momentum along the magnetic field $M_S = -1/2$, which is opposite in direction to the magnetic moment of the electron. This is due to the negative sign of the electron charge.

It is of use to repeat this analysis for nuclear spins. Let us consider the Zeeman effect for the nuclear magnetic moment (e.g., for a proton which is the nucleus of the hydrogen atom). According to (1.10), the operator of the magnetic moment of the proton has the form

$$\hat{\vec{\mu}}_p = g_p \mu_N \hat{\vec{I}},$$

with $I = 1/2$. Substituting this magnetic moment into Hamiltonian (1.22) we obtain the expression for the Zeeman energy of the proton magnetic moment in an external magnetic field in the form of

$$\hat{H}_{Zeem} = -g_p \mu_N \hat{\vec{I}} \cdot \vec{B}, \tag{1.30}$$

For $B \parallel z$: $B_z = B$, $B_x = B_y = 0$, the expression for the energy simplifies to

$$\hat{H}_{Zeem} = -g_p \mu_N \hat{I}_Z B. \tag{1.30a}$$

By analogy with the electron spin, we introduce the wave functions for the two projections of the angular momentum of the nuclear spin along the z axis as $|m_I\rangle = |+1/2\rangle$ and $|-1/2\rangle$ (angular momentum projections for nuclei, unlike electrons, are further written as lowercase m). By analogy with the electronic spin moment, we use the notation $|+1/2\rangle = |\alpha_n\rangle$ and $|-1/2\rangle = |\beta_n\rangle$ (index "n" is introduced to denote the nucleus). As a result, the eigenvalue equation with operator (1.30a) takes the form

$$\hat{H}_{Zeem}|\alpha_n\rangle = -g_p \mu_N \hat{I}_Z B|\alpha_n\rangle = -\frac{1}{2} g_S \mu_B B|\alpha_n\rangle \text{ and}$$
$$\hat{H}_{Zeem}|\beta_n\rangle = -g_p \mu_N \hat{I}_Z B|\beta_n\rangle = \frac{1}{2} g_S \mu_B B|\beta_n\rangle,$$

where 1/2 and −1/2 are the eigenvalues of the equations

$$\hat{I}_Z|\alpha_n\rangle = \frac{1}{2}|\alpha_n\rangle \text{ and } \hat{I}_Z|\beta_n\rangle = -\frac{1}{2}|\beta_n\rangle.$$

The eigenvalues of the Hamiltonian $\hat{H}_{Zeem}$ give the energy levels

$$E_{\alpha n} = -1/2 g_p \mu_N B \text{ and } E_{\beta n} = 1/2 g_p \mu_N B. \tag{1.31}$$

That is qualitatively implemented energy-level diagram shown in Fig. 1.3a for the electron spin. However, there is a significant difference due to the positive sign of the proton (Fig. 1.3b). For the nuclear magnetic moment of the proton in an external magnetic field, the situation is the opposite, i.e., the angular momentum and the magnetic moment of the proton have the same direction and the lowest energy state corresponds to the $m_I = +1/2$ (level $E_{\alpha n}$). The separation between the energy levels is given by

$$\Delta E = g_p \mu_N B. \tag{1.32}$$

and the resonant frequency, which corresponds to the nuclear magnetic resonance, is

$$\nu = g_p \mu_N B / h, \tag{1.33}$$

or, for the angular frequency,

$$\omega = g_p \mu_N B / \hbar. \tag{1.33a}$$

Because the nuclear magnetic moment is about three orders of magnitude smaller than the magnetic moment of the electron, the NMR frequency is about three orders of magnitude lower than the EPR frequency in the same external magnetic field (GHz in EPR and MHz in NMR).

The above consideration is suitable for any system with an angular momentum equal to 1/2, e.g., for the case $J = 1/2$, which occurs in subterm $^2P_{1/2}$. In this case, the g-factor is the Landé g-factor $g_J = 2/3$, rather than 2.0, as in the case of the spin magnetic moment. As a result, the splitting of energy levels in a magnetic field looks like

$$\Delta E = g_J \mu_B B. \tag{1.34}$$

The behavior of the energy levels of the upper subterm $^2P_{3/2}$, $J = 3/2$ is different from the case of $J = 1/2$, because in this case, the subterm is split in a magnetic field into four sublevels. The number of sublevels is given by $2J + 1$ (above, we used a similar formula for the multiplicity of the spin levels in a term as $2S + 1$). The problem solved is as easy as it is for a two-level system. It is necessary to make a table of matrix elements $\langle M_J | \hat{H}_{Zeem} | M'_J \rangle$ with wave functions for $J = 3/2$ in a magnetic field in the form of $|J, M_J\rangle$. Because the wave functions will only differ by M_J, we write them as $|M_J\rangle$, where M_J takes the values 3/2, 1/2, −1/2, −3/2.

$\langle M_J \vert \hat{H}_{Zeem} \vert M'_J \rangle$	$\vert 3/2\rangle$	$\vert 1/2\rangle$	$\vert -1/2\rangle$	$\vert -3/2\rangle$
$\langle 3/2 \vert$	$3/2 g_J \mu_B B - E$	0	0	0
$\langle 1/2 \vert$	0	$1/2 g_J \mu_B B - E$	0	0
$\langle -1/2 \vert$	0	0	$-1/2 g_J \mu_B B - E$	0
$\langle -3/2 \vert$	0	0	0	$-3/2 g_J \mu_B B - E$

and make a determinant of these.

There are only diagonal matrix elements, because the projections $|M_J\rangle$ are eigenfunctions of the Hamiltonian $\hat{H}_{Zeem}$, which is written as

$$\hat{H}_{Zeem} = g_J \mu_B \hat{J}_Z B. \tag{1.35}$$

The matrix elements were calculated with (1.15) and the orthogonality relation for the wave functions $\langle M_J | M_J' \rangle = 1$ at $M_J = M_J'$ and $\langle M_J | M_J' \rangle = 0$ at $M_J \neq M_J'$. In general, the energy level can be written as

$$E_{MJ} = M_J g_J \mu_B B, \quad (1.36)$$

and the separation between adjacent levels is given by

$$\Delta E = g_J \mu_B B, \quad (1.37)$$

This is analogous to the magnetic-resonance condition of (1.27). For the ground-state $^2P_{1/2}$ subterm, the Landé g-factor equal to 2/3 is smaller than the Landé g-factor of the excited-state $^2P_{3/2}$ subterm equal to 4/3, and the magnetic transitions correspond to a higher magnetic field for the same microwave frequency.

1.3.2 Interaction Between the Magnetic Dipoles

The magnetic dipole–dipole interaction (dipolar coupling) is a direct interaction between two magnetic dipoles. Let us consider the interaction of magnetic dipoles with each other in terms of the classical electrodynamics. The interaction of the magnetic moments can be represented as the interaction of the magnetic moment of a magnetic dipole with the magnetic field generated by the other magnetic dipole at the location of the first dipole (Fig. 1.4). The same result is provided by the inverse scheme, if the dipoles are reversed. The magnetic field generated by a magnetic dipole designated, to be specific, as $\boldsymbol{\mu}_1$ is expressed by (1.2),

$$\vec{B}_{\mu 1} = \frac{1}{r^3}\left[\frac{3(\vec{\mu}_1 \cdot \vec{r}) \cdot \vec{r}}{r^2} - \vec{\mu}_1\right].$$

The potential energy of the second magnetic dipole $\boldsymbol{\mu}_2$ placed in the magnetic field of the first dipole is written according to (1.21) in the form $E = -\boldsymbol{\mu}_2 \cdot \boldsymbol{B}_{\mu 1}$, which gives the interaction energy of two magnetic dipoles with each other in the form

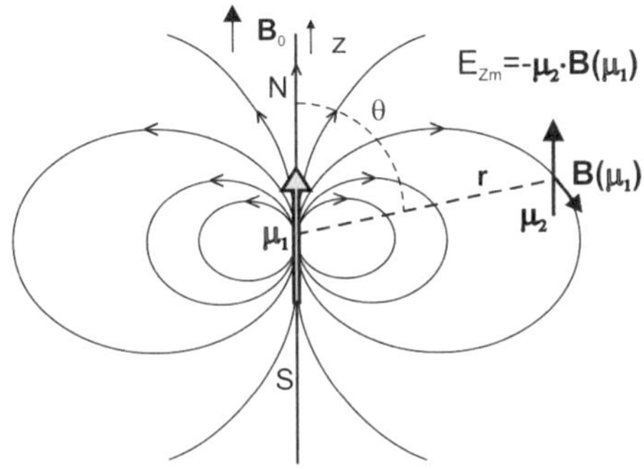

Fig. 1.4 External magnetic field produced by a magnetic dipole moment $\boldsymbol{\mu}_1$ and the position of the second magnetic dipole moment $\boldsymbol{\mu}_2$ in this field

$$E = \frac{1}{r^3}\left[\vec{\mu}_1 \cdot \vec{\mu}_2 - \frac{3(\vec{\mu}_1 \cdot \vec{r})(\vec{\mu}_2 \cdot \vec{r})}{r^2}\right]. \quad (1.38)$$

As shown below, similar relationships can be used for a quantum-mechanical description of the interaction of magnetic moments.

1.4 Populations of Energy Levels for Magnetic Moments in a Magnetic Field in Thermal Equilibrium

1.4.1 Magnetization of the Paramagnetic Materials, Magnetic Susceptibility

Let us consider a system of energy levels in a magnetic field. For simplicity, we consider a two-level system with levels a and b (see Fig. 1.5a), which is formed by placing a particle with an angular momentum 1/2 in a magnetic field. For definiteness, we again consider the spin $S = 1/2$ in a magnetic field (Fig. 1.5b). The spacing between the levels is given by $\Delta E = g_s \mu_B B$.

Consider the condition of thermal equilibrium. We denote the number of electrons in the upper energy level by N_b, and that in the lower energy level, by N_a. Then, in thermal equilibrium at temperature T, the population ratio of two states, provided by the *Boltzmann statistics*, will be:

$$N_b/N_a = \exp(-\Delta E/kT), \quad (1.39)$$

where k is the Boltzmann constant and T is the absolute temperature in Kelvin degrees. Rewrite this formula in the approximation $\Delta E \ll kT$, (which is usually the case in the EPR and NMR experiments in which the energy difference ΔE is small as compared with the average energy kT of thermal motion), i.e., $\Delta E/kT \ll 1$. Then

$$N_a/N_b = \exp(\Delta E/kT) \approx 1 + \Delta E/kT = 1 + (g_s \mu_B B)/kT. \quad (1.40)$$

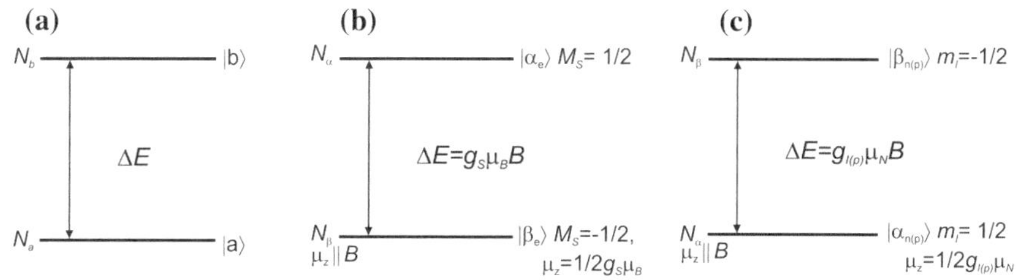

Fig. 1.5 **a** Two-level system for $S = 1/2$ ($J = 1/2$ or $I = 1/2$). **b** Two-level system for $S = ½$. **c** Two-level system for $I = ½$

We introduce a new notation that will simplify further calculations: the total number of N spins, with N_a spins in state a and N_b spins in state b, and the population difference n:

$$N = N_a + N_b \text{ and } n = N_a - N_b. \quad (1.41)$$

Then, we have

$$N_a = (N + n)/2 \text{ and } N_b = (N - n)/2. \quad (1.41a)$$

A combination of (1.40) and (1.41) yields

$$2n = (Ng_s\mu_B B)/kT + (ng_s\mu_B B)/kT.$$

Because $n \ll N$, only the first term can be left, with the expression for the population difference of two levels having the form

$$n = (Ng_s\mu_B B)/2kT. \quad (1.42)$$

In the case of the electron spin, the lower level is for the direction of the magnetic spin moment along the magnetic field, which corresponds to the projection of angular momentum in the opposite direction, $M_S = -1/2$, that is, the wave function $|\beta_e\rangle$. Since the projection of the magnetic moment along the magnetic field is proportional to the projection of the angular momentum, the magnetic moment is also quantized, as well as the angular momentum.

In a magnetic field, electron spin magnetic moments are oriented in accordance with the condition of thermodynamic equilibrium (1.42) with n electrons more oriented in the direction of the magnetic field, compared with that against the field. As a result, the system creates a magnetization M

$$M = \mu_z n, \quad (1.43)$$

where μ_z is a projection of the magnetic moment along the magnetic field for the lower energy level (i.e., along the field direction). The quantity μ_z is an eigenvalue of the projection of the spin magnetic moment. According to (1.8), $\boldsymbol{\mu}_S = -g_S\mu_B\boldsymbol{S}$ and, hence, we can write the projection of the magnetic moment operator as

$$\hat{\mu}_{Sz} = \hat{\mu}_z = -g_S\mu_B\hat{S}_z.$$

We apply this operator to the wave function $|\beta\rangle$ corresponding to the lower energy state

$$\hat{\mu}_z|\beta\rangle = -g_S\mu_B\hat{S}_z|\beta\rangle = \frac{1}{2}g_S\mu_B|\beta\rangle \text{ (we used } \hat{S}_z|\beta\rangle = -\frac{1}{2}|\beta\rangle).$$

As a result,

$$\mu_z = 1/2g_S\mu_B. \tag{1.44}$$

Thus, the magnetization M is given in accordance with (1.42) and (1.44) by

$$M = 1/2g_s\mu_B(Ng_s\mu_B B)/2kT,$$

and we finally obtain the expression for the magnetization in thermal equilibrium at temperature T in a magnetic field B (for the high-temperature approximation $g_s\mu_B B \ll kT$)

$$M = [(Ng_s^2\mu_B^2)/4kT]B. \tag{1.45}$$

The coefficient at the magnetic field B in (1.45) is named the static magnetic susceptibility and is denoted by χ

$$\chi = [(Ng_s^2\mu_B^2)/4kT] \text{ and } M = \chi B. \tag{1.46}$$

1.4.2 *Curie's Law*

It can be seen that, in the high-temperature approximation, the magnetic susceptibility of the paramagnetic materials is inversely proportional to temperature, which is known as Curie's law (French physicist Pierre Curie, 1895). In the absence of an external magnetic field, the magnetic moments $\boldsymbol{\mu}$ are randomly oriented. A magnetic field $\boldsymbol{B}$ orients the magnetic moments along the field, which is counteracted by the thermal motion of the particles. In strong magnetic fields and at low temperatures, the thermal motion does not disturb the orientation of the magnetic moments and the magnetization M tends to saturate, and Curie's law is inapplicable.

Magnetic resonance experiments measure the behavior of an ensemble of identical spins, typically 10^{10} electron spins and 10^{18} nuclear spins. The magnetization M is one of the important ensemble averages in the magnetic resonance, given by the equation Magnetization = Magnetic moment/Volume. For individual, uncoupled spins with fixed density, the magnetization can be easily calculated by using (1.46).

All the preceding expressions are derived for the magnetization in a two-level system, i.e., a system with angular momentum equal to 1/2. Curie's law is also applicable to the paramagnetism of nuclear magnetic moments. The nuclear spin of hydrogen (proton) is $I = 1/2$. Thus, the above (1.45) and (1.46) can be used with minor changes to describe the proton magnetic moment in a magnetic field (Fig. 1.5c).

Instead the Bohr magneton, the nuclear magneton can be written as μ_N (1.9), and the projection of the proton magnetic moment has a value $\mu_z = 1/2g_p\mu_N$ [see (1.44)]. In contrast to the case of an electron, the directions of the magnetic moment and angular momentum in the magnetic field coincide for the proton due to its positive charge. The expression for the magnetic susceptibility has the form

$$\chi_p = [(Ng_p^2\mu_N^2)/4kT] \text{ and } M = \chi_p B. \tag{1.47}$$

In the general case of an arbitrary angular momentum I, the expression for the magnetic susceptibility can be written as (without derivation of the formula)

$$\chi_I = [(Ng_I^2\mu_N^2)I(I+1)/3kT] \text{ and } M = \chi_I B. \tag{1.48}$$

For nuclear spins, the assumption of individual spins is valid to a very good approximation, even at very high densities. As an example, we calculate the susceptibility of water at room temperature, which has high density of protons with $I = 1/2$. From the density of water, we can calculate the density of protons to be of $6.7 \cdot 10^{28}$ m^{-3}. Using the proton magnetic moment $\mu_p = (h/2\pi)\gamma = 2.82 \cdot 10^{-26}$ J/T, we obtain $\chi(H_2O) = 4.1 \times 10^{-9}$. So the nuclear susceptibilities are very small compared to unity even at low temperatures!

For electrons, the susceptibilities are commonly substantially larger (by more than factor of 1000), due to the larger magnetic moment of the electron. It should be noted that the electron susceptibilities are usually more complicated to calculate because the electron spins are coupled at high electron concentrations by a strong exchange interaction and the assumption of individual spins is only valid for the low density of electron spins.

1.5 Magnetic Resonance Conditions

The magnetic resonance conditions can be satisfied by varying either the magnitude of the magnetic field or the frequency. It is more common to meet the resonance condition by varying the magnetic field strength while keeping the frequency fixed.

1.5.1 Probability of Transitions Between Levels for EPR (NMR)

To observe the EPR (NMR), is necessary to satisfy the following basic conditions.

1. Resonance condition.
2. A perturbation (electro-magnetic radiation field) is necessary that has non-zero matrix elements between the stationary states. Transitions from the lower to the upper energy level correspond to absorption of energy, and those in the reverse direction to emission of energy. Both transitions are associated with a spin orientation reversal and are equally probable.
3. The population difference of the energy levels should be non-zero. Due to the population excess in the lower level in thermal equilibrium, the absorption of the microwave energy from the irradiating field is the dominant process.

We consider a system that has two stationary energy levels a and b in a magnetic field (Zeeman levels) with wavefunctions $|a\rangle$ and $|b\rangle$ (Fig. 1.6a).

Let us consider the conditions under which transitions between energy levels are possible and those cases in which there will be EPR (NMR), i.e., a resonant absorption or emission of the microwave energy. The transitions between the stationary energy levels can be described by the nonstationary perturbation theory. Let the Hamiltonian of the system be represented by two terms

$$\hat{H} = \hat{H}_0 + \hat{H}'(t), \text{ where } \hat{H}'(t) \ll \hat{H}_0.$$

The time-independent part of the Hamiltonian, $\hat{H}_0$, substantially exceeds the time-dependent perturbation $\hat{H}'(t)$. We consider a system in which the function $\hat{H}'(t)$ can be represented as a product

$$\hat{H}'(t) = \hat{H}' \times f(t),$$

where $\hat{H}'$ is time-independent and the limited time dependence has the form of a periodic function $f(t) = a\cos(\omega t)$. Note that this is the form of perturbation considered below.

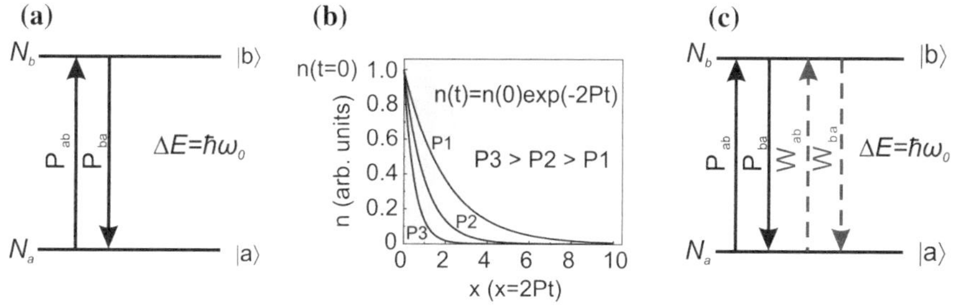

Fig. 1.6 **a** Two-level system for $S = 1/2$ ($I = 1/2$ or $J = 1/2$) with microwave transitions for absorption (P_{ab}) and emission (P_{ba}), $P_{ab} = P_{ba} = P$. **b** Dependence $n = n_0\exp(-2Pt)$. **c** Two-level system with microwave transitions (*solid line*) for absorption (P_{ab}) and emission (P_{ba}), $P_{ab} = P_{ba} = P$, and the spin-lattice relaxation transitions W_{ab} and W_{ba} (*dashed line*), $W_{ab} \neq W_{ba}$

Let there be a solution to Schrödinger's equation for the Hamiltonian in the absence of a perturbation $\hat{H}_0$

$$\hat{H}_0\Psi_n^0 = E_n^0\Psi_n^0$$

in the form of a set of eigenfunctions (wavefunctions) Ψ_n^0 and eigenvalues E_n^0 (we added '0' as a superscript to show that there is no perturbation). Then, at an instant of time $t = 0$, the electro-magnetic field is switched on and, at $t > 0$, the perturbation $\hat{H}'(t)$ reflects the interaction of the electro-magnetic field with a kind of electronic system, which will cause transitions between levels E_n^0 with an energy difference ΔE. Then, it is necessary to solve the time-dependent Schrödinger's equation

$$i\hbar\frac{\partial\Psi}{\partial t} = \hat{H}\Psi,$$

which determines how the function Ψ evolves with time. Instead of solving this problem exactly, one usually considers the term $\hat{H}'(t)$ a perturbation and solves the problem by using the time-dependent perturbation theory on the basis of the unperturbed eigenfunctions of the Hamiltonian $\hat{H}_0$ (it should be noted that, in many modern experiments, e.g., with the advent of lasers, $\hat{H}'(t)$ can no longer be considered a perturbation and the electric field in the light wave causing an electric-dipole transition may even be comparable with the intra-molecular field).

In accordance with the time-dependent perturbation theory, the probability of transitions between two adjacent levels n_1 and n_2 under the influence of a perturbation is constant in time and can be expressed as (given without proof)

$$P_{n_1,n_2} = \frac{2\pi}{\hbar^2}\left|\langle n_1|\hat{H}'|n_2\rangle\right|^2\delta(\omega - \omega_0), \tag{1.49}$$

where ω is the frequency of the electro-magnetic field, and $\omega_0 = \Delta E/\hbar$. The probability of transitions between the $|a\rangle$ and $|b\rangle$ levels (see Fig. 1.6a) is given by

$$P_{ab} = \frac{2\pi}{\hbar^2}\left|\langle a|\hat{H}'|b\rangle\right|^2\delta(\omega - \omega_0). \tag{1.49a}$$

Here, P_{ab} is the probability of the transition from level a to level b, equal to the number of transitions per second, its dimension being $[P_{ab}] = 1/\mathrm{s}$; $\left|\langle a|\hat{H}'|b\rangle\right|$ is the transition matrix element between levels; and $\delta(\omega - \omega_0)$ is the Dirac function having the form

$$\delta(\omega - \omega_0) = \infty \text{ if } \omega = \omega_0 \text{ and}$$
$$\delta(\omega - \omega_0) = 0 \text{ if } \omega \neq \omega_0,$$

whereas the integral over all the possible values of $\omega - \omega_0$, from $-\infty$ to ∞, is unity; and ω_0 is the resonant frequency in accordance with Planck's formula, (1.28)

$$\Delta E = \hbar\omega.$$

Here, ΔE is the splitting between the levels (see Fig. 1.13). Expression (1.49a) contains the first two conditions for observation of the EPR (NMR): (1) the transition matrix element must be non-zero: $\left|\langle a|\hat{H}'|b\rangle\right| \neq 0$; and (2) the resonance conditions must be satisfied because, in accordance with the definition of the Dirac δ-function, the frequency of the electro-magnetic field must be equal to the resonant frequency ω_0, determined by Planck's formula. That is the energy quantum $\hbar\omega$ must be equal to the energy difference ΔE between the levels, $\omega = \omega_0$. In the case of the system having the form of an electron spin in a magnetic field

$$\omega_0 = g_S\mu_B B/\hbar. \tag{1.27b}$$

Surely, the Dirac δ-function is a kind of idealization with a zero line width and infinite amplitude. In practice, the electro-magnetic radiation is not strictly monochromatic, but rather has a bandwidth; the amplitude and width of the lines are finite and can be written as a kind of a normalized form function $\Gamma(\omega)$ such as, e.g., the Lorentzian or Gaussian shape, or a combination thereof.

It is important to emphasize that, in accordance with (1.49a), the probabilities of up and down transitions are the same, i.e., $P_{ab} = P_{ba} = P$. Transitions between levels, caused by the interaction of the magnetic moment of the electron (nucleus) with an electro-magnetic field, are known as stimulated (in optics, the corresponding probabilities are named the Einstein coefficients).

Let us consider the interaction of electro-magnetic waves with the spin magnetic moment of the electron. The oscillating magnetic field generated by the magnetic component of an electro-magnetic wave affects the magnetic moment. Suppose that an oscillating magnetic field, designated $\boldsymbol{B}_1$ as opposed to the applied static magnetic field denoted by $\boldsymbol{B}_0$ (above, we only used a static magnetic field, so it was designated simply $\boldsymbol{B}$), has the form

$$\boldsymbol{B}_1(\mathrm{t}) = B_{1\mathrm{A}}2\cos(\omega t). \tag{1.50}$$

Here, $\boldsymbol{B}_{1\mathrm{A}}$ is the vector amplitude of the wave, and factor 2 is introduced for convenience in calculations. $\boldsymbol{B}_1(\mathrm{t})$ is the linearly polarized oscillating magnetic field that induces the resonance transitions and can be written as a sum of left- and right-rotating fields. The important point is that there will be only energy absorption by the field component that rotates in phase with the precession of $\boldsymbol{M}$ in the magnetic field. The total magnetic field acting on the spin magnetic moment is given by

$$\boldsymbol{B}_0 + \boldsymbol{B}_1(\mathrm{t}). \tag{1.50a}$$

As a result, the energy operator (Hamiltonian) of the magnetic moment in a magnetic field can be written, in accordance with the expression for the energy of the magnetic moment in a magnetic field (1.23), as

$$\hat{H} = \hat{H}_0 + \hat{H}'(t), \tag{1.51}$$

where

$$\hat{H}_0 = -\hat{\vec{\mu}}_S \cdot \vec{B}_0 = g_S\mu_B\hat{\vec{S}} \cdot \vec{B}_0 \tag{1.52}$$

and

$$H'(t) = -\hat{\vec{\mu}}_S \cdot \vec{B}_1(t) = g_S\mu_B\hat{\vec{S}} \cdot \vec{B}_1(t) = g_S\mu_B\hat{\vec{S}} \cdot \vec{B}_{1A}2\cos(\omega t). \tag{1.53}$$

The time-independent energy operator causes splitting of the energy levels of the spin magnetic moment in a static external magnetic field $\boldsymbol{B}_0$, (Zeeman splitting), whereas the time-dependent operator $\hat{H}'(t)$ must lead to transitions between these energy levels, which corresponds to a reorientation of the magnetic moment. As a rule, the amplitude of the oscillating magnetic field is substantially smaller than the strength of the static magnetic field, so that the condition $|\boldsymbol{B}_1(\mathrm{t})| \ll |\boldsymbol{B}_0|$ is usually satisfied. We now substitute $\hat{H}'(t) = \hat{H}' \times f(t)$ (where $H' = g_S\mu_B\hat{\vec{S}} \cdot \vec{B}_{1A}$ and $f(t) = 2\cos(\omega t)$) into the expression for the matrix element between the two levels $|a\rangle$ and $|b\rangle$ in (1.49a) and replace the corresponding wavefunctions with the projections of the electron spin $|\beta\rangle$ (lower level) and $|\alpha\rangle$ (upper level). Equation (1.49a) then becomes

$$P_{\beta\alpha} = \frac{2\pi}{\hbar^2}\left|\langle\beta|\hat{H}'|\alpha\rangle\right|^2\delta(\omega - \omega_0), \tag{1.54}$$

while the time-dependent part of (1.50) with a factor 2 is converted to a δ-function (without derivation of the formula), and only the amplitude of the oscillating magnetic field is present in the matrix element (in what follows, we simplify the notation for the amplitude by replacing $\boldsymbol{B}_{1\mathrm{A}}$ with $\boldsymbol{B}_1$). This leaves only the calculation of matrix elements of the type

$$\begin{aligned} &\left|\langle\beta|g_S\mu_B\hat{\vec{S}} \cdot \vec{B}_1|\alpha\rangle\right| \text{ up and} \\ &\left|\langle\alpha|g_S\mu_B\hat{\vec{S}} \cdot \vec{B}_1|\beta\rangle\right| \text{ down for the reverse transition.} \end{aligned} \tag{1.54a}$$

We direct the static magnetic field $\boldsymbol{B}_0$ along the z-axis, i.e., $\boldsymbol{B}_\mathbf{0} \parallel z$: $B_z = B_0$, $B_x = B_y = 0$, which does not lead to loss of generality by the problem. Then, the system gives rise to two spin energy levels with wavefunctions $|\alpha\rangle$ and $|\beta\rangle$, which are the eigenfunctions of the Hamiltonian

$$\hat{H}_0 = g_S \mu_B \hat{S}_Z B_0. \tag{1.23}$$

Now we add the resonance ($\omega = \omega_0$) oscillating magnetic field $\boldsymbol{B}_1(t)$ and calculate the matrix elements of the type (1.54). Because we have a preferred direction, the z-axis, associated with the direction of the $\boldsymbol{B}_0$, that of $\boldsymbol{B}_1$ should be chosen in a certain way with respect to the z, x and y axes. We are going to show that the transitions between levels in the system can be caused only by the component of the oscillating magnetic field that is directed perpendicular to the static magnetic field. Consider two cases.

(1) $\boldsymbol{B}_1 \parallel \boldsymbol{B}_0$, i.e., $\boldsymbol{B}_1$ is directed along the z-axis: $\boldsymbol{B}_1 \parallel z$, $B_{1z} = B_1$, $B_{1x} = B_{1y} = 0$. Then the expression for the matrix element takes the form (for definiteness, we consider the bottom-up transition) $\langle\beta|g_S \mu_B \hat{S}_Z B_1|\alpha\rangle$ or, factoring constant coefficients from the brackets, we have $g_S \mu_B B_1 \langle\beta|\hat{S}_Z|\alpha\rangle = 1/2 g_S \mu_B B_1 \langle\beta \mid \alpha\rangle = 0$ (the ratio $\hat{S}_Z|\alpha\rangle = 1/2|\alpha\rangle$ and the orthogonality of the wavefunctions $\langle\beta \mid \alpha\rangle = 0$ were used). It is also clear that the matrix element of the inverse transition is zero. Thus, the transition probability $P_{\beta\alpha} = P_{\alpha\beta} = 0$ if an oscillating magnetic field is directed along the static magnetic field, i.e., the first condition for the EPR is not satisfied.

(2) $\boldsymbol{B}_1 \perp \boldsymbol{B}_0$, i.e., $\boldsymbol{B}_1$ is directed perpendicularly to the static magnetic field (perpendicularly to the z axis): $\boldsymbol{B}_1 \perp z$, for definiteness, the oscillating magnetic field is directed along the x-axis, then $B_{1x} = B_1$, $B_{1z} = B_{1y} = 0$. The expression for the matrix element (1.54a) takes the form (for definiteness, we consider the bottom-up transition) $\langle\beta|g_S \mu_B \hat{S}_X B_1|\alpha\rangle$, or, factoring constant coefficients from the brackets, we obtain the expression $g_S \mu_B B_1 \langle\beta|\hat{S}_X|\alpha\rangle$. So far, we disregarded the situation in which the operator $\hat{S}_X$ (or $\hat{S}_Y$) acts on the eigenfunctions of $\hat{S}_Z$. In quantum mechanics, only certain values of $\vec{S}^2$ and $\hat{S}_Z$ (squared length of the angular momentum and one of its projections) can have a common set of eigenvectors (say, $\vec{S}^2$ and $\hat{S}_Z$ commute). That is, the eigenvalues $\vec{S}^2$ and $\hat{S}_Z$ can be written in this case, according to (1.15), as

$$\vec{S}^2|S, M_S\rangle = S(S+1)|S, M_S\rangle, \ \hat{S}_Z|S, M_S\rangle = M_S|S, M_S\rangle.$$

1.5.2 Step-up and Step-down Spin Operators

Let us consider the quantum-mechanical relations that can help us to calculate the matrix element for the second case, i.e., for $\langle\beta|\hat{S}_X|\alpha\rangle$ (or $\langle\beta|\hat{S}_Y|\alpha\rangle$). We write two operators as $\hat{S}_+$ and $\hat{S}_-$. They are named shift (or ladder) operators. These, very useful, operators are defined by

$$\hat{S}_+ = \hat{S}_X + i\hat{S}_Y,\ \hat{S}_- = \hat{S}_X - i\hat{S}_Y. \tag{1.55}$$

The theory gives the mode in which the shift operators $\hat{S}_+$ and $\hat{S}_-$ act upon the eigenfunctions $|M_S\rangle$. The following results are obtained:

$$\hat{S}_+|S, M_S\rangle = 1 \cdot |S, M_S + 1\rangle,\ \hat{S}_-|S, M_S\rangle = 1 \cdot |S, M_S - 1\rangle. \tag{1.56}$$

This is the reason for the term 'shift operator': $\hat{S}_+$ and $\hat{S}_-$ move one step up (step-up or raising operator $\hat{S}_+$), or down (step-down or lowering operator $\hat{S}_-$) the ladder of the eigenfunctions $|M_S\rangle$, i.e., they raise or reduce by unity the projection of the spin angular momentum. In other words, this leads to a reorientation of the angular momentum.

The inverse relations to (1.55), to be used in our calculation, are as follows:

$$\hat{S}_X = \frac{1}{2}(\hat{S}_+ + \hat{S}_-),\ \hat{S}_Y = \frac{1}{2i}(\hat{S}_+ - \hat{S}_-) \tag{1.57}$$

The factor 1 in (1.56) for $\hat{S}_+$ and $\hat{S}_-$ operators are implemented only for systems with the angular momentum of 1/2, e.g., for $S = 1/2$, $J = 1/2$ or $I = 1/2$

We do not prove this statement, but, in general case, the step-up and step-down operators are defined for a system with the angular momentum J and projection of the angular momentum M_J as

$$\hat{J}_+ = \hat{J}_X + i\hat{J}_Y,\ \hat{J}_- = \hat{J}_X - i\hat{J}_Y. \tag{1.58}$$

The shift operators $\hat{J}_+$ and $\hat{J}_-$ act upon the eigenfunctions $|J,M_J\rangle$ as follows

$$\begin{aligned} \hat{J}_+|J, M_J\rangle &= [J(J+1) - M_J(M_J+1)]^{1/2}|J, M_J+1\rangle, \\ \hat{J}_-|J, M_J\rangle &= [J(J+1) - M_J(M_J-1)]^{1/2}|J, M_J-1\rangle. \end{aligned} \tag{1.59}$$

In the case of a spin vector $|S,M_s\rangle$ for an electron, $S = 1/2$, while $M_S = 1/2$ or $-1/2$ are the only possible combinations. Just as mentioned above, it is convenient to use different labels for the spin states

$$|1/2, 1/2\rangle \equiv |\alpha\rangle \text{ and} |1/2, -1/2\rangle \equiv |\beta\rangle.$$

Each ladder has a top and a bottom beyond which it is impossible to go. This is true for the shift operators as well. For the wavefunctions, such as $|\alpha\rangle$ and $|\beta\rangle$ from (1.56), we obtain the relations

$$\hat{S}_{+}|\beta\rangle = |\alpha\rangle,\ \hat{S}_{-}|\alpha\rangle = |\beta\rangle,\ \hat{S}_{+}|\alpha\rangle = 0,\ \hat{S}_{-}|\beta\rangle = 0. \tag{1.60}$$

The last two equations mean that it is impossible to go off the top or the bottom.

Let us substitute $\hat{S}_X$ into the matrix element, using (1.57), (1.60) and the orthogonality of wavefunctions,

$$\langle\beta|\hat{S}_X|\alpha\rangle = \langle\beta|1/2(\hat{S}_{+} + \hat{S}_{-})|\alpha\rangle = 1/2\langle\beta \mid \beta\rangle = 1/2.$$

The result is

$$g_S\mu_B B_1\langle\beta|\hat{S}_X|\alpha\rangle = 1/2 g_S\mu_B B_1,$$

and, finally, the transition probability between the levels is given by

$$P_{\beta\alpha} = \frac{2\pi}{\hbar^2}(1/2 g_S\mu_B B_1)^2\delta(\omega - \omega_0) = \frac{\pi}{2\hbar^2} g_S^2\mu_B^2 B_1^2\delta(\omega - \omega_0),$$

where $P_{\beta\alpha} = P_{\alpha\beta} = P$. For real systems, the δ-function should be replaced with a distribution denoted as $\Gamma(\omega - \omega_0)$. Then,

$$P = \frac{\pi}{2\hbar^2} g_S^2\mu_B^2 B_1^2\Gamma(\omega - \omega_0). \tag{1.61}$$

1.5.3 Selection Rules

Thus, in the case of $\boldsymbol{B}_1 \perp \boldsymbol{B}_0$, the matrix element for the transition between the energy levels is non-zero, and, consequently, the transition probability for the resonance conditions is not zero. In this case, the M_S can only change by unity

$$\Delta M_S = \pm 1. \tag{1.62}$$

Relation (1.62) is named "selection rules". Physically, this means that microwave photons possess an intrinsic spin angular momentum of one unit of the angular momentum ($\hbar$). To conserve the angular momentum in a magnetic dipole transition with a photon is emitted (absorbed) from an atom, the atom must lose (gain) one angular momentum unit. It should be emphasized that all the above arguments are valid only under the condition that the static magnetic field splitting the levels is substantially stronger than the amplitude of the oscillating magnetic field that causes transitions between the levels. Otherwise, the wavefunctions of the types $|\alpha\rangle$ and $|\beta\rangle$ will not be the eigenfunctions of the total energy and the transitions between levels will be invoked as a parallel and perpendicular components of the oscillating magnetic field.

Frequently, the expression for the transition probability (1.61) is written in a different form by using the magnetogyric ratio γ. This ratio has the form

$$g_S^2\mu_B^2/\hbar^2 = \gamma^2.$$

Combining this ratio with (1.61), we obtain

$$P = (\pi/2)\gamma^2 B_1^2 \Gamma(\omega - \omega_0). \tag{1.61a}$$

Hence, the transition probability is the higher, the larger the magnetogyric ratio and the amplitude of the oscillating magnetic field.

The above analysis can be fully applied to NMR systems with $I = 1/2$ (e.g., protons). In this case, $g_S^2\mu_B^2$ is replaced in all formulas with $g_I^2\mu_N^2$, and all the operators with the electron spin $\boldsymbol{S}$ are replaced with the corresponding operators with the nuclear spin $\hat{\boldsymbol{I}}$. For the system constituted by a nuclear magnetic moment in a magnetic field, the NMR frequency is expressed as

$$\omega_0 = g_I\mu_N B_0/\hbar^2.$$

The Hamiltonian for the interaction of the nuclear magnetic moment with a static magnetic field $\boldsymbol{B}_0$ that splits the energy levels (Zeeman splitting) and an oscillating magnetic field $\boldsymbol{B}_1(t)$ ($|\boldsymbol{B}_1(t)| \ll |\boldsymbol{B}_0|$) that causes stimulated transitions between these levels can be written as

$$\hat{H}_0 = -\hat{\vec{\mu}}_I \cdot \vec{B}_0 = -g_I\mu_N\hat{\vec{I}} \cdot \vec{B}_0 \tag{1.52a}$$

and

$$H'(t) = -\hat{\vec{\mu}}_I \cdot \vec{B}_1(t) = -g_I\mu_N\hat{\vec{I}} \cdot \vec{B}_1(t) = -g_I\mu_N\hat{\vec{I}} \cdot \vec{B}_{1A}2\cos(\omega t), \tag{1.53a}$$

respectively. Let us substitute the operator of the time-dependent perturbation in the expression for the matrix element between the two levels $|a\rangle$ and $|b\rangle$ in (1.49), with the corresponding wavefunctions replaced with the projections of the nuclear spin $|\alpha_n\rangle$ (lower level) and $|\beta_n\rangle$ (upper level). Then, (1.61) is given by

$$P_{\alpha\beta} = \frac{2\pi}{\hbar^2}\left|\langle\alpha_n|g_I\mu_N\hat{\vec{I}} \cdot \vec{B}_{1A}|\beta_n\rangle\right|^2 \delta(\omega - \omega_0), \tag{1.61b}$$

The shift operators for the nuclear magnetic moment expressed in terms of operators $\hat{I}_x$ and $\hat{I}_y$ are defined as

$$\widehat{I}_{+} = \widehat{I}_{x} + i\widehat{I}_{y}, \widehat{I}_{+} = \widehat{I}_{x} - i\widehat{I}_{y}. \tag{1.63}$$

For nuclear magnetic moment as for the electron spin transitions are possible only for the case of $\boldsymbol{B}_1 \perp \boldsymbol{B_0}$. The shift operators $\hat{I}_+$ and $\hat{I}_+$ move one step up, or down the ladder of eigenfunctions $|m_I\rangle$, i.e., they raise or reduce by a single unit the projection of the nuclear angular momentum (reorient the angular momentum).

The results are as follows:

$$\widehat{I}_{+}|m_I\rangle = |m_I + 1\rangle, \widehat{I}_{-}|m_I\rangle = |m_I - 1\rangle, \tag{1.64}$$

or

$$\widehat{I}_{+}|\alpha_{\mathrm{n}}\rangle = 0, \widehat{I}_{+}|\beta_{\mathrm{n}}\rangle = |\alpha_{\mathrm{n}}\rangle, \widehat{I}_{-}|\alpha_{\mathrm{n}}\rangle = |\beta_{\mathrm{n}}\rangle, \widehat{I}_{-}|\beta_{\mathrm{n}}\rangle = 0. \tag{1.65}$$

Then, using $B_1 = B_{1\mathrm{A}}$, we have

$$P_{\alpha\beta} = \frac{\pi}{2\hbar^2} g_I^2 \mu_N^2 B_1^2 \delta(\omega - \omega_0), \tag{1.66}$$

where $P_{\alpha\beta} = P_{\beta\alpha} = P$.

For real systems

$$P = \frac{\pi}{2\hbar^2} g_I^2 \mu_N^2 B_1^2 \Gamma(\omega - \omega_0) = \frac{\pi}{2} \gamma_I^2 B_1^2 \Gamma(\omega - \omega_0). \tag{1.66a}$$

The selection rules for the NMR are

$$\Delta m_I = \pm 1. \tag{1.67}$$

1.5.4 Changing the Populations of Spin Levels by Resonant Microwave Field and Spin Relaxation; Absorption (Emission) of Electro-magnetic Energy in EPR (NMR) Experiments

Let us again consider a system of two energy levels (Fig. 1.6a). The rate at which the populations of energy levels vary under the influence of a perturbation that causes transitions between the levels is recorded in the form of the so-called 'rate equations'. In this system, the role of the perturbation is played by the interaction of the oscillating magnetic component of the EM field with the magnetic moment of the electron or nucleus, which unduces transitions between the Zeeman levels (frequently named spin sublevels) created by the static magnetic field.

Thus, changes in the population of the lower level can be written as the rate equation that determines the occupation of the two levels.

$$dN_a/dt = -N_a P_{ab} + N_b P_{ba} = -(N_a - N_b)P. \tag{1.68}$$

Taking that $P_{ab} = P_{ba} = P$ and substituting the expressions for n and N (1.41) instead of N_a and N_b, we obtain the equation for the time evolution of the population difference n

$$dn/dt = -2Pn. \tag{1.69}$$

The solution to (1.69) has the form

$$n(t) = n_0 \exp(-2Pt), \tag{1.70}$$

where n_0 is the population difference at the initial instant of time $t = 0$ (this may be the difference of the populations in thermal equilibrium).

The solution is the population difference $n(t)$ exponentially decaying with a time constant $\tau = 1/(2P)$. It is evident that, under the influence of the perturbations, the population difference between levels of n, initially equal to n_0, exponentially tends to zero, the rate of the decrease being the faster, the higher the transition probability P.

We write the expression for the EM energy absorption rate dE/dt (microwave for the EPR, or radio-frequency for the NMR). Because each photon has the energy

$$\Delta E = \hbar\omega,$$

the energy absorption by the spin system is given by

$$dE/dt = N_a P_{ab} \Delta E - N_b P_{ba} \Delta E = P\Delta E(N_a - N_b) = P\Delta E n. \tag{1.71}$$

Using (1.70), we obtain

$$dE/dt = P\Delta E n_0 \exp(-2Pt). \tag{1.71a}$$

It follows from (1.71a) that the EM energy absorption, proportional to the population difference n between the levels, also tends to zero similarly to the dependence in Fig. 1.6b. The absorbed power decays in the course of time with a time constant $\tau = 1/(2P)$. The EM energy absorption and, hence, the resonance (EPR or NMR) are only recorded when there is a population difference, and, therefore, it is also proportional to the total number of spins N (concentration in a sample). Under the influence of a perturbation, the populations of the levels become equal ($N_a = N_b$), and, as a result, the absorption and emission processes cancel each other out and no signal is observed. In this situation, named "saturation", the system becomes transparent to the microwave (in EPR) or radio-frequency (in NMR) radiation. There is no permanent power absorption!

The fact that the EPR and NMR can be observed and the population ratio in thermal equilibrium suggest that there must be more processes leading to transitions between two levels than only the interaction with the oscillating B_1-field. All these interaction processes are referred to as a coupling between the spin system and the environment, or "lattice". This mechanism of interaction between the spins (electronic or nuclear) and the surrounding environment (e.g., lattice vibrations) is named the spin-lattice interaction, and the relaxation is named the spin-lattice relaxation (term used for all systems, including non-crystalline materials). The heat capacity of the lattice is assumed to be infinitely large, compared to the spin-system capacity.

Let us consider a two-level system shown in Fig. 1.6c and denote in this figure the stimulated transitions due to the interaction with the electro-magnetic field by solid lines, and the transitions due to spin-lattice relaxation, by dashed lines. For stimulated transitions, the probabilities of the bottom-up and top-down transitions are the same, i.e., $P_{ab} = P_{ba}$. For the transitions caused by the relaxation, the probabilities (denoted by W) of the bottom-up and top-down transitions are not equal, $W_{ab} \neq W_{ba}$. W_{ab} corresponds to transitions from the lower to the upper energy level; W_{ba} corresponds to transitions in the reverse direction.

We now write the rate equations for the rate of change in the population of level a (N_a) under relaxation in the absence of an external electro-magnetic field This is the rate equation without transitions caused by the external alternating field $B_1(t)$.

$$\mathrm{d}N_a/\mathrm{d}t = -N_a \mathrm{W}_{\mathrm{ab}} + N_b \mathrm{W}_{\mathrm{ba}}. \tag{1.72}$$

Let us find an expression that relates W_{ab} and W_{ba}. In the steady state, $\mathrm{d}N_a/\mathrm{d}t = 0$ because the level populations do not change. In thermal equilibrium and in the steady state, we introduce the designations N_a^0 and N_b^0 and then (1.72) is written as

$$0 = -N_a^0 \mathrm{W}_{\mathrm{ab}} + N_b^0 \mathrm{W}_{\mathrm{ba}} \text{ or } N_a^0 \mathrm{W}_{\mathrm{ab}} = N_b^0 \mathrm{W}_{\mathrm{ba}}. \tag{1.73}$$

There exist generally applicable principles of thermodynamics. Using the Boltzmann ratio, we obtain

$$\mathrm{W}_{\mathrm{ab}}/\mathrm{W}_{\mathrm{ba}} = N_b^0/N_a^0 = \exp(-\Delta E/kT), \text{ and finally}$$

$$\mathrm{W}_{\mathrm{ab}}/\mathrm{W}_{\mathrm{ba}} = \exp(-\Delta E/kT). \tag{1.74}$$

Thus, we derived a fundamental relation indicating that, for the two-level system, the probability of bottom-up transitions is lower than that of top-down transitions: $W_{ab} < W_{ba}$. This circumstance determines the level populations in thermal equilibrium: the lower level is more populated than the upper one, and W_{ab} and W_{ba} depend on ΔE and temperature T. The population difference in thermal equilibrium can be expressed in terms of transition probabilities if we substitute the relations for N and n in (1.73). The steady state solution in thermal equilibrium is given by

$$[(N+n)/2]\mathrm{W}_{\mathrm{ab}} = [(N-n)/2]\mathrm{W}_{\mathrm{ba}}, \text{ or}$$

$$n = n_0 = N(\mathrm{W}_{\mathrm{ba}} - \mathrm{W}_{\mathrm{ab}})/(\mathrm{W}_{\mathrm{ba}} + \mathrm{W}_{\mathrm{ab}})$$

(hence follows that the population difference $n = 0$ at $\mathrm{W}_{\mathrm{ba}} = \mathrm{W}_{\mathrm{ab}}$).

We substitute relations for N and n in (1.72) and, after grouping, we obtain the expression

$$\mathrm{d}n/\mathrm{d}t = N(\mathrm{W}_{\mathrm{ba}} - \mathrm{W}_{\mathrm{ab}}) - n(\mathrm{W}_{\mathrm{ba}} + \mathrm{W}_{\mathrm{ab}}).$$

If we artificially multiply the right-hand side in the above expression by $(\mathrm{W}_{\mathrm{ba}} + \mathrm{W}_{\mathrm{ab}})/(\mathrm{W}_{\mathrm{ba}} + \mathrm{W}_{\mathrm{ab}})$ equal to unity (i.e., making no changes), the result is

$$\mathrm{d}n/\mathrm{d}t = [N(\mathrm{W}_{\mathrm{ba}} - \mathrm{W}_{\mathrm{ab}})/(\mathrm{W}_{\mathrm{ba}} + \mathrm{W}_{\mathrm{ab}})](\mathrm{W}_{\mathrm{ba}} + \mathrm{W}_{\mathrm{ab}}) - n(\mathrm{W}_{\mathrm{ba}} + \mathrm{W}_{\mathrm{ab}}).$$

The term in square brackets is equal to n_0. We introduce the notation

$$(\mathrm{W}_{\mathrm{ba}} + \mathrm{W}_{\mathrm{ab}}) = 1/T_1,$$

as the spin-lattice relaxation rate, where T_1 has dimension of time and is named the spin-lattice relaxation time. Upon the appropriate substitution, we have

$$\mathrm{d}n/\mathrm{d}t = -(n - n_0)/T_1. \tag{1.75}$$

Combining the effects of the microwave radiation (1.70) and the relaxation (1.75) on the level populations, we obtain

$$\mathrm{d}n/\mathrm{d}t = -2\mathrm{P}n - (n - n_0)/T_1. \tag{1.76}$$

Here, the first term corresponds to the microwave-induced transitions and the second represents the spin-lattice relaxation.

Under the action of these two effects in equilibrium, i.e., at $\mathrm{d}n/\mathrm{d}t = 0$, (1.76) gives a formula for the population difference between two levels in the form

$$n = n_0/(1 + 2\mathrm{P}T_1). \tag{1.77}$$

Thus, due to relaxation processes, the population difference can be maintained at a certain level, there will be no saturation (alignment of the level populations), and, therefore, the EPR (NMR) can be detected. To avoid saturation, it is necessary to reduce the product $\mathrm{P}T_1$ ($2\mathrm{P}T_1 \ll 1$), i.e., to make the EM radiation power (B_1 field) lower or the spin-lattice relaxation time T_1 shorter. At $2PT_1 \ll 1$, $n \to n_0$.

Formula (1.77) expresses the absorption of EM energy (1.71) in terms of relaxation

$$dE/dt = Pn\Delta E = n_0 \Delta E P/(1 + 2PT_1). \tag{1.78}$$

We now use the expression for the transition probabilities $P = (\pi/2)\gamma^2 B_1^2 \Gamma(\omega - \omega_0)$, which gives

$$dE/dt = n_0 \Delta E (\pi/2)\gamma^2 B_1^2 \Gamma(\omega - \omega_0)/[1 + \pi\gamma^2 B_1^2 \Gamma(\omega - \omega_0) T_1]. \tag{1.79}$$

1.6 Bloch Equations

1.6.1 Classical Behavior of the Magnetic Moment in a Magnetic Field

In the preceding sections, we concentrated on individual particles. In actual practice, experiments are performed with macroscopic samples, i.e., with an ensemble of particles. Let us consider the magnetic resonance from a macroscopic point of view by analyzing the total magnetic moment (or magnetization) $\boldsymbol{M}$ of a certain ensemble of magnetic moments (such as electron or nuclear spins in a volume of 1 cm^3) at a certain temperature. The EPR experiments are usually made with a large number of spins, at least 10^8–10^9 spins (for NMR, about 10^{18} or more nuclear spins). In the absence of an external magnetic field, there is no physical difference between the projections of the magnetic field M_X, M_Y and M_Z. In an external magnetic field, which is taken to be aligned with the Z-axis, the magnetic moments are oriented, and it was already shown (1.43) that the projection of the total magnetic moment on the Z-direction (magnetization M_Z) is given by $M_Z = \mu_z n$ (for uncoupled spins). According to (1.75), the rate at which the population difference n varies can be written as $dn/dt = -(n - n_0)/T_1$. Multiplying both sides of this equation by the projection of the magnetic moment of a single spin along the Z-axis, we obtain an expression for changing the projection of the total magnetic moment on the direction of the static magnetic field, which coincides with the Z-axis

$$dM_Z/dt = -(M_Z - M_0)/T_1, \tag{1.80}$$

Here, M_0 is the magnetization in a static magnetic field $\boldsymbol{B}_0$ in thermal equilibrium, which is written, in accordance with (1.46), in the form $M_0 = \chi_0 B_0$.

Equation (1.80) shows that, in the magnetic field, the projection of the total magnetic moment M_Z tends to the equilibrium value M_0 with a relaxation time T_1. When you turn off the magnetic field, the projection of the total magnetic moment M_z tends to the equilibrium value $M_Z = 0$ with a relaxation time T_1 in accordance with the equation

$$\mathrm{d}M_Z/\mathrm{d}t = -M_Z/T_1. \tag{1.81}$$

In the absence of a magnetic field, there is no physical difference between the *Z*, *X* and *Y* directions due to the isotropy of space, so the components M_X and M_Y are changing similarly to M_Z, i.e., they tend to zero.

While the transverse components M_X and M_Y still tend to zero in the general case with a magnetic field directed along the *Z*-axis, the corresponding relaxation time, we name it the transverse relaxation time T_2, is different from the spin-lattice relaxation time T_1, which is frequently named the longitudinal relaxation time. As a result, the equation for the time variation of the components M_X and M_Y can be written as

$$\begin{aligned}\mathrm{d}M_X/\mathrm{d}t &= -M_X/T_2,\\ \mathrm{d}M_Y/\mathrm{d}t &= -M_Y/T_2.\end{aligned} \tag{1.82}$$

Typically, T_1 and T_2 are not equal because of being caused by different processes. The change in the longitudinal magnetization is due to the energy exchange between the spin system and the surrounding environment (e.g., the crystal lattice). The change in the transverse components of magnetization is not related to the energy exchange between the spin system and the environment. Because the energy of interaction of the magnetic moment and the magnetic field is expressed as the inner product of these vectors, the interaction energy in their mutually perpendicular directions is zero.

According to classical mechanics, the equation of motion of the angular momentum $\boldsymbol{G}$ under the influence of the torque $\boldsymbol{F}_\mathrm{M}$ can be written as

$$\frac{d\vec{G}}{dt} = \vec{F}_M.$$

When a magnetic moment $\boldsymbol{M}$ is placed in a magnetic field $\boldsymbol{B}$, it experiences a torque that can be expressed in the form of a vector product. The torque $\boldsymbol{F}_\mathrm{M}$ for the magnetic moment $\boldsymbol{M}$ associated with the angular momentum $\boldsymbol{G}$ in a magnetic field is given by

$$\vec{F}_M = \gamma\vec{G} \times \vec{B} = \vec{M} \times \vec{B},$$

and the equation of motion of the angular momentum becomes

$$\frac{d\vec{G}}{dt} = \vec{M} \times \vec{B}.$$

Multiplying both sides of this equation by the magnetogyric ratio γ, we obtain the final equation of motion of the magnetic moment $\boldsymbol{M}$

$$\frac{d\vec{M}}{dt} = \gamma(\vec{M} \times \vec{B}). \tag{1.83}$$

These are the equations of motion for the magnetization $\boldsymbol{M}$ without dissipation. This equation can be written in terms of projections as (an external magnetic field is taken to be aligned with the Z-axis)

$$\begin{aligned} \frac{dM_X}{dt} &= \gamma(M_Y B_Z - M_Z B_Y) \Rightarrow \frac{dM_X}{dt} = \gamma M_Y B_0 \Rightarrow \frac{dM_X}{dt} = \omega_0 M_Y \\ \frac{dM_Y}{dt} &= \gamma(M_Z B_X - M_X B_Z) \Rightarrow \frac{dM_Y}{dt} = -\gamma M_X B_0 \Rightarrow \frac{dM_Y}{dt} = -\omega_0 M_X \, . \\ \frac{dM_Z}{dt} &= \gamma(M_X B_Y - M_Y B_X) \Rightarrow \frac{dM_Z}{dt} = 0 \end{aligned} \tag{1.83a}$$

Here, the designation $\omega_0 = \gamma B_0$ is introduced. The torque produces a change in the angular momentum that is perpendicular to the angular momentum, which results in the precession of the magnetic moment $\boldsymbol{M}$ around the direction of the magnetic field (Z) rather than in its settling along the magnetic field (Fig. 1.7a). Equation (1.83) describes the precession of the magnetic moment about the Z-axis at a frequency ω_0. These precession and frequency are named the Larmor precession and the Larmor frequency (after English physicist Joseph Larmor, 1895). This solution is analogous to that for a gyroscope (top) in a gravitational field.

The solution to these equations under the conditions in which the static magnetic field is directed along the Z-axis, i.e., $B_z = B_0$, $B_x = B_y = 0$, has the form

$$\begin{aligned} M_X &= M_\perp \cos \omega_0 t \\ M_Y &= -M_\perp \sin \omega_0 t \\ M_Z &= M_\parallel = \text{const}, \end{aligned} \tag{1.84}$$

where the transverse magnetization

$$M_\perp = \left[M_X^2 + M_Y^2\right]^{1/2}.$$

The rather complicated motion of $\boldsymbol{M}$, both during the application of a microwave field and in the subsequent relaxation, was analyzed by F. Bloch (Swiss physicist Felix Bloch, 1946).

It can be shown that the above classical description of the magnetic moment in a magnetic field can be made in accordance with the quantum mechanical consideration of a two-level system, i.e., a system with an angular momentum equal to 1/2. For definiteness, we consider the magnetic moment of the electron spin in a magnetic field. In accordance with Planck's formula for quantum-mechanical description, $\omega_0 = g_S \mu_B B_0/\hbar = \gamma B_0$ because $\gamma = g_S \mu_B/\hbar$. Thus, the frequency of the

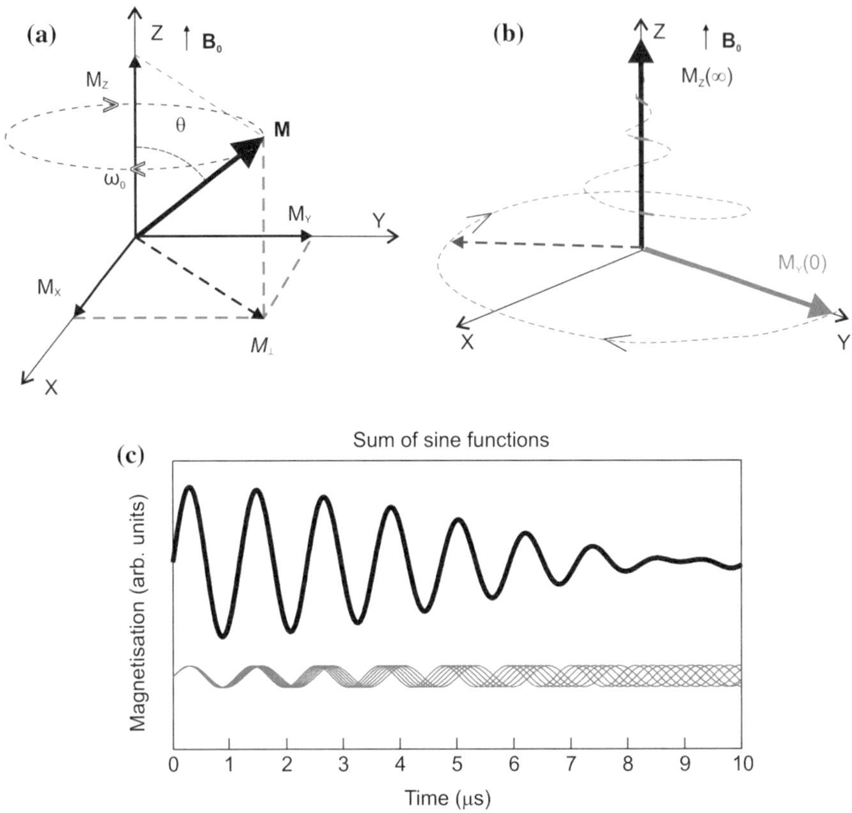

Fig. 1.7 **a** After being turned from its equilibrium orientation through an angle θ (e.g., by applying a microwave pulse), the macroscopic magnetization vector ***M*** precesses, like individual spins, about the direction of the magneic field (*Z*-axis) at the Larmor frequency ω_0. The vector ***M*** has the coordinates M_X, M_Y, and M_Z in the stationary coordinate system (*X*, *Y*, *Z*) at the instant of time t. The perpendicular component of the magnetic moment $M_\perp$ rotates with an angular velocity $\omega_0 = \gamma B_0$ in the *X*-*Y* plane. **b** The trajectory of the tip of the magnetization vector, which shows the combined regrowth of the longitudinal magnetization M_Z and the decay of the transverse magnetization components M_Y and M_X. The initial value was along the Y axis, M_Y (0), and the final value is along the Z axis, $M_Z(\infty)$; the reference frame is the laboratory. **c** Resultant magnetization, which is the sum of all sine functions, decays as described by the Bloch equations with the relaxation constant T_2

Larmor precession in the classical description of motion of the magnetic moment in the magnetic field coincides with the frequency of the resonant transition between the levels of the two-level system. This result is due to the fact that the quantum-mechanical expectation value for the magnetic moment of the magnetic dipole obeys the same equation of motion as that for a classical magnetic moment.

1.6.2 *Bloch Equations. Two Spin Relaxation Times Introduced for Longitudinal (T1) and Transverse (T2) Spin Relaxation*

The Bloch equations are a set of macroscopic equations used to calculate the magnetization ***M*** as a function of time when there are a longitudinal relaxation time T_1 and a transverse relaxation time T_2. These are phenomenological equations of motion for magnetization. The equations introduced by Bloch to interpret first magnetic resonance experiments used a combination of two different approaches. The interaction of the spin system with the applied magnetic field was described in terms of classical mechanics, whereas the decay to the magnetization in thermal equilibrium is accounted for by purely phenomenological relaxation terms. In addition to the equations of motion for the magnetization ***M*** without dissipation, (1.83), two relaxation rates $1/T_1$ and $1/T_2$ were introduced by Bloch in a phenomenological way, with T_1 named the longitudinal relaxation time and T_2 named the transverse relaxation time. The introduction of these terms associated with relaxation suggests that the magnetization ***M*** returns to the thermal equilibrium state by exponential decays. The great success in using the Bloch equations is due to the fact that the spin systems are usually rather isolated from their surrounding environment and the precession of the magnetization ***M*** about the applied magnetic field is much faster that the changes of ***M*** caused by interactions with the surrounding environment.

Addition to (1.83) of the terms describing the relaxation processes in accordance with expressions (1.80)–(1.82) leads to the equations named the Bloch equations

$$\begin{aligned}\frac{dM_X}{dt} &= \omega_0 M_Y - \frac{M_X}{T_2}\\ \frac{dM_Y}{dt} &= -\omega_0 M_X - \frac{M_Y}{T_2}\\ \frac{dM_Z}{dt} &= -\frac{M_Z - M_0}{T_1}\end{aligned}\quad . \qquad (1.85)$$

The Bloch equations describe in the classical form the behavior of the magnetic moment in a magnetic field in the presence of a relaxation. The static-magnetic-field solutions to the Bloch equations at $B \parallel Z$ ($\boldsymbol{B} = B_0\mathbf{k}$) are given by

$$\begin{aligned}M_X(t) &= e^{-t/T_2}[M_X(0)\cos\omega_0 t + M_Y(0)\sin\omega_0 t]\\ M_Y(t) &= e^{-t/T_2}[M_Y(0)\cos\omega_0 t + M_X(0)\sin\omega_0 t]\\ M_Z(t) &= M_Z(0)e^{-t/T_1} + M_0(1 - e^{-t/T_1})\end{aligned}\quad . \qquad (1.86)$$

The equilibrium or steady-state solutions are found by letting $t \to \infty$. In practice, the steady-state solutions to the Bloch equations are obtained when one ‘waits long enough', i.e., on a time scale several times longer than the relaxation times T_1 and T_2.

$$M_X(\infty) = M_Y(\infty) = 0,\ M_Z(\infty) = M_0. \tag{1.86a}$$

Figure 1.7b shows the evolution of the magnetization according to solution for Bloch equations in static magnetic field $\boldsymbol{B}$ along the Z axis according to (1.86) and (1.86a). In thermal equilibrium, the magnetization $\boldsymbol{M}$ will tend to align with the static magnetic field $\boldsymbol{B}$. According to (1.46), $\boldsymbol{M} = \chi\boldsymbol{B}$, where χ is the static magnetic susceptibility, $\chi = (Ng_s^2\mu_B^2)/4kT$. Thus, the decay of the transverse and longitudinal magnetization components is exponential; the decay of the Z component, on the one hand, and that of X, Y components, on the other hand, are described by different time constants T_1 and T_2. The relaxation is caused by interactions of the spin system with the surrounding environment and by mutual interactions of the spins. As mentioned earlier, there is a fundamental difference between the two relaxation mechanisms. In contrast to the longitudinal decay, the transverse decay conserves energy in the static magnetic field. When the perpendicular magnetization $M_\perp$ relaxes to zero, the energy of the spin system does not change. This relaxation is the result of interactions that change the resonance frequency of individual spins, e.g., the dipole-dipole interactions of the spins.

In general, when there are a static $\boldsymbol{B}_0$ and an oscillating $\boldsymbol{B}_1$ magnetic fields, where $\boldsymbol{B}_1 = B_1(\boldsymbol{i}\cos\omega t - \boldsymbol{j}\sin\omega t)$, the Bloch equations are written in the vector form as follows:

$$\frac{d\vec{M}}{dt} = \gamma(\vec{M} \times \vec{B}_0) + \gamma(\vec{M} \times \vec{B}_1) - \frac{\vec{i}M_X + \vec{j}M_Y}{T_2} - \frac{\vec{k}(M_Z - M_0)}{T_1}. \tag{1.87}$$

The resonance conditions are satisfied for $\omega = \omega_0$, whereas the expression for $\omega_0 = \gamma B_0$ coincides with the above quantum-mechanical expression for a two-level system.

If, instead of the stationary coordinate system X, Y, Z, we use a rotating coordinate system x, y, z that rotates at the Larmor frequency, the Bloch equations become much simpler because of no longer including the precession about the Z-axis. By defining the quantities ω_0, ω_1 and $\Delta\omega$, we can present the Bloch equations in a convenient form:

$\omega_0 = \gamma B_0$ is the Larmor frequency,
$\omega_1 = \gamma B_1$ is the Rabi frequency,
$\Delta\omega = \omega - \omega_0$ is the difference between the Larmor frequency and the rotation frequency of the reference frame.

Bloch equations in the rotating frame (the label "$\cap$" shows that one is in the rotating frame) of reference have the form:

$$\frac{d\widehat{M}_X}{dt} = -\frac{\widehat{M}_X}{T_2} + \Delta\omega\widehat{M}_Y$$
$$\frac{d\widehat{M}_Y}{dt} = -\frac{\widehat{M}_Y}{T_2} - \Delta\omega\widehat{M}_X - \omega_1\widehat{M}_Z \, , \tag{1.88}$$
$$\frac{d\widehat{M}_Z}{dt} = -\frac{\widehat{M}_Z - M_0}{T_1} + \omega_1\widehat{M}_Y$$

where M_0 is the magnetization in thermal equilibrium in the absence of microwave transitions, and $\boldsymbol{B}_1 \parallel X$.

The steady-state solutions to the Bloch equations (1.88) in the rotating frame of reference give the frequency response of the magnetization.

$$\frac{d\widehat{M}_X}{dt} = 0, \frac{d\widehat{M}_Y}{dt} = 0, \frac{d\widehat{M}_Z}{dt} = 0$$

$$\widehat{M}_X = \frac{\Delta\omega\omega_1 T_2^2}{1 + (\Delta\omega T_2)^2 + \omega_1^2 T_1 T_2} M_0$$
$$\widehat{M}_Y = \frac{\omega_1 T_2}{1 + (\Delta\omega T_2)^2 + \omega_1^2 T_1 T_2} M_0 \, . \tag{1.89}$$
$$\widehat{M}_Z = \frac{1 + (\Delta\omega T_2)^2}{1 + (\Delta\omega T_2)^2 + \omega_1^2 T_1 T_2} M_0$$

Note that these expressions apply to the rotating frame. A transformation back to the laboratory frame yields

$$M_X(t) = \widehat{M}_X \cos(\omega t) + \widehat{M}_Y \sin(\omega t)$$
$$M_Y(t) = -\widehat{M}_X \sin(\omega t) + \widehat{M}_Y \cos(\omega t) \, \cdot$$
$$M_Z(t) = \widehat{M}_Z$$

The power absorbed by the spin system from the microwave field is given by

$$\mathrm{d}E/\mathrm{d}t = \mathrm{d}/\mathrm{d}t(-\boldsymbol{M}\cdot\boldsymbol{B}) = -\boldsymbol{M}\cdot\mathrm{d}\boldsymbol{B}/\mathrm{d}t - \mathrm{d}\boldsymbol{M}/\mathrm{d}t\cdot\boldsymbol{B} = -\boldsymbol{M}\cdot\mathrm{d}\boldsymbol{B}/\mathrm{d}t$$

(The term $-\mathrm{d}\boldsymbol{M}/\mathrm{dt}\cdot\boldsymbol{B}$ is zero because the variation of $\boldsymbol{M}$ is always perpendicular to $\boldsymbol{B}$)

The solution to the Bloch equations provides a macroscopic description of the magnetic resonance absorption (EPR, NMR) and gives an absorption line $\Gamma(\omega - \omega_0)$ that is the so-called Lorentz lineshape. The Lorentz lineshape is obtained when the frequency of the applied microwave field slowly passes across the resonance condition. The important aspect is that the spin system is in thermal equilibrium at any instant of time during the passage. The width of this line in the small-microwave-power limit, $\omega_1^2 T_1 T_2 \ll 1$ (expressed in angular frequency) at half intensity $\Delta\Gamma = 2T_2^{-1}$, is the so-called homogeneneous linewidth. In the Bloch equation, T_2 is a measure of the decay rates of M_X and M_Y; in addition, T_2 also governs the linewidth of the resonant absorption curve, i.e., it determines the magnetic resonance damping.

In continuous-wave (cw) magnetic resonance spectroscopy, the detected signal is proportional to the perpendicular components of the magnetization (M_X and M_Y). The solution yields an expression for the absorption of electro-magnetic energy in the form that is similar to that obtained above by quantum-mechanical description of (1.79) with addition of the Lorentz lineshape.

$$\begin{aligned} dE/dt &= n_0 \Delta E(\pi/2)\gamma^2 B_1^2 \Gamma(\omega - \omega_0)/[1 + \pi\gamma^2 B_1^2 T_1 \Gamma(\omega - \omega_0)] \\ \Gamma(\omega - \omega_0) &= (T_2/\pi)[1/(1 + T_2^2(\omega - \omega_0)^2] \end{aligned} \tag{1.90}$$

is the Lorentz lineshape.

In practice, the linewidth is almost always larger (even at a low microwave power) than the homogeneneous linewidth and is commonly given by a Gaussian lineshape, rather than by that of the Lorentz type. This inhomogeneous linewidth is due to several factors.

The effect of power broadening is observed as a broadening of the absorption line with increasing microwave power (B_1 amplitude). In this case, M_Z deviates appreciably from M_0 and the longitudinal relaxation time T_1 appears in the expression for the linewidth.

Each spin experiences a "local" magnetic field from its neighbours, e.g., a proton magnetic moment induces a magnetic field of 0.1 mT at a distance of 0.2 nm. A random precession of different spins in this magnetic field will lead to a transverse spin relaxation with a time T_2 on the order of $1/\gamma B_{loc}$.

Figure 1.7c shows how the transverse spin relaxation or the dephasing works. The transverse relaxation (or dephasing) works as follows: transverse spin components of different electrons precess at different frequencies according to the sin (ωt) law. The resultant magnetization (the sum of all sine functions) rapidly decays as described by the Bloch equations with the relaxation constant T_2.

There is an additional magnetization dephasing introduced by external field inhomogeneities, and also by inhomogeneities of the spin ensemble. This reduction in the initial decay of the magnetization can be characterised by a separate decay time T_2'. Thus, the total decay rate will be given by:

$$1/T_2^* = 1/T_2 + 1/T_2'$$

It should be noted that the decay due to the field or ensemble inhomogeneities is reversible (phase relationship between spins is recoverable) in "spin-echo" experiments. The decay due to T_2 is irreversible.

To detect the magnetic resonance, the law of electro-magnetic induction (Faraday's law) can be used. According to Faraday's law, the electromotive force induced in any closed circuit is equal to the rate at which the magnetic flux through the circuit varies with time. Once the magnetization has a transverse component an electromotive force will be created in a coil, a consequence of Faraday's law.

The so-called free induction decay (FID) technique provides the simplest way to detect the magnetic resonance by using a coil in which the varying magnetic flux will produce an electromotive force. The way in which the FID (or a primary spin echo) signal is generated can be understood as follows.

Let us assume that the net magnetization vector of a large number of electron spins is initially oriented along the static external magnetic field $\boldsymbol{B}_0$ aligned with the z-axis. The electron spins are known to be characterized by two quantum-mechanical states, one with its magnetic moment parallel to $\boldsymbol{B}_0$, which has a lower energy, and the other antiparallel. In thermal equilibrium, there should be, according to the Boltzmann distribution, a net magnetization parallel to the z-axis. The electron spins are still precessing about the z-axis; however, their orientations are random in the x-y plane because there is no reason to prefer one direction to another. For a very large number of electron spins, the magnetic moment components in the x-y plane cancel each other out and the result is a stationary magnetization $\boldsymbol{M}_0$ aligned with $\boldsymbol{B}_0$.

The $\boldsymbol{B}_1$ vector is assumed to be oriented in the perpendicular plane, e.g., B_1 is parallel to the x-axis. There will be two rotations. The magnetic field will rotate the magnetization about the x-axis as long as microwaves are applied.

The angle by which the magnetization $\boldsymbol{M}$ is rotated, or the so-called tip angle, is equal to $\theta = \gamma B_1 \tau$, where τ is the pulse width. The tip angle is dependent on both the magnitude of B_1 and the pulse width. The $\pi/2$ pulse corresponds to a rotation of the magnetization $\boldsymbol{M}$ by $\pi/2$. The $\pi/2$ pulse results in magnetization along the y-axis (the rotating frame) rotating in the x-y plane at the Larmor frequency (the lab frame).

Let us describe the evolution of electron (nuclear) spins after a $\pi/2$-pulse. The motion of the spins will be independent of the oscillating field $\boldsymbol{B}_1$ and will only be determined by the static external field $\boldsymbol{B}_0$. The angle of rotation in the plane normal to $\boldsymbol{B}_0$ is given by: $\theta = \gamma B_0 t$, where $\omega_0 = \gamma B_0$ is the Larmor frequency. Note that the frequency $\omega_1 = \gamma B_1$ is named the Rabi frequency.

In a standard experiment aimed to detect the free induction decay, the field associated with a precessing magnetization sweeps past fixed receiving coils. Once the magnetization has a transverse component an electromotive force will be created in a coil, a consequence of Faraday's law. The time-dependent form of this current carries information that is transformed into the magnetic resonance signal. The free induction decay signal decreases in the course of time, and B_1 is only applied for a short time, this being an advantage of the FID method. Here, "free" means free of the oscillating field $\boldsymbol{B}_1$. These effects are considered in more detail below, in the Pulse EPR section.

1.7 Hydrogen Atom in a Magnetic Field

1.7.1 Hyperfine Interaction in the Ground State of a Hydrogen Atom

The hydrogen atom is the simplest system with an electron spin $S = 1/2$ and a nuclear spin $I = 1/2$. In general, the electronic-nuclear structure with $S = 1/2$ and $I = 1/2$ is widely used in calculations as a model system. The fundamental importance of the hydrogen atom follows from the possibility of its exact analytical calculation.

Each wavefunction that is a solution to the Schrödinger equation is known as an atomic orbital. Each orbital has a particular set of values of the following three quantum numbers: principal quantum number n whose value determines the energy $E_n \propto 1/n^2$; azimutal or orbital quantum number l, which furnishes information about the orbital angular momentum of the electron and, as result, about the shape of the orbital; magnetic quantum number m_l provides information about the number of orbitals with a given value of l and their behaviour in a magnetic or an electric field [1]. These three quantum numbers are used to name the orbital and reflect the way in which the orbitals spread in space.

The Schrödinger equation for the hydrogen atom has an exact solution in the form of a set of wavefunctions and eigenvalues of energy and angular momentum. The 1*s* atomic orbital has the lowest energy; thus, in the ground state, one electron occupies the 1*s* orbital and its wavefunction is

$$\Psi_{1s} = \left(\frac{1}{\pi a_0^3}\right)^{\frac{1}{2}} \exp\left(-\frac{r}{a_0}\right), \tag{1.91}$$

where a_0 is the Bohr radius, $a_0 = 0.529$ Å $= 0.529 \cdot 10^{-8}$ cm. The ground state of the hydrogen atom can be written as $^2S_{1/2}$, (subterm $^{2S+1}L_J$) because the orbital angular momentum $L = 0$ and the spin moment $S = 1/2$ (hence $J = 1/2$). All the *s* orbitals have the same spherical symmetry. The optical spectra of the hydrogen atom have been studied: the Lyman (ultraviolet), Balmer (visible), Paschen (infrared), Brackett (infrared), and Pfund (infrared) series of lines carry information about the excited states of the atom.

The radial parts of the wavefunctions (1*s*, 2*s*, 2*p*) are shown in Fig. 1.8a.

Let us consider the effect of an external magnetic field on the ground state of the hydrogen atom. Naturally, the magnetic field will affect the magnetic moments of the electron and the nucleus. The unpaired electron on the 1*s* orbital has a spin magnetic moment with the *g* factor nearly equal to $g_S = 2.0023$, it is

$$\hat{\vec{\mu}}_S = -g_S \mu_B \hat{\vec{S}}$$

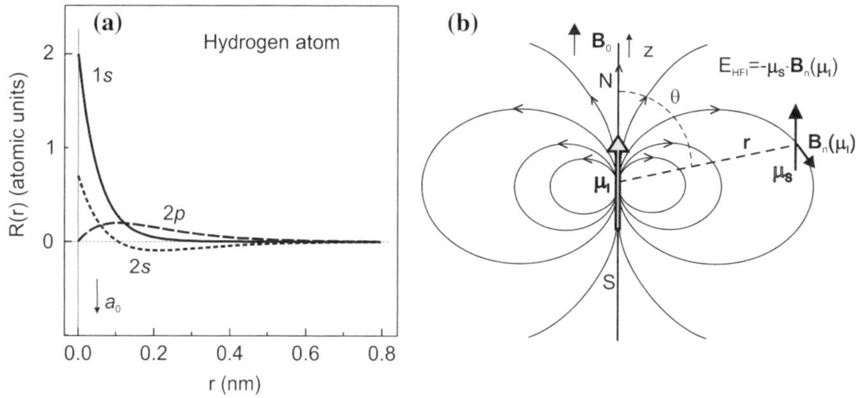

Fig. 1.8 **a** Diagrams for the radial wavefunctions R(r), for 1*s*, 2*s*, and 2*p* orbitals of hydrogen; the units of R(*r*) are $(1/a_0)^{3/2}$. **b** External magnetic field produced by a nuclear magnetic dipole moment $\boldsymbol{\mu}_I$ and the position of the second electron magnetic dipole moment $\boldsymbol{\mu}_S$ in this field

The nucleus of the hydrogen atom consists of one proton whose nuclear magnetic moment can be written as

$$\hat{\vec{\mu}}_I = g_I \mu_N \hat{\vec{I}},$$

where $I = 1/2$.

The Zeeman interaction of the electron and nuclear magnetic moments with an applied magnetic field $\boldsymbol{B}$ will be given by the Hamiltonian

$$\hat{H}_{Zeem} = g_S \mu_B \hat{\vec{S}} \cdot \vec{B} - g_p \mu_N \hat{\vec{I}} \cdot \vec{B}, \tag{1.92}$$

where the first and second terms describe the Zeeman energy of the electron and nucleus in an external magnetic field, respectively.

As before (when considering a magnetic moment in a magnetic field), we direct, without loss of generality, the magnetic field along an arbitrary axis z, i.e., $B \parallel z$ and $B_z = B$, $B_x = B_y = 0$.

For $B \parallel z$, the expression for the energy simplifies

$$\hat{H}_{Zeem} = g_S \mu_B \hat{S}_Z B - g_I \mu_N \hat{I}_Z B. \tag{1.92a}$$

In the next stage, we chose wavefunctions for a system of two spins that would be the eigenfunctions of the Hamiltonian (1.92a). The electron and nuclear spins have two projections on the external magnetic field, designated by us as $|+1/2\rangle = |\alpha_e\rangle$ and $|-1/2\rangle = |\beta_e\rangle$ for the electron spin and $|+1/2\rangle = |\alpha_n\rangle$ and $|-1/2\rangle = |\beta_n\rangle$ for the nuclear spin. Because the electron and nuclear wavefunctions in this approximation are independent of each other, the total wavefunction of the system can be written as a product of the wavefunction of the electron and nucleus. There will be four independent spin combinations: $|\alpha_e, \alpha_n\rangle$; $|\alpha_e, \beta_n\rangle$; $|\beta_e, \alpha_n\rangle$ and $|\beta_e, \beta_n\rangle$ (subscripts 'e' and 'n' denote the electron and nucleus, respectively). The energy matrix consists of the matrix elements of the Hamiltonian (1.92) between all the spin eigenfunctions

$$\langle M_S, m_I | \hat{H}_{Zeem} | M'_S, m'_I \rangle. \tag{1.7.3}$$

and the matrix of the Hamiltonian (1.92) is now just of the 4×4 type:

The matrix is diagonal because the wavefunctions we used are eigenfunctions of the Zeeman energy Hamiltonian $\hat{H}_{Zeem}$. The diagonal matrix elements are the energy levels (eigenvalues of energy) for the states described by the corresponding wavefunctions

$$\begin{aligned} E_{|\alpha e,\alpha n\rangle} &= +1/2 g_S \mu_B B - 1/2 g_I \mu_N B, \\ E_{|\alpha e,\beta n\rangle} &= +1/2 g_S \mu_B B + 1/2 g_I \mu_N B, \\ E_{|\beta e,\alpha n\rangle} &= -1/2 g_S \mu_B B - 1/2 g_I \mu_N B, \\ E_{|\beta e,\beta n\rangle} &= -1/2 g_S \mu_B B + 1/2 g_I \mu_N B. \end{aligned} \tag{1.93a}$$

However, these calculations do not correctly describe the energy levels of the hydrogen atom in a magnetic field, because the actually occurring interaction of the electron magnetic moment with the nuclear magnetic moment is disregarded. Let us consider this interaction, which is named the 'hyperfine interaction' (HF interaction). The hyperfine interaction is responsible for quite a number of important aspects of EPR and nuclear magnetic resonance (NMR) spectra. We already examined the interaction between two magnetic moments in the classical version, which is written as (1.38)

$$E = \frac{1}{r^3}\left[\vec{\mu}_1 \cdot \vec{\mu}_2 - \frac{3(\vec{\mu}_1 \cdot \vec{r})(\vec{\mu}_2 \cdot \vec{r})}{r^2}\right].$$

This expression can be obtained in two ways from the formula for the energy E of the magnetic moment $\boldsymbol{\mu}$ placed in an external magnetic field $\boldsymbol{B}$: $E = -\boldsymbol{\mu} \cdot \boldsymbol{B}$. The interaction between the magnetic moments can be regarded as an interaction of the first magnetic moment in a magnetic field produced by the second magnetic moment at the position of the first one, or vice versa; in both cases, the same result must be obtained, namely, expression (1.38).

As always, to proceed with the quantum-mechanical consideration, we replace the classical quantities in (1.38) with operators and obtain the Hamiltonian of the so-called dipole-dipole interaction in the form

$$\hat{H} = \frac{1}{r^3}\left[\hat{\mu}_1 \cdot \hat{\mu}_2 - \frac{3(\hat{\mu}_1 \cdot \vec{r})(\hat{\mu}_2 \cdot \vec{r})}{r^2}\right]. \tag{1.94}$$

The hyperfine interaction can be represented as the motion of the electron in the magnetic dipole field of the nucleus. The nuclear magnetic moment induces a nuclear magnetic field $\boldsymbol{B}_n(\boldsymbol{r})$ and the electron magnetic moment interacts with this nuclear magnetic field, with the interaction energy

$$\hat{H} = -\hat{\vec{\mu}}_S \cdot \vec{B}_n(\vec{r}).$$

To calculate the Hamiltonian for the energy of the hyperfine interaction, we replace the magnetic moments in (1.94) with the magnetic moment of the electron spin $\boldsymbol{\mu}_S$ and magnetic moment of the nuclear spin $\boldsymbol{\mu}_I$. As a result, we have

$$\begin{aligned} \hat{H}_{HFI} &= \frac{1}{r^3}\left[\hat{\mu}_S \cdot \hat{\mu}_I - \frac{3(\hat{\mu}_S \cdot \vec{r})(\hat{\mu}_I \cdot \vec{r})}{r^2}\right] \\ &= -g_S\mu_B g_I\mu_N \frac{1}{r^3}\left[\hat{S} \cdot \hat{I} - \frac{3(\hat{S} \cdot \vec{r})(\hat{I} \cdot \vec{r})}{r^2}\right] \end{aligned}. \tag{1.95}$$

In the atom, the nucleus is situated at the center, while the electron is distributed in space in accordance with the wavefunction for a particular state (Fig. 1.8b). In the ground state of the hydrogen atom, the electron is in the 1*s* orbital characterized by a spherically symmetric density distribution of the wavefunction. To find the HF interaction energy, it is necessary to average the position of an electron in the system, and, as a result, the HF interaction energy for the 1*s*-state (and, in general, for any *ns*-state) vanishes according to (1.95). Thus, the HF interaction must not be observed for the hydrogen atom in its ground state. But it is still there!

Consider the nature of this interaction, which was first analyzed by Fermi (Italian physicist Enrico Fermi, 1930), so this interaction is frequently named the Fermi-contact interaction. The contact interaction only occurs for *s*-electrons, for which it is responsible for the appearance of an isotropic hyperfine coupling. For definiteness, we consider the 1*s* wavefunction (1.91). As already noted, the integration over a sphere leads to the disappearance of the dipole-dipole interaction between the spin magnetic moment of the electron and the nuclear magnetic moment. Fermi pointed out that there is a nonzero wavefunction density within the nucleus for the *ns*-electron and it is approximately equal to the density of the wavefunction at $r = 0$.

Fermi examined the magnetic field inside the nucleus and obtained an uniform distribution of the magnetic moment of the electron within the nucleus (the classical counterpart is the magnetic field generated inside a uniformly magnetized sphere), which can be written as

$$\vec{B}_{\mu s} = \frac{8\pi}{3}\hat{\vec{\mu}}_s|\Psi_{1s}(0)|^2. \tag{1.96}$$

As a result, if we imagine that the nuclear magnetic moment is placed in a magnetic field $\boldsymbol{B}_{\mu S}$, the hyperfine Hamiltonian can be written as

$$\hat{H}_{HFI} = -\hat{\vec{\mu}}_I \vec{B}_{\mu s} = -\frac{8\pi}{3} \hat{\vec{\mu}}_s \cdot \hat{\vec{\mu}}_I |\Psi_{1s}(0)|^2$$

Substituting the magnetic moments expressed in terms of the electron and nuclear spin operators, we obtain the Hamiltonian

$$\hat{H}_{HFI} = \frac{8\pi}{3} g_s \mu_B g_I \mu_N |\Psi_{1s}(0)|^2 \hat{\vec{S}} \cdot \hat{\vec{I}} = A \hat{\vec{S}} \cdot \hat{\vec{I}}, \tag{1.97}$$

where the isotropic hyperfine interaction constant A is given by

$$A = \frac{8\pi}{3} g_s \mu_B g_I \mu_N |\Psi_{1s}(0)|^2. \tag{1.98}$$

For the ground state of the hydrogen atom with a $1s$ unpaired electron, the HF imteraction constant becomes

$$A = \frac{8\pi}{3} g_s \mu_B g_I \mu_N \frac{1}{\pi a_0^3} \cong 1420\,\text{MHz}.$$

(A = 1420.4057517662(3) MHz!)

This splitting corresponds to a wavelength of 21 cm at which science-fiction believers are trying to establish contact with extraterrestrial civilizations.

In general, for the unpaired electron situated in the ns-shell in a free atom or ion, the hyperfine interaction constant is given by

$$A = \frac{8\pi}{3} g_s \mu_B g_I \mu_N |\Psi_{ns}(0)|^2. \tag{1.98a}$$

It should be noted that, the hyperfine interaction is, as shown below, a sum of both the contact (isotropic) and dipole-dipole (anisotropic) contributions.

To summarize, there are two contributions into the HF interaction constant A:

(i) The wavefunction of the electron has an angular dependence and vanishes at the position of the nucleus (e.g. a p-, d-, f-function). For np-function $A = \frac{2}{5} g_s \mu_B g_I \mu_N \left\langle \frac{3\cos^2\theta - 1}{r^3} \right\rangle$, with averaging over the electron wavefunction.

(ii) The wavefunction of the electron does not vanish at the position of the nucleus (ns-function), $|\Psi_{ns}(0)|^2 \neq 0. A = \frac{8\pi}{3} g_s \mu_B g_I \mu_N |\Psi_{ns}(0)|^2$

1.7.2 Hamiltonian and Energy Levels for the Hydrogen Atom in a Magnetic Field (Breit-Rabi Formula); Selection Rules

We write the Hamiltonian for the hydrogen atom in a magnetic field with a hyperfine interaction as

$$\hat{H}_{HFI} = A\hat{\vec{S}} \cdot \hat{\vec{I}},$$

which is added to the Hamiltonian (1.92)

$$\hat{H} = g_S\mu_B\hat{\vec{S}} \cdot \vec{B} - g_I\mu_N\hat{\vec{I}} \cdot \vec{B} + A\hat{\vec{S}} \cdot \hat{\vec{I}}. \tag{1.99}$$

Here, the first and second terms describe, respectively, the Zeeman energy of the electron and nucleus in an external magnetic field, and the third term represents the isotropic hyperfine interaction (Fermi interaction). Without loss of generality by the problem, we direct the magnetic field along an arbitrary axis z, i.e., $B \parallel z$ and $B_z = B$, $B_x = B_y = 0$. As a result, we have the following expression for the Zeeman energy (with the inner product for the HF interaction also shown)

$$\hat{H} = g_S\mu_B\hat{S}_ZB - g_I\mu_N\hat{I}_ZB + A(\hat{S}_Z\hat{I}_Z + \hat{S}_X\hat{I}_X + \hat{S}_Y\hat{I}_Y). \tag{1.99a}$$

This Hamiltonian can be solved by diagonalisation to give the energy eigenvalues (energy levels) and the eigenstates (wavefunctions). As the base states for the quantum-mechanical treatment, we take product states between the electron spin states $|S, M_S\rangle$ and the nuclear spin states $|I, m_I\rangle$:

$$\Psi = |S, M_S\rangle|I, m_I\rangle$$

For the sake of simplicity, we omit S, I in the designations of the base states and write:

$$\Psi = |M_S\rangle|m_I\rangle.$$

In general, there are $(2S + 1)(2I + 1)$ base states; for the case $S = 1/2$, $I = 1/2$, we have four base states

$$|\alpha_e, \alpha_n\rangle; |\alpha_e, \beta_n\rangle; |\beta_e, \alpha_n\rangle \text{ and} |\beta_e, \beta_n\rangle$$

To calculate the matrix elements, it is of use to express the electron operators $\hat{S}_X$, $\hat{S}_Y$ and the nuclear operators $\hat{I}_X$, $\hat{I}_Y$ in (1.99a) in terms of the shift operators $\hat{S}_+$, $\hat{S}_-$, $\hat{I}_+$, $\hat{I}_-$ and, as a result, (1.99a) can be rearranged to

$$\hat{H}_{HFI} = A(\hat{S}_Z\hat{I}_Z + \hat{S}_X\hat{I}_X + \hat{S}_Y\hat{I}_Y) = A[\hat{S}_Z\hat{I}_Z + 1/2(\hat{S}_+\hat{I}_- + \hat{S}_-\hat{I}_+)]. \qquad (1.99b)$$

The operators $\hat{S}_Z$ and $\hat{I}_z$ are diagonal in the base states; $\hat{S}$ operates on the electron spin only, with the nuclear spin state unaffected; and $\hat{I}$ operates on the nuclear spin only, with the electron spin state unaffected. The matrix elements of the shift operators are easy to find: $\hat{S}_+$ shifts a state $|M_S\rangle$ to $|M_S + 1\rangle$, unless $|M_S\rangle$ is the highest state. $\hat{S}_-$ shifts $|M_S\rangle$ to $|M_S\text{-}1\rangle$, unless $|M_S\rangle$ is the lowest state.

The operators $\hat{S}_+\hat{I}_- + \hat{S}_-\hat{I}_+$ couple the states $|\alpha_e, \beta_n\rangle$ and $|\beta_e, \alpha_n\rangle$ with the matrix elements A/2.

Let us consider the states in the limit of strong magnetic fields, when the Zeeman interaction is substantially stronger than the hyperfine interaction. In the limit of strong magnetic fields, the base states $|\alpha_e, \alpha_n\rangle$, $|\alpha_e, \beta_n\rangle$, $|\beta_e, \alpha_n\rangle$, $|\beta_e, \beta_n\rangle$ are the eigenstates. Due to the hyperfine interaction, the states $|\alpha_e, \beta_n\rangle$ and $|\beta_e, \alpha_n\rangle$ have anti-parallel electron and nuclear spins and, as a result, they are lower in energy.

We choose as the zero approximation the first two terms of the Hamiltonian (1.99), i.e., we disregard the HF interaction. We use the zero-approximation wavefunctions to find the matrix elements in the block-diagonal matrix of (1.93a) (Table 1.1).

The matrix is not diagonal (block-diagonal) because the wavefunctions we used are not eigenfunctions of the total Hamiltonian due to the presence of the term describing the HF interaction. To obtain all the four energy levels (Fig. 1.9), we have to diagonalize the energy matrix of (1.100, see Table 1.2). The diagonalization must set to zero the resulting secular determinant. The four roots of the equation yield four energy levels corresponding to the four possible wavefunctions. These levels are given by (Table 1.2)

$$\begin{aligned}
E_{|\alpha e,\alpha n\rangle} &= +1/2g_S\mu_B B - 1/2g_I\mu_N B + A/4;\\
E_{|\beta e,\beta n\rangle} &= -1/2g_S\mu_B B + 1/2g_I\mu_N B + A/4;\\
E_{(|\alpha e,\beta n\rangle)} &= +1/2[(g_S\mu_B + g_I\mu_N)^2B^2 + A^2]^{1/2} - A/4;\\
E_{(|\beta e,\alpha n\rangle)} &= -1/2[(g_S\mu_B + g_I\mu_N)^2B^2 + A^2]^{1/2} - A/4.
\end{aligned} \qquad (1.101)$$

Table 1.1 Matrix elements of the Hamiltonian (1.92)

$\langle M_S,m_I\| H_{Zeem}\|M_S', m_I'\rangle$	$\|\alpha_e, \alpha_n\rangle$	$\|\alpha_e, \beta_n\rangle$	$\|\beta_e, \alpha_n\rangle$	$\|\beta_e, \beta_n\rangle$	
$\langle\alpha_e, \alpha_n\rangle\|$	$1/2g_S\mu_B B - 1/2g_I\mu_N B$	0	0	0	
$\langle\alpha_e, \beta_n\|$	0	$1/2g_S\mu_B B + 1/2g_I\mu_N B$	0	0	(1.93a)
$\langle\beta_e, \alpha_n\|$	0	0	$-1/2g_S\mu_B B - 1/2g_I\mu_N B$	0	
$\langle\beta_e, \beta_n\|$	0	0	0	$-1/2g_S\mu_B B + 1/2g_I\mu_N B$	

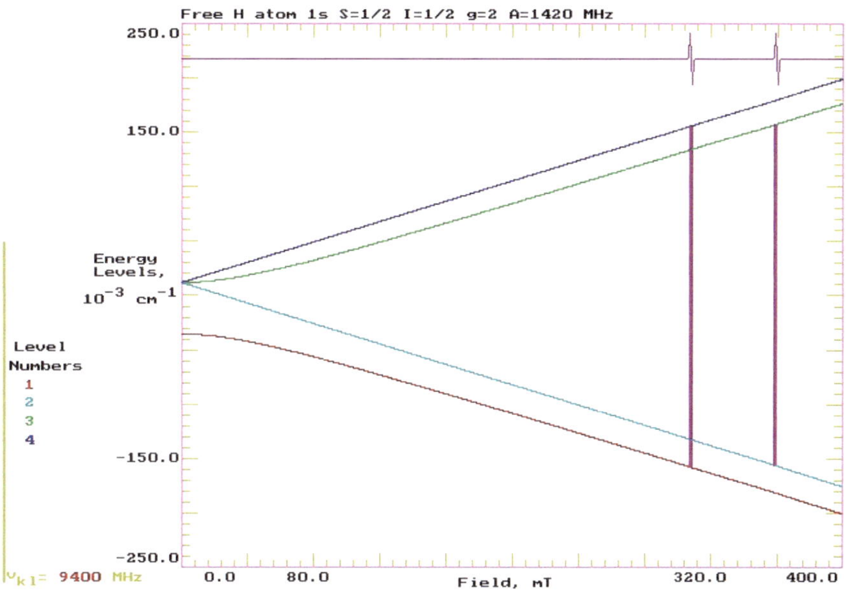

Fig. 1.9 Diagram representing the magnetic field dependence of the sublevel energies for the ground state of the free hydrogen atom H^0, known as the Breit-Rabi diagram. X-band EPR spectra are shown

Table 1.2 Matrix elements of Hamiltonian (1.99)

$\langle M_S,m_I\|\boldsymbol{H}\|M_S',m_I'\rangle$	$\|\alpha_e, \alpha_n\rangle$	$\|\alpha_e, \beta_n\rangle$	$\|\beta_e, \alpha_n\rangle$	$\|\beta_e, \beta_n\rangle$	
$\langle\alpha_e, \alpha_n\|$	$1/2g_S\mu_BB - 1/2g_I\mu_NB + A/4 - E$	0	0	0	
$\langle\alpha_e, \beta_n\|$	0	$1/2g_S\mu_BB + 1/2g_I\mu_NB - A/4 - E$	A/2	0	(1.100)
$\langle\beta_e, \alpha_n\|$	0	A/2	$-1/2g_S\mu_BB - 1/2g_I\mu_NB - A/4 - E$	0	
$\langle\beta_e, \beta_n\|$	0	0	0	$-1/2g_S\mu_BB + 1/2g_I\mu_NB + A/4 - E$	

The subscripts correspond to the designations of the wavefunction.

For two levels, $E_{(|\alpha e,\ \beta n\rangle)}$ and $E_{(|\beta e,\ \alpha n\rangle)}$, the subscripts are given in parentheses in order to emphasize that the wavefunctions within the parentheses describe the energy levels of the data only in the approximation of high magnetic fields in which the Zeeman energy substantially exceeds the hyperfine interactions, i.e., $B \gg A$. In intermediate magnetic fields, the parentheses indicate that there is a mixture of the states $|\alpha_e, \beta_n\rangle$ and $|\beta_e, \alpha_n\rangle$. Expressions (1.101) are known as the Breit-Rabi formulas (the simplest case for $I = ½$) [2]. This is the exact solution for a system with $S = 1/2$ and $I = 1/2$. It should be added that there may be more general exact solutions for systems with $S = 1/2$ and any nuclear angular momentum I (I can take for different atoms and their isotopes the values of 1/2, 1, 3/2, 2, 5/2, 3, 7/2, 4, 9/2, 5), which also are known as the Breit-Rabi formula (to be discussed below).

For high magnetic fields B_0, if the electron Zeeman energy is high compared to the hyperfine interaction, the base states are nearly the eigenstates of the Hamiltonian. Thus, at high fields, the states can be classified by $|\alpha_e, \alpha_n\rangle$, $|\alpha_e, \beta_n\rangle$, $|\beta_e, \alpha_n\rangle$, and $|\beta_e, \beta_n\rangle$. The Breit-Rabi diagram at a low magnetic field is discussed below.

The electron spin and the nuclear spin are quantized along the magnetic field $\boldsymbol{B}_0$ (z-axis). The hyperfine interaction is thus +A/4 or −A/4, depending on whether the spins are parallel or anti-parallel, respectively. In mathematical terms, this means that only the diagonal matrix elements of the operators are to be taken into account. The eigenstates are the base states $|\alpha_e, \alpha_n\rangle$, $|\alpha_e, \beta_n\rangle$, $|\beta_e, \alpha_n\rangle$, and $|\beta_e, \beta_n\rangle$.

Consider the interaction of electromagnetic waves with a hydrogen atom in a magnetic field. An oscillating magnetic field generated by the magnetic component of the electromagnetic wave acts on the magnetic moments of the electron and nucleus. As discussed above, we have a static magnetic field $\boldsymbol{B}_0$ and an oscillating magnetic field (1.50) $\boldsymbol{B}_1 = \boldsymbol{B}_{1A}2\cos(\omega t)$. Let us consider the resonance conditions for the electron spin resonance. In this case, the oscillating magnetic field interacts only with the magnetic moment of the electron, and then the Hamiltonian of a time-dependent perturbation can be written in the form:

$$\hat{H}'(t) = -\hat{\vec{\mu}}_S \cdot \vec{B}_1 = g_S\mu_B\hat{\vec{S}} \cdot \vec{B}_1 = g_S\mu_B\hat{\vec{S}} \cdot \vec{B}_{1A}2\cos(\omega t). \tag{1.53}$$

The operator of the Zeeman energy in a static magnetic field with the HFI of (1.99) causes splitting of energy levels in accordance with the Breit-Rabi formula (1.101). The operator $\hat{H}'(t)$ (1.53) must lead to transitions between these levels, which correspond to reorientation of magnetic moments of the electron and nucleus. We only consider the approximation of high magnetic fields, in which case, the Zeeman energy substantially exceeds the hyperfine interaction. Then we can use the four wavefunctions corresponding to the four energy levels: $|\alpha_e, \alpha_n\rangle$, $|\alpha_e, \beta_n\rangle$, $|\beta_e, \alpha_n\rangle$, and $|\beta_e, \beta_n\rangle$. We need to consider the matrix elements of the transitions between the levels, which have the form

$$\langle M_S, m_I|\hat{H}'(t)|M'_S, m'_I\rangle = g_S\mu_B\langle M_S, m_I|\hat{\vec{S}} \cdot \vec{B}_1|M'_S, m'_I\rangle. \tag{1.102}$$

Here, as above, we denote the amplitude $\boldsymbol{B}_{1A}$ in the matrix element as $\boldsymbol{B}_1$. Because the resonance conditions are satisfied only for the electron paramagnetic resonance, and, therefore, the oscillating magnetic field interacts only with the magnetic moment of the electron, we can rewrite (1.102) as a product

$$g_S\mu_B\langle M_S|\hat{\vec{S}}\cdot\vec{B}_1|M_S'\rangle\langle m_I \mid m_I'\rangle.$$

Owing to the orthogonality of the nuclear wavefunctions, the following relations can be written: $\langle m_I|m_I'\rangle = 0$ for $m_I \neq m_I'$ and $\langle m_I|m_I'\rangle = 1$ for $m_I = m_I'$. Thus, it makes sense to consider only the transitions with the selection rule $\Delta m_I = 0$, and the calculation of matrix elements (1.102) reduces to finding $g_S\mu_B\langle M_S|\hat{\vec{S}}\cdot\vec{B}_1|M_S'\rangle$ for the same wavefunction $|m_I\rangle$.

We direct a static magnetic field $\boldsymbol{B}_0$ along the z-axis, i.e., $\boldsymbol{B_0} \parallel z$ and $B_z = B_0$, $B_x = B_y = 0$, which does not lead to loss of generality by the problem. As a result, the system has four spin energy levels with the wavefunctions $|\alpha_e, \alpha_n\rangle$, $|\alpha_e, \beta_n\rangle$, $|\beta_e, \alpha_n\rangle$, and $|\beta_e, \beta_n\rangle$, which are the eigenfunctions of (1.92a). Now, we add the oscillating magnetic field corresponding to the EPR resonance condition and calculate the matrix elements (1.102). Because there is a preferred direction associated with the direction of $\boldsymbol{B}_0$ along the z-axis, the direction of $\boldsymbol{B}_1$ should be chosen in a certain way with respect to the z, x and y axes. We show that the transitions between the levels in the system are caused only by the component of the oscillating magnetic field that is directed perpendicularly to the static magnetic field. Let us consider the following two cases.

(1) $\boldsymbol{B}_1 \parallel \boldsymbol{B_0}$, i.e., $\boldsymbol{B}_1$ is directed along the axis z: $\boldsymbol{B}_1 \parallel z$, $B_{1z} = B_1$, $B_{1x} = B_{1y} = 0$. Then the expression for the matrix element (1.102) has the form $g_S\mu_B\langle\beta_e\alpha_n|\hat{S}_Z B_1|\alpha_e\alpha_n\rangle$ and $g_S\mu_B\langle\beta_e\beta_n|\hat{S}_Z B_1|\alpha_e\beta_n\rangle$ (for definiteness, we consider the bottom-up transitions), with allowance for the selection rules for the nuclear wavefunctions. In accordance with the equation $\hat{S}_Z|\alpha_e\rangle = 1/2|\alpha_e\rangle$, we obtain the matrix element $\langle\beta_e \mid \alpha_e\rangle = 0$, which vanishes as a consequence of the orthogonality of the wavefunctions. It can be shown that similar relations hold for the top-down transitions. Thus, all the matrix elements are zero, i.e., the transition probability is zero when the oscillating magnetic field is directed along the static magnetic field, so the first condition for the EPR (NMR) is not satisfied.

(2) $\boldsymbol{B}_1 \perp \boldsymbol{B_0}$, i.e., $\boldsymbol{B}_1$ is directed perpendicularly to the static magnetic field; for definiteness, the oscillating magnetic field is directed along the x-axis, then $B_{1x} = B_1$, $B_{1z} = B_{1y} = 0$. Consider the matrix element $g_S\mu_B\langle\beta_e\alpha_n|\hat{S}_X B_1|\alpha_e\alpha_n\rangle$ (matrix element $g_S\mu_B\langle\beta_e\beta_n|\hat{S}_X B_1|\alpha_e\beta_n\rangle$ has a similar form). The calculation reduces to finding the matrix element $\langle\beta_e|\hat{S}_X|\alpha_e\rangle$, as it was done for the two-level system. Using the up-step and down-step operators $\hat{S}_+$ and $\hat{S}_-$ and relations (1.55)–(1.57) we obtain the equation $\langle\beta_e|\hat{S}_X|\alpha_e\rangle = 1/2\langle\beta_e \mid \beta_e\rangle = 1/2$. Thus, in the case of $\boldsymbol{B}_1 \perp \boldsymbol{B}_0$, the transition matrix element between the levels is non-zero and, therefore, the probability of transitions between the levels is non-zero when the resonance conditions for the EPR are satisfied. In this case, the electron angular momentum can change only by unity as a result of such a transition,

whereas the angular momentum of the nucleus remains unchanged. Thus, the general selection rules for EPR transitions can be written as

$$\Delta M_S = \pm 1,\ \Delta m_I = 0. \tag{1.103}$$

It should be noted that, if the resonance condition for the nuclear magnetic resonance is satisfied, the selection rules look like

$$\Delta M_S = 0,\ \Delta m_I = \pm 1. \tag{1.104}$$

Thus $|\beta_e,\alpha_n\rangle \leftrightarrow |\alpha_e,\alpha_n\rangle$ and $|\beta_e,\beta_n\rangle \leftrightarrow |\alpha_e,\beta_n\rangle$ are the EPR transitions in the high-magnetic-field approximation (Fig. 1.9).

The frequency of the $|\beta_e,\alpha_n\rangle \leftrightarrow |\alpha_e,\alpha_n\rangle$ and $|\beta_e,\beta_n\rangle \leftrightarrow |\alpha_e,\beta_n\rangle$ transitions in constant (fixed) magnetic field with a varying microwave frequency according to the selection rules for the high-magnetic-field approximation [see (1.103)]

$$\begin{aligned} \nu_1 &= 1/h(E_{|\alpha_e,\alpha_n\rangle} - E_{(|\beta_e,a_n\rangle)});\\ \nu_2 &= 1/h(E_{(|ae,bn\rangle)} - E_{|be,bn\rangle}). \end{aligned} \tag{1.101a}$$

As a result, the difference

$$\nu_1 - \nu_2 = A$$

is exactly the hyperfine interaction constant (in Hz).

$$\begin{aligned} E_{|\alpha e,\alpha n\rangle} &= +1/2 g_S \mu_B B - 1/2 g_I \mu_N B + A/4;\\ E_{|\beta e,\beta n\rangle} &= -1/2 g_S \mu_B B + 1/2 g_I \mu_N B + A/4;\\ E_{(|\alpha e,\beta n\rangle)} &= +1/2[(g_S\mu_B + g_I\mu_N)^2 B^2 + A^2]^{1/2} - A/4;\\ E_{(|\beta e,\alpha n\rangle)} &= -1/2[(g_S\mu_B + g_I\mu_N)^2 B^2 + A^2]^{1/2} - A/4. \end{aligned} \tag{1.101}$$

The resonant magnetic fields at constant microwave frequency ν_0 (conventional arrangement for high frequency EPR experiments) can be also obtained from (1.101) and (1.101a). The resonant magnetic fields H_1 and H_2 will be taken for $|\beta_e,\alpha_n\rangle \leftrightarrow |\alpha_e,\alpha_n\rangle$ and $|\beta_e,\beta_n\rangle \leftrightarrow |\alpha_e,\beta_n\rangle$ transitions. For high frequency approximation (the Zeeman unteraction is much higher then hyperfine interaction, in addition, nuclear Zeeman interaction is neglected compared to the electron Zeeman interaction) can be written in the form [3]

$$H_{1/2} = h/(4 g_S \mu_B)[2\nu_0 - / + A + (4\nu_0^2 - / + 4A\nu_0 - 3A^2)^{1/2}]$$

The hyperfine splitting (in G or T): $H_2 - H_1 \neq hA/g_S\mu_B$ and only for $\nu_0 \to \infty$ $H_2 - H_1 \to hA/g_S\mu_B$.

The ESR signal consists of two lines, with the hyperfine splitting A measured in magnetic-field units. It should be noted that the nuclear Zeeman energy is of no importance in high magnetic fields because the nuclear spin is not "flipped" in an EPR transition. If the hyperfine interaction is resolved in the EPR spectrum, it can be directly seen in the spectrum. The hyperfine interaction is frequently small and unresolved and only results in the EPR-line broadening. The hyperfine coupling to many nuclei makes the EPR spectrum fairly complicated and very informative.

To observe EPR transitions in free atoms (e.g., H^0, N^0, O^0, F^0) with a conventional EPR spectrometer, an appropriate low-pressure gas is pumped through the microwave cavity on being subjected to an electric discharge. As a result, sharp EPR lines are observed, the g-values being fully consistent with the g-factors calculated by the Landé formula: for H^0 ($^2S_{1/2}$) and N^0 ($^4S_{3/2}$) $g_J = 2.00$; for O^0 ($^3P_2$) $g_J = 3/2$; for F^0 ($^2P_{3/2}$) $g_J = 4/3$.

1.7.3 Uncoupled and Coupled Bases for Angular Momenta

In the approximation of low magnetic fields, i.e., fields comparable in magnitude with the hyperfine interactions, the selection rules of (1.103) for the EPR transitions are violated. The fact that off-diagonal elements appear in the energy matrix of (1.99) means that the basis spin functions are not eigenfunctions of the Hamiltonian (1.99a).

We examined the wavefunctions for the angular momentum in the form of $|S, M_S, I, m_I\rangle$. Such a basis of the wavefunctions in the form of angular momenta and their projections for each particle (electron and nucleus) is known as the uncoupled basis. The hyperfine interaction

$$\hat{H}_{HFI} = A\hat{\vec{S}} \cdot \hat{\vec{I}} \tag{1.97}$$

couples the electron angular momentum and the nuclear angular momentum in accordance with the addition rule for the momenta, given by

$$\hat{\vec{F}} = \hat{\vec{S}} + \hat{\vec{I}}, \tag{1.105}$$

where the total angular momentum F can take on a range of possible values for fixed S and I, given by

$$F = S + I,\ S + I - 1,\ \ldots |S - I|. \tag{1.105a}$$

The wavefunctions for the total angular momentum F can be written as

$$|F, M_F, S, I\rangle. \tag{1.106}$$

This concept is named the coupled basis. The transformation from the uncoupled to the coupled bases is expressed in terms of the Clebsch-Gordan coefficients. In general, we can work with any of these bases, as also in any coordinate system, but, to solve specific problems, it is of use to choose a basis appropriate to the task. Thus, for the hydrogen atom and for similar systems in strong magnetic fields, in which the HF interaction has no significant effect on the wavefunctions and energy levels, i.e., the electron and nuclear magnetic moments are oriented in an external magnetic field independently of each other, it is convenient to work with an uncoupled basis $|S,M_S,I,m_I\rangle$ or more simply with $|M_S,m_I\rangle$. In weak magnetic fields, including zero magnetic field, when the HF interaction dominates, it is convenient to make analysis in a coupled basis, $|F,M_F,S,I\rangle$ or $|F,M_F\rangle$.

It should be noted that similar relations were just written for the spin-orbit interaction because the Hamiltonians for the isotropic hyperfine interaction (1.97) and the spin-orbit interaction are similar:

$$\hat{H}_{SO} = \lambda \hat{\vec{L}} \cdot \hat{\vec{S}}, \tag{1.13}$$

and

$$\hat{\vec{J}} = \hat{\vec{L}} + \hat{\vec{S}},$$

i.e., the wavefunctions have the form $|L,M_L,S,M_S\rangle$ for the uncoupled basis and $|J, M_J,L,S\rangle$ for the coupled basis. Due to this similarity, the Landé interval rule introduced for the spin-orbit interaction, in which the distance between the energy levels corresponding to the total angular momentum J and $(J-1)$ is expressed by $E_J - E_{J-1} = \lambda J$ (1.16a), can be applied to the HF interaction. Then, the distance between the hyperfine levels F and $(F-1)$ in zero magnetic field is given by

$$E_F - E_{F-1} = AF. \tag{1.107}$$

The hydrogen atom in its ground state in zero magnetic field has two levels, the distance between these levels being the HF interaction constant A. A magnetic field will split these levels, with the number of the resulting levels determined by the formula $(2F + 1)$. Thus, there will be three sublevels for $F = 1$ and one sublevel for $F = 0$; in general, there will be four sublevels.

Let us consider the states in the limit of weak magnetic fields ($B \rightarrow 0$). There are two solutions for the states with total spin F at $B = 0$: a singlet state with the total spin $F = 0$ and $|F,M_F\rangle = |0,0\rangle$:

$$|0,0\rangle = 1/\sqrt{2}(|\alpha_e, \beta_n\rangle - |\beta_e, \alpha_n\rangle)$$

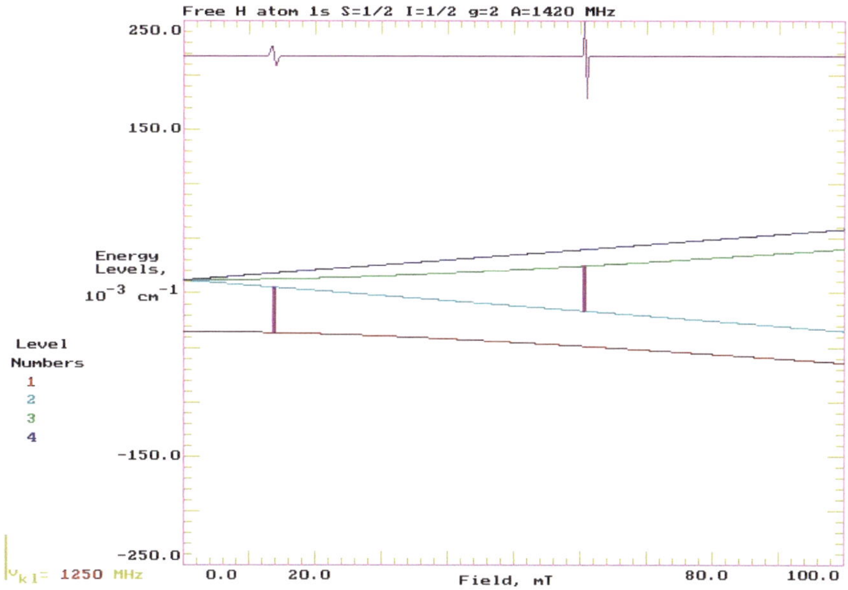

Fig. 1.10 Breit-Rabi diagram for the H^0 ground state representing the magnetic field dependence of the sublevel energies in low magnetic field. Low frequency (1.25 GHz) EPR transitions are shown

and a triplet state with the total spin $F = 1$ and $|F,M_F\rangle = |1,1\rangle$, $|1,0\rangle$, $|1,-1\rangle$,

$$|1,1\rangle = |\alpha_e,\alpha_n\rangle;\ |1,0\rangle = 1/\sqrt{2}(|\alpha_e,\beta_n\rangle + |\beta_e,\alpha_n\rangle);\ |1,-1\rangle = |\beta_e,\beta_n\rangle.$$

In week magnetic fields, the eigenstates $|\alpha_e,\alpha_n\rangle$, $|\alpha_e,\beta_n\rangle$, $|\beta_e, \alpha_n\rangle$ and $|\beta_e, \beta_n\rangle$ are no longer base states. Because the states are now a mixture of four base states, there may be, in principle, four magnetic resonance transitions, which depend on the microwave frequency used in the EPR experiment (Fig. 1.10). The states coupled by these transitions are mixtures of base states, and the magnetic dipole transitions apparently disobey the selection rules $\Delta M_S = \pm 1$, $\Delta m_I = 0$.

The eigenfunctions $|1,0\rangle$ and $|0,0\rangle$ are obtained by diagonalization of the matrix 2.7.10 (2×2 part) [3] and are expressed as

$$|1,0\rangle = \cos\omega|\alpha_e,\beta_n\rangle + \sin\omega|\beta_e,\alpha_n\rangle)$$
$$|0,0\rangle = -\sin\omega|\alpha_e,\beta_n\rangle + \cos\omega|\beta_e,\alpha_n\rangle),$$

where

$$\cos^2\omega = 1/2\{1 + [1 + (\sqrt{2}hA)^2/(g_S\mu_B B)^2]^{-1/2}\}$$
$$\sin^2\omega = 1/2\{1 - [1 + (\sqrt{2}hA)^2/(g_S\mu_B B)^2]^{-1/2}\}$$

For B $\to \infty$ $|1,0\rangle \to |\alpha_e,\beta_n\rangle$, $|0,0\rangle \to |\beta_e, \alpha_n\rangle$.

As a result of the mixing of states, four transitions can be observed at low magnetic fields and the relative intensities can be obtained by calculation of the matrix elements:

$$|0,0\rangle \leftrightarrow |1,-1\rangle \text{ relative intensity } \propto \sin^2\omega$$
$$|0,0\rangle \leftrightarrow |1,1\rangle \text{ relative intensity } \propto \cos^2\omega$$
$$|1,-1\rangle \leftrightarrow |1,0\rangle \text{ relative intensity } \propto \cos^2\omega$$
$$|1,0\rangle \leftrightarrow |1,1\rangle \text{ relative intensity } \propto \sin^2\omega$$

Figures 2.9 and 2.10 show the evolution of energy levels as a function of the applied magnetic field for the case when the nuclear spin quantum number I is ½. The zero-field states can be labeled with the total angular momentum ($\boldsymbol{F} = \boldsymbol{S} + \boldsymbol{I}$) quantum numbers: $F = 0$ for the singlet state and $F = 1$ for the triplet state (the zero energy is (3/4)A above the singlet state chosen for the approximation in the absence of a hyperfine interaction).

In general, when there is a system of two spins ($S = 1/2$ and I) whose associated magnetic moments are coupled by the hyperfine interaction, the energies of the magnetic sublevels depend nonlinearly on the magnetic field B. The diagram representing the magnetic field dependence of the sublevel energies is known as the Breit-Rabi diagram [2].

The energy eigenvalues for arbitrary I ($S = 1/2$) are summarized by the Breit-Rabi formula

$$E^{(F)}_{\pm,M_F} = -\frac{A}{4} - g_I\mu_N BM_F \pm (I+\frac{1}{2})\frac{A}{2}(1+\frac{2M_F}{I+1/2}\eta+\eta^2)^{1/2}, \qquad (1.101a)$$
$$\eta = [(g_e\mu_B + g_I\mu_N)/(I+1/2)A]B$$

where η is a dimensionless parameter, and $M_F = M_S + m_I$.

1.7.4 Energy Levels for the Deuterium Atom and Atoms and Ions with One Unpaired s-Electron in the Ground-State (2S1/2 State)

Hydrogen has the second stable isotope, deuterium, for which $I = 1$. Thus, in accordance with the rule of addition of angular momenta (1.105) in zero magnetic field, $F = 3/2$ and $F = 1/2$; these levels have a fourfold and twofold degeneracy, respectively. In a magnetic field, the degeneracy is lifted completely, the level $F = 3/2$ is split into four sublevels, and $F = 1/2$, into two sublevels (see Fig. 1.11), i.e., there will be six sublevels in general. In the approximation of a strong magnetic field, the EPR spectrum will be produced, in accordance with the selection rules, by

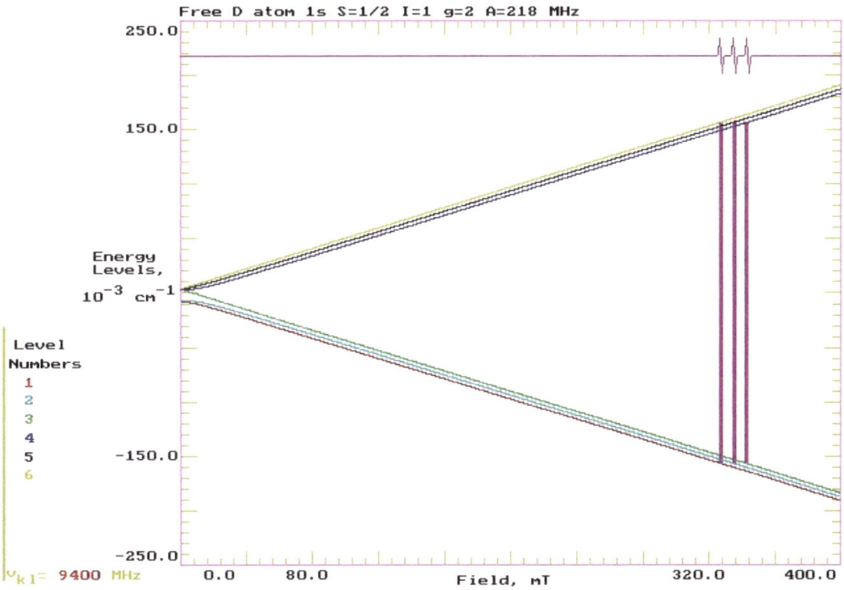

Fig. 1.11 Breit-Rabi diagram representing the magnetic field dependence of the sublevel energies and the EPR transitions for a free deuterium atom with $S = 1/2$, $I = 1$, HF interaction constant A for the X-band (9.4 GHz) is 0.218 GHz

three transitions, in contrast to the two transitions for the hydrogen atom. The HF interaction constant for deuterium, corresponding to the magnetic-field distance between adjacent lines will be substantially smaller than that for hydrogen because the nuclear magnetic moment of deuterium is much smaller than that of hydrogen. The 1s wavefunction (1.91) densities at the nucleus are, of course, identical in both cases.

Using the same principle and the Breit-Rabi formula (1.7.1
1), we can find the energy levels for other atoms and ions in the $^2S_{1/2}$ state: Li^0, Na^0, K^0, Rb^0, Cs^0, Cu^0, Ag^0, and Au^0 atoms; Zn^+, Cd^+, and Hg^+ singly charged positive ions; and Ga^{2+}, In^{2+}, Tl^{2+}, etc. doubly charged positive ions. Table 1.1 lists estimated parameters describing the isotropic HF interaction constans for a number of free atoms and ions, taken from [4]. The atomic parameters $|\Psi_{ns}(0)|^2$ and the isotropic hyperfine interaction constant A for a unit spin density in the corresponding ns orbital have been calculated for the most abundant nuclei of elements from Lithium to Lead from the Hartree-Fock-Slater atomic wavefunctions by Herman and Skillman. EPR spectra of these atoms and ions in the $^2S_{1/2}$ state have been observed in different solid state materials (ionic crystals, molecular crystals, glasses, etc.).

All calculations of the spin Hamiltonian parameters were performed using the program View EPR, written Grachev [5] (Figs. 1.12 and 1.13).

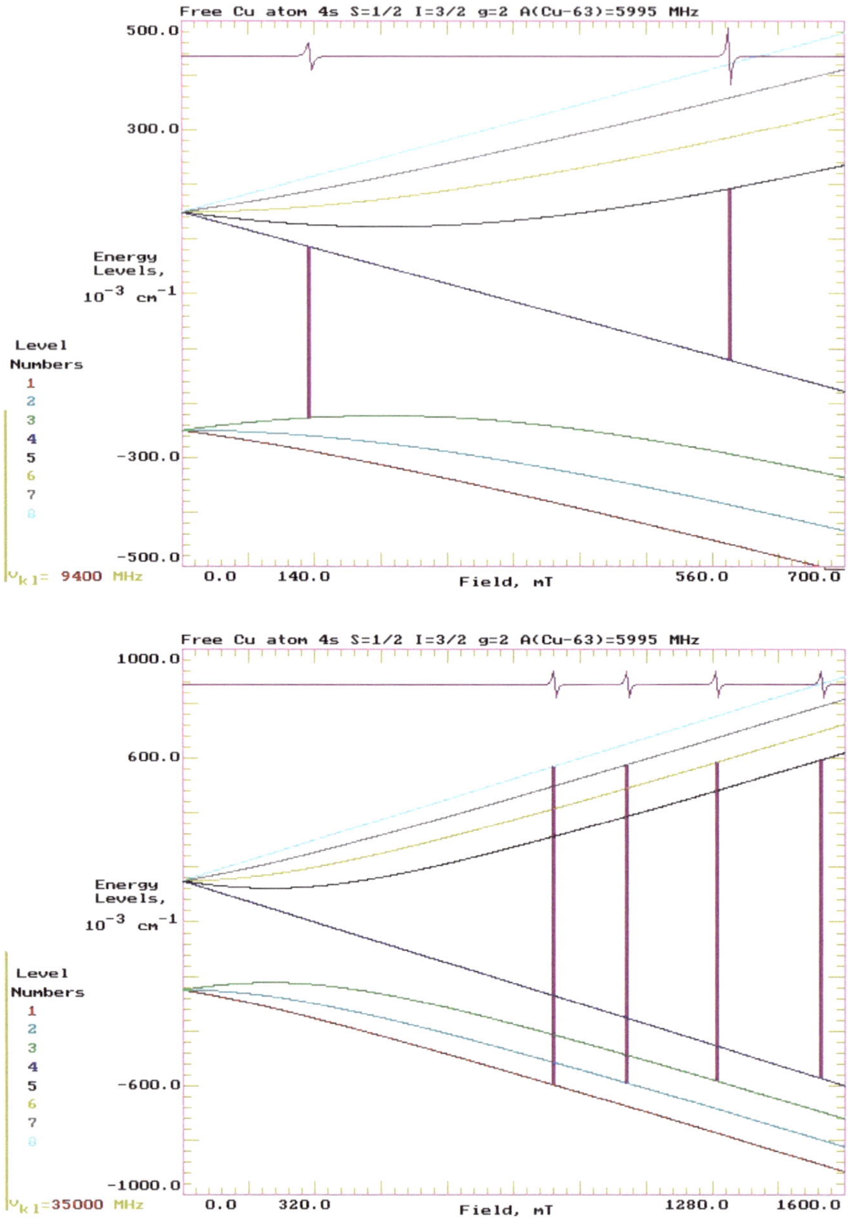

Fig. 1.12 Breit-Rabi diagram representing the magnetic field dependence of the sublevel energies and the EPR transitions for free Cu atoms with $S = 1/2$, $I = 3/2$, HF interaction constant $A = 5.995$ GHz: (*top panel*) for the X-band (9.4 GHz) and (*bottom panel*) for the Q-band (35 GHz)

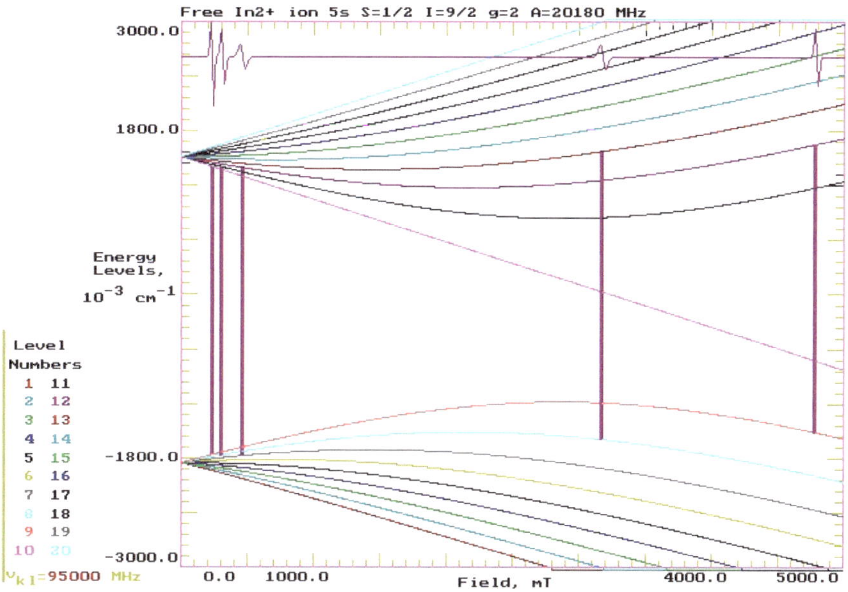

Fig. 1.13 Breit-Rabi diagram representing the magnetic field dependence of the sublevel energies and the EPR transitions for free In^{2+} ion with $S = 1/2$, $I = 9/2$, HF interaction constant $A = 20.180$ GHz for the W-band (95 GHz)

1.7.5 Hydrogen Atoms in Excited States. Spin-Orbit Interaction

Consider the excited states of a hydrogen atom. We restrict ourselves to the energy levels corresponding to the principal quantum number $n = 2$, i.e., to the levels formed by 2*s*- and 2*p*-electrons. In accordance with the solutions to the Schrödinger equation for the hydrogen atom, the energy levels depend only on the principal quantum number; in other words, the energy levels described by wavefunctions for the 2*s*- and 2*p*-electrons are degenerate. In fact, this is not the case, there being a certain splitting due to the relativistic corrections and, in particular, due to the spin-orbit interaction.

The wavefunction for the 2*s*-electron has the form

$$\Psi_{2s} = \left(\frac{1}{32\pi a_0^3}\right)^{\frac{1}{2}}\left(2 - \frac{r}{a_0}\right)\exp\left(-\frac{r}{2a_0}\right) \tag{1.108}$$

and the density of the wavefunction at the nucleus is given by

$$|\Psi_{2s}(0)|^2 = \frac{1}{8\pi a_0^3},$$

which is eight times smaller than the corresponding value for the 1s-electron, and thus, the hyperfine interaction constant for the 2s-electron A_{2s} = 1420/8 = 177.5 MHz. In a magnetic field, four energy sublevels are described by the Breit-Rabi formula of (1.101) with the corresponding HF interaction constant A_{2s}.

Above, we described the state of an electron with zero orbital angular momentum. Consider a state with a single 2p-electron having orbital and spin magnetic moments. In this example, we explain the nature of the spin-orbit interaction. The appropriate scheme of levels is illustrated by Fig. 1.27: there are two subterms for this system: $^2P_{1/2}$ and $^2P_{3/2}$, separated by the energy of the spin-orbit interaction, equal to $J\lambda = 3/2\lambda$.

Let us calculate the energy of the spin-orbit interaction for the 2p-electron. As already mentioned, the spin-orbit interaction is a purely relativistic effect caused by the motion of an electron in an orbit in the field of the positively charged nucleus. In the coordinate system associated with the electron, the nuclear charge is moving relative to the electron, thus creating a magnetic field at the position of the spin magnetic moment of the electron.

This magnetic field is given by

$$\vec{B}_l = -\frac{1}{c}(\vec{v} \times \vec{E}),$$

where $\vec{v}$ is the speed of an electron in the orbit, and the electric field generated by the core is given by

$$\vec{E} = -grad\varphi = -\frac{d\varphi}{dr}\frac{\vec{r}}{r}.$$

Then, the magnetic field $\boldsymbol{B}_l$ becomes

$$\vec{B}_l = -\frac{1}{cmr}\frac{d\varphi}{dr}(\vec{r} \times m\vec{v}) = -\frac{1}{cmr}\frac{d\varphi}{dr}\vec{G}_l,$$

where

$$\hat{\vec{G}}_l = \vec{r} \times \hat{\vec{p}}.$$

is the orbital angular momentum, which can be written as

$$\hat{\vec{G}}_l = \hbar\hat{\vec{l}}.$$

Thus, the Hamiltonian for the interaction of the magnetic moment of the electron spin with the magnetic field $\vec{B}_l$, which is the spin-orbit interaction, can be written as

$$\hat{H}_{SO} = -\hat{\vec{\mu}}_s \cdot \vec{B}_l = \frac{e\hbar^2}{m^2c^2r}\frac{d\varphi}{dr}\hat{\vec{l}} \cdot \hat{\vec{s}} = \lambda'\hat{\vec{l}} \cdot \hat{\vec{s}}. \tag{1.109}$$

Here, we used the expression for the Bohr magneton.

Equation (1.109) was derived in the approximation of the electron motion in an inertial reference system. Because the motion actually occurs in a non-inertial reference system, we should introduce a correction in the form of a factor ½ (given without proof). Then, the final expression for the spin-orbit coupling λ has the form

$$\lambda = \frac{e\hbar^2}{2m^2c^2r}\frac{d\varphi}{dr} = \frac{\hbar^2}{2m^2c^2}\frac{U'(r)}{r},$$

where $U(r)$ is the potential energy.

The total Hamiltonian in the Schrödinger equation for the hydrogen atom in an excited $2p$ state can be written with consideration for the spin-orbit interaction as

$$\hat{H} = -\frac{\hat{\vec{p}}^2}{2m} + U(r) + \frac{\hbar^2}{2m^2c^2}\frac{U'(r)}{r}\hat{\vec{l}} \cdot \hat{\vec{s}}. \tag{1.110}$$

Figure 1.14 shows the energy level scheme for hydrogen atoms in the ground $1s$ and the excited $2s$ and $2p$ electron configurations. The hyperfine structure for the $1s$ and $2s$ electron configurations and the energy level positions as a function of the magnetic field B are given. For the hydrogen atom in an excited state with the $2p$-

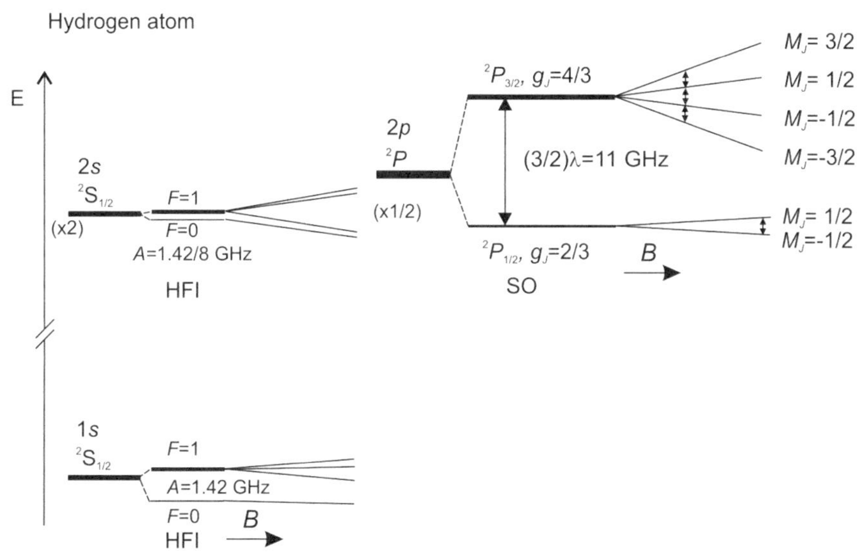

Fig. 1.14 Energy level scheme for hydrogen atoms in the ground $1s$ and excited $2s$ and $2p$ electron configurations. The hyperfine structure for the $1s$ and $2s$ electron configurations and the energy level positions as a function of the magnetic field B are shown

electron, the spin-orbit interaction constant is 7.3 GHz, i.e., in accordance with the Landé interval rule, the energy difference between the two subterms $^2P_{1/2}$ and $^2P_{3/2}$ is $1.5 \times 7.3 = 11$ GHz. According to experimental data, the $^2P_{1/2}$ level is 1.058 GHz below the 2*s*-electron level. In a magnetic field, the subterms $^2P_{1/2}$ and $^2P_{3/2}$ are split by the Zeeman effect into two and four sublevels, respectively. The Zeeman splitting is given by $\Delta E = g_J \mu_B B$, and the Landé *g*-factors are 2/3 and 4/3 for the $^2P_{1/2}$ and $^2P_{3/2}$ subterms, respectively. Thus, the energy levels of the hydrogen atom in the ground and excited states are calculated with high accuracy, and it is important that all these calculations are confirmed experimentally.

1.8 EPR in Condensed Matter

In the preceding sections, we considered atoms and ions in free states. When these atoms and ions are placed in a condensed matter, for example, in a crystalline or molecular matrix, they are exposed to the external influence of the environment. In a crystalline or molecular environment, the EPR spectra may either depend on the orientation of the sample in the magnetic field $\boldsymbol{B}$ (e.g., single crystals) or are superpositions of many different spectra of single crystals with random orientations (e.g., powder samples).

1.8.1 Atoms and Ions in the S-state (L = 0) in the Crystal Field

The interaction of magnetic moments with the surrounding environment usually significantly alters the paramagnetic properties of atoms and ions and their EPR spectra. The exceptions are the atoms and ions in the S-state, for which the interaction with the environment only slightly changes their paramagnetic properties because these atoms and ions have zero orbital angular momentum in the ground state. Atoms H^0, Li^0, Na^0, K^0, Rb^0, Cs^0, Cu^0, Ag^0, Au^0 of Group-I of the Periodic system; singly charged positive ions of Group-II elements, Zn^+, Cd^+, Hg^+; doubly charged positive ions of Group-III elements, Ga^{2+}, In^{2+}, Tl^{2+} (see Table 1.3), etc. have unpaired *ns*-electrons and are characterized by the $^2S_{1/2}$ state. Centers of this kind are rarely formed in as-grown crystals and can be produced by the trapping of irradiation-induced electrons or holes by non-paramagnetic impurities. The EPR spectra of the $^2S_{1/2}$ atoms and ions can be described, as in the free state, by the Hamiltonian (1.99) whose exact solution is given by the Breit-Rabi formula (1.101). An interaction with the environment leads to a slight change in the *g*-factor and in the hyperfine coupling constants because the wave function of the unpaired electron undergoes minor changes. The overlapping of the *ns*-wave function of the unpaired electron with wave functions of the ligand ions results in that the *ns*-wave

Table 1.3 Parameters obtained from the Hartree-Fock-Slater atomic wavefunctions by Herman and Skillman

Free atom (ion)	*ns* configuration	Isotope, nuclear spin *I*	Wavefunction density (a.u.)	Isotropic HFI constant (GHz)
Hydrogen (H^0)	1*s*, ground state 2*s*, excited state	H^1, 1/2 H^2 (D), 1 H^1, 1/2	$1/\pi = 0.318$ $1/\pi = 0.318$ $1/(8\pi)$	1.420 0.218 0.177
Lithium (Li^0)	2*s*	Li^7, 3/2	0.2101	0.365
Sodium (Na^0)	3*s*	Na^{23}, 3/2	0.780	0.927
Potassium (K^0)	4*s*	K^{39}, 3/2	1.066	0.228
Rubidium (Rb^0)	5*s*	Rb^{85}, 5/2	2.000	1.037
Cesium (Cs^0)	6*s*	Cs^{133}, 7/2	2.538	2.467
Copper (Cu^0)	4*s*	Cu^{63}, 3/2	4.617	5.995
Silver (Ag^0)	5*s*	Ag^{107}, 1/2	7.170	−1.831
Gold (Au^0)	6*s*	Au^{197}, 3/2	12.86	2.876
Zinc (Zn^+)	4*s*	Zn^{67}, 5/2	6.739	2.087
Cadmium (Cd^+)	5*s*	Cd^{111}, 1/2	10.03	−13.65
Mercury (Hg^+)	6*s*	Hg^{199}, 1/2	17.37	41.88
Gallium (Ga^{2+})	4*s*	Ga^{69}, 3/2	10.18	12.21
Indium (In^{2+})	5*s*	In^{115}, 9/2	14.06	20.18
Thallium (Tl^{2+})	6*s*	Tl^{205}, 1/2	22.97	183.8
Tin (Sn^{3+})	5*s*	Sn^{119}, 1/2	17.64	−43.921
Lead (Pb^{3+})	6*s*	Pb^{207}, 1/2	27.96	81.51

function density at the nucleus of the central atom or ion changes, which alters the isotropic hyperfine interaction constants in accordance with the equation

$$A = \frac{8\pi}{3} g_e \mu_B g_I \mu_N |\Psi_{ns}(0)|^2. \tag{1.111}$$

Interaction with the environment for the $^2S_{1/2}$-state leads to mixing of the wave functions of the *ns*-unpaired electron and those of the ligands carrying an orbital angular momentum, which causes a deviation of the *g*-factor of the impurity atom or ion (g_e) from the pure-spin *g*-factor $g_S = 2.00232$.

In addition, the S-state atoms and ions have half-filled electronic shells with p^3 ($^4S_{3/2}$ ground state), d^5 ($^6S_{5/2}$), and f^7 ($^8S_{7/2}$) configurations: e.g., the configurations $2p^3$, $3p^3$, $4p^3$, $3d^5$, $4d^5$, $5d^5$, $4f^7$, and $5f^7$. When these atoms or ions are placed in a crystal field, the ground state remains an S-state and only high-order perturbations, including the crystal field and spin-orbit interactions, can lift the four-, six- or eightfold degeneracy of the $^4S_{3/2}$, $^6S_{5/2}$ or $^8S_{7/2}$ ground states, respectively, and result in a small splitting of the ground state.

The anomalous isotropic hyperfine interactions have been observed for all the *S*-state atoms and ions with half-filled *p*-*d*-*f* shells, although these unpaired electrons must not show any isotropic hyperfine structure because of the spin density on the nucleus being zero. It has been shown that the hyperfine interaction originates from the effect of *s*-electrons or the so-called "core polarization", i.e., the magnetic polarization of an atomic or ionic core of closed n*s*-shells by an unfilled shell with a total spin *S* [6]. The electrostatic repulsion for an *s*-electron of a closed shell with a spin parallel to the total spin *S* of an atom or ion will not be the same as that for an electron with the opposite spin direction. This is a consequence of Pauli's exclusion principle, which prevents two electrons with parallel spins from occupying the same position in space.

Only a minor admixture of *ns*-unpaired electrons is necessary for explaining the hyperfine splitting. To find the hyperfine interaction splitting, the spin density at the origin should be calculated as [6]

$$\frac{4\pi}{2S}\sum_{n}\left[\left|\psi_{ns}^{\uparrow}(0)\right|^{2}-\left|\psi_{ns}^{\downarrow}(0)\right|^{2}\right]. \qquad (1.112)$$

In this equation, the arrow points in the direction of the *s*-electron spin in a closed shell, which coincides with (e.g., the upward arrow), or is opposite to the total spin of an atom or ion *S* [$S = 3/2$ for half-filled *p*-shell (p^3-configuration), $S = 5/2$ for half-filled *d*-shell (d^5-configuration) and $S = 7/2$ for half-filled *f*-shell (f^7-configuration)].

1.8.2 Transition Elements in Condensed Matter. Crystal Field Approach; Classification of Crystal Fields

A completely different scenario occurs for atoms and ions with an orbital angular momentum, e.g., transition elements [7]. Apart from having a half-filled shell, nd^5 (^{6}S term, $n = 3, 4, 5$), free ions are characterized by ground orbital states of the *nd*-shell that are either of the *D* or *F* type. In most cases, the crystal field has a predominately cubic symmetry with small distortions, in the sense that the splitting of the orbital states due to the cubic field exceeds that due to the terms of lower symmetry.

The transition-element ions have unfilled 3*d* electron shells (iron group or the first-transition series), unfilled 4*d* electron shells (palladium group or the second-transition series), unfilled 5*d* electron shells (platinum group or the third-transition series), unfilled 4*f* electron shells (rare-earth group), and unfilled 5*f* electron shells (actinide group).

When a free atom or ion is considered, the total Hamiltonian $\hat{H}_{Free}$ includes several basic interactions, the dominant term being the Coulomb interaction of electrons with the nucleus (potential energy due to the attraction between the electrons and the nucleus) and with each other (potential energy due to the repulsion between the electrons), $\hat{H}_{Coul}$. It is these interactions that form the terms ^{2S+1}L; then, there are the total orbital L and spin S moments (for not too heavy elements), with energies on the order of 10^5 cm^{-1}. In addition to the Coulomb interactions, there is a substantially weaker spin-orbit interaction $\hat{H}_{SO}$, which forms the subterms $^{2S+1}L_J$, i.e., it couples the orbital angular momentum L and the spin angular momentum S to give the total angular momentum J of the electron shell of a free atom (or ion); the energies of the spin-orbit interaction are on the order of 10^2–10^3 cm^{-1}. Significantly weaker interactions in moderate magnetic fields are the Zeeman interaction of electrons, $\hat{H}_{Zeem-e}$, and the hyperfine interaction $\hat{H}_{HFII}$; then follow the several orders of magnitude weaker nuclear Zeeman interaction $\hat{H}_{Zeem-n}$ and the nuclear quadrupole interaction $\hat{H}_Q$. As a result, we have

$$\hat{H}_{Free} = \hat{H}_{Coul} + \hat{H}_{SO} + \hat{H}_{Zeem-e} + \hat{H}_{HFI} + \hat{H}_{Zeem-n} + \hat{H}_Q. \tag{1.113}$$

The order in which the terms are introduced is determined by their relative strengths.

When the ions under consideration are placed in a condensed medium (e.g., in a crystal field), the crystal environment affects their electronic shells. This is usually named the interaction with the crystal field, $\hat{H}_{Crys}$, even when the environment is not a real crystal. The strength of this field may exceed the spin-orbit interaction, or be weaker. Since this crystal field has a major influence on changes in the EPR spectra in comparison with the spectra of free atoms and ions, the ratio between these two interactions is used for classifying the crystal fields.

We can consider the following three cases:

(1) The case of a weak crystal field, i.e., the interaction with the crystal field is weaker than the spin-orbit interaction, i.e., subterms can be formed and J is a 'good' quantum number.

$$\hat{H}_{Crys} < \hat{H}_{SO}. \tag{1.114}$$

In the case of the weak crystal field, the spin-orbit interaction couples $\boldsymbol{L}$ and $\boldsymbol{S}$ into a total angular momentum $\boldsymbol{J}$, which is then split by the crystal field.

(2) The case of an intermediate field, when the interaction with the crystal field is stronger then the spin-orbit interaction, but substantially weaker than the intra-Coulomb interactions.

$$\hat{H}_{Crys} > \hat{H}_{SO}. \tag{1.115}$$

In this case, the terms are formed, but subterms can not appear, i.e., J is not a 'good' quantum number. It is assumed that, first, the individual spins $\boldsymbol{s}$ couple to form $\boldsymbol{S}$ and the individual orbital angular momenta $\boldsymbol{l}$ couple to form $\boldsymbol{L}$ in accordance with Hund's rule, and then the interaction with the crystal field and, finally, the spin-orbit coupling are taken into account. This scheme is usually implemented for the transition elements with external 3*d*-shells.

(3) The case of a strong crystal field, when the interaction with the crystal field is substantially stronger than the spin-orbit interaction and is close to the Coulomb interactions. This interaction may even affect the formation of terms, i.e., the interaction with the crystal field may be comparable with the Coulomb interaction within the atom (ion). This case will not be considered here, we refer the reader to the authoritative sources for full details [see references to Chap. 1], because each particular situation requires an individual consideration and there is no general approach.

$$\hat{H}_{Crys} \gg \hat{H}_{SO}. \tag{1.116}$$

The ions of the 4d and 5d groups tend to form strong covalent bonds with neighboring ions, which gives rise to a strong crystal field. For d^n ions with $n \le 3$, the results are similar to those for the corresponding 3*d* ions, but for $n \ge 4$, there always exists the case of a strong crystal field.

We consider the situation with an intermediate field, which is the case for the transition atoms and ions with the unfilled *d*-shells, and that with a weak field, which is the case for atoms and ions of rare-earth elements with unfilled *f*-shells. In the case of the intermediate field, the spin-orbit interaction can not form a total angular momentum J (say, J is not a 'good' quantum number). Rare-earth ions imbedded in a medium are exposed to a weak crystal field and this perturbation is weaker than the spin-orbit interaction. This is so because the *f*-shells, being situated within the atom (ion), are weakly susceptible to the external crystal field (hence the name of the rare-earth scheme). So, we can assume that J is a good quantum number. The crystal field can not destroy the coupling between the spin and orbital angular momenta and will only split the subterm J into a number of Stark components.

1.9 The Case of Intermediate Crystal Field

1.9.1 Ground-State Terms for Transition Elements with Unpaired d-Electrons

For transition-metal ions in the condensed phase, the interactions responsible for the large splittings are the crystal-field splitting and the spin-orbit coupling. The combined effect of these two interactions lifts the orbital degeneracy of the energy levels for most transition metal ions completely, leaving a non-degenerate ground state, frequently with zero orbital angular momentum (quenching of the orbital angular momentum). It should be noted that only the lowest energy levels are populated with electrons if the energy splitting is much larger than kT, where k is the Boltzmann constant and T is temperature. In this case, EPR spectra can only be observed in the ground-state manifold of the paramagnetic system.

Let us consider the terms of free transition-element ions with unfilled d-shells; for definiteness, we take the elements with $3d$-electrons, although these schemes are also suitable, in general, for other d-shells. We construct a diagram in which the full orbital and spin angular momenta of the free atom (ion) are plotted along the vertical axis, and the horizontal axis shows the number of $3d$-electrons; as a result, we obtain the diagram in Fig. 1.28.

The ground-state terms are formed in accordance with Pauli's exclusion principle and Hund's rule, i.e., there can not be two states with identical quantum numbers and should be the maximum orbital angular momentum when the condition of the maximum spin moment is satisfied. The result is a "triangle" of Hund, because, for any configuration in the ground state, all the spin angular moments of $3d$-electrons are oriented in the same direction and the spin (multiplicity) is at a maximum.

Because J is not a 'good' quantum number, we do not consider in Fig. 1.15 the subterms of free atoms or ions. So, we have to introduce, instead of J and M_J, other designations for the levels associated with the orbital angular momentum. At the same time, the designations for the spin levels remain the same: S and M_S.

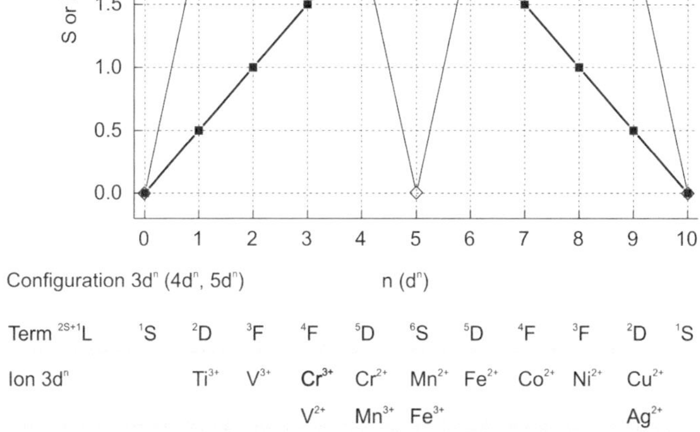

Fig. 1.15 Spectroscopic electronic properties (terms) of $3d^n$ ions in the ground state

Commonly, group-theoretic notations for the symmetry of the crystal field are introduced. It should be noted that the designations for the free atoms (ions) are also group-theoretic, used to describe the spherical symmetry.

The majority of semiconductors crystallize in the following four major structural forms. (i) The diamond structure, with each atom covalently bonded in a perfect tetrahedral fashion to its four neighbors, is adopted not only by diamond but also by silicon and germanium. (ii) The zincblende structure is related to the diamond structure in that it consists entirely of tetrahedrally-bonded atoms. The main difference is that, unlike diamond, each atom is bonded to four unlike atoms, with the result that the structure lacks an inversion center. (iii) The wurtzite structure also has entirely tetrahedrally bonded atoms; however, it is a hexagonal crystal system, unlike the cubic form of the zincblende lattice. (iv) There is a tendency for the zincblende and wurtzite structure to transform to the rock salt structure (NaCl) as covalent bonds become increasingly ionic.

We restrict our consideration to the case of a field with tetrahedral symmetry. In this book, we review the application of magnetic resonance methods to semiconductor materials that mainly crystallize into a tetrahedral lattice. In these materials, atoms of one type are surrounded tetrahedrally by atoms of the same or other type. In a field of this kind, the five orbital states of a *d*-electron split into a lower doublet (e) and an upper triplet (t_2), with the separation labeled 10Dq (Fig. 1.16).

For configurations containing more then a single *d*-electron, we have a choice of states for various electrons [8]. The crystal field energy is minimized by placing as many electrons as possible in the lower *e*-states, but we should take into account the strong coupling between the electron spins, expressed in the first Hund's rule according to which the ground state is that with the maximum spin. We begin by assuming that the spin coupling energy is higher than the crystal field energy; then, only two electrons can be placed in the *e*-orbitals with parallel spins because of Pauli's exclusion principle, but three further electrons can be accommodated in the t_2-orbitals with parallel spins. This maximum of five electrons with parallel spins corresponds to a half-filled shell with $S = 5/2$; since there is only a single configuration of this kind, the overall state is an orbital singlet with $L = 0$, or $^6S_{5/2}$. The ground state of any given configuration can be constructed by following these rules, whereby the crystal field energy is minimized subject to maximizing the spin and satisfying the restrictions imposed by Pauli's exclusion principle. The orbital multiplicity of the ground state (see Fig. 1.17) is given by the number of ways in which electrons can be distributed between orbitals of the same energy. A similar procedure can be carried out for an octahedral field by placing the triplet t_2-states below, rather than above, the doublet e-states.

In some compounds with ions of the 3*d* group and most of the 4*d* and 5*d* groups, the magnetic behaviour suggests that the crystal field energy is substantially higher and suffices to outweigh the spin coupling. We can construct the appropriate ground states for this situation by placing electrons as far as possible in the *e*-orbitals. For configurations containing up to two electrons, the ground states are the same as those in Fig. 1.18, but the remainder are different (with the exception of d^7, d^8 and d^9).

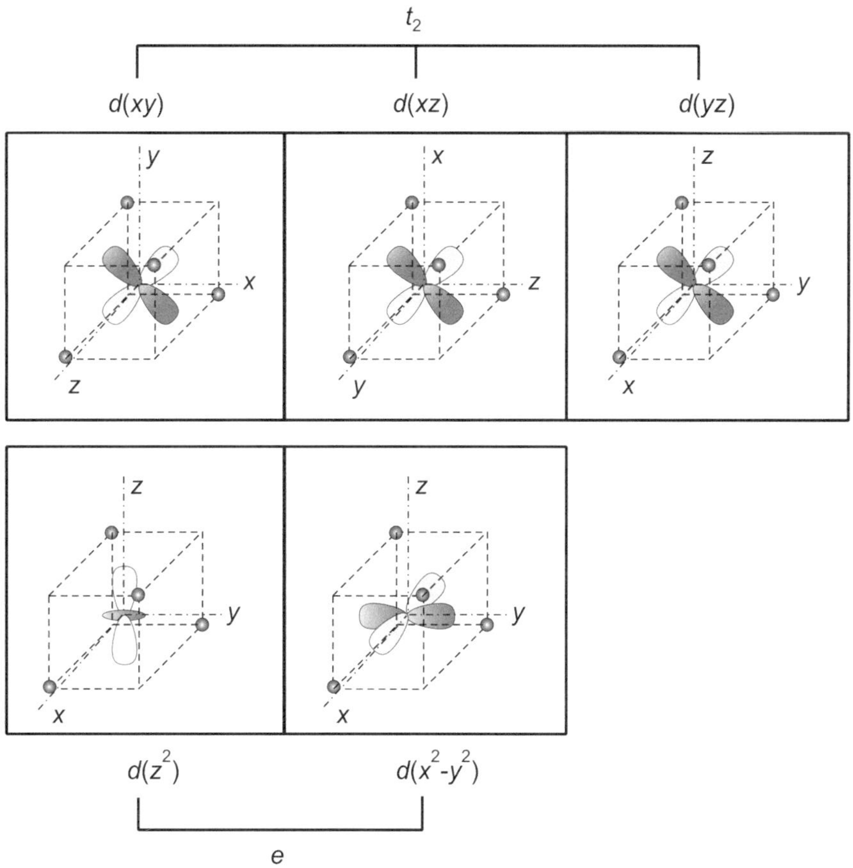

Fig. 1.16 Orbitals e and t_2 represent the energy levels corresponding to the twofold degenerate d-wave functions $(3z^2 - r^2)/r^2$, $(x^2 - y^2)/r^2$, and threefold degenerate wave functions xy/r^2, yz/r^2 and zx/r^2, in a tetrahedral crystal field. The geometrical arrangement of the ligands is shown schematically. A schematic representation of the central ion orbitals of symmetry e and t_2 in a tetrahedral crystal field is depicted

The effect of a crystal field on a transition-metal ion is illustrated by Fig. 1.18. The figure shows the schemes of lowest energy levels and the ground-state splitting for the d-configurations under the action of a tetrahedral (T_d) or axial (trigonal C_{3v}) crystal field, spin-orbit coupling ($\lambda \boldsymbol{LS}$), and an external magnetic field $\boldsymbol{B}$ parallel to the symmetry (z) axis, constructed using the intermediate crystal field approach, which assumes that the crystal field energy is stronger than the spin-orbit coupling. The inversion of the energy diagrams for d^3 and d^8, compared with those for d^7 and d^2, follows from the fact that the former can be regarded as a half-filled or completely occupied shell with two holes, and the latter, as a half-filled or empty shell plus two electrons [8]. Various d^n ions are discussed below in order of increasing number of unpaired d electrons.

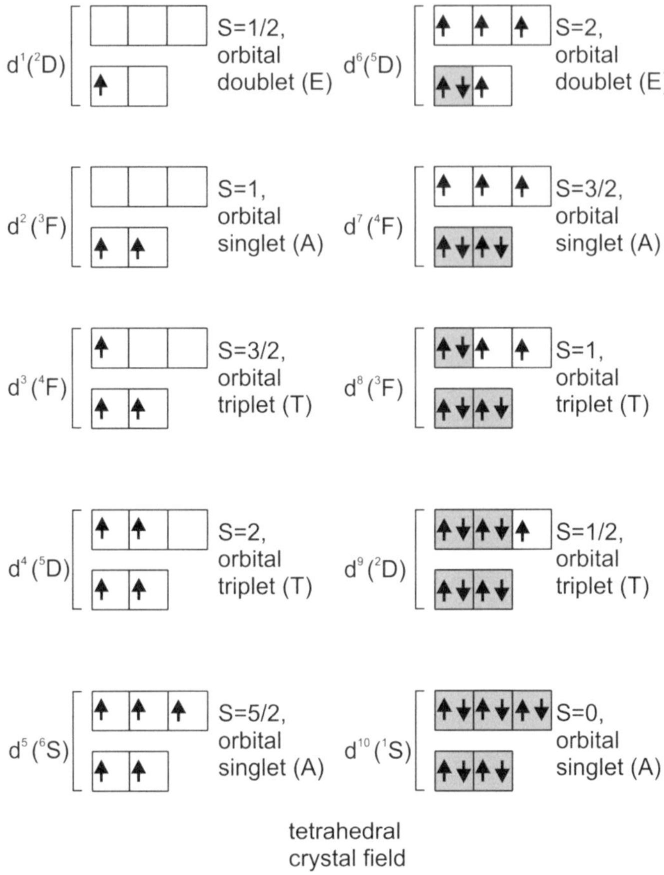

Fig. 1.17 Construction of the ground states of the *d*-configurations in a tetrahedral crystal field by using the intermediate crystal field approach, which assumes the spin-spin coupling to be stronger than the crystal field energy [8]

$3d^1$ The ground state for free $3d^1$ ions is ^{2}D, $S = 1/2$. The ^{2}D ground state is split by a tetrahedral field into an ^{2}E orbital doublet and ^{2}T orbital triplet, the ground state being the ^{2}E state, which can be further split by an axial field (or by the Jan-Teller effect) to give two nondegenerate orbital-singlet states.

$3d^2$ The ground state for free $3d^2$ ions is ^{3}F, $S = 1$. A tetrahedral field splits the ^{3}F ground state into two triply degenerate states and the orbitally nondegenerate ^{3}A singlet state, which is the lowest in energy. The threefold spin degeneracy of this orbital singlet is lifted by the combined action of the axial crystal field and the spin-orbit coupling.

$3d^3$ The ground state for free $3d^3$ ions is ^{4}F, $S = 3/2$. The ^{4}F ground state is split by a tetrahedral field into two triply degenerate states and the orbitally nondegenerate singlet state with an orbital triplet $^4T_1$ being the lowest in energy, which is

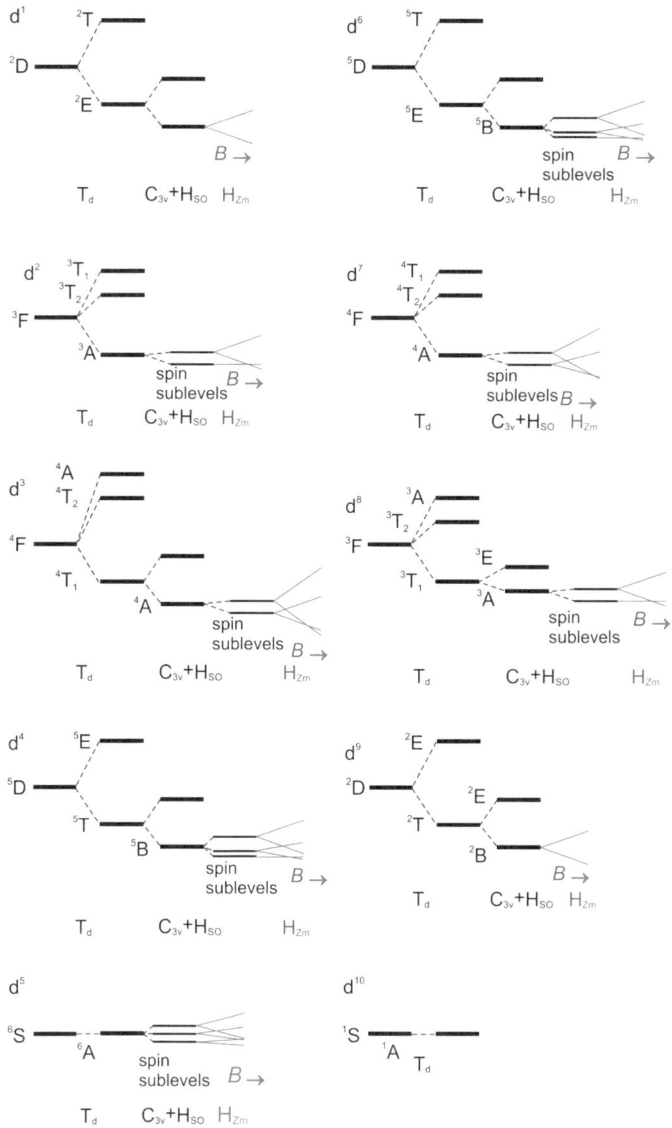

Fig. 1.18 Ground-state splitting for the *d*-configurations under the action of a tetrahedral (T_d) or axial (trigonal C_{3v}) crystal field, spin-orbit coupling ($\lambda \boldsymbol{LS}$), and an external magnetic field $\boldsymbol{B}$ parallel to the symmetry (z) axis), found by using the intermediate crystal field approach, which assumes the crystal field energy to be stronger than the spin-orbit coupling. The inversion of the energy diagrams for d^3 and d^8, compared with those for d^7 and d^2, follows from the fact that the former can be regarded as a half-filled or completely occupied shell with two holes, and the latter, as a half-filled or empty shell plus two electrons [8–10]

further split by an axial field to give a nondegenerate ^{4}A orbital-singlet state, which is the lowest in energy. The fourfold spin degeneracy with the total spin $S = 3/2$ is split by axial fields into two Kramers doublets.

$3d^4$ The ground state for free $3d^4$ ions is ^{5}D, $S = 2$. The ^{5}D ground state is split by a tetrahedral field into an ^{5}E orbital doublet and ^{5}T orbital triplet, the ground state being the ^{5}T state, which can be further split by an axial field (or by the Jan-Teller effect), resulting in a nondegenerate ^{2}B orbital-singlet state, which is the lowest in energy.

$3d^5$ The ground state for free $3d^5$ ions is ^{6}S, $S = 5/2$. When these ions are placed in a tetrahedral crystal field, the ground state is the nondegenerate ^{6}A orbital-singlet state. High-order perturbations, including simultaneously the crystal field and spin-orbit coupling, are necessary to split this state. The sixfold spin degeneracy with the total spin $S = 5/2$ is lifted by the combined action of the axial crystal field and the spin-orbit coupling and is split into three Kramers doublets.

$3d^6$ The ground state for free $3d^4$ ions is ^{5}D, $S = 2$. The ^{5}D ground state is split by a tetrahedral field into an ^{5}E orbital doublet and ^{5}T orbital triplet, the ground state being the ^{5}E state, which can be further split by an axial field (or by the Jan-Teller effect) into two nondegenerate orbital-singlet states.

$3d^7$ The ground state for free $3d^7$ ions is ^{4}F, $S = 3/2$. A tetrahedral field splits the ^{4}F ground state into two triply degenerate states and the orbitally nondegenerate ^{4}A singlet state, which is the lowest in energy. The fourfold spin degeneracy with the total spin $S = 3/2$ is lifted by the combined action of the axial crystal field and the spin-orbit coupling and is split into two Kramers doublets.

$3d^8$ The ground state for free $3d^8$ ions is ^{3}F, $S = 1$. The ^{3}F ground state is split by a tetrahedral field with an orbital triplet ^{3}T$_1$ being the lowest in energy, which is further split by an axial field to give a nondegenerate ^{3}A orbital-singlet state, which is the lowest in energy. The threefold spin degeneracy of this orbital singlet is lifted by the combined action of the axial crystal field and the spin-orbit coupling.

$3d^9$ The ground state for free $3d^9$ ions is ^{2}D, $S = 1/2$. The ^{2}D ground state is split by a tetrahedral crystal field into an ^{2}E orbital doublet and ^{2}T orbital triplet, the ground state being the ^{2}T state, which can be further split by an axial field (or by the Jan-Teller effect) to give a nondegenerate ^{2}B orbital-singlet state, which is the lowest in energy.

1.9.2 Quenching of the Orbital Angular Momentum in the Orbitally Nondegenerate Singlet State

Consider the example of the most extensively studied case of the Cr^{3+} ion (see Fig. 1.28) with the electron configuration $3d^3$. According to the rule of addition of angular momentum and Hund's rule, the maximum spin angular momentum corresponds to the maximum value of M_S, i.e., $S = 3/2$. For each d-electron, $s = 1/2$ and the projection $m_s = +1/2$, so the quantum numbers for electrons must be different from the projection of the orbital angular momentum. Each d-electron has $l = 2$ and a projection must be selected by the maximum value in accordance with Hund's rule. However, they can not be chosen the same as, e.g., $m_l = +2$, there may be only three options for three electrons $m_l = +2, +1, 0$. This gives a total of $M_L = 3$ and the total orbital angular momentum is $L = 3$, the result being a term ^4F. Likewise, you can get all the terms presented in the graph (Fig. 1.28). For example, for a half-filled $3d$ shell $3d^5$ with the maximum spin, we obtain $S = 5/2$ and each of the five electrons must have a different projection of the orbital angular momentum, that is we have to take all the values $m_l = +2, +1, 0, -1, -2$ and the resulting sum gives $M_L = 0$, i.e., $L = 0$, and the term has the form ^6S.

When an ion with an unfilled d-shell is placed in the crystal field (for definiteness, we consider the ^4F term of a Cr^{3+} ion with the $3d^3$ electron shell), the level positions and their splitting change. In the case of the intermediate crystal field, $\hat{H}_{Crys} > \hat{H}_{SO}$.

Figure 1.19a shows the energy level scheme for the $3d^3$ electronic configuration in an octahedral crystal field, in which the triplet t_2-state lies below the doublet e-state instead of being in the position in Fig. 1.29, shown for the tetrahedral crystal field. These energy levels (Fig. 1.19a) represent the so-called zero-order approximation, being the energy eigenvalues for the Hamiltonian

$$\hat{H}_0 = \hat{H}_{Coul} + \hat{H}_{Crys}. \tag{1.117}$$

For a free ion, the ^4F term will be multiply degenerate similarly to the orbital angular momentum ($2L + 1 = 7$) and spin angular momentum ($2S + 1 = 4$), i.e., there is a 7 * 4 = 28 fold degeneracy. In a cubic crystal field, the term ^4F is split into three orbital levels of two orbital triplets and one orbital singlet, as shown in Fig. 1.19a, while the spin state remains fourfold degenerate. As before in the case of free atoms and ions, we are interested only in the ground state, which is a non-degenerate orbital singlet in this system. We denote the wave function of this

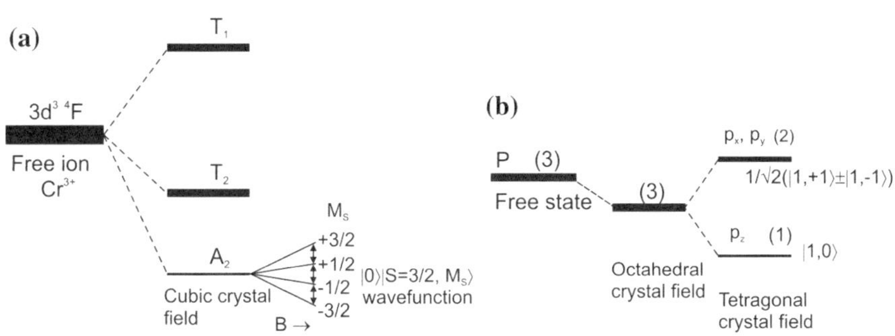

Fig. 1.19 **a** Energy level scheme for $3d^3$ electronic configuration in an octahedral crystal field, in which the triplet t^2-state lies below the doublet e-state instead of being in the position in Fig. 1.16, shown for a tetrahedral crystal field. **b** Energy level scheme for np electronic configuration in an octahedral crystal field with a tetragonal distortion

state as $|0\rangle$, although, in general, the wave function $|L,M_L,S,M_S\rangle$ is transformed into a kind of wave function of the type $|\Gamma,\gamma,S,M_S\rangle$, where the first two letters are used to indicate the levels according to irreducible representations of the cubic group (Γ takes the values A_2, T_2 and T_1 in the figure). We are not going to specifically discuss these designations, but, because the wave functions for the orbital motion and for the spin do not mix in our approximation, the eigenfunctions (wave functions) of H_0 can be written in general as a product of an orbital part, which we label $|n\rangle$, and a spin factor, labeled $|S,M_S\rangle$. It can be written for the ground state as a product of the wave function for the orbital motion in a cubic crystal field and spin wave function, i.e., as $|0\rangle|S,M_S\rangle$ or, because the spin S remains unchanged, as $|0\rangle|M_S\rangle$.

The lower state is a nondegenerate orbital singlet. We use the perturbation theory for nondegenerate states and regard as the perturbation the spin-orbit interaction and the Zeeman interaction of the electron, given by

$$\hat{H}' = \hat{H}_{SO} + \hat{H}_{Zeem-e}. \tag{1.118}$$

According to the perturbation theory for the nondegenerate states, the ground-state energy can be written, up to second-order perturbation, as

$$\begin{aligned} E_0 &= E_0^{(0)} + E_0^{(1)} + E_0^{(2)} \\ E_0 &= E_0^{(0)} + \langle 0|\hat{H}'|0\rangle + \sum\nolimits_n' \frac{\langle 0|\hat{H}'|n\rangle\langle n|\hat{H}'|0\rangle}{E_0^{(0)} - E_n^{(0)}}. \end{aligned} \tag{1.119}$$

The first term gives the energy of the ground state in the zero-order perturbation theory, which is derived from the Hamiltonian (1.117) and is shown in Fig. 1.19a. This term is not directly involved in the level splitting and can be regarded as a kind of zero point of reference for energy.

The second and third terms in (1.119) give the correction to energy in the first- and second-order perturbation theory, $E_0^{(1)}$ and $E_0^{(2)}$.

Consider the perturbation (1.118), in which

$$\hat{H}_{Zeem-e} = g_L \mu_B \hat{\vec{L}} \cdot \vec{B} + g_S \mu_B \hat{\vec{S}} \cdot \vec{B} = \mu_B(\hat{\vec{L}} + 2\hat{\vec{S}}) \cdot \vec{B},$$

where $g_L = 1$, $g_S = 2$ are used, and

$$\hat{H}_{SO} = \lambda \hat{\vec{L}} \cdot \hat{\vec{S}}.$$

We have as a result

$$\hat{H}' = \mu_B(\hat{\vec{L}} + 2\hat{\vec{S}}) \cdot \vec{B} + \lambda \hat{\vec{L}} \cdot \hat{\vec{S}}. \tag{1.120}$$

Substituting (1.120) into (1.119), we write the ground-state energy up to the second-order perturbation. We consider only the wave function of the ground and excited states for the orbital motion, with the spin wave functions disregarded in this approximation. When calculating the expected values, it is convenient to perform integration only over the orbital variables. As a result, the orbital angular momentum will be hidden in the parameters, leaving the spin-dependent part in the operator form. Then, as a result, we obtain the so-called "spin Hamiltonian", which depends only on the operators of the spin moments (for electron, and subsequently the nucleus) and on the magnetic field.

Substituting (1.120) into (1.119),we obtain the second term of (1.119) as

$$
\begin{aligned}
&\langle 0|\hat{H}'|0\rangle = \langle 0|\mu_B(\hat{\vec{L}}+2\hat{\vec{S}})\cdot\vec{B}+\lambda\hat{\vec{L}}\cdot\hat{\vec{S}}|0\rangle = \\
&\langle 0|\mu_B\hat{\vec{L}}\cdot\vec{B}|0\rangle + \langle 0|2\mu_B\hat{\vec{S}}\cdot\vec{B}|0\rangle + \langle 0|\lambda\hat{\vec{L}}\cdot\hat{\vec{S}}|0\rangle = \\
&\mu_B\vec{B}\cdot\langle 0|\hat{\vec{L}}|0\rangle + 2\mu_B\hat{\vec{S}}\cdot\vec{B}\langle 0\mid 0\rangle + \lambda\hat{\vec{S}}\cdot\langle 0|\hat{\vec{L}}|0\rangle
\end{aligned}
$$

It can be shown that the integral $\langle 0|\hat{L}|0\rangle$ vanishes for a nondegenerate level, i.e., $\langle 0|\hat{L}|0\rangle = 0$, also due to the orthogonality of the wave functions $\langle 0\mid 0\rangle = 1$. Then, the second term in (1.119), i.e., the expression for the energy in the first-order perturbation theory, has the form

$$
\hat{H}_0^{(1)} = \langle 0|\hat{H}'|0\rangle = g_S\mu_B\hat{\vec{S}}\cdot\vec{B}. \tag{1.121}
$$

Let us prove that $\langle 0|\hat{L}|0\rangle = 0$. The operator of the orbital angular momentum is a purely imaginary operator, e.g.,

$$
\hat{L}_Z = -i(x\frac{\partial}{\partial y} - y\frac{\partial}{\partial x}).
$$

This operator is a Hermitian operator, because, being measurable quantities, the eigenvalues must be real numbers,

$$
\hat{L}_Z|n\rangle = M_L|n\rangle. \tag{1.122}
$$

For the nondegenerate case, the wave function can always be expressed in the form of a real-valued function (any state that describes the complex wave function must be at least twice degenerate because there always is at least one more independent eigenfunction corresponding to the state with the same energy, that is the complex conjugate function). In accordance with (1.122), applying an imaginary operator to the real eigenfunction must lead to an imaginary or zero eigenvalue, then $M_L = 0$, as required. In this state, the expectation value of $\hat{L}$ is zero. In this case, the orbital angular momentum is said to be "quenched" and the magnetic moment is only caused by the spin. The quenching of the orbital angular momentum is one of the most important effects of the crystal field surrounding the paramagnetic ion. This is the reason why the EPR is frequently named Electron Spin Resonance (ESR). The electron being on the non-degenerate energy level, as it loses the ability to move.

1.9.3 The Spin Hamiltonian

Let us consider the second-order effects in the perturbation theory, which arise from matrix elements of the orbital angular momentum between the ground orbital state and the excited orbital states. The second-order effects on the manifold of the spin sublevels in the ground singlet orbital level can be written as

$$
\begin{aligned}
\hat{H}_0^{(2)} &= -\sum_n{}' \frac{\langle 0|\hat{H}'|n\rangle\langle n|\hat{H}'|0\rangle}{E_n^{(0)} - E_0^{(0)}} = \\
&-\sum_n{}' \frac{\langle 0|\mu_B(\hat{\vec{L}}+2\hat{\vec{S}})\cdot\vec{B}+\lambda\hat{\vec{L}}\cdot\hat{\vec{S}}|n\rangle\langle n|\mu_B(\hat{\vec{L}}+2\hat{\vec{S}})\cdot\vec{B}+\lambda\hat{\vec{L}}\cdot\hat{\vec{S}}|0\rangle}{E_n^{(0)} - E_0^{(0)}} = \\
&-\sum_n{}' \frac{\langle 0|\mu_B\hat{\vec{L}}\cdot\vec{B}+\lambda\hat{\vec{L}}\cdot\hat{\vec{S}}|n\rangle\langle n|\mu_B\hat{\vec{L}}\cdot\vec{B}+\lambda\hat{\vec{L}}\cdot\hat{\vec{S}}|0\rangle}{E_n^{(0)} - E_0^{(0)}} = \\
&-(\mu_B\vec{B}+\lambda\hat{\vec{S}})\sum_n{}' \frac{\langle 0|\hat{\vec{L}}|n\rangle\langle n|\hat{\vec{L}}|0\rangle}{E_n^{(0)} - E_0^{(0)}}(\mu_B\vec{B}+\lambda\hat{\vec{S}}).
\end{aligned}
\tag{1.123}
$$

The quantities

$$
\overleftrightarrow{\Lambda} = -\sum_n{}' \frac{\langle 0|\hat{\vec{L}}|n\rangle\langle n|\hat{\vec{L}}|0\rangle}{E_n^{(0)} - E_0^{(0)}}, \tag{1.124}
$$

with components in the form

$$
\Lambda_{ij} = -\sum_n{}' \frac{\langle 0|\hat{L}_i|n\rangle\langle n|\hat{L}_j|0\rangle}{E_n^{(0)} - E_0^{(0)}}, \tag{1.124a}
$$

where i and j take the values x, y, z, are tensor quantities formed from the matrix elements of L, which connect the ground orbital state $|0\rangle$ having energy $E_0^{(0)}$ with excited orbital states $|n\rangle$ having energy $E_n^{(0)}$ (see, e.g., Fig. 1.21).

As a result, formula (1.123) is written as

$$\hat{H}_0^{(2)} = \mu_B^2 \vec{B} \cdot \overleftrightarrow{\Lambda} \cdot \vec{B} + 2\lambda\mu_B \hat{\vec{S}} \cdot \overleftrightarrow{\Lambda} \cdot \vec{B} + \lambda^2 \hat{\vec{S}} \cdot \overleftrightarrow{\Lambda} \cdot \hat{\vec{S}}. \quad (1.125)$$

The first term is independent of spin, so it simply shifts all the levels and is of no interest for EPR, therefore, it is not to be considered. Combining the above two corrections to the energies, we obtain

$$\hat{H}_0^{(1)} + \hat{H}_0^{(2)} = g_S\mu_B \hat{\vec{S}} \cdot \vec{B} + 2\lambda\mu_B \hat{\vec{S}} \cdot \overleftrightarrow{\Lambda} \cdot \vec{B} + \lambda^2 \hat{\vec{S}} \cdot \overleftrightarrow{\Lambda} \cdot \hat{\vec{S}}. \quad (1.126)$$

The first two terms can be combined to form

$$g_S\mu_B \hat{\vec{S}} \cdot \vec{B} + 2\lambda\mu_B \hat{\vec{S}} \cdot \overleftrightarrow{\Lambda} \cdot \vec{B} = \mu_B \hat{\vec{S}} \cdot (g_S \overleftrightarrow{1} + 2\lambda \overleftrightarrow{\Lambda}) \cdot \vec{B} = \mu_B \hat{\vec{S}} \cdot \overleftrightarrow{g} \cdot \vec{B},$$

where $\overleftrightarrow{1}$ is the unit tensor (unit diagonal matrix) and

$$\overleftrightarrow{g} = g_S \overleftrightarrow{1} + 2\lambda \overleftrightarrow{\Lambda}.$$

As a result, we obtain the expression

$$\hat{H}_0^{(1)} + \hat{H}_0^{(2)} = \mu_B \hat{\vec{S}} \cdot \overleftrightarrow{g} \cdot \vec{B} + \hat{\vec{S}} \cdot \overleftrightarrow{D} \cdot \hat{\vec{S}},$$

where

$$\overleftrightarrow{D} = \lambda^2 \overleftrightarrow{\Lambda}.$$

Two new parameters in the form of tensors $\boldsymbol{g}$ and $\boldsymbol{D}$ were introduced, which include tensor $\overleftrightarrow{\Lambda}$. This tensor reflects two different effects based on a single phenomenon, which is the impurity of orbital excited states due to the spin-orbit interaction. As a result, we have a new Hamiltonian, named the spin Hamiltonian:

$$\hat{H} = \mu_B \hat{\vec{S}} \cdot \overleftrightarrow{g} \cdot \vec{B} + \hat{\vec{S}} \cdot \overleftrightarrow{D} \cdot \hat{\vec{S}}. \quad (1.127)$$

In this spin Hamiltonian, there is no orbital angular momentum, which is "hidden" in the following parameters: anisotropic g-factor in the form of a tensor $\boldsymbol{g}$, which is an analogue of the g factor of free atoms (ions), and an expression describing the so-called 'fine-structure splitting', where the tensor $\boldsymbol{D}$ describes this splitting. Equation (1.127) reflects that the orbital angular momentum includes a second-order contribution, which is not exactly zero. We see that the fine-structure splitting is independent of the magnetic field, i.e., it can be observed in zero magnetic field. As shown below, the fine structure can be observed only for systems with spin $S > 1/2$.

The first term represents the Zeeman energy of the system with an anisotropic g-factor in an external magnetic field. It is seen that the ground state due to the spin-orbit coupling is partly mixed with the excited states with orbital angular momenta. As a result, the magnetisation is now due not only to the spins and the **g** tensor differs from the free spin g-factor of 2.0. The main goal of the theory is to calculate the tensors **g** and **D**. $\hat{\vec{S}}$ in the spin Hamiltonian (1.127) is the operator corresponding to the effective spin, which does not necessarily match the actual spin. If the angular momentum of the system is due only to the spin, the **g** tensor must be an isotropic scalar quantity equal to $g_S = 2.00232\ldots$. Any anisotropy of the g-factor or its deviation from the purely spin g-factor is associated with the $\overleftrightarrow{\Lambda}$ tensor, which includes the contributions of the orbital angular momentum of excited states.

1.9.4 Application to an Orbital Triplet in the Ground State

The energy-level scheme for the $3d^3$ electronic configuration in a tetrahedral crystal field can be obtained by inversion of the energy diagrams for d^3 in an octahedral crystal field (Fig. 1.19a). As can be seen from Fig. 1.18, an orbital triplet for the $3d^3$ electronic configuration in a tetrahedral crystal field is the ground state T_1. In the tetrahedral symmetry, if we represent the orbital triplet as a fictitious angular momentum $\tilde{l} = 1$, then the $(2\tilde{l} + 1)(2S + 1)$ manifold of states splits under the action of the spin-orbit coupling into series of levels characterized by a fictitious total angular momentum [9]

$$\hat{\vec{\tilde{j}}} = \hat{\vec{\tilde{l}}} + \hat{\vec{S}} : \tilde{j} = S + 1, S, |S - 1|$$

The fictitious angular momentum $\tilde{j}$ is introduced here in a manner that demonstrates its resemblance to the real angular momentum $\boldsymbol{J} = \boldsymbol{L} + \boldsymbol{S}$ used in the ordinary atomic theory. The energy separations are just those that would be produced by an effective spin-orbit coupling parameter, which we write as $\tilde{\lambda}$, i.e., by

$$\hat{\tilde{H}}_{SO} = \tilde{\lambda}(\hat{\vec{\tilde{l}}} \cdot \hat{\vec{S}}), \tag{1.128}$$

so that they obey the interval rule known as the Landé interval rule for the real angular momentum

$$E(\tilde{j}) - E(\tilde{j} - 1) = \tilde{\lambda}\tilde{j}. \tag{1.129}$$

The Zeeman splitting becomes just

$$\hat{\tilde{H}}_{Zeem} = \tilde{g}_J \mu_B (\vec{B} \cdot \hat{\vec{\tilde{j}}}). \tag{1.130}$$

The effective orbital g-factor has the form of Landé g-factor for the real angular momentum

$$\tilde{g}_J = \frac{1 + \tilde{j}(\tilde{j}+1) + S(S+1) - \tilde{l}(\tilde{l}+1)}{2\tilde{j}(\tilde{j}+1)}, \tag{1.131}$$

where $\tilde{l} = 1$.

1.10 Anisotropic g-Factor

The orientation dependence of the g-factor reflects the orientation dependence of the Zeeman levels. Consider the first term in the spin Hamiltonian (1.127), which is an expression for the Zeeman energy

$$\hat{H}_{Zeem} = \mu_B \vec{B} \cdot \overleftrightarrow{g} \cdot \hat{\vec{S}}.$$

In general, this part of the spin Hamiltonian (1.127) can be written for arbitrarily chosen axes in the matrix form

$$\hat{H}_{Zeem} = \mu_B \left| B_x, B_y, B_z \right| \begin{vmatrix} g_{xx} & g_{xy} & g_{xz} \\ g_{yx} & g_{yy} & g_{yz} \\ g_{zx} & g_{zy} & g_{zz} \end{vmatrix} \begin{vmatrix} \hat{S}_x \\ \hat{S}_y \\ \hat{S}_z \end{vmatrix}. \tag{1.132}$$

The double indices, e.g., those in g_{yx}, reflect the contribution to the g factor along the y axis from the magnetic field applied along the x axis. The expression $\vec{B}\overleftrightarrow{g}$ can be regarded as a vector of a certain effective magnetic field in which the electron spin is placed. That is the external magnetic field is transformed in a crystal into an effective magnetic field. Equation (1.132) can be significantly simplified if the system is brought to the principal axes in which the tensor is diagonal. To do this, let us combine the x, y and z axes with the symmetry axes of the system under consideration in the form of a molecule, defect or crystal. Then we have

$$\hat{H}_{Zeem} = \mu_B \left| B_x, B_y, B_z \right| \begin{vmatrix} g_x & 0 & 0 \\ 0 & g_y & 0 \\ 0 & 0 & g_z \end{vmatrix} \begin{vmatrix} \hat{S}_x \\ \hat{S}_y \\ \hat{S}_z \end{vmatrix} = \mu_B (g_x B_x \hat{S}_x + g_y B_y \hat{S}_y + g_z B_z \hat{S}_z). \tag{1.132a}$$

In the case of the axial symmetry of the system, $g_z = g_{\|}$ and $g_x = g_y = g_{\perp}$ and, as a result, the spin Hamiltonian for the Zeeman interaction can be written as

$$\hat{H}_{Zeem} = \mu_B[g_{//}B_z\hat{S}_z + g_{\perp}(B_x\hat{S}_x + B_y\hat{S}_y)]. \tag{1.132b}$$

The effective value of the g-factor in any direction of the magnetic field is given by the expression

$$g_{eff}^2 = g_x^2\cos^2(\angle\vec{B},x) + g_y^2\cos^2(\angle\vec{B},y) + g_z^2\cos^2(\angle\vec{B},z), \tag{1.133}$$

transformed in the case of the axial symmetry into the formula

$$g_{eff}^2 = g_{//}^2\cos^2\theta + g_{\perp}^2\sin^2\theta, \tag{1.133a}$$

where θ is the angle between the magnetic field and the axial axis of the system (a defect in the crystal, molecule, etc.).

We consider as an example calculating the anisotropic g-factor of the simplest system in the form of a single unpaired p-electron, i.e., a system in the P state. The orbital angular momentum $L = 1$ and the wave eigenfunctions for the angular momentum operator are given by

$$|L, M_L\rangle = |1,0\rangle, |1,+1\rangle, |1,-1\rangle. \tag{1.134}$$

These functions can be expressed in terms of the real shape of the wave functions of p electrons as complex combinations of

$$\begin{aligned}|1,+1\rangle &= (2)^{-\frac{1}{2}}(p_x + ip_y),\\ |1,-1\rangle &= (2)^{-\frac{1}{2}}(p_x - ip_y),\\ |1,0\rangle &= p_z.\end{aligned} \tag{1.135}$$

Physically, $|1,+1\rangle$ and $|1,-1\rangle$ correspond to a clockwise rotation about the z-axis through p_x and p_y, and in the opposite direction, respectively, carrying angular momentum about this axis. For $|1,0\rangle$, there is no component of angular momentum about the z-axis (when viewed along the axis, the p_z-orbital looks like an s-orbital).

In the absence of a crystal field, there is a threefold degeneracy of the orbital angular momentum, as shown in parentheses in Fig. 1.19b. When the system is placed in a crystal field having an octahedral symmetry, the degeneracy is not lifted, only all the energy levels will experience a certain offset (Fig. 1.19b). If the symmetry is lowered further to that of the tetragonal type (ions on one of the axes of the octahedron are displaced symmetrically), there will be splitting into two levels characterized by wave functions of p_z for the lower non-degenerate level and p_x, p_y for the upper doubly degenerate level.

We find the components of the g-factor along the direction z aligned with the axial tetragonal axis, and along the perpendicular direction, which coincides with one of the axes x or y. In accordance with (1.124) and (1.125), we calculate the tensor components

$$\Lambda_{ij} = -\sum_n{}' \frac{\langle 0|\hat{L}_i|n\rangle\langle n|\hat{L}_j|0\rangle}{E_n^{(0)} - E_0^{(0)}},$$

namely, Λ_{zz} and Λ_{xx} or Λ_{yy}; the last two components are identical, so we just calculate Λ_{xx}. When calculating the matrix elements, we use the wave functions in the form

$$|L, M_L\rangle = |1,0\rangle, |1,+1\rangle, |1,-1\rangle,$$

shown in Fig. 1.19b.

$$g_{zz} = g_s + 2\lambda\Lambda_{zz} = g_s - 2\lambda\sum_n{}' \frac{\langle 0|\hat{L}_z|n\rangle\langle n|\hat{L}_z|0\rangle}{E_n^{(0)} - E_0^{(0)}} = g_s$$

because $\langle 0|\hat{L}_z|n\rangle = M_L\langle 0|n\rangle = 0$ due to the orthogonality of the wave functions

$$g_{xx} = g_s + 2\lambda\Lambda_{xx} = g_s - 2\lambda\sum_n{}' \frac{\langle 0|\hat{L}_x|n\rangle\langle n|\hat{L}_x|0\rangle}{E_n^{(0)} - E_0^{(0)}} = g_s - \frac{2\lambda}{\delta}, \tag{1.136}$$

where δ is the splitting between the ground- and excited-state energy levels in the tetragonal crystal field (Fig. 1.19b). This result was obtained by representing the operator $\hat{L}_x$ in the form of raising (up-step) and lowering (dow-step) operators (1.59), as it was done above for the spin $S = 1/2$ operators:

$$\hat{L}_x = \frac{1}{2}(\hat{L}_+ + \hat{L}_-).$$

For $L = 1$:

$$\hat{L}_+|L, M_L\rangle = \sqrt{2}|L, M_L + 1\rangle \text{ and } \hat{L}_-|L, M_L\rangle = \sqrt{2}|L, M_L - 1\rangle$$

The wave functions shown in Fig. 1.19b were used

$$\Lambda_{xx} = -\frac{\langle 0|\hat{L}_x|1\rangle\langle 1|\hat{L}_x|0\rangle + \langle 0|\hat{L}_x|-1\rangle\langle -1|\hat{L}_x|0\rangle}{\delta} = -\frac{1}{\delta}. \tag{1.137}$$

The result is an anisotropic g-factor with different components for the parallel and perpendicular directions with respect to the axial-symmetry axis:

$$g_{\parallel} = g_s;\ g_{\perp} = g_s - 2\lambda/\delta. \tag{1.138}$$

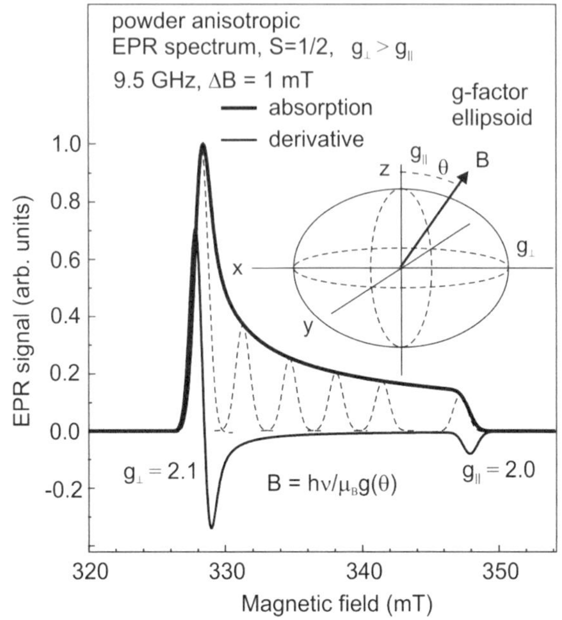

Fig. 1.20 The asymmetric EPR spectrum of paramagnetic centers with axial g-factor (absorption and derivative) simulated (at 9.5 GHz) for a powder sample assuming a large number of paramagnetic centers with their random orientation with respect to the static magnetic field B. An axial g factors with $g_\perp > g_\parallel$ are represented by a rotational ellipsoid. The spin packets with linewidth ΔB of 1 mT are schematically exemplified in the powder line shape

It should be noted that λ is positive for a single *p*-electron in the outer shell (*p*-shell is less than half-filled), but if there is one missing *p*-electron in the outer shell (*p*-shell is more than half-filled), e.g., in the p^5 electronic configuration (e.g., O^- ion), then $\lambda < 0$ and, as a result, $g_\perp > g_\parallel$.

For a powder sample the asymmetric EPR line shape can be observed (absorption and derivative) due to the fact that the number of spin packets contributing to the spectrum is much larger in the xy-plane than along the z-axis. In Fig. 1.20 an axial g factors with $g_\perp > g_\parallel$, represented by a rotational ellipsoid, and the line shape of the corresponding EPR spectrum are drawn, assuming a large number of paramagnetic centers with their random orientation with respect to the static magnetic field B. For a given magnetic field strength B, all spins fulfilling the resonance condition $B = h\nu/[g(\theta)\mu_B]$, i.e. all spins for which B makes an angle θ with the z-axis of the g ellipsoid, contribute to the spectrum and are considered to form a spin packet. These spin packets are schematically exemplified in the powder line shape in Fig. 1.20. The extreme positions of the powder spectrum are obtained by inserting $g_\parallel$ and $g_\perp$ into the resonance condition. It should be noted that for $g_\perp > g_\parallel$, one obtains $B(g_\perp) < B(g_\parallel)$ due to the inverse proportionality of g and B.

1.11 Fine Structure

1.11.1 Contribution of Dipole-Dipole Interaction Between Two Electron Spins to the Fine Structure

Consider the second term in the spin Hamiltonian (1.127), which is named the fine-structure (FS) interaction and gives rise to a fine-structure splitting of the spin sublevels even in zero-magnetic fields (so-called zero-field splitting)

$$\hat{H}_{FS} = \hat{\vec{S}} \cdot \overleftrightarrow{D} \cdot \hat{\vec{S}}. \tag{1.139}$$

It should be noted that this interaction can occur only for spins $S > 1/2$. In general, (1.139) can be written for arbitrarily chosen axes in the matrix form

$$\hat{H}_{FS} = \left|\hat{S}_x, \hat{S}_y, \hat{S}_z\right| \begin{vmatrix} D_{xx} & D_{xy} & D_{xz} \\ D_{yx} & D_{yy} & D_{yz} \\ D_{zx} & D_{zy} & D_{zz} \end{vmatrix} \begin{vmatrix} \hat{S}_x \\ \hat{S}_y \\ \hat{S}_z \end{vmatrix}. \tag{1.139a}$$

Expression (1.139a) is considerably simplified if the system is brought to the principal axes in which the tensor is diagonal, i.e., the axes x, y and z are combined with the symmetry axes of the system: a crystal, a molecule, a defect. Then, we have

$$\hat{H}_{FS} = \left|\hat{S}_x, \hat{S}_y, \hat{S}_z\right| \begin{vmatrix} D_{xx} & 0 & 0 \\ 0 & D_{yy} & 0 \\ 0 & 0 & D_{zz} \end{vmatrix} \begin{vmatrix} \hat{S}_x \\ \hat{S}_y \\ \hat{S}_z \end{vmatrix} = D_{xx}\hat{S}_x^2 + D_{yy}\hat{S}_y^2 + D_{zz}\hat{S}_z^2. \tag{1.139b}$$

$\overleftrightarrow{D}$ is the second-rank tensor shown to be reducible to the diagonal form. The trace of this tensor is invariant under the transformation, i.e., $D_{xx} + D_{yy} + D_{zz} = \text{const}$. The trace of the $\overleftrightarrow{D}$-tensor is usually set to zero ($D_{xx} + D_{yy} + D_{zz} = 0$), because the trace only shifts the total spin sublevels multiplet energy and does not appear in the EPR spectra. Therefore, only two independent diagonal parameters are needed to describe the fine structure, and the fine-structure spin Hamiltonian in (1.139) can be normally rewritten in the principal axes of $\overleftrightarrow{D}$ in terms of the zero-field parameters D and E

$$\hat{H}_{FS} = D_{xx}\hat{S}_x^2 + D_{yy}\hat{S}_y^2 + D_{zz}\hat{S}_z^2 = D[\hat{S}_z^2 - \frac{1}{3}S(S+1)] + E(\hat{S}_x^2 - \hat{S}_y^2). \tag{1.140}$$

Here, $D = D_{zz} - (D_{xx} + D_{yy})/2$ is the axially symmetric parameter and the parameter $E = (D_{xx} - D_{yy})/2$ reflects the deviation from the axial symmetry of the fine-structure interaction.

It can be seen that the fine-structure term in the spin Hamiltonian is not field dependent, but, in combination with the electron Zeeman interaction term, it leads to a field-dependent mixing of the electron spin eigenfunctions. When the electron Zeeman interaction is comparable in magnitude with the fine-structure splitting, this mixing requires complicated calculations to analyse EPR spectra. High-field (high-frequency) experiments (when the electron Zeeman interaction substantially exceeds in magnitude the fine-structure splitting) greatly simplify the analysis of EPR spectra.

The expression for the spin Hamiltonian (1.139) describing the fine structure was derived when considering the admixture of excited states due to the spin-orbit interaction. It turns out that there is another kind of interaction leading to an expression of a similar type, which is the interaction between the magnetic moments of electrons themselves or, as it is frequently named, "the magnetic dipole-dipole interaction".

The classical expression for the energy of interaction between two magnetic dipoles was written above in the form (1.27). In the quantum treatment, the interaction between the magnetic moments of two electrons (for definiteness, we consider the interaction of two spin magnetic moments of S_1 and S_2, each equal to 1/2) can be represented as a dipole-dipole interaction Hamiltonian $\hat{H}_{DD}$. With the spin magnetic moments given in the operator form, we have

$$\hat{H}_{DD} = \frac{1}{r^3}\left[\hat{\vec{\mu}}_1 \cdot \hat{\vec{\mu}}_2 - \frac{3(\hat{\vec{\mu}}_1 \cdot \vec{r})(\hat{\vec{\mu}}_2 \cdot \vec{r})}{r^2}\right] = g_s^2 \mu_B^2 \frac{1}{r^3}\left[\hat{\vec{S}}_1 \cdot \hat{\vec{S}}_2 - \frac{3(\hat{\vec{S}}_1 \cdot \vec{r})(\hat{\vec{S}}_2 \cdot \vec{r})}{r^2}\right]. \tag{1.141}$$

Expanding the scalar products in (1.141), we obtain the expression

$$\begin{aligned}\hat{H}_{DD} = g_s^2 \mu_B^2 \frac{1}{r^5}[&(r^2 - 3x^2)S_{1x}S_{2x} + (r^2 - 3y^2)S_{1y}S_{2y} + (r^2 - 3z^2)S_{1z}S_{2z} \\ &- 3xy(S_{1x}S_{2y} + S_{1y}S_{2x}) - 3xz(S_{1x}S_{2z} + S_{1z}S_{2x}) - 3yz(S_{1y}S_{2z} + S_{1z}S_{2y})].\end{aligned} \tag{1.142}$$

This expression can be written in the matrix form

$$\hat{H}_{DD} = g_s^2 \mu_B^2 \left|\hat{S}_{1x}, \hat{S}_{1y}, \hat{S}_{1z}\right| \begin{vmatrix} \frac{(r^2-3x^2)}{r^5} & \frac{-3xy}{r^5} & \frac{-3xz}{r^5} \\ \frac{-3xy}{r^5} & \frac{(r^2-3y^2)}{r^5} & \frac{-3yz}{r^5} \\ \frac{-3xz}{r^5} & \frac{-3yz}{r^5} & \frac{(r^2-3z^2)}{r^5} \end{vmatrix} \begin{vmatrix} \hat{S}_{2x} \\ \hat{S}_{2y} \\ \hat{S}_{2z} \end{vmatrix}. \tag{1.143}$$

The interaction of two spin moments is frequently described by using the concept of the total spin, i.e., two spin moments are formed in correspondence with the quantum mechanical rule of addition of angular moments $\hat{\vec{S}} = \hat{\vec{S}}_1 + \hat{\vec{S}}_2$, the result is $S = 1$ and $S = 0$. Consider the case where the lower energy level corresponds to the state $S = 1$ (triplet state), and the energy level corresponding to the state $S = 0$ (singlet state) lies substantially higher than the triplet level and has no effect on the properties of the triplet state. Expressing the total spin angular momentum via the spin moments of the individual electrons and making several transformations [11], we obtain

$$\hat{H}_{DD} = \frac{1}{2} g_s^2 \mu_B^2 \left| \hat{S}_x, \hat{S}_y, \hat{S}_z \right| \begin{vmatrix} \frac{(r^2-3x^2)}{r^5} & \frac{-3xy}{r^5} & \frac{-3xz}{r^5} \\ \frac{-3xy}{r^5} & \frac{(r^2-3y^2)}{r^5} & \frac{-3yz}{r^5} \\ \frac{-3xz}{r^5} & \frac{-3yz}{r^5} & \frac{(r^2-3z^2)}{r^5} \end{vmatrix} \begin{vmatrix} \hat{S}_x \\ \hat{S}_y \\ \hat{S}_z \end{vmatrix}. \quad (1.144)$$

This expression can be represented as

$$\hat{H}_{DD} = \hat{\vec{S}} \cdot \overleftrightarrow{D} \cdot \hat{\vec{S}},$$

i.e., it can be written in the same form as (1.139). To obtain the final expression for the dipole-dipole interaction, the tensor components should be averaged over the electron wave function because the relative positions of two electrons vary with the wave function of the system of two electrons, e.g.,

$$D_{xx} = \frac{1}{2} g_s^2 \mu_B^2 \left\langle \frac{r^2 - 3x^2}{r^5} \right\rangle, \text{ etc.} \quad (1.145)$$

The tensor $\overleftrightarrow{D}$ is a symmetric second-rank tensor with zero trace (which is evident from expression (1.144)). This tensor can be reduced to a diagonal form by combining the coordinate axes with the principal axes of symmetry of the system. Then, expression (1.144) with (1.145) reduces to

$$\hat{H}_{DD} = \left| \hat{S}_x, \hat{S}_y, \hat{S}_z \right| \begin{vmatrix} D_{xx} & 0 & 0 \\ 0 & D_{yy} & 0 \\ 0 & 0 & D_{zz} \end{vmatrix} \begin{vmatrix} \hat{S}_x \\ \hat{S}_y \\ \hat{S}_z \end{vmatrix} = D_{xx}\hat{S}_x^2 + D_{yy}\hat{S}_y^2 + D_{zz}\hat{S}_z^2. \quad (1.144a)$$

We see that formulas (1.139) and (1.144a) are absolutely identical, and both equations describe the fine structure, although being of totally different nature. Experimentally, these two contributions to the fine structure are commonly difficult to separate. The first contribution, associated with the spin-orbit interaction, is predominant for heavy elements. The second contribution is the case for organic molecular systems dominated by light elements, for which the spin-orbit interaction is relatively weak, whereas the dipole-dipole interaction makes a major contribution to the fine structure.

1.11.2 Energy Levels in Magnetic Field of Systems with Half-Integer and Integer Spins. Kramers Doublets

Consider a few examples of how the spin Hamiltonians (1.127) and (1.140) are used to calculate systems with total spin $S = 1$ (triplet state) and $S = 3/2$.

$$\hat{H} = \mu_B \vec{B} \cdot \overleftrightarrow{g} \cdot \hat{\vec{S}} + \hat{\vec{S}} \cdot \overleftrightarrow{D} \cdot \hat{\vec{S}} = \mu_B \vec{B} \cdot \overleftrightarrow{g} \cdot \hat{\vec{S}} + D[\hat{S}_z^2 - \frac{1}{3}S(S+1)] + E(\hat{S}_x^2 - \hat{S}_y^2). \quad (1.146)$$

For simplicity, we consider a system with an axial symmetry and an isotropic g-factor, and the magnetic field is directed along the z-axis coinciding with the axial axis of the system, i.e., $B = B_z$. Then, the spin Hamiltonian can be written as

$$\hat{H} = \mu_B g B \hat{S}_z + D[\hat{S}_z^2 - \frac{1}{3}S(S+1)]. \quad (1.146a)$$

For the triplet state in the approximation of a strong magnetic field, i.e., when the Zeeman energy is substantially higher than the energy of the fine structure, the eigenfunctions have the form $|S, M_S\rangle = |1,0\rangle, |1,+1\rangle, |1,-1\rangle$. To find the energy levels of the system, we have to construct a 3 × 3 matrix with the matrix elements $\langle S, M_S|\hat{H}|S, M_S'\rangle$. For the problem at hand, this matrix will be diagonal because the eigenfunctions of the spin Hamiltonian (1.146a) are used (Fig. 1.21).

$$\begin{vmatrix} \langle M_S|\hat{H}|M_S'\rangle & |+1\rangle & |0\rangle & |-1\rangle \\ \langle +1| & \mu_B g B + \frac{1}{3}D & 0 & 0 \\ \langle 0| & 0 & -\frac{2}{3}D & 0 \\ \langle -1| & 0 & 0 & -\mu_B g B + \frac{1}{3}D \end{vmatrix}. \quad (1.147)$$

Thus, the energy levels for this system have the form (see Fig. 1.21)

$$E_{|+1\rangle} = \mu_B g B + \frac{1}{3}D,\ E_{|-1\rangle} = -\mu_B g B + \frac{1}{3}D,\ E_{|0\rangle} = -\frac{2}{3}D. \quad (1.147a)$$

It should be noted that, if the magnetic field is directed at an angle θ to the axial axis of the system, the matrix is not diagonal because the selected functions will no longer be eigenfunctions of the spin Hamiltonian, and, in this case, it is necessary to diagonalize the matrix to calculate the energy levels. Figure 1.22 shows the energy levels and allowed X-band EPR transitions for nitrogen-vacancy (NV) defects in diamond with $S = 1$, $g = 2.0028$, $D = 961 \times 10^{-4}$ cm^{-1}: $\theta = 54.5°$ (top) and $\theta = 90°$ (bottom) where θ is the angle between the orientation of the magnetic field and the z-axis of the fine-structure tensor.

Figure 1.23 shows the angular variation of two allowed X-band (9.3 GHz) EPR transitions for a single type of the axial NV-defect in diamond in the triplet ground state with $S = 1$ (top), and four $\langle 111 \rangle$ oriented NV-defects (bottom) where θ is the

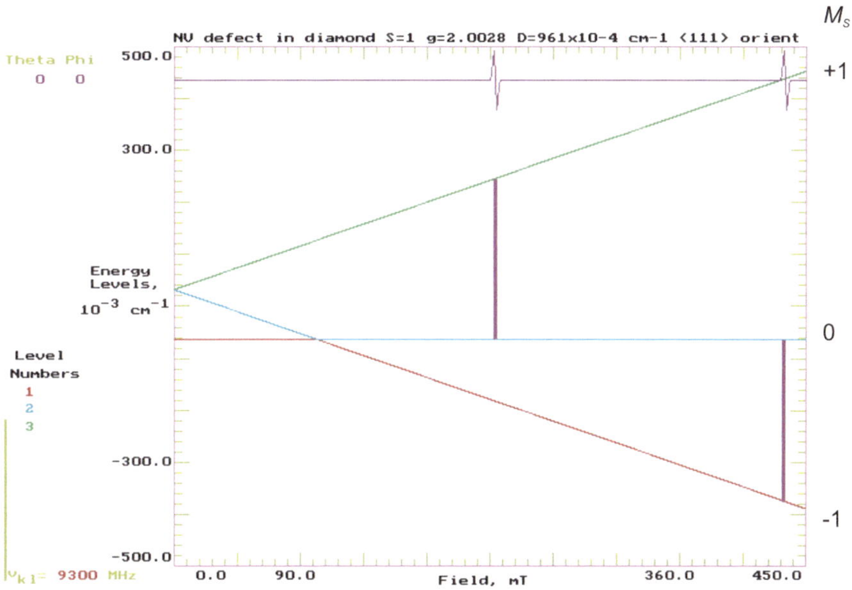

Fig. 1.21 Energy levels and allowed X-band EPR transitions for the spin Hamiltonian (1.146a) in a static magnetic field for the triplet system with $S = 1$ and $\boldsymbol{B}$ aligned with the z-axis. The transitions are shown for NV-defects in diamond. The field difference between two EPR transitions in this orientation is almost equal to $2D/(g\mu_B)$

angle between the orientation of the magnetic field and the z-axis of the fine-structure tensor. The splitting varies as $(3\cos^2\theta - 1)$ in the approximation with $D \ll g\mu_B B$. The field differences between two EPR transitions in the orientations $\theta = 0°$ and $\theta = 90°$ are $2D/(g\mu_B)$ and $D/(g\mu_B)$, respectively.

For the total spin $S = 3/2$, the wave functions can be written in the form

$$|S, M_S\rangle = \left|\frac{3}{2}, +\frac{3}{2}\right\rangle, \left|\frac{3}{2}, +\frac{1}{2}\right\rangle, \left|\frac{3}{2}, -\frac{1}{2}\right\rangle, \left|\frac{3}{2}, -\frac{3}{2}\right\rangle,$$

The matrix will be diagonal because the eigenfunctions of the spin Hamiltonian (1.146a) are used.

$$\begin{vmatrix} \langle M_S|/|M'_S\rangle & \left|+\frac{3}{2}\right\rangle & \left|+\frac{1}{2}\right\rangle & \left|-\frac{1}{2}\right\rangle & \left|-\frac{3}{2}\right\rangle \\ \left\langle +\frac{3}{2}\right| & \frac{3}{2}\mu_B gB + D & 0 & 0 & 0 \\ \left\langle +\frac{1}{2}\right| & 0 & \frac{1}{2}\mu_B gB - D & 0 & 0 \\ \left\langle -\frac{1}{2}\right| & 0 & 0 & -\frac{1}{2}\mu_B gB - D & 0 \\ \left\langle -\frac{3}{2}\right| & 0 & 0 & 0 & -\frac{3}{2}\mu_B gB + D \end{vmatrix}. \tag{1.148}$$

The energy levels corresponding to the diagonal elements of the 4×4 matrix constructed with these wave functions have the form (see Fig. 1.24):

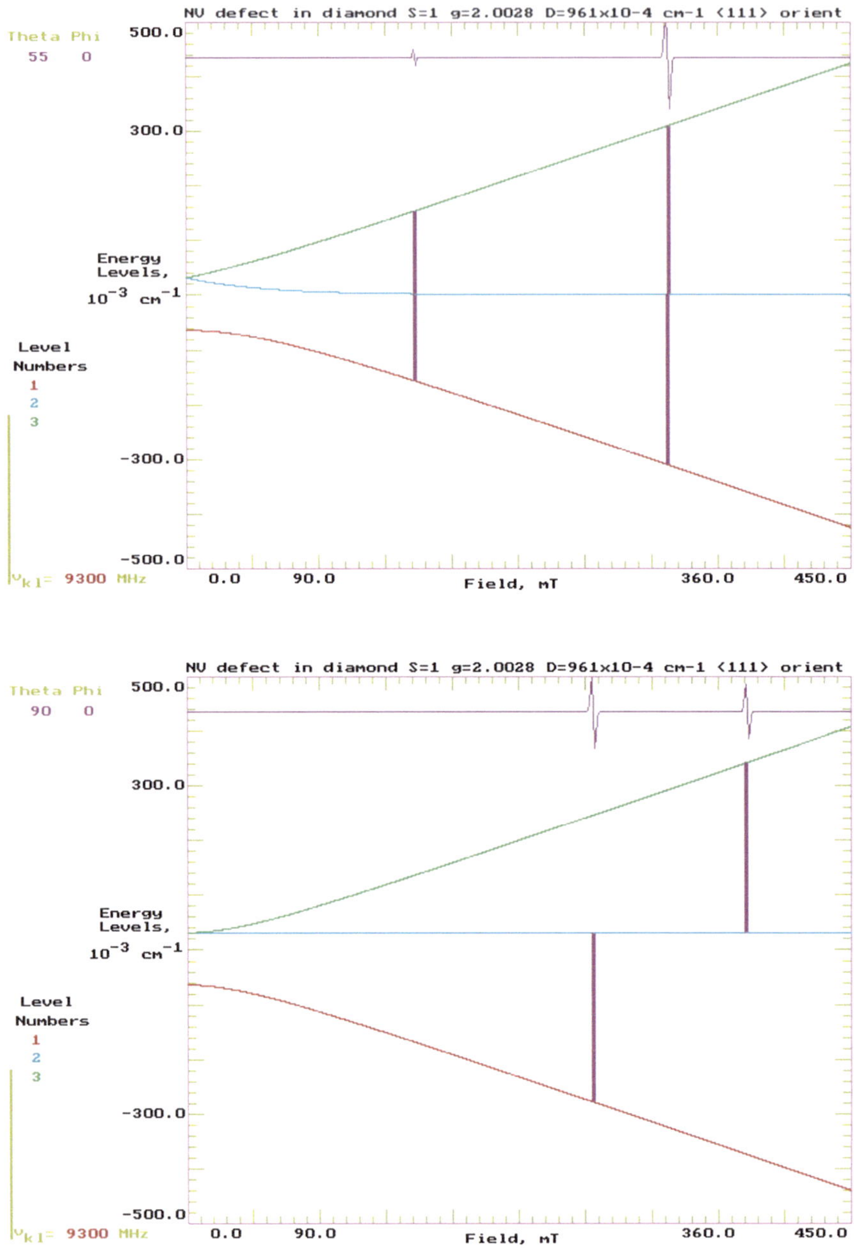

Fig. 1.22 Energy levels and allowed X-band EPR transitions for NV-defects in diamond with $S = 1$: $\theta = 54.5°$ (*top*) and $\theta = 90°$ (*bottom*) where θ is the angle between the orientation of the magnetic field and the z-axis of the fine-structure tensor

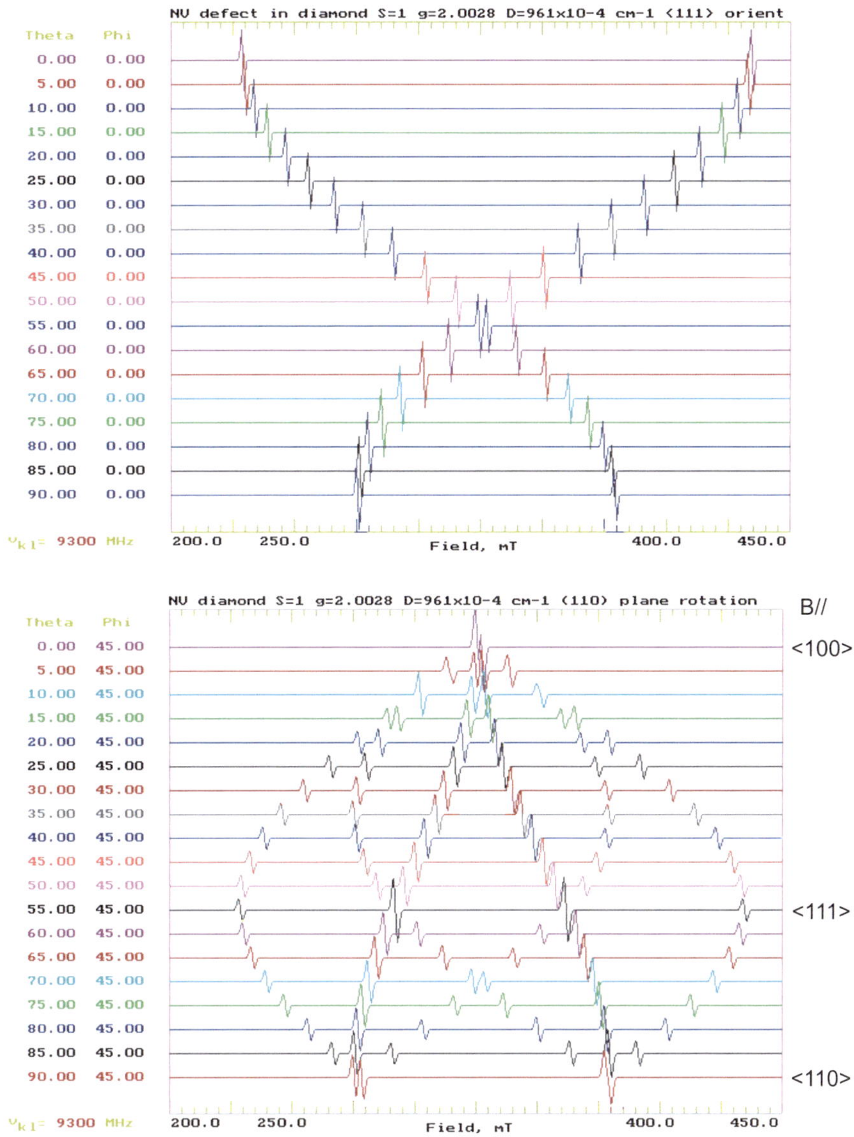

Fig. 1.23 Angular variation of two allowed X-band (9.3 GHz) EPR transitions for $S = 1$ for a single NV-defect in diamond (*top*), and four $\langle 111 \rangle$ oriented NV-defects (*bottom*) where θ is the angle between the orientation of magnetic field and the *z*-axis of the fine-structure tensor. The splitting varies as $(3\cos^2\theta - 1)$ in approximation $D \ll g\mu_B B$. The field differences between two EPR transitions in the orientations $\theta = 0°$ and $\theta = 90°$ are $2D/(g\mu_B)$ and $D/(g\mu_B)$, respectively

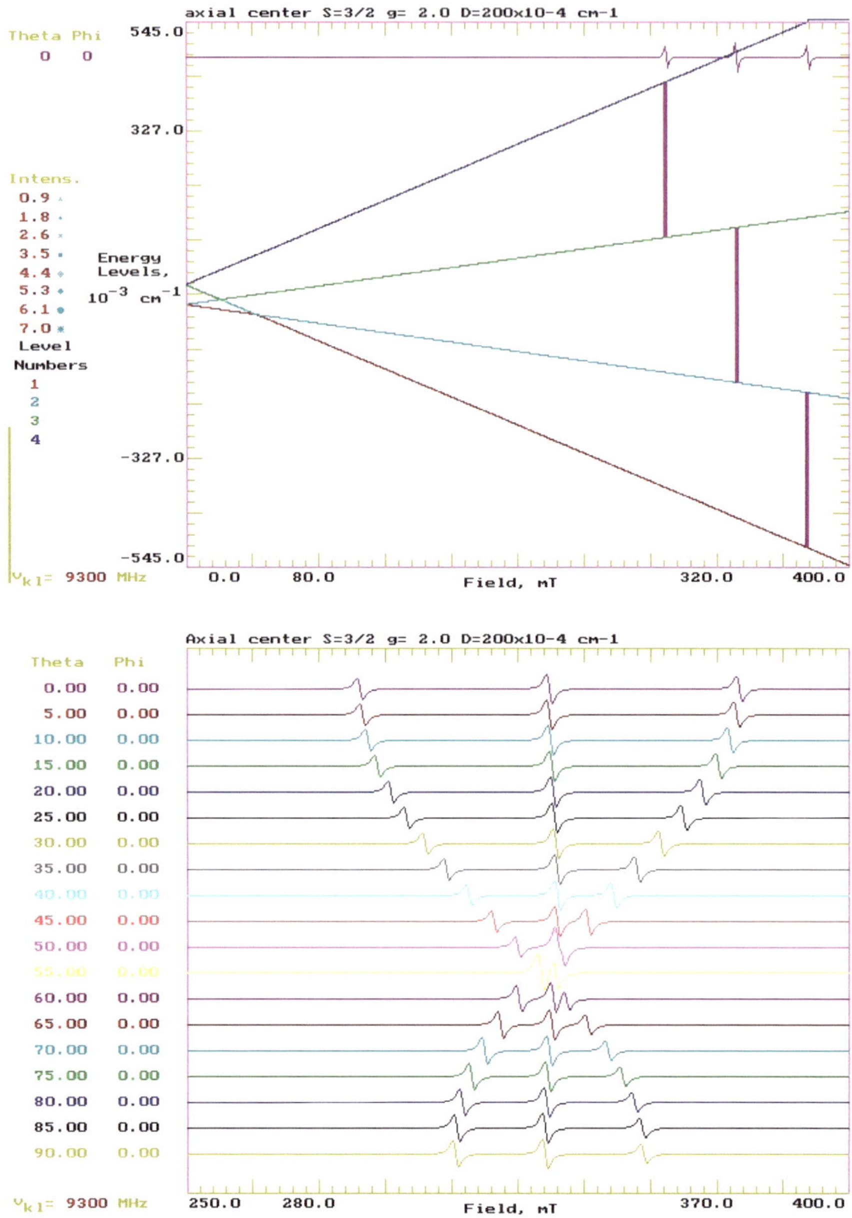

Fig. 1.24 (*Top*) Energy levels and allowed X-band EPR transitions for the spin Hamiltonian (2.146a) in a static magnetic field for a quartet system with $S = 3/2$ and $\boldsymbol{B}$ along the z-axis. The field difference between two adjacent EPR transitions in the orientations $\theta = 0^0$ is $2D/(g\mu_B)$. (*bottom*) Angular variation at X-band (9.3 GHz) of the three allowed EPR transitions for quartet system with $S = 3/2$, where θ is the angle between the orientation of the magnetic field and the z-axis of the fine-structure tensor

$$E_{\left|+\frac{3}{2}\right\rangle}=\frac{3}{2}\mu_B gB+D,\ E_{\left|+\frac{1}{2}\right\rangle}=\frac{1}{2}\mu_B gB-D,\ E_{\left|-\frac{1}{2}\right\rangle}=-\frac{1}{2}\mu_B gB-D,\ E_{\left|-\frac{3}{2}\right\rangle}=-\frac{3}{2}\mu_B gB+D. \tag{1.148a}$$

If the magnetic field is directed at an angle θ to the axial axis of the system, the matrix is not diagonal because the selected functions will no longer be eigenfunctions of the spin Hamiltonian, and, to calculate the energy levels in this case, it is necessary to diagonalize the matrix. Figure 1.24 (bottom) shows the angular dependence of the EPR lines for the quartet system with $S = 3/2$. The field differences between two adjacent EPR transitions in the orientations $\theta = 0°$ and $\theta = 90°$ are $2D/(g\mu_B)$ and $D/(g\mu_B)$, respectively.

It is seen that the fine-structure interaction leads to splitting of the levels even in the absence of a magnetic field, this splitting is frequently named zero-field splitting. The splitting of the fine structure is observed only for systems with $S > 1/2$. In systems with an integer spin ($S = 1, 2, 3$, etc.), complete lifting of degeneracy in zero magnetic field can be observed. For systems with a half-integer spin ($S = 3/2$, 5/2, 7/2, etc.), an at least twofold degeneracy of the levels in zero magnetic field always remains. For example, in a system with $S = 3/2$, the energy levels for the states $\left|+\frac{3}{2}\right\rangle,\left|-\frac{3}{2}\right\rangle$ and $\left|+\frac{1}{2}\right\rangle,\left|-\frac{1}{2}\right\rangle$ in zero magnetic field are the same (see Fig. 1.24).

There is the so-called "Kramers theorem" (Dutch physicist Hendrick Anton Kramers, 1930), which maintains that, in the absence of an external magnetic field, the electron states of any system (molecules, defects) with an odd number of electrons remain at least doubly degenerate (Kramers degeneracy). In other words, all the interactions (such as the crystal field, spin-orbit interaction, electron spin-spin interaction) can not lift this degeneracy. The Kramers degeneracy is a consequence of the time-reversal invariance of the electron Hamiltonian, and this degeneracy is lifted by placing the system in an external magnetic field.

Figure 1.25 shows the EPR spectrum simulated (at 9.5 GHz) for a powder sample (absorption and derivative) for a triplet state (S = 1) of the axial center with fine structure parameter D = 10 mT. The form of the spectrum is a result of the fact that the number of spin packets contributing to the spectrum is much larger in the xy-plane than along the axiall z-axis. EPR spectrum is calculated assuming a large number of paramagnetic centers with their random orientation with respect to the static magnetic field B. The spin packets with linewidth ΔB of 1 mT are schematically exemplified in the powder line shape in Fig. 1.25.

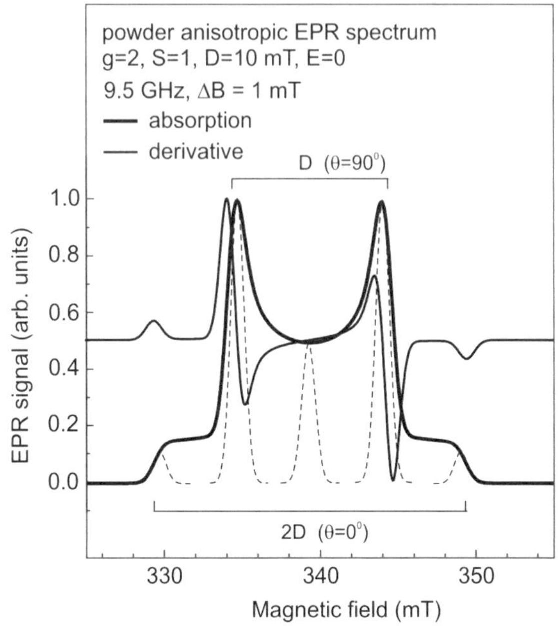

Fig. 1.25 The EPR spectrum simulated (at 9.5 GHz) for a powder sample (absorption and derivative) for a triplet state (S = 1) of the axial center with fine structure parameter D = 10 mT. EPR spectrum is calculated assuming a large number of paramagnetic centers with their random orientation with respect to the static magnetic field B. The spin packets with linewidth ΔB of 1 mT are schematically exemplified in the powder line shape

1.12 Anisotropic Hyperfine Interaction

The interaction between the magnetic moment of the electron and the nuclear magnetic moment, or the hyperfine (HF) interaction of free atoms or ions, can be written as

$$\hat{H}_{HFI} = A\hat{\vec{J}} \cdot \hat{\vec{I}}, \tag{1.149}$$

where J is the total angular momentum of the electron shell. The anisotropic g-factor in a crystal field reflects the presence of an anisotropic electron magnetic moment. The hyperfine interaction, which is the interaction of the electron and the nuclear magnetic moments, will also be anisotropic due to the anisotropy of the electron magnetic moment. This interaction can be described by the spin Hamiltonian (1.127) with addition of the appropriate term describing the HF interaction.

$$\hat{H}_{HFI} = \frac{1}{r^3}\left[\hat{\vec{\mu}}_s \cdot \hat{\vec{\mu}}_I - \frac{3(\hat{\vec{\mu}}_s \cdot \vec{r})(\hat{\vec{\mu}}_I \cdot \vec{r})}{r^2}\right] = -g_s\mu_B g_I \mu_N \frac{1}{r^3}\left[\hat{\vec{S}} \cdot \hat{\vec{I}} - \frac{3(\hat{\vec{S}} \cdot \vec{r})(\hat{\vec{I}} \cdot \vec{r})}{r^2}\right]. \tag{1.150}$$

As in the case of the dipole-dipole interaction between two electronic magnetic moments, discussed above, expression (1.150) can be represented in the matrix form

$$\hat{H}_{HFI} = -g_s\mu_B g_I\mu_N\left|\hat{S}_x, \hat{S}_y, \hat{S}_z\right|\begin{vmatrix} \frac{(r^2-3x^2)}{r^5} & \frac{-3xy}{r^5} & \frac{-3xz}{r^5} \\ \frac{-3xy}{r^5} & \frac{(r^2-3y^2)}{r^5} & \frac{-3yz}{r^5} \\ \frac{-3xz}{r^5} & \frac{-3yz}{r^5} & \frac{(r^2-3z^2)}{r^5} \end{vmatrix}\begin{vmatrix} \hat{I}_x \\ \hat{I}_y \\ \hat{I}_z \end{vmatrix}. \tag{1.151}$$

As a result, we obtain an expression for the spin Hamiltonian describing the anisotropic hyperfine interaction

$$\hat{H}_{HFI} = \hat{\vec{S}} \cdot \overleftrightarrow{T} \cdot \hat{\vec{I}}, \tag{1.152}$$

where the tensor components $\overleftrightarrow{T}$ are averaged over the wave functions of unpaired electrons because the relative positions of electrons and the nucleus vary with the wave function of the system. For example,

$$T_{xx} = -g_s\mu_B g_I\mu_N\left\langle\frac{r^2 - 3x^2}{r^5}\right\rangle, T_{xy} = -g_s\mu_B g_I\mu_N\left\langle\frac{-3xy}{r^5}\right\rangle, \text{ etc.} \tag{1.152a}$$

The tensor $\overleftrightarrow{T}$ is a symmetric second-rank tensor with zero trace. This tensor can be reduced to the diagonal form by combining the coordinate axes with the principal axes of symmetry of the system. Then, expression (1.151) becomes

$$\hat{H}_{HFI} = \left|\hat{S}_x, \hat{S}_y, \hat{S}_z\right|\begin{vmatrix} T_{xx} & 0 & 0 \\ 0 & T_{yy} & 0 \\ 0 & 0 & T_{zz} \end{vmatrix}\begin{vmatrix} \hat{I}_x \\ \hat{I}_y \\ \hat{I}_z \end{vmatrix} = \hat{S}_x T_{xx}\hat{I}_x + \hat{S}_y T_{yy}\hat{I}_y + \hat{S}_z T_{zz}\hat{I}_z. \tag{1.153}$$

As shown above for the S state, the isotropic hyperfine interaction is observed. The isotropic HF interaction is caused by the interaction of the unpaired s-electrons and the nucleus due to the presence of a nonzero density of the unpaired electron wave function within the nucleus. The Hamiltonian of this interaction (e.g., for hydrogen) was written as

$$\hat{H}_{HFI(iso)} = A_{iso}\hat{\vec{S}} \cdot \hat{\vec{I}}, \tag{1.97}$$

where

$$A_{iso} = \frac{8\pi}{3} g_s\mu_B g_I\mu_N|\Psi_{ns}(0)|^2 \tag{1.98}$$

is the constant of the isotropic HF interaction.

Equations (1.97) and (1.152) can be combined into one general formula, with the constant of the isotropic hyperfine interaction represented as a second-rank tensor $A_{iso}\overleftrightarrow{1}$, where $\overleftrightarrow{1}$ is the unit tensor

$$\overleftrightarrow{1} = \begin{vmatrix} 1 & 0 & 0 \\ 0 & 1 & 0 \\ 0 & 0 & 1 \end{vmatrix}.$$

Then, we can add two tensor quantities, so that the hyperfine interaction tensor is written in the form

$$\overleftrightarrow{A} = A_{iso}\overleftrightarrow{1} + \overleftrightarrow{T}, \tag{1.154}$$

which yields the final formula for the HF interaction in the condensed medium, e.g., in a crystal: given by

$$\hat{H}_{HFI} = \hat{\vec{S}} \cdot (A_{iso}\overleftrightarrow{1} + \overleftrightarrow{T}) \cdot \hat{\vec{I}} = \hat{\vec{S}} \cdot \overleftrightarrow{A} \cdot \hat{\vec{I}}, \tag{1.155}$$

or, in the matrix form,

$$\hat{H}_{HFI} = \left|\hat{S}_x, \hat{S}_y, \hat{S}_z\right| \begin{vmatrix} A_{xx} & A_{xy} & A_{xz} \\ A_{yx} & A_{yy} & A_{yz} \\ A_{zx} & A_{zy} & A_{zz} \end{vmatrix} \begin{vmatrix} \hat{I}_x \\ \hat{I}_y \\ \hat{I}_z \end{vmatrix}. \tag{1.155a}$$

Equation (1.155a) can be greatly simplified if the system is brought to the principal axes in which the tensor is diagonal, i.e., if the coordinate axes x, y and z are combined with the symmetry axes of the system under consideration (crystal, molecule, defect). Then, we have

$$\hat{H}_{HFI} = \left|\hat{S}_x, \hat{S}_y, \hat{S}_z\right| \begin{vmatrix} A_x & 0 & 0 \\ 0 & A_y & 0 \\ 0 & 0 & A_z \end{vmatrix} \begin{vmatrix} \hat{I}_x \\ \hat{I}_y \\ \hat{I}_z \end{vmatrix} = A_x\hat{S}_x\hat{I}_x + A_y\hat{S}_y\hat{I}_y + A_z\hat{S}_z\hat{I}_z. \tag{1.155b}$$

In the case of the axial symmetry of the system, $A_z = A_{\|}$, $A_x = A_y = A_{\perp}$, and the spin Hamiltonian for the HF interaction can be written as

$$\hat{H}_{HFI} = A_{//}\hat{S}_z\hat{I}_z + A_{\perp}(\hat{S}_y\hat{I}_y + \hat{S}_z\hat{I}_z). \tag{1.155c}$$

The effective value of A for any direction of the magnetic field is given by

$$A_{eff}^2 = A_x^2\cos^2(\angle\vec{B}, x) + A_y^2\cos^2(\angle\vec{B}, y) + A_z^2\cos^2(\angle\vec{B}, z), \tag{1.156}$$

and for the axial symmetry, it has the form

$$A_{eff}^2 = A_{//}^2\cos^2\theta + A_{\perp}^2\sin^2\theta, \tag{1.156a}$$

where θ is the angle between the magnetic field and the axial axis of the system (a defect in a crystal, a molecule, etc.).

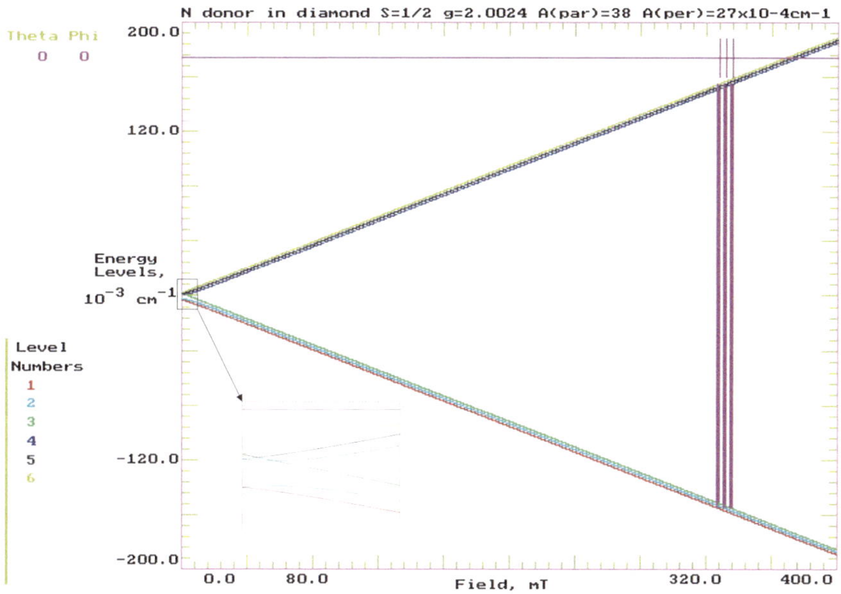

Fig. 1.26 Energy levels and the three allowed X-band EPR transitions for the axial system with $S = 1/2$, $I = 1$, $\theta = 0$. The transitions are shown for one axial deep N-donor in diamond. The field difference between the three EPR transitions in this orientation is equal to the HF structure constant $A_{\|}/(g\mu_B)$. The *inset* shows the low magnetic field range

Figure 1.26 shows the energy levels and the three allowed X-band EPR transitions for the axial system with $S = 1/2$, $I = 1$, $\theta = 0$, $A_{\|} = 38 \times 10^{-4}$ cm^{-1}, $A_{\perp} = 27 \times 10^{-4}$ cm^{-1}. The transitions are shown for one axial nitrogen (N) deep-donor in diamond. The field difference between the three EPR transitions in this orientation is equal to the HF structure constant $A_{\|}/(g\mu_B)$. The angular variation of the three allowed EPR transitions for the axial system with $S = 1/2$ and $I = 1$ (deep N-donors in diamond) is presented in Fig. 1.27: (top) for one axial deep N-donor in diamond; (bottom) for four $\langle 111 \rangle$ oriented N-donors where θ is the angle between the orientation of the magnetic field and the *z*-axis of the hyperfine interaction (structure) tensor. It can be deduced from the number of lines that the unpaired electron of the center can be localized on any of the four N-C bonds, and the symmetry axis of the HF interaction with the nitrogen ^{14}N nucleus is parallel to any of the four $\langle 111 \rangle$ bond directions.

Figure 1.28 shows the EPR spectrum of paramagnetic centers with S = ½, I = ½ having an axial HF structure with $A_{\|} > A_{\perp}$ and isotropic g-factor g = 2.0 (absorption and derivative) simulated (at 35.2 GHz) for a powder sample. The EPR line shape is a result of the fact that the number of spin packets contributing to the spectrum is much larger in the xy-plane than along the axial symmetry z-axis. In Fig. 1.28 an axial HF structure with $A_{\|} > A_{\perp}$ represented by a rotational ellipsoid, and the line shape of the corresponding EPR spectrum are drawn, assuming a large

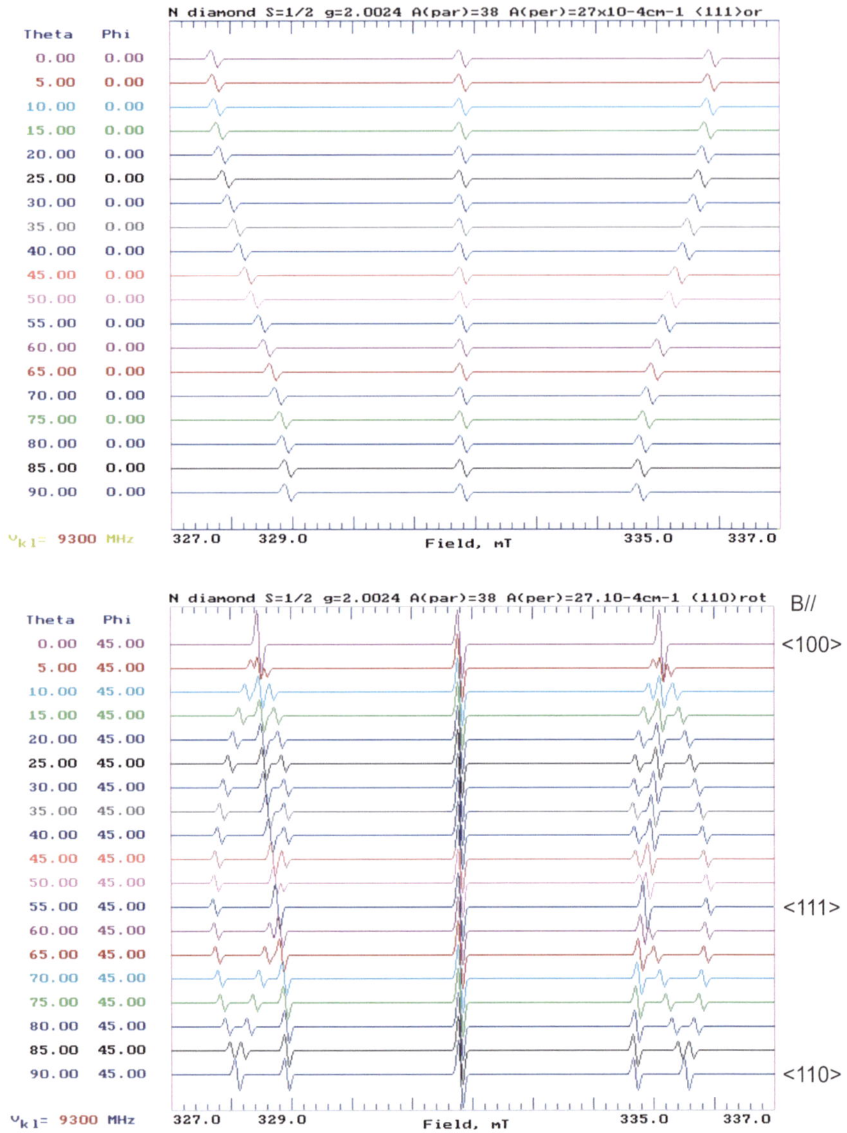

Fig. 1.27 Angular variation of the three allowed X-band (9.3 GHz) EPR transitions for the axial system with $S = 1/2$, $I = 1$ for one axial deep N-donor in diamond (*top*), and four $\langle 111 \rangle$ oriented N-donors (*bottom*) where θ is the angle between the orientation of the magnetic field and the z-axis of the hyperfine interaction (structure) tensor

number of paramagnetic centers with their random orientation with respect to the static magnetic field B. For a given magnetic field strength B, all spins fulfilling the resonance condition contribute to the spectrum and are considered to form a spin

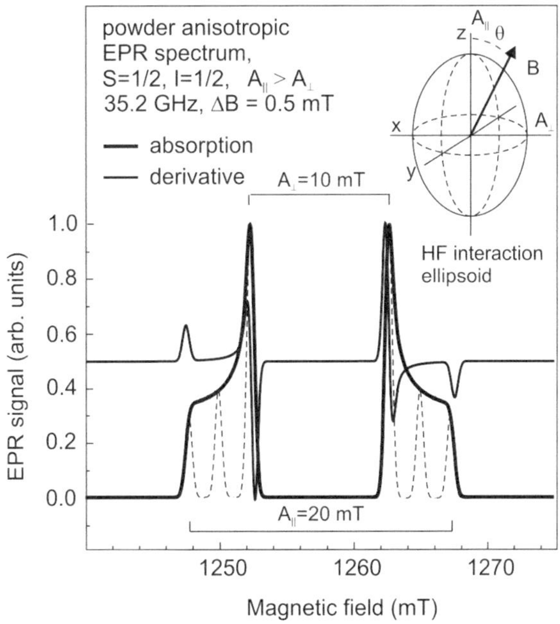

Fig. 1.28 The EPR spectrum of paramagnetic centers with S = ½, I = ½ and with axial HF structure $A_{\|} > A_{\perp}$ and isotropic g-factor (absorption and derivative) simulated (at 35.2 GHz) for a powder sample assuming a large number of paramagnetic centers with their random orientation with respect to the static magnetic field B. An axial HF structure with $A_{\|} > A_{\perp}$ is represented by a rotational ellipsoid. The spin packets with linewidth ΔB of 0.5 mT are schematically exemplified in the powder line shape

packet. These spin packets with linewidth ΔB of 0.5 mT are schematically exemplified in the powder line shape in Fig. 1.28. The extreme positions of the powder spectrum are obtained by inserting $A_{\|}$ and $A_{\perp}$ into the resonance condition.

Finally, the full spin Hamiltonian can be written as

$$\hat{H} = \mu_B \vec{B} \cdot \overleftrightarrow{g} \cdot \hat{\vec{S}} + \hat{\vec{S}} \cdot \overleftrightarrow{D} \cdot \hat{\vec{S}} + \hat{\vec{S}} \cdot \overleftrightarrow{A} \cdot \hat{\vec{I}}, \tag{1.157}$$

where the first term describes the electron Zeeman interaction with g-factors that differ from the free-spin value and may be anisotropic; the second term is the fine structure interaction, and the last term describes the hyperfine interaction which may be anisotropic.

In general, it is necessary to add to the spin Hamiltonian (1.157) two additional terms describing the nuclear Zeeman interaction

$$\hat{H}_{Zeem-n} = -g_I \mu_N \vec{B} \cdot \hat{\vec{I}}, \tag{1.158}$$

and the nuclear quadrupole interaction

$$\hat{H}_Q = \hat{\vec{I}} \cdot \overleftrightarrow{Q} \cdot \hat{\vec{I}}, \tag{1.159}$$

where $\overleftrightarrow{Q}$ is the quadrupole-interaction tensor.

As a result, we have the final spin Hamiltonian describing all the interactions

$$\hat{H} = \mu_B \vec{B} \cdot \overleftrightarrow{g} \cdot \hat{\vec{S}} + \hat{\vec{S}} \cdot \overleftrightarrow{D} \cdot \hat{\vec{S}} + \hat{\vec{S}} \cdot \overleftrightarrow{A} \cdot \hat{\vec{I}} - g_I \mu_N \vec{B} \cdot \hat{\vec{I}} + \hat{\vec{I}} \cdot \overleftrightarrow{Q} \cdot \hat{\vec{I}}. \tag{1.160}$$

It can be seen from the spin Hamiltonian (1.160) that some interactions are magnetic field dependent (electron and nuclear Zeeman interactions), while others are not (fine structure interaction, hyperfine interaction, nuclear quadrupole interaction). To separate these interactions from each other, it is frequently necessary to make EPR measurements with various magnetic field/frequency settings.

1.13 Case of a Weak Crystal Field or the Rare-Earth Arrangement

1.13.1 Terms and Subterms of the Ground States of Rare-Earth Elements with Unpaired f-Electrons

Let us consider the case of a weak crystal field. Then, the interaction with the crystal field is weaker than the spin-orbit interaction, i.e.,

$$\hat{H}_{Crys} < \hat{H}_{SO}.$$

As a result, subterms can be generated and J is a "good" quantum number.

We have a weak crystal field for atoms and ions of rare-earth elements, e.g., the lanthanide (4*f*) group or actinide (5*f*) group. The main feature of the lanthanide and actinide groups is the filling of the 4*f* or 5*f* shells, respectively. Because the 4*f* (5*f*) electrons form an "inner shell", the unpaired electrons are but little affected by the surroundings of the ions in a crystal, i.e., they are rather well shielded from the effect of the crystal field (hence, the name "rare-earth scheme").

Consider the group of lanthanides. Choose positive triply charged ions. The total Hamiltonian of the free ion is composed of interactions described in the preceding sections, which are arranged in order of decreasing energy

$$\hat{H}_{Free} = \hat{H}_{Coul} + \hat{H}_{SO} + \hat{H}_{Zeem-e} + \hat{H}_{HFI} + \hat{H}_{Zeem-n} + \hat{H}_{Q}. \tag{1.113}$$

The Coulomb interaction $\hat{H}_{Coul}$ in the Hamiltonian (1.113) forms terms ($L, S \rightarrow {}^{2S+1}L$); upon addition of the spin-orbit coupling $\hat{H}_{SO} = \lambda \hat{\vec{L}} \cdot \hat{\vec{S}}$, the subterms ($L, S, J \rightarrow {}^{2S+1}L_J$) are generated. The ground-state terms are formed in accordance with Pauli's principle and Hund's rule, i.e., there should not be two states with identical quantum numbers and should be the maximum orbital angular momentum when the condition of the maximum spin momentum is satisfied. The result is a "Hund's triangle", because, for any configuration in the ground state, all the spin angular moments of the electrons are aligned and the spin is at a maximum. According to the spectroscopic alphabet, the values of $L = 0, 1, 2, 3, 4, 5, 6$ correspond to the capital letters *S, P, D, F, G, H, I*, respectively. The spin-orbit coupling leads to splitting of the term into subterms differing in the total angular momentum of a free atom (or ion) $\boldsymbol{J}$, $\boldsymbol{J} = \boldsymbol{L} + \boldsymbol{S}$, and a subterm is written as $^{2S+1}L_J$. In accordance with Hund's second rule, the subterm with the minimum J has the lowest energy if the electron shell is less than half-filled and the subterm with the maximum J has the lowest energy if the electron shell is more than half-filled.

Let us consider the terms and subterms of rare-earth ions with shells under construction. For definiteness, we take the elements with 4*f*-electrons, although these schemes are, in general, also suitable for other *f*-shells. For the tripositive ions, the configuration is $4f^n$, where $n = Z - 57$ and Z is the atomic number of lanthanum. Unlike the scheme of the terms discussed above for the transition elements, rare-earth elements are to be considered as terms and subterms because J is a "good" quantum number.

We construct a diagram (Fig. 1.29) in which the vertical axis shows the total electron orbital and spin moments of the free ion and the horizontal axis reflects the number of *f*-electrons.

As an example we will consider the configuration $4f^3$. First, taking into account the Pauli exclusion principle and Hund's rule, we find the term corresponding to the minimum energy. We write down the corresponding quantum numbers for the maximum spin angular momentum, given that this should be the maximum orbital angular momentum:

for the first electron: $l_1 = 3$, $m_{l1} = 3$; $s_1 = 1/2$; $m_{s1} = 1/2$;
for the second electron: $l_2 = 3$, $m_{l2} = 2$; $s_2 = 1/2$; $m_{s2} = 1/2$;
for the third electron: $l_2 = 3$, $m_{l2} = 1$; $s_2 = 1/2$; $m_{s2} = 1/2$.

The total orbital angular momentum L is equal to the maximum projection of the total orbital angular momentum, as $m_{l1} + m_{l2} + m_{l3} = 3 + 2 + 1 = 6$, $L = 6$, for the spin moment $m_{s1} + m_{s2} + m_{s3} = 3/2$, $S = 3/2$. Thus, the lower term has the form ^4I. As a result of the spin-orbit interaction (in line with the rules of addition of angular momenta) the term is split into four subterms $^4I_{9/2}$, $^4I_{11/2}$, $^4I_{13/2}$, and $^4I_{15/2}$. In line with Hund's second rule, subterm $^4I_{9/2}$ is the lowest energy sublevel, the splitting between the levels equal to $11/2\lambda$, $13/2\lambda$ and $15/2\lambda$. Note that for the half-filled shell ($4f^7$) the total orbital angular momentum is zero ($L = 0$) and for $4f^{11}$ configuration in line with Hund's second rule, subterm $^4I_{15/2}$ is the lowest energy sublevel.

In the case of a weak crystal field, when the interaction with the crystal field $\hat{H}_{Crys}$ is weaker than the spin-orbit interaction $\hat{H}_{SO}$, i.e., $\hat{H}_{Crys} \lhd \hat{H}_{SO}$, the sum of the first two terms of the Hamiltonian (1.113) will be considered a zero approximation

$$\hat{H}_0 = \hat{H}_{Coul} + \hat{H}_{SO}. \tag{1.161}$$

(Compare with (1.117) $\hat{H}_0 = \hat{H}_{Coul} + \hat{H}_{Crys}$ for the case of an intermediate crystal field).

The Hamiltonian $\hat{H}$ of an ion in a crystal field is then assumed to have the form

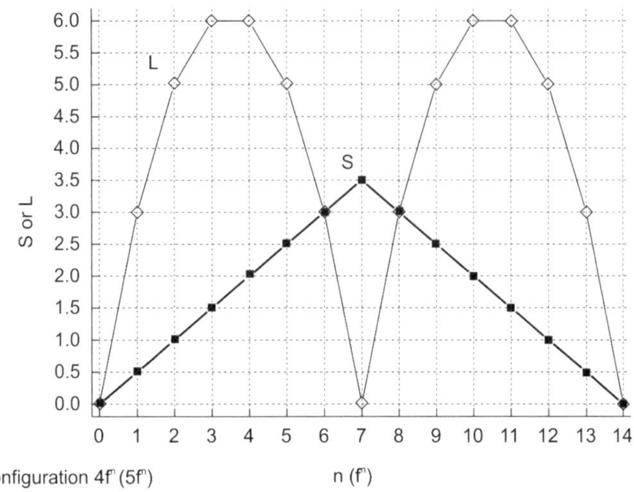

	0	1	2	3	4	5	6	7	8	9	10	11	12	13	14
Subterm $^{2S+1}L_J$	^1S	$^2F_{5/2}$	$^3H_4$	$^4I_{9/2}$	$^5I_4$	$^6H_{5/2}$	$^7F_0$	$^8S_{7/2}$	$^7F_6$	$^6H_{15/2}$	$^5I_8$	$^4I_{15/2}$	$^3H_6$	$^2F_{7/2}$	^1S
Lande *g*-factor g_J		6/7	4/5	8/11	3/5	2/7	-	2	3/2	4/3	5/4	6/5	7/6	8/7	
λ (cm^{-1})		640	800	900	1070	1200	1410	1540	-1770	-1860	-2000	-2350	-2660	-2940	
Ion $4f^n$ ($5s^25p^6$)	La^{3+}	Ce^{3+}	Pr^{3+}	Nd^{3+}	Pm^{3+}	Sm^{3+}	Eu^{3+}	Gd^{3+}	Tb^{3+}	Dy^{3+}	Ho^{3+}	Er^{3+}	Tm^{3+}	Yb^{3+}	

Fig. 1.29 Spectroscopic electronic properties of the lanthanide-group ($4f^n$) free ions in the ground state (low energy subterms) in accordance with Pauli's principle and Hund's rules. The free-ion spin-orbit interaction constants λ (cm^{-1}) are presented

$$\hat{H} = \hat{H}_0 + \hat{H}'$$

where $\hat{H}'$ is an electrostatic potential energy that has the symmetry of the surroundings of the rare-earth ion.

We use a perturbation theory for degenerate states.The interaction with the crystal field $\hat{H}_{Crys}$ is considered to be the perturbation $\hat{H}'$. Direct calculation of the crystal field potential at a rare-earth ion site is a difficult task, but this potential must reflect the symmetry of the crystal at the ion site. The symmetry of the site allows selection of the non-zero matrix elements within the manifold of $4f$-states.

1.13.2 Energy Levels and Wave Functions for the Ground State of Rare-Earth Ions in a Magnetic Field

For ions with half-integer values of J (odd number of electrons, ions with Kramers degeneracy or the so-called Kramers ions), the crystal field lifts the degeneracy except for the necessary at least a pair of degenerate levels in accordance with the Kramers theorem. Application of a magnetic field will lift this degeneracy (Zeeman spltting). The description of the EPR for these ions follows the common pattern. The ground-state multiplet J of a free ion (see Fig. 1.29) will be split by the low-symmetry crystal field into $J + 1/2$ doublets.

We consider only the first-order perturbation approach in which matrix elements within the manifold of a given J are considered (the second-order perturbation approach includes matrix elements between states with different J). To calculate the Zeeman splitting, it is necessary to know the wave-functions of the ground-state doublet. These wave-functions would allow us to calculate the g-tensor and the hyperfine interaction tensor (for ions with a nuclear spin).

To illustrate the method for calculation of the energy levels, we consider a simple system with the minimum number of levels, which is the case for Ce^{3+} ions. The ground configuration of Ce^{3+} is $4f^1$ so the ground term is ^2F. The spin-orbit coupling splits this term into a sixfold-degenerate $^2F_{5/2}$ subterm and eightfold-degenerate $^2F_{7/2}$ subterm. The separation between these levels is $7/2\lambda$, which is about 2200 cm^{-1} for the free ion. The energy levels of the lower subterm of the Ce^{3+} ion is $^2F_{5/2}$. To a first approximation, J is assumed to be a good quantum number $J = 5/2$. In a crystal field, the sixfold-degeneracy will be lifted (number of levels is given by $2J + 1$, we already used a similar formula for the multiplicity of the spin levels in a term: $2S + 1$). The separation between the adjacent J levels due to the spin-orbit coupling exceeds the splitting of the ground-state levels due to the crystal field by at least an order of magnitude.

The eigenfunctions that can be obtained by diagonalization of the Hamiltonian $H_{SO} = \lambda \boldsymbol{L} \cdot \boldsymbol{S}$ in the product basis $|M_L, M_S\rangle$ (uncoupled basis) are eigenfunctions of the set $\boldsymbol{J}^2$, J_z, $\boldsymbol{L}^2$, and $\boldsymbol{S}^2$ $|J, M_J\rangle$ (coupled basis). In terms of the products $|M_L, M_S\rangle$, the eigenfunctions $|J = 5/2, M_J\rangle$ are given by [12]:

$$\begin{aligned}
\left|\frac{5}{2}, \pm\frac{1}{2}\right\rangle &= \left(\frac{4}{7}\right)^{1/2}\left|\pm1, \mp\frac{1}{2}\right\rangle - \left(\frac{3}{7}\right)^{1/2}\left|0, \pm\frac{1}{2}\right\rangle \\
\left|\frac{5}{2}, \pm\frac{3}{2}\right\rangle &= \left(\frac{5}{7}\right)^{1/2}\left|\pm2, \mp\frac{1}{2}\right\rangle - \left(\frac{2}{7}\right)^{1/2}\left|\pm1, \pm\frac{1}{2}\right\rangle \\
\left|\frac{5}{2}, \pm\frac{5}{2}\right\rangle &= \left(\frac{6}{7}\right)^{1/2}\left|\pm3, \mp\frac{1}{2}\right\rangle - \left(\frac{1}{7}\right)^{1/2}\left|\pm2, \pm\frac{1}{2}\right\rangle.
\end{aligned} \tag{1.162}$$

According to the Kramers theorem, the crystal field can at most split the sixfold degeneracy into three doublets. Thus, if the cerium ion is placed in a crystal field of low symmetry, the limit can be split into three pairs of levels (three Kramers doublets). Commonly, the lower energy level is of interest in EPR experiments because the excited levels are weakly populated and their EPR spectra are rarely observed. Because the wave functions will differ only by M_J, we write these functions as $|M_J\rangle$, where M_J takes the values 5/2, 3/2, 1/2, −1/2, −3/2, −5/2. Thus, with the vector $|J,M_J\rangle$ for one subterm, it is convenient to use simply $|M_J\rangle$

$$\begin{aligned} \left|\pm\frac{1}{2}\right\rangle &= \left(\frac{4}{7}\right)^{1/2}\left|\pm1,\mp\frac{1}{2}\right\rangle - \left(\frac{3}{7}\right)^{1/2}\left|0,\pm\frac{1}{2}\right\rangle \\ \left|\pm\frac{3}{2}\right\rangle &= \left(\frac{5}{7}\right)^{1/2}\left|\pm2,\mp\frac{1}{2}\right\rangle - \left(\frac{2}{7}\right)^{1/2}\left|\pm1,\pm\frac{1}{2}\right\rangle \\ \left|\pm\frac{5}{2}\right\rangle &= \left(\frac{6}{7}\right)^{1/2}\left|\pm3,\mp\frac{1}{2}\right\rangle - \left(\frac{1}{7}\right)^{1/2}\left|\pm2,\pm\frac{1}{2}\right\rangle. \end{aligned} \quad (1.162a)$$

Let us suggest that the $M_J = \pm 1/2$ doublet is the lowest in energy. An applied magnetic field lifts the degeneracy. We use the Zeeman Hamiltonian

$$\hat{H}_{Zeem-e} = g_J\mu_B\hat{\vec{J}}\cdot\vec{B} = \mu_B(\hat{\vec{L}}+2\hat{\vec{S}})\cdot\vec{B}$$

($g_S = 2$) to calculate the splitting in the magnetic field, g_J is the Landé g-factor (1.2.9)

$$g_J = 1 + \frac{J(J+1)+S(S+1)-L(L+1)}{2J(J+1)}$$

In the basis of functions $|M_J = \pm 1/2\rangle$, the matrix of the Zeeman Hamiltonian has the form [12]:

	$\|+1/2\rangle$	$\|-1/2\rangle$
$\langle+1/2\|$	$3/7\mu_B B_z$	$9/7\mu_B(B_x - iB_y)$
$\langle-1/2\|$	$9/7\mu_B(B_x + iB_y)$	$-3/7\mu_B B_z$

and, as a result, we obtain $g_{\|} = 6/7$ and $g_{\perp} = 18/7$.

In the same way, we can calculate the splitting and, hence, the g-factors for the other two Kramers doublets with $M_J = \pm 3/2$ and $M_J = \pm 5/2$ (1.162a). We have

$$\begin{aligned} M_J &= \pm 3/2:\ g_{\|} = 18/7\,\text{and}\,g_{\perp} = 0 \\ M_J &= \pm 5/2:\ g_{\|} = 30/7\,\text{and}\,g_{\perp} = 0 \end{aligned}$$

In general, the procedure for calculation of the energy levels consists of several steps. In the first stage of the calculation, we construct a matrix $\langle M_J|\hat{H}_{Crys}|M'_J\rangle$ and then diagonalize it and find the wave-functions of various doublets. The crystal field mixes states with different M_J to give a resultant state of the type

$$|\psi_i\rangle = \sum_M k_M |J, M_J\rangle, \tag{1.163}$$

with

$$\sum_M k_M^2 = 1$$

to satisfy the normalization condition. The real energy levels and the relative admixtures of the wave-functions depend on parameters of the crystal field. The Kramers doublets are commonly separated in energy by about 20–100 cm^{-1}, and, as a result, EPR transitions are observable only between the components of the lowest doublet.

For the threefold symmetry, the wave-functions contain all values of M differing by 3. We choose a low-symmetry crystal field appropriate to a site of D_2 symmetry. For a site of D_2 symmetry and for $J = 5/2$, we obtained three doublets with the wave-functions in the form of combinations of wave-functions of the zeroth approximation in which the wave-functions contain all values of M_J differing by 2

$$\begin{aligned} |\alpha_i\rangle &= a_i|5/2\rangle + b_i|1/2\rangle + c_i|-3/2\rangle, \\ |\beta_i\rangle &= a_i|-5/2\rangle + b_i|-1/2\rangle + c_i|3/2\rangle. \end{aligned} \tag{1.164}$$

The lower doublet is written as

$$\begin{aligned} |\alpha_1\rangle &= \alpha_1|5/2\rangle + b_1|1/2\rangle + c_1|-3/2\rangle, \\ |\beta_1\rangle &= a_1|-5/2\; n + b_1|-1/2\; n + c_1|3/2\; n. \end{aligned} \tag{1.165}$$

Next, consider the effect of the magnetic field on the lower Kramers doublet. To find the EPR conditions, we have to calculate the Zeeman splitting for each doublet (as a rule, for the lowest doublet). The Hamiltonian of the Zeeman interaction can be again written as

$$\hat{H}_{Zeem-e} = g_J \mu_B \hat{\vec{J}} \cdot \vec{B}.$$

The calculation of the Zeeman levels is reduced to finding the matrix elements of $\langle\psi|\hat{H}_{Zeem-e}|\psi'\rangle$ or finally the matrix elements of the operators $\hat{J}_Z$, $\hat{J}_X$ and $\hat{J}_Y$ within each doublet. In general, the Zeeman splitting within each doublet is described by a spin Hamiltonian with an effective spin $S' = 1/2$ and an anisotropic g-factor. For the axial symmetry, $\hat{J}_X$ and $\hat{J}_Y$ are equal and the anisotropic effective g-factor has components $g_{effZ} = g_{eff\|}$ and $g_{effX} = g_{effY} = g_{eff\perp}$. As a result, the spin Hamiltonian for the Zeeman interaction can be written as

$$\hat{H}_{Zeem} = \mu_B[g_{eff//}B_z\hat{S}'_z + g_{eff\perp}(B_x\hat{S}'_x + B_y\hat{S}'_y)].$$

For $B \| z$, the Zeeman Hamiltonian simplifies

$$\hat{H}_{Zeem-e} = g_J \mu_B \hat{J}_Z B.$$

The energy matrix $\langle \psi | \hat{H}_{Zeem-e} | \psi' \rangle$ for the lower doublet has the form

$\langle \psi \| \hat{H}_{Zeem-e} \| \psi' \rangle$	$\|\alpha_1\rangle$	$\|\beta_1\rangle$
$\langle \alpha_1 \|$	$g_J \mu_B B(5/2\, a_1^2 + 1/2 b_1^2 - 3/2 c_1^2)$	0
$\langle \beta_1 \|$	0	$-g_J \mu_B B(5/2 a_1^2 + 1/2 b_1^2 - 3/2 c_1^2)$

(1.166)

Thus, there will be the usual linear dependence of the magnetic field splitting

$$\Delta E = g_{\text{eff}\|} \mu_B B \tag{1.167}$$

with an effective g-factor

$$g_{\text{eff}\|} = g_J \left(5a_1^2 + b_1^2 - 3c_1^2\right). \tag{1.168}$$

The effective g-factor depends on the Landé g-factor and on the coefficients in the wave-functions, expressed in terms of the characteristics of the crystal field potential.

The scheme of energy levels for the lower doublet in a magnetic field (for $B \| z$) has the form.

For $B \| z$, the energy difference $\Delta E = g_{\text{eff}\|} \mu_B B$. According to Planck's formula, $\Delta E = hv$, and, if microwave radiation is applied to a system with these levels, there will be resonant transitions between these levels, with a reorientation of the electron spin in a magnetic field (shown by the arrow in Fig. 1.30). Planck's formula will be written by using the angular frequency $\omega = \nu/2\pi$, in the form $\Delta E = \hbar\omega$,

$$\hbar\omega = g_{eff\|} \mu_B B. \tag{1.169}$$

It is straightforward to calculate $g_{\text{eff}\perp}$ by using these results (see authoritative sources for full details [12]).

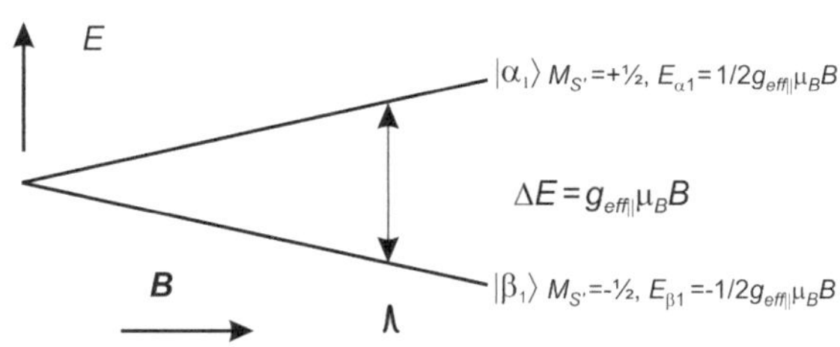

Fig. 1.30 Scheme of energy levels for the lower doublet in a magnetic field (for $B \| z$)

References

1. Matthews, P.S.C.: Quantum Chemistry of Atoms and Molecules. Cambridge University Press (1986)
2. Breit, G., Rabi, I.I.: Measurement of nuclear spin. Phys. Rev. **38**, 2082–2083 (1931)
3. Wertz, J.E., Bolton, J.R.: Electron Spin Resonance: Elementary Theory and Practical Applications. McGraw-Hill, New York (1972); Wertz, J.E., Bolton, J.R.: Electron Spin Resonance: Elementary Theory and Practical Applications. Chapman and Hall, London (1986)
4. Morton, J.R., Preston, K.F.: Atomic parameters for paramagnetic resonance data. J. Magn. Reson. **30**, 577–583 (1978)
5. Grachev, V.: View EPR/ENDOR program. http://www.physics.montana.edu/faculty/grachev
6. Abragam, A., Bleaney, B.: Electron Paramagnetic Resonance of Transition Ions, p. 702. Clarendon Press, Oxford (1970)
7. Abragam, A., Bleaney, B.: Electron Paramagnetic Resonance of Transition Ions, p. 372. Clarendon Press, Oxford (1970)
8. Abragam, A., Bleaney, B.: Electron Paramagnetic Resonance of Transition Ions, p. 376. Clarendon Press, Oxford (1970)
9. Sushil, K. (ed.): Misra Multifrequency Electron Paramagnetic Resonance: Theory and Applications. Wiley (2011)
10. Abragam, A., Bleaney, B.: Electron Paramagnetic Resonance of Transition Ions, p. 403. Clarendon Press, Oxford (1970)
11. Carrington, A., McLachlan, A.D.: Introduction to Magnetic Resonance with Applications to Chemistry and Chemical Physics. Harper & Row, Publishers (1967)
12. Atherton, N.M.: Electron Spin Resonance—Theory and Applications, p. 164. Wiley, New York (1973)

Chapter 2
Fundamentals of EPR Related Methods

2.1 Basics of Pulse Magnetic Resonance Spectroscopy

We present a brief historical introduction to the field of pulse magnetic resonance (see, e.g., [1]). The first EPR phenomenon was observed by Zavoisky in 1944 [2]. In 1946 Bloch [3] and Purcell, Torrey and Pound [4] reported the first nuclear magnetic resonance (NMR) experiments. The first nuclear spin echo was described by Hahn in 1950 [5]. In 1958 the first observation of electron spin echo (ESE) was reported by Blume [6] and pulse electron-nuclear double resonance (ENDOR) was developed by Mims in 1965 [7]. The first echo modulation experiments were reported in 1968 by Novosibirsk group [8]. The basic physical principles of electron spin echo and nuclear spin echo (NSE) are essentially the same. The difference between ESE and NSE experiments are related to the technical problems of scaling from the radiofrequency range (MHz) to the microwave range (GHz) [9–14].

Let us consider the difference between conventional continuous wave (CW) and pulse magnetic resonance (EPR) spectroscopy. CW and pulse EPR are complementary and the application of both gives a more reliable picture of the spin phenomena under investigation. In the CW EPR, the magnitude of the magnetic field B_0 (static magnetic field) is swept, while the amplitude of the microwave field B_1 is constant with time. The great potential of the EPR spectroscopy cannot be fully realized with only CW EPR due to limitations in time resolution. In the pulse EPR experiments (time-resolved experiments) a time-dependent microwave pulse B_1 is applied in addition to a static magnetic field B_0. The short microwave pulses are used to deflect the magnetization and monitor their return to thermodynamic equilibrium. In the pulse EPR spectroscopy relaxation times can be directly measured by monitoring the magnetization on the same timescale in which relaxation occurs.

One can compare CW and pulse techniques by using a well known analogy (see, e.g., [15]) which relates them to tuning a bell (or other musical instrument) by measuring the frequency spectrum of acoustic resonances in the bell. In CW

P.G. Baranov et al., *Magnetic Resonance of Semiconductors and Their Nanostructures*, Springer Series in Materials Science 253,
DOI 10.1007/978-3-7091-1157-4_2

experiment one slowly sweeps the frequency and detect resonances in the bell (similarly magnetic resonances in the sample), while in the pulse experiment one strikes the bell to cause its acoustic resonances and as a result the bell will resonate acoustically at the own multiple frequencies. The frequency spectrum of acoustic resonances can be obtained by the Fourier transformation (FT) of the digitized sound signal.

2.1.1 Free Induction Decay (FID) and the Electron Spin-Echo (ESE) Phenomenon

The quantum mechanical description of pulse magnetic resonance experiments uses the density matrix formalism [9–12]. This description is not enough illustrative therefore, we will use classical treatment for pulse phenomena. The important aspect is that the classical precession frequency ω_0, so-called Larmor frequency (Joseph Larmor, British physicist) is equal to the resonance condition for two level system that was derived according to a quantum mechanical description. The description of spin-echo phenomenon is based on the concept of an inhomogeneously broadened resonance line which is composed of homogeneous spin packets (Fig. 2.1). The lineshape of homogeneously broadened line is determined by the relaxation time and therefore this line has so-called Lorentz lineshape. The magnetic resonance signal of homogeneously broadened line is the sum of a large number of lines (spin packet $\mathbf{M_i}$) each of which is characterized by the same Larmor frequency and linewidth (Fig. 2.1). The magnetic resonance signal of inhomogeneously broadened line is the sum of a large number of homogeneously broadened line each of which is shifted in frequency with respect to each other. The

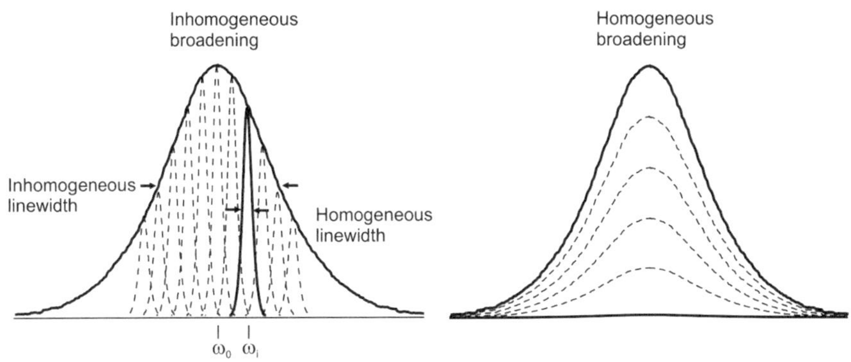

Fig. 2.1 Schematic diagram showing an inhomogeneously broadened EPR line of a paramagnetic center as the sum of a large number of separate narrow homogeneously broadened lines that are each shifted in frequency with respect to each other (*left*). A homogeneously broadened line shown as the superposition of a number of similar spin packets; the individual spin packet width is here the same as the width of the overall line (*right*)

inhomogeneously broadened line is usually described by a Gaussian lineshape. This inhomogeneous linewidth is mainly caused by effect of the local magnetic field inhomogeneity in the sample which results in a large number of spin packets characterized by different Larmor frequencies.

The time dependence of the magnetisation associated with each spin packet $\mathbf{M_i}$ is described by the precession equation [13, 14]

$$\frac{d\vec{M}_i}{dt} = -\gamma \vec{B}_i \times \vec{M}_i, \tag{2.1}$$

here $\mathbf{B_i}$ is the static magnetic field experienced by the spin packets and γ is the gyromagnetic ratio (also known as the magnetogyric ratio) which gives the proportionality constant between the magnetic moment and the angular momentum. Thus $\mathbf{M_i}$ precesses around $\mathbf{B_i}$ with the Larmor frequency $\omega_i = \gamma B_i$. Due to the inhomogeneous broadening, the magnetic resonance line is a distribution of ω_i around the center frequency $\omega_0 = \gamma B_0$, where B_0 is the static external magnetic field. In the laboratory frame one uses the magnetic field, B_0, which is parallel to the z axis, the oscillation magnetic field, B_1, is parallel to the x axis, and the y axis is orthogonal to the x and z axes. If we just introduce a frame rotating with frequency ω_0 around B_0 then the rotation frequencies in this frame will be given by $\omega_i' = \omega_0 - \gamma B_i$. All spins in a packet precess with the same frequency but different phase and therefore the resultant magnetic moment of all spin packets in steady state lies along B_0. Let us now apply microwave radiation with frequency ω_0, polarized along, e.g., the x-direction, and decompose this oscillating field with amplitude $2B_1$ into two counter-rotating components with amplitude B_1. One component is stationary in the rotating frame but the other component is off resonance by $2\omega_0$ and neglected in this approximation. If this radiation is strong and γB_1 larger than the spread in Larmor frequencies, the magnetizations of the spin packets will precess around B_1 with frequency $\omega_1 = \gamma B_1$ (so-called, "Rabi frequency"). In the case when the duration of the microwave pulse is just enough to flip the magnetizations $\mathbf{M_i}$ into the xy plane, the pulse is defined as a "$\pi/2$-pulse". When the duration of the microwave pulse is enough to invert the magnetization (the equilibrium magnetization M_0 is reversed) the pulse is defined as a "π-pulse".

When the magnetization flips into the xy plane, the magnetization along the z-axis decreases to zero, in other words, the population difference between the levels tends to zero. The π-pulse (or inversion pulse) will rotate the magnetization one step further—into the spin state which is anti-parallel to the z-axis.

Figure 2.2 shows populations of quantum states in thermal equilibrium and after $\pi/2$ (90°) and π (180°) pulses. In thermal equilibrium Boltzmann distribution between spin states is realized and M is directed along the z axis. As a result of a $\pi/2$ pulse (so called "saturating pulse") the magnetization along the z axis goes to zero since the population difference between parallel and antiparallel states goes to zero (Fig. 2.2). A $\pi/2$ pulse creates electron coherence which corresponds to in-plane rotating electron magnetization M. Coherence is a condition in which two (or more) signals maintain a fixed phase relationship relative to each other and therefore have identical frequencies (or sets of frequencies). A π pulse (or an inversion pulse) tips the magnetization 180°, exchanging the populations of the quantum states. A vector of the magnetic moment M making a finite angle with static magnetic field B_0 precesses with angular frequency of magnitude $\omega_0 = \gamma B_0$ about the direction of B_0. In magnetic resonance studies it is usual to describe this precession in the rotating reference frame in which the static magnetic field is completely or partially eliminated. In the rotating frame the coordinate system is rotated at an angular velocity equal to the applied microwave frequency; for magnetic resonance condition the magnetic components precessing at the Larmor frequency and B_1 is stationary. This makes the visualization and mathematical treatment of the magnetic behavior easier. If one is off magnetic resonance, the Larmor frequency is not exactly equal to the applied microwave frequency, then the magnetization vector will rotate in the x-y plane at the resonance offset $\Delta\omega = \omega - \omega_{mw}$. This is usually the case for inhomogeneously broadened EPR line because of different resonance fields and B_1 inhomogeneities, which is due to the fact that different parts of the sample can experience slightly different magnetic fields. Nutation of the net magnetic moment as viewed from the frame of reference rotating at the precessional frequency $\omega_L = \gamma B_0$ is shown in Fig. 2.2, bottom.

Figure 2.3 shows the formation of so-called the free induction decay (FID) and an electron spin echo (ESE). An electron spin echo can be generated with the two-pulse microwave sequence shown in the upper part of the figure. Pulse is an abrupt application of microwave radiation which is characterized by its strength, duration, and frequency range.

So B_1 is defined along the $-X'$-axis of the rotating frame, first one gives a pulse with a duration $t_{p1} = (\pi/2)/(\gamma B_1)$ so that the magnetization of the spin packets is brought along Y', a so-called $\pi/2$-pulse. As there is a spread in offset frequencies ω_i', the different spin packets will start to diphase and as a result one has the free induction decay.

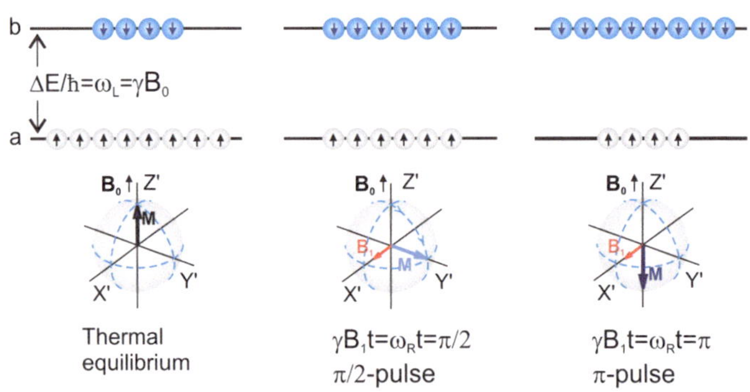

Fig. 2.2 Populations of quantum states in thermal equilibrium and after $\pi/2$ and π pulses (*top*). Nutation of the net magnetic moment as viewed from the frame of reference rotating at the precessional frequency $\omega_L = \gamma B_0$ (*bottom*)

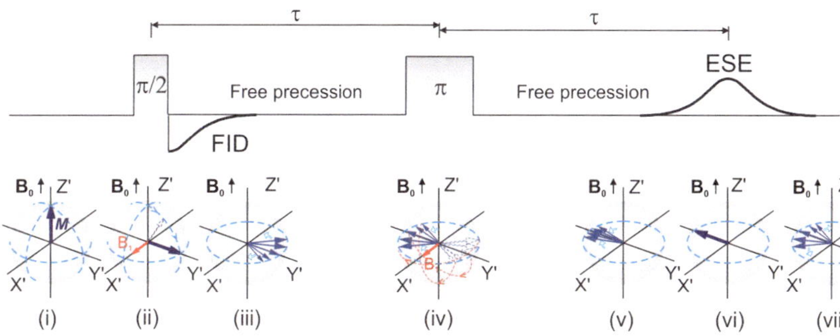

Fig. 2.3 The formation of a free induction decay (*FID*) and an electron spin echo (*ESE*) for the two-pulse sequence. Nutation of the net magnetic moment as viewed from the frame of reference rotating at the precessional frequency (*bottom*)

A single π/2 (90°)-pulse produces an FID, and by the Fourier transformation of the FID one will get the frequency domain which is an EPR spectrum. One can not collect FID signal immediately after the pulse is turned off because the receiver would be destroyed by the MW pulse high power. There is a lag between the end of the MW pulse and when the signal can be measured, so called "dead time" which is typically ~80 ns at X-band, depending on the quality factor (Q) of the cavity and MW frequency. The response function of the cavity in the time domain [16]

$$f = \exp(-t/T_R) \quad \text{with} \quad T_R = 2Q/\omega_0,$$

where T_R is the ringing time of the EPR cavity. The ringing time T_R of the EPR cavity is inversely proportional to the MW frequency of the EPR spectrometer. The ringing time gives the physical limit of the EPR instrument response time. Fixing a typical loaded quality value Q = 2000 of a single-mode MW cavity, T_R = 67 and 6.7 ns is calculated for X-band (9.5 GHz) and W-band (95 GHz) [16], respectively. Obviously, the time resolution of X-band EPR can be improved by reducing the cavity Q-factor. However, decreasing the cavity Q-factor leads to a lowering of the sensitivity of the experiment at X-band. Thus, the high-frequency EPR is the method of choice for pulse experiments.

EPR spectrum is usually inhomogeneously broadened. The homogeneous broadening is result of random and irreversible events, unlike inhomogeneous broadening which is reversible and static. If we apply the second pulse at a time τ after the first π/2-pulse when dephasing will start, then the electron spin echo signal will occur at time τ after the second pulse. Thus the second pulse produces an electron spin echo signal after the dead time, and the echo shape compares two back-to-back FID pattern.

Thus, an electron spin echo can be generated by giving a π-pulse with duration $t_{p2} = \pi/(\gamma B_1)$ at a time τ after the first pulse. As the B_0 field has not changed its direction after the pulse, the magnetizations of the spin packets will continue to dephase in the same direction and at the same speed. As the whole X′Y′-plane is turned around the X′-axis by the π-pulse, the magnetizations of the spin packets will come together again along the –Y′-axis at time 2τ, generating a two-pulse spin echo.

One should note that any pulse sequence with rotation θ_1 and θ_2 will cause a spin echo, however the maximum effect occurs for $\theta_1 = \pi/2$ and $\theta_2 = \pi$. The spin echo has found wide-spread application to measure the relaxation times T_1 and T_2.

The refocussing of the magnetization of the individual spin packets only works if the "fast", "middle" and "slow" spins remain fast, middle and slow spins during all the time 2τ, respectively. If there are random jumps in the resonance frequency of the spin packets in the time interval 2τ than this effect will be observed as a decay of the echo intensity with increasing time 2τ. The decay function of the echo intensity as a function of 2τ gives the rate of irreversible loss of phase coherence, i.e. the relaxation time T_2. In other words, whereas the decay of the FID after a π/2-pulse is determined by T_2*, i.e., by the total inhomogeneous linewidth, the decay of the spin echo signal as a function of 2t is determined by T_2, i.e., by the homogeneous linewidth.

In comparison to the FID decay, there are more contributions to the spin echo decay than simply transverse relaxation that affect the signal intensity and duration. The time constant for spin echo decay is often called the phase memory time, or T_M. Many processes contribute to the phase memory time, such as spin-spin relaxation, spectral diffusion, spin diffusion, and instantaneous diffusion. Spectral diffusion is a process by which the frequency of a spin packet changes with time and can be caused by molecular motion, exchange interactions, nuclear spin flip-flops, or electron-nuclear cross-relaxation. In spin diffusion process, spins undergo the continuous exchange of energy via spin flip-flops, causing energy to be dissipated throughout the sample rather than being lost to the lattice. This effect reduces polarization differences between quantum states to a timescale which is much shorter than the intrinsic relaxation rate. Instantaneous diffusion takes place in the case wherein the distance between spins is small at high spin concentrations or in aggregates, a quickly inverted spin changes the local field at a nearby spin and causes its frequency to shift and also results in a phase shift and shorten the phase memory time. This frequency change affects the ability of the spin packet to refocus properly, and thus the spin echo intensity decreases.

As shown in Fig. 2.3 initially the total (net) magnetic moment vector M is in the equilibrium position (i) parallel to the direction of the strong external static magnetic field **B** along Z′-axis (B_z). The next step is the application of an alternating microwave (mw) magnetic field $\mathbf{B_1}$ along the *X*′-axis. In order to describe this process, it is useful to visualize the magnetization vector M in the rotating reference frame X′Y′Z′ (Fig. 2.3). In the rotating reference frame the net magnetic moment rotates about B_1 and at the end of a π/2 (90°) pulse the net magnetic moment M is in the equatorial plane (ii). During the relatively long period of time following the removal of B_1, the separate incremental moment vectors begin to fan out slowly (iii). This is mainly caused by effect of the local magnetic field inhomogeneity (the variation in B_z) over the sample which results in a large number of spin packets characterized by different Larmor frequencies (see Fig. 2.1). At time t = τ, the mw field B_1 is again applied and the moments (iv) begin to rotate about the direction of B_1 long enough to satisfy the π (180°) pulse condition. After π pulse, all the separate incremental vectors (magnetization of spin packets) are again in the equatorial plane and the inverted magnetization of spin packets still rotates in the same direction with the same speed (v) and as a result the magnetization eventually refocuses at $t = 2\tau$ to form an electron spin echo (vi). The echo signal then begins to decay as the incremental vectors again fan out (vii) [17].

All the previous arguments can be repeated for clarity on the example of the stadium with motorcyclists that just will serve as the end of the magnetization vector (see Fig. 2.4). Two cases are shown. (a) If we look down the z axis at the stadium, e.g., XY plane (Fig. 2.4a), there will be faster moving motorcyclists and slower moving motorcyclists, depending on their sports training. (b) If we look

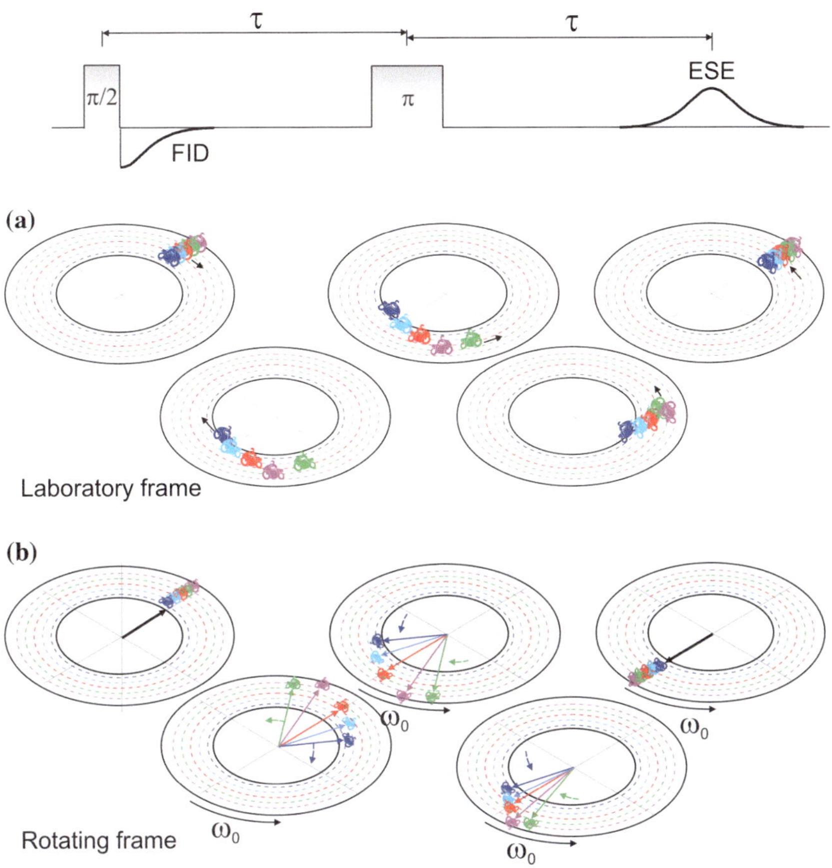

Fig. 2.4 The pulse sequence in a two-pulse electron spin echo. The refocusing of spin packets represented by faster, normal and slower motorcyclists when viewed from the laboratory frame (**a**) and from the rotating frame (**b**)

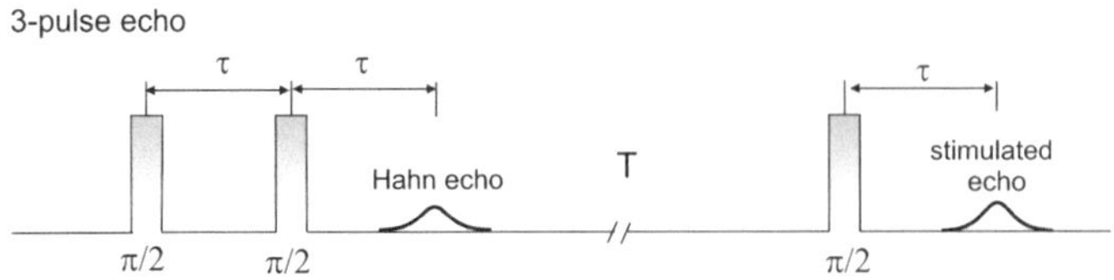

Fig. 2.5 The three-pulse sequence. In this sequence the π-pulse is split into two π/2 pulses. After the first of these π/2 pulses, the +Y′ components of the dephased magnetization pattern are stored along the –Z′-axis where they remain during the waiting time T. The third microwave pulse brings the M_Z components back into the X′Y′-plane where they continue their time evolution and give rise to a stimulated echo at time τ after the third pulse

down the z axis at the stadium, which rotates area with an average angular velocity corresponding to the precession of the magnetic moment. When one applies a π pulse, this corresponds the case, when instead of moving away from the slower motorcyclists, the faster moving motorcyclists move towards the slower motorcyclists. The motorcyclists distribution eventually refocuses to form what is called an echo at time 2τ.

The three-pulse sequence is often used in ESE spectroscopy which is sketched in Fig. 2.5. In this sequence the π-pulse is so to say split into two π/2 pulses. After the first of these π/2 pulses, the +Y′ components of the dephased magnetization pattern are stored along the −Z′-axis where they remain during the waiting time T. The X′-components generate a Hahn echo after time τ (Fig. 2.5). The third microwave pulse brings the M_Z components back into the X′Y′-plane where they continue their time evolution and give rise to a so-called "stimulated" spin echo at time τ after the third pulse (Fig. 2.5). Both spin echoes have half the intensity of the two-pulse spin echo. The characteristic time of the stimulated spin echo decay as a function of the waiting time, T, is much longer than the phase memory time, T_M, since the "phase information" is stored along the Z′-axis where it can only decay via spin-lattice relaxation processes or via spin diffusion. In general, this relaxation time has a value between T_1 and T_M.

The free-induction decay (FID), primary two-pulse spin echo and the stimulated spin echo experiments are the main tools in the pulse spectroscopy. In all the experiments the nuclear modulation effects could be observed. These effects are important for measuring hyperfine interactions which can not be resolved in EPR spectra.

2.1.2 *The ESE as a Spectroscopic Tool*

With pulse EPR spectroscopy, an EPR spectrum can be observed by performing an ESE field-swept experiment. In this method, a simple two-pulse spin echo sequence is applied to the spin system as a function of the magnetic field B. Since the selection rules for EPR transitions are $\Delta M_S = \pm 1$ and $\Delta M_{Ii} = 0$ for all nuclei, for N nuclei this gives $\prod_N (2I_N + 1)$ possible transitions. To detect an EPR spectrum one sweeps the magnetic field and keeps the microwave frequency constant, e.g., at 9.5 GHz (X-band) or 95 GHz (W-band) and the cavity is tuned to the microwave frequency. If the magnetic field passes through resonance, an electron spin echo is generated and detected, e.g., with a boxcar integrator and as result the ESE-detected EPR spectrum is recorded.

The ESE signal can be used to obtain the EPR spectrum (pulse EPR) by recording the echo height as a function of swept magnetic field. As s result one has absorption shape of line. Such ESE-detected EPR spectrum is similar to the conventional continuous-wave EPR spectrum (CW-EPR) with derivative shape of line.

From the EPR spectra of single crystals, one can find the symmetry of the defect or the molecule since the position of the EPR lines depends on the direction of the magnetic field with respect to the crystal axis of the defect (molecule) due to anisotropic g-factor, fine structure or hyperfine structure. So the different possible orientations of the defect (molecule) in the crystal are separated in the EPR spectrum as they are at resonance at different magnetic field values, because their orientations are at different angles with respect to the external magnetic field.

An useful application of ESE techniques is the study of short-lived paramagnetic species, e.g. free radicals and photo-excited triplets generated by a laser flash prior to the ESE experiment. By varying the delay between the laser flash and the microwave pulses, the population and depopulation behaviour of the triplet sublevels can be studied very accurately.

The following chapters will examine many examples of EPR spectra using electron spin echo.

2.1.3 The ESE as a Direct Way for Measuring Relaxation Times

A pulse EPR experiment can be employed to directly measure the electron spin relaxation rates of a spin system on the same timescale in which relaxation occurs. After a microwave pulse, the magnetization M will interact with its surroundings and return to equilibrium. This process which is called "relaxation", is characterized by two time constants, T_1 and T_2. T_1 is a characteristic of the spin-lattice relaxation time. T_1 describes how quickly the magnetization recovers its longitudinal component along the Z-axis, that is, time interval in which the energy absorbed from the microwave pulse is dissipated to the crystal lattice as the system returns to equilibrium.

In CW experiment the microwave field is switch on for a long period. When the microwave field is only applied during a time such that $\omega_R t = \pi$ (ω_R – Rabi frequency) then a non-equilibrium state is created and the result of such a π (180°)-pulse is that according to the equation $M_Z(t) = M_0 \cos\omega_R t$ the equilibrium magnetization M_0 is reversed. This situation is in contradiction with thermodynamic equilibrium, thus after this pulse, spin-lattice relaxation process in the system will take place which lead the system back to thermodynamic equilibrium. According to Bloch this is an exponential process in the form

$$(dM_Z/dt) = -(M_Z - M_0)/T_1.$$

After integration one has

$$(M_Z - M_0) = C\exp\{-t/T_1\},$$

where the constant C represents the deviation of the magnetization from equilibrium at $t = 0$, e.g., after π (180°)-pulse $C = 2M_0$. The constant T_1 was introduced by Bloch phenomenologically and is commonly called the longitudinal or spin-lattice relaxation time (see Chap. 1 of this book).

The designation "spin-lattice relaxation" is due to the fact that the relaxation is related to the exchange of energy between the spin system and the "bath". This name "spin-lattice" is even used to describe the relaxation in liquids or other non-crystal medium. According to the equation for the energy $W = -M_zB$ ($B = B_Z$) if M_z changes, magnetic energy is transferred to thermal energy and vice versa. For nuclear spins the value T_1 is usually in the second range, for electron spins the value T_1 is in the ms and μs range.

As was mentioned before, the spin echo has found application to measure the relaxation times T_1 and T_2.

Spin-lattice (longitudinal) relaxation time T_1. One such method involves the inversion of the electron spin magnetization of the spin system and monitoring its recovery to equilibrium via the spin echo approach. This method is called inversion-recovery and consists of the following pulse sequence:

$$180°\text{---}T\text{---}90°\text{---}\tau\text{---}180°\text{---}\tau\text{---echo}.$$

The first pulse inverts the electron spin magnetization, while the recovery time T is varied stepwise until the magnetization returns to equilibrium. The relaxation of the inverted magnetization to the equilibrium value under the influence of spin-lattice relaxation is shown in Fig. 2.6 (left). The ESE amplitude is plotted versus the time t. One can see that the rate constant for magnetization, M_Z, recovering to the thermal equilibrium M_0 is the spin-lattice relaxation time T_1. The M_Z recovery after a microwave pulse is described by the following expression: (i) after a 90° pulse, $M_Z = M_0[1 - \exp\{-t/T_1\}]$; (ii) after a 180° pulse, $M_Z = M_0[1 - 2\exp\{-t/T_1\}]$.

In spin-lattice electron relaxation the energy of the spin system changes and due to energy conservation law the same energy is absorbed by the environment. We start with the phenomenon of spontaneous emission. As follows from the so-called Einstein coefficients for absorption and emission, the rate of spontaneous emission W by a magnetic dipole contains a factor ω^3 ($W \propto \omega^3$). As a result, spontaneous emission may be neglected in EPR experiments where the frequency employed is of the order of

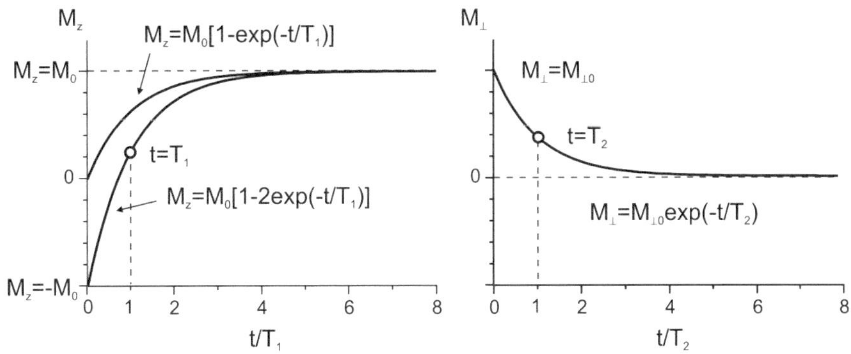

Fig. 2.6 (*Left*) Recovery of the magnetization M_Z after 90° or 180° microwave pulse. One carries the magnetization **M** out of the equilibrium, such that the system is saturated ($M_Z = 0$) or the magnetization is inversed, so-called "inversion-recovery experiment" ($M_Z = -M_0$). The system returns to the equilibrium with characteristic time T_1 (spin-lattice or longitudinal relaxation time). (*Right*) The relaxation of the transverse magnetization $M_\perp$ under the influence of the spin-spin relaxation. One turns the magnetization in the XY-plane after 90° microwave pulse. Magnetization $M_\perp$ is perpendicular to B_0 and rotates with Larmor frequency ω_L. At thermal equilibrium the magnetization in the XY-plane should be zero and the system will returns to this position with characteristic time T_2 (transversal or spin-spin relaxation time)

1–1000 GHz (about 10^{-8} s^{-1} for 1000 GHz that corresponds $T_1 \sim 10^8$ s). Thus, spontaneous emission can not be the reason for the spin-lattice electron relaxation.

The main source of spin-lattice electron relaxation is the thermal motion of the environment which in solids is usually described by phonons and in liquids by molecular motion. The energy transfer between the electron spin system and the lattice is interposed by local magnetic field fluctuations induced by phonons or molecular motion, which, e.g., will modulate the spin-orbit coupling.

The simplest spin-flip process is the direct process, where one phonon is absorbed or emitted by the spin system. $T_1 \propto B_0^{-4}T^{-1}$ law has been found for the direct process in the high-field (B_0) and high-temperature (T) approximation. The most efficient is so-called "Raman process", when the spin system absorbs a phonon of higher frequency and emits a phonon with smaller frequency. The dependence $T_1 \propto T^{-9}$ was found for half-integer electron spin (Kramers systems) and as $T_1 \propto T^{-7}$ for integer electron spin (non-Kramers systems). Depending on temperature and magnetic field, either the direct or the Raman process may dominate.

In some systems there is a very efficient process for spin-lattice electron relaxation, which is described by the relation $T_1 \propto [\exp(-\Delta/k_BT) - 1]^{-1}$. This is so-called "Orbach process" can be observed in the systems with a low-lying (Δ) excited spin state.

Spin-spin (transversal) relaxation time T_2. The 90° microwave pulse creates a non-equilibrium situation when the magnetization M, which in equilibrium is pointing along the Z-axis, is transferred into direction parallel to the Y′-axis of the rotating frame such that after the pulse $M_X' = 0$, $M_Y' = M_0$ and $M_Z' = 0$ (B_1 is directed along the X′-axis). The transformation to the laboratory axis system (X, Y, Z) results in $M_X(t) = M_0\sin\omega_L t$, $M_Y(t) = M_0\cos\omega_L t$, $M_Z(t) = 0$.

The conclusion is that after the 90°-pulse a transverse magnetization is present which oscillates with the Larmor frequency ω_L, and which decays to zero because $M_X \neq 0$ and $M_Y \neq 0$ do not correspond to an equilibrium magnetization conditions. The characteristic time constant T_2 with which the transverse magnetization disappears is called the transverse or spin-spin relaxation time. The relaxation of the transverse magnetization $M_\perp$ under the influence of the spin-spin relaxation is described by the following expression: $M_\perp = M_{\perp 0}\exp(-t/T_2)$.

It should be noted that in two cases considered one observes an exponential decay to an equilibrium state on the one hand $M_Z \rightarrow M_0$, on the other hand $M_\perp \rightarrow 0$. Relaxation is caused by interactions of the spin system with the surrounding and by mutual interactions of the spins. However, there is a fundamental difference between the two cases due to different relaxation mechanisms. After the 180°-pulse the energy of the spin system changes from $+M_0B_0$ to $-M_0B_0$. The loss of energy is transferred from the spin system to the surrounding (to the lattice). In contrast, when the perpendicular magnetization $M_\perp$ relaxes to zero the energy of the spin system does not change. This relaxation is the result of interactions which change the resonance frequency of the individual spins, e.g., by the mutual dipole-dipole interactions of spins.

It should be noted that a spin flip which result in longitudinal relaxation also contributes to transverse relaxation, however an energy conserving flip-flop process of two spins is twice as effective as a single spin flip in destroying coherence [12].

Figure 2.7 shows the decay of the electron spin-echo signal as a function of 2τ. The pattern represents the result of a series of two-pulse ESE experiments with various values of τ, the time interval between π/2 and π pulses. The decay of the electron spin-echo signal intensity as a function of 2τ gives the rate of irreversible loss of phase coherence, i.e. the relaxation time T_2. It should be noted, whereas the decay of the FID after a π/2 is determined by T_2^*, i.e., by total inhomogeneous

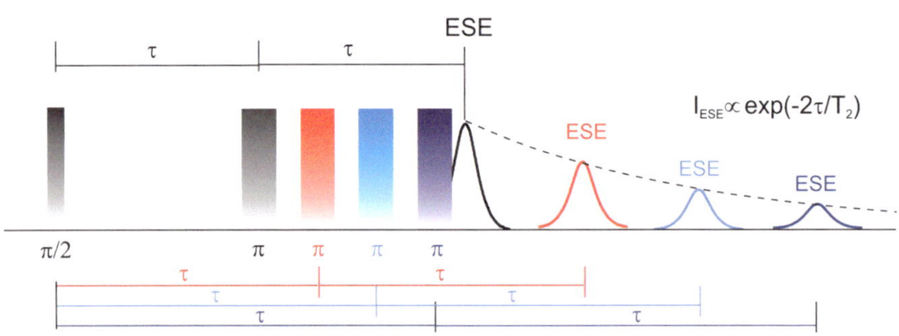

Fig. 2.7 The decay of the electron spin-echo signal as a function of 2τ. The pattern represents the result of a series of two-pulse ESE experiments with various values of τ, the time interval between π/2 and π pulses

linewidth, the decay of the spin echo signal intensity as a function of 2τ is determined by T_2, i.e., by the homogeneous linewidth.

Monitoring the two-pulse echo intensity as a function of inter-pulse distance τ offers a method to obtain the phase-memory time T_M of the spin system which is related to the homogeneous linewidth of the spin packets. T_M is the time during which a spin can 'remember' its position in the dephased pattern after the first pulse. In other words: T_M represents the contribution of irreversible mechanisms to the time T_2, associated with the total (homogeneous plus inhomogeneous) linewidth. The difference in T_M for different paramagnetic species provides an opportunity to separate the spectra of spin systems that would overlap when using conventional EPR techniques.

2.1.4 *Electron Spin Echo Envelope Modulation (ESEEM) Spectroscopy*

Electron Spin Echo Envelope Modulation (ESEEM) Spectroscopy was developed by Mims [18]. Similar to ENDOR, ESEEM spectroscopy detects the nuclear spin transitions of NMR-"active" nuclei that are coupled to a paramagnetic center. The ESEEM technique is an excellent spectroscopic tool in detecting nuclei that are weakly-coupled to a paramagnetic center. Advantages of ESEEM technique are the lower technical effort compared to ENDOR. Simulations of the spectral data can indicate the number, identity, and radial distance of the weakly-coupled nuclei from the paramagnetic center.

When a paramagnetic centre is coupled to surrounding nuclear spins via anisotropic hyperfine interaction which are of the order of magnitude of the nuclear Zeeman energy, the nuclear spin transitions frequencies (NMR transitions) can be observed as modulations on the ESE decay generated in both, two-pulse and three-pulse experiments.

Let us consider a spin system with one electron S = ½ and one nucleus I = ½ coupled via an anisotropic hyperfine (HF) interaction [13, 14]. Figure 2.8 shows the energy level diagram consisting of four levels for this system. For isotropic HF interaction only allowed EPR transitions 1–3 and 2–4 can be observed, however, for anisotropic HF interaction the nuclear spin states are mixed and, therefore, also the transitions 1–4 and 2–3, involving a simultaneous nuclear spin transitions, are allowed to some extent. In case of short microwave pulses (large frequency spread), all transitions are excited coherently. The process that the microwaves can induce transitions starting at one level but ending at different ones is called "branching of transitions" which is responsible for the nuclear modulation phenomenon. To illustrate this process we will follow resonance transitions 1–3 and 2–3 which correspond to the spin packets with the frequencies ω_{13} and ω_{23}. For two-pulse sequence we assume that ω_{13} spin packet is at resonance and it is fixed in the rotating frame. After the $\pi/2$-pulse both ω_{13} and ω_{23} spin packets are in the X′Y′

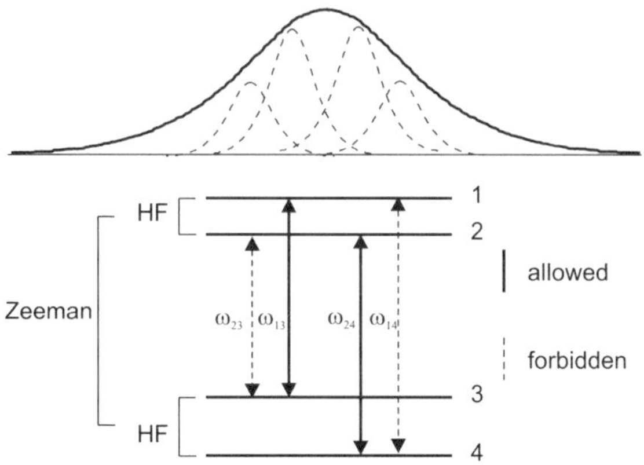

Fig. 2.8 Energy-level diagram and the corresponding spin packets for ω_{23}, ω_{13}, ω_{24}, and ω_{14} for an electron S = ½ and a nucleus I = ½

plane and ω_{23} spin packet start to diphase. After π-pulse ω_{13} spin packet is inverted along the $-Y'$-axis. Because of the branching of transitions also a component with frequency ω_{23} is inverted. The ω_{13} and ω_{23} spin packets will refocus at different times. The maximum contribution is obtained only if during the refocussing process (in time τ) it precesses an integral number of times in the rotating frame i.i. $(\omega_{23} - \omega_{13})\tau = n \cdot 2\pi$. Therefore, the spin echo envelope as a function of τ is modulated by a function $(1 - k\cos(\omega_{23} - \omega_{13})\tau)$. The same holds for all the other combinations of spin packets formed by branching of transitions. Therefore, one can expect all hyperfine frequencies to occur in the echo envelope. By Fourier transform of the echo-envelope an ENDOR-like spectrum is obtained. The intensities of the ENDOR-like lines are determined by the probabilities of the forbidden EPR transitions.

There are several different pulse sequences that are used in ESEEM spectroscopy. We are going to focus on two of them.

Two-pulse ESEEM: Consists of the following pulse sequence:

$$90°\text{---}\tau\text{---}180°\text{---}\tau\text{---echo.}$$

The amplitude of the ESE arising from a Hahn echo is modulated by the type of nuclei that are weakly coupled to the paramagnetic center. The spin echo amplitude is monitored as a function of the time interval between pulses τ. A Fouier Transform (FT) is used to show the corresponding nuclear frequencies. The time domain data is limited by the length of the T_2 of the spin system. T_2 can be measured as well.

Three-pulse ESEEM: Consists of the following pulse sequence:

$$90°\text{---}\tau\text{---}90°\text{---T---}90°\text{---}\tau\text{---echo,}$$

where the spin echo amplitude is varied as a function of T and a FT of the time domain data reveals the corresponding nuclear frequencies. Since the inter pulse time T is varied after the second 90° pulse (which transfers the magnetization to the

Z plane) the modulation patterns are limited by T_1 instead of T_2. Since, $T_1 > T_2$ the modulation patterns can be detected out further in time (when compared to the two-pulse ESEEM experiment) in time. Thus, low-frequency (long modulation period) transitions are easier to resolve when compared to the two-pulse experiment. Also, the inter-pulse time τ can be properly selected to remove certain nuclei from the ESEEM spectrum.

The time domain free induction decay and the EPR spectrum are related by the Fourier transform. The idea behind this transform, for example, can be demonstrated in music: a sound is represented as a note, an ear is sensing oscillations in air density, and the brain recognizes what frequency it is. The louder the note, the larger the amplitude of the sound wave, the higher the note, the higher the frequency of sound oscillations.

In pulse magnetic resonance the Fourier transform is a complex integral which is used to switch data between frequency-domain and time-domain representations, in other words, how to mathematically relate some oscillating signal (like sound) and transform it into a plot of amplitude versus frequency, that is

$$I(\omega) = \int_{-\infty}^{\infty} I(t)e^{-i\omega t}dt.$$

One integrates I(t) over all time for each ω to get $I(\omega)$. I(t) and $I(\omega)$ represent the time domain and frequency domain signals, respectively. The inverse Fourier transform

$$I(t) = \frac{1}{2\pi}\int_{-\infty}^{\infty} I(\omega)e^{i\omega t}d\omega$$

takes one from the frequency domain signal to the time domain signal. I(t) and $I(\omega)$ form a Fourier transform pair.

Figure 2.9a shows an example of an ESEEM spectrum in the time domain. An nuclear modulation effect was observed in the stimulated echo decay at 9.3 GHz following a three $\pi/2$-pulse sequence by keeping τ fixed and varying the interval T between the second and the third $\pi/2$ pulse. In Fig. 2.9b the related frequency-domain spectrum is displayed as obtained by a fast Fourier transform (FFT). The spectrum was measured on the quasicubic part of the EPR line for deep B in 6H-SiC (see Chap. 4, 4.22, p. 264). At 7.5 MHz we see the ^{11}B HF line belonging to the manifold above the ^{11}B nuclear Zeeman frequency of the hexagonal site. At 9 MHz we see the HF line of ^{11}B in the quasicubic site (consisting of three components, which are not visible in the picture), the line at 18 MHz is its second harmonic and the one at 27 MHz its third. The ^{11}B HF lines around 1 MHz belong to the manifold below the ^{11}B nuclear Zeeman frequency of the ^{11}B sites.

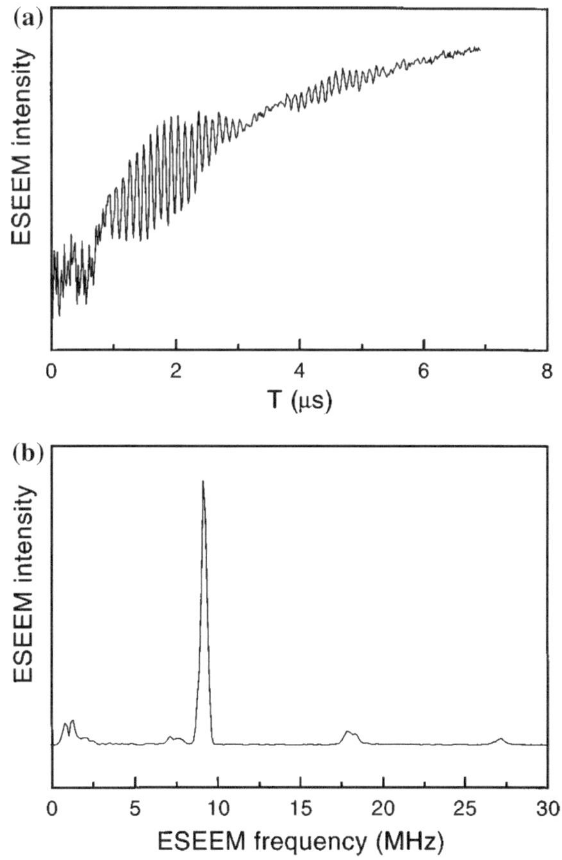

Fig. 2.9 **a** An example of an ESEEM spectrum in the time domain. **b** The spectrum of (**a**) after a fast Fourier transform

2.1.5 *Transient Nutation and the Rotary Echo*

The essence of the transient nutation and the rotary echo experiments are shown in Fig. 2.10. The transient nutation is presented in Fig. 2.10a. The equilibrium magnetization **M**, which at time t = 0 is parallel to the constant magnetic field $\mathbf{B}_0$ starts a precession in the Z′Y′ plane under the influence of the resonant microwave field with amplitude B_1 in a frame rotation at the resonance frequency. The result is that the magnetization M_Z oscillates in time according to $M_Z(t) = M_0 \cos\omega_R \tau$ with Rabi frequency of $\omega_R = |\gamma| B_1$ which depends on microwave power [13, 14]. Inset shows the Fourier transform (FT).

The oscillation of M_Z damps as a result of the spread in B_1–values over the sample because of the inhomogeneity of the microwave field. This leads to a fanning out of the ensemble of spins in Z′Y′ plane. When this fanning out covers the whole disc he resulting value of $M_Z = 0$. One will conclude that the transitions

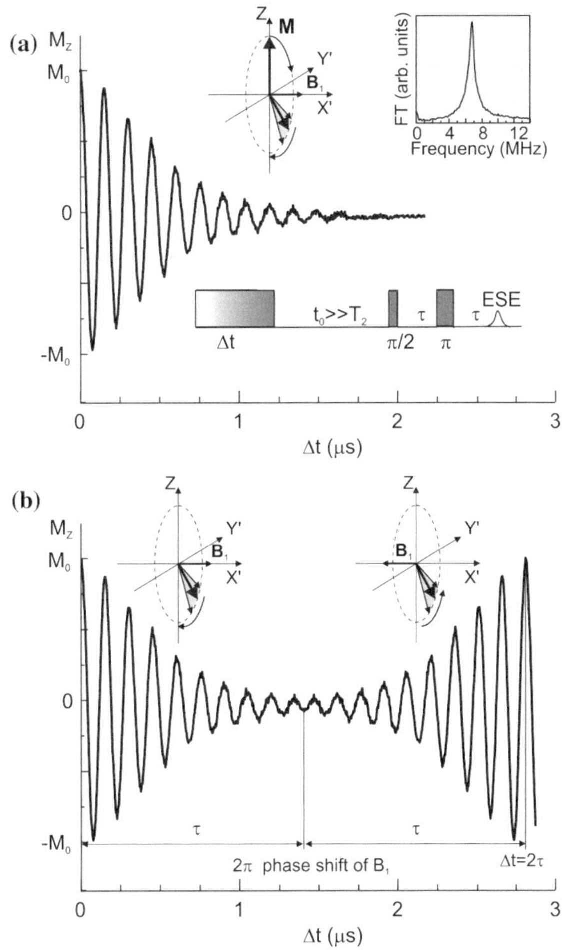

Fig. 2.10 **a** The transition nutation of the magnetization M_0 under the influence of a resonant microwave field with amplitude B_1 in a frame rotation at the resonance frequency. The transition nutation decays as a result of the spread in values of B_1 over the sample. Rabi oscillations were detected monitoring of the ESE intensity versus nutation pulse length (Δt). The *lower part* of the figure shows the pulse sequence used to measure Rabi oscillations. *Inset* depicts the Fourier transform (FT). It is shown the real Δt scale for NV-defect in diamond (see the following chapters). **b** The reversal of the loss of phase coherence by a phase shift of 180° applied at time $\Delta t = \tau$ and the refocusing of the magnetization at time $\Delta t = 2\tau$ leading to the rotary echo

is saturated because, macroscopically seen, the two spin states have the same population.

Figure 2.10a shows an example of Rabi oscillations measured as an ESE intensity versus nutation pulse length (Δt). Rabi oscillations decay with a characteristic time constant. The pulse sequence used to measure Rabi oscillations is presented in the lower part of Fig. 2.10a.

Figure 2.11 demonstrates another example of Rabi oscillations measured as EPR intensity versus nutation pulse length Δt. The observed oscillatory behavior demonstrates that the probed spin center ensemble in SiC (see also section of this book) can be prepared in a coherent superposition of the spin states at resonant magnetic fields at room temperature. The population difference of spin states becomes modulated in time with the Rabi frequency given by ω_R. The Fourier transforms corresponding to observed oscillations are presented in the inset in

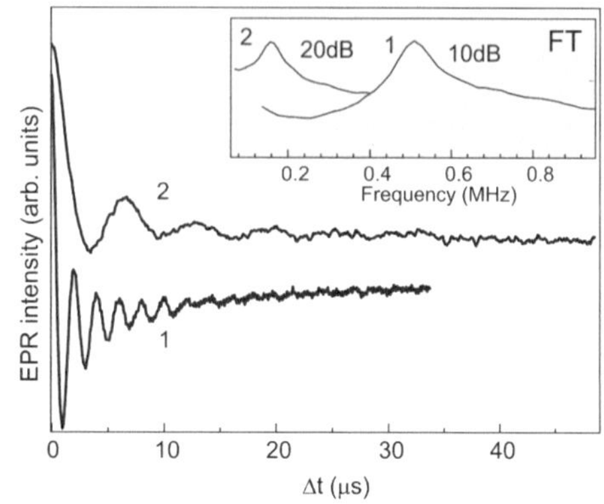

Fig. 2.11 Transient nutations for the spin center in silicon carbide at room temperature are shown for two values of microwave power P. Absolute microwave power value is estimated as ca. 5 mW at 10 dB attenuation. (*Inset*) Corresponding fast Fourier transform

Fig. 2.11. The Rabi frequencies are of 0.16 and 0.5 MHz at microwave power P = 20 and 10 dB, respectively. Rabi oscillations decay with a characteristic time constant that depends on the microwave power.

One can eliminate the loss in phase of the different spin packets with their different values of $\omega_R = |\gamma|B_1$ by applying a 180° phase shift to the microwave field at time τ, when the nutation has decayed to zero. There will be a re-phasing at time $t = 2\tau$ and a macroscopic magnetization will be re-established leading to the re-appearance of the transition nutation (Fig. 2.10b). This effect is called the rotary spin echo [13, 14, 19, 20]. The phase shift of 180° in the laboratory frame leads to a reversal of B_1 in the rotation frame and thus to a reversal of the transition nutation. The spin packets that were ahead at time $t = \tau$ are suddenly lagging behind. In the second period after $t = \tau$ they compensate exactly the loss in phase acquired in the first period.

If the spin packets maintain a fixed nutation frequency caused by the spread in B_1 fields over the sample, then the loss of phase in the first period τ will always be compensated in the second period τ and the rotary spin echo at $t = 2\tau$ will always have the same value. This is an example of reversible phase coherence. If however during the period 2τ there are random jumps in the resonance frequency, this will lead to irreversible loss of phase coherence and the rotary echo will decay when the time interval 2τ is made longer. The possibility to measure the irreversible loss of phase coherence makes this pulse magnetic resonance technique very well suited for measuring of dynamical processes in the materials.

The rotary echo can be also detected optically, e.g., as a modulation of the intensity of phosphorescence emitted by the photo-excited triplet state of an aromatic molecules [20].

2.2 Basics of Double Resonance Spectroscopy

One of the main tasks of modern spectroscopy is to increase the sensitivity up to the registration of individual quantum states: single photons, single spins, single defects and molecules. A promising solution to this problem is the use of double resonances. In this case high-energy transitions serve as highly sensitive detector for the low-energy transitions.

Double resonance is a "trigger detection" in that the absorption of a resonance low-energy photon triggers a change in emission of an high-energy photon. The detection of photons is thus displaced from the low-energy range to the far more sensitive high-energy range by means of quantum transformation (see Fig. 2.12).

In this chapter, we briefly introduce several important double-resonance techniques that represent elaborations of standard continuous-wave (CW) and pulse EPR spectroscopy. The experimental aspects and instrumentation of these methods are described in the literature (see, for instance, [21–23]), including special applications, presented in the following chapters.

2.2.1 *Electron Nuclear Double Resonance (ENDOR)*

Electron nuclear double resonance (ENDOR) is widely used for the measurement of small hyperfine and nuclear quadrupole interactions that are not resolved in inhomogeneously broadened EPR spectra. In the original continuous-wave ENDOR (CW ENDOR) technique, introduced by Feher in 1956 [24], an EPR transition is partially saturated with an intense CW microwave irradiation field. A second irradiation field in the radiofrequency (RF) region induces nuclear magnetic resonance (NMR) transitions which can be detected as a desaturation of the EPR signal.

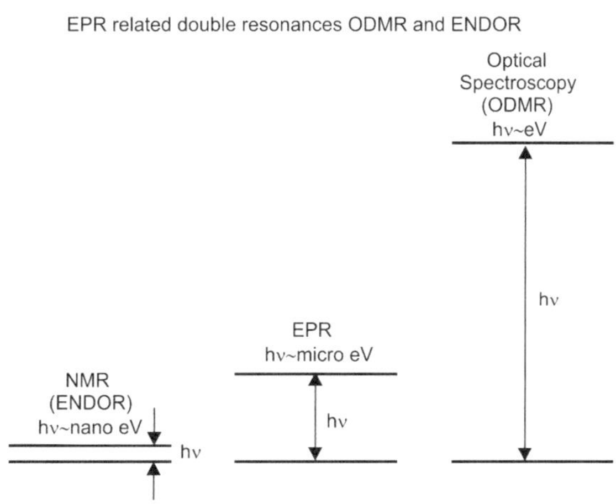

Fig. 2.12 Schematic representation of the energy ratio in the double resonances

The CW ENDOR effect depends on the delicate balance between the effective electron and nuclear spin relaxation rates which, in practice, restricts the observation of CW ENDOR to narrow temperature region. In contrast, in pulse ENDOR the applied pulse sequences can usually be made short enough to suppress the effect of relaxation. As a result, pulse ENDOR can be used at almost any temperature, provided that an electron spin echo can be detected. Two standard ENDOR pulse sequences have been introduced by Mims [7, 25] in 1965 and Davies [26] in 1974. Along with the rapid development in pulse EPR spectroscopy in the last several years [12, 27, 28], a variety of pulse ENDOR techniques exists nowadays which are almost all based on the Mims and Davies pulse sequences. These more advanced methods can be used to enhance the ENDOR efficiency or to simplify and unravel complicated spectra [27, 29–31].

Recently, an increasing interest exists in performing EPR spectroscopy at high magnetic fields and at microwave frequencies much higher than the conventional range of 9.5 (X-band) to 35 GHz (Q-band). The main advantages are the high spectral resolution and the high absolute sensitivity that can be obtained not only for EPR but also for ENDOR spectroscopy [16, 32–35].

In ENDOR, the sample is irradiated simultaneously by two electromagnetic fields, a microwave (MW) field (to drive EPR transitions with the selection rules $\Delta M_S = \pm 1$, $\Delta m_I = 0$) and a radio-frequency field (to drive NMR transitions $\Delta M_S = 0$, $\Delta m_I = \pm 1$). Under appropriate experimental conditions, which are different for CW and pulse irradiation schemes ENDOR signals are observed by monitoring the changes of EPR line amplitudes when sweeping the RF field through the nuclear resonance frequencies. Thus, every group of equivalent nuclei, no matter how many nuclei are involved and what their spin is, contributes only two ENDOR lines because, within an M_S manifold, the hyperfine levels are equidistant to first order. The gain in resolution of ENDOR versus EPR, therefore, becomes very drastic for low-symmetry systems because, with increasing number of groups of nuclei, the number of ENDOR lines increases only in an additive way. This gain in resolution is particularly pronounced when nuclei with different magnetic moments are involved. Their ENDOR lines appear in different frequency ranges, and from their Larmor frequencies these nuclei can be immediately identified. In the case of accidental overlap of ENDOR lines from different nuclei they can be separated when working at higher MW frequencies and Zeeman fields.

ENDOR is a variant of nuclear magnetic resonance on paramagnetic systems, the unpaired electron serving as highly sensitive detector for the NMR transitions. Double resonance excitation thus offers the advantage of detecting low-energy RF transitions by high-energy MW transitions, i.e., by means of quantum transformation. The sensitivity of ENDOR is orders of magnitude greater than the direct observation of the same number of nuclei by nuclear magnetic resonance and in favorable cases approaches the sensitivity of EPR.

2.2.1.1 Continuous-Wave ENDOR Spectroscopy

In the conventional continuous-wave ENDOR experiment, a sample is placed in a constant magnetic field and irradiated sequentially with a microwave followed by radio frequency. The changes are then detected by monitoring variations in the intensity of the saturated EPR transition. The stationary ENDOR technique consists of the following steps. Keeping the microwave frequency constant, the magnetic field is adjusted to the center of EPR line and maintained constant B_0. The microwave power is set at a level high enough to saturate partially the electron transition, and the ENDOR radio-frequency is swept through the frequency region of interest. An ENDOR spectrometer is essentially an EPR spectrometer in which supplies are made to apply to the sample an RF magnetic field $\mathbf{B}_2$ perpendicular to the constant magnetic field $\mathbf{B}_0$ and to microwave magnetic field $\mathbf{B}_1$.

Figure 2.13a illustrates the energy diagram of the simplest spin system with S = ½ and I = ½ where A is the isotropic hyperfine interaction constant (more complex spin systems will be discussed in the following chapters). The diagram indicates the electron Zeeman, nuclear Zeeman and hyperfine splittings. In a steady state ENDOR experiment, an EPR transition (1–4) obey the EPR selection rules $\Delta M_S = \pm 1$ and $\Delta M_I = 0$, is partly saturated by microwave radiation of amplitude B_1 while a radio frequency field of amplitude B_2, induces nuclear transitions. Transitions happen at frequencies f1 and f2 and obey the NMR selection rules $\Delta M_S = 0$ and $\Delta M_I = \pm 1$. It is these NMR transitions that are detected by ENDOR via the intensity changes to the simultaneously irradiated EPR transition. It is important to realize that both the hyperfine interaction constant (A) and the nuclear Larmor frequencies (f_L) are determined when using the ENDOR method: $f(1,2) = |f_L \pm A/2|$.

Figure 2.13b shows ENDOR lines measured as a change in the EPR signal amplitude, for a system with S = ½, I = 1/2 (top) and S = ½, I = 3/2 (bottom) as the radiofrequency is scanned through the region including the NMR frequencies (f1 and f2 for the S = ½, I = ½ system, see Fig. 2.13a). These are separated by the hyperfine coupling A (to first order), and are symmetrically spaced about the Larmor frequency (nuclear magnetic resonance frequency in the selected magnetic field) f_L of the nucleus for the magnetic field at which the microwave saturation is being carried out. For S = ½, I = 3/2 system, there is an additional quadrupole splitting with the number of lines equal to 2I.

The spin Hamiltonian can be described as (see Chap. 1)

$$\mathbf{H}_0 = \mathbf{H}_{eZ} + \mathbf{H}_{nZ} + \mathbf{H}_{HFS} + \mathbf{H}_Q$$

The terms in this equation describe the electron Zeeman interaction ($\mathbf{H}_{eZ}$), the nuclear Zeeman interaction ($\mathbf{H}_{nZ}$), the hyperfine interaction ($\mathbf{H}_{HFS}$), and the nuclear quadrupole interaction ($\mathbf{H}_Q$), the last term is added only for the case I > 1/2. The electron Zeeman interaction describes the interaction between an electron spin and the applied magnetic field. The nuclear Zeeman interaction is the interaction of the nuclear magnetic moment (e.g., of the proton) with an applied magnetic field. The

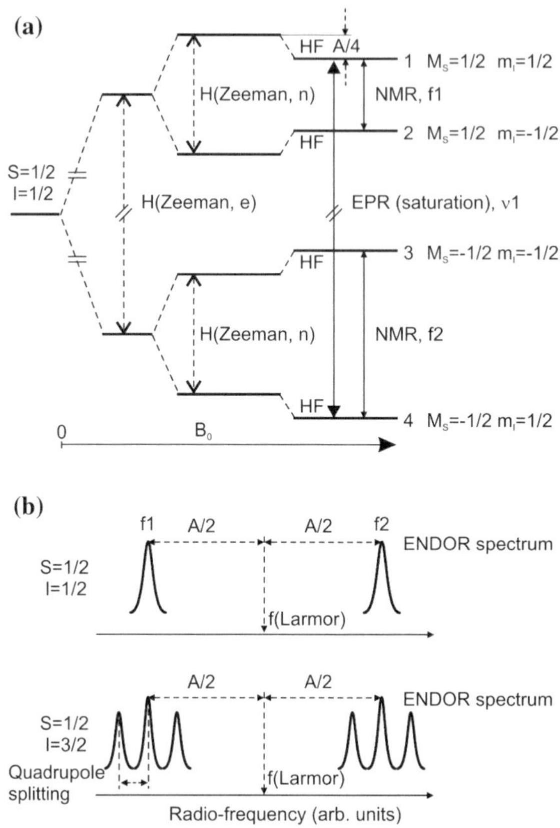

Fig. 2.13 **a** Energy level diagram for the ENDOR of the simplest spin system with S = ½ and I = ½ where A is the isotropic hyperfine interaction constant. **b** ENDOR lines measured as a change in the EPR signal amplitude, for a system with S = ½, I = 1/2 (*top*) and S = ½, I = 3/2 (*bottom*) as the radiofrequency is scanned through the region including the NMR frequencies (f1 and f2 for the S = ½, I = ½ system, see Fig. 2.2a). These are separated by the hyperfine coupling A (to first order), and are symmetrically spaced about the Larmor frequency (nuclear magnetic resonance frequency in the selected magnetic field) f_L of the nucleus for the magnetic field at which the microwave saturation is being carried out. For S = ½, I = 3/2 system, there is an additional quadrupole splitting with the number of lines equal to 2I

hyperfine interaction is the interaction between the electron spin and the nuclear spin.

The ENDOR spectra contain information on the type of nuclei in the vicinity of the unpaired electron, on the spin density distribution and on the electric field gradient at the nuclei.

The requirement for ENDOR is the partial saturation of both the EPR and the NMR transitions defined by $\gamma_e^2 B_1^2 T_{1e} T_{2e} \geq 1$ and $\gamma_n^2 B_2^2 T_{1n} T_{2n} \geq 1$. Here γ_e and γ_n are the gyromagnetic ratio of the electron and the nucleus respectively; B_1 and B_2

are the magnetic field of the microwave and radiofrequency radiation, respectively; T_{1e} and T_{1n} are the spin-lattice relaxation time for the electron and the nucleus, respectively; T_{2e} and T_{2n} are the spin-spin relaxation time for the electron and the nucleus, respectively.

Let us consider the population of the four levels in Fig. 2.13a under various conditions. In the zero magnetic field, $B_0 = 0$, the population of each of degenerate levels (ignoring hyperfine interaction) would be N/4, where N is the total number of unpaired electrons. In the presence of a constant magnetic field, B_0 (ignoring hyperfine and nuclear Zeeman interactions), the populations of each of the states 1 and 2 ($M_S = +1/2$) are $N_{+1/2} \cong 1/4\ N[1 - g_e\mu_B B_0/(2kT)]$ and the populations of each of the states 3 and 4 ($M_S = -1/2$) are $N_{-1/2} \cong 1/4\ N[1 + g_e\mu_B B_0/(2kT)]$ (see Chap. 1). It is to be noted that we neglect the population difference between the states 1 and 2 (or between the states 3 and 4) in the presence of a constant magnetic field, B_0, and in the absence of microwave saturation, since this value is of the order $g_n\mu_N B_0/(2kT)$ and too small in comparison with the value of $g_e\mu_B B_0/(2kT)$.

If the 4–1 EPR transition (Fig. 2.13) is excited by the microwave field, the only effective relaxation path is 1–4 with electron spin-lattice relaxation time of T_{1e}. The nuclear spin-lattice relaxation time T_{1n} is ineffective, since is much longer.

Let us consider qualitatively the appearance of the ENDOR signal. For complete saturation of the transition between states 4 and 1 with $m_I = +1/2$, 1 and 2 states differ in population by $g_e\mu_B B_0/(2kT)$, whereas in the absence of microwave saturation they would have differed by $g_n\mu_N B_0/(2kT)$. RF saturation of transition between states 1 and 2 stimulated effective relaxation path between states 2 and 4 that gives rise to an ENDOR line at frequency f_1. Similarly, saturation of the EPR transition 3–2 with $m_I = -1/2$ (not shown in Fig. 2.13) and NMR transition 4–3 will result in ENDOR line f_2.

2.2.1.2 Pulse ENDOR Spectroscopy

The ENDOR technique can also be used in pulse EPR spectroscopy. One of the most important application of ESE is ESE-detected ENDOR technique. To detect ENDOR spectrum a microwave pulse-sequence is combined with radio-frequency pulse. The ENDOR spectrum is detected by recording the stimulated electron spin echo height as a function of the radio frequency of the RF pulse. Thus the ESE serves as a detector of the ENDOR effect induced by the radio-frequency pulses. The ENDOR spectrum is obtained by measuring the ESE intensity as a function of the radio-frequency. Various sequences of pulse microwave and RF excitations can be used. At present, Mims ENDOR and Davies ENDOR (Fig. 2.14) have a most preferred application.

In the ESE-ENDOR techniques based on the stimulated echo sequence (Mims), or the inversion recovery sequence (Davies) a population difference between hyperfine sublevels is established during a preparation period. When an NMR-transition is hit by a subsequent RF-pulse a magnetization exchange occurs within the EPR-spectrum. The resulting change of the stimulated echo amplitude

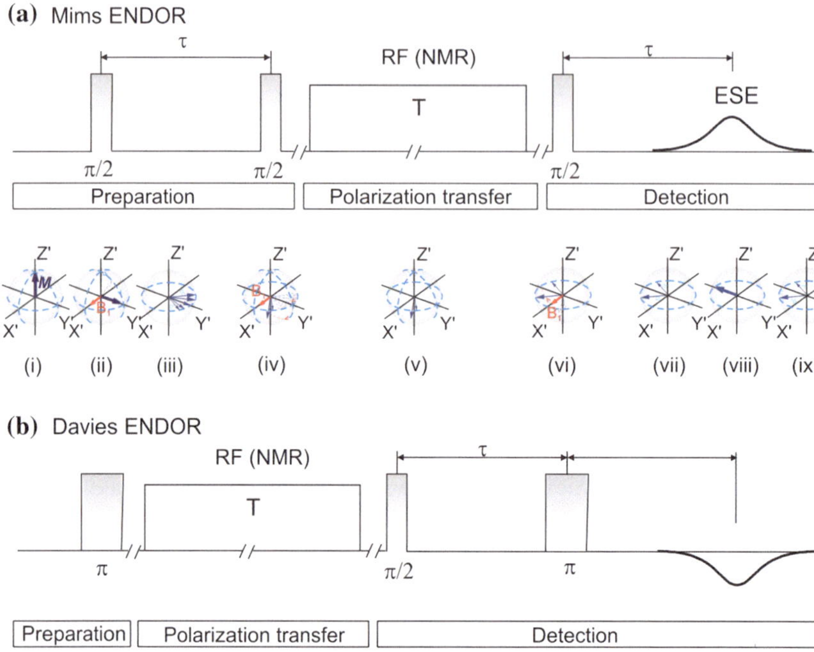

Fig. 2.14 ESE-ENDOR 3-pulse sequence consisting of a preparation period, a polarization transfer (mixing) period and a detection period. **a** The Mims-type pulse ENDOR sequence 90°—RF pulse-90°—echo. **b** The Davies-type pulse ENDOR sequence 180°—RF pulse-90°—180°—echo

(Mims) respectively the 2-pulse-echo amplitude (Davies), is detected as the ENDOR-signal (Fig. 2.14). A qualitative understanding can be obtained as follows [13, 14]. *Mims-type-ENDOR.* Mims-type-ENDOR is well suited for studying nuclei with small hyperfine interactions and small nuclear Zeeman splittings. With this technique the population difference of the hyperfine sublevels is established by the $\pi/2 - \tau - \pi/2$ microwave sequence applied at the EPR-transition (see Fig. 2.14a). In the frequency domain the preparation pulses produce a periodic pattern $M_{zi}(\Delta\omega) = M_0\cos(\Delta\omega_i\tau)$. The M_z-component of a spin packet 'i' depends on how its precessing frequency in the rotating frame ($\Delta\omega_i$) "fits" in the waiting time τ. For example, spin packets at resonance will be oriented along negative z-axis whereas spin packets with a frequency offset $\Delta\omega_i = \pi/\tau$ will be oriented along the positive z-axis. In the standard stimulated echo the whole pattern refocusses at time τ after the third pulse. If the RF-pulse, inserted between the second and the last microwave pulse, is resonant with a transition corresponding to a hyperfine interaction A, the polarization transfer shifts the whole M_z pattern up or down in frequency by an amount of A. Therefore the pattern of M_z components will be blurred and as a result the intensity of the stimulated echo will be reduced. Only when $A = n/\tau$

(n = 0, 1, 2, ...) the pattern is retained and the stimulated echo amplitude is unaffected. In other words the echo amplitude will be modulated by a factor cos $(2\pi A\tau)$. The "blind spot" for $n/\tau = A$ are the only drawback of this technique. One can circumvent this problem by taking ENDOR spectra at different τ-values.

Davies-type-ENDOR. Consider again an electron spin (S = 1/2) coupled to a nearby nuclear spin (I = 1/2) by the hyperfine interaction *A* which results in the four-level energy scheme sketched in Fig. 2.13 and an EPR-spectrum consisting of two lines which are separated by *A*. With the first microwave π-pulse applied to one of the EPR-transitions (see Fig. 2.14b), e.g. 4–1, a large population difference of the NMR sublevels is established because the population of levels 4 and 1 are interchanged. Neglecting relaxation effects the two-pulse-echo—excited at a time T after the inversion π-pulse is now inverted with respect to an initial equilibrium state. When the RF-pulse applied during the waiting time T, is resonant with one of the NMR-transitions, i.e. 4–3 or 1–2, the population change of level 4 respectively 1 is detected as an increase of the echo amplitude toward equilibrium. The maximum effect is obtained when the sublevel populations of transition 4–3 or 1–2 are inverted, i.e. for nuclear flip angle $\beta = \omega_R t_p = \pi$, where ω_R is the effective nuclear Rabi-frequency and t_p is the RF pulse length.

In the EPR-spectrum the change in amplitude of the inverted line, induced by the RF-pulse, is accompanied by an opposite change of the noninverted line. When the echo signal is obtained via a pulse sequence that excites both components of the EPR spectrum, i.e. the detection is non selective, the resulting ENDOR effect will not be observable. Similarly if the preparation pulse is non-selective, i.e. $\omega_1 \gg A$, then both components are inverted and no appreciable population difference of the hyperfine levels is created.

In the situation that is most interesting for ENDOR experiments the hyperfine splitting is not resolved in the EPR spectrum and the preparation pulse burns a hole with a width of $\omega_1 = \gamma_e B_1$ in the inhomogeneously broadened EPR line. Thus, the Davies type ENDOR signal in a frequency range ω_1 around the nuclear Larmor Frequency are suppressed.

Let's look at the advantages offered by high-field EPR and ENDOR spectroscopy.

(i) High spectral resolution due to high magnetic fields (see Fig. 2.15) and high orientation selectivity for powder samples, disordered systems, biological systems.
(ii) Accessibility of high-spin systems with large fine-structure (zero-field) and hyperfine structure splittings owing to the large microwave energy quantum.
(iii) Enhancement of detection sensitivity (particularly for small samples and pulse EPR). If one assumes that the equipment is frequency independent in their noise and performance, the expression for the frequency dependence of the minimum number of detectable electron spins N_{min} can be obtained [21].

   (a) $N_{min}/V_S \propto \omega_0^{-3/2}$ for the minimum detectable spin concentration, when the sample size is scaled by the same factor as the cavity dimensions.

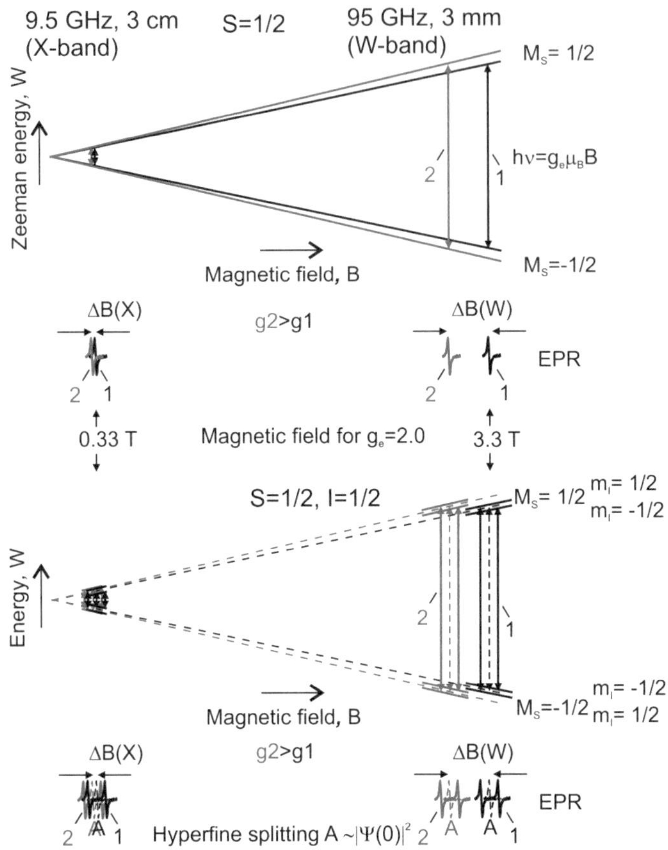

Fig. 2.15 Comparison of resolution for standard (X-band) and high-frequency (W-band) EPR for system with S = ½ (*top*) and system S = ½, I = ½ with hyperfine interaction A (*bottom*)

(b) $N_{min} \propto \omega_0^{-9/2}$ for the minimum number of detectable spins when the sample size is not varied, i.e. when the sample volume V_S is constant and the filling factor is proportional to V_S/V_C with the cavity volume V_C being proportional to ω_0^{-3}.

(iv) High spectral resolution of ENDOR spectra from different nuclei (see Fig. 2.16) and enhancement of ENDOR sensitivity for low gyromagnetic ratio nuclei

(v) Suppression of second order effects and decoupling of spin exchange.

We now turn to high-field/high-frequency EPR and ENDOR experiments in more detail to show what can be additionally learned about paramagnetic systems when going beyond conventional X-band EPR/ENDOR. From the conventional spin Hamiltonian (see Chap. 1) one sees that some interactions are magnetic field dependent (the Zeeman interactions), while others are not (fine structure interaction,

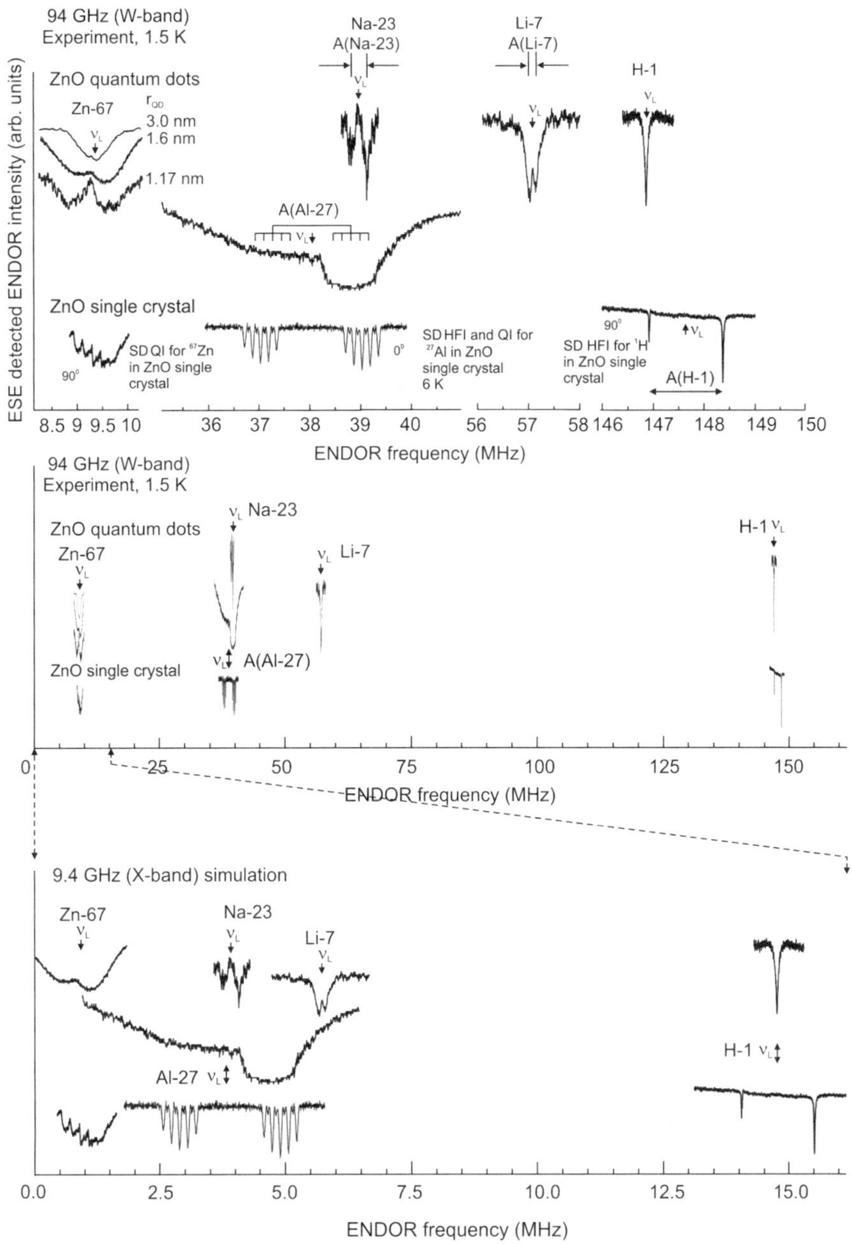

Fig. 2.16 Comparison of ESE-detected ENDOR spectra resolution for measuring at standard X-band and high-frequency W-band in the example of shallow donors in ZnO single crystal and ZnO quantum dots. Gain in ENDOR resolution with increasing microwave frequency and Zeeman field are presented. ENDOR lines of different nuclei for shallow donors in ZnO, largely overlapping at the traditional frequency of 9.5 GHz (X-band), become almost completely separated at 95 GHz (W-band)

the hyperfine interactions, quadrupole interaction). Obviously, in complex systems it will be necessary to measure at various field/frequency settings in order to separate these interactions from each other. Operating frequencies from 9.5 GHz to almost 300 GHz are used.

Figure 2.15 demonstrates comparison of resolution in the EPR spectra for standard X-band (9.5 GHz) and high-frequency W-band (95 GHz) experiments. The EPR line shifts in two bands when changing g-factor for system with S = ½ are schematically shown in Fig. 2.15. It is seen that a shift line in the high frequency band ten times larger as compared to the low frequency band. Therefore signals with nearby values of g-factors can be resolved. Figure 2.15 (bottom) shows the EPR signal in the case of the system S = ½, I = ½ with hyperfine interaction A which does not depend on the magnetic field. Again, it is clear that high-frequency spectra are well resolved and their analysis is much easier.

To illustrate the power of the high-frequency/high-field ENDOR method, shallow donors in ZnO single crystals and ZnO quantum dots were chosen as example (Fig. 2.16).

See also the following chapters.

2.2.2 *Optically Detected Magnetic Resonance (ODMR)*

Optically detected magnetic resonance (ODMR) is a double resonance method that combines EPR spectroscopy with optical measurements. The population redistribution among magnetic sublevels, in passage through magnetic resonance in the ground or excited state of a paramagnetic center, produces a change of either emitted or absorbed light associated with the center. ODMR is a "trigger detection" in that the absorption of a resonance microwave photon triggers a change in emission of an optical photon. The detection of photons is thus displaced from the microwave range to the optical range. As a result, the change in light acts as the indicator of magnetic resonance rather then the direct observation of microwave power absorption by the paramagnetic system. The scaling up of the magnetic resonance detection from the microwave to the optical region performs these optical magnetic resonance methods extremely sensitive. Therefore, a strong motivation for using these techniques has been to study the EPR spectra in sparsely populated excited states or in ground states with low spin concentration, e.g., nanostructures, which impossible to do by the conventional magnetic resonance methods. Note, the typical sensitivity of conventional magnetic resonance methods is limited to and 10^{10}–10^{14} spins for EPR and 10^{16}–10^{18} spins for NMR experiments, while ODMR allows us to record the magnetic resonance on a single spin (see Chap. 6, p. 435).

One must distinguish between "forced" polarization of the spin sublevels by optical pumping with polarized light (see, e.g., [36]) and the natural spin polarization by unpolarized light due to the existence spin-dependent recombination channels. The selective feeding of magnetic sublevels is operative in nonradiative decay from different sublevels in semiconductor systems. At the root of this selective feeding are spin selection rules for the optical transitions and spin selection rules for the radiationless decay. The selectivity could be also connected with ground-state polarization (e.g., due to the Boltzmann distribution) as a result of spin memory in optical excitation cycle.

The discovery of spin-dependent recombination processes goes back to the first ODMR experiments carried out by Geschwind et al. in 1959 [37, 38]. In these experiments, spin configurations of excited electronic states were manipulated with EPR, which led to a change of the decay rate that could be observed by photoluminescence (PL). ODMR has become a successful technique for studying the emission processes in a large number of different materials. The technique, which has been applied mainly to the investigation of insulating materials, has been reviewed by Geschwind [39]. Initially, ODMR was carried out as continuous wave (CW) experiments, later first transient ODMR experiments were made which allowed the measurement of spin coherence, revealing information such as coherence times and therefore transition probabilities and spin-spin interactions within spin pairs [40]. Soon, transient ODMR became a frequently utilized method for spin-dependent reaction analysis with development of commercially available pulse EPR spectrometers, optically detected electron spin-echo and Rabi oscillations techniques [41–45].

The basic apparatus required for ODMR: (i) microwave oscillator with amplitude modulated to enhance the sensitivity by means of a lock-in amplifier; (ii) optical excitation source (lamp or laser beam); (iii) optical detector (photomultiplier or photodiode); (iv) a liquid helium cryostat, with optical and microwave access; (v) PC interfaced to the set-up for recording and handling of the data. A simple scheme of the set-up, optimized for high-frequency EPR/ODMR research, is shown in Fig. 2.17, where line of microwave units developed for EPR/ODMR/ODCR studies in Ioffe Institute (St. Petersburg) is presented [46].

The use of a high frequency (94 GHz, 140 GHz) for detecting magnetic resonance makes it possible to increase significantly the energy resolution of spectra and apply a quasi-optical (rather than waveguide) channel to form a microwave field on the sample. As a result, microwave power can be supplied to the sample directly through the optical cryostat window. Figure 2.17b shows a block diagram of the noncavity W-band ODMR spectrometer. 94 GHz microwave generator provides output power up to 100 mW and ultrahigh frequency stability (5×10^{-7}). To obtain a frequency of 94 GHz, A highly stable solid-state oscillator at 7.23 GHz and a frequency multiplier (with a multiplication factor of 13) has been used; the microwave radiation from the multiplier output was fed to a power amplifier. The oscillator can operate in both cw and pulse modes. The oscillator output is loaded by horn antenna, radiation of which is focused by Teflon lenses to pass through the quartz window of a magneto-optical cryostat with superconducting magnet and arrive at the sample. A specific feature of the proposed microwave scheme, along with the presence of highly stable oscillator, is the use of copper insert in the cryostat (microwave field concentrator). Sample is placed in the center of the magnetic system; it can be rotated with respect to the vertical axis. The

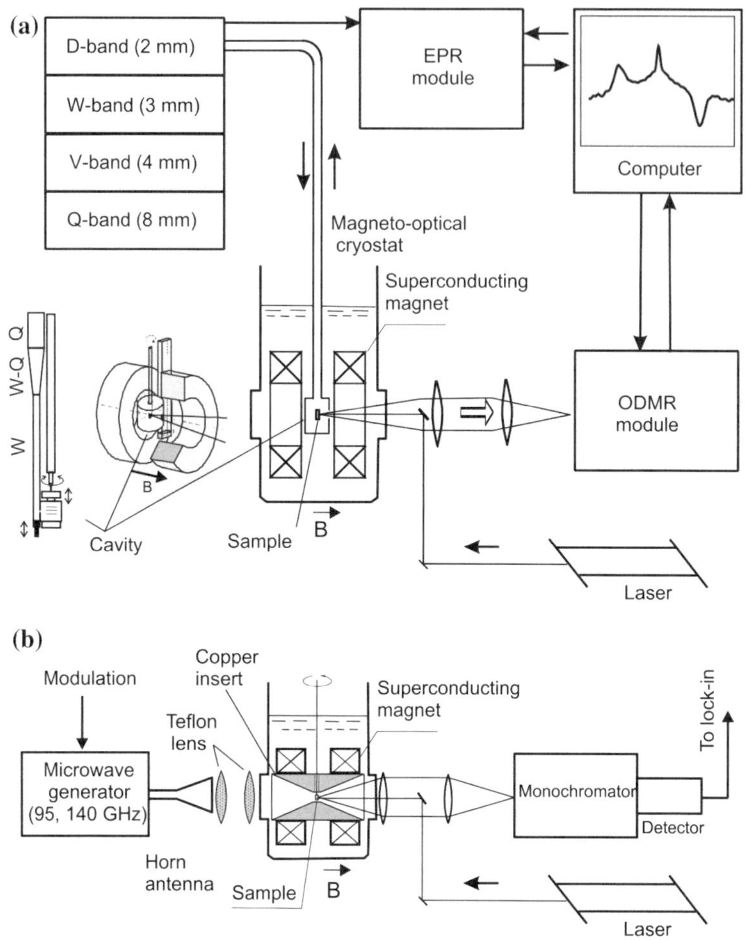

Fig. 2.17 **a** Basic block scheme of the high-frequency EPR/ODMR set-up. Only the photoluminescence detection ODMR line is shown. **b** ODMR spectrometer using a quasi-optical microwave channel with a noncavity scheme

luminescence of the sample is excited by focused laser beam and recorded by a grating monochromator with photodetector (for example, photoelectronic multiplier or photodiode). The concentrator insert is made of copper in the form of a polished cone (horn antenna) connected with the same cone through a segment of round waveguide. The second cone is a matched load. Application of this field concentrator allows one to amplify the field B_1 by no less than an order of magnitude and extend the range of the objects under study in comparison with the quasi-optical scheme, where only one transmitting horn antenna is used. Samples with sizes considerably exceeding those used in 3 mm spectrometers with a microwave cavity

(0.3–0.5 mm) can be analyzed. This is important for local diagnostics of nanostructures with spatial resolution over the sample surface.

In semiconductors ODMR has been shown to be a powerful method of obtaining information about both the recombination mechanisms and the identity of the emitting centres. This aspect has been reviewed by Cavenett [47] and by Nicholls et al. [48].

Let us consider the mechanism by which the ODMR is observed. The spin states of an electron (e) on the donor and of a hole (h) on the acceptor are involved in a recombination event. The spin levels of the excited state are shown in Fig. 2.18, the final ground state being a singlet is formed by the recombination of the electron and the hole. The upper part of the figure shows the levels which are drawn for the case of a vanishingly small electron-hole exchange interaction, so that the allowed

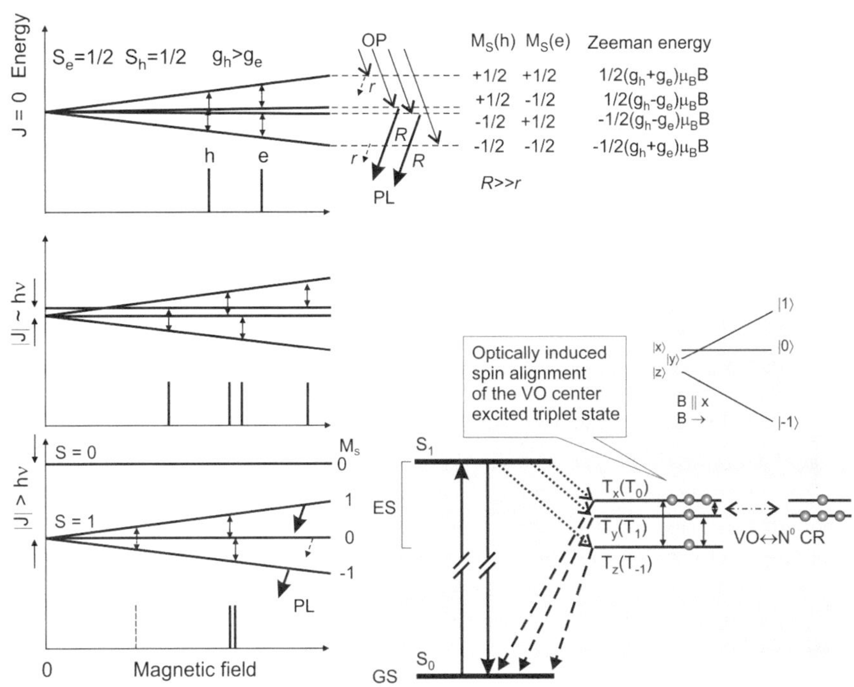

Fig. 2.18 Zeeman splitting of the energy levels of an electron-hole (donor-acceptor) pair for three cases of the exchange interaction J between two spins $S_e = S_h = \frac{1}{2}$: $|J| \sim 0$ (*upper part*), $|J| \sim h\nu$ (*middle part*) and $|J| > h\nu$ (*lower part*). Microwave resonance transitions and recombination transitions (rate constant: R and r) are indicated by *arrows*, OP—optical pumping rate. (*Inset*) Five energy level system representing the singlet ground state S_0, the first excited singlet (S_1) and the triplet (T_1, T_0, T_{-1}) states. (*Top inset*) Schematic energy level diagram of the triplet sublevels with the external magnetic field B parallel to the x center axis. $|x\rangle$, $|y\rangle$, and $|z\rangle$ are the wave functions of the triplet sublevels at B = 0, and $|1\rangle$, $|0\rangle$, and $|-1\rangle$ are those with $B > |D|/g\mu_B$. $|1\rangle$ and $|-1\rangle$, are linear combinations of $|y\rangle$, and $|z\rangle$

magnetic resonance transitions are indicated at magnetic fields of $h\nu/g_e\mu_B$ (electron) and $h\nu/g_h\mu_B$ (hole).

Energy levels can be described by a simple spin Hamiltonians, if we consider the case with isotropic g-factors and isotropic exchange interactions

$$H = g_e\mu_B\mathbf{S}_e \cdot \mathbf{B} + g_h\mu_B\mathbf{S}_h \cdot \mathbf{B} + \mathbf{J}\mathbf{S}_e \cdot \mathbf{S}_h. \quad (2.2)$$

Here the first and the second terms are electron Zeeman interaction and hole Zeeman interaction, respectively; the third term is exchange interaction between two spins $S_e = S_h = ½$.

The properties of system depend on the magnitude of J. If it is zero the two spins behave completely independently. At the other extreme, when J is big and the singlet level lies far above or far below (depends on sign of the isotropic exchange interactions) the triplet, the magnetic resonance properties are determined solely by the interactions within the triplet manifold. If J is of the same order as the HF interactions then the singlet and triplet states become mixed and complicated spectra are onserved.

Increasing the singlet-triplet splitting from a low value leads to a regime where their mixing is negligible. We analyze the EPR behaviour in the low-J region in some detail using a system containing two unpaired electrons as an example.

If the exchange interaction were small or comparable to the microwave quantum each line would be split into two: thus a small exchange coupling which varied from pair to pair would result in a broadening of the resonance signals (Fig. 2.18, middle part). It is to this that we ascribe the fact that the lines in ODMR experiments, as a rule, are broader than their counterparts observed using conventional EPR.

Recombination occurs from the outer two spin levels with a rate constant r and from the inner two with R (Fig. 2.18, upper part). The recombination rates for the strong emission are expected to be mainly radiative. In general, R and r will be different and this is a crucial factor in making the experiment possible. If the recombination processes are such as to conserve spin we expect that $R \gg r$. If the recombination rates are fast compared with the spin-lattice relaxation rates, the spin populations will not thermalise and will be determined by the relative magnitudes of R and r [48]. Since the distance between donor and acceptor varies from pair to pair, the recombination rate constants R and r will have a range of values.

Quite likely that the more short-lived pairs have insufficient time to thermalise. In considering the magnetic behaviour we shall therefore consider two limits: first, donor-acceptor pairs that are sufficiently long lived for thermalisation to occur (for which $T \gg R, r$), and, secondly, those that recombine so quickly that thermalisation is ineffective ($R, r \ll T$).

A major contribution in ODMR signal was concluded in [48] must arise from electron-hole (donor-acceptor) pairs which are not thermalised, thermalised pairs are unlikely to produce intensity changes of sufficient strength to account for the observed ODMR signals.

Table 2.1 Maximum equilibrium changes in emission intensity (ΔI/I) produced by saturation of the electron or hole resonance for different conditions: thermalised pairs and non-thermalised pairs

		Low optical excitation	High optical excitation (saturation)
Thermalised pairs	R, r both radiative	Zero	2ΔF (if $r \ll R$)
	R radiative and r non-radiative	Zero (if $r \ll R$) 4ΔF (if $r \gg R$)	2ΔF (for all values of r/R)
Non-thermalised pairs	R, r both radiative	Zero	R/4r (if $r \ll R$)
	R radiative and r non-radiative	100% (if $r \ll R$)	R/2r (if $r \ll R$)

ΔF = 0.6% for the conditions of the spin-dependent donor-acceptor pair recombination in ZnS crystals [48]. It is expected that $r \ll R$

The conclusions in [48] are summarised in Table 2.1. Two limiting cases were considered: (i) recombination of donor-acceptor pairs that have sufficient time to thermalise; (ii) recombination of donor-acceptor pairs that have not thermalised. Pairs in thermal equilibrium can contribute to the magnetic resonance signal in two ways. The first requires the emission to be optically saturated: microwave transitions increase the recombination rate and the emission intensity becomes greater at resonance. The second mechanism requires a significant non-radiative recombination rate from the outer two spin levels (Fig. 2.18, upper part): in this case the effect of microwaves is to alter the relative numbers of pairs that recombine with and without the emission of light.

In contrast, for pairs that are non-thermalised, the relative spin populations in the excited state are determined by the recombination rate constants R and r (Fig. 2.18, upper part). For example, if R is radiative and r non-radiative, with $R \gg r$, the inner two levels are much less heavily populated than the outer two. Application of microwaves at resonance tends to equalise the spin populations and to increase the emission intensity. Note, that the idealised conditions lead however to signals that are much larger than those observed experimentally. The signals can be produced either by desaturation (in the optical sense) or by alterations in the ratio of radiative to non-radiative recombination. For the latter mechanism to be effective one requires a strict spin selection rule for radiative decay (emission from the centre two spin levels only) and a small non-radiative recombination probability.

Since there are donor-acceptor pairs of different separations, the recombination rates R and r will have a wide range of values. Thus some pairs will have sufficient time to thermalise, others not. The observed signal will therefore be composed of contributions from both types: the large fractional changes predicted for non-thermalised pairs will add to the smaller changes from those that have attained a Boltzmann distribution.

We turn now to the case where the exchange interaction is large compared with the microwave quantum, i.e., consider the lower part of Fig. 2.18. We consider here only briefly triplet states because we pay attention to these studies in the following chapters.

Triplet states are formed as excited states of two-electron systems by way of strongly coupled electron and holes in semiconductors after a band-to-band optical excitation. The triplet sublevels are usually filled after an optical excitation from singlet ground state into a singlet excited state and subsequent intersystem crossing (ISC).

Electronic transitions can be approximated by a five-level system, including ground S_0 and excited S_1 singlet electronic states and the photoexcited triplet state T (see Fig. 2.18, inset). In a typical experiment, the sample is illuminated with laser light in resonance with the $S_0 \rightarrow S_1$ transition and Stokes-shifted fluorescence emission is detected. After being excited in the S_1 state the system can either relax back to the ground state via fluorescence emission and internal conversion, or can be trapped in the triplet state via the intersystem crossing process. In the inset of the Fig. 2.18, the sublevels of the triplet state related to zero-field splitting are shown.

A radiative transition to the singlet ground state is partially allowed for the $|+1\rangle$ and $|-1\rangle$ levels, but is not allowed for $|0\rangle$ level. Therefore, the $|0\rangle$ level will have a higher population compared to the $|+1\rangle$ and $|-1\rangle$ levels, provided spin-lattice relaxation time T_1 is larger than the radiative lifetime, in which case no thermalization of the electron occurs. The $|+1\rangle$ and $|-1\rangle$ levels are weakly radiative due to admixtures of higher singlet states with the selection rule, that from the $|+1\rangle$ state σ_+ light is emitted, and from the $|-1\rangle$ state σ_- light is emitted. EPR transitions will enhance the emitted light intensity by population redistribution from the $|0\rangle$ level into the $|+1\rangle$ and $|-1\rangle$ levels [49].

Let us consider radiative properties of a photoexcited triplet state [50]. Electronic dipole transitions between a photoexcited triplet state and a singlet ground state are forbidden due to the fact that spin angular momentum must be conserved during the transition ($\Delta S = 0$). However, a phosphorescence due to a spin-forbidden transition can be observed. The reason for this is that spin-orbit couplings result in mixing between states of different spin multiplicity, opening up the possibility of radiative decay to the ground state.

The triplet eigenfunctions for the canonical orientations x, y, z of fine-structure tensor **D** (see Chap. 1) can be in the first order to be represented by

$$^3\psi_i^1 = {}^3\psi_i^0 + \sum_n \frac{\langle {}^1\psi_n^0 | \hat{H}_{SO} | {}^3\psi_i^0 \rangle}{{}^3E_i^0 - {}^1E_n^0} {}^1\psi_n^0, \quad i = x, y, z$$

in which $^3\psi_i^0$ is an unperturbed triplet spin component, $^1\psi_n^0$ denote the perturbing singlet eigenfunctions and $^3E_i^0$ and $^1E_n^0$ denote their respective energies, H_{SO} is the operator of the spin-orbit interaction. Although the amount of admixed singlet character is small, it still allows for processes such as phosphorescence and ISC to take place.

The quantity of singlet character mixed into the triplet state strongly depends on the symmetric representation of the triplet sublevel due to the spin-orbit operator transform properties. Therefore, the populating and depopulating rates of the triplet sublevels will be different, and as a result, this effect can be utilized for ODMR.

In ODMR resonant microwaves are applied which give rise to transfer of population between triplet sublevels. As the radiative properties of these sublevels generally will different, a change of the phosphorescence intensity arises: the greater the difference in radiative decay rates, the stronger the ODMR effect will be.

Figure 2.18 (inset) shows five energy level system representing the singlet ground state S_0, the first excited singlet (S_1) and the triplet (T_1, T_0, T_{-1}) states. To clarify the relationship between the different energy levels and corresponding wave functions a schematic energy level diagram of the triplet sublevels with the external magnetic field B parallel to the x center axis is given in Fig. 2.18 (inset). $|x\rangle$, $|y\rangle$, and $|z\rangle$ are the wave functions of the triplet sublevels at B = 0, and $|1\rangle$, $|0\rangle$, and $|-1\rangle$ are those with $B > |D|/g\mu_B$. $|1\rangle$ and $|-1\rangle$, are linear combinations of $|y\rangle$, and $|z\rangle$ (after [51]).

The ODMR for two types of defects in diamond in the ground and excited triplet states can be regarded as model systems. An example of a well studied defect with triplet ground state is NV-center, which consists of a substitutional nitrogen atom next to a carbon vacancy. The ODMR study of this center will be presented in Chap. 7. Here we will consider a diamond center, so called 2.818 eV center, possessing a photoexcited triplet state [52–54].

Photoexcitation of a brown diamond at 364 nm gives rise to a long-lived luminescence with a zero phonon line peaking at 441 nm (2.818 eV). The lifetime of the emissive state of a few milliseconds is suggestive of a spin-forbidden transition to the ground state. ODMR experiments confirmed the triplet state nature of the excited luminescent state. Figure 2.19 shows the ODMR spectrum observed in zero magnetic field. After the first triplet-state ODMR experiments in zero magnetic field reported in 1968 by Schmidt and van der Waals (see [50]), the number of double resonance studies on excited triplet states grew rapidly. The basic principles of this technique are described in many publications [50]. Two zero-field transitions are observed, the first peaking at 396 MHz and the second peaking at 1122 MHz. In a magnetic field both transitions give rise to Zeeman splittings.

By means of optically detected magnetic resonance spectroscopy, it was shown that the emission originates from an excited triplet state characterized by S = 1, g = 2.00, |D| = 924(2) MHz, and |E| = 198(2) MHz.

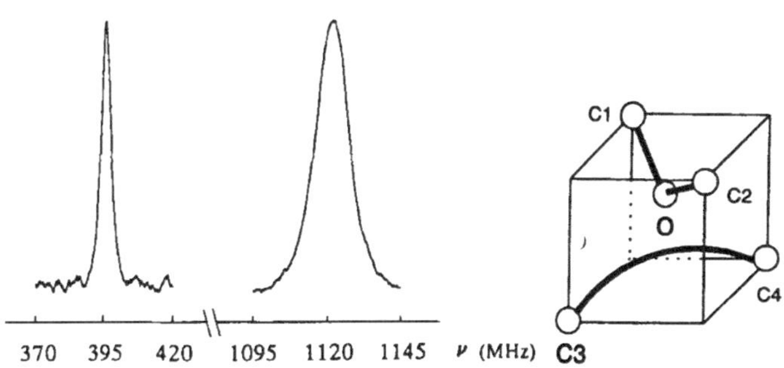

Fig. 2.19 Zero-field ODMR spectrum for the VO center in diamond at 1.4 K (*left*). Schematic structure of the VO center in diamond (*right*). After [54]

The angular dependences of the ODMR transitions in a magnetic field could be fitted using a spin Hamiltonian of the form

$$H = g_e\mu_B \mathbf{S} \cdot \mathbf{B} + D\left[S_z^2 - 1/3S(S+1)\right] + E\left(S_x^2 - S_y^2\right). \quad (2.3)$$

The magnetic main axes of the VO center are along the crystallographic [100], [011], and [0-11] directions. The defect excited triplet state ODMR spectrum does not show hyperfine interactions. The defect has been proposed to involve an oxygen impurity center. Figure 2.19 (right) shows a schematic structure of the VO center in diamond. The electrons in the dangling bonds of carbon atoms 3 and 4 give rise to an electron spin triplet excitation.

The principal axes of the zero-field fine-structure tensor for the VO center were found to be along the [100], [011], and [01-1) axes of the diamond crystal, showing the presence of a defect of rhombic symmetry. At 1.4 K, the lifetimes of the triplet-state sublevels, T_x, T_y, and T_z are 0.5, 1.8, and 23 ms, respectively. The T_x level is the most emissive substate, whereas T_z, is almost nonemissive. Under continuous optical excitation, a steady-state spin alignment is produced: the population of the T_x level is larger than for the other two sublevels [Fig. 2.18 (inset)]. Furthermore, the absence of the ODMR signal for the $|D| - |E|$ transition, i.e., the transition between the T_y and the T_z levels, indicates that the populations of these two sublevels are nearly equal.

It is reported cross-relaxation (CR) and level anticrossing (LAC) effects in the phosphorescence intensity of the 2.818-eV center for suitable values and orientations of an externally applied magnetic field. The basis for the experiments is that spin alignment becomes relaxed for those magnetic field strengths for which the triplet spins become resonant with other triplet or doublet defect spins, which are characterized by a spin temperature different from that of the probed 2.818-eV center triplet spins.

Figure 2.20 shows angular variation of the CR signals for rotation of the diamond crystal about the [011] axis perpendicular to the magnetic field. Experimental

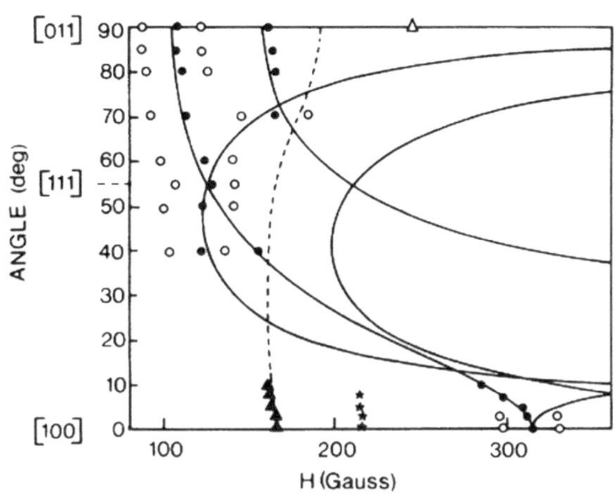

Fig. 2.20 Angular variation of the CR signals of the VO center in diamond for rotation of the crystal about the [0–11] axis perpendicular to the magnetic field. The data points denoted by ● and ○ derive from CR with the P1 center, ▲ derives from N-V centers, Δ reflects level-anticrossing and * is due to CR with an as yet unknown defect. After [54]

data include signals representative of CR between the 2.818-eV center and g = 2 doublet spins (filled circle) including hyperfine splittings (open circle) originated from the P1 center, CR between the 2.818-eV center, and NV center triplet spins (filled triangle), LAC (open triangle). The origin of a weak signal represented by stars is unknown. Drawn curves represent the calculated angular variation for the CR with the N° g = 2.00 spin doublet (solid line) and the CR with the NV center ground state triplet (dashed line) [54].

One of the common methods of ODMR is to use magnetic circular dichroism (MCD). MCD is the differential absorption for left (σ+) and right ($\sigma-$) circularly polarized lights, induced by a magnetic field B which is applied along the direction of propagation of the beam. We will not consider the MCD ODMR techniques in this book, and refer the reader to a number of reviews, where a detailed description of this method and its use for the study of magnetic resonance in semiconductors are given (see [22, 23] and references therein). Just note the very important achievements of this technique in the study of antisite defects in semiconductors, e.g., so called EL2 defect in GaAs [22].

2.2.3 *Electrically Detected Magnetic Resonance (EDMR) in Semiconductors*

The sensitivity of magnetic resonance was shown in previous section can be enhanced by shifting the detection of the magnetic resonance effect from microwave into the optical domain. In ODMR approach, the spin state is transferred to a photon state. Another approach to increase the sensitivity is to transfer the spin state to a charge state, this is so called electrically detected magnetic resonance (EDMR). EDMR is a powerful and sensitive method for investigating spin dependent processes including the recombination between different centers in semiconductors. The change in conductivity and photoconductivity of silicon under magnetic resonance of phosphorus (P^0) donor centers was observed and investigated many years ago [55]. It was argued that the processes of spin-dependent scattering of the conducting electrons from paramagnetic donor centers [56] and spin dependent capture of electrons by neutral shallow donors [57] are responsible for the change in conductivity under magnetic resonance. The spin-to-charge transfer is typically achieved in spin-dependent photoconductivity via a spin-dependent process governed by the Pauli principle involving two paramagnetic states. Suggested mechanisms of spin dependent conductivity require high electron spin polarization achieved at low temperatures below 4.2 K and strong magnetic fields used in standard electron paramagnetic resonance (EPR) spectrometer. Since the observation of SDR effect in silicon at room temperature, reported by Lepine [58], many experimental and theoretical investigations of SDR were performed. The most significant feature of SDR is the independence or weak dependence of EDMR signals on the magnetic field strength. Independence of the EDMR signals on the strength of magnetic field was explained by the model of SDR [59] taking intoaccount the exchange-interaction coupled electron-hole pairs in the triplet spin $S = 1$ state. A similar SDR model considering exchange-interaction coupled donor-acceptor pairs was suggested to explain the broadening of magnetic resonance lines detected optically [60]. This weak dependence of EDMR signals on magnetic field allows one to observe magnetic resonance transitions of paramagnetic recombination centers in weak magnetic fields at low frequencies. The first low frequency observation of EDMR spectra of P0 centers and of the excitedspin S = 1 states of the neutral A-centers (oxygen + vacancy complex) in low dos eirradiated silicon has been reported in [61]. An additional line with g-factor g = 2 was observed but not identified at that time. Similar EDMR spectra were observed in irradiated and post annealed samples later [23, 62] and it was pointed out that no EDMR spectra of shallow donors were observed in as-grownn-type silicon and that the g = 2 line originated from thermal donors and from A-centers. EDMR method using detection of conductivity by applying electrical contacts provide the opportunity to observe EPR signals of recombination centers in small samples with electrical contacts. In addition, complementary information concerning the properties of paramagnetic centers and SDR processes can be obtained from the EPR spectra in weak magnetic fields because the additional EDMR lines due to the mixing of spin states, magnetic level crossing, and anticrossing can be observed.

Spin-dependent recombination between phosphorus donors in silicon and Si/SiO_2 interface states was investigated with pulse electrically detected electron double resonance in [63].

Lepine's model. Let us consider mechanisms of spin dependent recombination (SDR) and will start with Lepine's model [58]. Spin-dependent recombination of excess carriers can be explained in the following simple model. Electrons and holes in the conduction and valence bands, respectively, generated in excess recombine through an intermediate recombination center, which is supposed to have an unpaired electron, before trapping a conduction electron.

The essence of the model lies in the fact that the capture of the conduction electron by the recombination center depends on the relative orientation of the spins of the conduction electron and the recombination center. Parallel or antiparallel spins of the conduction electron and the recombination center give rise to triplet or singlet final states for the system, conduction electron plus center.

We suppose that the only possible intermediate state in the recombination process is a singlet state (antiparallel spins of the conduction electron and the recombination center) and the capture cross section for the conduction electron $\sum$ should present the following form:

$$\sum = \sum_0 (1 - P_e P_{rc}), \tag{2.4}$$

where P_e and P_{rc} are the spin polarizations of the conduction electrons and of the recombination centers; $\sum_0$ is the spin independent part of the capture cross-section.

The recombination time of the excess carriers τ is a function of $\sum$ and a relative variation of $\sum$ will approximately result in an equal and opposite variation of τ: $\delta\tau/\tau \propto -\delta\sum/\sum$.

Equation (2.4) shows that such a variation can be induced by a change of the spin polarization of the carriers or of the recombination centers. This can be achieved by applying a resonant microwave frequency for the carriers or the recombination centers. The resonant microwave field induces transitions between the Zeeman spin levels and tends to equalize their populations. Hence the spin polarization of the system is reduced and, in the saturation limit, vanishes.

The steady-state value of the excess carrier concentration Δn is related to the generation rate G and to the recombination time τ by [58]

$$\Delta n = G\tau.$$

The transient behavior of Δn is described by

$$d\Delta n/dt = G - \Delta n/\tau.$$

The photoconductivity $\Delta\sigma$ is proportional to the excess carrier concentration: $\Delta\sigma = \Delta n e\mu$, where μ is the sum of the mobilities of both types of carriers. As a result the photoconductivity decreases when the magnetization of the carriers or of the recombination centers is reduced. The complete saturation of either system, $P_e = 0$ or $P_{rc} = 0$, results in the maximum effect and the relative decrease of the photoconductivity $\delta\Delta\sigma/\Delta\sigma$ is then equal to the product of the equilibrium spin polarizations: $\delta\Delta\sigma/\Delta\sigma = P_e P_{rc} \approx 10^{-6}$ at T = 300 K in a static magnetic field of $\approx$300 mT for spins with a g factor of g $\approx$ 2.0. The response time of the photoconductivity to a sudden change of P_e or P_{rc} is equal to the recombination time τ.

Lepine's model predicts square dependence of the SDR signal amplitude on the magnetic field

$$\Delta\sigma/\sigma \approx P_e P_P \approx B^2/T^2$$

It should be noted, that experimental value of $\Delta\sigma/\sigma$ was 10–100 times higher than predicted by Lepine's model and very weak dependence of $\Delta\sigma/\sigma$ on B was found experimentally.

Kaplan-Solomon-Mott (KSM) model. This discrepancy was eliminated by Kaplan-Solomon-Mott (KSM) model [64]. KSM model was developed for explanation SDR effects in amorphous semiconductors having high density of the localized states in gap. It was suggested that photo excited electrons and holes form the weakly coupled e–h pairs and SDR takes place through the excited spin S = 1 states of defects. In contrast to D-A recombination in this case recombination occurs through the single center. Specific features of the excited triplet states of defects are:

(i) Nonequilibrium distribution of population between magnetic sublevels with $M_S = +1$, 0, and -1.
(ii) These centers are metastable and have relatively long lifetime, form energy level in the forbidden gap, participate in the generation-recombination processes.
(iii) Excitation of magnetic resonance between Zeeman levels changes the recombination rate of photo-excited carriers and, consequently, the conductivity of samples.

EPR spectra of the excited triplet states can be detected electrically using electrical contacts or by contact-free method based on microwave photoconductivity [61]. The method of detection of microwave photoconductivity is based on sensitivity of the EPR spectrometer cavity to the conductivity of sample.

Electrically detected magnetic resonance of defects due to spin dependent recombination with triplet centers in irradiated silicon was observed [65]. EDMR spectra of phosphorus P^0 in silicon was detected in weak magnetic fields at low resonance frequencies of 200–400 MHz before and after irradiation of samples by γ-rays. EDMR spectra were detected by measuring dc-photoconductivity of samples under band-gap illumination. Phosphorus (P^0) EDMR lines are accompanied always with the single line (S-line) with g factor ≈2.01 originated most likely from the surface recombination centers. Strong, about 10 times, increase of the P^0 and S signals was found in the same samples after irradiation with the doses of $(3–6) \times 10^{15}$ γ/cm^2. The EPR transition between entangled states of phosphorous formed at low magnetic field were observed in the irradiated samples. New EDMR lines emerged after irradiation were observed due to the spin dependent recombination through the photo excited triplet states of oxygen + vacancy complexes (so called A-centers in silicon).

SDR detection of the recombination centers in silicon p-n junctions were presented in [66]. It was reported the results of investigation the radiation defects produced by γ-irradiation in silicon tracking detectors used now in Large Hadron Collider (LHC) analyzing the particles created under high energy proton collision.

The contact free method for detection spin-dependent photoconductivity is based on the sensitivity of microwave cavity to the free carrier density in illuminated semiconductor sample even in the case when the sample is placed in maximum of magnetic component [67]. Figure 2.21a shows the dependence of the intensities of EDMR and usual EPR spectra of the photo excited triplet states of (O + V) centers in silicon (spectrum Si-SL1) on the dose of γ-irradiation. Taking into account the production rate of (O + V) centers under γ-irradiation $k \approx 0.01\ cm^{-1}$, the minimal concentration of defects, $N \approx k\Phi$, detectable by SDR-EPR method is about 10^9–$10^{10}\ cm^{-3}$. Traditional EPR technique can be applied for investigation the samples containing four orders higher defects concentration. Absorption of the electrical component of microwave field by free carriers reduces the Q-factor of cavity as shown schematically in Fig. 2.21b. Maximal response of cavity is achieved at the maximal slope of dependence shown in Fig. 2.21b (point II). Therefore, the detection of microwave photoconductivity requires the optimization of the

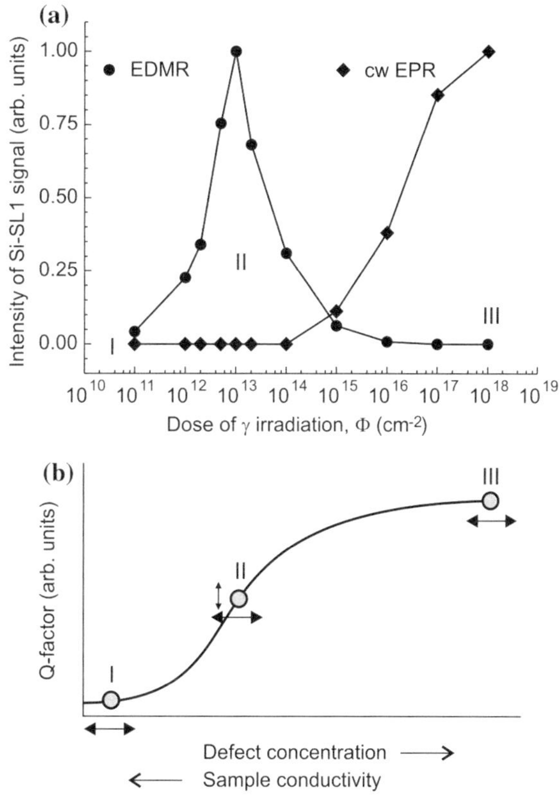

Fig. 2.21 **a** Dependence of the intensity of the Si-SL1 spectrum on the dose of γ-irradiation detected by EDMR due to SDR and convenient EPR methods at T = 20 K. **b** Schematic dependence of cavity Q-factor on the photoconductivity of sample. *Horizontal arrows* show small photoconductivity change under magnetic resonance. *Vertical arrow* is the response of Q-factor which is maximal at point II. After [67]

experimental conditions, such as a light intensity, sample temperature, and concentration of the investigated recombination centers. At high concentration of defects photoconductivity is low because of short lifetime of photoexcited carriers (point III). At low concentration of defects samples have high photoconductivity (point I).

It should be noted that electrical and magnetic components of microwave field play the different roles in SDR-EPR method. Electrical component is used for detection of microwave photoconductivity which changes under magnetic resonance excited by magnetic component. It allows to observe EPR spectra at different resonance frequencies applying the additional resonance magnetic field, for example to detect spectra at low resonance frequencies in weak magnetic fields [61, 68]. Furthermore, the change in photoconductivity can be observed even without magnetic resonance at the points of anticrossing of magnetic sublevels of spin S = 1 centers as well as at specific magnetic field values for which the Zeeman splitting of the different paramagnetic recombination centers become equal [69].

Electrical detection of cross relaxation. The electrical detection of cross relaxation processes in phosphorus-doped γ-irradiated silicon was reported in [69], where the dipolar-coupled electron spins of phosphorus and oxygen-vacancy

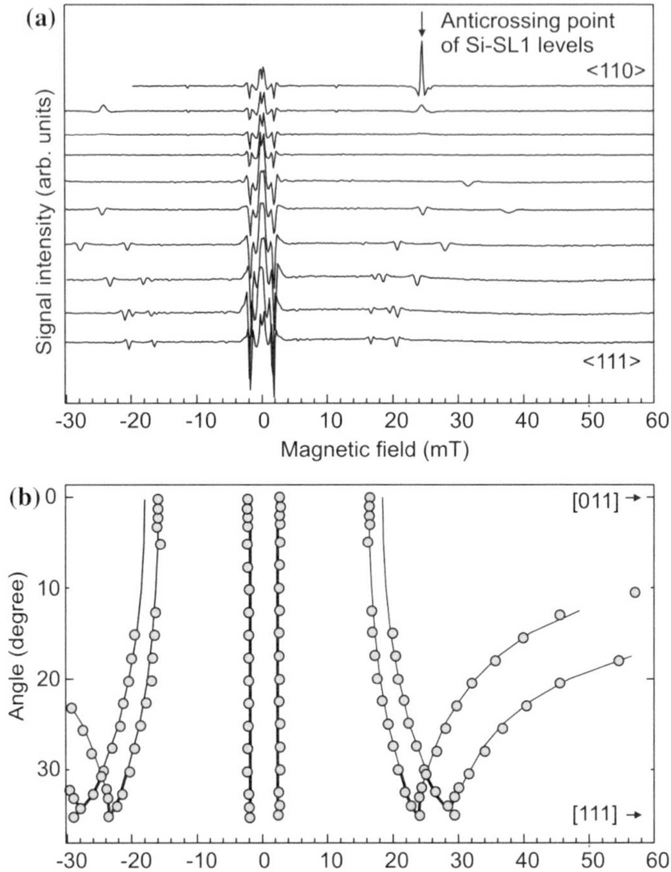

Fig. 2.22 **a** Cross relaxation signals detected by the microwave photoconductivity technique and **b** angular dependencies of their positions revealed by the microwave-photoconductivity technique. *Dots* represent the experimentally obtained line positions of cross relaxation signals and *curves* represent the calculated positions of Zeeman energy crossing points between phosphorus and SL1 centers

complex (Si-SL1 center) undergo spin flip-flop transitions at specific magnetic field values for which the Zeeman splitting of the two centers become equal. Such cross relaxation signals are observed as the change in the sample photoconductivity at theoretically predicted magnetic fields without application of resonance frequency (Fig. 2.22). This electrical detection of cross relaxation is a very simple and sensitive method for detecting paramagnetic centers in semiconductors. Effects of cross-relaxation of the SDR spectra were observed without resonant field in γ-irradiated Si(P), T = 20 K. Cross-relaxation change of photoconductivity caused by interaction between phosphorus and triplet Si-SL1 centers.

In the presence of external magnetic fields, the spin population among magnetic sublevels approaches the Boltzmann distribution in a time frame known as spin-lattice relaxation time (T_1). When two different kinds of spins coexist in solid, they can achieve identical spin temperature through energy conserving flip-flop transitions, provided that their Zeeman energies are made nearly equal by tuning the magnetic field strength. This phenomenon is known as cross relaxation and occurs if the exchange of energy between the two different spins is significantly faster than the exchange with the lattice, i.e., $\tau_{CR} < T_1$, where τ_{CR} is the cross relaxation time. Optical detection of cross relaxation has been studied in solids (see Sect. 2.2.2 and following chapters), where the change in the luminescence intensity from one of two different dipolar coupled paramagnetic centers is monitored as their Zeeman splittings are brought into resonance by appropriate tuning of the magnetic field. However, the optical method can be used only if recombination through one or both the paramagnetic centers are radiative.

In [69] the electrical method for detecting cross relaxation probes the change in photoconductivity when two different spin systems are brought into resonance by tuning the magnetic field. Unlike the case of EPR and EDMR spectroscopy, electrically-detected cross relaxation does not require external irradiation to induce transitions between the magnetic sublevels because two different centers that are coupled by magnetic dipolar interactions undergo energy-conserving flip-flop transitions. Therefore, electrically detected cross relaxation measurement is as simple as monitoring photoconductivity under a scanning magnetic field and applicable for detecting both radiative and nonradiative centers.

The cross relaxation between electron spins of phosphorus and oxygen-vacancy centers (A centers) in a γ-ray irradiated Czochralski (CZ)-grown, phosphorus (P)-doped silicon single crystal was studied [69]. A-centers can be easily transformed by bandgap illumination into excited triplet states (electron spin $S = 1$) that lead to well-known Si-SL1 EPR spectra. Thus, under illumination the sample contains predominantly two kinds of paramagnetic centers: phosphorus ($S = 1/2$ and ^{31}P nuclear spin $I = 1/2$) and Si-SL1 centers ($S = 1$). Cross relaxations are expected when Zeeman splittings of phosphorus and SL1 centers are made equal by tuning the magnetic field.

In [70] the authors present the results of electrically detected magnetic resonance experiments on ion-implanted Si:P nanostructures at 5 K, consisting of high-dose implanted metallic leads with a square gap, in which phosphorus is implanted at a nonmetallic dose corresponding to 10^{17} cm^{-3}. By restricting this secondary implant to a 100×100 nm^2 region, the EDMR signal from less than 100 donors is detected. This technique provides a pathway to the study of single donor spins in semiconductors, which is relevant to a number of proposals for quantum information processing. It should be noted, this technique is not restricted to P, but can be extended to other dopants.

The linewidth of the low-field electrically detected magnetic resonance of phosphorus electrons in silicon was investigated using samples with various ^{29}Si nuclear spin fractions and is compared to that of X-band electron paramagnetic resonance [71]. The linewidths of low-field electrically detected magnetic resonance and EPR are the same even though EDMR signals are obtained based on spin-dependent recombination, suggesting that the interaction between electron spins of phosphorus and recombination centers is strong enough for the EDMR detection but weak enough not to affect the linewidths. This favorable balance makes EDMR an attractive method to elucidate the low-field behavior of paramagnetic defects in semiconductors. The narrowest linewidth only limited by the inhomogeneity in the external field, is determined as $\Delta B_{1/2} = 0.004$ mT [71], which is already sufficient for the investigation of a variety of paramagnetic centers in solids, and could be improved by the use of a magnet with better homogeneity.

Pulsed electrically detected magnetic resonance. Spin-dependent recombination between phosphorus donors in silicon and Si/SiO_2 interface states was investigated with pulse electrically detected electron double resonance [72]. The dominant spin-dependent recombination transition was demonstrated to occur between phosphorus donors and Si/SiO_2 interface states. Combining pulses at different microwave frequencies allowed to selectively address the two spin subsystems participating in the recombination process and to coherently manipulate and detect the relative spin orientation of the two recombination partners. The spin-dependent process monitored in these experiments can be attributed to the transition from the ^{31}P donor to the dangling bond states P_{b0} at the Si/SiO_2 interface.

The theoretical foundation of pulse EDMR presented in [73] provides a broad base for the quantitative and qualitative investigation of various electronic processes in different materials and new insights into the nature of charge carrier recombination in bulk semiconductors, semiconductor interfaces as well as semiconductors devices such as thin film transistors and solar cells. Pulse EDMR is based on the transient measurement of electrical currents in semiconductors after a coherent manipulation of paramagnetic centers with pulse electron paramagnetic resonance. A model of spin-dependent recombination is introduced combining features of previous models (Lepine's model, Kaplan-Solomon-Mott (KSM) model etc.) into one general picture that takes influences by spin-relaxation, singlet and triplet recombination as well as spin-spin interactions within recombining charge carrier pairs into account. Based thereon, predictions for excess charge carrier currents after short coherent pulse EPR excitations are made which show that spin coherence in semiconductors can be observed by means of current measurements and hence, microscopic, quantitative information about charge carrier recombination dynamics by means of pulse EDMR is attainable.

Electrical detection of the spin resonance of a single electron. Electrical detection of the spin resonance of a single electron in semiconductors was demonstrated in [74, 75]. In quantum dot devices, single electron charges are easily measured. Spin states in quantum dots, however, have only been studied by measuring the average signal from a large ensemble of electron spins. Ensembles of many spins have found diverse applications ranging from magnetic resonance imaging to magneto-electronic devices, while individual spins are considered as carriers for quantum information. The experiment presented in [74] aims at a single-shot measurement of the spin orientation (parallel or antiparallel to the field) of a particular electron; only one copy of the electron is available, so no averaging is possible. The quantum dot under investigation was formed in the two-dimensional electron gas (2DEG) of a GaAs/AlGaAs heterostructure by applying negative voltages to the metal surface gates. The authors of [74] used spin-to-charge conversion of a single electron confined in the dot, and detected the single electron charge using a quantum point contact; the spin measurement visibility was about 65%. In addition, very long single-spin energy relaxation times (up to, 0.85 ms at a magnetic field of 8 T) was observed, which are encouraging for the use of electron spins as carriers of quantum information.

Electrical detection of the spin resonance of a single electron in a silicon field-effect transistor was made in [75]. The ability to manipulate and monitor a single-electron spin using electron spin resonance was demonstrated. Such control would be invaluable for nanoscopic spin electronics, quantum information processing using individual electron spin qubits and magnetic resonance imaging of single molecules. Several examples of magnetic resonance detection of a single-electron spin in solids will be considered in Chap. 7. It was demonstrated electrical sensing of the magnetic resonance spin-flips of a single electron paramagnetic spin centre, formed by a defect in the gate oxide of a standard silicon transistor [75]. The spin orientation is converted to electric charge, which was measured as a change in the source/drain channel current. Our set-up may facilitate the direct study of the physics of spin decoherence, and has the practical advantage of being composed of test transistors in a conventional, commercial, silicon integrated circuit. For a small transistor, there might be only one isolated trap state that is within a tunnelling distance of the channel, and that has a charging energy close to the Fermi level. When a defect is present, the source/drain channel current can experience random telegraph signal, jumping between two discrete current values. These arise from two possible trapped electric charge states of the defect. The two charge states can correspond to the two spin orientations of a trapped electron. Field effect transistor (FET) current senses electrostatic charge (by definition), and can thus sense single-electron spin resonance.

Electrically detected magnetic resonance in silicon carbide (SiC). The main studies in the field of electrically detected magnetic resonance were performed on silicon [see also 76–79]. In [80] EDMR techniques was applied to ion implanted 4H-SiC pn junctions. Nitrogen implantation was shown to create a high density of recombination centers in SiC which can degrade the performance of ion implanted pn junctions. Spin dependent recombination was used to identify deep level defects associated with these centers. A dominating EDMR spectrum was found to be a defect complex involving nitrogen.

In a later publication [81] electrically detected magnetic resonance was applied to investigate defects observed in nitrogen implanted silicon carbide 4H-SiC. Nitrogen implantation is a high energy process that gives rise to a high defect concentration. The majority of these defects were removed during the dopant activation anneal, shifting the interstitial nitrogen to the desired substitutional lattice sites, where they act as shallow donors. EDMR shows that a deep-level defect persists after the dopant activation anneal. This defect is characterized as having a $g_{\|} = 2.0054$ and $g_{\perp} = 2.0006$, with pronounced hyperfine shoulder peaks with a 1.3 mT peak to peak separation. The model as the nitrogen at a carbon site next to a silicon vacancy center was suggested and this deep-level defect responsible for the observed EDMR signal and the associated dopant deactivation. The defect identification was based upon the defect symmetry and the form of the hyperfine structure. In addition, defect spectra are identified by comparing EDMR measurements with extensive ab initio calculations [81].

In [82] the electrically-detected EPR spectra of point centers were measured inside of a sandwich nanostructure with embedded microcavities, which consists of an ultra-narrow p-type quantum well, confined by δ-barriers, heavily doped with boron, on an n-6H-SiC surface. This procedure is based on the assumption that an experimental sample with Hall geometry of contacts can be used as an EPR spectrometer.

2.2.4 Optically Detected Cyclotron Resonance (ODCR)

In addition to the magnetic characteristics of carriers (electrons and holes) such as g-factors discussed in the previous sections, the effective masses of carriers are the basic parameters for semiconductors and semiconductor nanostructures. The cyclotron resonance (CR) is a method of choice to determine carrier properties (effective masses and scattering times) with high precision and this technique is widely used for evaluation of the fundamental parameters of nanostructures. Cyclotron resonance refers to the resonant absorption of microwave or far-infrared radiation by electrons in static magnetic field. It was first observed by Dresselhaus et al. [83] in germanium and silicon single crystals. The resonance cyclotron frequency is given by $\omega_c = eB/m$, where e is the elementary charge, B the magnetic field, and m the mass of the charge carrier. In semiconductors this is effective mass, m*, ($\omega_c = eB/m^*$) which obtains from the band structure of the material: its inverse reflects the curvature of the dispersion relation E(**k**). Information on scattering times and the free-carrier concentration can be deduced from the CR linewidth and from the integrated absorption strength, respectively.

The basic theory of cyclotron resonance in semiconductors, can be built from both a classical and a quantum mechanical point of view [84]. The formulas describing the cyclotron resonance frequency and absorption can be derived from the classical equation of motion of a carrier with effective mass m*, and −e (electron) or +e (hole) charge under the influence of crossed magnetic (**B**) and electric (**E**) fields. The equation of motion for the drift velocity **v** (classical Drude model) is

$$m^* d\mathbf{v}/dt + m^* \mathbf{v}/\tau = -e\mathbf{E} - e(\mathbf{v} \times \mathbf{B}). \qquad (2.5)$$

Here, **v** is the electron drift velocity, m* is the electron effective mass, τ a phenomenological scattering time (the time between two scattering events), and e the elementary charge. The right side contains the driving terms: **B** is the static magnetic field, which assumed to be directed along the z-axis, and **E** is an alternating electric field polarized in the xy plane, corresponding to a plane wave propagating along the z-axis. This equation can be solved in steady state for the velocity components. A carrier will move in a spiral orbit about the magnetic field, with the angular rotation frequency (cyclotron frequency) equal to

$$\omega_c = \pm eB/m^*,$$

$\pm$ sign indicates that electrons and holes rotate in opposite directions. The conductivity σ can be defined via Ohm's law and the current density $\mathbf{j} = ne\mathbf{v} = \sigma\mathbf{E}$, where n is the electron density. Assuming an orientation of the magnetic field in z-direction $B_z = B$, $B_x = B_y = 0$ and the oscillatory components of the electric field in xy plane $\mathbf{E} = \mathbf{E}_0\exp(i\omega t)$ one finds the frequency dependent conductivity for circular polarized microwave radiation by solving (2.5). According to classical electrodynamics the power absorption is proportional to the real part of the conductivity $\mathrm{Re}\,\sigma$ which is given by

$$\mathrm{Re}\,\sigma = \sigma_0[1 + (\omega\tau)^2 + (\omega_c\tau)^2]/\{[(1 + (\omega_c\tau)^2 - (\omega\tau)^2]^2 + 4(\omega\tau)^2]\}, \qquad (2.6)$$

where $\sigma_0 = (ne^2\tau)/m^*$ is the steady state conductivity, n is the carrier concentration. The resonance is quite well defined for cyclotron frequency ω_c when $\omega_c\tau > 1$. The scattering is often expressed by the means of the electron mobility $\mu = e\tau/m^*$. Let us state the condition $\omega_c\tau > 1$ for the observation of cyclotron resonance expressed by the means of the electron mobility. To have $\omega_c\tau > 1$ requires $\mu > e/(\omega_c m^*)$, thus to observe the cyclotron resonance are essential a high mobility and appropriate average effective mass. Since the radius of the cyclotron orbit is $r = v/\omega_c$, the mean radius for carriers in a Maxwellian velocity distribution at temperature T is $\langle r\rangle = [(8kT)/(\pi m^*)]^{1/2}(1/\omega_c)$ as $\langle v\rangle = [(8kT)/(\pi m^*)]^{1/2}$. Example: for $T = 4$ K, $\langle v\rangle \cong 4 \times 10^4$ m/s, $\omega_c = 1.5 \times 10^{11}\ \mathrm{s}^{-1}$, one has $\langle r\rangle \cong 3 \times 10^{-7}$ m [83].

The width of the cyclotron-resonance line contains information about the electron scattering time τ. The half-width of the half maximum (HWHM) $\Delta B/2$ of the cyclotron active conductivity component $\mathrm{Re}\,\sigma(B)$ is related to the electron mobility μ and to the electron scattering time: $\Delta B/2 = 1/\mu = m^*/(e\tau)$.

Note the transition probability in cyclotron resonance is proportional to the square of the electric dipole moment; in electron paramagnetic resonance the transition probability between Zeeman levels is proportional to the square of the magnetic moment. As the maximum electric field in a resonant cavity is of the same order of magnitude as the maximum magnetic field, the ratio of the transition probabilities for cyclotron and for spin resonance will be the order of $(e\langle r\rangle)^2/\mu_B^2 \approx 10^{12}$. The substantial advantage favoring the detection of cyclotron resonance is partly lost because of the low carrier concentrations used in cyclotron resonance compared with EPR.

A description of cyclotron resonance based on the simplest quantum mechanical theory leads to the formulas which describe an absorption process between Landau levels, described by Fermi's Golden Rule. The influence of a crystal periodic potential U(r) can be described by the effective mass approximation. In an isotropic crystal the effective mass is a scalar, whereas in an anisotropic crystal the effective mass is a tensor. For sufficiently small **k** (so called **kp** perturbation theory) crystal electrons act like free electrons, having an effective mass m*. The crystal periodic potential is condensed in the effective mass only and the band electrons can be approximated as freely moving particles with an effective mass, and as a result the Hamiltonian can be transformed

$$\widehat{\mathrm{H}} = \mathrm{p}^2/(2\mathrm{m}_0) + \mathrm{U(r)} \rightarrow \widehat{\mathrm{H}}^* = \mathrm{p}^2/(2\mathrm{m}^*). \tag{2.7}$$

The effective mass is commonly expressed in units of the vacuum mass m_0 (e.g., in GaAs $m* = 0.067m_0!$). The real Schrödinger equation is replaced by the so called effective Schrödinger equation, which does not contain the crystal potential and the free carrier masses are substituted by the effective masses. The effective Schrödinger equation (with, e.g., Coulomb interactions, interactions with external magnetic fields or confinement potentials) coincides with the corresponding equation in the free space, its eigenfunctions do not contain lattice periodic parameters.

In the effective mass approximation the interaction with an external magnetic field is introduced by performing the standard substitution p → p + eA:

$$\widehat{\mathrm{H}}_0 = [1/(2\mathrm{m}^*)](\mathbf{p} + \mathrm{e}\mathbf{A})^2, \tag{2.8}$$

where **A** is the vector potential of the magnetic field **B** = rot**A**, which is given by **A** = $(0,B_x,0)$ in the asymmetric (Landau) gauge, when B has only a z-component **B** = (0,0,B).

One has to solve the Schrödinger equation $\hat{H}_0\Psi_N = E_N\Psi_N$ (or $\hat{H}_0|N\rangle = E_N|N\rangle$) and the solution will lead to the Landau level eigenvalues E_N and eigenfunction (wavefunctions) $|N\rangle$. The corresponding Schrödinger equation can be written in the form of a harmonic oscillator with eigenvalues (energy levels) (see, e.g. [84])

$$\mathrm{E_N} = \hbar\omega_c(\mathrm{N} + 1/2), \tag{2.9}$$

where $\omega_c = eB/m^*$ the cyclotron frequency. These levels are known as Landau levels. For a finite magnetic field, the motion in the xy-plane is quantized and the energy eigenvalues E_N are equally distanced by the cyclotron energy $\hbar\omega_c$, while ω_c itself increases linearly with magnetic field B.

The analog to the classical orbital motion of electrons is reflected by the wavefunctions Ψ_N. From the radial probability density one can find for each N (Ψ_N) only one maximum at $\langle r\rangle = l\sqrt{(2N + 1)}$, where l is the magnetic length which is a natural length scale for the expansion of the electron wavefunction in an external magnetic field and defines the radius of the cyclotron orbit $l = \sqrt{\hbar/(m * \omega_c)} = \sqrt{\hbar/(eB)}$.

For a bulk three-dimensional semiconductor, there is a series of Landau levels. The free motion of the electrons remains possible along the z-direction but xy motion now is condensed into the Landau levels. In a quantum well (two-dimensional electron system), the z-motion is quantized by a quantum well potential, and application of a magnetic field results in a completely quantized system.

So, we have a system of Landau energy levels in a static magnetic field of the moving charge in the same way as we have a system of Zeeman energy levels for a spin in a static magnetic field. In order to calculate the absorption between Landau levels according Fermi's Golden Rule, we need the interaction Hamiltonian of the electromagnetic field acting on the Landau levels. We will use the time dependent vector potential of the electromagnetic microwave field $\mathbf{A}_{mw}$, taking into consideration that the static magnetic field is described by time invariant vector potential $\mathbf{A}$. The Hamiltonian is then given by [84]

$$\widehat{H} = [1/(2m^*)][\mathbf{p} + e(\mathbf{A} + \mathbf{A}_{mw})]^2 \approx \widehat{H}_0 + (e/m^*)(\mathbf{p} + \mathbf{A}) \cdot \mathbf{A}_{mw} = \widehat{H}_0 + \widehat{H}', \tag{2.10}$$

where the second term we consider as the time dependent perturbation $\hat{H}' = (e/m^*)(\mathbf{p} + \mathbf{A}) \cdot \mathbf{A}_{mw}$, which results in the resonant promotion of electrons to higher Landau levels, i.e. from an energy E_N to $E_{N'}$. This transition is excited by a weak MW field $\mathbf{A}_{mw}$, which is a superposition of two oppositely circular polarized fields. We will consider for simplicity only one circular polarized component, which rotates in the same direction as the charge: $\mathbf{A}_{mw} = A_0(\mathbf{e}_x + \mathbf{e}_y)\exp(i\omega t)$.

According to the Fermi's Golden Rule for a periodic perturbation the transition rate W between Landau levels can be given by

$$W_{N,N'} = (2\pi/\hbar)\delta(E_N - E_{N'} - \hbar\omega)|\left\langle N'|\widehat{H}'|N\right\rangle|^2. \tag{2.11}$$

The operator $\hat{H}'$ can be presented as raising or lowering operator for Landau level wavefunction $|N\rangle$ with the selection rule of $\Delta N = \pm 1$. The δ-function only permits MW-quantum energies $\Delta E = \hbar\omega_c = \hbar eB/m^*$ coinciding with the Landau level energy splitting.

In summary, for fixed microwave frequency ω_{mw} and tunable magnetic field a cyclotron resonance at $B_{cr} = m^* \omega_{mw}/e$ will be observed. By measuring B_{cr} and ω_{mw} the effective electron (hole) mass can be determined according to equation

$$m^* = e\, B_{cr}/\omega_{mw}.$$

For more extensive descriptions of conventional cyclotron resonance, the reader is referred, e.g., to the book [84].

One of the obstacles that prevent using of the conventional cyclotron resonance technique is that the carrier density has to be large enough to produce a appreciable change in the absorption of microwave radiation. This limitation very often does not allow to measure carrier effective masses in undoped systems. It has been overcome by the invention of the optically detected cyclotron resonance (ODCR) technique. The ODCR technique is based on variation of the optical properties, such as the photoluminescence (PL) intensity under absorption of microwaves or FIR radiation by free carriers. Baranov et al. [85] were the first to demonstrate the ODCR method in 1977 for study of semiconductors, they observed microwave-induced changes in the photoluminescence intensity of condensed excitons in germanium single crystal. Later ODCR was detected in GaAs and CdTe [86] and in GaSb [87], see also review [88] and references therein.

In this section we will discuss the optically detected cyclotron resonance in semiconductor single crystals and based nanostructures. ODCR is analogous to the more known optically detected magnetic resonance (ODMR), reviewed in Sect. 2.2.2.

It was pointed out that ODCR technique may have several advantages over conventional absorption techniques at microwave frequencies.

(i) The optical excitation of the photoluminescence provides the required free carriers in the conduction and valence band without the necessity of doping the material (which itself can provide scattering centres and thus destroy the cyclotron resonance signal).
(ii) The optical experiment allows the selective investigation of inhomogeneous samples such as epitaxial layers and nanostructures.
(iii) The study of ODCR in GaP and ZnTe [89] and in silicon [90] highlights the possibilities of this technique as a probe of luminescence processes.
(iv) The condition $\omega_C\tau \geq 1$ is satisfied at lower fields owing to the light-induced impurity neutralization (the large orbits at lower field being of particular interest in quantum wells and superlattices as reported in [91, 92].

As an example of ODMR application, Fig. 2.23 shows the ODMR signals of electrons and light and heavy holes in silicon single crystal registered at two frequencies of the microwave field (35.2 and 95 GHz) simultaneously applied to the sample by monitoring the exciton emission intensity. The electron energy surfaces in silicon near the band edge are prolate spheroids oriented along $\langle 100 \rangle$ axes with longitudinal mass parameter $m_l = 0.98m_e$ and transverse mass parameter $m_t = 0.19m_e$ are shown in inset. Note, that the electron energy surfaces in germanium near the band edge are prolate spheroids oriented along $\langle 111 \rangle$ axes with $m_l = 1.58m_e$ and $m_t = 0.082m_e$.

ODCR measurements [87] give a value for the conduction electron effective mass in GaSb of $0.039m_e \pm 0.005m_e$ in agreement with previously reported values and a lineshape analysis of the resonance reveals an electron momentum relaxation time in close agreement with theory. Remarkable sensitivity is possible with ODCR even at low microvawe frequency. The experiments in [87] were performed on the GaSb layer with the highest observed hole peak mobility. The sample was mounted on a rotating sample holder in the electric field maximum of a 16.5 GHz resonator, T = 2 K. Free carriers were photoexcited by the 514.5 nm line of an Ar^+ ion laser with a power density of about 10 W cm^{-2}; the microwave power was in the range 0.1–0.5 W and was modulated at a frequency of 1 kHz. The ODCR signals were detected via the total luminescence emission using a Ge p-i-n diode detector.

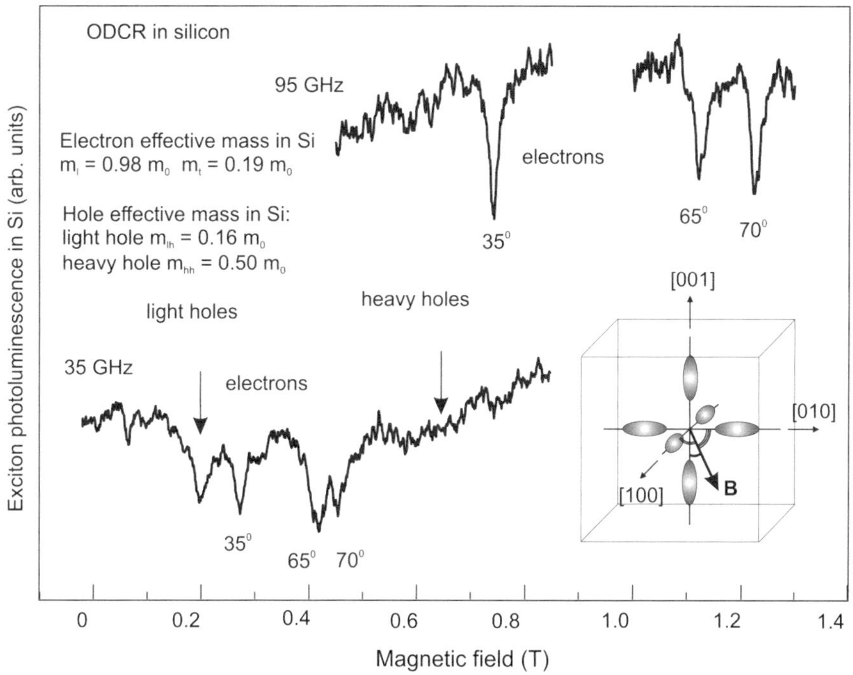

Fig. 2.23 Optically-detected cyclotron resonance signal of electrons and light and heavy holes in silicon single crystal registered at two frequencies of the microwave field (35.2 GHz and 95 GHz) simultaneously applied to the sample by monitoring the exciton emission intensity. (*Inset*) The electron energy surfaces in silicon near the band edge are prolate spheroids oriented along $\langle 100 \rangle$ axes with longitudinal mass parameter $m_l = 0.98m_e$ and transverse mass parameter $m_t = 0.19m_e$ (For comparison, the electron energy surfaces in germanium near the band edge are prolate spheroids oriented along $\langle 111 \rangle$ axes with $m_l = 1.58m_e$ and $m_t = 0.082m_e$)

Measurements were later extended to the far-infrared (FIR) where much higher resolution of the electron and hole masses was possible.

Thus, ODCR has proved to be extremely sensitive and has been successfully used to measure the effective masses of electrons and holes in bulk III–V compounds GaAs, InP, II–VI compound CdTe, [86, 93, 94] and IV–IV compound SiC [95–101].

Optically detected cyclotron resonance at X-band frequency in high-purity 6H-SiC epilayers has been observed [95]. The electron effective mass values for 6H-SiC were obtained as $m_{\perp}^* = 0.42 \pm 0.02m_0$ and $m_{\|}^* = 2.0 \pm 0.02m_0$. The electron mobility at 6 K in the basal plane was determined to be $\mu_{\perp} \approx 1.1 \times 10^5$ cm^2/V s. The anisotropy of the electron mobilities in 6H-SiC can be explained by the corresponding anisotropy of the effective masses.

The results on 4H-SiC were presented in [96], ODCR has been observed in high-purity 4H SiC CVD epilayers at both *X*-band and *Q*-band MW frequencies. For the first time, the electron effective masses have been directly determined for this polytype: $m_\perp{}^* = 0.42m_0$ and $m_\parallel{}^* = 0.29m_0$. A scattering time $\tau_\perp \approx 4.3 \times 10^{-11}$ s was obtained for the carriers in the basal plane, and hence the corresponding electron mobility: $\mu_\perp \approx 1.8 \times 10^5$ cm^2/V s. The results from this work clearly show that the electron mobility in 4H-SiC is higher than that in 6H-SiC and is also much less anisotropic.

Experimental and theoretical results from studies of electron effective masses in 4*H* SiC were presented in [97]. Three principal values of the mass tensor were experimentally resolved by ODMR, and were determined as m(ML) = 0.33 ± $0.01m_0$, m(MΓ) = 0.58 ± $0.01m_0$, and m(MK) = 0.31 ± $0.01m_0$. These values are in good agreement with m(ML) = $0.31m_0$, m(MΓ) = $0.57m_0$, and m(MK) = $0.28m_0$, obtained from band-structure calculations based on the local density approximation to the density-functional theory using the linearized augmented plane-wave method. The conduction-band minimum was found to be at the *M* point of the Brillouin zone.

The effects of microwave fields on recombination processes, which are responsible for the ODCR in 4H- and 6H-SiC epitaxial layers, have been investigated [99] and experimental evidence indicating that the dominant mechanism of ODCR in SiC was microwave-induced lattice heating under the cyclotron resonance conditions were presented. The results also show that at low temperatures and low microwave power the dominant scattering mechanism is impurity scattering, while carrier scattering by lattice phonons dominates under high microwave power conditions.

Experimental data on the band-structure and high-mobility transport properties of 6*H* and 4*H*-SiC epitaxial films based on ODMR investigations were presented in [100]. From the orientational dependence of the electron effective mass in 6*H*-SiC direct evidence for the camels back nature of the conduction band between the *M* and *L* points were obtained. The broadening of the resonance signal in 4*H*-SiC as a function of temperature was used to extract information on electron mobilities and to conclude on the role of the different scattering mechanisms.

ODCR was used to study hole effective masses in 4*H* SiC [101]. In the vicinity of the maximum of the uppermost valence band, the constant energy surface was concluded can be considered as an ellipsoid with the principal axis along the *c* axis and the effective masses of the holes were determined as $m_{h\perp} = 0.66m_0$ and $m_{h\parallel} = 1.75m_0$. The influence of the polaron coupling effect on the effective mass values in 4*H* SiC was discussed.

The authors in [102] report the first direct measurements of the conduction band electron effective mass in MBE grown $A1_{0.48}In_{0.52}As$ on InP by the technique of optically detected cyclotron resonance. The effective mass value derived was $m^* = 0.1m_e \pm 0.01m_e$. A value for the carrier relaxation time was also deduced, indicating a lattice-limited mobility for this material of the order of 10^5 cm^2 V^{-1} s^{-1}.

Semiconductor micro- and nanostructures prepared for the study of low dimensional phenomena consist in many cases of rather complex layer structures. Investigations on such structures require highly selective characterization methods which have the ability to distinguish between the different structures present in one sample. Cyclotron resonance can be used, but performing conventional CR experiments is problematic since the absorption of the microwave (MW) or far-infrared (FIR) power of the whole sample is measured. It can be advantageous to use optically detected cyclotron resonance where the effect of the applied MW- or FIR-power on specific photoluminescence bands is measured. Thus it may allow to separate the carrier properties of different structures present in one sample. Furthermore this method has the potential to obtain enhanced optical resolution compared to photoluminescence (PL) measurements. The MW- or FIR-power absorption of carriers under cyclotron resonance conditions can lead to a selective enhancement or quenching of different recombination channels being unresolved in PL [103–107]. Energy transfer can occur through the following processes: (i) impact ionization of bound or localized excitons by free carriers accelerated in the MW field [105, 107–109], or (ii) heating the crystal lattice caused by the thermal coupling of the accelerated by microwave field free carriers which effects the PL [85, 88, 110]. A way to distinguish between these two processes is to measure the response time of the signals on the applied resonance MW field. The optically detected impact ionization process is expected to be "fast", i.e., in the order of the lifetime of the recombination (typically 1 ns). The temperature modulation effect should be comparable "slow", in the μsec range or below.

In [103] GaAs/$Al_xGa_{1-x}As$ quantum wells and quantum wires were studied by optically detected cyclotron resonance. It was shown that the microwave modulation of the photoluminescence signal is enhanced under cyclotron resonance conditions of the electrons. The energy transfer to the luminescence is thermal by heating the crystal lattice. The ODCR experiments allowed to define selectively the effective masses and mobilities of electrons confined in an 80 Å quantum well and in nominally 80-nm-wide quantum wires.

Cyclotron resonance has been optically detected in a GaAs/$Ga_{0.67}Al_{0.33}As$ superlattice with the wells doped with Si to 6×10^{16} cm^{-3} [92]. The sign and magnitude of the signal depend on the emission monitored, and an effective mass of $0.062m_0$ was obtained for the electrons. The doping of the GaAs wells results in both free- and bound-exciton emission, and the different sign of the microwave effect on each distinguishes these transitions. Because of the low magnetic field' needed at 22 GHz for cyclotron resonance, the confinement of the orbits is readily observed. Cyclotron resonance at microwave frequencies has allowed to examine the case where the cyclotron orbit is much larger than the well width, and deviations from the expected angular dependence were found.

An important application of the method is the study of 2D electron states and internal transitions of neutral and charged magnetoexcitons in GaAs/(Al, Ga)As heterostructures [111–124], which are the best known experimental realizations of a 2D system. The electrons are located in a potential notch in the vicinity of the interface plane and along this plane they can move freely, while the motion perpendicular to the interface is confined, so-called "two-dimensional-electron gas" (2DEG). High electron mobilities are of crucial importance for the technical application in a high-electron-mobility transistor (HEMT) where high operation frequencies and a low noise level can be realized due to high mobility 2DEGs. In addition, these dynamical properties play an essential role for the electron-hole separation process in a solar cell or for the carrier transport to the p-n junction in a light emitting diode. These properties of 2DEG are extremely important for basic research, as high mobility 2DEGs in GaAs/AlGaAs heterojunctions and low carrier scattering is of interest to investigate the electronic band structure. A number of publications investigated properties of GaAs/AlGaAs heterojunctions by optically detected far infrared cyclotron resonance (FIR-ODCR) which is widely recognized as a is a powerful tool to investigate electronic properties in semiconductors.

A triangular quantum well forms at the AlGaAs/GaAs interface, referred to as the heterointerface (see Fig. 2.24a). Often, only the quantum mechanical ground state in the triangular well is populated (band E_0 in Fig. 2.24a), at low temperatures, T < 100 K. As long as the dopants are removed from the lower GaAs/AlGaAs interface, it is referred to as modulation doping. Usually, Si is used as the dopant. It

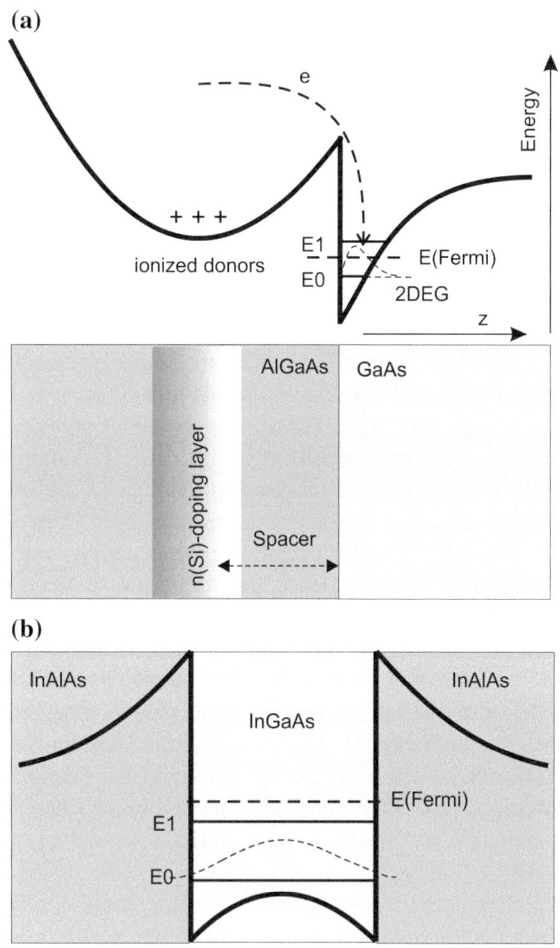

Fig. 2.24 **a** Growth profile and bandstructure of typical GaAs/GaAlAs heterostructure. Schematic of electronic structure and wavefunctions of the triangular well; the envelope wavefunction of the first subband E0 is shown. **b** Growth profile and bandstructure of InGaAs/InAlAs square well; the envelope wavefunctions of the first subband is shown

only goes into the doping region, all the other regions are intrinsic semiconductors. Only a fraction of the donor atoms are ionized. Part of that fraction goes into surface states, and part into the quantum/triangular well.

Very large mobilities reaching 30×10^6 $cm^2/(V\ s)$ corresponding to a mean free backscattering path of about 300 μm have been achieved. The ionized donors, which are a significant source of scattering, are spatially well separated from the 2DEG, usually between 20 and 120 nm. By controlling the Al content the z-dependence of the band gap/band structure can be custom engineered.

Optically detected cyclotron resonance studies of multisubband $In_{0.52}Al_{0.48}As/In_{0.53}Ga_{0.47}As$ quantum wells were performed in [125]. Modulation-doped quantum wells (MDQWs) are a unique system for use both in studying fundamental physics as well as in potential device applications. Due to the spatial separation of the dopant atoms and the mobile electrons, MDQW based devices have considerable technological importance. ODCR measurements on two subband occupied electronic systems confined in InAlAs/InGaAs triangular and square quantum wells were reported (see Fig. 2.24). The ODCR measurements were demonstrated to be a useful tool for the understanding of the luminescence processes, in addition to the characterization of carrier concentration, effective mass and Fermi energy. It was found that only one broad photoluminescence line locates below the energy gap for the triangular well, which is attributed to donor-to-acceptor recombinations. For the square well, two lines with emission energies higher than the bandgap are observed. Each line width is consistent with the Fermi energy of the first and second subband. This suggests that the emissions involve electron transitions from the first and second subbands at all occupied *k* states to localized holes. The effective mass of electrons in the first subband (E_0 level) was shown to be heavier than that of electrons in the second subband (E_1 level) in the triangular well due to the effect of nonparabolicity and the electron distribution, while the reverse is observed in the square well. The obtained effective masses in the square well are $m^*_0 = 0.050m_e$ and $m^*_1 = 0.060m_e$, which shows the reverse behaviour compared with the result of the triangular well sample. This is explained in part by the nonparabolicity and, significantly, by the barrier leakage of the electron wavefunction. Furthermore, it was found that the single particle relaxation time obtained from magnetoresistance measurements and the scattering time from cyclotron resonance are longer for electrons in the upper subband for both triangular and square wells, and the relaxation time is smaller than the corresponding scattering time, consistent with previous transport results. In addition, quantum oscillations in the ODCR spectra due to the effect of the crossing between the Landau levels and the Fermi level have observed, from which useful information of the properties of the 2DEG was obtained.

The properties of a two-dimensional electron gas confined at an $In_{0.53}Ga_{0.47}As/In_{0.52}Al_{0.48}As$ interface were investigated by optically detected cyclotron resonance [126] and spin-splitting crossing between subbands was demonstrated. ODCR spectra structured by quantum oscillations and subharmonic cyclotron resonances have been observed. The ODCR spectra reveal a very strong and sharp peak with several fine structures. This observation was shown to be due to the energy-level crossing of the spin splitting between the first and second subbands of the two-dimensional electron gas. The strong and sharp absorption is made possible only under conditions such that the energy separation of Landau levels is equal to the incident far-infrared photon energy, the Fermi energy lies in the center of a Landau level, and the uppermost occupied states of the two subbands have the same energy and spin. This behavior provides a precise determination of the effective mass and g factor.

Chen et al. [127] have directly measured for the first time the electron effective mass of the 2DEG in InAlAs/InGaAs heterojunction bipolar transistors (HBTs) using the ODCR technique at 2 K. They reported an investigation of optical and electronic properties in InAlAs/InGaAs heterojunction bipolar transistor layers. Strong ODCR spectra structured by quantum oscillations have been observed, from which the effective mass and the carrier concentration of the two-dimensional electron gas can be obtained. In addition to the observation of a broad cyclotron resonance, the ODCR spectrum is structured by quantum oscillations due to the effect induced when the Landau levels of the 2DEG move through the Fermi level. The crossover between the Landau levels and the Fermi level modulates the populations and the width of the Landau levels of the 2DEG active in CR transitions, which results in oscillations in the CR amplitude. The measured cyclotron mass is heavier than the conduction-band-edge mass in bulk InGaAs. The carrier concentration was found to increase with the spacer thickness and with decreasing the carrier concentration.

Optically detected cyclotron resonance has been observed in doped GaAs quantum wells [128]. A variety of far-infrared resonances were measured, including the transition from the ground state to the first excited state in neutral donors, and singlet and triplet transitions of negative donor ions (D^-), as well as electron-cyclotron resonance, in well-center-doped GaAs quantum wells. The power of ODCR technique for studying impurity states in confined systems is clearly revealed. Results provide evidence for the existence of D^- centers under optical excitation in multiple-quantum-well structures doped only in the wells.

ODCR technique was applied to determine composition dependence of the in-plane conduction band effective mass in strained lattice-mismatched $Ga_{1-x}In_xAs$/InP single quantum wells [129, 130]. ODCR results show a strong increase of the in-plane effective mass of electrons with increasing quantum confinement. The results are in agreement with a self-consistent calculation taking into account effects due to nonparabolicity, confinement, strain, and finite two-dimensional carrier densities. InAs/GaAs self-assembled quantum dots [131] and InSb/GaSb quantum dot structures [132] were investigated by ODCR technique. Microwave and far-infrared induced optically detected cyclotron resonance in epitaxial InP and GaAs were studied in [133]. Cyclotron-resonance-induced impact ionization of shallow donors and bound excitons was concluded to be the basic mechanism for the observation by photoluminescence.

Another advantage of the ODCR technique is related to its spectral selectivity, which allows for selecting the signal from different quantum wells grown in the same structure by analyzing the corresponding photoluminescence emission lines.

Therefore, the ODCR technique is very well suited for measurements of the electron effective masses in undoped CdTe-based QW's of different widths [134].

Optically detected cyclotron resonance of two-dimensional electrons has been studied in nominally undoped CdTe/(Cd,Mn)Te quantum wells [134]. The enhancement of carrier quantum confinement results in an increase of the electron cyclotron mass from $0.099m_e$ to $0.112m_e$ with well width decreasing from 30 down to 3.6 nm. Comparison with model calculations performed for this material system highlights two contributions to the mass increase, the first one determined by band structure parameters and the second one due to the polaron effect modified by reducing the dimensionality of the electronic system.

The electronic properties of self-organized InAs/GaAs QDs are attracting great attention because of their device applications for 1.3 μm laser diodes emitting at the optical fiber window [135–137]. InGaAsN/GaAs QWs with low nitrogen content have been suggested as a novel material for the realization of 1.3 μm lasers [138]. The incorporation of even small portion of nitrogen is suffient to reduce the band-gap energy of the alloy. The interaction between the conduction band and a band formed by nitrogen states was supposed to account for the reduction of the band gap energy in the quaternary materials InGaAsN.

The study of InGaAsN/GaAs quantum wells with low nitrogen content and InAs/GaAs quantum dots using ODMR and ODCR was reported in [139]. ODMR and ODCR were applied to study two types of nanostructures emitting around 1.3 μm: quantum wells with low nitrogen content and InAs/GaAs quantum dots (both isolated and vertically-coupled).

Samples. The InGaAsN/GaAs multiple quantum well (MQW) structures with low nitrogen content were grown by molecular-beam epitaxy (MBE) on [001] oriented GaAs substrates [139]. The sample contained five $In_{0.36}Ga_{0.64}As_{0.98}N_{0.02}$ 6 nm-width quantum wells capped by 300 nm nominally-undoped GaAs. To obtain high luminescence efficiency a part of InGaAsN/GaAs MQW samples was annealed at a temperature 700 °C. The InAs/GaAs QDs structures were MBE grown using self-organized effects. Two types of the InAs/GaAs structures were investigated: QDs formed by a single cycle InAs deposition (sample #1) and vertically-coupled quantum dot (VCQD) structures (sample #2). The VCQD structures were obtained by MBE on GaAs (100) substrate by sixfold deposition of 2.0 monolayers of InAs separated by 4 nm-thick GaAs spacers. The lateral dimension of the lower island in InAs VCQDs is about 16 nm and the lateral dimension of the island in each row gradually increases, reaching 26 nm for the upper island. Each InAs/GaAs VCQD structure consists of six InAs islands separated by a narrow 4 nm-thick GaAs spacer. The lateral dimensions of the dot were measured using transmission electron microscopy.

The wave functions of carriers of all vertically-coupled islands overlap and the electronic properties of the VCQD can be considered as a single object. Luminescence was excited far above the band gap by an Ar ion laser and analyzed with a grating monochromator and InGaAs detector diode. ODMR spectrometer operating at 35 GHz and providing magnetic field B up to 4.5 T was used in the experiments. The microwave power up to 400 mW was applied. Microwaves were modulated at audio frequency and the microwave-induced variations of the luminescence intensity were detected with a lock-in amplifier.

ODMR with the axial symmetry (effective g factor $|g_{\|}| = 3.61$ and $|g_{\perp}| = 0.7$) was found in unannealed InGaAsN/GaAs multiple-quantum well structures and ascribed to electrons in the quantum wells; the sign of g factor is suggested to be negative. There is evidence that before annealing the InGaAsN/GaAs structure with low nitrogen content has properties, which are typical for quantum wells, but after annealing the structure is completely-changed and resembles a quantum-dot-like structure. The effect of cyclotron resonance on the luminescence of InAs quantum dots was found.

The observed ODCR belongs to a two-dimensional system, such as a heterointerface InAs/GaAs. Figure 2.25a shows the optically-detected cyclotron resonance signal (similar in the two samples) recorded at 35.2 GHz with different angles θ between the magnetic field and the [001] axis by monitoring the whole luminescence. In Fig. 2.25b angular dependence of ODCR in sample #1 (open circle) and #2 (black circle) are depicted; the *solid line* shows the calculated angular dependence for $m^* = 0.07m_0$.

Before annealing the luminescence properties are consistent with QWs structures. The results on the effect of the microwaves on luminescence of InAs/GaAs QDs give confirmation of QD-like structures in the annealed sample. This effect contains two contributions: (i) the first is independent on the magnetic field; (ii) the

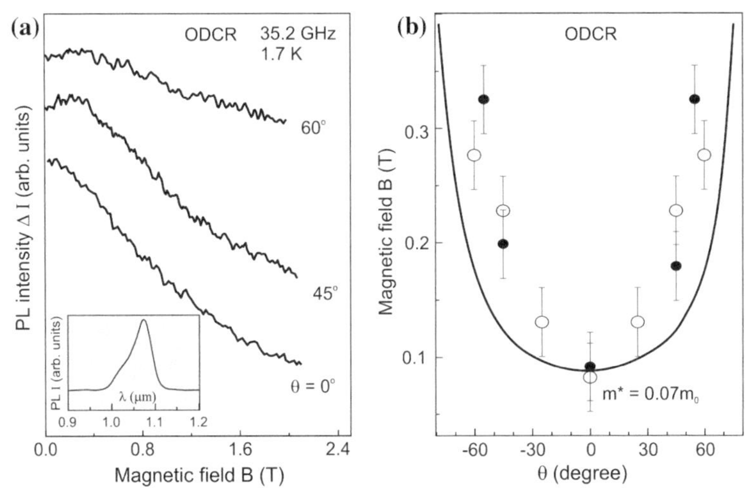

Fig. 2.25 **a** The optically-detected cyclotron resonance signal in InGaAsN/GaAs multiple quantum well structures with low nitrogen content (similar in the two samples: #1—one row of InAs/GaAs QDs; #2—six rows of vertically coupled quantum dots) recorded at 35.2 GHz with different angles θ between the magnetic field and the [001] axis by monitoring the whole photoluminescence (offset for clarity); (*inset*) photoluminescence spectra in the samples #2 at B = 0. **b** Angular dependence of ODCR in sample #1 (*open circle*) and sample #2 (*black circle*); the *solid line* shows the calculated angular dependence for $m = 0{:}07m_0$

second is a broad anisotropic resonance signal typical for cyclotron resonance in a 2D electron gas. The ODCR experiments can allow for a direct identification of 2D related recombination because for Q-band (35 GHz) microwaves the CR orbit is in the range or larger than a typical size of the quantum well. This is why a strong anisotropy of the CR signal is expected for the 2D electrons: the resonance is produced by a magnetic field B which is normal to the 2D plane: B = Bcosθ. For the CR measured with the magnetic field perpendicular to the interface the whole cyclotron orbit is localized in the QW. However, for parallel (in-plane) orientation of the field with respect to the interface, the CR orbit is subjected to confinement effects.

Thus, ODCR have been studied in two types of InAs/GaAs quantum dots structures: QDs formed by a single cycle InAs deposition and vertically-coupled quantum dot structures. An anisotropic low magnetic field signal was ascribed to cyclotron resonance corresponding to an effective mass of $0.07m_0$ of the electrons with cyclotron orbits confined in the 2D system, such as a heterointerface InAs/GaAs.

The basic problem is to clarify the origin of the cyclotron resonance detected on the luminescence of QDs. In two-dimensional (2D) systems, such as heterointerfaces or quantum wells the electron mass is divided into a mass perpendicular with respect to the growth plane, and a mass parallel to it or in-plane mass. It has been shown that the in-plane effective mass of electrons strongly increases as the thickness of the QW decreases. The in-plane effective mass $m^* = 0.07m_e$ found in the InAs/GaAs QW samples is close to that of electrons in GaAs QWs. Since the ODCR signal observed in [139] is the same in both structures under investigation there is a small probability that all these signals belonged to the cyclotron resonance inside QDs, and this signal was suggested to be observed at the interface on the heterostructure wetting layer. In addition, the cyclotron orbit seems to be much larger than a typical size of QDs and at least the effective mass should be strongly dependent on the size of the QDs and their arrangement. In [140] the anisotropic lowfield signal observed in a different InAs/GaAs structure with shallowly-formed QDs was assigned to the ODCR of electrons with the effective mass $0.059m_e$ inside InAs/GaAs QDs. Later a new interpretation of their result was presented in [141] in which the signal was ascribed to cyclotron resonance of the electron in the two-dimensional wetting layer.

The cyclotron resonance in semiconductor nanostructures was electrically detected without an external cavity, a source, and a detector of microwave radiation [142, 143]. An ultra-narrow p-Si quantum well on an n-Si (100) surface confined by heavily boron doped δ-shaped barriers is used as the object of investigation and provides microwave generation within the framework of the non-stationary Josephson effect. The cyclotron resonance is detected upon the presence of a microcavity, which is incorporated into the quantumwell plane, by measuring the longitudinal magnetoresistance under conditions of stabilization of the source–drain current. The cyclotron resonance spectra and their angular dependences measured in a low magnetic field identify small values of the effective mass of light and heavy holes in various 2D subbands due to the presence of edge channels with a high mobility of carriers.

References

1. Eaton, G.R., Eaton, S.S., Salikhov, K.M. (eds.): Foundations of Modern EPR. World Scientific, Singapore (1998)
2. Zavoisky, E.J.: Relaxation of liquid solutions for perpendicular fields. J. Phys. (USSR) **9**, 211–216 (1945)
3. Bloch, F., Hansen, W.W., Packard, M.: The nuclear induction experiment. Phys. Rev. **70**, 474–485 (1946)
4. Purcell, E.M., Torrey, H.C., Pound, R.V.: Resonance absorption by nuclear magnetic moments in solids. Phys. Rev. **69**, 37–38 (1946)
5. Hahn, E.L.: Spin echoes. Phys. Rev. **80**, 580–594 (1950)
6. Blume, R.J.: Electron spin relaxation times in sodium-ammonia solutions. Phys. Rev. **109**, 1867–1873 (1958)
7. Mims, W.B.: Pulsed ENDOR experiments. Proc. R. Soc. A **283**, 452–457 (1965)
8. Zhidomirov, G.M., Salikhov, K.M., Tsvetkov, YuD, Yudanov, V.F., Raitsimring, A.M.: A study of the interaction between paramagnetic species and the magnetic nuclei of the surrounding molecules by electron spin echo spectroscopy. J. Struct. Chem. **9**, 704–708 (1968)
9. Salikhov, K.M., Semenov, A.G., Tsvetkov, Yu.D.: Electron Spin Echoes and its Applications. Nauka, Novosibirsk (1976)
10. Dikanov, S.A., Tsvetkov, Yu.D.: Electron Spin Echo Envelope Modulation (ESEEM) Spectroscopy. CRC Press, Boca Raton (1992)
11. Mims, W.B.: Electron spin echoes. In: Geschwind, S. (ed.) Electron Paramagnetic Resonance, pp. 263–351. Plenum Press, New York (1972)
12. Schweiger, A., Jeschke, G.: Principles of Pulse Electron Paramagnetic Resonance. Oxford University Press, London (2001)
13. Keijzers, C.P., Reijerse, E.L., Schmidt, J.: Pulse EPR: A New Field of Applications, pp. 15–42. North Holland, Amsterdam/Oxford/New York/Tokyo (1989)
14. Schmidt, J.: Interaction of Two-Level Systems with Coherent Radiation Fields. Leiden University (2000)
15. Weber, R.T.: Pulsed EPR Primer. Bruker Instruments, Inc., Billerica, MA USA Biospin (2000)
16. Möbius, K., Savitsky, A.: High-Field EPR Spectroscopy on Proteins and their Model Systems: Characterization of Transient Paramagnetic States. Published by the Royal Society of Chemistry (2008)
17. Carr, H.Y., Purcell, E.M.: Effects of diffusion on free precession in nuclear magnetic resonance experiments. Phys. Rev. **94**, 630–638 (1954)
18. Mims, W.B.: Envelope modulation in spin-echo experiments. Phys. Rev. B **5**, 2409–2419 (1972)
19. Solomon, I.: Rotary spin echoes. Phys. Rev. Lett. **2**, 301–302 (1959)
20. Schmidt, J., van der Waals, J.H.: ESE in triplet states: In: Kevan, L., Schwarz, H. (eds.) Modern Pulsed and CW EPR Techniques. New York (1979)
21. Poole Jr., C.P.: Electron Spin Resonance: A Comprehensive Treatise on Experimental Techniques, 2nd edn. Wiley, New York (1983)
22. Spaeth, J.-M., Niklas, J.R., Bartram, R.H.: Structural analysis of point defects in solids: an introduction to multiple magnetic resonance spectroscopy. In: von Klitzing, K., Merlin, R.,

Queisser, H.-J., Keimer, B. (eds.) Springer Series in Solid-State Sciences, vol. 43. Springer, Berlin, Heidelberg (1992)
23. Spaeth, J.-M., Overhof, H.: Point defects in semiconductors and insulators: determination of atomic and electronic structure from paramagnetic hyperfine interactions. In: Hull, R., Jagadish, C., Kawazoe, Y., Osgood, R.M., Parisi, J., Seong, T.-Y., Uchida, S.-I., Wang, Z. M. (eds.) Springer Series in Materials Science, vol. 51. Springer, Berlin, Heidelberg (2003)
24. Feher, G.: Observation of nuclear magnetic resonances via the electron spin resonance line. Phys. Rev. **103**, 834–835 (1956)
25. Mims, W.B.: Pulsed ENDOR experiments. Proc. R. Soc. Lond. A **283**, 452 (1965)
26. Davies, E.R.: A new pulse ENDOR technique. Phys. Lett. A **47**, 1–2 (1974)
27. Schweiger, A.: Creation and detection of coherence and polarization in pulsed EPR. J. Chem. Soc. Faraday Trans. **91**, 177–190 (1995)
28. Freed, J.H.: The Bruker lecture. Modern techniques in electron paramagnetic resonance spectroscopy. J. Chem. Soc. Faraday Trans. **86**, 3173–3180 (1990)
29. Gemperle, C., Schweiger, A.: Pulsed electron-nuclear double resonance methodology. Chem. Rev. **91**, 1481–1505 (1991)
30. Jeschke, G., Schweiger, A.: Time-domain chirp electron nuclear double resonance spectroscopy in one and two dimensions. J. Chem. Phys. **103**, 8329–8337 (1995)
31. Jeschke, G., Schweiger, A.: Hyperfine-correlated electron nuclear double resonance spectroscopy. Chem. Phys. Lett. **246**, 431–438 (1995)
32. Disselhorst, J.A.J.M., van der Meer, H., Poluektov, O.G., Schmidt, J.: A pulsed EPR and ENDOR spectrometer operating at 95 GHz. J. Magn. Reson. A **115**, 183–188 (1995)
33. Prisner, T.F., Rohrer, M., Moebius, K.: Pulsed 95 GHz high-field EPR heterodyne spectrometer with high spectral and time resolution. Appl. Magn. Reson. **7**, 167–183 (1994)
34. Weil, J.A., Bolton, J.R.: Electron Paramagnetic Resonance, Chapter 12, 2nd edn. Wiley (2007)
35. Mehring, M., Hoefer, P., Grupp, A.: Bloch-Siegert shift, Rabi oscillation, and spinor behaviour in pulsed electron-nuclear double-resonance experiments. Phys. Rev. A **33**, 3523–3526 (1986)
36. Zakharchenya, B.P., Meier, F.: Optical Orientation. Elsevier, Amsterdam (1984)
37. Geschwind, S., Collins, R.J., Schawlow, A.L.: Optical detection of paramagnetic resonance in an excited state of Cr^{3+} in Al_2O_3. Phys. Rev. Lett. **3**, 545–548 (1959)
38. Brossel, J., Geschwind, S., Schawlow, A.L.: Optical detection of paramagnetic resonance in crystals at low temperatures. Phys. Rev. Lett. **3**, 548–549 (1959)
39. Geschwind, S.: Optical detection of paramagnetic resonance. In: Geschwind, S. (ed.) Electron Paramagnetic Resonance, Ch. 5. Plenum Press, New York (1972)
40. Klein, J., Voltz, R.: Time-resolved optical detection of coherent spin motion for organic-radical-ion pairs in solution. Phys. Rev. Lett. **36**, 1214–1217 (1976)
41. Hoff, A.J., Lous, E.J., Vreeken, R.: In: Keijzers, C.P., Reijerse, E.J., Schmidt, J. (eds.) Pulsed EPR A New Field of Application, Chap. 19, pp. 219–226. North-Holland, Amsterdam (1989)
42. Weis, V., Möbius, K., Prisner, T.: Optically detected electron spin echo envelope modulation on a photoexcited triplet state in zero magnetic field—A comparison between the zero-field and high-field limits. J. Magn. Reson. **131**, 17–24 (1998)
43. Purvis, K.L., Wiemelt, S.P., Maras, T., Blue, M., Melkonian, V., Ashby, P.D., Riley, S.A., Fifield, L.S., Martin, K.A., Nishimura, A.M.: Spin-echo in the phosphorescent triplet state of crystalline 2-indanone. J. Lumin. **71**, 199–205 (1997)
44. van Oort, E, Glasbeek, M.: In: Keijzers, C.P., Reijerse, E.J., Schmidt, J. (eds.) Pulsed EPR A New Field of Application, Chap. 19, pp. 227–231. North-Holland, Amsterdam (1989)
45. Tadjikov, B.M., Astashkin, A.V., Sakaguchi, Y.: Quantum beats of the reaction yield induced by a pulsed microwave field. Chem. Phys. Lett. **283**, 179–186 (1998)
46. Babunts, R.A., Badalyan, A.G., Romanov, N.G., Gurin, A.S., Tolmachev, D.O., Baranov, P. G.: A noncavity scheme of optical detection of high-frequency magnetic and cyclotron resonances in semiconductors and nanostructures. Tech. Phys. Lett. **38**, 887–890 (2012)

47. Cavenett, B.C.: Optically detected magnetic resonance (ODMR) investigations of recombination processes in semiconductors. Adv. Phys. **30**, 475–538 (1981)
48. Nicholls, J.E., Davies, J.J., Cavenett, B.C., James, J.R., Dunstan, D.J.: Spin-dependent donor-acceptor pair recombination in ZnS crystals showing the self-activated emission. J. Phys. C: Solid State Phys. **12**, 361–379 (1979)
49. Hiromitsu, I., Kevan, L.: An improved analysis of transient ESR signals of photoexcited triplet states: Application to chlorophyll-a in a glassy matrix. J. Chem. Phys. **88**, 691–695 (1988)
50. Schmidt, J., van der Waals, J.H.: Transient ESR studies of molecular triplet states in zero-field. In: Kevan, L., Schwartz, R.N. (eds.) Time Domain Electron Spin Resonance, pp. 343–398. Wiley, New York (1979)
51. Kevan, L., Kispert, L.D.: Electron Spin Double Resonance Spectroscopy. Wiley Interscience, New York (1976)
52. Westra, J., Sitters, R., Glasbeek, M.: Optical detection of magnetic resonance in the photoexcited triplet state of a deep center in diamond. Phys. Rev. B **45**, 5699–5702 (1992)
53. Hiromitsu, I., Westra, J., Glasbeek, M.: Cross-relaxation effects in the 2.818 eV zero-phonon emission in brown diamond. Phys. Rev. B **46**, 5303–5310 (1992)
54. Glasbeek, M.: Cross-relaxation of localized triplet states in diamond. Appl. Magn. Reson. **7**, 479–494 (1994)
55. Shmidt, J., Solomon, I.: Compt. Rend. **263**, 169 (1966)
56. Honig, A.: Neutral-impurity scattering and impurity Zeeman spectroscopy in semiconductors using highly spin-polarized carriers. Phys. Rev. Lett. **17**, 186–188 (1966)
57. Thorntont, D.D., Honig, A.: Shallow-donor negative ions and spin-polarized electron transport in silicon. Phys. Rev. Lett. **30**, 909–912 (1973)
58. Lepine, D.J.: Spin-dependent recombination on silicon surface. Phys. Rev. B **6**, 436–441 (1972)
59. Kaplan, D., Solomon, I., Mott, N.F.: Explanation of the large spin-dependent recombination effect in semiconductors. J. Phys. Lett. **39**, L51–L54 (1978)
60. Cox, R.T., Block, D., Herve, A., Picard, R., Santier, C., Helbig, R.: Exchange broadened, optically detected ESR spectra for luminescent donor-acceptor pairs in Li doped ZnO. Solid State Commun. **25**, 77–80 (1978)
61. Vlasenko, L.S., Khramtsov, V.A.: Sov. Phys. Semicond. **20**, 688 (1986)
62. Stich, B., Greulich-Weber, S., Spaeth, J.-M.: Electrical detection of electron paramagnetic resonance: new possibilities for the study of point defects. J. Appl. Phys. **77**, 1546–1553 (1994)
63. Hoehne, F., Huebl, H., Galler, B., Stutzmann, M., Brandt, M.S.: Spin-dependent recombination between phosphorus donors in silicon and Si/SiO_2 interface states investigated with pulsed electrically detected electron double resonance. Phys. Rev. Lett. **104**, 046402 (2010)
64. Mortemouque, P.A., Sekiguchi, T., Culan, C., Vlasenko, M.P., Elliman, R.G., Vlasenko, L. S., Itoh, K.M.: Spin dependent recombination based magnetic resonance spectroscopy of bismuth donor spins in silicon at low magnetic fields. Appl. Phys. Lett. **101**, 082409 (2012)
65. Akhtar, W., Morishita, H., Vlasenko, L.S., Poloskin, D.S., Itoh, K.M.: Electrically detected magnetic resonance of phosphorous due to spin dependent recombination with triplet centers in γ-irradiated silicon. Phys. B: Condens. Matter **404**, 4583–4585 (2009)
66. Eremin, V., Poloskin, D.S., Verbitskaya, E., Vlasenko, M.P., Vlasenko, L.S., Laiho, R., Niinikoski, T.O.: Spin-dependent recombination electron paramagnetic resonance spectroscopy of defects in irradiated silicon detectors. J. Appl. Phys. **93**, 9659–9664 (2003)
67. Vlasenko, L.S.: Effects of spin dependent recombination and EPR spectroscopy of the excited triplet states of point defects in silicon. Appl. Magn. Reson. **47**, 813–822 (2016)
68. Franke, D.P., Otsuka, M., Matsuoka, T., Vlasenko, L.S., Vlasenko, M.P., Brandt, M.S., Itoh, K.M.: Spin-dependent recombination at arsenic donors in ion-implanted silicon. Appl. Phys. Lett. **105**, 112111 (2014)

69. Akhtar, W., Morishita, H., Sawano, K., Shiraki, Y., Vlasenko, L.S., Itoh, K.M.: Electrical detection of cross relaxation between electron spins of phosphorus and oxygen-vacancy centers in silicon. Phys. Rev. B **84**, 045204 (2011)
70. McCamey, D.R., Huebl, H., Brandt, M.S., Hutchison, W.D., McCallum, J.C., Clark, R.G., Hamilton, A.R.: Electrically detected magnetic resonance in ion-implanted Si: P nanostructures. Appl. Phys. Lett. **89**, 182115 (2006)
71. Morishita, H., Abe, E., Akhtar, W., Vlasenko, L.S., Fujimoto, A., Sawano, K., Shiraki, Ya., Dreher, L., Riemann, H., Abrosimov, N.V., Becker, P., Pohl, H.-J., Thewalt, M.L.W., Brandt, M.S., Itoh, K.M.: Linewidth of low-field electrically detected magnetic resonance of phosphorus in isotopically controlled silicon. Appl. Phys. Express **4**, 021302 (2011)
72. Hrubesch, F., Braunbeck, G., Voss, A., Stutzmann, M., Brandt, M.S.: Broadband electrically detected magnetic resonance using adiabatic pulses. J. Magn. Reson. **254**, 62 (2015)
73. Boehme, C., Lips, K.: Theory of time-domain measurement of spin-dependent recombination with pulsed electrically detected magnetic resonance. Phys. Rev. B **68**, 245105 (2003)
74. Elzerman, J.M., Hanson, R., Willems van Beveren, L.H., Witkamp, B., Vandersypen, L.M. K., Kouwenhoven, L.P.: Single-shot read-out of an individual electron spin in a quantum dot. Nature **430**, 431–435 (2004)
75. Xiao, M., Martin, I., Yablonovitch, E., Jiang, H.W.: Electrical detection of the spin resonance of a single electron in a silicon field-effect transistor. Nature **430**, 435–439 (2004)
76. Morishita, H., Vlasenko, L.S., Tanaka, H., Semba, K., Sawano, K., Shiraki, Y., Eto, M., Itoh, K.M.: Electrical detection and magnetic-field control of spin states in phosphorus-doped silicon. Phys. Rev. B **80**, 205206 (2009)
77. Akhtar, W., Filidou, V., Sekiguchi, T., Kawakami, E., Itahashi, T., Vlasenko, L.S., Morton, J.J.L., Itoh, K.M.: Coherent storage of photoexcited triplet states using ^{29}Si nuclear spins in silicon. Phys. Rev. Lett. **108**, 097601 (2012)
78. Matsuoka, T., Vlasenko, L.S., Vlasenko, M.P., Sekiguchi, T., Itoh, K.M.: Identification of a paramagnetic recombination center in silicon/silicondioxide interface. Appl. Phys. Lett. **100**, 152107 (2012)
79. Mortemouque, P.A., Sekiguchi, T., Culan, C., Vlasenko, M.P., Elliman, R.G., Vlasenko, L. S., Itoh, K.M.: Spin dependent recombination based magnetic resonance spectroscopy of bismuth donor spins in silicon at low magnetic fields. Appl. Phys. Lett. **101**, 082409 (2012)
80. Aichinger, T., Lenahan, P.M., Tuttle, B.R., Peters, D.: A nitrogen-related deep level defect in ion implanted 4H-SiC pn junctions—A spin dependent recombination study. Appl. Phys. Lett. **100**, 112113 (2012)
81. Cottom, J., Gruber, G., Hadley, P., Koch, M., Pobegen, G., Aichinger, T., Shluger, A.: Recombination centers in 4H-SiC investigated by electrically detected magnetic resonance and ab initio modeling. J. Appl. Phys. **119**, 181507 (2016)
82. Bagraev, N.T., Gets, D.S., Kalabukhova, E.N., Klyachkin, L.E., Malyarenko, A.M., Mashkov, V.A., Savchenko, D.V., Shanina, B.D.: Electrically detected electron paramagnetic resonance of point centers in 6H-SiC nanostructures. Semiconductors **48**, 1467–1480 (2014)
83. Dresselhaus, G., Kip, A.F., Kittel, C.: Cyclotron resonance of electrons and holes in silicon and germanium crystals. Phys. Rev. **98**, 368–384 (1955)
84. Drachenko, O., Helm, M.: Cyclotron resonance spectroscopy. In: Patane, A., Balkan, N. (eds.) Semiconductor Research, vol. 150, pp. 283–307. Springer, Berlin, Heidelberg (2012)
85. Baranov, P.G., Veshchunov, YuP, Zitnikov, R.A., Romanov, N.G., Schreter, Yu.G.: Optical detection of microwave resonance in germanium by means of luminescence of electron hole drops. JETP Lett. **26**, 249–252 (1977)
86. Romestain, R., Weisbuch, C.: Optical detection of cyclotron resonance in semiconductors. Phys. Rev. Lett. **45**, 2067–2070 (1980)
87. Johnson, G.R., Cavenett, B.C., Kerr, T.M., Kirby, P.B., Wood, C.E.C.: Optical, Hall and cyclotron resonance measurements of GaSb grown by molecular beam epitaxy. Semicond. Sci. Technol. **3**, 1157–1165 (1988)

88. Godlewski, M., Chen, W., Monemar, B.: Optical detection of cyclotron resonance for characterization of recombination processes in semiconductors. Crit. Rev. Solid State Mater. Sci. **19**, 241–301 (1994)
89. Booth, I.J., Schwerdtfeger, C.F.: Optical detection of cyclotron resonance in GaP and ZnTe. Solid State Commun. **55**, 817–822 (1985)
90. Pakulis, E.J., Northrop, G.A.: Optically detected electron cyclotron resonance in silicon. Appl. Phys. Lett. **50**, 1672–1674 (1987)
91. Cavenett, B.C.: Optically detected magnetic resonance (ODMR) investigations of recombination processes in semiconductors. Adv. Phys. **30**, 475–538 (1981)
92. Cavenett, B.C., Pakulis, E.J.: Optically detected cyclotron resonance in a GaAs/Ga0.67A10.33As superlattice. Phys. Rev. B **32**, 8449–8451 (1985)
93. Moll, A., Wetzel, C., Meyer, B.K., Omling, P., Scholz, F.: Microwave and far-infrared induced optically detected cyclotron resonance in epitaxial InP and GaAs. Phys. Rev. B **45**, 1504–1506 (1992)
94. Michels, J.G., Warburton, R.J., Nicholas, R.J., Stanley, C.R.: An optically detected cyclotron resonance study of bulk GaAs. Semicond. Sci. Technol. **9**, 198–206 (1994)
95. Son, N.T., Kordina, O., Konstantinov, A.O., Chen, W.M., Sörman, E., Monemar, B., Janzén, E.: Electron effective masses and mobilities in high-purity 6H-SiC chemical vapor deposition layers. Appl. Phys. Lett. **65**, 3209–3211 (1994)
96. Son, N.T., Chen, W.M., Kordina, O., Konstantinov, A.O., Monemar, B., Janzén, E., Hofman, D.M., Volm, D., Drechsler, M., Meyer, B.K.: Electron effective masses in 4H SiC. Appl. Phys. Lett. **66**, 1074–1076 (1995)
97. Volm, D., Meyer, B.K., Hofmann, D.M., Chen, W.M., Son, N.T., Persson, C., Lindefelt, U., Kordina, O., Sörman, E., Konstantinov, A.O., Monemar, B., Janzén, E.: Determination of the electron effective-mass tensor in 4H SiC. Phys. Rev. B **53**, 15409–15412 (1996)
98. Chen, W.M., Son, N.T., Janzén, E., Hofmann, D.M., Meyer, B.K.: Effective masses in SiC determined by cyclotron resonance experiments. Phys. Status Solidi A **162**, 79–93 (1997)
99. Son, N.T., Sörman, E., Chen, W.M., Bergman, J.P., Hallin, C., Kordina, O., Konstantinov, A.O., Monemar, B., Janzén, E., Hofmann, D.M., Volm, D., Meyer, B.K.: Effects of microwave fields on recombination processes in 4H and 6H SiC. J. Appl. Phys. **81**, 1929–1932 (1997)
100. Meyer, B.K., Hofmann, D.M., Volm, D., Chen, W.M., Son, N.T., Janzén, E.: Optically detected cyclotron resonance investigations on 4H and 6H SiC: Band-structure and transport properties. Phys. Rev. B **61**, 4844–4849 (2000)
101. Son, N.T., Hai, P.N., Chen, W.M., Hallin, C., Monemar, B., Janzén, E.: Hole effective masses in 4H SiC. Phys. Rev. B **61**, R10544–R10546 (2000)
102. Wright, M.G., Kanáah, A., Cavenett, B.C., Johnson, G.R., Davey, S.T.: Optically detected cyclotron resonance measurements in Al0.48In0.52As MBE layers. Semicond. Sci. Technol. **4**, 590–592 (1989)
103. Hofmann, D.M., Drechsler, M., Wetzel, C., Meyer, B.K., Hirler, F., Strenz, R., Abstreiter, G., Böhm, G., Weimann, G.: Optically detected cyclotron resonance on GaAs/$Al_xGa_{1-x}As$ quantum wells and quantum wires. Phys. Rev. B **52**, 11313–11318 (1995)
104. Ashkinadze, B.M., Belkov, V.V., Krasinskaya, A.G.: Effects of hot-electrons on the luminescence of GaAs. Sov. Phys. Semicond. **24**, 555 (1990)
105. Godlewski, M., Prone, K., Gajewska, M., Chen, W.M., Monemar, B.: Optically detected impact ionization related chaotic oscillations in GaInAs. Phys. Rev. B **44**, 8357–8360 (1991)
106. Ashkinadze, B.M., Cohen, E., Ron, A., Pfeiffer, L.: Microwave modulation of exciton luminescence in GaAs/$Al_xGa_{1-x}As$ quantum wells. Phys. Rev. B **47**, 10613–10618 (1993)
107. Wang, F.P., Monemar, B., Ahlstroem, M.: Mechanisms for the optically detected magnetic resonance background signal in epitaxial GaAs. Phys. Rev. B **39**, 11195–11198 (1989)
108. Weman, H., Godlewski, M., Monemar, B.: Optical detection of microwave-induced impact ionization of bound excitons in silicon. Phys. Rev. B **38**, 12525–12530 (1988)
109. Dedulewicz, D., Godlewski, M.: Carrier heating efficiency in optically detected cyclotron resonance experiment. Acta Phys. Pol. A **84**, 535–537 (1993)

110. Delong, M.C., Viohl, I., Ohlsen, W.D., Taylor, P.C., Olson, J.M.: Microwave thermal modulation of photoluminescence in III-V semiconductors. Phys. Rev. B **43**, 1510–1519 (1991)
111. Gubarev, S.I., Dremin, A.A., Kukushkin, I.V., Malyavkin, A.V., Tyazhlov, M.G., von Klitzing, K.: Optical detection of cyclotron resonance at a GaAs-GaAlAs heterojunction. JETP Lett. **54**, 355–359 (1991)
112. Ahmed, N., Agool, I.R., Wright, M.G., Mitchell, K., Koohian, A., Adams, S.J.A., Pidgeon, C.R., Cavenett, P.C., Stanley, C.R., Kean, A.H.: Far-infrared optically detected cyclotron resonance in GaAs layers and low-dimensional structures. Semicond. Sci. Technol. **7**, 357–363 (1992)
113. Warburton, R.J., Michels, J.G., Nicholas, R.J., Harris, J.J., Foxon, C.T.: Optically detected cyclotron resonance of GaAs quantum wells: effective-mass measurements and offset effects. Phys. Rev. B **46**, 13394–13399 (1992)
114. Nickel, H.A., Kioseoglou, G., Yeo, T., Cheong, H.D., Petrou, A., McCombe, B.D., Broido, D., Bajaj, K.K., Lewis, R.A.: Internal transitions of confined neutral magnetoexcitons in $GaAs/Al_xGa_{1-x}As$ quantum wells. Phys. Rev. B **62**, 2773–2779 (2000)
115. Nickel, H.A., Yeo, T.M., Dzyubenko, A.B., McCombe, B.D., Petrou, A., Sivachenko, AYu., Schaff, W., Umansky, V.: Internal transitions of negatively charged magnetoexcitons and many body effects in a two-dimensional electron gas. Phys. Rev. Lett. **88**, 056801 (2002)
116. Meining, C.J., Nickel, H.A., Dzyubenko, A.B., Petrou, A., Furis, M., Yakovlev, D.R., McCombe, B.D.: Many body effects and internal transitions of confined excitons in GaAs and CdTe quantum wells. Solid State Commun. **127**, 821–827 (2003)
117. Michels, J., Warburton, R., Nicholas, R., Harris, J., Foxon, C.: Optically detected cyclotron resonance of GaAs quantum wells: effective mass measurements and offset effects. Phys. B: Condens. Matter **184**, 159–163 (1993)
118. Wetzel, C., Meyer, B.K., Omling, P.: Electron effective mass in direct-band-gap $GaAs_{1-x}P_x$ alloys. Phys. Rev. B **47**, 15588–15592 (1993)
119. Hopkins, M.A., Nicholas, R.J., Brummell, M.A., Harris, J.J., Foxon, C.T.: Cyclotron-resonance study of nonparabolicity and screening in $GaAs$-$Ga_{1-x}Al_xAs$ heterojunctions. Phys. Rev. B **36**, 4789–4795 (1987)
120. Chou, M.J., Tsui, D.C., Weimann, G.: Cyclotron resonance of high mobility two-dimensional electrons at extremely low densities. Phys. Rev. B **37**, 848–854 (1988)
121. Herold, G.S., Nickel, H.A., Tischler, J.G., Weinstein, B.A., Mc-Combe, B.D.: Full-spectrum optically detected resonance (ODR) spectroscopy of GaAs/AlGaAs quantum wells. Phys. E: Low Dimens. Syst. Nanostruct. **2**, 39–43 (1998)
122. Nickel, H., Herold, G., Salib, M., Kioseoglou, G., Petrou, A., McCombe, B., Broido, D.: Internal transitions of excitons and hole cyclotron resonance in undoped GaAs/AlGaAs quantum wells by optically detected resonance spectroscopy. Physica B: Condens. Matter **249–251**, 598–602 (1998)
123. Nickel, H.A., Yeo, T., Meining, C.J., Yakovlev, D.R., Furis, M., Dzyubenko, A.B., McCombe, B.D., Petrou, A.: Interaction of an electron gas with photoexcited electron-hole pairs in modulation-doped GaAs and CdTe quantum wells. Phys. E: Low Dimens. Syst. Nanostruct. **12**, 499–502 (2002)
124. Hopkins, M.A., Nicholas, R.J., Barnes, D.J., Brummell, M.A., Harris, J.J., Foxon, C.T.: Temperature dependence of the cyclotron-resonance linewidth in $GaAs$-$Ga_{1-x}Al_xAs$ heterojunctions. Phys. Rev. B **39**, 13302–13309 (1989)
125. Dai, Y.T., Chang, Y.H., Lee, T.F., Chen, Y.F., Fang, F.F., Wang, W.I.: Optically detected cyclotron resonance studies of multisubband $In_{0.52}Al_{0.48}As/In_{0.53}Ga_{0.47}As$ quantum wells. J. Phys. D Appl. Phys. **29**, 3089–3095 (1996)
126. Chen, Y.F., Shen, J.L., Dai, Y.D., Fang, F.F.: Observation of spin-splitting crossing between subbands in the optically detected cyclotron-resonance spectra of $In_{0.53}Ga_{0.47}As/In_{0.52}Al_{0.48}As$ heterojunctions. Phys. Rev. B **52**, 4692–4695 (1995)

127. Chen, Y.F., Shen, J.L., Dai, Y.D., Jan, G.J., Lin, H.H.: Study of InAlAs/InGaAs heterojunction bipolar transistor layers by optically detected cyclotron resonance. Phys. Lett. **66**, 2543–2545 (1995)
128. Kono, J., Lee, S.T., Salib, M.S., Herold, G.S., Petrou, A., McCombe, B.D.: Optically detected far-infrared resonances in doped GaAs quantum wells. Phys. Rev. B **52**, R8654–R8657 (1995)
129. Meyer, B.K., Drechsler, M., Wetzel, C., Linke, H., Omling, P., Sobkowicz, P.: Composition dependence of the in-plane effective mass in lattice-mismatched, strained Ga1-xInxAs/InP single quantum wells. Appl. Phys. Lett. **63**, 657–659 (1993)
130. Wetzel, C., Efros, A.L., Moll, A., Meyer, B.K., Omling, P., Sobkowicz, P.: Dependence on quantum confinement of the in-plane effective mass in $Ga_{0.47}In_{0.53}As$/InP quantum wells. Phys. Rev. B **45**, 14052–14056 (1992)
131. Murdin, B.N., Hollingworth, A.R., Barker, J.A., Clarke, D.G., Findlay, P.C., Pidgeon, C.R., Wells, J.-P.R., Bradley, I.V., Malik, S., Murray, R.: Double-resonance spectroscopy of InAs/GaAs self-assembled quantum dots. Phys. Rev. B **62**, R7755–R7758 (2000)
132. Child, R.A., Nicholas, R.J., Mason, N.J., Shields, P.A., Wells, J.-P.R., Bradley, I.V., Phillips, J., Murdin, B.N.: Far-infrared modulated photoluminescence spectroscopy of InSb/GaSb quantum dot structures. Phys. Rev. B **68**, 165307 (2003)
133. Booth, I.J., Schwerdtfeger, C.F.: Optically detected cyclotron resonance in AgBr. Phys. Status Solidi B **130**(749–756), 749–756 (1985)
134. Dremin, A.A., Yakovlev, D.R., Sirenko, A.A., Gubarev, S.I., Shabelsky, O.P., Waag, A., Bayer, M.: Electron cyclotron mass in undoped CdTe/CdMnTe quantum wells. Phys. Rev. B **72**, 195337 (2005)
135. Bimberg, D., Grundmann, M., Ledentsov, N.N.: Quantum Dot Heterostructures. Wiley, Chichester (1998)
136. Ledentsov, N.N., Ustinov, V.M., Shchukin, V.A., Kop'ev, P.S., Alferov, Zh.I., Bimberg, D.: Quantum dot heterostructures: fabrication, properties, lasers (Review). Semiconductors **32**, 343–365 (1998)
137. Ledentsov, N.N., Shchukin, V.A., Grundmann, M., Kirstaedter, N., Bohrer, J., Schmidt, O., Bimberg, D., Ustinov, V.M., Egorov, A.Yu., Zhukov, A.E., Kop'ev, P.S., Zaitsev, S.V., Gordeev, N.Yu., Alferov, Zh.I., Borovkov, A.I., Kosogov, A.O., Ruvimov, S.S., Werner, P., Gosele, U., Heydenrech, J.: Direct formation of vertically coupled quantum dots in Stranski-Krastanow growth. Phys. Rev. B **54**, 8743–8750 (1996)
138. Egorov, A.Yu., Bernklau, D., Livshits, D., Ustinov, V., Alferov, Zh.I., Riechert, H.: High power CW operation of InGaAsN lasers at 1.3 μm. Electron. Lett. **35**, 1643–1644 (1999)
139. Baranov, P.G., Romanov, N.G., Preobrazhenski, V.L., Egorov, A.Yu., Ustinov, V.M., Sobolev, M.M.: Optically-detected microwave resonance in InGaAsN/GaAs quantum wells and InAs/GaAs quantum dots emitting around 1.3 μm. Int. J. Nanosci. **2**, 469–478 (2003)
140. Zurauskiene, N., Janssen, G., Goovaerts, E., Bouwen, A., Schoemaker, D., Koenraad, P.M., Wolter, J.H.: Optically detected microwave resonance at 95 GHz of exciton states in InAs/GaAs quantum dots. Phys. Status Solidi B **224**, 551–554 (2001)
141. Janssen, G., Goovaerts, E., Bouwen, A., Partoens, B., Van Daele, B., Zurauskiene, N., Koenraad, P.M., Wolter, J.H.: Observation of cyclotron resonance in an InAs/GaAs wetting layer with shallowly formed quantum dots. Phys. Rev. B **68**, 045329 (2003)
142. Bagraev, N.T., Gets, D.S., Danilovsky, E.Y., Klyachkin, L.E., Malyarenko, A.M.: On the electrically detected cyclotron resonance of holes in silicon nanostructures. Semiconductors **47**, 525–531 (2013)
143. Bagraev, N.T., Mashkov, V.A., Danilovsky, E.Yu., Gehlhoff, W., Gets, D.S., Klyachkin, L. E., Kudryavtsev, A.A., Kuzmin, R.V., Malyarenko, A.M., Romanov, V.V.: EDESR and ODMR of impurity centers in nanostructures inserted in silicon microcavities. Appl. Magn. Reson. **39**, 113–135 (2010)

Chapter 3
Retrospectives: Magnetic Resonance Studies of Intrinsic Defects in Semiconductors

3.1 Introduction

Point defects of intrinsic nature in semiconductors influence the electronic and optical properties of the main technologically important semiconductors such as diamond, silicon, silicon carbide and the 3–5 compounds. The most simple defects are vacancies and interstitials in elemental semiconductors as well as antisite defects in the compound materials. Often, these defects can form complexes at elevated temperatures at which they become mobile or even at lower temperature at irradiation conditions which might equally induce an thermal mobility. Intrinsic defects are present in many as-grown epitaxial layers and bulk samples due to the difficulty in realizing "perfect" growth conditions. They can also purposely be introduced by non stoichiometric growth conditions; well known examples are the arsenic antisite defects in GaAs [1–3] which allow the growth of high resistive bulk substrates or the carbon vacancy defect in silicon carbide [4] which allows also the growth of high resistive material due to its deep level character.

Exposure to particle irradiation introduces equally intrinsic point defects and degrades the electronic properties of microelectronic devices. The associated radiation hardness of the different semiconductors is material dependent and depends on the nature of the irradiation damage (X-rays, γ-rays, electrons, protons, alpha particles or swift heavy ions) to cite the most common cases and the thresholds for the displacement of lattice atoms; further, annealing stages and Fermi level position in the semiconductors are important parameters which will mediate the radiation damage.

In addition to the modification of the electronic properties (conductivity changes), or optical properties (introduction of sub bandgap absorption bands) individual intrinsic defects can also be used selectively as qubits in quantum systems for future use in advanced information and communication technologies. Important

---

The original version of this chapter was revised: Incorrect cross-citations of the references have been corrected. The correction to this chapter is available at https://doi.org/10.1007/978-3-7091-1157-4_7

P.G. Baranov et al., *Magnetic Resonance of Semiconductors and Their Nanostructures*, Springer Series in Materials Science 253,
DOI 10.1007/978-3-7091-1157-4_3

examples are the NV centre in diamond [5] or more recently the divacancy centres in SiC [6]. This is a new and rapidly developing subject since early 2013.

Most of these intrinsic defects can be studied by the electron paramagnetic resonance technique. Both "classical" microwave absorption techniques (EPR, ENDOR) and optically detected magnetic resonance (ODMR, ODENDOR, MCDA) have proven to be useful for their study. In fact, EPR and to a less extent ODMR are the main techniques for the identification of the microscopic and electronic structure of intrinsic point defects via their spin Hamiltonian parameters: the electron spin S, the g-tensor, the crystal field parameters and the hyperfine and superhyperfine interactions.

In the following we will review some of these defects in the main semiconductor materials C, Si, SiC, GaAs and GaN. Of particular importance are also intrinsic defects at interfaces such as the Si dangling bond centres at the Si/dielectric interfaces [7]. Their control is a prerequisite for many semiconductor devices based on the FET structure. It is instructive to follow the properties of these defects in different semiconductors and we have chosen this approach in their presentation.

The modeling of the spin Hamiltonian parameters is of prime importance for defect identification as "intuitive" models have often lead to wrong assignments in materials with complex hyperfine interactions [8, 9]. Modern calculation tools are now at hand for the modeling not only of HF interactions but also of g-tensors [10] and fine structure splitting [11].

The labeling of the intrinsic centres is often confusing as in the initial stage of their study they were labeled by letters A-center, E-center, later according to the laboratory where these studies were performed: G for General Electrics, NL for the Ammerlaan laboratory in Amsterdam, Ky for the Institute of Semiconductors in Kiev etc.

The intrinsic defects in bulk semiconductors and epitaxial layers have been studies by various magnetic resonance techniques. Whereas mostly "classical" CW EPR at X-, K- and Q-band have been applied, optically detection via PL–ODMR and MCDA have equally shown to be useful. Whenever possible, ENDOR has equally been used both in connection with CW EPR and optical detection; the results obtained have contributed dominantly to the defect identification. Various authors have reviewed these techniques. For a recent review see, for example [12].

In the case of nanometer sized samples optical detection of the magnetic resonance of intrinsic defects is the appropriate approach as CW EPR lacks the required sensitivity. Recently, ODMR has been applied with great success in the case of NV centres (carbon vacancy-nitrogen donor associates) in nm sized individual diamond particles [13].

3.2 Experimental Results

3.2.1 Vacancy Defects

3.2.1.1 Diamond

In diamond the ground states for the vacancy predicted in a many electron LCAO model are respectively V^+: $^2T_2$ (S = 1/2), V^0: $^1E_1$ (S = 0) and V^-: $^4A_2$ (S = 3/2). The optical and magnetic properties of the neutral and negatively charged monovavcancy have been studied in detail (Table 3.1). In the ground state V^0 is diamagnetic whereas V^- is paramagnetic with S = 3/2 (Fig. 3.1). V^0 has however an excited paramagnetic state $^5A_2$, which has been studied by EPR and ENDOR [14] (van Wyk et al., PRB52, 12657 (1995)). The positively charged vacancy has not yet been observed. In the T_d symmetry of the diamond lattice the distinction between S = 1/2 and S > 1/2 states is not possible from the EPR measurement alone as in the absence of zero field splittings a single line is observed in all cases. The identification of the S = 3/2 state of the V-related EPR spectrum has been obtained from ENDOR measurements as the ENDOR transitions depend on the M_S value of spin state [15]. The EPR spectrum of V^- has initially been attributed by Baldwin in 1963 [16] to the "+" charge state of V_C (S1 centre).

The spin Hamiltonian parameters of V^- are presented in Table 3.1.

Table 3.1 EPR parameters of the negatively charged vacancy in diamond

Label	Model	Point symmetry	g-tensor	CHF (10^{-4} cm^{-1})	LHF (10^{-4} cm^{-1})	T (K)
S1 [16]	V_C^-S = 3/2	T_d	2.0023 isotropic	$A_\parallel$ = 47.2 $A_\perp$ = 27.3	$A_\parallel$ = 4.44 $A_\perp$ = 3.1	300
[15]	V_C^-S = 3/2	T_d		$A_\parallel$ = 47.2 $A_\perp$ = 27.3	$A_\parallel$ = 4.5 $A_\perp$ = 3.2	77
				$A_\parallel$ = 47.3 $A_\perp$ = 27.3	A_1 = 4.48 A_2 = 3.14 A_3 = 3.08	4
NV	$(V_C\text{-}N_C)^-$ S = 1D = 961 × 10^{-4} cm^{-1}	C_{3V}	2.0028	$A_{//}(^{13}C)$ = 66 $A_\perp(^{13}C)$ = 40	$A(^{14}N)$ = 0.7	300

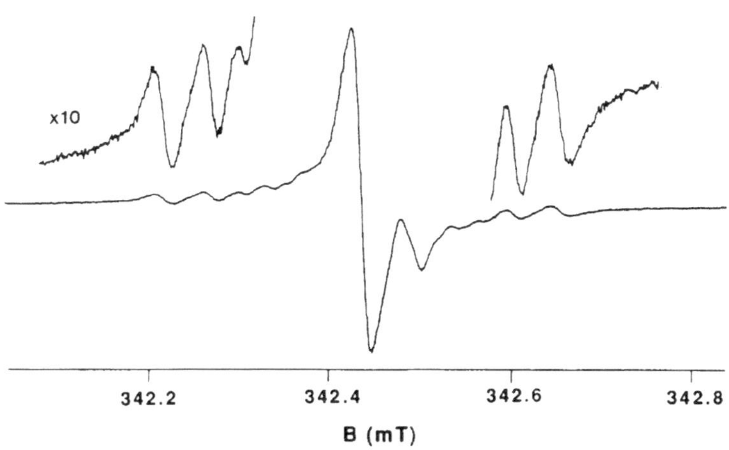

Fig. 3.1 X-band EPR spectrum at 4 K of the V_C^- centre in diamond showing the HF interaction with the 12 NNN C atoms, B II [110]; after [15]

The level positions of the vacancy have also been determined: −/0 at $E_C - 2.5$ eV, 0/+ at $E_V + 0.6$ eV [17]. The closely related defect, a Frenkel pair V_C-C_i, which can be generated by electron irradiation at energies close to the displacement threshold, has not been observed up to now.

During thermal annealing at 850 °C the V_C become mobile and form associates with N_C donor atoms to form the so-called NV center which in its negative charge state has found various applications. The ZPL of the intracenter transition $^3A_2 \rightarrow {}^3E$ of the NV centre is situated in the red (1.945 eV) and can be observed up to room temperature. It is characterized by intense phonon sidebands (intensity of the zero phonon line is only 4%) due to strong coupling to phonons (see Chap. 6). Early ensemble studies show the strong inhomogeneous broadening of optical transitions with a linewidth of 1000 GHz for the zero phonon line [18] (Redman, Brown, and Rand, 1992). Nevertheless the intensity of the ZPL has been shown to be sufficient to allow single defect spectroscopy by ODMR techniques.

3.2.1.2 Silicon

The vacancy defects in silicon are one of the first defects having been studied by EPR in semiconductors. They can be conveniently introduced in monocrystalline bulk samples by high energy electron irradiation at low temperature. Electrons with kinetic energies of >100 keV are sufficient to transmit in elastic collisions sufficient energy to the lattice atoms to displace them from their lattice sites. The primary defects generated are Frenkel pairs (close vacancy-interstitial complexes) or separated vacancies and interstitial defects depending on the energy transmitted. The particularity of silicon is the existence of low temperature annealing stages, which for the generally applied room temperature irradiation leads to the formation of more complex associated defects. The interaction of the primary radiation defects (V_{Si}, Si_i) with impurity or dopant related centres is thus of great importance in irradiation studies of Si. The study of isolated monovacancies requires special setups allowing combined low temperature irradiation and EPR measurements. Watkins and Corbett, who published the pioneering results on this subject, have used this approach in the early studies [19–21].

The monovacancy in Si (V) can exist in four different charge states (Fig. 3.2). Which charge state is observed in an EPR experiment depends on the Fermi level position; of the four possible charge states, "2+", "+", "0", two are paramagnetic "+", "−" and can thus be studied by EPR.

In Table 3.2 we present the spin Hamiltonian parameters of these centres.

For V^+ the HF interaction tensor is oriented close to the [111] directions with a 7° tilt towards the [100]. Thus, each EPR line is accompanied by four ^{29}Si (I = 1/2) HF doublets. In spite of several attempts no ENDOR spectra could be obtained for this defect. The evaluation of the HF tensor in a simple, dangling bond model predicts values of $A_{\parallel} = 49 \times 10^{-4}$ cm^{-1} and $A_{\perp} = 39 \times 10^{-4}$ cm^{-1}in fair agreement with the experimental findings. In the case of the negatively charged vacancy the electron is localized only on two of the four nearest Si neighbours in agreement

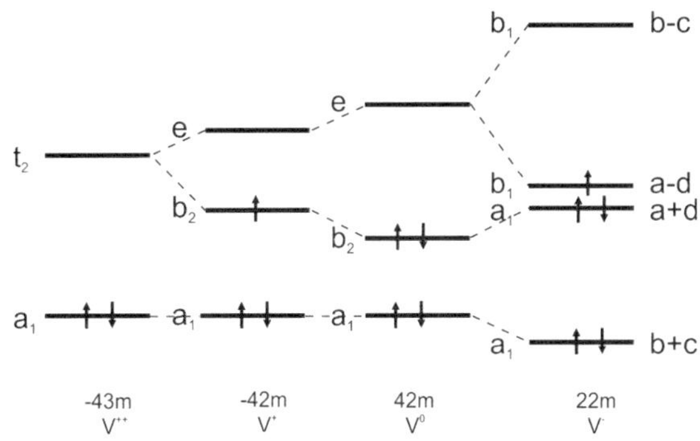

Fig. 3.2 Energy levels of the vacancy in a one electron molecular orbital treatment. The various symmetry-lowering Jahn-Teller distortions are indicated, after M. Sprenger et al. (1987) [22]

Table 3.2 EPR parameters of the positively and negatively charged vacancies in silicon

Label	Model	Point symmetry	g_1	g-tensor g_2	g_3	Annealing T (K)
G1 [20]	V^+ S = 1/2	42 mD_{2d}	2.0087 [100]	1.9989	1.9989	150
				CHF 10^{-4} cm^{-1}		
			A_1 = 43.9 [111]	A_2 = 29.8	A_3 = 29.8	
G2 [19]	V^- S = 1/2	2 mm	2.0151	2.0028	2.0038	60 K
				CHF 10^{-4} cm^{-1}		
			$A_{\parallel}$ = 44.4 [111]	A_2 = 37.0	A_3 = 37.3	

with the bond formation predicted by the LCAO models. For the 1-charge state extensive EPR (Fig. 3.3) and ENDOR measurements have been performed [22].

Interestingly the vacancy in Si behaves as a negative U centre with the 1+ charge state being a metastable state. The +/2+ state is at Ev + 0.13 eV and the 0/+ charge transition level at Ev + 0.05 eV.

In Fig. 3.2 the level scheme in the one electron molecular orbital scheme is given. The splitting of the t_2 level is due to several charge state dependent Jahn-Teller distortions.

3.2.1.3 Silicon Carbide

Among the many polytypes of silicon carbide only three are technologically relevant and have been studied by EPR. They are the cubic 3C and the hexagonal 4H and 6H polytypes. The 3C polytype is the closest relative to diamond and Si from a structural point of view. Whereas in 3C-SiC all Si or C lattice sites are equivalent this is no longer the case for 4H and 6H polytypes where we have respectively one so-called quasi cubic site (k) and one hexagonal site (h) in 4H or two quasi cubic

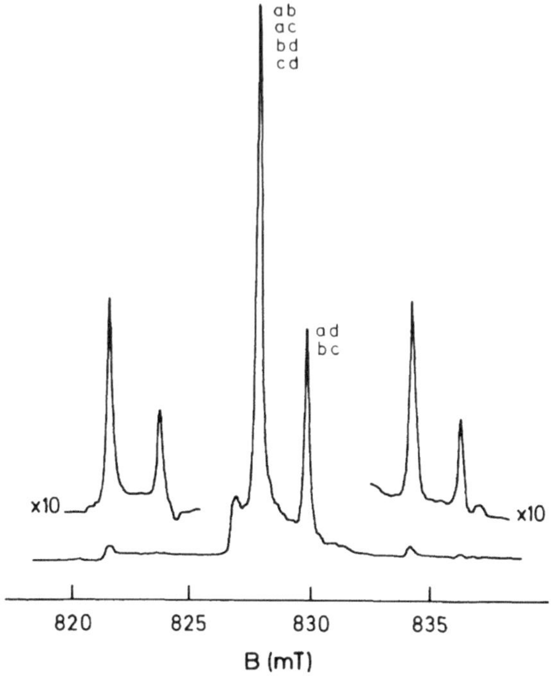

Fig. 3.3 K-band EPR spectrum of the negatively charged silicon vacancy ($V_{Si}-$) for B || [100]; the HF lines due to interaction with two nearest neighbour (NN) Si atoms are directly resolved; after [22]

sites (k1, k2) and one hexagonal site (h) in 6H. In principle, defects situated on k and h sites have different electronic properties and should and do show distinctive defect signatures (Table 3.3).

For the 4H polytype n-type, p-type and semi-insulating bulk materials are commercially available. 6H-SiC is "naturally" n-type conductive due to nitrogen contamination but can be rendered p-type by Al co-doping. Bulk 3C-SiC is only available as n-type material.

Both silicon and carbon vacancies defects can be easily introduced by particle irradiation and have been studied by EPR. Due to the large band gap of SiC different charge states are possible for each defect.

Silicon Vacancy. In SiC the following paramagnetic charge states of the silicon monovacancy can be expected: V_{Si}^{+}: S = 1/2, V_{Si}^{0}: S = 1 or S = 0, V_{Si}^{-}: S = 3/2 and V_{Si}^{2-}: S = 1. Whereas the negatively charged V_{Si}^{-} has been clearly observed in many EPR studies of irradiated SiC the identification of the other charge states has been widely and controversially disputed.

In initially n-type SiC the negatively charged Si vacancy is one of the most prominent defects after high-energy particle irradiation. Contrary to the case of Si it has a high spin S = 3/2 $^4A_1$ ground state; in the case of 3C-SiC [23] its EPR spectrum should not show any zero field splitting contrary to the hexagonal polytypes 4H and 6H, where a zero filed splitting is a priori expected. But surprisingly, in both 4H [24] and 6H [27] polytypes no zero field splitting was

Table 3.3 EPR parameters of the silicon vacancy in 3C-, 4H- and 6H-SiC

Label	Model	Point symmetry	g-tensor	D	T (K)	Comment
T1 [23]	V_{Si}^- S = 3/2	T_d	2.0028	0	4– 300	Epi n, p on Si e^-, p^+ irradiation
			4H			
V_{Si} [24] V_{Si}(I) [25] V_{Si}(II) [25] [26]	V_{Si}^- V_{Si}^-(h) S = 3/2 V_{Si}^- (k) S = 3/2 V_{Si}^- (k, h) S = 3/2	C_{3V} C_{3V} C_{3V}	2.00342.0028 2.0028 $g_{\parallel}(h) - g_{\parallel}(k) = 0.00004$ $g_{\perp}(h) - g_{\perp}(k) =$ 0.00002	0 0 0	300 300 300	n° Wimbauer et al. Bulk n e^- 3 meV n° irrad. 95 GHz
			6H			
V_{Si} [27] V_{Si}(I) V_{Si}(II) [25]	V_{Si}^- V_{Si}^-(h) V_{Si}^-(k1) V_{Si}^-(k2) S = 3/2	C_{3V} C_{3V} C_{3V} C_{3V}	2.003 2.0028 2.0028	0 0 0	300 300 300	n° Bulk n e^- 3 meV

observed in the early EPR studies of the V_{Si}-defect; less surprisingly, due to the small spin orbit coupling in SiC, an anisotropy and site dependence of the g-tensor could neither be resolved in X-band EPR. In 2003 Mizuochi et al. [25] published a high resolution EPR study of V_{Si}^- in 4H and 6H material which showed the C_{3V} symmetry of this defect from its central hyperfine structure with the four non equivalent nearest C neighbours. They showed further by pulsed EPR [28] in 4H that the central line contains all three transition of the S = 3/2 state and concluded that the zero field splitting is zero, i.e. D = 0. Orlinski et al. [26] applied high frequency EPR (95 GHz) and were able to resolve a weak g-tensor anisotropy for the h lattice site in neutron irradiated 4H. Representative EPR spectra are shown in Figs. 3.4 and 3.5.

The central hyperfine interaction is with 4C nearest neighbours in 3C and 1 axial and 3 basal C nearest neighbours in 4H and 6H (Table 3.4).

The central hyperfine interaction is with 4C nearest neighbours in 3C and 1 axial and 3 basal C nearest neighbours in 4H and 6H (Table 3.4).

There is still some controversy on the zero field splitting of the $V_{Si}-$ centre. In 2009 Janzen et al. [29] claimed that the isolated V_{Si} of S = 3/2 with no zero field splitting does not exist. From room temperature measurements in 4H they claimed that the central line observed in all previous studies is in fact only the $M_S = 1/2 - 1/2$ transition of V_{Si}^-; the additional lines, labeled T_{v1a}, T_{v2a}, previously attributed to a different charge state (V_{Si}^0) of the silicon vacancy, were reinterpreted by these authors as the $M_S = 3/2 - 1/2$ transitions of V_{Si}^-. The zero field splittings of 11.4 and 53.1×10^{-8} eV for these centres deduced previously assuming an S = 1 ground state have in this new model of S = 3/2 to be divided by two. As calculations of the

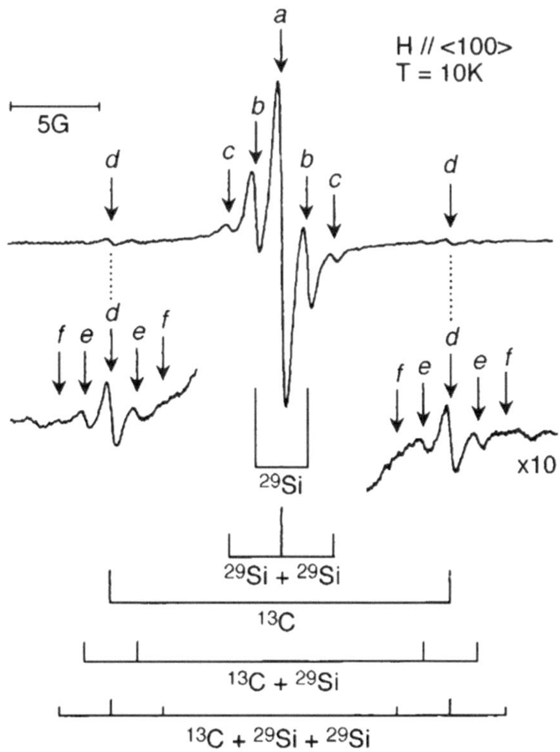

Fig. 3.4 EPR spectrum of the negatively charged silicon vacancy V_{Si}^{-} in 3C-SiC; T = 10 K and B || [100]; the HF interaction with the 4 NN C atoms and the 12 NNN Si atoms are directly resolved; after [8]

expected zero field splitting for the V_{Si} centre in the two charge states 0/− are not available, this issue remained open.

In addition to isolated Si monovacancies slightly distorted vacancies have equally been reported. The case of the so-called T_{v2a} centre is particular, as different authors have assigned it controversially. Initially it has been assigned in 2000 by Sörman et al. [30] based on ODMR studies to a S = 1 excited state of the neutral Si vacancy. Later it has been shown by EPR that it is a groundstate, which can be observed at 4 K without any photoexcitation [26]. In 2002 Mizuochi et al. [28] concluded from pulsed EPR measurements that it is not a S = 1 but a S = 3/2 state and thus should be a $V_{Si}-$ related centre with a distorted local environment. In 2003 Son et al. [31] claimed to have shown on the contrary that it is a S = 1 groundstate. In a different W-band EPR study in 2003 Orlinski et al. [26] claim equally to have confirmed the S = 1 character of the T_{v2a}, T_{v2b} centres and attribute them to the neutral charge state of the V_{Si}. To further complicate the discussion, Son et al. more recently questioned even its assignment to the V_{Si} as it seemed incompatible with the observed high thermal stability (1600 °C).

In p-type 6H-SiC irradiated with electrons of energy close to the displacement threshold for Si atoms the negatively charged V_{Si} with a resolved zero field splitting has been observed (von Bardeleben et al. [32]). The assignment to the V_{Si} centre is

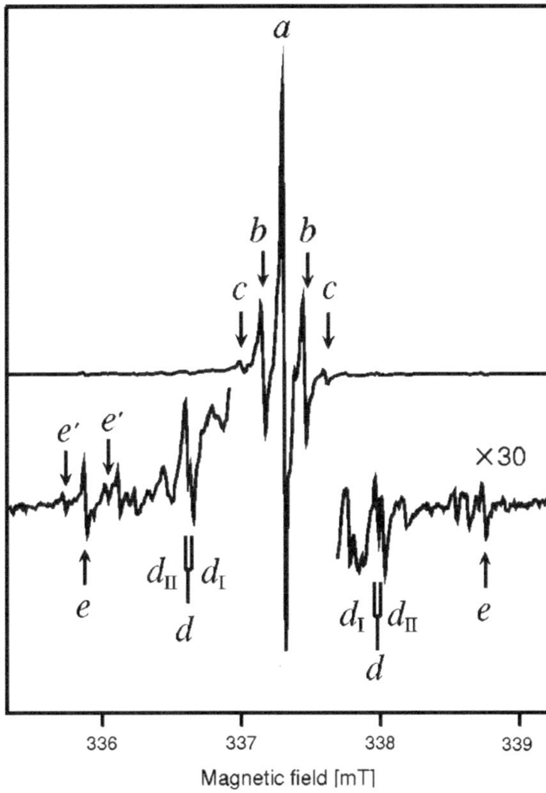

Fig. 3.5 EPR spectrum of the negatively charged silicon vacancy V_{Si}^- in 4H-SiC; T = 300 K and B || [0001]; the HF interaction with the (1 + 3) NN C atoms and the 12 NNN Si atoms are directly resolved; after [25]

Table 3.4 Hyperfine interaction parameters of the negatively charged silicon vacancy on cubic, quasicubic and hexagonal sites in 3C-, 4H-, 6H-SiC

3C A (10^{-4} cm^{-1})	4H A (10^{-4} cm^{-1})	6H A (10^{-4} cm^{-1})
$A_{\parallel}$ = 26.8 [111] $A_{\perp}$ = 11.0	k $A_{\parallel}$ = 26.7 (axial) $A_{\perp}$ = 11.3 A_x = 26.5 A_y = 10.5 A_z = 10.4	k1 $A_{\parallel}$ = 26.7 (axial) $A_{\perp}$ = 10.9 A_x = 25.3 A_y = 9.5 A_z = 9.4
	h $A_{\parallel}$ = 26.8 (axial) $A_{\perp}$ = 11.1 A_x = 25.5 A_y = 9.4 A_z = 9.4	k2 $A_{\parallel}$ = 26.7 (axial) $A_{\perp}$ = 10.9 A_x = 26.8 A_y = 10.5 A_z = 10.5
		h

based on the CHF structure with (1 + 3) nearest neighbour (NN) C atoms and 12 NNN Si atoms (Figs. 3.6 and 3.7). Its S = 3/2 ground state follows directly from the observation of the expected three electronic transitions and the second order angular

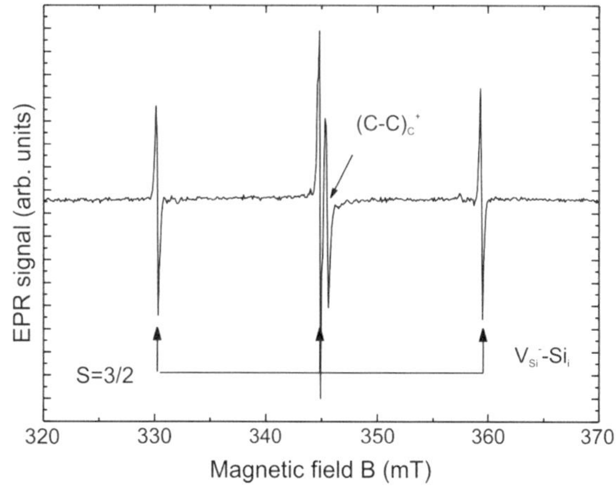

Fig. 3.6 EPR spectrum of p-type electron irradiated 6H-SiC; T = 300 K, B ‖ [0001]; the S = 3/2 three line spectrum has been attributed to the V_{Si}^--Si_i Frenkel pair after [34]

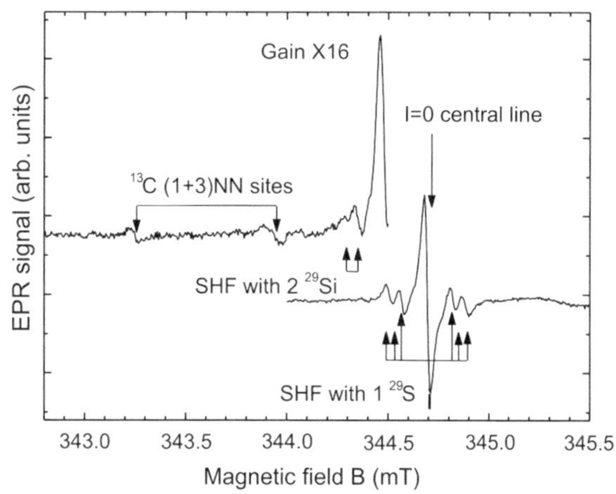

Fig. 3.7 EPR spectrum of the V_{Si}–Si_i centre in 6H-SiC showing the HF interaction with (1 + 3) C and (6 + 6) Si neighbours; after [34]

variation of the central $M_S = -1/2$ <–> $+1/2$ line. For irradiation with slightly higher electron energy an additional V_{Si}^- spectrum with reduced symmetry is observed. Both spectra have been attributed to Frenkel pairs, i.e. close V_{Si}-Si_i associated defects in different configurations (Table 3.5). The observation of the negatively charged Si monovacancy in p-type electron irradiated material is surprising and does not fit with the idea that the stable configuration of the VSi in p-type material is the antisite vacancy pair C_{Si}-V_C [33]. The reason might be an important thermal barrier for this defect transformation $V_{Si} - C_{Si}V_C$.

The issues of spin S = 1 or S = 3/2 and the neutral versus the negative charge state of the silicon vacancy have been largely disputed in the last 10 years. Early theoretical papers had both predicted a spin S = 0 diamagnetic and a spin S = 1

Table 3.5 EPR parameters of the negatively charged silicon vacancy in p-type 6H-SiC

Polytype	Model	Spin	g-tensor	D, E (10^{-4} cm^{-1})	CHF (10^{-4} cm^{-1})
6H [32]	V_{Si}^{-}-Si_i Frenkel pair	3/2	$g_{\parallel c} = 2.0032$ $g_{\perp c} = 2.0028$	D = 68.7 E = 0	$A_{\parallel} = 26.6$ (axial) $A_{\perp} = 16.1$ $A_{\parallel} = 26.7$ (basal) $A_{\perp} = 12.4$
6H [32]	V_{Si}^{-}-Si_i Frenkel pair	3/2	$g_{xx} = 2.0015$ $g_{yy} = 2.0039$ $g_{zz} = 2.0035$	D = 76 E = 19	

paramagnetic ground state for the neutral silicon vacancy [35, 36]; thus the S = 1 model of V_{Si}^{0} could not be excluded a priori. The difficulty in the assignment from standard CW X-band EPR measurements of the S = 1 or S = 3/2 character of a centre in near T_d symmetry with quasi isotropic g-values is not evident; due to the superposition of the EPR lines at the same field corresponding to g = 2.0028 this distinction is rendered very difficult. As the debate is not yet closed it is useful to recall the previously published results of the S = 1 "V_{Si}^{0}" centre obtained by ODMR and EPR (Table 3.6). If this assignment turns out to be incorrect, the zero-field splitting parameters have to be reinterpreted.

No clear results for the remaining two paramagnetic charge states "+" and "2−" of the Si monovacancy have been published (Fig. 3.8).

Silicon Vacancy-Nitrogen Donor Complex. At temperatures where silicon vacancies can diffuse, association with donor impurities can be formed. An important example vacancy-nitrogen donor complex is the NV centre in diamond; the NV center is a nearest neighbor V_C-N_C complex. In 6H-SiC such a defect has been claimed to be observed already many years ago [39]. This defect, the so-called P12 centre, has spin S = 1/2, axial symmetry and shows a weak HF interaction with one nitrogen atom. In the absence of modeling the attribution to a V_{Si}-N_C centre was speculative at that time. More recently, Gerstmann et al. [40], have studied theoretically the electronic structure and formation of such centres in 3C-SiC and found the P12 centre properties compatible with those predicted for the neutral $(V_{Si}\text{-}N_C)^0$ centre. Very recently we [41] have observed in 4H-SiC (Figs. 3.9 and 3.10) and later in 3C, 6H different spin S = 1 centres which based on the excellent agreement with the calculated Spin Hamiltonian parameters could be assigned to of the negatively charged, c-axis aligned V_{Si}-N_C centres. The NV centers in all three SiC polytypes have interesting optical properties with a narrow intracenter ZPL similar to the NV center in diamond which should give rise to numerous applications. A main difference is the reduced energy of the intracenter transitions which in diamond is situated in the visible and is shifted in SiC to the near infrared (1300 nm). A different centre with very different zero field splitting was reported in

Table 3.6 EPR parameters of the silicon vacancy centres in 4H- and 6H-SiC initially attributed to spin S = 1 centres but later shown in 4H-SiC to be S = 3/2 centers; thus the zero field splitting parameter has to be divided by 2

Polytype label	Model	Technique	g-tensor	D 10^{-4} cm^{-1}	T (K)	References	Comment
4H							
T_{v2a}	V_{Si}^{-} (hex) S = 3/2 C_{3V}	ODMR	g_{iso} = 2.004	23.2/2 = 11.6	2	[30]	epi n e− 2.5 meV
T_{v2b}	V_{Si}^{-} (c) S = 3/2 C_{3V}	ODMR	g_{iso} = 2.004	12.1/2 = 6.1	2	[30]	epi n e− 2.5 meV
	V_{Si}^{-} (hex) S = 3/2 C_{3V}	EPR	g_{iso} = 2.0032	22/2 = 11	300	[34]	Bulk n p+ 12 meV
	V_{Si}^{-} (c) S = 3/2 C_{3V}	EPR	g_{iso} = 2.0032	13/2 = 6.5	300	[34]	Bulk n p+ 12 meV
T_{v2a}	V_{Si}^{-} S = 3/2 C_{3V}	Pulsed EPR		11.7	160	[28]	Bulk n e− 3 meV
T_{v2b}	V_{Si}^{-} S = 3/2 C_{3V}	Pulsed EPR		6.67	160	[28]	Bulk n e− 3 meV
6H							
T_{v1a}	V_{Si}^{0} (c1) S = 1	ODMR	g_{iso} = 2.0035	9.2	2	[30]	epi n e− 2.5 meV
T_{v2a}	V_{Si}^{0} (hex) S = 1	ODMR	g_{iso} = 2.0035	42.8	2	[30]	epi n e− 2.5 meV
T_{v3a}	V_{Si}^{0} (c2) S = 1	ODMR	g_{iso} = 2.0035	9.2	2	[30]	epi n e− 2.5 meV
P3	V_{Si}-V_C S = 1	EPR	$g_{\parallel c}$ = 2.0026 $g_{\perp}$ = 2.0031	43	77	[37, 38]	Bulk n thermal quench
P5	V_{Si}-V_C S = 1	EPR	$g_{\parallel c}$ = 2.0026 $g_{\perp}$ = 2.0031	9	77	[37, 38]	Bulk n thermal quench
	V_{Si}^{0} (hex) S = 1	EPR	g_{iso} = 2.0032	42.8	300	[34]	Bulk n p+ 12 meV
	V_{Si}^{0} (c1,c2) S = 1	EPR	g_{iso} = 2.0032	9	300	[34]	Bulk n p+ 12 meV

This should also apply to 6H-SiC and the model of V_{Si}° with S = 1 should be most probably changed to V_{Si}^{-} with S = 3/2 and D/2

6H-SiC nanostructures; it has equally been attributed to a NV centre related defect [42] but this model should be revised. Their parameters are shown in Table 3.7.

Carbon Vacancy. In the 3C, 4H and 6H polytypes three paramagnetic charge states of the carbon vacancy are expected to be observable by EPR: V_C^{+}: S = 1/2, V_C^{0}: S = 1 and V_C^{-}: S = 3/2. The carbon vacancy has also been predicted to be a negative

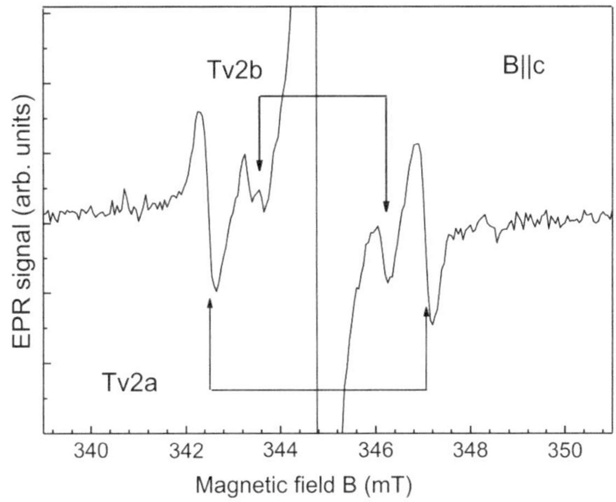

Fig. 3.8 EPR spectrum of the T_{v2a} and T_{v2b} centres in 4H-SiC; T = 300 K and B ∥ [0001]; after [34]

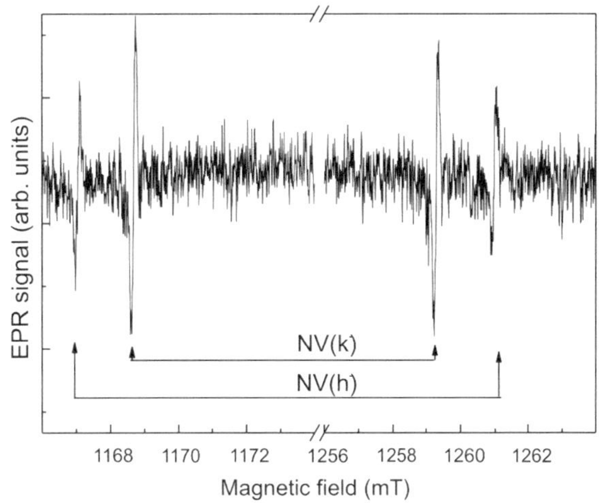

Fig. 3.9 EPR spectrum of the NV^- centers (k,h) in 4H-SiC for B//c

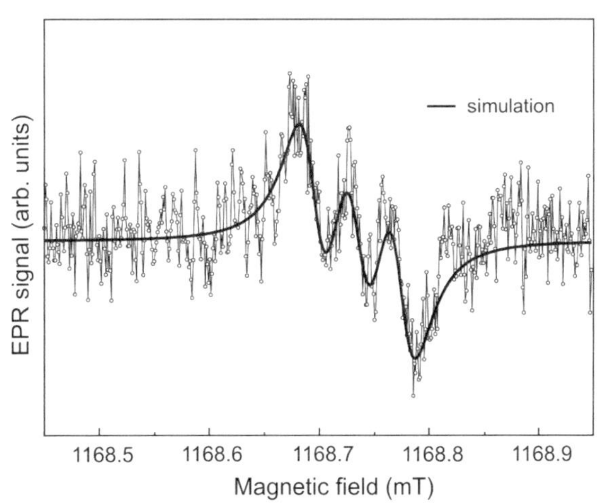

Fig. 3.10 High resolution EPR spectrum of the low field line of the axial $NV(k)^-$ center, displaying the resolved HF splitting with 1 ^{14}N neighbor

Table 3.7 Spin Hamiltonian parameters of silicon vacancy-nitrogen complexes in SiC

Polytype label	Model	Spin S	g-factors	HF 10^{-4} cm^{-1}	D 10^{-4} cm^{-1}	T (K)	References

6H
P12
	$(V_{Si}\text{-}N_C)^0$	1/2	$g_{//} = 2.0063$
			$g_{\perp} = 2.0044$
				$A_{//} = 14$
				$A_{\perp} = 0.74$
				(1C)
				$A_{//} = 0.94$
				$A_{\perp} = 0.74$
				(1 N)
					435	20	Vainer et al. (1981) [39]

3C	$(V_{Si}\text{-}N_C)^-$	1	$g_{//} = 2.004$ $g_{//} = 2.003$	A = 0.42 (1 N)	435		von Bardeleben (2016) to be published

4H	$(V_{Si}\text{-}N_C)^-$(kk)
	$(V_{Si}\text{-}N_C)^-$(hh)
		1
		1
			$g_{//} = 2.004$
			$g_{//} = 2.003$
			$g_{//} = 2.004$
			$g_{//} = 2.003$
				A = 0.37
				(1 N)
				A = 0.41
				(1 N)
					424
					438
						4
						4
							von Bardeleben (2015) [41] and to be published

6H	$(V_{Si}\text{-}N_C)^-$(k1)
	$(V_{Si}\text{-}N_C)^-$(k2)
	$(V_{Si}\text{-}N_C)^-$(h)
		1
		1
		1
			$g_{//} = 2.004$
			$g_{//} = 2.003$
			$g_{//} = 2.004$
			$g_{//} = 2.003$
			$g_{//} = 2.004$
			$g_{//} = 2.003$
				A = 0.44
				(1 N)
				A = 0.42
				(1 N)
				A = 0.37
				(1 N)
					426
					452
					443
						4
						4
						4
							von Bardeleben (2016) to be published

6H	$(V_{Si}\text{-}N_C)$ has to be revised	1	$g_{//} = 1.9700$ $g_{\perp} = 1.9964$	$A_{//} = 0.51$ $A_{\perp} = 0.42$ (1 N)	1140	5	Kalabukhova et al. (2013) [42]

U system [43]. The first observation of the positively charged V_C in 3C has been reported many years ago by Itoh et al. [8]; its EPR spectrum has been labeled T5 by these authors, but it turned out later that this assignment was erroneous [44]: the T5 centre is now definitely identified as the positively charged carbon split interstitial $(C\text{-}C)_C^+$. Since then, no further results concerning the carbon vacancy in 3C-SiC have been published. Very recently the positively charged carbon vacancy have been observed [45], which have been identified based on its spin, symmetry, g-tensor and HF values (Table 3.8).

In the 4H and 6H polytypes the situation is different. The positively charged carbon vacancy has first been observed and identified in 6H [44] and later in 4H. Bratus et al. [44] observed three centres labeled Ky1–Ky3 with very similar properties in 6H (Fig. 3.11). Based on model calculations these authors were able to assign these three spectra to the positively charged carbon vacancy on the three different lattice sites and questioned at the same time the previous assignment by Itoh et al. [8]. These important results lead also to the identification of the carbon monovacancy in the 4H polytype. The corresponding centres in 4H are labeled EI5 and EI6. EI6 has initially been assigned by Son et al. to the Si antisite [50] but this model has been revised in 2004 by Umeda et al. [47].

Table 3.8 EPR parameters of the carbon vacancy defects in 3C, 4H- and 6H-SiC

Polytype label	Model	Point symmetry	g-tensor	CHF 10^{-4} cm^{-1}	T (K)	References
3C						
	V_C^+ S = 1/2	D_{2d}	$g_\parallel = 2.0064$ $g_\perp = 2.0024$		4	[45]
4H						
EI5	$V_C^+(k)$ S = 1/2	C_{1h}	$g_1 = 2.0056$ $g_2 = 2.0048$ $g_3 = 2.0030$	$A_1 = 41.5$ $A_2 = 40.8$ $A_3 = 6.0$ (Si1)	5	[46]
EI6	$V_C^+(h)$ S = 1/2	C_{3V}	$g_\parallel = 2.0026$ $g_\perp = 2.0052$	$A_\parallel = 144.7$ $A_\perp = 99.2$	10	[47]
	$V_C^-(k)$	C_{1h}	$g_1 = 2.0027$ $g_2 = 2.0038$ $g_3 = 2.0054$	$A_1 = 93.85$ $A_2 = 94.88$ $A_3 = 121.4$ (Si1,2) $A_1 = 8.13$ $A_2 = 7.95$ $A_3 = 10.5$ (Si3,4)	30	[48]
HEI1	$V_C^-(h)$ S = 1/2	C_{1h}	$g_1 = 2.00287$ $g_2 = 2.00407$ $g_3 = 2.00459$	$A_1 = 72.5$ $A_2 = 72.5$ $A_3 = 94.1$ (Si1) $A_1 = 110.1$ $A_2 = 109.1$ $A_3 = 142.0$ (Si2)		[49]
6H						
Ky1	$V_C^+(k1)$ S = 1/2	C_S	$g_1 = 2.0025$ $g_2 = 2.0026$ $g_3 = 2.0060$	$a_{iso} = -42.5$ $b = -11.2$ (Si1) $a_{iso} = -40.8$ $b = -10.6$ (Si2) $a_{iso} = -41.1$ $b = -10.8$ (Si3) $a_{iso} = -41.1$ $b = -10.8$ (Si4)	4	[44]
Ky2	$V_C^+(k2)$ S = 1/2	C_S	$g_1 = 2.0023$ $g_2 = 2.0040$ $g_3 = 2.0050$	$a_{iso} = -47.1$ $b = -13.1$ (Si1) $a_{iso} = -33.6$ $b = -9.4$ (Si2) $a_{iso} = -41.2$ $b = -10.4$ (Si3) $a_{iso} = -41.2$ $b = -10.4$ (Si4)	4	[44]

(continued)

Table 3.8 (continued)

Polytype label	Model	Point symmetry	g-tensor	CHF 10^{-4} cm^{-1}	T (K)	References
Ky3	$V_C^+(h)$ S = 1/2	C_{3V}	$g_\parallel = 2.0020$ $g_\perp = 2.0046$	$a_{iso} = 116.9$ b = 15.2 (1Si)	15	[44]

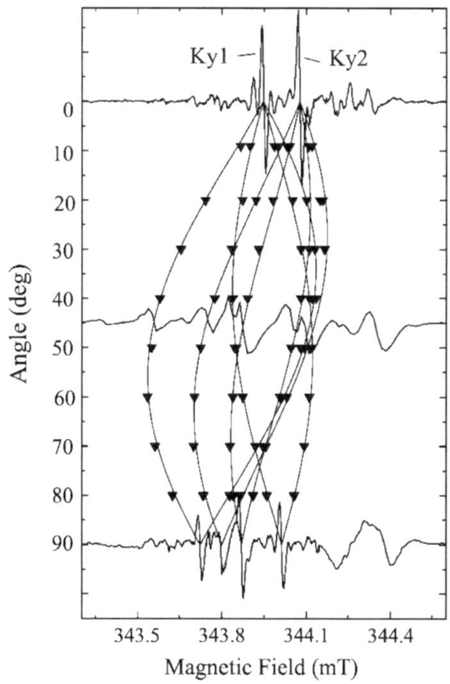

Fig. 3.11 EPR spectra (Ky1, Ky2) of V_C^+ on the (k1,k2) quasicubic sites in 6H-SiC; T = 4 K; after [44]

Divacancies. The formation of divacancies can be achieved by annealing at temperatures where the monovacancies become mobile. High energy particle irradiation can also lead to direct divacancy formation by double displacements. In diamond monovacancies become mobile at temperatures above 800 K but in the case of silicon this occurs at temperatures below room temperature already. For both materials two paramagnetic charge states of the "ideal" [111] oriented nearest neighbour (V-V) defect have been studied by EPR: VV^0 and VV^-. Due to opposing distortions in silicon and diamond the localization of the spin density is very different [51]. Split divacancies, V-C-V or [110] oriented divacancies have not been observed in Si and C.

Diamond. In electron irradiated diamond, annealed at temperatures above 700 °C, a paramagnetic defect labeled R4 or W6 has been observed by different authors (Fig. 3.12). This defect is attributed to the neutral divacancy VV^0 (Table 3.9).

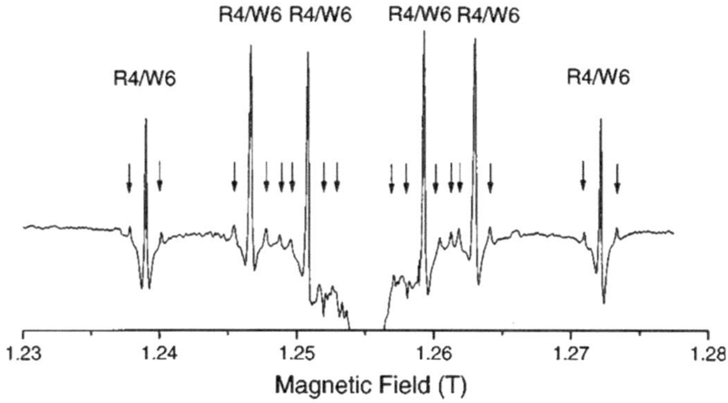

Fig. 3.12 Q-band EPR spectrum of the neutral divacancy in diamond; T = 30 K and B || [111]; after [51]

Table 3.9 EPR parameters of the divacancy centres in diamond

Label	Model	Point symmetry	g-tensor	D 10^{-4} cm^{-1}	CHF $10^{-4}cm^{-1}$	T (K)
R4/W6 [51]	VV^0 S = 1	C_{2h}	g_1 = 2.0022 [90°,315°] g_2 = 2.0026 [141°,45°] g_3 = 2.0013 [51°,45°]	D_1 = +34.4 [90°,315°] D_2 = +68.7 [144°,45°] D_3 = −34.4 [54°,45°]	$A_{\parallel}$ = 38.0 $A_{\perp}$ = 18.7 [55°,315°]	30
W29 [52]	VV^- S = 3/2		g_1 = 2.0019 g_2 = 2.0020 g_3 = 2.0024	D_1 = 99.0 D_2 = 52.1 D_3 = 151.2		100

The level position of the divacancy, a deep acceptor, has also been determined: it is: −/0: Ev + 1.7 eV [53].

Silicon. Divacancy defects in Si produced by high-energy electron irradiation have been studied in great detail. Initial studies were published by Corbett and Watkins and later on by the Amerlaan group. They are one of the most important defects in Si submitted to room temperature irradiation. They are formed by direct double displacements of two Si lattice atoms but also by the annealing of monovacancies. As in the case of diamond only [111] oriented nearest neighbour configurations have been observed (Fig. 3.13). In spite of the small band gap of Si 4 stable charge states can exist 2−, −, 0, +. Two of them (+, −) are paramagnetic and have been studied by EPR and ENDOR. They have been labeled G6 and G7 respectively (Table 3.10).

The charge transition levels of the di-vacancy are 2−/−: E_v + 0.4 eV, 0/+: E_v + 0.25 eV.

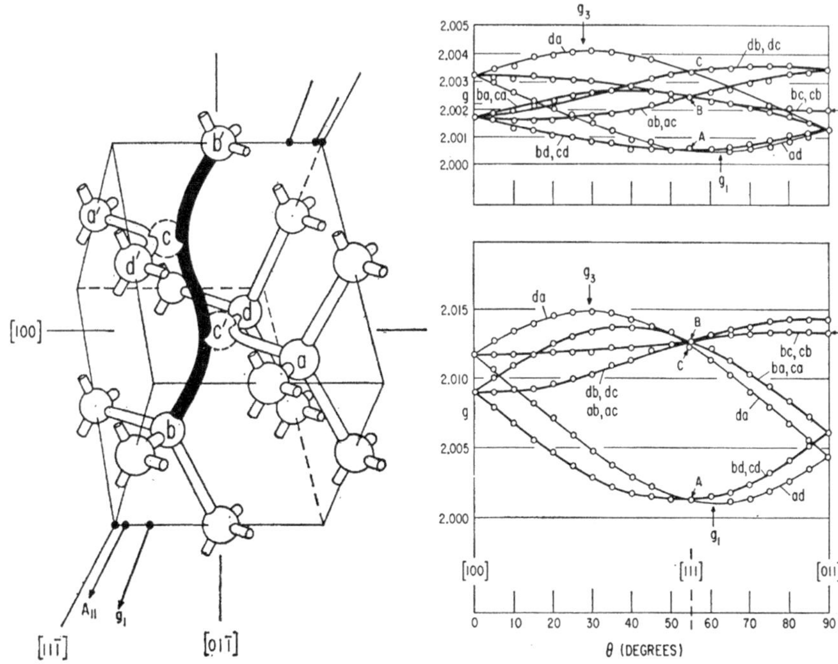

Fig. 3.13 Model of the negatively charged divacancy in Si (*left*) and the angular variation of the VV^+ (*top*) and VV^- (*bottom*) centres in silicon (*right*); after [54]

Table 3.10 EPR parameters of the divacancy centres in silicon

Label	Model	Point symmetry	g-tensor	CHF $10^{-4}cm^{-1}$	T (K)
G6 [54–56]	VV^+ S = 1/2	C_{2h}	g_{xx} = 2.0004 g_{yy} = 2.0020 g_{zz} = 2.0041	$A_{\parallel}$ = 67.8 $A_{\perp}$ = 40.0 (2 Si)	20
G7 [55, 57]	VV^- S = 1/2	C_{2h}	g_{xx} = 2.0012 g_{yy} = 2.0135 g_{zz} = 2.0150	$A_{\parallel}$ = 79 $A_{\perp}$ = 56 (2Si)	10

Silicon Carbide. Di-vacancy defects in SiC have been reported already in the early EPR studies of SiC [38]. Vainer et al. observed a large number of different spin S = 1 centres in heat-treated 6H-SiC. In particular they assigned two centres labelled P6, P7 to close pair vacancy centres (V_C^--V_{Si}^-) in axial (P6) and basal (P7) coordination. No distinction of lattice sites (k1, k2, h) has been made in that study. Baranov et al. [58] confirmed the assignment of the P6, P7 centres in 6H-SiC to the divacancy —a model questioned at that time by Lingner et al. [59] who assigned P6 to a carbon antisite carbon vacancy defect (C_{Si}-V_C); Baranov et al. demonstrated further the

Table 3.11 EPR parameters of the divacancy centres in 4H- and 6H-SiC

Polytype label	Model	Point symmetry	g-tensor	D, E 10^{-4} cm^{-1}	CHF 10^{-4} cm^{-1}	Reference
4H						
P6b	$(V_C\text{-}V_{Si})^0$ (k-k) S = 1	C_{3V}	2.003	D = 436	A_{xx} = 15.7 A_{yy} = 15.0 A_{zz} = 34.7 (3xC)	[62]
P6b	$(V_C\text{-}V_{Si})^0$ (h-h) S = 1	C_{3V}	2.003	D = 447	A_{xx} = 17.7 A_{yy} = 16.7 A_{zz} = 36.7 (3xC)	[62]
P7b	$(V_C\text{-}V_{Si})^0$ (h-k) S = 1	C_{1h}	2.003	D = 408 E = 10	A_{xx} = 17.3 A_{yy} = 17.3 A_{zz} = 36.7 (1xC)	[62]
P7b	$(V_C\text{-}V_{Si})^0$ (k-h) S = 1	C_{1h}	2.003	D = 447 E = 90	A_{xx} = 17.0 A_{yy} = 17.3 A_{zz} = 39.4 (1xC)	[62]
6H						
P6	$V_C\text{-}V_{Si}$ (h) S = 1	C_{3V}	$g_{\parallel}$ = 2.0023 $g_{\perp}$ = 2.0024	D = 449	A_{iso} = 6.6 (3 Si)	[38]
P7	$V_C\text{-}V_{Si}$ (basal) S = 1	C_{1h}	g_1 = 2.0033 g_2 = 2.0025 θ = 71°	D = −442 E = −35	A_{iso} = 6.8 (3 Si)	[38]

ground state character of the P6, P7 centres. This model is now fully reconfirmed (Table 3.11). Son et al. had attributed a different EPR centre (SI5) to the divacancy [60] but have reassigned it later to the carbon antisite carbon vacancy pair [61].

3.2.2 *Interstitial Defects*

3.2.2.1 Diamond

As vacancy defects are the primary irradiation induced centres in diamond the question of the fate of the associated carbon interstitials arises. It has been shown by Twitchen et al. [64] that carbon interstitials form [100] oriented split interstitial centres $(C\text{-}C)_C$, the so-called R2 centre (Fig. 3.12), under irradiation below room temperature and combine in di-split interstitial associates (R1) when irradiated to room temperature (Table 3.12).

Table 3.12 EPR parameters of the split interstitial carbon defects in diamond

Label	Model	Point symmetry	g-tensor	D, E 10^{-4} cm^{-1}	CHF 10^{-4} cm^{-1}	T (K)
R2 [64]	$(C\text{-}C)_C^0$ S = 1	D_{2d}	$g_{\parallel} = 2.0021$ $g_{\perp} = 2.0019$	D = 1392		300
R1 [63]	$(C\text{-}C)_C\text{-}(C\text{-}C)_C$ S = 1	C_{1h}	$g_1 = 2.0018$ $g_2 = 2.0019$ $g_3 = 2.0025$	$D_1 = +469.7$ $D_2 = -935.8$ $D_3 = +465.9$	$A_{\parallel} = 41.0$ $A_{\perp} = 4.0$	300

3.2.2.2 Silicon

Isolated silicon interstitial defects have not convincingly been observed in Si by magnetic resonance spectroscopy even though Si interstitials are easily formed by electron irradiation. Even for irradiation and in-situ EPR detection at temperatures as low as T = 4 K paramagnetic Si interstitial defects were not observed. This "non" observation might be related to their high reactivity to form complexes with other defects such as substitutional acceptors: carbon, boron, aluminium as well as molecular hydrogen, and interstitial oxygen even at cryogenic temperatures [65]. Two paramagnetic centres AA12 and P6 have been tentatively attributed to Si interstitial defects (Table 3.13) but further information is required to confirm this assignments.

3.2.2.3 Silicon Carbide

Similar to the case of diamond but different from that of silicon, interstitial defects have been detected by EPR in all SiC polytypes. As it is a binary compound and both type of vacancies are easily created under particle irradiation both Si and C interstitials are generated in the irradiation process. Whereas C interstitials have been evidenced clearly by EPR (Fig. 3.15), the case of Si interstitials seems still open. Apparently isolated interstitials at high symmetry lattice sites are not stable; instead complexes with themselves or complexes with other dopants are the predominant configuration at room temperature [70].

Concerning the carbon interstitial, in all three polytypes of SiC the [100] oriented split interstitial $(C\text{-}C)_C$ is the dominant defect configuration (Table 3.14). It has been observed in two paramagnetic charge states: 1+ with S = 1/2 and 0 with S = 1. Its defect configuration has many similitudes with the R2 centre in diamond, its direct counterpart.

In 3C-SiC the positively charged carbon split interstitial has first been observed by Itoh et al. [8]. In their initial paper they attributed this defect erroneously to the carbon vacancy defect V_C^+. In 2002 Petrenko et al. [68] have shown that it should be assigned to $(C\text{-}C)_C^+$. The neutral charge state of this defect have also been observed in the 3C polytype [71]. In the 4H and 6H polytypes these omnipresent defects in

Table 3.13 EPR parameters of silicon interstitial defects in silicon

Label	Model	Point symmetry	g-tensor	CHF 10^{-4} cm^{-1}	T (K)	Reference
AA12	Si_i		1.9998	A = 15.0 MHz (1 Si)		[66]
P6	Si_i-Si-Si_i S = 1/2	C_2	g_1 = 2.0040 g_2 = 2.0062 g_3 = 2.0010		200	[67]

Table 3.14 Carbon split interstitial defects in 3C, 4H- and 6H-SiC

Polytype label	Model	Point symmetry	g-tensor	D, E 10^{-4} cm^{-1}	CHF 10^{-4} cm^{-1}	T (K)
3C						
T5 [8] p-type	$(C\text{-}C)_C^+$ S = 1/2 [63]	D_2	g_1 = 2.0020 g_2 = 2.0007 g_3 = 1.9951		$A_\parallel$ = 18.9 $A_\perp$ = 13.8	<100
– [68] n-type	$(C\text{-}C)_C^0$ S = 1	D_{2d}	g = 2.003	D = 551		4
4H						
EI1 [69]	$(C\text{-}C)_C^+$ S = 1/2 [63]		g_x = 1.9962 g_y = 2.0019 g_z = 2.0015 θ = 41°			
EI3 [69]	$(C\text{-}C)_C^0$ S = 1		g = 2.0063	D = 552 θ = 46°		
EI1 [69]	$(C\text{-}C)_C^+$ S = 1/2		g_x = 1.9962 g_y = 2.0019 g_z = 2.0015 θ = 41°			
EI3 [69]	$(C\text{-}C)_C^0$ S = 1		g = 2.0063	D = 559 θ = 45.5°		

particle irradiated SiC have also been observed very early but not assigned correctly [50]. The initial assignments were also based on the carbon vacancy model [45]. Bratus et al. [44] published detailed modelling and experimental results, which have clarified this issue.

3.2.2.4 Gallium Nitride (h-GaN)

The first interstitial defect observed in GaN was the Ga interstitial (Table 3.15). It is detected after low temperature (4 K) electron irradiation and was observed by ODMR [72]. In these ODMR experiments photoexcitation at 364 nm (near band to band excitation) is used and the PL is detected in a broad band at 0.95 eV. The calculated HF values [73] for this defect ($A\|$ = 2960 and $A_\perp$ = 2830 MHz) are in

Table 3.15 EPR parameters of the nitrogen split interstitial and the gallium interstitial in h-GaN

Label	Model	Point symmetry	g-tensor	CHF 10^{-4} cm^{-1}	T (K)	Reference
			N interstitial			
	$(N\text{-}N)_N^0$ S = 1/2	C_1	g_{xx} = 1.9895 g_{yy} = 2.0016 g_{zz} = 2.0036	N1 A_{xx} = 3.1 A_{yy} = 3.2 A_{zz} = 24.8 N2 A_{xx} = 3.4 A_{yy} = 3.6 A_{zz} = 25.5 Ga1 A_{xx} = 26.1 A_{yy} = 26.9 A_{zz} = 40.1 Ga2 A_{xx} = 25.2 A_{yy} = 25.7 A_{zz} = 37.2 Ga3 A_{xx} = 13.4 A_{yy} = 13.9 A_{zz} = 21.9 Ga4 A_{xx} = 18.0 A_{yy} = 18.6 A_{zz} = 28.4	4	[75, 73]
D2	V_{Ga}-O S = 1/2	C_{3V}	$g_{\parallel}$ = 2.001 $g_{\perp}$ = 1.999	A_{iso} = 44.0 (3 N)	77	[76]
			Ga interstitial			
L5	Ga_i^{2+} S = 1/2 O site	–	2.000(1)	$A_{\parallel}$ = 1331 $A_{\perp}$ = 1258 (1 Ga)	4	[72]
Ga-1	Ga_i-X S = 1/2	C_{3V}	$g_{\parallel}$ = 2.008 $g_{\perp}$ = 2.001	$A_{\parallel}$ = 253.5 $A_{\perp}$ = 186.8	5	[74]

fair agreement with the model of a Ga interstitial at a octahedral lattice site; the calculated HF interactions exclude an assignment to a split interstitial configuration. This defect is not stable at room temperature and thus is not expected to be observable in room temperature irradiated samples. In p-type as grown GaN a native Ga interstitial related defect has equally been reported by Hai et al. [74].

The nitrogen interstitial has only been identified very recently by CW EPR [75]. It can be introduced by room temperature irradiation of n-type GaN with various particles (e−, p+, ions). It is stable at room temperature and anneals at 300 °C. An analysis of its spin Hamiltonian parameters requiring the use of very high frequency EPR (>300 GHz) combined with extensive modelling has allowed to attribute this centre to the neutral split interstitial $(N\text{-}N)_N^0$ [75, 73]. Above T = 40 K this defect

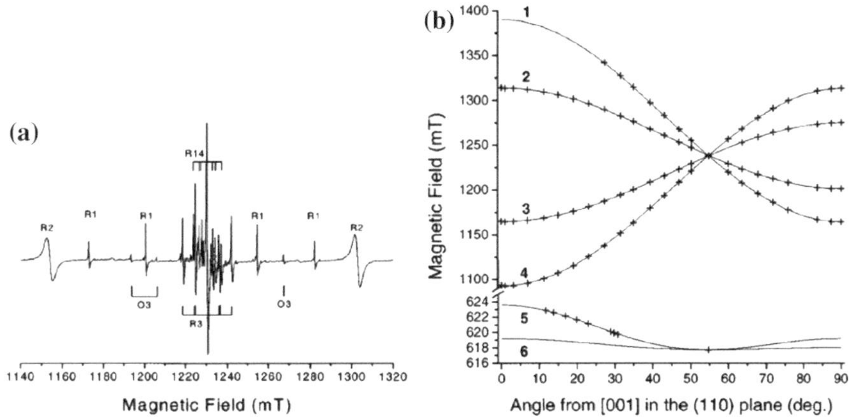

Fig. 3.14 Q-band EPR spectra (*left*) and angular variation (*right*) of the neutral $(C\text{-}C)_C^0$ centre in diamond; after [64]

displays at X-band a configuration of higher symmetry induced by thermal reorientation (Fig. 3.14). In its high temperature configuration this centre has been reported before by Son et al. [76] but assigned to the Ga vacancy defect, a model not supported by our recent calculations [75, 73].

3.2.3 *Antisite Defects*

Antisite defects have been evidenced by EPR many years ago in the 3–5 compounds GaAs, GaP and InP. In these materials they occur as growth related native defects but they can also be introduced by high-energy electron irradiation or even plastic deformation. One of the most studied antisite defect is the so-called EL2 defect, a double donor which gives rise to the semi-insulating character of Czochralski grown (LEC) GaAs substrates. The studies of the EL2 related As_{Ga} antiste have shown that actually various configurations of the antisite can occur; in addition to the isolated As_{Ga} antisite, complexes with vacancies or interstitials and even other antisite may be formed.

3.2.3.1 Silicon Carbide

In SiC different antisite defects have been claimed to be detected: they are Si_C, $(Si_C\text{-}C_{Si})$ (Table 3.16) and the $C_{Si}\text{-}V_C$ complex. However, a clear identification has only been achieved for the nearest neighbour antisite pairs SiC-C_{Si}, which seem to

Table 3.16 EPR parameters of the antisite pair defects and the isolated carbon antisite in SiC

Polytype label	Model	Point symmetry	g-tensor	CHF 10^{-4} cm^{-1}	T (K)	Reference
3C						
P1	$(Si_C\text{-}C_{Si})^+$ S = 1/2	C_{1h}	g_{xx} = 2.0049 g_{yy} = 2.0210 g_{zz} = 2.0497 θ = 69°		30	[77]
4H						
P1	$(Si_C\text{-}C_{Si})^+$ k-h basal S = 1/2	C_{1h}	g_{xx} = 2.0030 g_{yy} = 2.0161 g_{zz} = 2.0407 θ = 63°		30	[77]
P2	$(Si_C\text{-}C_{Si})^+$ h-k basal S = 1/2	C_{1h}	g_{xx} = 2.0030 g_{yy} = 2.0161 g_{zz} = 2.0407 θ = 63°		30	
6H						
P1	$(Si_C\text{-}C_{Si})^+$ k1-h basal S = 1/2	C_{1h}	g_{xx} = 2.0041 g_{yy} = 2.0161 g_{zz} = 2.0407 θ = 63°		30	[77]
P2	$(Si_C\text{-}C_{Si})^+$ k2-k2 basal S = 1/2	C_{1h}	g_{xx} = 2.0030 g_{yy} = 2.0139 g_{zz} = 2.0323 θ = 50°		30	
P3	$(Si_C\text{-}C_{Si})^+$ h-k1 basal S = 1/2	C_{1h}	g_{xx} = 2.0060 g_{yy} = 2.0196 g_{zz} = 2.0582 θ = 68°		30	
	C_{Si} S = 1/2	C_{3V}	$g_{\parallel}$ = 2.0045 $g_{\perp}$ = 2.0055	$A_{\parallel}$ = 75.8 $A_{\perp}$ = 28.0 (1 C)	25	[78]

form easily under room temperature particle irradiation (Fig. 3.15) [77]. This defect has the particularity that it may exist in great concentrations without modifying the overall stoichiometry. It is considered to be an important precursor defect in the amorphisation process under high dose particle irradiation (Fig. 3.16).

The complex defect C_{Si}-V_{Si} is considered as the stable form of the silicon vacancy in p-type material (Fig. 3.17).

The carbon antisite-silicon vacancy complexes (Fig. 3.18) have been observed in two different charge states + and −. Their properties are shown in Table 3.17.

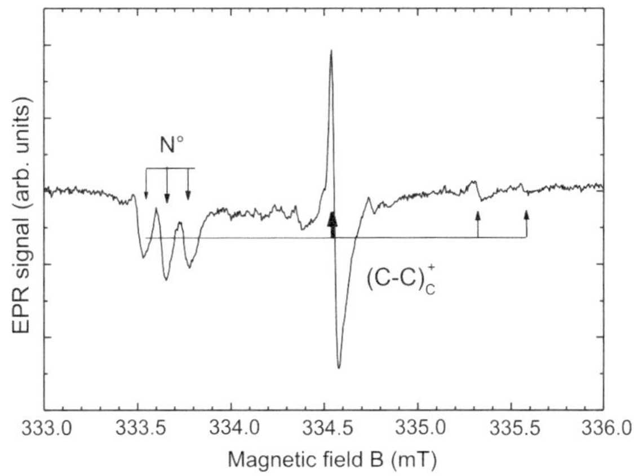

Fig. 3.15 EPR spectrum of the positively charged carbon split interstitial $(C\text{-}C)_C^+$ in n-type electron irradiated 3C-SiC displaying the CHF with (1 + 3) Si NN neighbours; T = 4 K and B ‖ [111]; a weak spectrum from the neutral nitrogen donor N° is also seen

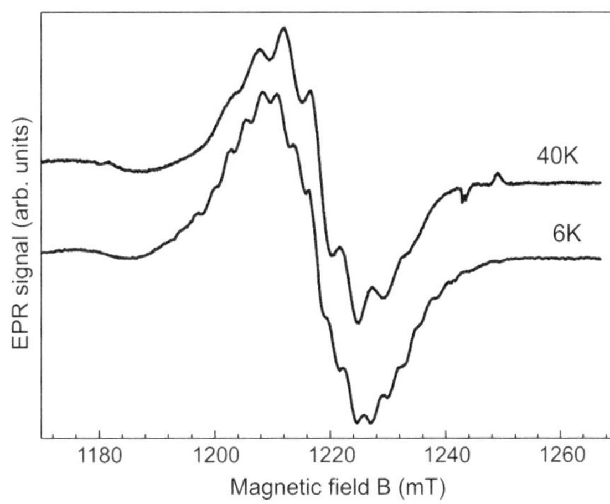

Fig. 3.16 Q-band EPR spectra of the neutral nitrogen split interstitial $(N\text{-}N)_N^0$ in h-GaN at T = 6 K and T = 40 K displaying HF interaction with two central N atoms and four non equivalent Ga neighbors; B ‖ c; after [75]

3.2.3.2 GaAs

Arsenic antisites: As_{Ga}. Arsenic antisite defects have been observed in GaAs as native, growth related defects and after particle irradiation (Table 3.18). The numerous results obtained showed that the simple picture of a simple As_{Ga}-As_4 centre had to be enlarged and actually a variety of defects exist with subtle differences as concerns their g-values, hyperfine interactions and optical properties. The most studied case is the one of the native As_{Ga} defect in Cz grown bulk GaAs, EL2 (Fig. 3.19), which is responsible for its semi-insulating properties (Fig. 3.20).

The EL2 related As_{Ga} defect is a deep double donor in GaAs with two charge transition levels 0/+ and +/2+. It is paramagnetic with a spin S = 1/2 in the "1+" charge state and diamagnetic in the two others. It is characterized by a strong central HF interaction, generally resolved in X-band EPR, and weaker SHF interactions,

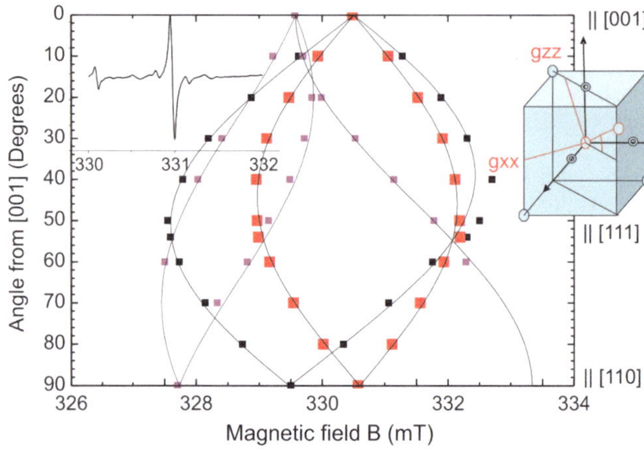

Fig. 3.17 EPR spectrum for B || [100] (*inset*) and angular variation of the antisite pair defect $(Si_C\text{-}C_{Si})^+$ in 3C-SiC; after [77]

Fig. 3.18 Atomic models for 4 $C_{Si}\text{-}V_C$ pairs in 4H-SiC; after [79]

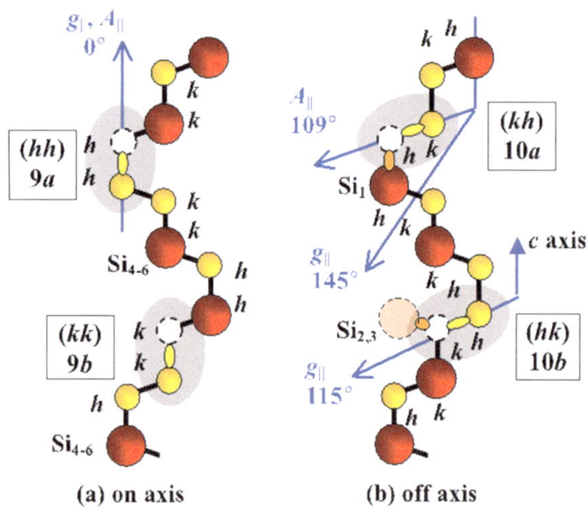

which have been studied by ENDOR spectroscopy. For the basic defect configuration, the isolated As_{Ga}, detailed modeling results have been reported.

By electron irradiation in n-type GaAs a different arsenic antisite with a reduced central hyperfine interaction is generated [82, 83]. Its introduction rate is high, comparable to that of primary defects. Contrary to EL2, the paramagnetic charge state of this defect does not show any optically induced metastability.

Gallium antisite Ga_{As}. The Ga antisite has only been detected by EPR in electron irradiated p-type GaAs. Its Spin Hamiltonian parameters are given in Table 3.19.

Table 3.17 EPR parameters of the carbon antisite silicon vacancy pairs in 4H-SiC

Label	Model	Point symmetry	g-tensor	CHF 10^{-4} cm^{-1}	T (K)	Reference
HEI9a	$(C_{Si}\text{-}V_{Si})^+$	C_{3V}	$g_\parallel = 2.00227$	$A_\parallel = 77.1$	30	[79]
HEI9b	S = 1/2	C_{3V}	$g_\perp =$	$A_\perp = 21.2$		
HEI10Ia	h-h	C_{1h}	2.00408	$A_\parallel = 93.0$		
HEI10b	k-k	C_{1h}	$g_\parallel = 2.00195$	$A_\perp = 34.7$		
	k-h		$g_\perp = 2.00379$	$A_{xx} = 24.3$		
	h-k		$g_{xx} = 2.00339$	$A_{yy} = 24.8$		
			$g_{yy} = 2.00258$	$A_{zz} = 81.8$		
			$g_{zz} = 2.00226$	$A_{xx} = 21.6$		
			$g_{xx} = 2.00399$	$A_{yy} = 22.9$		
			$g_{yy} = 2.00345$	$A_{zz} = 78.9$		
			$g_{zz} = 2.00263$	(1 C)		
SI5	$(C_{Si}\text{-}V_{Si})^-$ S = 1/2	C_{1h}	$g_x = 2.00372$ $g_y = 2.00259$ $g_z = 2.00534$	$A_x = 95.0$ $A_y = 93.9$ $A_z = 121.4$	30	[61]

Table 3.18 EPR parameters of the arsenic antisite defects in GaAs

Model	Spin	Point symmetry	g-factor	CHF 10^{-4} cm^{-1}	T (K)	Reference
$As^+_{Ga}\text{-}As_4$	S = 1/2	T_d	2.04 ± 0.01	900	4–300	[80]
$As^+_{Ga}\text{-}As_4$	S = 1/2	T_d	2.0037	867		[81]
$As^+_{Ga}\text{-}V^-_{As}As_3$	S = 1/2	C_{3V}	1.97	680		[82, 83]
$As^+_{Ga}\text{-}As_4\text{-}As_i$ (EL2)	S = 1/2	T_d	2.0047	900		[1]

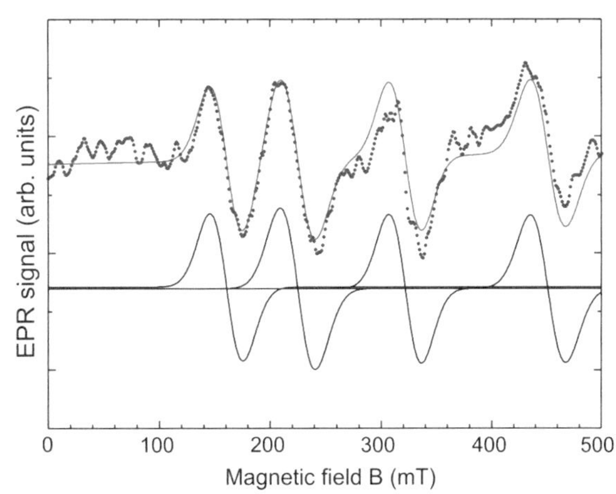

Fig. 3.19 EPR spectrum of the EL2 related arsenic antisite defect $As_{Ga}^+\text{-}As_i$ in LEC-GaAs; T = 4 K and B ∥ [100] (*top*) and its simulation (*bottom*) with parameters given in table; after [1]

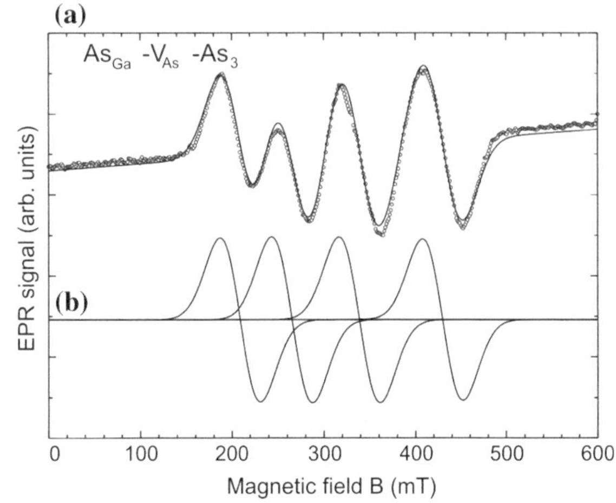

Fig. 3.20 EPR spectrum at T = 4 K of the As^{+}_{Ga}-$V_{As}As_3$ defect in electron irradiated n-type GaAs (*top*) and the simulation (*b*) with parameters given in table [82]

Table 3.19 EPR parameters of the gallium antisite defect in GaAs

Model	Spin	Point symmetry	g-factor	CHF 10^{-4} cm^{-1}	T (K)	Reference
Ga^{-}_{As}	S = 1/2	C_{3V}	$g_{\parallel} = 2.00$ $g_{\perp} = 2.05$	$A_{\parallel} = 240$ $A_{\perp} = 175$ (1 Ga)	4	[84]

3.2.3.3 GaP

Phosphorous antisite P_{Ga}. Phosphorous antisite defects have been observed in neutron irradiated GaP bulk materials. Due to the large central and ligand hyperfine interactions their microscopic structure could be deduced directly from the resolved HF structure of the EPR spectra (Fig. 3.21). As in the case of GaAs, antisite defects with different ligand configuration have been detected (Table 3.20).

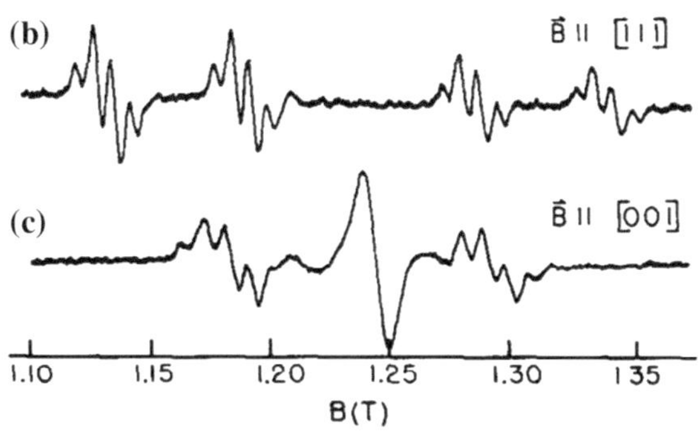

Fig. 3.21 ODMR spectra of the phosphorous antisite spectra P_{Ga}-P_3(Y) (**b**) and P_{Ga}-P_4 (**c**) in p-type GaP; after [85]

Table 3.20 EPR parameters of the phosphorous antisite defects in GaP

Label	Model	Spin	D 10^{-4} cm^{-1}	g-factor	HF 10^{-4} cm^{-1}	Reference
	P_{Ga}-P_4	S = 1/2		2.008	A_{CHF} = 966 (1P) A_{LHF} = 82 (4P)	[86]
	P_{Ga}-P_3(X)	S = 1/2		2.006	A_{CHF} = 704 (1P) A_{LHF} = 117 (3P)	[87]
	P_{Ga}-P_3(Y)	S = 1	717	2.007	A_{CHF} = 530 (1P) A_{LHF} = 67 (3P)	[85]

3.3 Outlook

Magnetic resonance spectroscopy has been shown to be the technique of choice for the study of the microscopic and electronic structure of intrinsic defects in the main semiconductor materials. These defects can be seen as radiation damage, purposely growth induced or treatment related defects. Via their electrical activity they can trap charges and modify the electrical and optical properties of these materials. More recently, the fact that most of them carry also a spin (often >1/2) has been shown to be interesting for optical selective manipulation of individual spins; this finds application in the context of qubits for quantum information in diamond and silicon carbide as the most important examples; this last development is only at the on-set and is expected to give strong further impulse in correlated microwave/optical spin manipulation experiments based on the use of intrinsic defects.

References

1. von Bardeleben, H.J., Stiévenard, D., Deresmes, D., Huber, A., Bourgoin, J.C.: Identification of a defect in a semiconductor: EL2 in GaAs. Phys. Rev. B **34**, 7192–7202 (1986)
2. Dabrowski, J., Scheffler, M.: Isolated arsenic-antisite defect in GaAs and the properties of EL2. Phys. Rev. B **40**, 10391–10401 (1989)
3. Krambrock, K., Spaeth, J.-M., Delerue, C., Allan, G., Lannoo, M.: Identification of the isolated arsenic antisite defect in electron-irradiated gallium arsenide and its relation to the *EL2* defect. Phys. Rev. B **45**, 1481(R)–1484(R) (1992)
4. Son, N.T., Carlson, P., Gällström, A., Magnusson, B., Janzén, E.: Prominent defects in semi-insulating SiC substrates. Phys. B: Condens. Mat. **401–402**, 67–72 (2007)
5. Loubser, J.H.N., van Wyk, J.A.: Electron spin resonance in annealed type 1b diamond. Diam. Res. **11**, 4–7 (1977)

6. Falk, A.L., Buckley, B.B., Calusine, G., Koehl, W.F., Dobrovitski, V, Politi, A., Zorman, C. A., Feng, P.X., Awschalom, D.D.: Polytype control of spin qubits in silicon carbide. Nat. Commun. **4**:1819 (2013)
7. Cantin, J.L., Schoisswohl, M., von Bardeleben, H.J., Hadj Zoubir, N., Vergnat, M.: Electron-paramagnetic-resonance study of the microscopic structure of the Si(001)-SiO_2 interface. Phys. Rev. B **52**, R11599–R11602 (1995)
8. Itoh, H., Yoshikawa, M., Nashiyama, I., Misawa, S., Okumura, H., Yoshida, S.: Electron spin resonance study of defects in CVD grown 3C-SiC irradiated with 2 MeV protons. J. Electron. Mater. **21**, 707–710 (1992)
9. Son, N.T., Hai, P.N., Janzén, E.: Intrinsic defects in silicon carbide polytypes. Mater. Sci. Forum **353–356**, 499–504 (2001)
10. Pickard, C.J., Mauri, F.: First-principles theory of the EPR g tensor in solids: defects in quartz. Phys. Rev. Lett. **88**, 086403 (2002)
11. Szasz, K., Ivady, V., Erik Janzen, E., Gali, A.: First principles investigation of divacancy in SiC polytypes for solid state qubit application. Mater. Sci. Forum **778–780**, 499 (2014)
12. Spaeth, J.M., Overhof, H.: Point defects in semiconductors and insulators. Springer, Berlin (2003)
13. Gruber, A., Draebenstedt, A., Tietz, C., Fleury, L., Wrachtrup, J., von Borczyskowski, C.: Scanning confocal optical microscopy and magnetic resonance on single defect centers. Science **276**, 2012–2014 (1997)
14. Van Wyk, J.A., Tucker, O.D., Newton, M.E., Baker, J.M., Woods, G.S., Spear, P.: Magnetic resonance measurements on the $^5A_2$ excited state of the neutral vacancy in diamond. Phys. Rev. **B52**, 12657 (1995)
15. Isoya, J., Kanda, H., Uchida, Y., Lawson, S.C., Yamasaki, S., Itoh, H., Morita, Y.: EPR identification of the negatively charged vacancy in diamond. Phys. Rev. B **45**, 1436–1439 (1992)
16. Baldwin, J.A.: Electron paramagnetic resonance investigation of the vacancy in diamond. Phys. Rev. Lett. **10**, 220–222 (1963)
17. Deak, P., Aradi, B., Kaviani, M., Frauenheim, T., Gali, A.: Formation of NV centers in diamond: a theoretical study based on calculated transitions and migration of nitrogen and vacancy related defects. Phys. Rev. **B89**, 075203 (2014)
18. Redman, D.A., Brown, S., Rand, S.C.: Origin of persistent hole burning of NV centers in diamond. J. Opt. Soc. Am. **B9**, 768 (1992)
19. Watkins, G.D.: Radiation Damage in Semiconductors. Academic Press, New York (1965)
20. Watkins, G.W., Corbett, J.W.: An EPR study of the lattice vacancy in silicon. J. Phys. Soc. Jpn., Suppl. II, **18**, 22–27 (1963)
21. Corbett, J.W.: Electron Radiation Damage in Semiconductors and Metals (Solid State Physics, Suppl. 7). Academic Press, New York (1965)
22. Sprenger, M., Muller, S.H., Sieverts, E.G., Ammerlaan, C.A.J.: Vacancy in silicon: hyperfine interactions from electron-nuclear double resonance measurements. Phys. Rev. B **35**, 1566–1581 (1987)
23. Itoh, H., Masahito, Y., Nashiyama, I., Misawa, Sh., Okumura, H., Yoshida, S.: Radiation induced defects in CVD-grown 3C-SiC. IEEE Trans. Nucl. Sci. **37**, 1732–1738 (1990)
24. Wimbauer, T., Meyer, B.K., Hofstaetter, A., Scharmann, A., Overhof, H.: Negatively charged Si vacancy in 4H-SiC: a comparison between theory and experiment. Phys. Rev. B **56**, 7384–7388 (1997)
25. Mizuochi, N., Yamasaki, S., Takizawa, H., Morishita, N., Ohshima, T., Itoh, H., Isoya, J.: EPR studies of the isolated negatively charged silicon vacancies in n-type 4H- and 6H-SiC: identification of C_{3V} symmetry and silicon sites. Phys. Rev. B **68**, 165206 (2003)
26. Orlinski, S.B., Schmidt, J., Mokhov, E.N., Baranov, P.G.: Silicon and carbon vacancies in neutron-irradiated SiC: a high-field electron paramagnetic resonance study. Phys. Rev. B **67**, 125207 (2003)
27. Schneider, J., Maier, K.: Point defects in silicon carbide. Phys. B **185**, 199–206 (1993)

28. Mizuochi, N., Yamasaki, S., Takizawa, H., Morishita, N., Ohshima, T., Itoh, H., Isoya, J.: Continuous-wave and pulsed EPR study of the negatively charged silicon vacancy with S=3/2 and C_{3v} symmetry in n-type 4H−SiC. Phys. Rev. B **66**, 235202 (2002)
29. Janzén, E., Gali, A., Carlsson, P., Gällström, A., Magnusson, B., Son, N.T.: The silicon vacancy in SiC. Phys. B **404**, 4354–4358 (2009)
30. Sörman, E., Son, N.T., Chen, W.M., Kordina, O., Hallin, C., Janzén, E.: Silicon vacancy related defect in 4H and 6H SiC. Phys. Rev. B **61**, 2613–2620 (2000)
31. Son, N.T., Zolnai, Z., Janzén, E.: Silicon vacancy related TV2a centre in 4H-SiC. Phys. Rev. B **68**, 205211 (2003)
32. von Bardeleben, H.J., Cantin, J.L., Henry, L., Barthe, M.F.: Vacancy defects in p-type 6H-SiC created by low-energy electron irradiation. Phys. Rev. B **62**, 10841–10846 (2000)
33. Bockstedte, M., Mattausch, A., Pankratov, O.: Ab initio study of the migration of intrinsic defects in 3C−SiC. Phys. ReV. B **68**, 205201 (2003)
34. von Bardeleben, H.J., Cantin, J.L., Vickridge, I., Battistig, G.: Proton-implantation-induced defects in n-type 6H- and 4H−SiC: an electron paramagnetic resonance study. Phys. Rev. B **62**, 10126–10134 (2000)
35. Torpo, L., Nieminen, R.M., Laasonen, K.E., Poykko, S.: Silicon vacancy in SiC: a high-spin state defect. Appl. Phys. Lett. **74**, 221–223 (1999)
36. Deak, P., Miro, J., Gali, A., Udvardi, L., Overhof, H.: The spin state of the neutral silicon vacancy in 3C–SiC. Appl. Phys. Lett. **75**, 2103–2105 (1999)
37. Vainer, V.S., Veinger, V.I., Il'in, V.A., Tsvetkov, V.S.: Electron spin resonance associated wiyh the triplet state of secondary thermal defects in 6H-SiC, Sov. Phys. Solid State **22**, 2011 (1980)
38. Vainer, V.S., Il'in, V.A.: Electron spin resonance of exchange-coupled vacancy pairs in hexagonal silicon carbide. Sov. Phys. Solid State **23**, 2126 (1981)
39. Vainer, V.S., Il'in, V.A.: Sov. Phys. Solid State **23**, 1432 (1981)
40. Gerstmann, U., Rauls, E., Frauenheim, Th., Overhof, H.: Formation and annealing of nitrogen-related complexes in SiC. Phys. Rev. B **67**, 205202 (2003)
41. von Bardeleben, H.J., Cantin, J.L., Rauls, E., Gerstmann, U.: Identification and magneto-optical properties of the NV centre in 4H−SiC. Phys. Rev. B **92**, 064104 (2015) and von Bardeleben, H.J., Cantin, J.L., Csoré, A., Gali, A., Rauls, E., Gerstmann, U.: NV Centers in 3C, 4H and 6H Silicon Carbide: a novel platform for solid state qubits and nanosensors. Phys. Rev. **B94**, 121202(R) (2016)
42. Kalabukhova, E.N., Savchenko, D., Shanina, B., Bagraev, N.T., Klyachkin, L., Malyarenko, A.: EPR study of the nitrogen containing defect centre created in self-assembled 6H SiC nanostructure. Mater. Sci. Forum **740–742**, 389–392 (2013)
43. Zywietz, A., Furthmüller, J., Bechstedt, F.: Vacancies in SiC: influence of Jahn-Teller distortions, spin effects, and crystal structure. Phys. Rev. B **59**, 15166–15180 (1999)
44. Bratus', V.Ya., Petrenko, T.T., Okulov, S.M., Petrenko, T.L.: Positively charged carbon vacancy in three inequivalent lattice sites of 6H-SiC: combined EPR and density functional theory study. Phys. Rev. B **71**, 125202 (2005)
45. von Bardeleben, H.J., Cantin, J.L., Gerstmann, U.: (to be published)
46. Umeda, T., Isoya, J., Morishita, N., Ohshima, T., Kamiya, T.: EPR identification of two types of carbon vacancies in 4H-SiC. Phys. Rev. B **69**, 121201 (2004)
47. Umeda, T., Isoya, J., Morishita, N., Ohshima, T., Kamiya, T., Gali, A., Deák, P., Son, N.T., Janzén, E.: EPR and theoretical studies of positively charged carbon vacancy in 4H-SiC. Phys. Rev. B **70**, 235212 (2004)
48. Umeda, T., Ishitsuka, Y., Isoya, J., Son, N.T., Janzén, E., Morishita, N., Ohshima, T., Itoh, H., Gali, A.: EPR and theoretical studies of negatively charged carbon vacancy in 4H−SiC. Phys. Rev. B **71**, 193202 (2005)
49. Umeda, T., Morishita, N., Ohshima, T., Itoh, H., Isoya, J.: Photo-EPR study of vacancy type defects in irradiated n-type 4H-SiC. Materials Sci. Forum **600**, 409–412 (2009)
50. Son, N.T., Hai, P.N., Janzén, E.: Silicon antisite in 4H-SiC. Phys. Rev. Lett. **87**, 045502 (2002)

51. Coomer, B.J., Resende, A., Goss, J.P., Jones, R., Öberg, S., Briddon, P.R.: The divacancy in silicon and diamond. Phys. B **273–274**, 520–523 (1999)
52. Twitchen, D.J., Newton, M.E., Baker, J.M., Anthony, T.R., Banholzer, W.F.: Electron paramagnetic resonance measurements on the divacancy defect center R4/W6 in diamond. Phys. Rev. **B59**, 12900–12910 (1999)
53. Pu, A., Avalos, V., Dannefaer, S.: Negative charging of mono-and divacancies in IIa diamonds by monochromatic illumination. Diam. Relat. Mater. **10**, 585–587 (2000)
54. Corbett, J.W., Watkins, G.D.: Silicon divacancy and its direct production by electron irradiation. Phys. Rev. Lett. **7**, 314–316 (1961)
55. Watkins, G.D., Corbett, J.W.: Defects in irradiated silicon: electron paramagnetic resonance of the divacancy. Phys. Rev. **138**, 543–555 (1965)
56. De Wit, J.G., Sieverts, E.G., Ammerlaan, C.A.J.: Divacancy in silicon: hyperfine interactions from electron-nuclear double resonance measurements. Phys. Rev. B **14**, 3494–3503 (1976)
57. Sieverts, E.G., Muller, S.H., Ammerlaan, C.A.J.: Divacancy in silicon: Hyperfine interactions from electron-nuclear double-resonance measurements. II. Phys. Rev. B **18**, 6834 (1978)
58. Baranov, P.G., Il'in, I.V., Mokhov, E.N., Muzafarova, M.V., Orlinskii, S.B., Schmidt, J.: EPR identification of the triplet ground state and photoinduced population inversion for a Si-C divacancy in silicon carbide. JETP Lett. **82**, 441–443 (2005)
59. Lingner, Th., Greulich-Weber, S., Spaeth, J.-M., Gerstmann, U., Rauls, E., Hajnal, Z., Frauenheim, Th., Overhof, H.: Structure of the silicon vacancy in 6H−SiC after annealing identified as the carbon vacancy–carbon antisite pair. Phys. Rev. B **64**, 245212 (2001)
60. Son, N.T., Magnusson, B., Zolnai, Z., Ellison, A., Janzén, E.: Defects in high-purity semi-insulating SiC. Mater. Sci. Forum **457–460**, 437–442 (2004)
61. Umeda, T., Son, N.T., Isoya, J., Janzén, E., Ohshima, T., Morishita, N., Itoh, H., Gali, A., Bockstedte, M.: Identification of the carbon antisite-vacancy pair in 4H-SiC. Phys. Rev. Lett. **96**, 145501 (2006)
62. Son, N.T., Carlsson, P., ul Hassan, J., Janzén, E., Umeda, T., Isoya, J., Gali, A., Bockstedte, M., Morishita, N., Ohshima, T., Itoh, H.: Divacancy in 4H-SiC. Phys. Rev. Lett. **96**, 055501 (2006)
63. Hunt, D.C., Twitchen, D.J., Newton, M.E., Baker, J.M., Anthony, T.R., Banholzer, W.F., Vagarali, S.S.: Identification of the neutral carbon ⟨100⟩ split interstitial in diamond. Phys. Rev. B **61**, 3863–3876 (2000)
64. Twitchen, D.J., Newton, M.E., Baker, J.M., Tucker, O.D., Anthony, T.R., Banholzer, W.F.: Electron-paramagnetic-resonance measurements on the di-⟨001⟩-split interstitial centre (*R*1) in diamond. Phys. Rev. B **54**, 6988–6998 (1996)
65. Watkins, G.D.: Intrinsic defects in silicon. Mater. Sci. Semicond. Process. **3**, 227–235 (2000)
66. Abdullin, K.A., Mukashev, B.N., Gorelkinskii, YuV: Metastable oxygen-silicon interstitial complex in crystalline silicon. Semicond. Sci. Tech. **11**, 1696–1703 (1996)
67. Lee, Y.H., Corbett, J.W.: EPR evidence of self interstitials in neutron irradiated silicon. Solid State Commun. **15**, 1781–1784 (1974)
68. Petrenko, T.T., Petrenko, T.L., Bratus', V.Ya.: The carbon <100> split interstitial in SiC. J. Phys.: Condens. Matter **14**, 12433–12440 (2002)
69. Son, N.T., Chen, W.M., Lindström, J.L., Monemar, B., Janzén, E.: Carbon-vacancy related defects in 4H- and 6H-SiC. Mat. Sci. Eng.: B **61**, 202–206 (1999)
70. Lento, J.M., Torpo, L., Staab, T.E.M., Nieminen, R.N.: Self interstitials in 3C-SiC. J. Phys. Cond. Matter **16**, 1053–1060 (2004)
71. von Bardeleben, H.J., Cantin, J.L., Gerstmann, U., to be published
72. Chow, K.H., Watkins, G.D., Usui, A., Mizuta, M.: Detection of interstitial Ga in GaN. Phys. Rev. Lett. **85**, 2761–2764 (2000)
73. Gerstmann, U., Seitsonen, A.P., Mauri, F.: Ga self-interstitials in GaN investigated by *ab-initio* calculations of the electronic *g*-tensor. Phys. Status Solidi (b) **245**, 924–926 (2008)
74. Hai, P.N., Chen, W.M., Buyanova, I.A., Monemar, B., Amano, H., Akasaki, I.: Ga-related defect in as-grown Zn-doped GaN: an optically detected magnetic resonance study. Phys. Rev. B **62**, R10607–R10609 (2000)

75. von Bardeleben, H.J., Cantin, J.L., Gerstmann, U., Scholle, A., Greulich-Weber, S., Rauls, E., Landmann, M., Schmidt, W.G., Gentils, A., Botsoa, J., Barthe, M.F.: Identification of the nitrogen split interstitial $(N-N)_N$ in GaN. Phys. Rev. Lett. **109**, 206402 (2012)
76. Son, N.T., Hemmingson, C.G. et al.: Identification of the vacancy oxygen pair defect in GaN. Phys. Rev. **B80**, 153202 (2009)
77. Gerstmann, U., Seitsonen, A.P., Ceresoli, D., Mauri, F., von Bardeleben, H.J., Cantin, J.L.: Garcia Lopez, J.: Si_CC_{Si} antisite pairs in SiC identified as paramagnetic defects with strongly anisotropic orbital quenching. Phys. Rev. B **81**, 195208 (2010)
78. Baranov, P.G., Ilyin, I.V., Soltamova, A.A., Mokhov, E.N.: Identification of the carbon antisite in SiC: EPR of ^{13}C enriched crystals. Phys. Rev. B **77**, 085120 (2008)
79. Umeda, T., Ishoya, J., Ohshima, T., Morishita, N., Itoh, H., Gali, A.: Identification of positively charged carbon antisite-vacancy pairs in 4H-SiC. Phys. Rev. B **75**, 245202 (2007)
80. Wagner, R.J., Krebs, J.J., Stauss, G.H.: Submillimeter EPR evidence for the As antisite defect in GaAs. Solid State Commun. **36**, 15–17 (1980)
81. Goswami, N.K., Newman, R.C., Whitehouse, J.E.: The observation of high concentrations of arsenic anti-site defects in electron irradiated n-type GaAs by X-band EPR. Solid State Commun. **40**, 473–477 (1981)
82. von Bardeleben, H.J., Bourgoin, J.C., Miret, A.: Identification of the arsenic-antisite-arsenic-vacancy complex in electron-irradiated GaAs. Phys. Rev. B **34**, 1360(R)–1362(R) (1986)
83. Koschnick, F.K., Wietzke, K.-H., Spaeth, J.-M.: Optically detected magnetic resonance study of an arsenic-antisite–arsenic-vacancy complex in GaAs. Phys. Rev. B **58**, 7707–7716 (1998)
84. Jia, Y.Q., von Bardeleben, H.J., Stiévenard, D., Delerue, Ch.: Intrinsic defects in electron irradiated p-type GaAs. Mat. Sci. Forum **83–87**, 965–970 (1992)
85. Kennedy, T.A.: Defects and radiation effects in semiconductors. Inst. Phys. Conf. Ser. **46**, 375 (1978)
86. Kaufmann, U., Schneider, J., Räuber, A.: ESR detection of antisite lattice defects in GaP, $CdSiP_2$, and $ZnGeP_2$. Appl. Phys. Lett. **29**, 312–313 (1976)
87. O'Donnell, K.P., Lee, K.M., Watkins, G.D.: ODMR studies of antisite-related luminescence in GaP. Solid State Commun. **44**, 1015–1018 (1982)

Chapter 4
State-of-Art: High-Frequency EPR, ESE, ENDOR and ODMR in Wide-Band-Gap Semiconductors

4.1 Excitons, Shallow Electrons and Holes in Ionic-Covalent Silver-Halide Crystals: High-Frequency EPR, ESE, ENDOR and ODMR Studies

Silver halides have unique features in solid state physics because their properties are considered to be of borderline nature between ionic and covalent bonding. In AgCl the self-trapped hole (STH) is centred and partly trapped in the cationic sublattice, forming an Ag^{2+} ion inside of a $(AgCl_6)^{4-}$ complex as the result of the Jahn-Teller distortion. The STH in AgCl can capture an electron from the conduction band forming the self-trapped exciton (STE). The properties of the self-trapped excitons, such as exchange coupling, the ordering of the triplet and singlet sublevels, the dynamical properties of the singlet and triplet states, and the hyperfine interaction with the Ag and Cl (Br) nuclei are discussed. Recent results of a study of STE by means of high-frequency electron paramagnetic resonance, electron spin echo, electron–nuclear double resonance (ENDOR) and optically detected magnetic resonance (ODMR) are reviewed. The properties of the STE in AgCl crystals, such as exchange coupling, the ordering of the triplet and singlet sublevels, the dynamical properties of the singlet and triplet states, and the hyperfine interaction with the Ag and Cl (Br) nuclei are discussed. Direct information about the spatial distribution of the wave function of STE unpaired electrons was obtained by ENDOR. From a comparison with the results of an ENDOR study of the shallow electron centre and STH, it is concluded that the electron is mainly contained in a hydrogen-like 1*s* orbital with a Bohr radius of 1.51 ± 0.06 nm, but near its centre the electron density reflects the charge distribution of the hole. The hole of the STE is virtually identical to an isolated STH centre.

P.G. Baranov et al., *Magnetic Resonance of Semiconductors and Their Nanostructures*, Springer Series in Materials Science 253,
DOI 10.1007/978-3-7091-1157-4_4

4.1.1 Introduction

Self-trapping of charge carriers in the deformable lattice of a solid was predicted by Landau in 1933 for electrons [1] but among the best-known examples are the self-trapped holes (STHs) found in alkali halides and later in silver halides [2–9]. Silver halides have some unique features in solid state physics because their properties are considered to be of borderline nature between ionic and covalent bonding, that is, they are located close to the boundary corresponding to the critical value of ionicity 0.785, namely, 0.856 and 0.850 for AgCl and AgBr, respectively [10]. In contrast to the alkali halides where the self-trapped hole (V_K centre) is formed in the anionic sublattice and resides on two halide ions, in AgCl the hole is centered and partly trapped in the cationic sublattice, forming an Ag^{2+} ion inside of a $(AgCl_6)^{4-}$ complex. The ground state of the Ag^{2+} ($4d^9$) ion is a ^{2}D configuration and the corresponding energy level has fivefold degeneracy. In a cubic crystal field the level is split into a twofold (E_g) and a threefold (T_{2g}) degenerate level. These levels are split further due to a static Jahn-Teller distortion, which lowers the local symmetry to D_{4h} by elongation of the complex along a cubic axis.

Under ultraviolet (UV) light irradiation of AgCl and AgBr an electron is excited from the valence band into the conduction band and a hole is left in the valence band. In AgCl, a hole is self-trapped on a Ag^+ ion in a $d(x^2-y^2)$-type orbital as the result of a Jahn-Teller distortion of the Ag-ion coordination sphere along a cubic axis. The free electron can be captured by a Coulombic core to form shallow electron centres (SEC), which are believed to play an important role in the latent image formation process, or by STH to form self-trapped exciton (STE).

Although the STH is centered and partly located at a Ag^+ ion, its wave function also contains contributions from the $3s$ and $3p$ orbitals of the neighboring Cl-ions in the plane perpendicular to the distortion axis. In forming the STE, the hole is expected to weakly bind an electron, because of the high dielectric constant of AgCl. Recombination of the STE contributes to the broad luminescence band of AgCl crystals at low temperatures, which peaks at about 500 nm, as demonstrated by studies using the method of optically detected magnetic resonance (ODMR). The properties of the excited states of this STE have been the subject of many ODMR investigations [11–19]. However, a number of important questions remained unclear: (1) the structure of the energy levels for STE; (2) the exchange interaction between electron and hole in STE, the ordering of the singlet and triplet sublevels, (3) the dynamical properties of the excited singlet and triplet states, (4) the hyperfine (HF) interaction with the Ag and Cl nuclei; (5) the space distribution of the unpaired electron wave function; (6) the proofs of intrinsic nature of STH and STE; (7) the structure of STE as an STH coupled with an SEC. These problems cannot be solved with ODMR techniques at conventional microwave frequencies, because many resonance lines overlap (or partly overlap), and further because it is difficult to extract the dynamical properties from the continuous-wave (cw) ODMR spectra.

In this chapter the results of recent studies of bulk silver halides by means of ODMR, time-resolved electron paramagnetic resonance (EPR), electron spin echo (ESE), and electron-nuclear double resonance (ENDOR) at high frequency 95 GHz and multiquantum resonance studies are presented. Owing to the high-energy resolution that is achieved at such high frequencies and the time-resolved character of the experiment, it was possible to obtain detailed new information about the STE, STH and SEC in the silver chloride crystals.

4.1.2 *Self-trapped Excitons*

4.1.2.1 The Energy Levels of Self-trapped Excitons in AgCl

The existence of a singlet state of STE in AgCl separated by 1–25 cm^{-1} from the triplet state was assumed in [15]. In [16], the singlet state and the negative sign of the fine structure parameter D of the triplet state were suggested from the analysis of the dependence of ODMR intensity on the microwave chopping frequency. The singlet-to-triplet splitting was estimated to be 6–7 cm^{-1} in [16]. Up to 1992 only the triplet STE state was studied by ODMR. Two different experiments allowed to confirm the existence of the singlet state and to measure very precisely the triplet to singlet exchange splitting, namely, W-band (95 GHz) ODMR [20, 21] and multi-quantum Q-band (35 GHz) ODMR [22, 23].

In Fig. 4.1a the optically detected EPR spectrum is shown, which was observed at 95 GHz in the total luminescence of the AgCl crystal upon UV irradiation, with the magnetic field B parallel to one of the cubic axes $\langle 100 \rangle$ of the AgCl crystal. The transitions at 3.1572 and 3.3250 T correspond to resonances of the STH, with B_0 parallel to the distortion z-axis and B ‖ x,y: g values $g_{\parallel STH} = 2.147$ and $g_{\perp STH} = 2.040$, respectively, in agreement with previous studies [7–9]. The two lines at 3.3541 and 3.3865 T correspond to the low-field and high-field transitions of the STE, with B_0 parallel to the distortion z-axis: they are characterized by a $g_{\parallel STE} = 2.014$. The two lines at 3.4460 and 3.4782 T are the low-field and high-field transitions of the other two sites of the STE with B ‖ x,y: they are characterized by a g value $g_{\perp STE} = 1.960$. The resonance of the SEC at 3.6061 T has an isotropic g value $g_{SEC} = 1.881$.

The ODMR spectra of AgCl demonstrate the advantage of working at high microwave frequencies. The resonance lines of the STH, the STE, and the SEC in AgCl are separated so well that it is easy to follow their dependence on the orientation of B with respect to the crystal axes.

It was shown in [20] that the positions and the angular dependence of 95 GHz ODMR lines of the STE considerably deviate from those calculated with a simplified $S = 1$ spin Hamiltonian, which was used to treat the ODMR data at conventional X- and Q-band frequencies. At the high magnetic field of 3.3–3.5 T, at which the experiments at W-band have been performed, the triplet state of the STE cannot be considered as isolated from its accompanying singlet state. As a result,

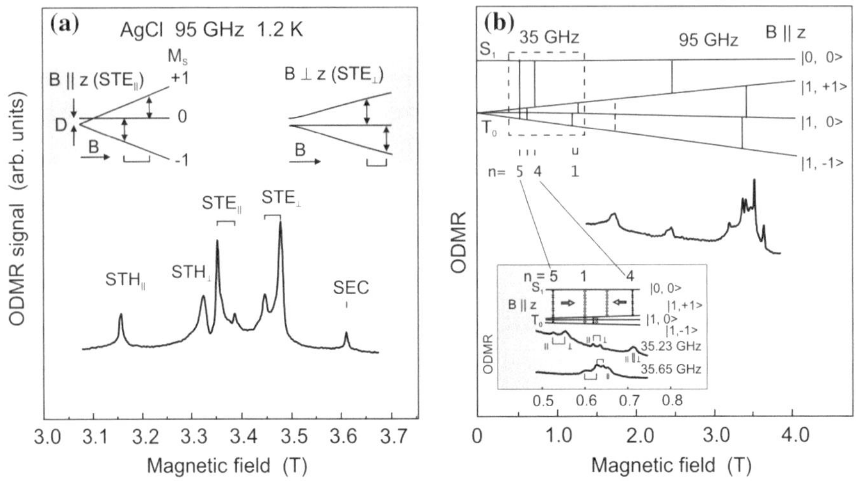

Fig. 4.1 **a** ODMR spectrum recorded at 95 GHz in the total luminescence of the AgCl crystal upon UV irradiation with the magnetic field B parallel to one of the cubic axes $\langle 100 \rangle$. The microwave was amplitude modulated at $f_m = 70$ Hz. T = 1.2 K. **b** 95 GHz ODMR obtained by sweeping the magnetic field over a broad range. The energy level scheme for STE and the observed transitions are shown in the *upper part* of the figure. *Inset* shows energy level scheme for STE in AgCl with the observed multiquantum transitions (n denotes the number of microwave quanta) and the ODMR signals measured at high microwave power (900 mW) and at two different microwave frequencies: 35.23 GHz (*black arrows* on the energy scheme) and 35.65 GHz (*open arrows*). $f_m = 80$ Hz, T = 1.8 K, B || [100]

the $M_S = 0$ sublevel of the triplet state is mixed with the singlet state. This leads, first of all, to a shift of the EPR transitions of the triplet state, which can be used for a determination of the value of the singlet-triplet splitting.

To explain the observed pattern of the resonance lines of the STE a more general spin Hamiltonian should be considered

$$\hat{H} = \mu_B \vec{B} \cdot \overleftrightarrow{g}_e \cdot \hat{\vec{S}}_e + \mu_B \vec{B} \cdot \overleftrightarrow{g}_h \cdot \hat{\vec{S}}_h + \hat{\vec{S}}_e \cdot \overleftrightarrow{D} \cdot \hat{\vec{S}}_h + J \hat{\vec{S}}_e \cdot \hat{\vec{S}}_h, \tag{4.1}$$

Here, the first two terms represent the Zeeman interaction of the electron and the hole forming STE. The third term describes the electron-hole spin-spin interaction and the last term describes their exchange interaction. The principal values of g-tensor for the hole are $g_{hz} = g_{h||}$ and $g_{hx} = g_{hy} = g_{h\perp}$, the g-factor of the electron g_e is isotropic. Hyperfine interactions are not included in (4.1).

In the first high-frequency ODMR experiments the exchange splitting J was derived, indirectly, via an analysis of the resonance fields of the ODMR transitions by taking into account the magnetic-field-induced mixing between the singlet state S_1 ($S = 0$) and the triplet state T_0 ($M_S = 0$) sublevel of the triplet state $S = 1$ [20].

Such a mixing makes the transition between the $M_S = 1$ triplet sublevel and S_1 slightly allowed. Indeed, a careful search revealed a resonance line in the ODMR spectrum which was attributed to this transition, as can be seen in Fig. 4.1b. Figure 4.1b shows the ODMR spectrum observed in the total luminescence of the AgCl crystal upon UV irradiation. The microwave power at 95 GHz was amplitude modulated at 70 Hz for lock-in detection and the magnetic field was directed parallel to the crystal [100] axis. Figure 4.1b clearly shows an ODMR line at the magnetic field 2.41 T. This line corresponds to the "forbidden" transition between the singlet state and the $M_S = +1$ sublevel of the triplet state of the STE. From its position the exchange coupling $J = -5.37 \pm 0.01\ \text{cm}^{-1}$ was derived [21]. A line at 1.7 T which is assigned to the $\Delta M_S = \pm 2$ transition.

Numerous multiquantum transitions corresponding to the absorption of up to seven microwave quanta (total energy 7*35 = 245 GHz) were found in the ODMR spectra of AgCl recorder in a wide magnetic field range from 0.5 to 4 T at high (>500 mW) microwave power [22].

Some of the observed 35 GHz multiquantum transitions are marked in the energy level scheme shown in Fig. 4.1b, n denotes the number of microwave quanta. The assignment of the ODMR lines to the multiquantum singlet-to-triplet transitions of the STE was unambiguously proved by the measurements of ODMR at slightly different microwave frequencies, i.e., with the different energy of the microwave quanta. Figure 4.1b (inset) shows a part of the ODMR spectrum recorded in AgCl at the frequencies of 35.23 and 35.63 GHz, the microwave power of 900 mW and T = 1.8 K, $B \parallel [100]$. The energy levels for $B \parallel z$ and the calculated positions of the EPR transitions for these two frequencies are shown in the upper part of the figure. One can see that the directions of the line shifts are different for different transitions and the value of the line shift is proportional to n in complete agreement with calculations. Observations of the singlet-to-triplet multiquantum transitions for STE in AgCl allowed one to measure J with extremely high accuracy: $J = 5.370 \pm 0.002\ \text{cm}^{-1}$ [22, 23].

Although multiquantum ODMR has been observed by different groups in several systems (see [22] and references therein), a complete understanding of this effect is still missing. One of possible explanations takes into account multiquantum transitions via real intermediate electronic states which may appear as a result of vibronic interaction in a quasi-degenerate system of electronic states of the centre. According to this mechanism both the electric (E) and magnetic (H) components of the microwave field can be active in the multiquantum transitions including an interference effect of E and H.

The sign of the zero-field splitting parameter D of the STE in AgCl was shown to be negative. This result confirms that the zero-field splitting in the triplet state is dominated by contributions of second-order spin-orbit coupling and that spin-spin (dipole-dipole) interactions are negligible.

4.1.2.2 The Dynamical Properties of the Singlet and Triplet States of the STE

The dynamical properties of the excited singlet and triplet state of the STE were studied in [21]. For this purpose, electron-spin-echo (ESE) technique in combination with pulsed laser excitation was used. The populating, depopulating, and spin-lattice relaxation processes have been studied. In addition, the results of ODMR experiments in magnetic field as well as in zero-field which support the structure of STE as the STH-captured shallow electron were presented.

The ESE-detected EPR spectrum of the AgCl crystal at 94.9 GHz with B ∥ [001] is presented in Fig. 4.2. The crystal is first excited by a laser flash of the XeCl excimer laser at 308 nm (duration 10 ns) and then subjected to a $(\pi/2) - \tau - \pi$ microwave pulse sequence starting at a time τ_d after the laser flash. The $\pi/2$- and π-pulse lengths are 30 and 60 ns, respectively, $\tau = 4$ μs, and $\tau_d = 5$ ms. The spectrum is obtained by monitoring the echo height at a fixed value of τ_d and τ and by sweeping the magnetic field. Two signals of the STH, four of the STE, and one of the SEC which are similar those in ODMR spectrum (Fig. 4.1) are recognized in the spectrum. We will concentrate on the signals of the STE.

First, the dephasing time T_2 of the signals in the spectrum displayed in Fig. 4.2 was measured. The results are listed in Table 4.1. The remarkable finding is that the value of T_2, for the STE is about five times shorter than that for the STH and the SEC. In addition, the spin-lattice relaxation time T_1 of all ground-state

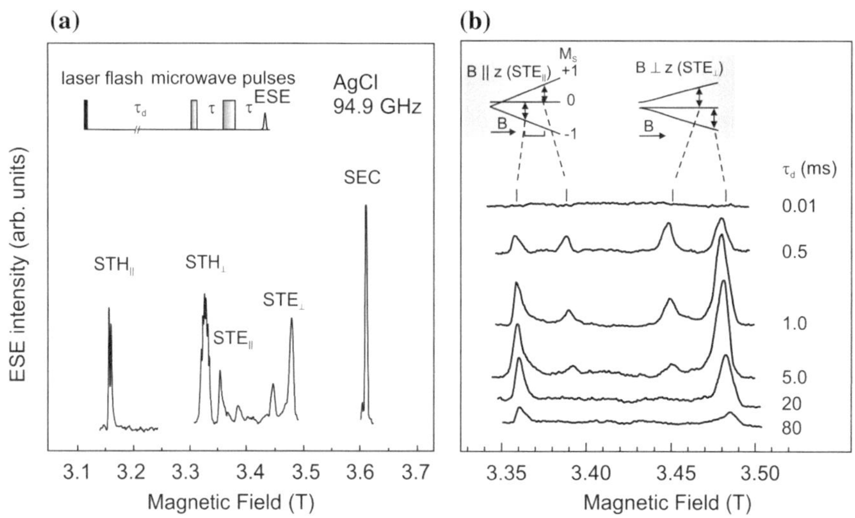

Fig. 4.2 **a** ESE-detected EPR spectrum of the AgCl crystal at 94.9 GHz with B ∥ [001] upon excitation by a laser flash at 308 nm (duration 10 ns). The $(\pi/2)$-τ-π microwave pulse sequence starts at a time τ_d after the laser flash. The $\pi/2$- and π-pulse lengths are 30 and 60 ns, respectively, $\tau = 4$ μs, and $\tau_d = 5$ ms, $T = 1.2$ K. **b** The ESE-detected EPR spectrum of the STE as a function of the delay τ_d after the laser flash

paramagnetic centres was estimated via the variation of the signal intensity with the repetition rate. These values are also given in Table 4.1.

Second, the evolution of the spectrum of the STE by varying the delay time τ_d after the laser flash was recorded for the delay times varying from 0.003 to 80 ms (see Fig. 4.2b). The striking aspect is that at the shorter delay time the signals, which are proportional to the population differences of the triplet sublevels connected by the microwave pulses, are zero. This means that within the experimental accuracy the populating rates of the sublevels are equal. The evolution of the signals is determined by the combined effect of decay and relaxation processes.

As mentioned already, the laser flash populates the sublevels M_S =+1, 0 and −1 equally. Then in the first 0.5 ms, the high-field and low-field signals of the $STE_{\parallel}$ and of the $STE_{\perp}$ start to develop equal absorptive intensities at the similar rates. This effect can only be understood by assuming that a dominant relaxation is present which transfers population from the $M_S = 1$ to the $M_S = -1$ sublevels. It is interesting to note that this rate is temperature independent, which is in agreement with the idea that this transfer corresponds to a direct process in which a phonon of energy ~190 GHz is created.

In the time interval between 1 and 5 ms the low-field signal of the $STE_{\parallel}$ and high-field signal of the $STE_{\perp}$ is further increasing but the high-field signal of the $STE_{\parallel}$ and the low-field signal of the $STE_{\perp}$ decreases to zero with the same rate. In the time interval between 5 and 100 ms, only the low-field $STE_{\parallel}$ and high-field $STE_{\perp}$ signals remain and start to decay with different rates. These rates become temperature independent below 1.3 K.

The time-resolved ESE and ODMR experiments have allowed unraveling the populating and decaying processes of the triplet spin sublevels. The relative populating rates in zero fields as well as in magnetic field were found to be equal. Apparently, upon excitation over the band-gap, pairs of electrons and holes are formed and the probabilities of generating the singlet state or the three sublevels of the triplet state of the STE are equal. The remarkable observation was the presence of the dominant and temperature-independent relaxation rate from the $M_S = 1$ to the $M_S = -1$ magnetic sublevels. It was proposed to be caused by a tunneling process in which the elongation of the STE changes direction from one cubic axis to another one. The consequence of this tunneling is that the expression for the zero-field Hamiltonian becomes time dependent and varies between $DS_z^2 - 1/3S(S+1)$ and $DS_x^2 - 1/3S(S+1)$ or $DS_y^2 - 1/3S(S+1)$. The two latter forms contain the bilinear operators S_x^2: and S_y^2, which are capable of inducing selectively transitions between the $M_S = 1$ and the $M_S = -1$ sublevels.

Table 4.1 The spin-spin relaxation time T_2 and the spin-lattice relaxation time T_1 of the STH, STE and SEC in the AgCl crystal at T = 1.2 K

Center	T_2 (μs)	T_1 (ms)
$STH_{\parallel}$	110 ± 5	10 ± 2
$STH_{\perp}$	110 ± 5	10 ± 2
$STE_{\parallel}$	17.5 ± 1.5	16.0 ± 0.4 (2 K)
$STE_{\perp}$	23.0 ± 0.8	20.4 ± 0.5 (2 K)
SEC	90 ± 5	10 ± 2

The value of T_2 shortens between 1.2 and 2.0 K. This indicates that the tunneling process is not a purely coherent process but that a thermally activated contribution is present. Further, it should be noted that T_2 for the parallel orientation is somewhat shorter than for the perpendicular one. This difference can be understood by the fact that in the parallel (z) orientation, energy jumps (to x or y) lead to a change in resonance frequency and thus to a dephasing of the triplet spins. In contrast in the perpendicular orientation, only half of the jumps (to z) destroy the phase coherence.

4.1.2.3 Hyperfine Interactions for the STE in AgCl

EPR and ODMR of the STE do not provide information on HF interactions since no resolved HF structure is observed. We, therefore, performed ENDOR spectroscopy at 94.9 GHz using a method which is based on the ESE-detected ENDOR, and in addition, the ODMR measurements in the zero magnetic fields.

Figure 4.3 shows the zero-field ODMR spectrum of the STE obtained at T = 1.2 K by scanning amplitude-modulated microwaves through resonance while detecting synchronously in the optical emission. The spectrum allows the observation of two components of the T_{0z}–$T_{0x,y}$ transitions between the triplet sublevels in zero magnetic field with a separation of 45 ± 4 MHz. We attribute the dominant contribution to this splitting to the HF-interaction term $A_{zz}S_zI_z$ of the triplet spin with the ^{107}Ag and ^{109}Ag nuclear spins. This term $A_{zz}S_zI_z$ gives matrix elements between T_{0x} and T_{0y} but since these two levels are degenerate it will lead to a first-order splitting of these levels equal to A_{ZZ}. The terms $A_{xx}S_xI_x$ and $A_{yy}S_yI_y$ only give second-order shifts of the order of 2 MHz. Further, for the Cl nuclei, the term

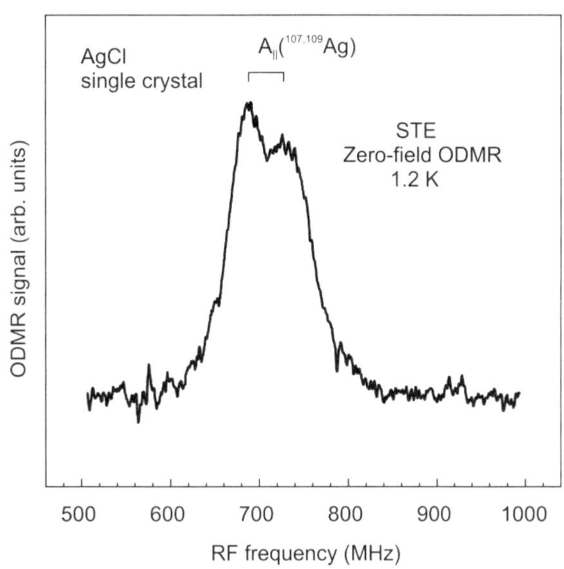

Fig. 4.3 Zero-field ODMR spectrum of the STE recorded by scanning the frequency of amplitude-modulated microwave field and lock-in detection of the emitted light. f_m = 43 Hz, T = 1.2 K

$A_{zz}S_zI_z$ is zero and the terms $A_{xx}S_xI_x$ and $A_{yy}S_yI_y$ also can give only second-order shifts of about 2 MHz. It is important to note that for the STH, a hyperfine interaction with the Ag nuclear spins $A_{zz} = 93 \pm 6$ MHz has been observed, i.e., almost exactly twice the value found for the STE. As a result it was concluded that this HF interaction in the STE is dominated by the electron spin of the self-trapped hole and that the contribution of the spin of the bound electron is small since the electron in the STE is very delocalized and the density of its wave function at the position of the Ag nucleus is too small to be detected. Thus, the STH is proved to reflect the hole part of the STE.

Figure 4.4a shows the ESE-detected EPR spectrum of STE and SEC in undoped AgCl crystal recorded with the magnetic field oriented along one of the cubic axes ($B \parallel [001]$) at 94.9 GHz and 1.2 K, under continuous UV irradiation. The spectrum is obtained by monitoring the height of the ESE signal detected at time τ after a $(\pi/2) - \tau - \pi$ microwave pulse sequence. The π/2 pulse length is 100 ns, τ = 650 ns, and the repetition rate of the pulse sequence is 33 Hz. The energy levels diagrams for the two orientations of STE in magnetic field with B parallel ($STE_{\parallel}$) and perpendicular ($STE_{\perp}$) to the distortion z-axis of the STE and for the SEC are presented on the top.

In Fig. 4.4b, two ENDOR spectra of the STE triplet state in AgCl are presented. For comparison the silver ENDOR spectrum of the shallow electron centre is shown in Fig. 4.4c. The spectra were recorded at 94.9 GHz and 1.2 K with B along a cubic axis, perpendicular to the distortion axis of the STE ($STE_{\perp}$): the upper spectrum in Fig. 4.4b was obtained by monitoring the high-field EPR transition of $STE_{\perp}$, which corresponds to a transition between the $M_S = 0$ and $M_S = -1$ sublevels of the triplet and the lower spectrum was recorded via the low-field transition of $STE_{\perp}$, which corresponds to the $M_S = 0 \leftrightarrow M_S = +1$ transition. Here a $(\pi/2) - \tau - \pi/2 - T - \pi/2$ microwave pulse sequence is applied resonant with the part of the EPR signal of STE presented in the insets. Two groups of lines can be distinguished which both contain nuclear transitions of ^{107}Ag and ^{109}Ag nuclei. The first group covers a range of a few MHz and is positioned in Fig. 4.4b above the nuclear Zeeman frequencies of ^{107}Ag and ^{109}Ag for the upper spectrum and below these frequencies for lower spectrum. The second group covers a range of only a few tenths of a MHz and is located below the nuclear Zeeman frequencies in both spectra. The larger part of the silver ENDOR spectrum is isotropic, but in some spectral regions an anisotropic behavior is observed. A similarity of the angular dependences for STE with the angular dependences for the STH, which will be described in Sect. 5.4, was observed.

The ^{35}Cl and ^{37}Cl ENDOR spectra of the STE show essentially the same features as that of silver; however, the interpretation of the chlorine spectra is more complicated owing to the appearance of additional quadrupole lines.

To describe HF interactions, new terms were added to the general Hamiltonian (4.1)

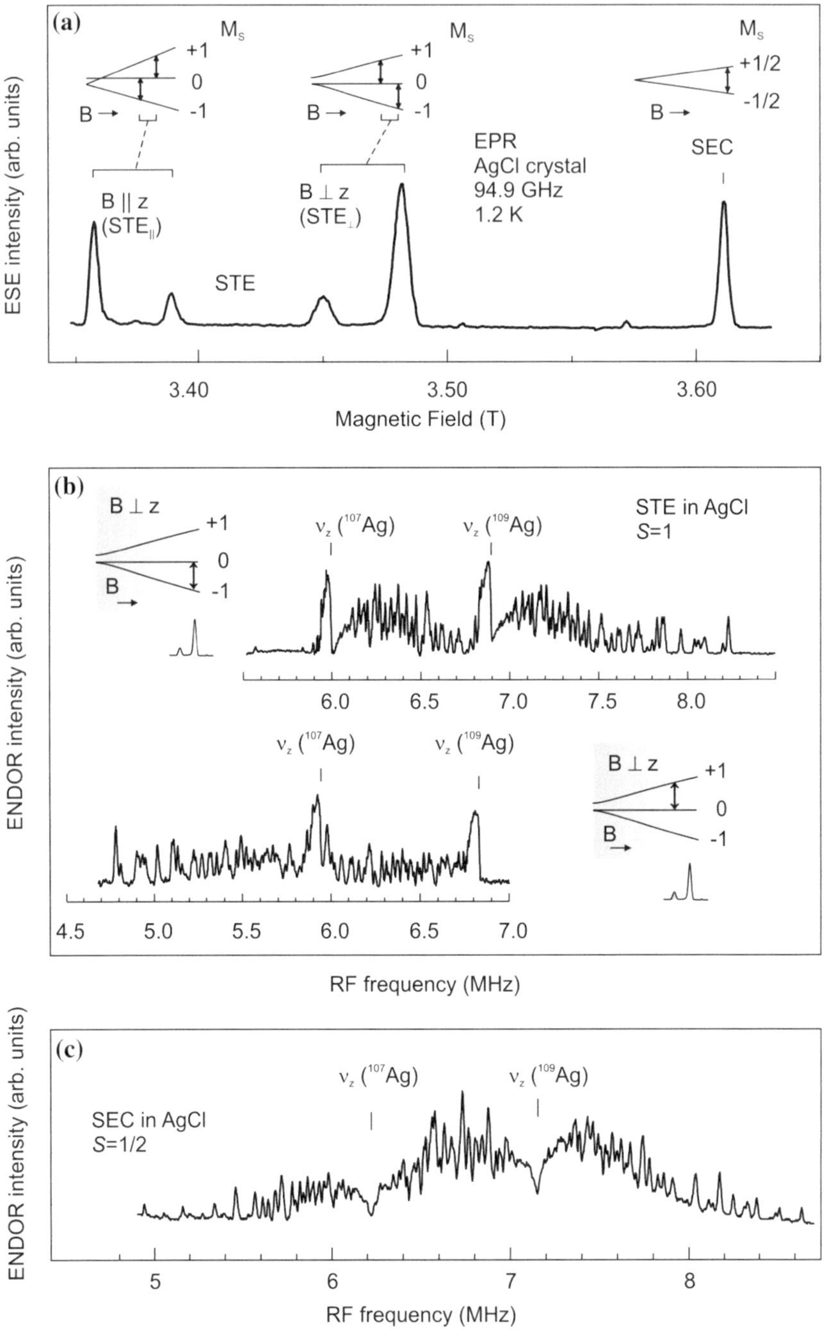
(a)
M_S
+1
0
-1
B →
M_S
+1
0
-1
B →
M_S
+1/2
-1/2
B →
EPR
AgCl crystal
94.9 GHz
1.2 K
SEC
B || z
(STE||)
B ⊥ z
(STE⊥)
STE
ESE intensity (arb. units)
3.40
3.50
3.60
Magnetic Field (T)
(b)
B ⊥ z
+1
0
-1
B
ν_z (107Ag)
ν_z (109Ag)
STE in AgCl
S=1
6.0
6.5
7.0
7.5
8.0
ν_z (107Ag)
ν_z (109Ag)
B ⊥ z
+1
0
-1
B
4.5
5.0
5.5
6.0
6.5
7.0
ENDOR intensity (arb. units)
RF frequency (MHz)
(c)
ν_z (107Ag)
ν_z (109Ag)
SEC in AgCl
S=1/2
ENDOR intensity (arb. units)
5
6
7
8
RF frequency (MHz)

◂**Fig. 4.4** **a** ESE-detected EPR spectrum of STE and SEC in undoped AgCl crystal recorded with B ∥ [001] at 94.9 GHz and 1.2 K, during continuous UV irradiation. The spectrum is obtained by monitoring the height of the ESE signal created at time τ after a (π/2) − τ − π microwave pulse sequence. The π/2 pulse length is 100 ns, τ = 650 ns, and the repetition rate of the pulse sequence is 33 Hz. The energy levels diagrams for the two orientations of the STE in magnetic field with *B* parallel ($STE_{\|}$) and perpendicular ($STE_{\perp}$) to the distortion *z*-axis and for SEC, are presented in the *top*. **b** Ag ENDOR spectra of the triplet state of the STE. The spectra were recorded at 94.9 GHz and 1.2 K with *B* along the cubic axis, which is perpendicular to the distortion axis of the STE ($STE_{\perp}$): the upper spectrum was obtained by monitoring the high-field EPR transition of $STE_{\perp}$, which corresponds to a transition between the $M_S = 0$ and $M_S = -1$ sublevels of the triplet and the lower spectrum was recorded via the low-field transition of $STE_{\perp}$, which corresponds to the transition between $M_S = 0$ and $M_S = +1$. **c** Silver ENDOR spectrum of the shallow electron centre at 94.9 GHz and 1.2 K. Typical pulse lengths: π/2 = 100 ns, τ between 400 and 900 ns, *T* = 700 μs and the length of radio-frequency (RF) pulse = 600 μs. Here a (π/2) − τ − π/2 − T − π/2 microwave pulse sequence is applied resonant with the part of the EPR signal of STE presented in (**b**)

$$-\mu_N g_n \vec{B}\cdot\hat{\vec{I}} + \hat{\vec{S}}_h\cdot\overleftrightarrow{A}_h\cdot\hat{\vec{I}} + \hat{\vec{S}}_e\cdot\overleftrightarrow{A}_e\cdot\hat{\vec{I}} + \hat{\vec{I}}\cdot\overleftrightarrow{Q}\cdot\hat{\vec{I}}. \tag{4.2}$$

Here, the first term describes the Zeeman interaction of a nucleus with spin $\hat{\vec{I}}$. The HF interaction of the nucleus with the hole and electron is given by the second and the third terms, respectively. The last terms describes the quadrupole interaction (only for Cl nuclei). The principal values of ***A*** tensor for hole are A_{hz}, A_{hx} and A_{hy}. The HF interaction for the electron is assumed to be isotropic, which is typical for the HF interaction of a shallow donor, thus ***A*** tensor is equal to A_e.

The frequencies for the $\Delta M_S = 0$, $\Delta m_I = \pm 1$ ENDOR transitions within the triplet sublevels are derived from (4.2) and are given by the following expressions [24]:

$$\begin{aligned}
&\text{for } M_S = +1,\ \nu_{\text{ENDOR}} = h^{-1}\left|-g_n\mu_N B_0 + \left(A_h^{\text{eff}} + A_e\right)/2\right|,\\
&\text{for } M_S = -1,\ \nu_{\text{ENDOR}} = h^{-1}\left|-g_n\mu_N B_0 - \left(A_h^{\text{eff}} + A_e\right)/2\right|,\\
&\text{for } M_S = 0,\ \nu_{\text{ENDOR}} = h^{-1}\left|-g_n\mu_N B_0 + J^{-1}\mu_B B_0\left(g_h^{\text{eff}} - g_e\right)\left(A_h^{\text{eff}} - A_e\right)/2\right|.
\end{aligned} \tag{4.3}$$

Here, $g_h^{\text{eff}} = (g_{hz} + g_{hx} + g_{hy})/3$ and $A_h^{\text{eff}} = (A_{hz} + A_{hx} + A_{hy})/3$.

On the basis of the expressions for the ENDOR frequencies, one can qualitatively understand the ENDOR spectra shown in Fig. 4.4b, which were recorded by monitoring the $M_S = 0 \leftrightarrow M_S = -1$ EPR transition. Since the STE is expected to consist of a delocalized electron trapped in the Coulombic field of a strongly localized hole, hyperfine constants A_h^{eff} are negligible for the majority of nuclei that contribute to the ENDOR spectrum. According to (4.3), the ENDOR transition in the $M_S = -1$ level will then lie at a frequency 1/2 $h^{-1}A_e$ above or below the nuclear Zeeman frequency $h^{-1}g_n\mu_N B_0$, depending on the sign of the hyperfine constant A_e. For the $M_S = 0$ level, (4.3) predicts a transition at the frequency $h^{-1}J^{-1}\mu_B B_0(g_h^{\text{eff}} - g_e)A_e/2$ above or below the nuclear Zeeman frequency, depending on the signs of *J* and A_e.

The ENDOR spectrum of the $M_S = 0$ level is compressed by a factor of ~ 20, as compared to the spectrum of the $M_S = -1$ level. The group of lines covering the broader frequency region in Fig. 4.4b is thus related to ENDOR transitions in the $M_S = -1$ level, whereas the compressed group corresponds to transitions in the $M_S = 0$ level. From the observation that the first group is positioned above the silver Zeeman frequencies and the fact that for silver g_n has a negative sign, we can derive from (4.3) that A_e has a negative sign. Because the compressed $M_S = 0$ ENDOR spectrum in Fig. 4.4b is located below the silver Zeeman frequencies and $A_e < 0$, the sign of the exchange energy J has to be negative according to (4.3). This is in agreement with previous results. The ENDOR spectrum shown in Fig. 4.4b, which was recorded by monitoring the $M_S = 0 \leftrightarrow M_S = +1$ EPR transition, confirms this qualitative analysis. Here, both the $M_S = +1$ spectrum and the compressed $M_S = 0$ spectrum are located below the silver Zeeman frequencies, just as predicted by (4.3) based on $A_h^{\text{eff}} \approx 0$ and the negative signs of A_e and J.

For those nuclei where $A_h^{\text{eff}} \neq 0$, the corresponding ENDOR lines are anisotropic. The dependences were obtained by monitoring the EPR transition of which the resonance field strength B_0 depends on the orientation of the magnetic field [24]. For a quantitative analysis of the ENDOR spectra, the lattice nuclei were grouped into shells. A shell contains all nuclei that have the same distance r from the central hole.

4.1.2.4 The Spatial Distribution of the Wave-Function of STE in AgCl

The ENDOR study has demonstrated that the lowest triplet state of the STE in AgCl consists of a very diffuse electron attracted by a strongly localized self-trapped hole [24]. The spatial distribution of the STE can roughly be divided into three regions. At distances larger than about 1.8 nm, the wave function of the STE is completely determined by the shallow electron, which occupies a hydrogen-like 1s orbital. At distances between 1.8 and 0.9 nm, the electron is still dominant but its density distribution deviates from the spherical symmetry and reflects the D_{4h} symmetry of the central STH. At distances shorter than about 0.9 nm, both the central hole and the shallow electron have a contribution to the wave function of the STE. From a comparison with the results of an ENDOR study of the STH, it is concluded that the central hole of the STE is virtually identical to an isolated STH.

For r > 1.8 nm the ENDOR frequencies are purely determined by HF interaction of the shallow electron and in the case of the $M_S = \pm 1$ sublevels are given by $\nu_{\text{ENDOR}} = h^{-1}|-g_n\mu_N B_0 \pm A_e/2|$. The ENDOR lines for Ag nuclei are isotropic, a property which is typical for the HF interaction between a nucleus and a very delocalized electron. This part of the ENDOR spectrum related to the remote nuclei resembles very much that of the intrinsic shallow electron centre in AgCl which is shown for Ag ENDOR in the bottom of Fig. 4.4b and will be discussed in the next section. The isotropic HF constant which reflects the spin density of the shallow electron wave function (Ψ) at the site of the nucleus (r_i) is given by

$$A_i = (8\pi/3)g_e\mu_B g_n\mu_N|\Psi(r_i)|^2,$$

where g_{ni} is the g-factor of nucleus i.

For a quantitative analysis of the observed isotropic HF constants, it is necessary to determine the spin density $|\Psi(r_i)|^2$ on the ligand nucleus. This analysis is based on the prediction of the effective mass theory that the ground state of a shallow donor in semiconductors can be described by a hydrogen like 1s wave function $\Phi(r) \sim \exp(-r/r_0)$ [25], where r_0 is effective Bohr radius of the shallow electron centre. Since in the STE the shallow electron is bound to a central hole with a Coulombic centre at a Ag ion, one expects that $\Phi(r)$ is centred on a Ag lattice position. Based on this information, it is possible to assign the lines in the silver ENDOR spectrum to the various shells of the remote class up to the 90th silver shell at a distance of 3.83 nm. In the remote region, the electron occupies a spherically symmetrical orbital with an exponential radial dependence, i.e., a hydrogen-like 1s orbital.

At smaller radii in the intermediate and nearby regions, the unpaired electrons spatial distribution differs in two ways from that in the remote region. First of all, the radial dependence deviates from the exponential behavior. This has been observed before in case of SECs and was interpreted as the breakdown of the oversimplifying hydrogen model [26]. Second, the angular distribution deviates from the spherical symmetry, thereby causing nuclei within a shell to become inequivalent. One may expect that at small radii, the bound electron becomes sensitive for the charge distribution of the binding core, leading to a redistribution of the electron density. Since the central hole is maintained in a $d(x^2 - y^2)$-type orbital, its charge distribution possesses D_{4h} symmetry. Therefore, the angular dependence of the electronic wave function is expected to reflect D_{4h} symmetry as well. Indeed this seems to be the case as suggested by the characteristic splitting of the intermediate ENDOR lines, and by the assignment of the nearby ENDOR lines. The analysis of the nearby silver nuclei showed that we are able to determine the signs of the HF-interactions of the central hole and the diffuse electron. The HF interactions of the hole and the electron were shown to have opposite sign. The HF interaction of the electron with nearby silver nuclei has a negative sign, just as for the remote silver nuclei, which implies that the shallow electron induces a positive electron spin density on silver nuclei (silver nuclei posses a negative magnetogyric ratio). Similarly, the positive sign of the HF interaction of the hole leads to the conclusion that nearby silver nuclei experience a negative hole spin density. From the remote part of the chlorine ENDOR spectrum, it follows that the shallow electron also induces a positive electron-spin density on the chlorine nuclei.

The central hole appears to be strongly localized, and the present results allow us to determine an upper limit of its spatial extension. The spin density of the central hole can only be detected up to the fifth silver shell at a radius of 0.87 nm. Since the electron is very diffuse, the overlap of the wave functions of the electron and hole building up the STE is very small, which accounts for the small value of the singlet-triplet splitting J.

The HF interactions with silver nuclei for the central hole of STE closely resemble those for STH, which will be discussed in detail further. This implies that the electronic structure of the STH is almost identical to that of the central hole of the STE. Apparently, the bound electron is so diffuse that it does not significantly influence the ionic equilibrium positions of the central hole. A small effect of the electron can be expected, since the activation energy for STE diffusion is lower than for STH diffusion and, moreover, the presence of the electron is believed to lower the barrier height between the three equivalent configurations of the elongated $(AgCl_6)^{4-}$ octahedron in the STE compared with that in the STH. In [21], a tunneling rate of $10^5\ s^{-1}$ at 1.2 K was derived, which should yield a lifetime-limited line width of about 30 kHz for the ENDOR transitions. The fact that in the present study line widths are observed down to 3 kHz suggests that the upper limit of the tunneling rate is rather in the order of $10^4\ s^{-1}$.

The similarity between the central hole of the STE and the STH obtained from the ENDOR results indicates that the STH is not accompanied by a nearby vacancy. Indeed, since such a STH-vacancy complex is charge-neutral, one does not expect this complex to attract an electron in a diffuse orbital to form the STE.

Circumstantial experimental evidence exists suggesting that Frenkel pairs are formed in the cationic sublattice of silver halides at liquid-helium temperatures (LHeT) upon UV irradiation. It was proposed that such a pair consists of a shallow electron centre at a silver interstitial and an STH centre at or near a silver vacancy. Recent experiments suggest that the shallow electron centre is in fact formed at a split-interstitial silver pair [26, 27], which is supported by the results of recent Hartree-Fock calculations [28]. In alkali halides, it is well known that Frenkel pairs are created in the anionic sublattice by the nonradiative decay of STEs. In contrast to AgCl where the STH centre resembles a $(AgCl_6)^{4-}$ molecular ion, the STH centre in alkali halides consists of a dimer X_2^- molecular ion, centred between two neighbouring anionic positions on the [110] axis (the V_k centre, X represents a halide ion). The V_k centre can capture an electron and form a STE which, after recombination, can produce a Frenkel pair consisting of a molecular X_2^- ionic hole centre at a single anionic lattice position (the H centre) and an electron trapped at an anion vacancy (the F centre). At LHeT the production of Frenkel pairs occurs most efficiently in those alkali halides where the lowest triplet state of the STE shows off-center relaxation, i.e., after the attraction of an electron the X_2^- dimer is shifted along the [110] axis. This off-center relaxation is believed to be essential for the Frenkel pair formation at LHeT. In case of the lowest triplet state of the STE in AgCl, no clear evidence for relaxation of the central hole or even of nearby silver ions is observed. Thus, there is no evidence that the suggested production of Frenkel pairs in AgCl at LHeT follows a similar path as in alkali halides. Moreover, the lack of observation of STHs accompanied by a vacancy in crystals, where no vacancies are deliberately introduced before irradiation, seems to indicate that the Frenkel pair formation at low temperatures is not efficient in AgCl.

Thus, the spatial distribution of the STE which will be compared with that of the SEC in Sect. 5.2 can roughly be divided into three regions: remote, intermediate and nearby as shown in Fig. 4.6b. At distances larger than about 1.8 nm, the wave function of the STE is completely determined by the shallow electron, which occupies a hydrogen-like 1*s* orbital. At distances between 1.8 and 0.9 nm, the electron is still dominant but its density distribution deviates from the spherical symmetry and reflects the D_{4h} symmetry of the central STH. At distances shorter than about 0.9 nm, both the central hole and the shallow electron have a contribution to the wave function of the STE. From a comparison with the results of an ENDOR study of the STH, one can conclude that the central hole of the STE is virtually identical to an isolated STH. Since the STE in AgCl consists of a very diffuse electron attracted by a strongly localized STH, it is of importance to consider separately the SEC and the STH.

4.1.3 Shallow Electron Centres

In [26, 27], the first direct reconstruction of the wave function of the intrinsic SEC in AgCl and AgBr was presented and a model of SEC was suggested in which an electron was shallowly trapped by two adjacent silver ions on a single cationic site.

In Fig. 4.4a, the ESE detected EPR spectrum of the SEC is shown. The ENDOR spectrum of the SEC is presented in Figs. 5.4c and 5.5a–c. The ENDOR transitions of silver nuclei are shown in Fig. 4.4c where the nuclear Zeeman frequencies of ^{107}Ag and ^{109}Ag are observable as dips at 6.224 and 7.156 MHz, respectively. Figure 4.5a shows the chlorine ENDOR transitions and the dips at 12.537 and 15.057 MHz indicate the nuclear Zeeman frequencies of ^{35}Cl (I = 3/2, 76%) and ^{37}CI (I = 3/2, 24%), respectively. The expanded spectra are shown in Fig. 4.5b, c for the high-frequency part of the ^{109}Ag and for the low-frequency part of ^{37}Cl, respectively.

The ENDOR spectra proved to be isotropic apart from a few lines in the chlorine spectrum which exhibit a quadruple splitting. The intensity of these few lines was so weak that the orientational dependences could not be resolved. In case of an electron spin (S = 1/2) coupled to a single silver nucleus (I = 1/2) the ENDOR transitions have the following frequencies:

$$\nu_{\mathrm{ENDOR}} = h^{-1}|-g_{\mathrm{n}}\mu_{\mathrm{N}}B_0 \pm A_{\mathrm{e}})/2|.$$

For a chlorine nucleus (I = 3/2) the same expression holds when the quadrupole interaction is neglected. Therefore, (4.3) predicts that each nucleus will give rise to two ENDOR transitions symmetrically placed around their nuclear Zeeman frequency. This behavior is indeed observed in the recorded spectra of Figs. 5.4c and 5.5. The fact that a multitude of lines is present indicates that we are dealing with a delocalized electron which interacts with a large number of Ag and C1 nuclei.

For a quantitative analysis of the HF constants, it is necessary to determine the spin density $|\Psi(r_i)|^2$ on each nucleus i. This problem has already been studied for the case of colour centres in alkali halides, particularly the F centre. It was shown [29] that, by orthogonalizing a suitable envelope function $\Phi(r)$ to the cores of the

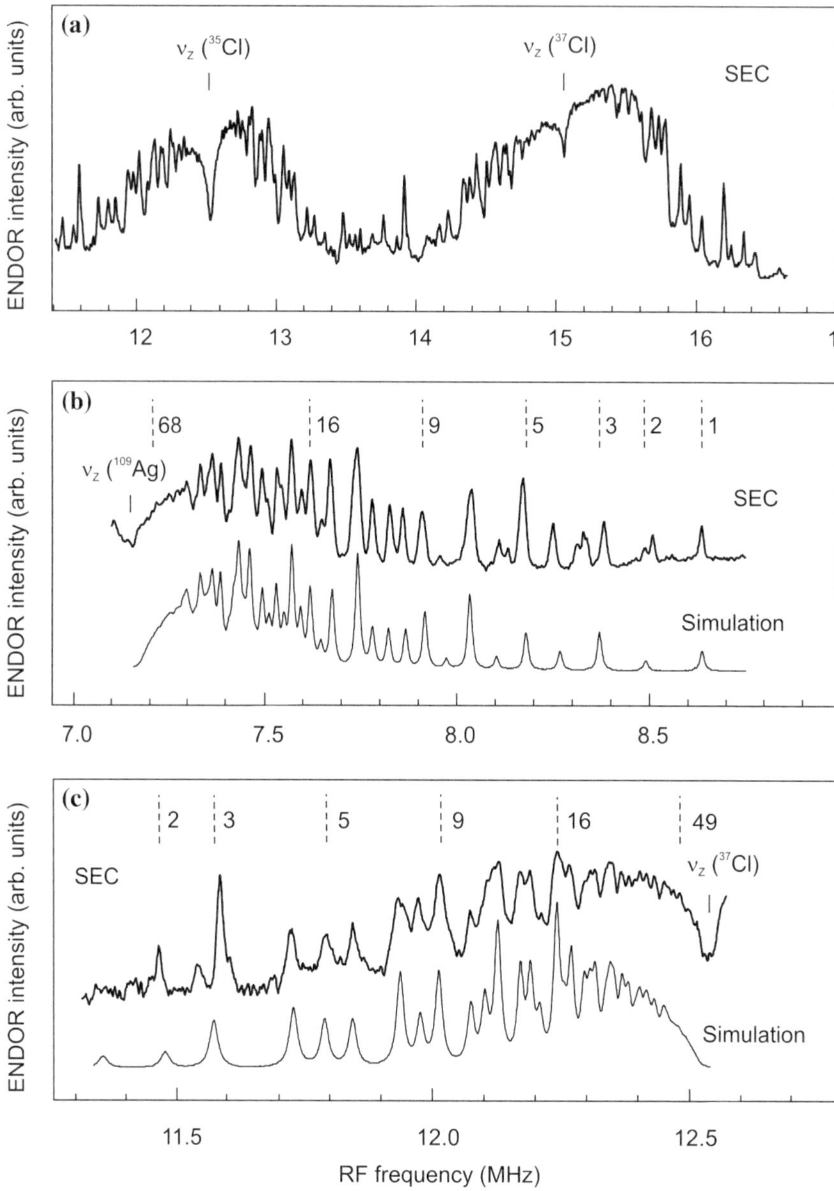

Fig. 4.5 **a** Chlorine ENDOR spectra of the shallow electron centre (S = 1/2, g = 1.878) recorded under the same conditions as in Fig. 4.4c. **b** Comparison between the recorded (*upper curve*) and simulated (*lower curve*) high-frequency part of the ^{109}Ag ENDOR spectrum of the SEC. **c** Similar comparison for the low-frequency part of the ^{37}C1 ENDOR spectrum. The first Cl shell is not observed in the ^{37}C1 ENDOR spectrum but is present in the ^{35}Cl spectrum. Lorentzian line shape was used to simulate the ENDOR spectrum

lattice ions in order to allow for the Pauli principle, the spin density on nucleus i may be written as an "amplification factor" K_i, times the density of the envelope function $\Phi(r)$ on that nucleus. Thus $|\Psi(r_i)|^2 = K_i|\Phi(r_i)|^2$. If the envelope function remains approximately constant within each ion core, the value of K_i will only depend on the species of ion i and not on its position in the lattice.

We used a trial-and-error procedure in which we calculated $|\Phi(r_i)|^2$ on a large number of Ag and Cl positions in the cubic AgCl lattice, trying various centre positions of Φ and by optimizing the amplification factor values for Ag and Cl and the Bohr radius r_0 to match the observed HF constants.

The result of our analysis is shown in Fig. 4.6a where the density of Φ is plotted as a function of r. This result is based on the assumption that Φ is centered on the Ag^+ lattice position. One has been able to derive the density of the envelope function on a large number of Ag and Cl shells and some of the Ag shells are indicated in Fig. 4.6a. It turned out that for shells with a radius larger than about 1.2 nm, the derived electron densities indeed obey the expected exponential form with a Bohr radius $r_0 = 1.66 \pm 0.08$ nm (dashed line in Fig. 4.6a); however, for nearby shells, there is a clear deviation. This results from the neglect of the influence of the chemical nature of the binding centre and it illustrates the need of the so-called central cell correction in the effective mass theory [25]. The derived densities can be very well described by the following normalized monotonically decreasing function [26]

$$|\Phi(r_i)|^2 == (7\pi r_1)^{-1}(1 + r/r_1)^2 \exp(-2r/r_1), \tag{4.4}$$

which for comparison is plotted as a solid line in Fig. 4.6a using amplification factors $K_{Ag} = 2450$, $K_{Cl} = 1060$ and $r_1 = 0.994$ nm.

The experimental ENDOR spectrum was simulated by calculating the frequency of the ENDOR transition using the result of such a simulation of the high-frequency part of the ^{109}Ag ENDOR spectrum shown in Fig. 4.5b, where it is compared to the recorded spectrum. The figure shows a good overall agreement between the recorded and the simulated spectrum. The simulation, however, does not account for all features observed at frequencies above 8.1 MHz. The ENDOR lines in this region correspond to shells that lie close to the centre of Φ and suggest the presence of the lattice distortion in the direct surrounding of the binding core. One indicated the ENDOR transitions of some of the Ag shells. One can see that even the contribution of the 68th silver shell can be resolved. Figure 4.5c compares the recorded spectrum of the low-frequency part of the ^{37}Cl ENDOR to its corresponding simulation. Again a good overall agreement is obtained and the contributions up to the 49th chlorine shell can be observed.

The results depicted in Fig. 4.5b, c were obtained by placing the centre of the envelope function Φ on the Ag^+ lattice position and this turned out to be the only position for which we could obtain a satisfactory analysis of the Ag and Cl ENDOR spectra. A displacement of the interatomic distance by only 4% in any direction would already worsen the agreement between the simulated and the recorded spectra.

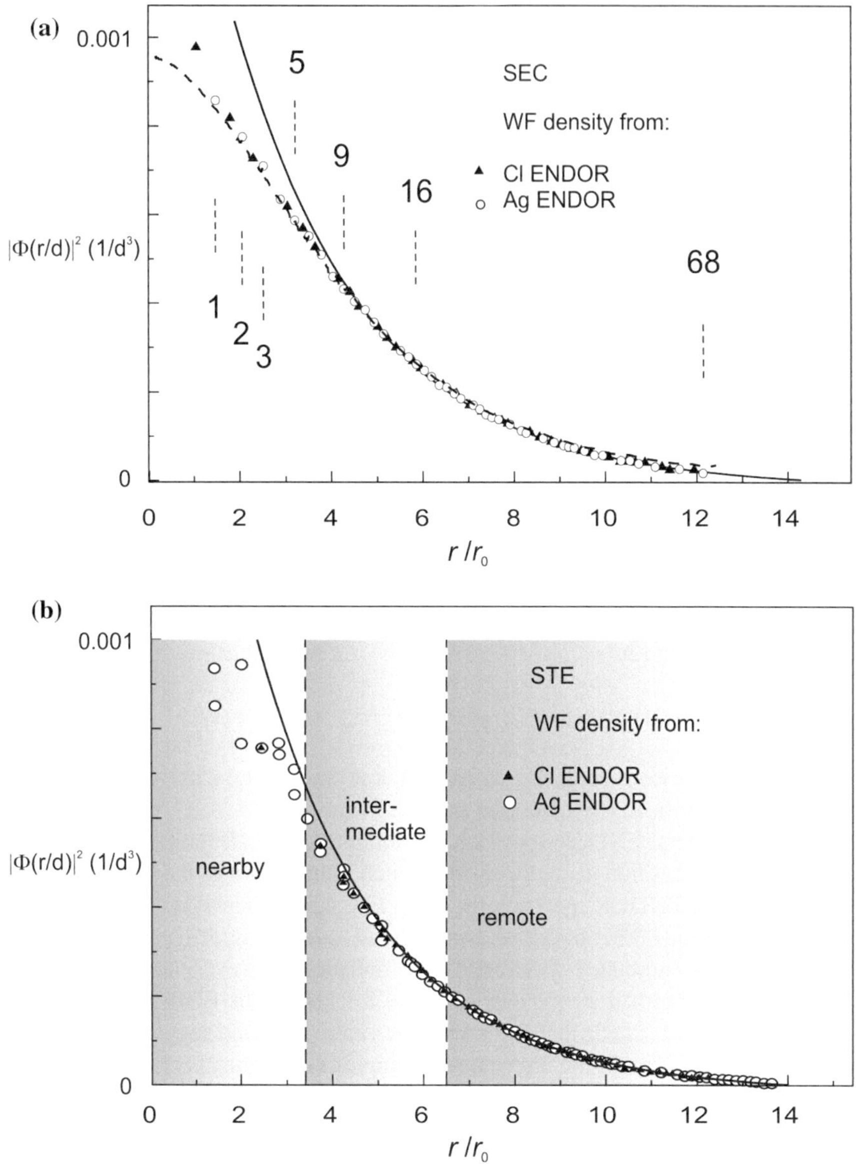

Fig. 4.6 Density of the envelope function $|\Phi(r)|^2$ of the SEC (**a**) and STE (**b**) as a function of (r/d) with the interionic distance $d = 0.2753$ nm. *Open circles* and *triangles* denote the densities derived from the Ag and Cl ENDOR spectra, respectively. Some neighboring silver shells are indicated. *Solid lines* illustrate the exponential radial dependence in the remote region corresponding to a Bohr radius $r_0 = 1.66$ nm for SEC and $r_0 = 1.51$ nm for STE. *Dashed line* in **d** is a result of calculation using (4.4)

Information concerning the charge of the intrinsic SEC can be obtained from ENDOR on the STE. The STE consists of an electron loosely bound to a STH and can therefore be considered as a special case of an SEC. It is possible to derive the spatial distribution of the electronic part of the STE, in a similar way as described here for the intrinsic SEC. The density of Φ for the STE is plotted as a function of r in Fig. 4.6b. One can found that the shallowly trapped electron of the STE also behaves very much like a hydrogen 1s electron, centered on the Ag^+ lattice position, with a Bohr radius $r_0 = 1.51 \pm 0.06$ nm [24]. The close agreement of this value with the one derived for the intrinsic SEC ($r_0 = 1.66 \pm 0.08$ nm) [27] and the fact that the electron of the STE is shallowly bound by the Coulombic field of an STH indicates that the SEC has the same Coulombic charge. For SEC in AgBr, $r_0 = 2.48 \pm 0.23$ nm [27].

Since these results show that the intrinsic SEC is located on the Ag^+ lattice position, the previously suggested model of an interstitial Ag^0 "atom", which consists of an electron loosely bound to a single interstitial Ag^+ ion, was rejected. Such molecular Ag_2^+ ions have been observed in KC1 crystals doped with silver after X-ray irradiation at room temperature, however, in KC1 they form, in contrast to the present situation, deep electron traps [30].

In conclusion, the presented ENDOR results prove that the $g = 1.878$ paramagnetic centre in undoped AgCl (previously observed in [31]) originates from an SEC. It is shown that the centre is located on the Ag lattice position within the accuracy of 4% of the interatomic distance and on the basis of the comparison of the ENDOR results with the optical data [32], the centre is concluded to be of intrinsic origin. The electron is suggested to be shallowly trapped in the Coulombic field of two adjacent Ag^+ ions symmetrically placed on a single cationic site.

4.1.4 Self-trapped Holes

In Fig. 4.7a, the low-temperature ESE-detected EPR spectrum of the STH in AgCl crystal at 94.9 GHz recorded with the magnetic field oriented along a cubic axis $\langle 100 \rangle$ and under continuous UV irradiation is shown [33]. The EPR signals of the STH are indicated in the figure. The transitions labelled with the symbol $\|$ belong to the site for which the magnetic field B_0 is oriented parallel to the distortion z-axis, whereas the transitions indicated by $\perp$ result from the two sites with B_0 perpendicular to this axis. Since the STH complex is Jahn-Teller distorted along one of the cubic axes, the resonances of three sites are observed simultaneously.

To analyse the EPR and subsequent ENDOR spectra of the STH one will use the Hamiltonian

$$\hat{H} = \mu_B \vec{B} \cdot \overleftrightarrow{g}_h \cdot \hat{\vec{S}}_h + \sum_i (\hat{\vec{S}}_h \cdot \overleftrightarrow{A}_{hi} \cdot \hat{\vec{I}}_i \cdot + \hat{\vec{I}}_i \cdot \overleftrightarrow{Q} \cdot \hat{\vec{I}}_i - g_n \mu_N \vec{B} \cdot \hat{\vec{I}}_i). \quad (4.5)$$

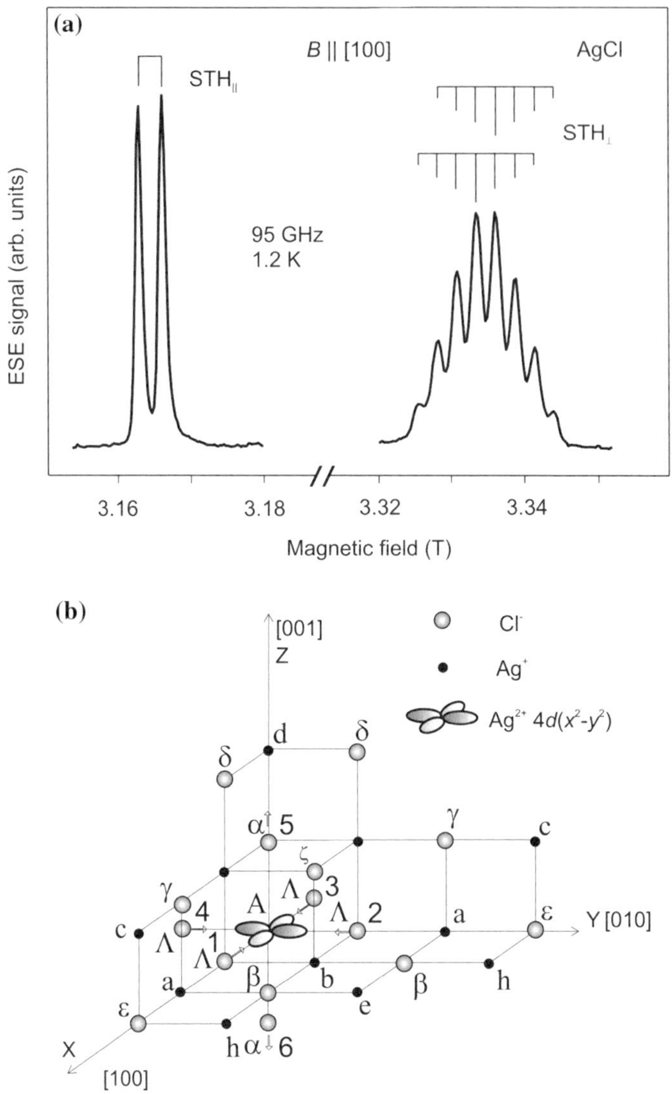

Fig. 4.7 **a** ESE-detected EPR spectrum of the STH in AgCl single crystal recorded under continuous ultraviolet irradiation at 94.9 GHz and 1.2 K, with the magnetic field oriented along a ⟨100⟩ cubic axis. **b** STH in the AgCl lattice. *X*, *Y* and *Z* are the symmetry axes of the centre, *Z* is taken along the distortion axis. The principal axis system of the HF tensor of the central silver ion is identical to that of the **g**-tensor. The hole is well localized inside the shells formed by Ag ions labeled as Latin characters A, a, b, c, d, e, h and by Cl ions labeled as Greek characters Λ, α, β, γ, δ, ε, ζ for which the HF and quadrupole interactions were resolved by ENDOR. The local principal *Z*′-axis for the HF tensor of each ligand ion is directed along the line interconnecting the nucleus under consideration and the central silver. The angle Θ is defined as the angle between the *Z*-axis and the magnetic field, the angle φ is defined as the angle between the *X*-axis and the magnetic field that is in the plane perpendicular to the *Z*-axis

Here, $\hat{\vec{S}}_h$ represents the electron spin operator of the hole with $S = 1/2$ and $\hat{\vec{I}}_i$ are nuclear spin operators. The terms with $\hat{\vec{I}}_i$ are summed over the nuclei of Ag and Cl that interact with the hole. The $\vec{g}_h$ tensor reflects axial symmetry around the distortion axis of the STH with principal values $g_{\|} = 2.147 \pm 0.002$ and $g_{\perp} = 2.040 \pm 0.004$. The second, third and the fourth terms reflect the HF, quadrupole (only for chlorine) and nuclear Zeeman interactions, respectively. The HF interaction parameters can be given in terms of an isotropic part, represented by a, and a (traceless) anisotropic part, represented by b and b', which are related to the principal HF-tensor axis system ($X'Y'Z'$) by $A_{X'X'} = a - b + b'$, $A_{Y'Y'} = a - b - b'$, $A_{Z'Z'} = a + 2b$. Similarly, one defines the diagonal matrix elements describing the nuclear quadrupole interaction in principal axis system, as follows $Q_{X'X'} = -q + q'$, $Q_{Y'Y'} = -q - q'$, $Q_{Z'Z'} = 2q$. The parameters b' and q' denote the deviation from the axial symmetry. The third term in (4.5), which reflects the quadrupole interaction of the chlorine nuclei, can be written in the principal axis system as $P[\hat{I}^2_{Z'} - 1/3I(I+1)] + q'(\hat{I}^2_{X'} - \hat{I}^2_{Y'})$. Here $P = 3/2Q_{Z'Z'} = 3q$ and $q' = ½(Q_{X'X'} - Q_{Y'Y'})$.

To explain the EPR spectrum of the STH presented in Fig. 4.7a it was assumed that only the central silver nucleus and the four equatorial chlorine nuclei in $\langle 100 \rangle$ positions in the plane perpendicular to the elongation axis affect the EPR spectrum, as their HF-tensor values are much larger then those of the other nuclei. The HF-interactions with other Ag and Cl nuclei were not known prior to [24, 33], and no information was available about quadrupole interactions with the Cl ions. The surrounding ions of the STH centre for which the HF and quadrupole interactions were resolved by EPR and ENDOR are presented in Fig. 4.7b. The hole is well localized inside the shell formed by Ag ions labelled with Latin characters as A, a, b, c, d, e, h and by Cl ions labelled with Greek characters as Λ, α, β, γ, δ, ε, ζ. The density outside this shell must be very small. The origin of the axis system coincides with the centre of the STH (a silver Ag^{2+} ion position labelled by A). The following orientations for the principal axes system X', Y', Z' of the equatorial chlorine HF-tensor with respect to the distortion axis X, Y and Z of the STH shown in Fig. 4.7b, were used: the Z' principal axis of HF-tensor is directed along the line interconnecting the chlorine nucleus 1, 2, 3 and 4 and the central silver, X' lies in the XY plane and Y' is perpendicular to the XY plane.

The doublet structure of the $STH_{\|}$ transition in Fig. 4.7a indicates that, when B_0 is oriented along the distortion axis, the HF coupling with the central silver nucleus is dominant compared to the HF couplings with four equatorial chlorine ligands labelled as Cl_{Λ}, and HF couplings with other surrounding nuclei. From the observed splitting, a value of $A_{\|}$ was previously derived (Table 4.2). In the case of the $STH_{\perp}$ transition, where B_0 is oriented perpendicular to the distortion axis, the observed eightfold structure results from the dominant HF interaction of the hole with the central silver nucleus and with the chlorine nuclei of two equivalent equatorial chlorine ligands, for which the HF-interaction parameters were previously derived and presented in Table 4.2.

Table 4.2 HF interaction parameters of the central silver ($^{109}Ag_A$) and HF and quadrupole interaction parameters of the equatorial chlorine ions (Cl_Λ) of the STH

Central silver (Ag_A)	$A_\parallel$ (MHz)	$A_\perp$ (MHz)	Equatorial chlorine ions (Cl_Λ)	$A_{Z'Z'}$ (MHz)	$A_{Y'Y'}$ (MHz)	$2\|P\|$ (MHz)	Ref.
^{109}Ag	(−) 100.0 (1)	(−) 65.8 (1)	^{35}Cl	(−) 81.4 (1)	(+) 6.4 (1)	10.3 (1)	[24]
^{109}Ag	96 ± 6	63 ± 6	^{35}Cl	86 ± 6			[7]

Only the absolute value has been obtained for the quadrupole interaction. The uncertainty in the last digit of these constants is indicated in brackets. The signs of the HF interactions (in brackets) were determined from the 'anomalous' ENDOR effect [24, 33]. Note that in [19, 21] only absolute values of the HF parameters were obtained

From the ENDOR experiment, the HF and quadrupole interactions with ligands that are not resolved in EPR were obtained. Also, much more accurate values for the central Ag_A and equatorial Cl_Λ HF interactions could be measured. Figure 4.8 shows the low-frequency part (Fig. 4.8a) and the high-frequency part (Fig. 4.8b, c) of the ENDOR spectra recorded with the magnetic field oriented along the Z (Fig. 4.8b) and X (Fig. 4.8c) axes of the $\boldsymbol{g}$-tensor by monitoring the EPR transitions $STH_\parallel$ and $STH_\perp$ shown in Fig. 4.7a. The frequencies of the ENDOR transitions in these spectra can be analysed by the first-order solution of the Hamiltonian (4.5), with the selection rules $\Delta M_S = 0$, $\Delta m_I = \pm 1$, which yields

$$\nu_{\mathrm{ENDOR}}(M_S) = h^{-1}|AM_S - g_n\mu_N B_0 + m_q(2P)|, \tag{4.6}$$

where A is the HF coupling with the central silver (Ag_A). The HF interaction of the four equatorial chlorine ligands are labelled as Cl_Λ and the HF interactions for remote silver and chlorine nuclei are labelled by a to h and α to ζ, respectively. P is the quadrupole interaction parameter for chlorine nuclei and the variable m_q is given by $m_q = 1/2\ (m_I + m_I^*)$, where m_I and m_I^* are the chlorine nuclear spin states involved in the transition. For the central silver, $|A(Ag_A)| \gg |g_n(Ag)\mu_B B_0|$ and it is clear that for each silver isotope, two transitions are expected which must be separated by two times the nuclear Zeeman frequency $\nu_Z(Ag) = |g_n(Ag)\mu_B B_0|$. These transitions are easily recognized in the spectra of Figs. 5.8b, c, and the extracted values are listed in Table 4.2. These values are in agreement with the ones found by EPR, which are less accurate and moreover reflect the averaged value of ^{107}Ag and ^{109}Ag interactions.

The HF couplings with chlorine isotope of the four equatorial chlorine ligands (Cl_Λ) are in the limit that $|A(Cl_\Lambda)| \gg |g_n(Cl)\mu_B B_0|$. This again leads to the detection of two lines with a separation of two times the nuclear Zeeman frequency $\nu_Z(Cl) = |g_n(Cl)\mu_B B_0|$, corresponding to the transitions for which $m_q = 0$. However, each $m_q = 0$ 'HF' line is accompanied by two additional lines at $\Delta\nu = \pm 2P$, which corresponds to $m_q = \pm 1$ transitions. In Fig. 4.8c one has marked all six ENDOR transitions of the ^{37}Cl and ^{35}Cl isotopes and attached the m_q value in the subscript.

The corresponding HF interactions $A_{Z'Z'}(Cl_\Lambda)$ are listed in Table 4.2 and agree with those obtained from EPR, which are again less accurate and reflect the

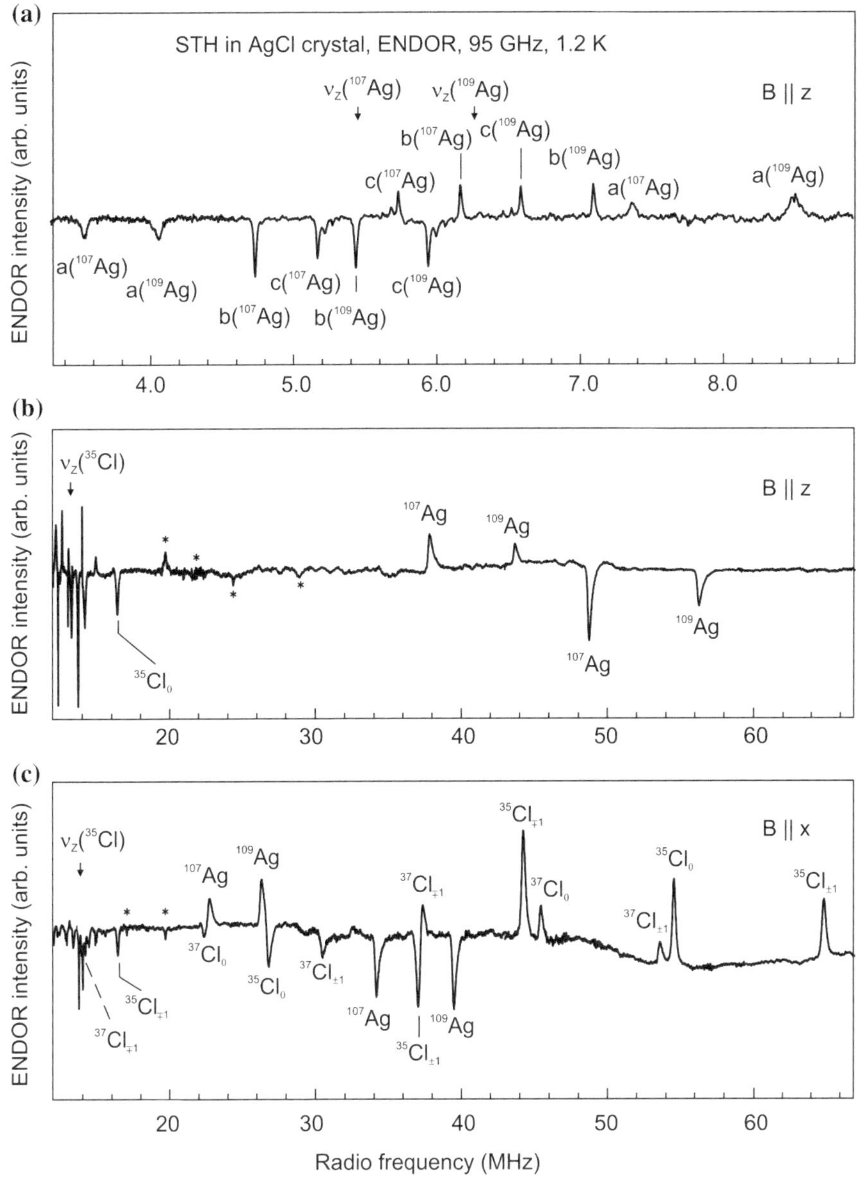

Fig. 4.8 Low-frequency (**a**) and high-frequency (**b**) parts of the ENDOR spectra of the STH, recorded with $B_0 \parallel Z$ and high-frequency part (**c**) recorded with $B_0 \parallel X$ by monitoring the EPR transitions indicated in Fig. 4.7. The ^{107}Ag and ^{109}Ag resonances of the central silver ion and the ^{37}Cl and ^{35}Cl resonances of the four equatorial chlorine ligands are indicated. In **a** $\nu_Z(^{107}\mathrm{Ag})$ = 5.4 MHz, $\nu_Z(^{109}\mathrm{Ag})$ = 6.3 MHz and $\nu_Z(^{35}\mathrm{Cl})$ = 13.2 MHz, whereas in **b** $\nu_Z(^{107}\mathrm{Ag})$ = 5.7 MHz, $\nu_Z(^{109}\mathrm{Ag})$ = 6.6 MHz, $\nu_Z(^{37}\mathrm{Cl})$ = 11.6 MHz and $\nu_Z(^{35}\mathrm{Cl})$ = 13.9 MHz. The transitions marked with *asterisks* are higher harmonics of the central silver ENDOR transition due to nonlinearity of the radio-frequency amplifier

averaged value of ^{35}Cl and ^{37}Cl interactions. Also, the magnitude of the quadrupole interaction $P_{Z'Z'}(\mathrm{Cl}_\Lambda)$ is included in Table 4.2. Figure 4.8c is recorded with $B_0 \parallel X$ and according to the definition of the principal axes of the HF tensor in Fig. 4.8b, the derived values apply for Cl_Λ ligands 1 and 3. For this orientation, Cl_Λ ligands 2 and 4 should give rise to resonances from which the value of $A_{X'X'}(\mathrm{Cl}_\Lambda)$ can be obtained. Since the EPR spectrum already indicates that the magnitude of $A_{X'X'}(\mathrm{Cl}_\Lambda)$ is small, the resonances of ligands 2 and 4 will probably lie close to or even coincide with those of the surrounding chlorine neighbours. In principle, the low-frequency Cl_Λ HF resonances might be identified from the angular dependence of the high-frequency ENDOR lines of Fig. 4.8c, when rotating the magnetic field B_0 from the X- to the Y-axis. Unfortunately, when B_0 is rotated over more than 10° from X, the high-frequency Cl_Λ HF resonances disappear. A similar observation is made when B_0 is rotated from X to Z, and consequently it is not immediately clear where the Cl_Λ HF-transitions are positioned for $B_0 \parallel Z$. The only candidate for the Cl_Λ HF-transition in this latter orientation, which is left after most of the chlorine HF-lines have been assigned, is the line at 16.4 MHz in Fig. 4.8b. If we assume that this line corresponds to the $m_q = 0$ 'HF' transition, this yields $|A_{Y'Y'}(^{35}\mathrm{Cl}_\Lambda)| = 6.4 \pm 0.1$ MHz. This value is in reasonable agreement with the one estimated from the EPR line width of the $\mathrm{STH}_\parallel$ transition $|A_{Y'Y'}(^{35}\mathrm{Cl}_\Lambda)| = 5.7$ MHz.

The spectra can be interpreted by (4.6) in the limit that the nuclear Zeeman frequencies of the silver and chlorine isotopes are large. Thus, two transitions are expected for each isotope at a distance $A/2$ above and below ν_Z. For the $m_q = 0$ 'hyperfine' transitions of the chlorine isotopes, a similar distribution is valid though the presence of the $m_q = \pm 1$ 'quadrupole' lines, which makes the chlorine spectrum less transparent.

The interaction parameters of the silver and chlorine HF-transitions, can be extracted from the angular dependence of the line positions. The results of the analysis of the silver HF interaction are presented in Table 4.2. Most of the HF tensors are assigned to nuclei assuming that the Z' principal axis of the tensor is directed along the line interconnecting the central silver ion and the involved nucleus. The listed values have been obtained from the best fit of the ENDOR angular dependences by the computer package 'Visual EPR' by Grachev [34].

The analysis of the chlorine ENDOR angular dependence is more complicated then the analysis of the Ag ENDOR spectra due to the presence of the $m_q = \pm 1$ 'quadrupole' lines (see Table 4.3). The obtained HF parameters are split into the isotropic and anisotropic parts. The assignment of the lines labeled with Greek characters is based on a simplifying assumption that the Z' principal axis of both the HF and quadrupole tensor is directed along the line interconnecting the central silver ion and the respective chlorine neighbour. Note that the chlorine HF results should be treated with some caution owing to the many overlapping ENDOR angular dependences. This especially applies for lines 'β' and 'γ', which can only be recognized in the XY plane and for which the derived parameters are based on fits that include a small number of points.

In general, the analysis of the ENDOR angular dependences only yields the relative signs of the HF parameters a, b, and b'. For the silver HF interaction, the absolute signs can be obtained from the ENDOR study of the lowest triplet state of

Table 4.3 ^{35}Cl HF constants a, b, and b' and quadrupole constant q for the STH

Line	^{35}Cl HF and quadrupole constants (MHz)						Assignment		Spin density per nucleus	
	a	b	b'	$\|q\|$	$\|q'\|$	q calc.	Position	Number of nuclei	s-orbital $\eta^2\alpha^2$	p-orbital $\eta^2\beta^2$
Λ	(−) 26.0 (2)	(−) 27.7 (2)	(−) 4.7 (4)	1.69 (2)	1.03 (1)	−2.2	[±100] + [0 ± 10]	4	0.005	0.158
α	(−) 1.85 (1)	(+) 0.11 (1)		0.157 (2)		−0.52	[00 ± 1]	2	0.0003	0.0006
β	(+) 1.39 (6)	(+) 0.62 (5)		0.32 (1)		−0.22	[±2±10] + [±1 ± 20]	8	0.0002	0.0035
γ	(+) 1.09 (5)	(+) 0.73 (5)		0.17 (1)		−0.23	[±20 ± 1] + [0 ± 2 ± 1]	8	0.0002	0.004
δ	(+) 0.50 (2)	(+) 0.42 (2)		0.104 (3)		−0.17	[±10 ± 2] + [±10 ± 2]	8	0.0001	0.0024
ε	(+) 0.420 (6)	(+) 0.358 (9)	+0.15 (2)	0.058 (3)	0.01 (2)	−0.12	[±300] + [0 ± 30]	4	0.0001	0.002
ζ	(−) 0.285 (3)	(−) 0.03 (2)				−0.23	[±1 ± 1 ± 1]	8	0.00005	0.0002

The asterisk indicates that the parameters derived for lines 'β' and 'γ' are questionable due to the complexity of the chlorine ENDOR angular dependence. The signs of the HF interactions (in brackets) were determined from the 'anomalous' ENDOR effect. The q and q′ quadrupole constants have the same sign. The last two columns contain the estimated distribution of the spin density per nucleus of the STH for 3s- and 3p-orbitals of chlorine ions in terms of $\eta^2\alpha^2$ and $\eta^2\beta^2$ as deduced from the HF interactions. The uncertainty in the last digit of these constants is indicated in brackets

Table 4.4 ^{109}Ag HF constants *a*, *b*, and *b′* for the STH

Line	^{109}Ag HF constants (MHz)			Assignment		Spin density per nucleus	
	a	*b*	*b′*	Position	Number of nuclei	*s*-orbital $\eta^2\alpha^2$	*d*-orbital $\eta^2\beta^2$
A	(−) 77.2 (4)	(−) 11.3 (2)		central	1	0.041	0.193
a	+3.02 (1)	−0.81 (1)		[±200] + [0 ± 20]	4	0.0016	0.014
b	+1.396 (3)	−0.033 (3)	−0.014 (3)	[±1 ± 10]	4	0.0008	0.0006
c	+0.531 (3)	−0.030 (3)	0.013 (5)	[±30 ± 1] + [0 ± 3 ± 1]	8	0.0003	0.0005
d	+0.364 (3)	+0.050		[00 ± 2]	2	0.0002	0.0009
e	+0.329 (3)	−0.011 (3)	−0.018 (3)	[±2 ± 20]	4	0.0002	0.0002
h	+0.10 (3)	0.01 (3)	0.01 (3)	[±3 ± 10] + [±1 ± 30]	8	0.00005	0.0001

The signs were obtained from the ENDOR study of the lowest triplet state of the STE in AgCl. The signs of the HF interaction for the central silver Ag_A (in brackets) was determined from the 'anomalous' ENDOR effect. The last two columns contain the estimated distribution of the spin density per nucleus of the STH for *s*- and *d*-orbitals of silver ions in terms of $\eta^2\alpha^2$ and $\eta^2\beta^2$ as deduced from the HF interactions. A possible contribution from the core polarization and point-dipole-dipole interaction was not considered. The uncertainty in the last digit is indicated in brackets

the STE in AgCl [27]. It is well established that the STE consists of an electron, which is shallowly trapped in the Coulombic field of an STH, and both the electron and the hole contribute to the observed ENDOR frequencies. The absolute signs of HF coupling for the STH complex are included in Table 4.4. Unfortunately, the absolute signs of the central silver and chlorine HF interactions cannot be verified from the study of the STE because the ENDOR resonances of the central silver and the equatorial chlorines were not observed for the STE. Since it is questionable whether the chlorine angular dependences in the very dense spectra of the more complex STE system would be resolved and because the recording of such dependences is very time-consuming, no attempt was undertaken to do so.

The spin density distribution of the STH complex in AgCl was found by using one-electron wave function of the unpaired electron bound to the STH which was constructed as a linear combination of atomic orbitals (LCAO) centered on silver and chlorine sites in the vicinity of the STH in line with theoretical calculations [35, 36]. It was shown that the wave function of STH is mainly distributed over five nuclei and not only located on the central nucleus. About 19% of the spin density is located in the $4^d(x^2-y^2)$ orbital on the central silver Ag_A, and about 65% of the spin density is located in the 3*s* and 3*p* orbitals of the four equatorial chlorines of Cl_A. The remainder of the wave function (~16%) is spread mostly over the chlorine and the silver shells that are mainly near the *XY* plane perpendicular to the elongation axis. This includes up to 7% of the spin density which is localized on the ions situated at the *X* and *Y* axes outside the central square-planar $(AgCl_4)^{2-}$ unit.

Thus, the hole is essentially located in the plane perpendicular to the elongation axis. In general, these results show that the hole is distributed on Ag ($\sim$30%) and Cl ($\sim$70%) sublattices and this finding confirms the $4d(Ag^+)$ and $3p(Cl^-)$ orbital admixture at the maximum of the valence band.

The quadrupole interactions have been determined from ENDOR analysis that gave the electrical field gradient distribution at the chlorine sites. It was shown that the largest quadrupole interactions were observed near the plane perpendicular to the elongation axis and that they correlate with the values of the anisotropic HF interactions. The large deviation from axial symmetry is found for quadrupole interactions with the four equatorial chlorine ions and for interaction with chlorine ions that are positioned along the X and Y axes outside of the central square-planar $(AgCl_4)^{2-}$ unit.

The results of the low-temperature ENDOR study of the STH complex in AgCl confirm that the trapping process is of intrinsic nature, i.e., a Jahn-Teller distortion not accompanied by a charged vacancy or impurity. This conclusion is based on the finding that the ENDOR angular dependences of neighbouring silver and chlorine nuclei still reflect a D_{4h} local symmetry. This eliminates the possible involvement of an uncharged impurity or intrinsic defect unless it is located at a distant position for which the role in the trapping process can only be of secondary importance. The results of calculations of the quadrupole interactions are also in line with the suggestion that the STH complex in AgCl is of intrinsic nature.

It was previously accepted that no self-localization of holes takes place in the bulk AgBr (see, for example [2, 16]). However, one can not exclude that the holes in AgBr can be self-localized as well. In contrast to the situation in AgCl, the dynamic Jahn–Teller effect may take place in AgBr, which leads to isotropic g value, as observed in experiment. The g value of holes in AgBr $g = 2.08$ is close to an average g value of the STH in AgCl. According to this approach, the bound exciton in AgBr may possess qualitatively the same structure as the STE in AgCl in which the wave function of an electron trapped by an STH is close to the wave function of a SEC. A smaller magnitude of the singlet–triplet splitting observed in AgBr could be explained by a more strongly delocalized wave function of the electron part of the exciton.

In conclusion, the results of the study of STE, SEC and STH in silver halide crystals by means of high-frequency EPR, ESE, ENDOR and ODMR were discussed. Application of high-frequency (95 GHz) and multiquantum (35 GHz) ODMR allowed to reveal the ordering of the singlet and triplet levels of the STE in AgCl and to measure their splitting with very high accuracy: $J = 5.370 \pm 0.002$ cm^{-1}.

The dynamical properties of the excited singlet and triplet states of the STE were studied by ESE technique in combination with pulsed laser excitation. The populating, depopulating and spin–lattice relaxation processes have been investigated.

The presented ENDOR study demonstrates that the lowest triplet state of the STE in AgCl consists of a very diffuse electron attracted by a strongly localized STH. It was found that the spatial distribution of the STE can roughly be divided into three regions. At distances larger than about 1.8 nm, the wave function of the STE is completely determined by the shallow electron, which occupies a hydrogen-like 1s orbital with a Bohr radius of 1.51 ± 0.06 nm. At distances between 1.8 and 0.9 nm, the electron is still dominant but its density distribution deviates from the spherical symmetry and reflects the D_{4h} symmetry of the central STH. At distances shorter than about 0.9 nm, both the central hole and the shallow electron have a contribution to the wave function of the STE.

The ENDOR studies allowed the direct determination of the wave function of the intrinsic SEC and STH in silver chloride. A model of SECs is suggested in which the electron is shallowly trapped in the Coulombic field of two adjacent Ag^+ ions, symmetrically placed on a single cationic site. The ENDOR studies of STH provide direct information about the spatial distribution of the unpaired electron at the silver and chlorine sites and of the electrical field gradient distribution at the chlorine sites of the STH. From a comparison of the ENDOR results for STE and STH, it is concluded that the central hole of the STE is virtually identical to an isolated STH centre.

4.2 Electronic Structure of Shallow Donors and Shallow Acceptors in Silicon Carbide

Silicon Carbide (SiC) is a promising wide-band-gap semiconductor for applications in high-frequency, high-temperature and high-power electronic devices. For this purpose n- and p-type SiC is grown by incorporation of donor impurities, like N, or acceptor impurities, like B, Al and Ga. To further the development of such semiconductor devices a good understanding of the electronic and geometric properties of the created donor and acceptor centres is imperative. A complicating factor in such studies is that SiC can occur in different polytypes, with greatly different band structures, and also that the donor and acceptor impurities seem to occur at different sites in the SiC polytypes.

SiC is the only chemically stable form of Si and C. Each Si (C) atom is surrounded by four C (Si) atoms in tetrahedral sp^3 bonds. The crystal structure consists of the close-packed stacking of layers containing Si and C atoms along the c-axis. By changing this stacking sequence different polytypes can be formed, alone cubic polytype denoted 3C-SiC, a great number of hexagonal polytypes denoted nH-SiC (n = 2, 4, 6, etc.) and rhombohedral polytypes denoted mR-SiC (m = 15, 21, 27, etc.) are identified. All SiC polytypes are indirect semiconductors. As much as 170 different polytypes have been observed. These polytypes are semiconductors with a varying band gap and one can consider them as natural short-period superlattices. For this reason they attract considerable interest for artificially grown nanostructures. The most common polytypes are 3C-SiC, which has a cubic symmetry, 4H-, 6H-SiC have hexagonal symmetry and 15R-SiC has rhombic symmetry.

In 4H-SiC the difference in the stacking sequence leads to the formation of 2 non-equivalent crystallographic positions, one hexagonal and one quasi-cubic site, called h and k, respectively. In 6H-SiC three non-equivalent positions are formed, one hexagonal and two quasi-cubic ones, called h, k1 and k2. The SiC crystal is built up of tetrahedrons, with four bonds for every atom. Two of the bonds lie in the (11–20) plane and parallel to the direction of the c-axis a staircase pattern is formed. For 6H-SiC this pattern gives rise to three inequivalent sites, k1, k2 and h, for

4H-SiC the staircase is shorter and there are only two sites, k and h. The difference between the h site and the two quasi-cubic sites is due to a difference in the position of the atoms in the second coordination sphere. In 6H-SiC a difference is found between the two quasi-cubic sites, k1 and k2, when the third coordination sphere is also considered. The same site is repeated along the c axis due to its stair-like structure, only rotated by 60°. This, together with the threefold symmetry around the c-axis leads to the existence of six subsites for every site.

An important issue is the spatial delocalization of the electronic wave function of the donor and acceptor centres in semiconductors. The method of choice to obtain this information is electron nuclear double resonance (ENDOR) spectroscopy developed by Feher [37], which has been applied to donor impurities in Si [37–39] and papers referenced therein, and to acceptor impurities in SiC [40]. In these experiments the hyperfine (HF) interaction between the unpaired electron spin of the donor and the nuclear spin of the surrounding atoms is determined, which is then translated into the spin density of the electronic wave function at the various atomic positions.

4.2.1 Nitrogen and Phosphorus Donors with Shallow Levels

4.2.1.1 Nitrogen Donors

In this section the results of an EPR and ENDOR study of the N-donor centre in 4H-SiC and 6H-SiC will be presented. The N donor in SiC has been studied extensively using optical absorption and emission spectroscopy [41] and using Raman experiments [42]. The first EPR measurements on the N-donor in SiC were done by Woodbury and Ludwig [43] and the first ENDOR measurements at 9.5 GHz in [44, 45]. High-frequency EPR measurements at 142 GHz by Kalabukhova et al. [46] allowed to separate overlapping EPR lines, owing to the high spectral resolution, and to assign the various EPR lines to specific sites in the SiC lattice. Additional ENDOR measurements were reported in [47] and an overview of the electronic properties of the N-donor, obtained by EPR and ENDOR at 9.5 GHz is found in [48].

The interpretation of the ENDOR data on the N-donor in SiC presents a considerable problem because the assignment of the ENDOR lines to specific nuclei is complicated. This is related to the fact that the N-donor electronic wave function is a linear combination of the wave functions in the six minima of the conduction band. As a result interference effects occur, which makes that the overall density of the wave function does not decay monotonically with the distance to the N donor. As yet we have not been able to come to an unambiguous assignment of the observed ENDOR lines to specific nuclei. Nevertheless, by comparing the general aspects of the ENDOR spectra with theoretical predictions of the electronic spin density distribution based on the Kohn-Luttinger theory [49, 50] for describing donor states, tentative explanations for the differences in the behaviour of the N donor at the k site in 4H-SiC and the h, k1 and k2 site in 6H-SiC were presented in [51]. Moreover for the k-site N donor in 4H-SiC an assignment of the first five

ENDOR lines to shells, using the similarity of this site with the P (As, Sb) donors in Si [37] was presented in [51].

The important aspect of the ENDOR investigation [51] is that it has been performed at a microwave frequency of 95 GHz, i.e., ten times higher than the conventional frequency of 9.5 GHz at which all ENDOR measurements reported so far have been carried out. The great advantage of this high microwave frequency is that the high spectral resolution not only allows to distinguish the various sites in the EPR spectra, but that it also leads to a separation of the ENDOR signal of the ^{13}C (natural abundance 1.11%, I = 1/2) and ^{29}Si (natural abundance 4.67%, I = 1/2) nuclei. Thus the HF interaction of the unpaired electron spin with the surrounding nuclear spins can be determined in great detail supplying information from which the spatial delocalisation of the electronic wave function can be obtained.

Theoretical and EPR studies of hyperfine interaction in nitrogen doped 4H and 6H SiC crystals have been recently presented in [52]. The hyperfine tensors within the framework of density functional theory have been calculated and results of [51, 53–55] have been analyzed.

Probing of the N shallow donor wave functions in silicon carbide through an EPR study of crystals with a modified isotopic composition was studied in [56].

Recent progress in the investigation of the electronic structure of the shallow nitrogen and phosphorus donors in 3C–, 4H– and 6H–SiC has been reviewed with focus on the applications of magnetic resonance (EPR, ESE, ESE-detected ENDOR, electron spin-echo envelope modulation and two-dimensional EPR [57].

EPR and ENDOR studies of the ^{29}Si and ^{13}C hyperfine interactions of the shallow N donors and their spin localization in the lattice were discussed. The use of high-frequency EPR in combination with other pulsed magnetic resonance techniques for determination of the valley–orbit splitting of the shallow N and P donors were presented and discussed.

Figure 4.9 shows the ESE-detected EPR spectra of the N donor in non-enriched and ^{13}C-enriched 6H-SiC (upper panel) and 4H-SiC (lower panel) as measured at 1.2 K and 95 GHz for two different extreme orientations of the magnetic field in the crystal: the magnetic field is parallel to the c-axis (B ‖ c) and the magnetic field lies in the plane perpendicular to the c-axis (B ⊥ c). Due to the HF interaction of the unpaired electron spin with the N-donor nucleus the EPR lines are split into three lines (isotope ^{14}N, natural abundance 99.63%, I = 1). For the h sites the splitting is too small to be resolved in the EPR spectra. For the Nk site in 4H-SiC the splitting is 1.9 mT, and for the Nk1 and Nk2 sites in 6H-SiC the splitting is roughly 1.23 and 1.25 mT, respectively. The central HF lines of the different EPR signals are marked using the following abbreviations. Nk for the N donor with N substituting on a quasi-cubic site in the 4H-SiC crystal and Nk1 and Nk2 for the N donor with N substituting on the quasi-cubic k1 or k2 site in 6H-SiC crystal respectively. Nh indicates the signal related to a N substituting on a hexagonal site in the 4H-SiC or 6H-SiC crystal (see inset). The experimental g- and HF interaction values for the N donor are in agreement with the values found in the literature.

From Fig. 4.9 it is clear that the EPR spectra changes due to the ^{13}C enrichment are very different for the two polytypes. In 4H-SiC a slight broadening of the EPR line is

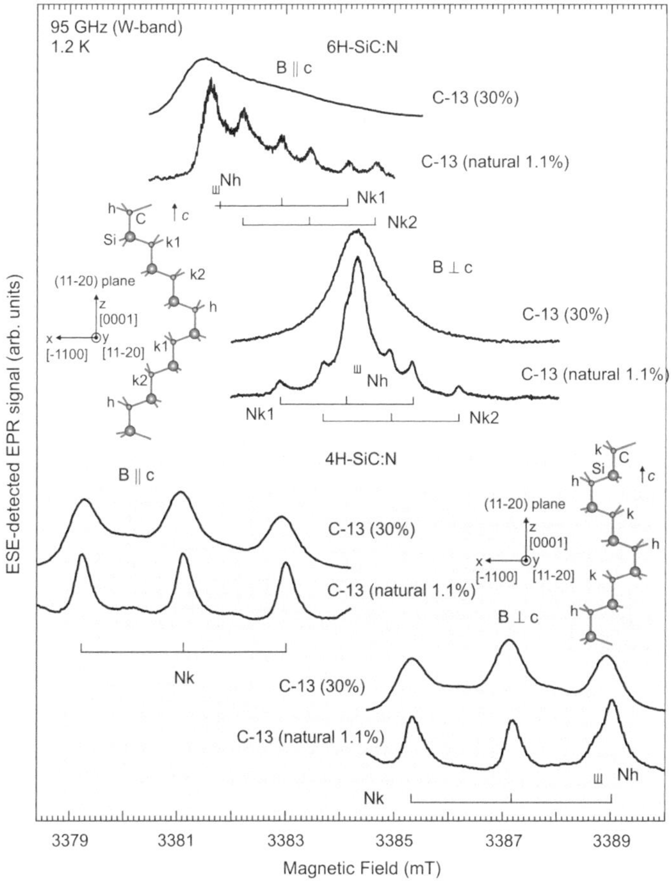

Fig. 4.9 The EPR spectra of the N donor in non-enriched and ^{13}C-enriched 4H-SiC and 6H-SiC as measured at 1.2 K and 95 GHz for two different extreme orientations of the magnetic field in the crystal. In the *upper panels* the magnetic field is parallel to the c-axis (B ∥ c), in the *lower panels* the magnetic field is perpendicular to the c-axis (B ⊥ c). In each panel the *upper spectrum* is connected to the ^{13}C-enriched sample and the lower one to the non-enriched sample. The linewidth is indicated for every spectrum. Of the three EPR HF lines connected to the quasi-cubic sites the central one is indicated by Nk (4H-SiC), Nk1 or Nk2 (6H-SiC). The HF splitting of the h site EPR line is too small to be resolved in EPR. The one line connected to the h site is indicated by Nh, but is hardly visible in the spectrum. (*Insets*) A schematic representation of the 4H- and 6H-SiC crystal structure and definition of the laboratory axis system within the crystallographic axis system as used throughout this chapter

observed. In 6H-SiC however, the line width is increased more than 3 times in the spectrum for B ∥ c. In EPR measurements at 9.5 GHz, a similar increase in line width is observed from 0.14 mT in non-enriched 6H-SiC to 0.55 mT in ^{13}C-enriched 6H-SiC. From the EPR results we conclude that in 4H-SiC the main part of the spin density resides on Si atoms because the 30% ^{13}C enrichment leads to a small line broadening. In contrast, in 6H-SiC we conclude that the main part of the spin density of the N donor is located on the C atoms, because of the large line broadening upon 30% ^{13}C enrichment. In the latter sample the line width of the non-enriched sample is slightly less than in 4H-SiC, suggesting that there is less spin density on the Si atoms. In the next section we will show that with the available ^{13}C and ^{29}Si ENDOR data a reasonable value for the observed linewidth is found.

The ^{29}Si and ^{13}C ESE-detected ENDOR spectra of the sites in 4H-Si^{13}C and 6H-Si^{13}C for B ∥ c at 1.2 K and 95 GHz are shown in Fig. 4.10a. Only the spectrum on the high-frequency side of the Zeeman frequency is shown. The upper spectrum, 4H, k, belongs to the N donor substituting on a k site in 4H-SiC. The positions of the lines have been marked by diamonds. The middle spectrum was measured at position k1 in the EPR spectrum, B = 3382.2 mT, the lower spectrum at position k2, B = 3381.5 mT in 6H-SiC. No distinction can be made between the two quasi-cubic sites, due to the low signal-to-noise ratio. The ^{29}Si k1, k2 site lines are both marked by squares, the positions of the h site lines are marked by circles. The ^{13}C positions of the h site lines are marked by open circles, those of the k1 site lines by squares and those of the k2 site by triangles. The ^{29}Si and ^{13}C nuclear Zeeman frequency are indicated by arrows.

From the comparison between the ^{29}Si and ^{13}C ENDOR spectra for the different sites in 4H-Si^{13}C and 6H-Si^{13}C in Fig. 4.10a it is clear that the spin density distribution of the N-donor electron over the ^{13}C and ^{29}Si nuclei differs between 4H-SiC and 6H-SiC. Whereas the main part of the spin density is localised on the ^{29}Si atoms for the k site in 4H-SiC (as expected for a donor electron) it is localised mainly on the ^{13}C atoms for the h and quasi-cubic sites in 6H-SiC. Note also, that the spectrum of 6H-SiC contains more lines around the ^{13}C nuclear Zeeman frequency (even though this spectrum shows the lines of three sites) than the spectrum of 4H-SiC around either the ^{13}C or ^{29}Si nuclear Zeeman frequency.

The EPR and ENDOR data for the N donors in SiC can be described by a spin Hamiltonian of the following form [51]:

$$\hat{H} = \mu_B \hat{\vec{S}} \cdot \vec{g} \cdot \vec{B} + (\hat{\vec{S}} \cdot \overleftrightarrow{A}_N - \gamma_N \vec{B}) \cdot \hat{\vec{I}}_N + \hat{\vec{I}}_N \cdot \overleftrightarrow{P}_N \cdot \hat{\vec{I}}_N + \sum_K (\hat{\vec{S}} \cdot \overleftrightarrow{A}_K - \gamma_K \vec{B}) \cdot \hat{\vec{I}}_K, \quad (4.7)$$

where $\overleftrightarrow{A}_N$ represents the hyperfine tensors of the ^{14}N (I = 1), and $\overleftrightarrow{A}_K$ represent the hyperfine tensors of the ^{13}C (I = 1/2) and ^{29}Si (I = 1/2) nuclear spins, $\overleftrightarrow{P}_N$ the quadrupole tensor of the ^{14}N spins; γ_N is the magnetogyric ratio for the N nucleus and γ_K the magnetogyric ratio for the C and Si nuclei. Assuming that the hyperfine tensors have nearly axial symmetry we can write the tensor as $A_{zz} = a + 2b$, $A_{yy} = a - b - b'$ and $A_{xx} = a - b + b'$ [58]. For the quadrupole interaction we can write, in the principal axis system and assuming axial symmetry, $P_N(I_z^2 - 1/3I^2)$ with $P_N = 3/2P_{zz} = 3q$ and $P_{xx} = P_{yy} = -q$. The deviation from axial symmetry is described by $q' = 1/2(P_{xx} - P_{yy})$, $P_{xx} = -q + q'$ and $P_{yy} = -q - q'$.

The angle θ is defined as the angle between the magnetic field and the c-axis, ϕ is the angle in the plane perpendicular to the c-axis. The laboratory axis system is oriented as follows. The (11–20) plane is equivalent to the xz plane. The z-axis

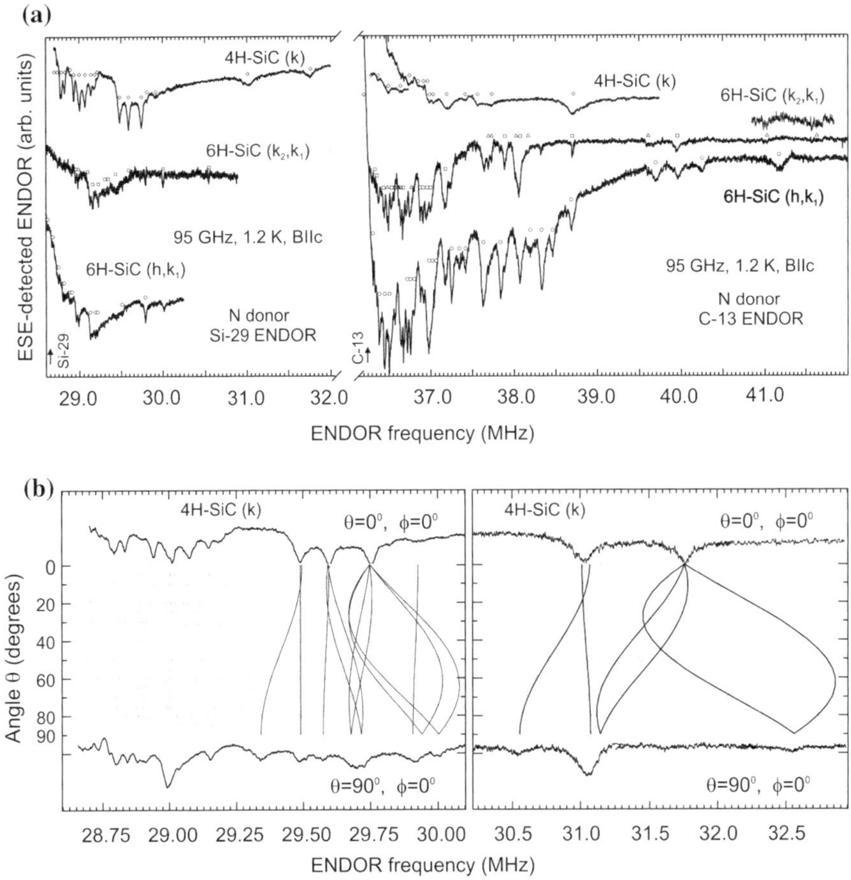

Fig. 4.10 **a** The ^{29}Si and ^{13}C ENDOR spectra of the sites in 4H-Si^{13}C and 6H-Si^{13}C for B ‖ c at 1.2 K and 95 GHz. Only the spectrum on the high-frequency side of the Zeeman frequency is shown. The *upper spectrum*, 4H, k, belongs to the N donor substituting on a k site in 4H-SiC. The positions of the lines have been marked by *diamonds*. The *middle spectrum* was measured at position k1 in the EPR spectrum (B = 3382.2 mT, see Fig. 4.9), the *lower spectrum* at position k2 (B = 3381.5 mT) in 6H-SiC. No distinction can be made between the two quasi-cubic sites, due to the low signal-to-noise ratio. The ^{29}Si k1, k2 site lines are both marked by *squares*, the positions of the h site lines are marked by *circles*. The ^{13}C positions of the h site lines are marked by *open circles*, those of the k1 site lines by *squares* and those of the k2 site by *triangles*. The ^{29}Si and ^{13}C nuclear Zeeman frequency are indicated by *arrows*. **b** The orientational dependence of the ^{29}Si ENDOR lines (experimental data points marked by *diamonds*) of the k site in 4H-SiC in the (11–20) or (zx) plane. The spectra have been corrected for the shift of the nuclear Zeeman frequency with respect to the orientation B ‖ c (θ = 0)

([0001] axis) is parallel to the crystallographic c-axis, and θ = 0° and ϕ = 0°. The x-axis ([−1100] axis) lies in the (11–20) plane perpendicular to the c-axis (θ = 90° and ϕ = 0°). The y-axis ([11–20] axis) stands perpendicular to the x-axis in the plane perpendicular to the c-axis (θ = 90° and ϕ = 90°).

In Fig. 4.10b the ^{29}Si ENDOR orientational dependence is presented of the k site in 4H-SiC in the (11–20) (zx) plane. In Fig. 4.10b the spectra have been corrected for the shift of the nuclear Zeeman frequency with respect to the orientation B ∥ c. This shift is due to the change in the magnetic field position of the EPR line for different orientations, which is connected to the anisotropy of the g-tensor. Figure 4.10b shows the experimental data points for all measured orientations of the k site, and the fits to these points as made using spin Hamiltonian (4.7). In total sixteen patterns have been found, belonging to sixteen different groups of nuclei, which are presented in [51], where the HF tensor principal values and the Euler angles of the HF tensor were calculated.

After the addition of the results of new studies [53–57] where strong HF interactions with the two Si and C shells were directly observed in the EPR spectra, number of shells that have been identified in ENDOR studies should be revised. So the maximum constants, which are registered in the ENDOR spectra should be attributed to the third shell, instead of the first one.

First we will discuss the results of the EPR and ENDOR investigations of the isotropic (a) and anisotropic (b) HF and quadrupole (q) interactions with the ^{14}N nucleus for the N donors in the main SiC polytypes 3C, 4H and 6H which are presented in Table 4.5. The values are taken from the [48, 51]. Only for the h site in 6H-SiC we observed a small difference in the parameters (Table 4.5). The other values are the same within experimental error and are not presented in the table. Table 4.5 also shows the experimental values of the ionization energy of the N donors, valley-orbit splitting, the g factors and the s and p spin densities corresponding to the a and b HF interaction parameters of the unpaired electron with the ^{14}N nuclei in 3C-, 4H- and 6H-SiC. For comparison the same parameters are

Table 4.5 The experimental values of shallow N donors occupying different positions in 3C-SiC, 4H-SiC and 6H-SiC from the [48, 51, 56]: ionization energy, valley-orbit splitting, g factors, isotropic (a) and anisotropic (b) HF interaction constants and their corresponding s and p spin densities on N, quadrupole parameters q

SiC polytype	Site	Ionization energy E_g/valley-orbit splitting (meV)	$g_{\parallel}$	$g_{\perp}$	a (MHz)	b (MHz)	q (MHz)	s (%)	p (%)
3C-SiC	k	54/8.37	2.0050	2.0050	3.5			0.19	
4H-SiC	h	52.1/7.6	2.0055	2.0043	2.9	0.080	…	0.16	0.14
	k	91.8/45.5	2.0010	2.0013	50.97	0.004	~0	2.8	0.007
6H-SiC	h	81/12.6	2.0048	2.0028	2.52	0.12	0.019	0.14	0.22
	k1	137.6/60.3	2.0040	2.0026	33.221	0.004	0.007	1.83	0.007
	k2	142.4/62.6	2.0037	2.0030	33.564	0.009	0.007	1.85	0.016
Si									
P		44/11.7	1.9985		117.53			0.9	
As		49/21.2	1.9984		198.35			1.35	

For comparison analogous parameters are also presented for shallow P and As donors in silicon after [37]

presented for the P and As shallow donors in Si after [37]. There the g-factor and HF interaction are isotropic. From Table 4.5 one can see that in general the value of the ionization energy for different sites does not correlate with the HF interactions, i.e., with the spin densities on the N nuclei. For instance, the ionization energies for the k site in 4H-SiC and the h site in 6H-SiC are approximately equal, but the nitrogen donor HF interaction for the k site in 4H-SiC is about 20 times larger than that for the h site in 6H-SiC.

Figure 4.11a shows EPR spectra of shallow nitrogen donors in three 6H-SiC crystals: with natural isotopic abundance, enriched in ^{28}Si (<0.5% ^{29}Si), and enriched in ^{13}C ($\sim$25%). The spectra were measured at 40 K with a magnetic field

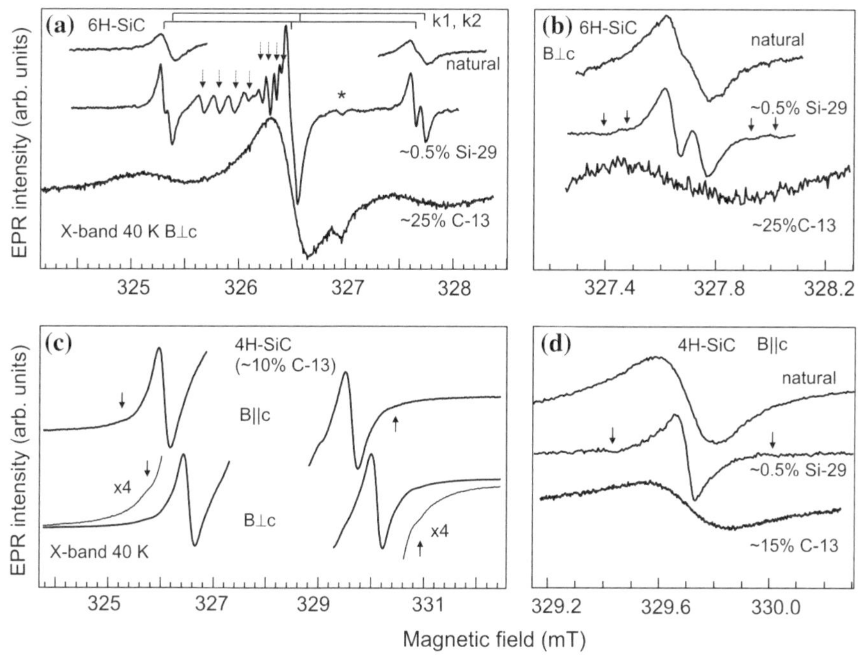

Fig. 4.11 **a** EPR spectra of shallow nitrogen donors in three 6H-SiC crystals: with natural isotopic abundance, enriched in ^{28}Si (<0.5% 29Si), and enriched in ^{13}C ($\sim$25%). The spectra were measured at 40 K with a magnetic field B ⊥ c. The ^{29}Si-depleted sample was used to obtain the EPR spectra after UV interband optical pumping. The *arrows* identify HF transitions for shallow boron acceptors, induced by UV. The *asterisk*-labeled reference signal belongs to quartz. **b** Expanded scale high-field components of the EPR spectra of shallow nitrogen donors in 6H-SiC in positions k1 and k2 shown in Fig. 4.11a for the B ⊥ c orientation. The *arrows* indicate the satellites in the spectrum of the ^{29}Si-depleted sample. **c** EPR spectra of shallow nitrogen donors in ^{13}C-enriched 4H-SiC for individual nitrogen HF-structure components measured for B ∥ c and B ⊥ c orientations. The satellite line pairs deriving from the HF interaction with one ^{29}Si nucleus are identified with *arrows*. **d** Expanded scale high-field components of the EPR spectra of shallow nitrogen donors in 4H-SiC in the k position for B ∥ c orientation. The *arrows* specify the satellites in the EPR spectrum deriving from the HF interaction with one ^{13}C nucleus of the ^{29}Si-depleted crystal

B ⊥ c. In the ^{29}Si-depleted sample the EPR spectra were measured after UV interband optical pumping and as a result, EPR spectra of shallow boron acceptors are observed (the arrows identify HF transitions for shallow boron acceptors, see the next section).

Nitrogen donors occupying positions k1 and k2 produce three EPR lines each, because nitrogen has only one isotope (^{14}N, with an abundance of 99.63%) with a nonzero nuclear spin, I = 1 (the number of lines is 2I + 1). Since the signals due to N donors in positions k1 and k2 differ in terms of their parameters, line splitting occurs. The vertical bars in Fig. 4.11a specify the nitrogen EPR signals for the k1 and k2 positions measured for the B ⊥ c orientation.

The high-field EPR components of shallow nitrogen donors in the k1 and k2 positions in 6H-SiC presented in Fig. 4.11a are displayed in Fig. 4.11b in an expanded scale for the B ⊥ c orientation. The slight narrowing of EPR lines resulting from a decrease of the ^{29}Si isotope concentration by an order of magnitude is seen to bring about a better resolution of the k1 and k2 signals. In opposite, enrichment in the ^{13}C isotope causes a substantial broadening of the EPR lines. The EPR linewidths measured in crystals with a natural and modified isotope composition are listed in Table 4.6. The spectrum of the ^{29}Si-depleted crystal shown in Fig. 4.11b contains additional lines as satellites, which are arranged symmetrically about the central lines and identified by arrows. Such satellites are seen to be present for each component of the hyperfine structure of nitrogen in positions k1 and k2, with the line splitting, 0.5 mT (14 MHz), being practically independent of crystal orientation. Because these satellites are observed in crystals depleted in the ^{29}Si isotope, they can only be due to the HF interaction with carbon. In accordance with the natural ^{13}C abundance, the intensity ratio attests to interaction with four practically equivalent carbon atoms; this interaction is practically isotropic to within experimental error and is the same for positions k1 and k2.

Figure 5.11c, d illustrate the results of a study of shallow nitrogen donors in 4H-SiC crystals, which are similar to those shown in Fig. 5.11a, b for the 6H-SiC polytype. Figure 5.11d shows expanded-scale high-field components of the EPR

Table 4.6 Based on ENDOR data calculated, ΔB(Si), ΔB(C), and W-band ESE measured EPR linewidths ΔBexp (in mT) for the different sites, as well as the total calculated linewidth ΔB (total) for the natural as well as the ^{13}C-enriched 4H-SiC and 6H-SiC samples

	ΔB(Si)	ΔB(C)	ΔB(C)	ΔB(total)	ΔB(total)	ΔB(exp)	ΔB(exp)
Abundance	Natural	Natural	Enriched	Natural	Enriched	Natural	Enriched
4H-SiC, Nk	0.194	0.084	0.43	0.28	0.62	0.32	0.57
6H-SiC, Nh	0.063	0.176	0.99	0.26	1.05	0.30	~1
6H-SiC, Nk1	0.066	0.102	0.58	0.17	0.64	0.25	~1
6H-SiC, Nk2	0.092	0.165	0.88	0.25	0.97	0.30	~1

All values are in mT. The different abundances are indicated in the text. In the enriched 6H-SiC sample it was not possible to distinguish between the experimental line widths, ΔBexp [mT], of the three sites. The estimated value of 1 mT is given for all three sites. The estimated error in the other experimental linewidths is 0.05 mT spectra of shallow nitrogen donors in 4H-SiC in the k position for B ∥ c orientation. The arrows specify the satellites in the EPR spectrum deriving from the HF interaction with one ^{13}C nucleus of the ^{29}Si-depleted crystal. The EPR spectra in Fig. 4.11d relate to shallow nitrogen donors obtained on three 4H-SiC crystals: (i) with natural isotopic abundance; (ii) enriched in ^{28}Si and, thus, having a low content (less than 0.5%) of the ^{29}Si isotope; and (iii) enriched in ^{13}C ($\sim$15%).

A decrease of the ^{29}Si isotope concentration by an order of magnitude is seen to result in a substantial narrowing of the EPR lines. However, enrichment by ^{13}C did not bring about noticeable EPR line broadening. This implies that the variation of the EPR linewidth in crystals with a modified isotopic composition in 4H-SiC differs markedly from what was observed in 6H-SiC. The ^{29}Si-depleted crystal reveals two additional lines located symmetrically about the central line. Similar satellites were observed with other k-nitrogen lines and there was practically no change in their relative intensity and line separation as the crystal orientation in the magnetic field was changed, although all lines change position because of the nitrogen donor g factor being slightly anisotropic. Because these lines were seen in crystals with low ^{29}Si content, just as in the case with 6H-SiC, they certainly cannot originate from HF interaction with ^{29}Si nuclei. It stands to reason that these satellites are caused by HF interaction with ^{13}C; for crystals with a natural abundance of the ^{13}C isotope, the satellites are masked by the broader nitrogen donor lines. As in the case with the 6H-SiC crystal, the intensity ratio suggests that they derive from interaction with four or five carbon atoms, with the strength of this interaction being 0.6 mT (16.8 MHz). The method of isotope composition modification used in [56] provides compelling evidence for the HF interaction being with carbon alone. This conclusion fits the results of the study of ^{13}C-rich crystals. As seen from Fig. 4.11d, the EPR line shape for the crystal with natural isotopic abundance differs from that obtained on the ^{13}C-rich crystal. Indeed, the wings of the EPR line in the latter case fall off substantially slower. The observed line shape reflects the presence of satellites whose intensities are substantially higher in crystals enriched with the ^{13}C isotope. The 6H-SiC crystals with natural isotopic abundance exhibited satellites corresponding to a practically isotropic HF interaction with one Si atom, whose strength was 0.96 mT (26.9 MHz). No such satellites were observed in crystals with a low ^{29}Si content.

Figure 4.11c displays EPR recorded in ^{13}C-enriched 4H-SiC for individual nitrogen HF-structure components measured for B ∥ c and B ⊥ c orientations. The arrows related to the ^{13}C-enriched crystal specify the outer satellite-pair lines assigned to HF interaction with one ^{29}Si nucleus, estimated as 1.46 mT (41 MHz). The central line of the nitrogen HF structure is omitted.

The EPR spectra of shallow nitrogen donors occupying positions k in the 4H-SiC and 6H-SiC polytypes and having the deepest levels (Table 4.5) exhibit, in addition to the unresolved HF structure determining the EPR linewidths, HF interactions with nearest neighbor coordination shells. These interactions can be resolved as satellites in the EPR spectra (see Fig. 5.11b–d). Satellites due to the HF interaction with silicon were observed in [51, 59, 60] and satellites deriving from interaction with carbon were observed in [51]. These additional EPR lines were unambiguously identified by studying crystals with a modified isotopic composition (indeed, this permitted observation of interaction with carbon due to EPR line narrowing [51]) and by ENDOR [60].

Being an element of Group V, nitrogen donor atom has five valence electrons, with four of them forming valence bonds with the nearest lattice atoms and the fifth being acted upon by the Coulomb field of the remaining pos-itive charge. In the effective mass (EM) approximation [49, 50], a weakly bound electron is treated as a hydrogen-like atom in which the Coulomb attraction of the donor nucleus is reduced by the semiconductor dielectric permittivity ε. It is also assumed that the electron moving in its orbit has an effective mass of a conduction electron. Under these assumptions, the wave function of a localized donor electron can be written as a product of the solution to the Schrödinger equation for a hydrogen-like atom formed by the donor and a weakly coupled electron and a Bloch function for an electron in the conduction band. Said otherwise, the Bloch function describing conduction band electrons is modulated by an envelope function which is a solution to the corresponding hydrogen-like Schrödinger equation. The result is the formation of a bound donor state with an ionization energy on the order of tens of millielectronvolts.

Silicon carbide similar to silicon belong to Group IV of the periodic table and, therefore, besides some substantial differences, possess qualitatively similar energy level structures. These crystals are indirect-gap semiconductors, because the conduction band has several minima shifted relative to the centre of the Brillouin zone. The pattern of this shift depends on the nature of the semiconductor material. Indeed, silicon has six minima displaced in the [100] directions and in SiC the character of the conduction band depends on the polytype. Because the conduction band has a many-valley character, shallow donor levels (which may be considered to be split off from the conduction band) are degenerate according to the number of minima; in silicon, for instance, there are six minima and they correspond to the A1, E, or T2 states. The EM approximation fails near the donor impurity, and the degeneracy is lifted. The level splitting, which is called valley–orbit splitting (see Table 4.5), is caused by the differences in electron distribution among the A1, E, and T2 states near the donor impurity. This difference is strongest in the energy of the A1 singlet state, because in this case the wave function reaches its maximum amplitude on the donor atom.

The donor electron wave function (for instance, for silicon or the 3C-SiC cubic polytype) can be written as

$$\psi(\vec{r}) = \sum_{j=1}^{6} \alpha_j F_j(\vec{r}) \varphi_j(\vec{r}), \tag{4.8}$$

where $\varphi_j(\vec{r}) = u_j(\vec{r}) e^{i\vec{k}_j \cdot \vec{r}}$ is the Bloch function at the jth minimum, which is located at point $\mathbf{k}_j$, and $u_j(\boldsymbol{r})$ is a periodic function. The coefficients α_j characterize the relative contribution of each valley and, thus, specify different combinations of the wave functions describing the regions near each minimum in the conduction band.

The function $F_j(\boldsymbol{r})$ is the hydrogen-like solution to the Schrödinger equation for the donor electron. Because the actual type of ground state is a priori unknown, the existence of HF interaction in the donor atom strongly suggests that the ground state is the A1 singlet; indeed, only in this state has the wave function a finite density at the donor nucleus site. In this case, the wave function is completely symmetric and the electron is spread uniformly over all the valleys; i.e., for Si or SiC, the wave function can be written as

$$\psi(\vec{r}) = \frac{1}{\sqrt{6}} \sum_{j=1}^{6} F_j(\vec{r}) u_j(\vec{r}) e^{i\vec{k}_j \cdot \vec{r}_j}. \tag{4.9}$$

In SiC, the degeneracy and further splitting depend on the polytype. In the cubic polytype 3C-SiC, we have a situation similar to that in silicon; i.e., the conduction band minima lie along the $\langle 100 \rangle$ axes, with the constant-energy surfaces in the proximity of the minima being shaped like ellipsoids extended along the $\langle 100 \rangle$ directions. The effective masses along and perpendicular to the ellipsoid axis are $m_l = 0.677m_0$ and $m_t = 0.247m_0$, respectively, where m_0 is the free electron mass. The hexagonal polytypes feature a more complex pattern. The conduction band minima in the 4H-SiC and 6H-SiC polytypes are in essentially different positions, and this is what gives rise in the final count to differences between their electronic properties.

The conduction band minimum in the 4H-SiC polytype is at point M, whereas that in 6H-SiC is located between points M and L, at a relative distance of approximately 60% from point M (see inset Fig. 4.12). The band structures are such that the effective masses in the plane perpendicular to the c axis are approximately the same for both polytypes ($m_\perp = 0.445m_0$ for 4H-SiC and $m_\perp = 0.43m_0$ for 6H-SiC), while differing noticeably along the c axis. Indeed, for 4H-SiC, the effective mass along the c axis is $m_\parallel = 0.32m_0$, whereas for 6H-SiC we have $m_\parallel = 1.7m_0$. Thus, for the 4H-SiC polytype (and 3C-SiC), the electron wave function is very nearly spherically symmetric, while for 6H-SiC the electron wave function is contracted noticeably along the c axis, i.e., is pancake-shaped. It appears reasonable to use averaged effective masses m* for electrons in the conduction bands. For 3C-SiC, $m^* = (m_l m_t m_t)^{1/3} = 0.345m_0$; for 4H-SiC, $m^* = (m_x m_y m_z)^{1/3} = 0.4m_0$; and for 6H-SiC, $m^* = 0.74m_0$. The effective Bohr radius of the shallow donor wave function is $a^* = [\varepsilon/(m^*/m_0)]a_0$, where $a_0 = 0.529$ Å is the Bohr radius. The shallow donor ionization energy is $E_{eff} = [(m^*/m_0)/\varepsilon^2]E_0$, where E_0 is the Rydberg energy. Thus, the effective Bohr radius of the electron wave function is $a^* = 15$ Å for 3C-SiC, $a^* = 13$ Å for 4H-SiC, and $a^* = 7.2$ Å for 6H-SiC and the corresponding shallow-donor ionization energies E_{eff} are 47 meV for 3C-SiC, 54 meV for 4H-SiC, and 101 meV for 6H-SiC.

It should be noted, we are dealing here with the wave function of a donor electron which is not localized in any specific position, k or h, of the polytype under study; it appears, however, only natural that the properties of donor electrons should contain information on the free-electron wave function. Thus, the pronounced difference

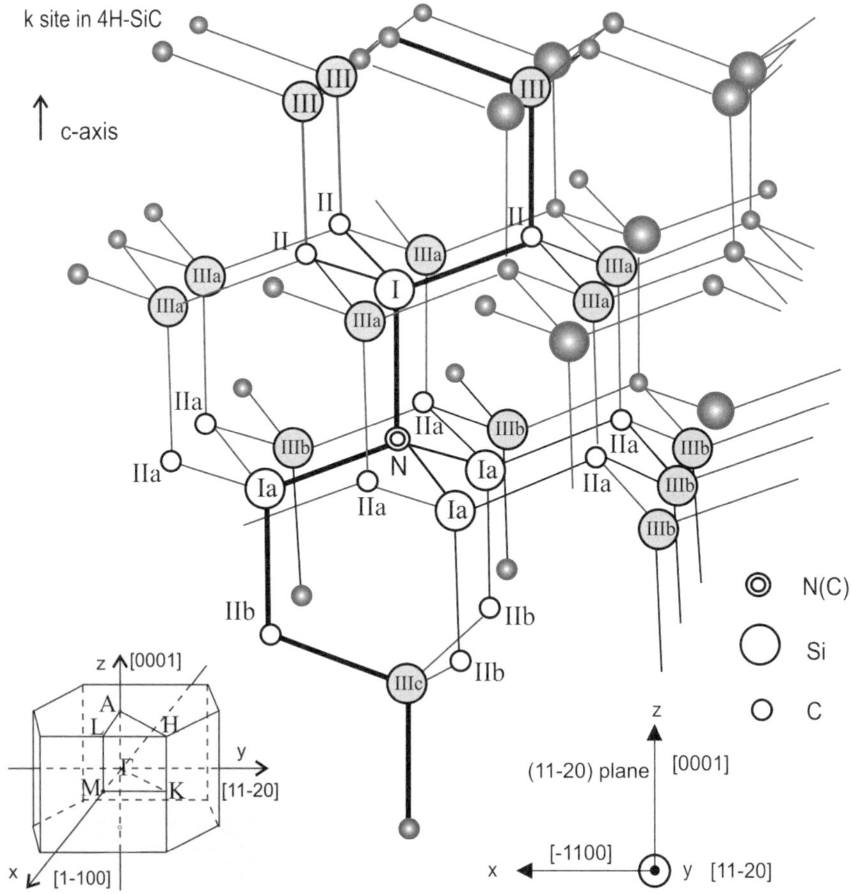

Fig. 4.12 A schematic representation of the crystal structure around the k site in 4H-SiC. The direction of the c-axis is shown, the staircase drawn with the heavy lines. In the bottom the orientation of the x, y, z axes are indicated. The {11–20} plane, contain the c-axis (z), x-axis and run parallel to the surface of the paper. The N atom in C position is indicated. The Roman numbers placed near the nuclei correspond to the numbers indicating the group in table. The surrounding of the k1-site in 6H-SiC is exactly the same as the surrounding of this k site in 4H-SiC. (*Inset*) The reciprocal lattice of 4H-SiC and 6H-SiC with lattice parameters $4\pi/(\sqrt{3}a)$ (distance from the centre of the hexagon to a corner) and $2\pi/c$ (hight of the hexagon). The lattice is rotated by 60° around the c-axis with respect to the lattice in real space. The positions of the special symmetry points are shown, together with the orientation of the hexagon with respect to the crystallographic axis system, x, y, and z in real space. The Γ point is the centre of the hexagon. The M point lies in the (11–20) plane, the K point in the (−1100) plane and the L point lies above the M point along the [0001] direction

between the wave function properties of nitrogen donors in the k or h positions (which does not depend on the polytype) apparently derives only from the local symmetry of these positions and does not have direct bearing on the general wave function of a band electron. These properties should be reflected in HF interactions with C and Si atoms in various coordination shells, which are proportional to the density of the donor wave function (5) $|\psi(\boldsymbol{r}_l)|^2$ at nucleus site l. Estimates of the wave function density on a nucleus are usually performed with a dimensionless quantity called the gain, which characterizes the degree of localization of the wave function near the nucleus; i.e., the gain is the density ratio of the actual wave function on the nucleus to the wave function envelope at the same site. The gain for the regions where the envelope wave function varies slowly (in accordance with EM theory and disregarding interference effects) can be expressed through a dimensionless quantity $\eta \equiv |u_j(\boldsymbol{r}_l)|^2/\langle u_j(\boldsymbol{r})\rangle^2_{av}$, where the denominator contains the cell-averaged Bloch function. This quantity is independent of the lattice site occupied by an atom but is dependent on the type of atom involved, Si or C.

As seen from Table 4.5, the donor level energies for 3C-SiC and for the hexagonal positions are comparable to the energies calculated in the EM approximation, whereas the levels for the k positions are substantially in excess of these values. The fact that the HF structure has not been observed in SiC for shallow donor levels in 3C-SiC and for hexagonal positions may suggest that the wave function distribution of donor electrons in these cases differs substantially from that of donors in quasi-cubic positions and, therefore, the EPR HF structure in this case is not resolved.

There are three major factors determining HF interactions for shallow donors in Group IV crystals, which are indirect-gap semiconductors: (i) smoothly decreasing modulation of HF interactions with increasing distance from the donor [in the EM approximation, this modulation is described by the envelope function $F(\boldsymbol{r})$]; (ii) the spin density localization on atoms, which is characterized by dimensionless gain coefficients and depends on the actual kind of atom and crystal involved; and (iii) interference effects originating from the existence of several valleys in these semiconductors [the interference destroys the smooth falloff of HF interactions on atoms or groups of atoms with distance, which is described by the function $F(\boldsymbol{r})$]. Incidentally, this interference effect considerably complicates the interpretation of EPR and ENDOR spectra in indirect-gap semiconductors, unlike in direct-gap semiconductors (ZnO, AgCl, AgBr), in which HF and SHF interactions have been assigned with a high degree of confidence [Sect. 5.1, this book].

The almost isotropic HF structure for nitrogen donors in quasi-cubic positions (k) due to the HF interaction with one Si atom was directly observed in EPR spectra in 4H-SiC and 6H-SiC crystals. The surrounding ions of the Nk shallow donor in 4H-SiC (the surrounding of the k1-site in 6H-SiC is exactly the same as the surrounding of this k site in 4H-SiC.), for which the HF interactions resolved by EPR and ENDOR and were analyzed in this section, are presented in Fig. 4.12. Figure 4.12 shows a schematic representation of the crystal structure around the k site in 4H-SiC. The direction of the c-axis is shown, the staircase drawn with the heavy lines. In the bottom the orientation of the x, y, z axes are indicated. The {11–20} plane, contain the c-axis (z), x-axis and run parallel to the surface of the paper. The N atom in C position is indicated. The Roman numbers placed near the nuclei correspond to the numbers indicating the different groups.

We will only consider the shells formed by Si and C atoms, labelled by I, Ia, II, IIa, III, IIIa, IIIb and IIIc. The HF structure of the first shell (one Si atom labelled I) and three carbon atoms of the second shell (labelled IIb) were resolved in the EPR spectra. The HF structure and unpaired electron density outside these shells were analyzed in [51]. Almost isotropic hyperfine splitting, caused by interaction with one Si atom in the first shell, labeled I Fig. 4.12, is the next: 1.46 mT (41 MHz) in 4H-SiC and 0.96 mT (26.9 MHz) in 6H-SiC.

The isotropic HF splitting due to interaction with ^{13}C which was observed in the EPR spectra as satellites in ^{29}Si depleted crystals for Nk1 and Nk2 in 6H-SiC and Nk in 4H-SiC (Fig. 4.11). We refer these EPR signals to the interaction with three carbon atoms, labelled by IIb, that bound along c-axis to three silicon nucleus labelled by Ia (see Fig. 4.12). This isotropic HF splitting is 0.6 mT (16.8 MHz) for 4H-SiC and 0.5 mT (14 MHz) for 6H-SiC. The observed isotropic HF interactions make it possible to estimate the unpaired-electron density in the *s*-orbitals of Si and C and, thus, calculate the degree of electron localization in these orbitals. The corresponding degrees of localization of the unpaired electron (spin density) on one Si atom in 4H-SiC and 6H-SiC are 0.89 and 0.6%, respectively, and those on each of the three C atoms in 4H-SiC and 6H-SiC are 0.44 and 0.37%, respectively. Significantly, the spin density ratio on one Si atom for the 4H-SiC and 6H-SiC polytypes, which is 1.5, coincides, to within experimental error, with that on the nitrogen atom (Table 4.5). This suggests that N centres in positions k in 4H-SiC and 6H-SiC have identical structure and that one Si atom occupies the position closest to the N atom; the wave function density decreases smoothly to about one-third its value as we go from N to Si. The spin density ratio on each of the C atoms of the 4H-SiC and 6H-SiC polytypes, which is 1.2, is noticeably smaller than that on the N atom, which implies that these atoms are located farther away from the nitrogen atom. Thus, the results of our study support the conclusion that nitrogen in positions k substitutes for C. It stands to reason that the degree of localization on the remaining three nearest Si atoms (Ia in Fig. 4.12) is of the same order of magnitude as that on one Si atom, for which isotropic HF splitting in the EPR spectrum was observed. The localization takes place in the s and p orbitals, thus implying that the HF interaction with the three Si atoms of the first coordination shell is anisotropic, which would account for its being not resolved in EPR spectra, because the isotropic HF interaction constant should decrease strongly if even a small part of the unpaired electron distribution extends to the p orbital. Anisotropic HF interaction with three silicon atoms in the first shell located in the basal plane which are labeled Ia from ENDOR data [51] are: $a = 6.54$ MHz, $b = 1.26$ MHz, $b' = 0.75$ MHz, s = 0.14%, p = 1.1%. Thus, in 4H-SiC, the total degree of localization of the unpaired electron on the central N atom and in the first coordination shell (containing four Si atoms I and Ia) should be 7.4%. We should add to this about 2% for the assumed three C atoms of the second shell. Thus, the total spin density within the first two coordination shells is approximately 9.5%. For 6H-SiC, the respective quantities should be about 1.5 times lower.

These values should be added to those obtained in an ENDOR study [51], which must be corrected by the including of the strong HF interactions for the nitrogen k positions, which were resolved in the EPR spectra. In 4H-SiC, according to ENDOR studies there are strongest HF interactions with C atoms in two groups: with $a = 5.02$ MHz (s = 0.13%) and with $a = 4.4$ MHz and b = 0.23 MHz (s = 0.12% and p = 0.22%). These HF interactions can be attributed to the C atoms labeled as II (3 atoms) and IIa (6 atoms) in Fig. 4.12. In 4H-SiC, according to ENDOR studies the strongest HF interactions with Si atoms in two groups: with $a = 4.8$ MHz (s = 0.1%), b = 0.04 MHz (p = 0.04%); and with $a = 4.14$ MHz and b = 0.34 MHz (s = 0.09% and p = 0.3%) with bonds along c-axis can be attributed to the Si atoms labeled as III and IIIc in Fig. 4.12. For IIIa and IIIb Si groups (Fig. 4.12) located in the basal plane HF interactions are the next: $a = 2.3$ MHz (s = 0.05%), b = 0.31 MHz (p = 0.27%) and $a = 2.0$ MHz (s = 0.04%), b = 0.12 MHz (p = 0.11%).

In the similar way the ENDOR data [51] for 6H-SiC can attributed to different Si and C shells presented in Fig. 4.12 since the surrounding of the k1-site in 6H-SiC is similar to the surrounding of k site in 4H-SiC.

Because the EPR linewidths reflect the density distribution of the unpaired electron of the shallow donor over coordination shells in the case of N positions k in SiC (with the exception of strong HF interactions, which were resolved in the EPR spectra), to which the EM approximation applies, we consider the HF interactions with these coordination shells in more detail.

The FWHM of an unresolved EPR line ΔB for the case where this width is dominated by HF interactions is given by [37]

$$\Delta B^2 = 2\left(\sum_i^R f(Si)n_i(Si)\left(\frac{a_i(Si)}{2}\right)^2 + \sum_j^M f(C)n_j(C)\left(\frac{a_j(C)}{2}\right)^2\right), \tag{4.10}$$

where n_i and n_j are the number of equivalent sites in the *i*th Si shell and the *j*th C shell. a_i and a_j are the HF interactions for equivalent sites in the *i*th Si shell and in the *j*th C shell. f(Si) and f(C) are the fractional abundances of ^{29}Si and ^{13}C (f(Si) = 0.047 and f(C) = 0.011 in natural abundance samples) f(Si)n_i(Si) and f(C) n_j(C) are the probabilities of having the *i*th and *j*th lattice sites occupied with a ^{29}Si or ^{13}C, respectively. The summations extend over the R silicon shells and the M carbon shells which contribute to EPR line width. Interpretation of the ^{13}C and ^{29}Si ENDOR data and calculation of the EPR linewidth were presented in [51]. First linewidth of the Nk donor in 4H-SiC in the crystal 4H-SiC was discussed because the ENDOR spectra are more informative for this centre than for the N donors in 6H-SiC. The Euler angles of the HF tensors give symmetry information about the shells to which they belong. Moreover, there seems to be a remarkable similarity between the Nk donor in 4H-SiC on the one hand and the shallow P (As, Sb) donors in Si on the other hand, when the HF interactions with the impurity nucleus and the surrounding ^{29}Si nuclei are compared. On the basis of this similarity an assignment of some of the ^{29}Si ENDOR lines to nuclei surrounding the Nk donor is proposed in [51]. The HF interaction with the central impurity atom leads in both cases to an s-like spin density and that it is of the same order of magnitude. The three nearest Si neighbours of the N atom (in carbon position) in the plane perpendicular to the c-axis show a remarkably large anisotropic HF interaction, which is almost of the same size as the isotropic one and is similar to the shell with (111) symmetry for the P donor in silicon. In the calculation only those shells that have an isotropic HF interaction that is larger than 1 MHz were included. The authors in [51] used six for the amount of nuclei in a shell. In the case of the isotopically enriched ^{13}C sample the number of ^{13}C nuclei with which the donor electron interacts is greatly enhanced and thus the linewidth increases. The results of the calculation as well as the experimental values, ΔB_{exp}, are shown in Table 4.6. The line width due to the C atoms, $\Delta B(C)$, and the total linewidth, ΔB_{total}, is given for the non-enriched (natural) as well as for the enriched case. It is clear that the calculations agree well with the experimentally found line width and thus it can be concluded that we did not miss large HF interactions with neighboring ^{29}Si or ^{13}C atoms with the exception of those for which resolved HF structure was observed in the EPR spectra.

According to (4.10), the EPR linewidth for shallow N donors in SiC has a complex pattern because of the presence of two types of atoms, Si and C. The actual value of the linewidth depends on the contribution of the HF interaction with the ^{29}Si and ^{13}C nuclei to the linewidth. Assuming the linewidth to be completely determined by the HF interaction with ^{13}C, the linewidth ratio for 6H-SiC enriched in ^{13}C to 25% to that of a crystal with natural isotopic abundance is calculated to be $(0.25/0.011)^{1/2} = 4.8$ and the EPR linewidth should be (based on the experimental linewidth in a natural-abundance crystal) 0.62 mT. The analogous ratio for 4H-SiC enriched in ^{13}C to 15% is (0.15/0.011)1/2 = 3.7, and the EPR linewidth should be 0.78 mT. Assuming now that the linewidth derives fully from the HF interaction with ^{29}Si, calculating the linewidth ratio for 6H-SiC with 0.5% ^{29}Si (the crystal is enriched in ^{28}Si) to a natural abundance crystal yields (0.005/0.047)1/2 = 0.33; so the EPR linewidth should be 0.04 mT. Similarly, the EPR linewidth for 4H-SiC can be found to be 0.07 mT.

A comparison of these figures with EPR experiments at X-band [56] (Table 4.7) shows that the first case (where the linewidth is fully determined by the HF interaction with ^{13}C) very nearly corresponds to the 6H-SiC polytype, whereas the second case should be identified with 4H-SiC, where the linewidth is primarily determined by the HF interaction with ^{29}Si. An increase in the ^{13}C content in the 6H-SiC polytype by approximately 20 times broadens the EPR linewidth nearly fourfold, while a decrease in the ^{29}Si content by approximately 10 times has practically no effect on the linewidth. By contrast, in 4H-SiC, a similar decrease in ^{29}Si content resulted in EPR line narrowing to one-third its previous width, with no noticeable broadening produced by a substantial increase in the ^{13}C content. Thus, the EPR study suggests the conclusion that the nitrogen-donor unpaired electron wave function distributions in the 4H-SiC and 4H-SiC polytypes are substantially different. In 4H-SiC, the wave function is primarily localized on silicon atoms, while in the 6H-SiC polytype it is localized mainly on carbon. This conclusion is in agreement with ENDOR studies [15], which also revealed a marked difference in the spatial distribution of the shallow N donor wave function between the 4H-SiC and 6H-SiC polytypes. It was demonstrated that, within the coordination shells the

Table 4.7 X-band experimental and calculated EPR linewidths (in mT) for shallow nitrogen donors in the k position in 4H-SiC and in the k1 position in 6H-SiC with different ^{13}C and ^{29}Si isotope abundances: natural, enriched in ^{13}C (25% for 6H-SiC and 15% for 4H-SiC), and depleted in ^{29}Si (0.5%)

	Experiment			Calculation		
	ΔB (natural)	ΔB (C-13)	ΔB (Si-29)	ΔB (natural)	ΔB (C-13)	ΔB (Si-29)
4H-SiC	0.21	0.35	0.07	0.21	0.29	0.08
6H-SiC	0.13	0.5	0.12	0.12	0.55	0.11

In each experiment, the content of only one isotope was varied

interactions with which account for the EPR linewidth, the unpaired electron is predominantly localized on silicon atoms in 4H-SiC and on carbon atoms in 6H-SiC. Also, the unpaired electron localized on silicon is distributed approximately evenly between the s and p orbitals, whereas on carbon the unpaired electron occupies predominantly the s orbitals. In other words, the gain coefficients for Si and C in these polytypes are essentially different. Indeed, judging from the maximum isotropic constants of HF interaction with ^{13}C, which in 4H-SiC and 6H-SiC are 5.02 and 10.75 MHz (k2), respectively [51], the gain coefficient for carbon in 6H-SiC is approximately twofold that for 4H-SiC. However, the distribution of the unpaired electron on silicon s orbitals reveals the reverse; namely, the HF interaction constants are 6.54 and 3.86 MHz for 4H-SiC and 6H-SiC, respectively. This means that the gain coefficient for the isotropic spin density distribution on silicon in 4H-SiC is about 1.7 times that in 6H-SiC. Note that a similar gain coefficient for the unpaired electron distribution over the p orbitals cannot be introduced, because one can speak here only about the density of the envelope wave function on the Si or C nuclei. Nevertheless, one has to bear in mind that, while in the 4H-SiC and 6H-SiC polytypes the unpaired electron density in the p orbitals of C is small and about the same, the maximum constant of anisotropic HF interaction b in the p orbitals of Si in 4H-SiC (1.26 MHz [51]) exceeds that for 6H-SiC by about 30 times.

It should be stressed that we do not compare the HF interaction constants for the first two coordination shells, including for which EPR spectra revealed a split structure. These quantities have practically no bearing on the properties of the donor electron that are described by EM theory and are characteristic of a band electron. These constants determine the depth of a state, and, for comparatively deep states, the difference between the polytypes, as a rule, disappears.

To obtain more information about the N-donor centre and the related electronic wave function we consider the distribution of s and p character. With the table of Morton and Preston [61] the observed HF interactions can be translated into spin density. As mentioned earlier the isotropic HF interaction (a) gives a measure of the s spin density (s), whereas the anisotropic HF interaction (b) is a measure for the p density (p). The results were presented in [51]. It is found that for the k site in 4H-SiC the ratio s:p is 1:4 for the density on the Si atoms and 2.5:1 for the density on the C atoms. For the h, k1 and k2 sites in 6H-SiC the ratio s:p is 1:1 (Si) and 19:1 (C); 3.5:1 (Si) and 6.8:1 (C); and 1.6:1 (Si) and 2:1 (C) respectively. Thus the spin density corresponding to the observed ENDOR lines is p like in character and located mainly on the Si atoms for the k site in 4H-SiC, whereas for the three sites in 6H-SiC the spin density is s-like in character and located mainly on the C atoms.

A possible explanation for the difference in the electronic wave function of the N donor in 4H-SiC and 6H-SiC can be found in the large difference in the band structure of the two polytypes and in the position of the minima in the Brillouin Zone. As a result the linear combination of atomic orbitals describing the wave function of the donor electron is different. Thus the wave function might have a completely different symmetry and a different distribution of s and p character on the Si and C atoms. From band structure calculations it is not clear whether Si or C bands lie lowest in the minima in 4H-SiC and 6H-SiC. The ENDOR and EPR results are consistent with mainly Si-like conduction-band minima in 4H-SiC (bottom of the conduction bands) and C-like conduction-band minima in 6H-SiC.

A large part of the spin density is related to the non-resolved ENDOR signal around the Si and C nuclear Zeeman frequencies, which consists of a superposition of a large number of ENDOR lines with a very small HF interaction [51]. In delocalised centres this signal is expected to be very pronounced, in localised centres it is almost not present. Inspection of the ENDOR spectra [51] shows that the non-resolved signal is indeed very pronounced, with an estimated width of 1.1 MHz. The amount of spin density incorporated in the non-resolved signal can be estimated using the approximation that the HF interaction A(r) is given by $A(r) = A_0\exp(-r/r_B)$, with A_0 the spin density at the centre and r_B the Bohr radius of the wave function. The amount of spin density in a shell δr at distance r from the centre is given by the product of the spin density ρ at r, calculated from A(r), and the number of atoms in δr. $\rho(r)$ has a maximum at twice the Bohr radius. Using for example $r_B = 7.2$ Å and $A_0 = 4$ MHz, 60% of the spin density turns out to originate from nuclei located between 9 and 30 Å (A(9 Å) < 0.03 mT). Increasing r_B even further it is possible to account for an even larger amount of spin density. Thus it seems reasonable to assume that the main part of the spin density is located in the crystal at distances of twice the Bohr radius and beyond.

Finally, we compare the properties of nitrogen donors in SiC with those in carbon (diamond) and silicon crystals. Nitrogen donors were first studied by EPR in diamond [62] and silicon [63, 64]. In both cases, the nitrogen donors had deep levels and their structure differed strongly from that of other donors in silicon belonging to Group V of the periodic table (P, As, Sb), which have shallow levels and whose electronic properties can be described in terms of the effective-mass approximation. EPR measurements show that nitrogen centres in diamond and silicon undergo strong trigonal distortions along one of the four $\langle 111 \rangle$ axes, which is caused by the nitrogen being in an off-center position. The donor electron occupies an antibonding orbital and is localized partially on the nitrogen atom in a substitutional position and partially on one of the four nearest neighbor carbon (in diamond) or Si (in silicon) atoms. In diamond, the unpaired donor electron is localized to 54.9% on one carbon atom (6% in the s orbital and 94% in the p orbital) and to 34% on the nitrogen atom (19% in the s orbital and 81% in the p orbital) and the donor level is ~2-eV deep. In silicon, the unpaired donor electron is localized to 72% on one Si atom (12% in the s orbital and 88% in the p orbital) and only to 9% on the nitrogen atom (28% in the s orbital and 72% in the p orbital) and the donor level depth is ~0.3 eV. A theoretical analysis of the formation of deep donor levels of nitrogen in silicon and diamond was performed in [65–68] and included the Jahn–Teller effect, pseudo-Jahn–Teller effect, and chemical rebonding. Because the issue of the formation of deep nitrogen donor levels in silicon and diamond still remains unsolved, there has not been, as far as we know, any comparative consideration of the differences in the nitrogen behavior between silicon and diamond, on the one hand, and SiC, on the other.

A radically different situation is observed for nitrogen donors in SiC, which does not compare, even qualitatively, to the behavior of deep donors in Si and C. In SiC, nitrogen produces donors with relatively shallow levels and, most remarkably, no bonds are formed with the nearest lattice atoms and no noticeable lowering of the symmetry occurs, unlike in Si and C crystals. While, in the case of the hexagonal polytypes of SiC, the axial symmetry of the crystal may be suggested to play a certain role (the g factor and the hyperfine structure constant exhibit approximately axial symmetry with respect to the crystal c axis) and the off-center nitrogen atom in positions k may manifest itself, the cubic modification of SiC (3C-SiC) has the same symmetry as diamond or silicon. Thus, there are firm grounds for developing a general theory which would account for the radical difference in the behavior of nitrogen donor impurities between silicon and diamond, on the one hand, and SiC, on the other.

4.2.1.2 Phosphorus Donors

A n-type conductivity of SiC crystals is generally achieved by nitrogen doping. Nitrogen introduces shallow donor states and is believed to substitute for carbon atoms. In the 6H-SiC polytype with three non equivalent lattice sites, two quasicubic (k1, k2) and one hexagonal (h) site, three different shallow nitrogen centres have been identified. Their thermal ionization energies are 81, 137.6 and 142.4 meV, respectively, with the hexagonal-site donor being the shallowest. Phosphorous, is an alternative shallow donor to nitrogen in SiC and it is considered to have advantages over nitrogen in the high doping range, where the electrical conductivity in N-doped SiC is being saturated.

Phosphorous can be introduced in SiC during the growth or after growth by either ion implantation or neutron transmutation doping (NTD) [57, 69–79]. In the cases of ion implantation and NTD, subsequent high-temperature thermal annealing is required to electrically activate the phosphorous impurity and to anneal the irradiation-related deep defects. In analogy to the case of substitutional N, three different P centres corresponding to P on the k1, k2, h sites are expected to coexist. This picture is confirmed by electrical measurements on P implanted, thermally annealed 6H-SiC layers [72, 73]. They have shown presence of the two P-related centres with ionization energies of (80 ± 5) and (110 ± 5) meV; these were attributed to the isolated P donor on the h and the (nondistinguished) k1, k2 states, respectively.

The microscopic and electronic structure of the phosphorous-related centres in SiC were studied by EPR and ENDOR spectroscopy. These measurements have been performed in SiC crystals doped with P by NTD [47, 69, 70, 74, 79], by P ion implantation [76] and during chemical vapor deposition (CVD) growth [57, 77]. The results indicated much more complex situation as compared to the case of N donors.

In the first publications [69, 70], instead of the expected three P centres on the k1, k2, h sites five different P-related centres were observed at 77 and 4.2 K in 6H-SiC crystals doped with P by NTD followed by annealing at about 2000 °C. Their interpretation has given rise to some controversy. The first X-band EPR results were obtained in [69] and three different doublet spectra with large, almost isotropic HF interaction and with small g-factor anisotropy observed at 77 K were labeled as P1, P2, P-V. The P1, P2 spectra have been attributed to the isolated P_{Si} donor on the hexagonal and the unresolved k1 and k2 Si sites; the third spectrum (with a smaller hyperfine interaction), P-V, was attributed to another P-related complex centre involving P atom and a vacancy.

Later, the EPR measurements were extended to the 4 K temperature range in [70]. The authors observed at 4 K two different P-related centres labeled as I1, I2 with strongly reduced and anisotropic central hyperfine interactions. As these centres could only be observed at temperatures below 15 K, an inverted level system for the P donor was proposed [47] with an E ground state corresponding to the I1, I2 spectra and an A1 excited donor state giving rise to spectra P1 and P2 spectral features. ENDOR measurements [47] at 4.2 K confirmed the phosphorus was the origin of the observed HF splitting of the I1 and I2 centres.

In a later EPR study [74], both sets of low- and high-temperature P-related spectra were simultaneously observed at 4.2 K, excluding the possibility that the P1 and P2 centres are related to the excited states. In [74] the P1, P2 and P-V centres were reassigned to the ground state of the shallow P donors at the quasicubic and hexagonal Si sites and labeled sP_{c1}, sP_{c2}, and sP_h, respectively (we reserve the original index "c" for quasi-cubic positions, in this review we will use index "k"). Also, the I1 and I2 spectra were suggested to be composed of three overlapping spectra, labeled dP_{c1}, dP_{c2}, and dP_h.

The striking result is that two types of the EPR spectra of P-related donor centres were observed only in SiC crystals doped with P by neutron transmutation. In SiC crystals doped with P by ion implantation [76] and during CVD growth [57, 77] only one type of P-related centres has been observed: sP_{c1}, sP_{c2}, and sP_h in the first case and the low temperature EPR spectra with parameters close to those of I1 and I2 (or dP_{c1}, dP_{c2}, and dP_h) in the second case. It should be noted that according to the recent theoretical study [78], despite higher formation energies the incorporation of phosphorous at the carbon sublattice is favored by kinetic effects during the annealing process after the neutron transmutation.

The EPR experiments with NTD of SiC crystals enriched with ^{30}Si were very promising because much higher concentration of P donors could be reached with much smaller n-irradiation dose and as a result the annealing temperature could be sufficiently decreased [79].

Figure 4.13a presents the EPR spectra for 6H-SiC crystal enriched with the ^{30}Si isotope subjected to neutron irradiation with thermal neutrons to a dose of $\sim 1 \times 10^{20}$ cm^{-2} measured in the B$\|$c orientation at 50 K. Before the irradiation, the crystal was n-type with a nitrogen donor concentration of $\sim 1 \times 10^{18}$ cm^{-3}. The spectra were measured after annealing at 1500 °C.

Before the annealing, the spectrum had a very strong single isotropic line with $g \approx 2$ that belongs to silicon vacancies and other deep-level defects of a high concentration with strong exchange coupling between centres. Upon annealing at 1500 °C, the EPR response changes drastically, the spectrum now becoming

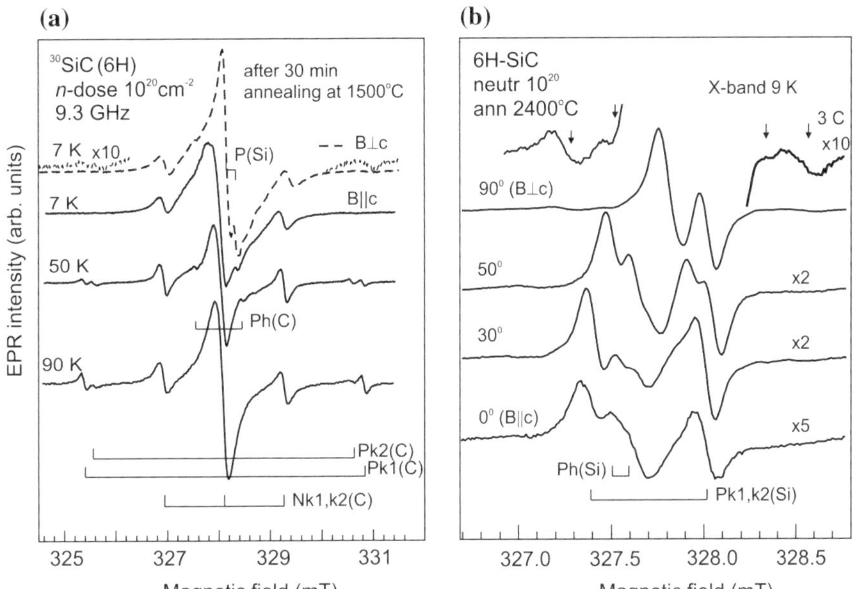

Fig. 4.13 **a** X-band EPR spectra measured at 7, 50 and 90 K in 6H-SiC crystal enriched with ^{30}Si and subjected to neutron irradiation at a dose of 1×10^{20} cm^{-2} and the annealing at 1500 °C. The double EPR lines of shallow phosphorus donors in two quasi-cubic positions Pk1(C), Pk2(C) and in hexagonal positions Ph(C) and three lines belonging to shallow nitrogen donors in the two quasi-cubic positions Nk1 and Nk2 are indicated by *bars*. The spectra were measured in the orientation B$\|$c (the *dashed line* shows the low temperature spectrum at 7 K for the orientation B$\perp$c). **b** X-band low-temperature EPR spectra of the superposed three P donor centres for different field orientations measured at 9 K in 6H-SiC crystal with natural abundance doped by neutron transmutation and annealed at 1900 °C. The spectra are double EPR lines of shallow phosphorus donors in two quasi-cubic Si positions Pk1(Si), Pk2(Si) and in hexagonal positions Ph(Si) (are indicated by *bars* for the orientation B$\|$c). Hyperfine interaction lines with 3 carbon atoms in the first shell are indicated by *arrows* for the orientation B$\perp$c

entirely dominated by the N-donor and P-related donor patterns that appear simultaneously. In the central part a strong three-line spectrum originates from N shallow donor spectra for two quasicubic sites labeled Nk1, Nk2, which are not resolved under these conditions. The phosphorous spectrum consists of three slightly anisotropic doublet spectra with axial symmetry along c-axis named Pk1, Pk2, and Ph (each doublet due to the ^{31}P hyperfine structure, I = 1/2) that correspond to the shallow P donors occupying the two quasi-cubic and hexagonal C lattice sites, respectively. For this reason, the carbon position of the phosphorus in the SiC crystal lattice is shown in parentheses (C).

The EPR spectra of P donors can be described using the conventional spin Hamiltonian (4.7). The Ph(C) lines partly overlap with the central line of the N shallow donors. The parameters of these centres at 70 K are almost identical to that in [55]: Pk1(C): $g_{\parallel} = 2.0038$, $g_{\perp} = 2.0028$, $A_{\parallel} = 5.50$ mT, $A_{\perp} = 5.42$ mT; Pk2(C): $g_{\parallel} = 2.0038$, $g_{\perp} = 2.0025$, $A_{\parallel} = 5.10$ mT, $A_{\perp} = 5.10$ mT and Ph(C): $g_{\parallel} = 2.0041$, $g_{\perp} = 2.0022$, $A_{\parallel} = 0.90$ mT, $A_{\perp} = 0.72$ mT. The value of HF interaction, g-factors and the weak localization of the wave function for shallow P donors in C lattice sites are very similar to the case of the shallow N donors (which occupy only C lattice sites) including a strongly reduced HF interaction for the hexagonal site of the N donor [46]. These similarities give the reason to conclude that EPR spectra belong to shallow P donor.

The study of the EPR signal temperature dependences of Pk1(C), Pk2(C), and Ph(C) centres allowed determining the thermal ionization energies that are also similar to shallow N donors in different C crystal sites. High concentration of the P donors in ^{30}Si enriched SiC allows measuring the EPR spectra in wide temperature range starting from 4 K. The EPR spectra of the P-related centres and the correlation with spectra of shallow N donors measured for 1500 °C annealed sample at several temperatures in the B||c orientation are presented in Fig. 4.13. The disappearance of Ph(C), Pk2(C), and Pk1(C) centres due to thermal ionization was observed at 60, 90 and 110 K, respectively. Thus, as in the case of the N donors, the P donor with the smaller HF interaction corresponds to that with the smaller thermal ionization energy.

The slight narrowing of the EPR line (about 10%) of shallow P donors P(P) was observed in 6H-SiC crystals with a reduced content of ^{29}Si with a nonzero nuclear magnetic moment, compared with natural 6H-SiC. A similar small narrowing was also observed for N shallow donors in 6H-SiC (see previous section). The wave function spatial distribution of N donor electrons in SiC was shown to depend substantially on the polytype; indeed, in 6H-SiC, the unpaired electrons occupy primarily the carbon s orbitals, whereas in 4H-SiC these electrons reside primarily in the silicon s and p orbitals. As a result, the linewidth of nitrogen donors in 6H-SiC with natural isotopic abundance was 0.13 mT and, ^{29}Si-depleted 6H-SiC (about 0.5% of ^{29}Si) showed 0.12 mT. In contrast, the linewidth for nitrogen donors in 4H-SiC with natural isotopic abundance was 0.21 mT, and in ^{29}Si-depleted 6H-SiC (about 0.5% of ^{29}Si) it was 0.07 mT. The data are presented for quasi-cubic positions. Thus, the wave function spatial distribution of unpaired electrons from N shallow donors and P shallow donors P(C) in 6H-polytype are similar, and these electrons reside primarily in C orbitals. This finding supports the idea that phosphorus for these centres occupies the carbon position, similarly to nitrogen.

In addition, the EPR spectrum measured at 7 K in the B⊥c orientation is presented in Fig. 4.13a (dashed line). For this orientation, the EPR spectra of another set of P-related donors with small anisotropic HF interactions exhibiting low temperature EPR spectra are more prominent. These centres appear after NTD at the same annealing temperature as the Pk1(C), Pk2(C), and Ph(C) shallow P donors. The parameters of these centres are almost identical to those presented in [74] and labeled as dP centres.

In Fig. 4.13b typical EPR spectra for different magnetic-field orientations are shown. The analysis of the total EPR spectrum for all magnetic-field orientations shows the presence of three partially overlapping doublet spectra [74]. The presence of three spectra is most easily seen at intermediate angles between the two orientations B‖c and B⊥c where all three are of comparable intensities. The complete analysis of the angular variation shows the presence of three spin $S = 1/2$ spectra with axial symmetry for the g tensor and the central phosphorous hyperfine interaction. By analogy with the P donors occupied carbon sites, these three centres were assumed to occupy positions of silicon. We will mark them as the Pk1(Si), Pk2(Si), and Ph(Si) shallow P donors. In the works, which have been cited in the beginning, these centres have different designations and slightly different parameters. We will use the parameters of [74] and simplify the notation, making them obvious. Of course, the k1 and k2 can be interchanged.

The parameters of these centres at 9 K: Pk1(Si): $g_{\|} = 2.0037$, $g_{\perp} = 2.0025$, $A_{\|} = 0.62$ mT, $A_{\perp} = 0.38$ mT; Pk2(Si): $g_{\|} = 2.0037$, $g_{\perp} = 2.0024$, $A_{\|} = 0.48$ mT, $A_{\perp} = 0.19$ mT, and Ph(Si): $g_{\|} = 2.0044$, $g_{\perp} = 2.0027$, $A_{\|} = 0.10$ mT, $A_{\perp} = 0.05$ mT.

The P(Si) spectra show additional small intensity lines due to hyperfine interaction with the ligands. They can easily be seen for B⊥c at the low- and high-field sides of the central set of lines (Fig. 4.13b). For this orientation, where the Pk2(Si) and Ph(Si) spectra are dominant in intensity, one observes two doublets centered at the Pk2(Si) and Ph(Si) lines, respectively, with splittings of 1.02 and 1.12 mT. For B‖c a value of 0.88 mT is obtained for Pk1(Si). The intensity ratio of the HF ligand lines to the corresponding central line has been evaluated to about 3%, which corresponds to the interaction with three equivalent C atoms (see [57, 77]). The C HF structure is observable for all orientations; it is approximately isotropic with values in the 1-mT range: Pk1(Si): A(3C) = 0.88 mT; Pk2(Si): A(3C) = 1.02 mT, and Ph(Si): A(3C) = 1.12 mT.

With the successful in situ P-doping during CVD growth [57, 77], the shallow P donors in 3C–, 4H– and 6H–SiC have been studied by magnetic resonance without interference from the normally strong signals of the residual N donors as in the cases of P-implanted or neutron-transmutation P-doping materials. The low-temperature P-related EPR spectra have been confirmed to be related to the shallow P occupying the Si lattice site [Pk1(Si), Pk2(Si), and Ph(Si) in this review], which are shown to be very different from the large HF splitting P donors in P-implanted or neutron-transmutation doping 6H–SiC samples: Pk1(C), Pk2(C), and Ph(C) in this review.

4.2.2 *Acceptors with Shallow and Deep Levels*

The group-III elements B, Al and Ga are the most important acceptor impurities in SiC. These impurities can be introduced during the growth or afterwards by diffusion or by implantation. A substitutional atom of a group-III element normally acts as an acceptor in SiC since there is a deficit of one valence electron to complete the normal tetrahedral bonding. At sufficiently low temperatures the hole is localized near the acceptor atom; at high temperatures it can ionize and give rise to p-type conduction. The group-III element acceptors in SiC were studied with using Hall effect, DLTS, optical spectroscopy measurements [80]. B creates in SiC band gap shallow and deep levels [80]. The B centre with a shallow acceptor level (shallow B) with activation energy 0.30–0.39 eV were usually observed in Hall effect studies and admittance spectroscopy [81, 82] measurements. The second B centre with a deep level (deep B) was first detected by photoluminescence measurements [83, 84]. It was suggested that the characteristic high-temperature bright-yellow luminescence in B doped 6H-SiC crystal (which can be observed up to room temperature) is due to pair recombination via the N donor and deep B centre. Based on this suggestion, the ionization energy of deep B centres was estimated to be more than 0.65 eV. From DLTS investigation on B diffused $p + n$ junctions in 6H-SiC Anikin et al. [82] observed a deep centre with a doublet structure at Ev + 0.63 eV and Ev + 0.73 eV. Only one level Ev + 0.58 eV due to deep B centre was observed on B-implanted aluminium-doped 6H-SiC epitaxial layers [85]. Two acceptor levels corresponding to two kinds of B centres (shallow and deep) were detected by capacitance methods in [86]. From Arrhenius plots and photocapacitance measurements the activation energies of the centres were estimated to be 0.22 and 0.35 eV for shallow B and 0.55 and 0.75 eV for deep B.

Shallow Al introduces acceptor levels in 6H-SiC at Ev + 0.239 eV for the hexagonal lattice site (h) and at Ev + 0.2485 eV for two cubic-like lattice sites (k1, k2); shallow Ga introduces acceptor levels at Ev + 0.317 eV (h) and Ev + 0.333 eV (k1, k2) [87]. The experimental energy levels of group-III acceptors are summarized in Fig. 4.14a. The ionization energy of group III impurity acceptors is weakly sensitive to a particular polytype and to the different lattice sites because the maximum of the valence band is located at the Γ-point and as a result there is no interference effect. In contrast to most acceptors, the differences in the ionization energies for donors (N, P) are much more sensitive to a particular polytype and to the different lattice sites that are mainly caused by the interference effect of the donor electron wave function, which is composed of the wave functions of the different conduction band minima.

4.2.2.1 Group-III Acceptors with Shallow Levels

EPR of acceptors in cubic semiconductors was believed can only be observed after uniaxial stress is applied to the crystal to remove the fourfold degeneracy of the

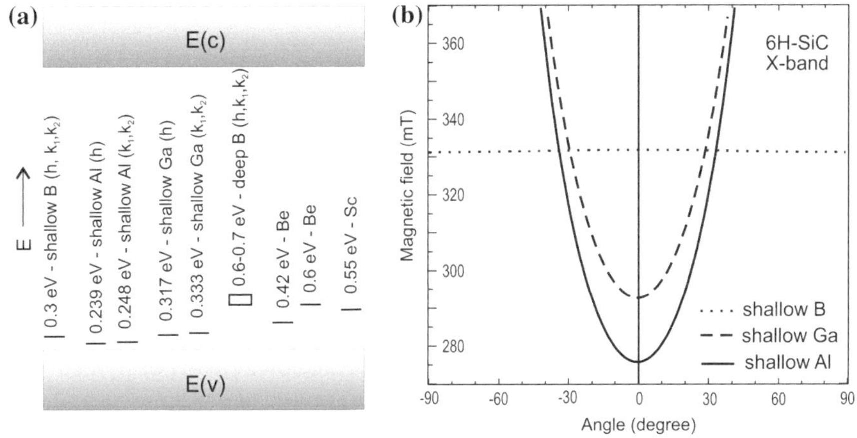

Fig. 4.14 **a** Schematic survey of acceptor levels observed in the band gap of 6H-SiC. **b** Angular dependencies of group-III shallow acceptors at X-band in 6H-SiC crystals. Only h-sites are presented

acceptor ground state, which is very sensitive to random crystalline strains. In the noncubic polytype of SiC this degeneracy is already lifted and EPR of acceptors can therefore be readily observed.

Let us now discuss the g-factors of the acceptors in SiC. First we will discuss Si to compare with SiC. The minimum hole energy in Si is at k = 0. The degeneracy of the valence bands at k = 0 is related to the symmetry of the lattice, and can be lifted (except for spin) by subjecting the crystal to an uniaxial stress. The binding energies of holes to the group III acceptors B, Al, Ga and In in Si range from about 0.05 to 0.16 eV. Since the spin-orbit splitting of the J = 3/2 and 1/2 bands is 0.035 eV [88] the ground state is fourfold degenerate. A single resonance line in B, Al. Ga and In-doped Si crystals subjected to an uniaxial stress was observed in [89] The effective g-factor was expressed as $g^2 = g_{\parallel}^2\cos^2\theta + g_{\perp}^2\sin^2\theta$, where θ is the angle between the stress axis (a cubic [001] direction) and the magnetic field. The acceptor EPR spectra were interpreted in terms of a strain-induced splitting of the J = 3/2 state into $M_S = \pm 1/2$ and $M_S = \pm 3/2$ doublets. The resonance transition is between the $M_S = -1/2$ and $M_S = +1/2$ states, which are split by the magnetic field. The $M_S = -3/2$ to $M_S = +3/2$ transition is forbidden for large zero field splitting and $M_S = \pm 1/2$ to $M_S = \pm 3/2$ transitions are strain sensitive and would require an energy greater than the microwave quantum at X- and Q-bands ($2|D| \gg h\nu$). The failures to detect resonance of shallow acceptors in the absence of applied uniaxial strain were attributed to the broadening of the EPR lines by random strains and a short spin-lattice relaxation time. For free hole one can show theoretically that $g_{\perp} \cong 2\ g_{\parallel}$, since J = 3/2. The best agreement with this ratio would be expected for B acceptors with the smallest binding energy, a fact which was confirmed experimentally and which indicates an effective-mass-like characters for the B acceptors. The experimentally observed features of the EPR spectra of Al, Ga and In acceptors in Si are identical in most details to those exhibited by B. Quantitative differences have only been observed in the spin Hamiltonian parameters of EPR spectra of B, Al, Ga and In acceptors in Si.

In hexagonal SiC the situation is different compared with Si. 6H-SiC hexagonal crystal field is larger than the spin-orbit splitting. Therefore J is no longer a good quantum number. The uppermost valence-band state is the doublet and the irreducible representation of this state is Γ_9 for C^4_{6v} crystal group of the 6H polytype. For the case of the free hole in Γ_9-state $g_{\|} = 4.0$ and $g_{\perp} = 0$. The best agreement with this ratio would be expected for acceptors with the smallest binding energy, a fact which was confirmed experimentally only for Al and Ga in SiC.

The first EPR observation of acceptors in SiC were made by Woodbury and Ludwig [43], who examined the EPR spectrum of shallow B in 6H-SiC. Since then the group-III impurities, B, Al and Ga with shallow levels in SiC have been studied using EPR [90–97], ODMR [98–102], pulsed EPR\ENDOR [40, 103–105] and conventional ENDOR [106–111] techniques. The main result was that the behaviour for shallow B is strikingly different from those observed for the other group-III impurities, Al and Ga. All other experimental finding should be consider as a consequence of this main result. The resonance properties of Al and Ga shallow acceptors are determined largely by the band structure of the host lattice. It was indicated in [96, 98–101] an effective-mass-like character for these acceptors, reflecting the uppermost Γ_9 valence-band hole state but with reduced orbital-momentum contributions resulting from the localisation. In contrast to Al and Ga, no effective-mass-like behaviour was observed for B shallow acceptors, and what is more, EPR of shallow B in contrast to shallow Al and Ga could also observe in cubic 3C-SiC without any uniaxial stress applied to the crystal.

Shallow B. EPR spectra of shallow B were studied in 3C-, 4H- and 6H polytypes of B-doped SiC in the temperature range 1.2–300 K [40, 43, 90, 92, 103]. The results in 4H have been used to aid the site assignment for EPR in 6H polytype. Hence, it is sufficient to summarize the results in 6H and 3C polytypes.

B has two stable isotopes, ^{10}B (natural abundance 19.8%) and ^{11}B (80.2%), having nuclear spins I of 3 and 3/2, respectively. The nuclear g-factors ratio $g_I(^{11}B)/g_I(^{10}B) = 3$, where $g_I(^{11}B)$ and $g_I(^{10}B)$ are the nuclear g-factors of ^{11}B and ^{10}B atoms, respectively. Since ^{11}B is about 80% abundant, one normally expects a spectrum consisting of one or more sets of four equally intense lines. The ^{10}B will produce a background of seven weaker lines (usually unresolved) whose spacing is 3 times smaller (according to the ratio of nuclear g-factors of ^{11}B–^{10}B atoms) as compared with those for the ^{11}B isotope.

In principal, EPR spectra of ^{11}B shallow acceptors should be resolved into three sets of four lines which belong to different lattice sites (hexagonal-like (h) and two cubic-like (k1, k2) crystallographic inequivalent lattice sites) and can be described by the conventional spin Hamiltonian. For h-centre, the principal axis of the g-tensor is parallel to c-axis. We can designate by the term 'axial' the defect corresponding to h-site. For k1 (or k2) centre, the principal axis of the g-tensor lies in a {11–20} symmetry plane of the crystal and axis z is parallel to one of the nonaxial Si-C directions (there are six such directions because there are two types of tetrahedrons in the wurtzite structure). We will term the corresponding defect the "nonaxial" defect. These results show that h-site corresponds to paramagnetic centre having C_{3v} symmetry. Cubic-like site (k1 or k2) corresponds to a centre having C_s symmetry, but the measurements show a very small deviation of the g-tensor and A-tensor from cylindrical symmetry.

The values of the g-tensor along the principal axes were measured in different polytypes of SiC [61, 112] to be: g_z ($g_{\|}$) = 2.0020 (±0.0002) and g_x, g_y ($g_{\perp}$) = 2.006 (±0.0005), where the local z-axis coincides with one of the direction Si-C (for h-site it is c-axis of the crystal, for k1 or k2-site it is one of six nonequivalent Si-C directions, which is not coincide with c-axis, for cubic 3C-SiC it is one of $\langle 111 \rangle$ directions of the crystal). x and y axes are in a plane perpendicular to the z axis and x axis lies in one of the {11–20} planes in 6H-SiC (4H-SiC). Thus, the g-tensor of shallow B centres is almost axial around the local z-axis for all shallow B centres in cubic and hexagonal polytypes of SiC. g-factor along z-axis was shown to be practically equal to the value g of the free electron spin. There is some influence of axial crystal field along c-axis in hexagonal polytype of SiC for cubic-like sites ("nonaxial" defect), but it the second order effect. The deviation from axial symmetry for the anisotropic g tensor were estimated to be $|g_x - g_y| <$ 0.0005 for qubic-like sites.

The HF interaction tensor A reflects the axial symmetry of the shallow B centre along the local z-axis (for cubic-like positions in hexagonal polytypes of SiC ("nonaxial" defect) there is small deviation from axial symmetry, which only reflects the influence of crystal field along c-axis and seems to be not important for the understanding of the structure of shallow B). At 4.2 K for 3C-SiC: $A_{\|} = \pm 1.0 \times 10^{-4}$ cm^{-1}, $A_{\perp} = -(\pm)1.9 \times 10^{-4}$ cm^{-1}; for 6H-SiC $A_{\|} = \pm 1.8 \times 10^{-4}$ cm^{-1}, $A_{\perp} = -(\pm)1.3 \times 10^{-4}$ cm^{-1} (h-site); $A_{\|} = \pm 1.7 \times 10^{-4}$ cm^{-1}, $A_{\perp} = -(\pm)0.9 \times 10^{-4}$ cm^{-1} (k1-site); $A_{\|} = \pm 2.0 \times 10^{-4}$ cm^{-1}, $A_{\perp} = -(\pm)1.1 \times 10^{-4}$ cm^{-1} (k2-site). The striking result is that the contact a and the dipole-dipole b HF interaction constant have opposite signs ($A_{\|} = a + 2b$, $A_{\perp} = a - b$). It was shown by EPR measurements in [43, 92] and later was confirmed by ENDOR [40, 100, 106–111].

With magnetic field parallel to the c-axis, additional HF lines have been observed. The intensity is approximately what one expect if the 4.7% abundant isotope ^{29}Si is randomly distributed in three equivalent positions near B site. The value of the HF interaction with these Si nuclei (or superhyperfine (SHF) interaction) was measured to be almost isotropic with $a \cong 9 \times 10^{-4}$ cm^{-1} for 3C and 6H (4H) polytypes.

The natural abundance of ^{13}C is only 1.1% and C SHF structure would be almost impossible to see. In a specially ^{13}C-enriched 6H-SiC:^{11}B sample, however, the ^{13}C isotope had been enriched to about 30% [40, 90, 103]. Compared with a sample with natural abundance of carbon and under the same detection conditions, the SHF interaction with one carbon atom for shallow B acceptor was clearly observed. This interaction was measured in different polytypes to be (56–58) × 10^{-4} cm^{-1} for magnetic field parallel to z axis of shallow B and (15–17) × 10^{-4} cm^{-1} for B perpendicular to z axis. For the shallow B acceptor substituting for Si the four neighbouring carbon atoms should be equivalent in the lattice but as one can see from EPR results they are not equivalent with respect to the spin distribution.

The EPR spectra of shallow B were observed [90, 92] to change drastically in the temperature region higher than 50 K. A thermally activated jumping takes place over the three displaced positions of shallow B in cubic-like positions or over the four displaced positions of shallow B in cubic SiC, which leads to motional averaging of the spectra in the higher temperature range.

A high-frequency (95 GHz) pulsed electron paramagnetic resonance and electron-nuclear-double-resonance study has been carried out on the shallow boron acceptor in natural and ^{13}C isotope enriched 4H-SiC and 6H-SiC. From the hyperfine interaction of the unpaired electron spin with the ^{13}C (I = 1/2) nuclei the spatial distribution of the electronic wave function has been established. It is found that there are subtle differences in the degree of localization of this wave function between the different sites—quasicubic (or cubic-like) and hexagonal in the two polytypes 4H-SiC and 6H-SiC. In particular it is found that the spatial distribution is highly anisotropic. This anisotropy can be rationalized by considering the anisotropy of the hole mass, i.e., by assuming that effective-mass theory is (partly) valid in describing the remote part of the spatial distribution of the electronic wave function.

Figure 4.15a shows three 95-GHz EPR spectra of the sB acceptor in SiC with B ∥ c axis. The upper spectrum is obtained in 4H-SiC single crystal (^{13}C enriched) and consists of two lines; one is related to a B atom on a quasicubic position (k) in the crystal, the other is related to a B atom on a hexagonal position (h). The latter line is only visible after magnifying the spectrum by a factor 10. The weakness of the hex signal is probably caused by the position of the Fermi level. The weak satellites in the magnified spectrum are caused by the hyperfine (HF) interaction with the nearest-neighboring ^{13}C nucleus. The middle spectrum was measured on 4H-SiC single crystal with ^{13}C natural abundance. Here, in addition to the two EPR signals of the sB acceptor, the signal of the k-site N donor, present in the central, uncompensated part of the sample, is observed, with a HF splitting in three components characteristic of 4H-SiC and caused by the ^{14}N (I = 1) nuclear spin. We remark that, owing to the small linewidth of the sB signals and the fact that this sample is not ^{13}C enriched, the HF interaction with the B nuclear spin is visible in this spectrum on the sB line related to the h site. The lowest spectrum is obtained under similar conditions on 6H-SiC crystal (^{13}C enriched) and ^{11}B isotope doped. The three lines marked with k1, k2, and h correspond to the two quasicubic sites and the hexagonal site of the sB acceptor. The broad line marked with N originates from the k1, k2 and h sites of N donors, present in the central part of this sample, which is not totally compensated with ^{11}B diffusion.

Figure 4.15b shows an overview of the total ENDOR spectrum as measured on the 4H-Si^{13}C sample with a natural abundance of B with B ∥ c. The figure is a nice example of the spectral resolution that can be obtained at high-frequency (95 GHz) ENDOR. The orientational dependence of the ENDOR spectra is obtained by tuning the magnetic field to the EPR line of the site studied. In the following we will concentrate on the one parts of the spectrum (for more information, see [40]).

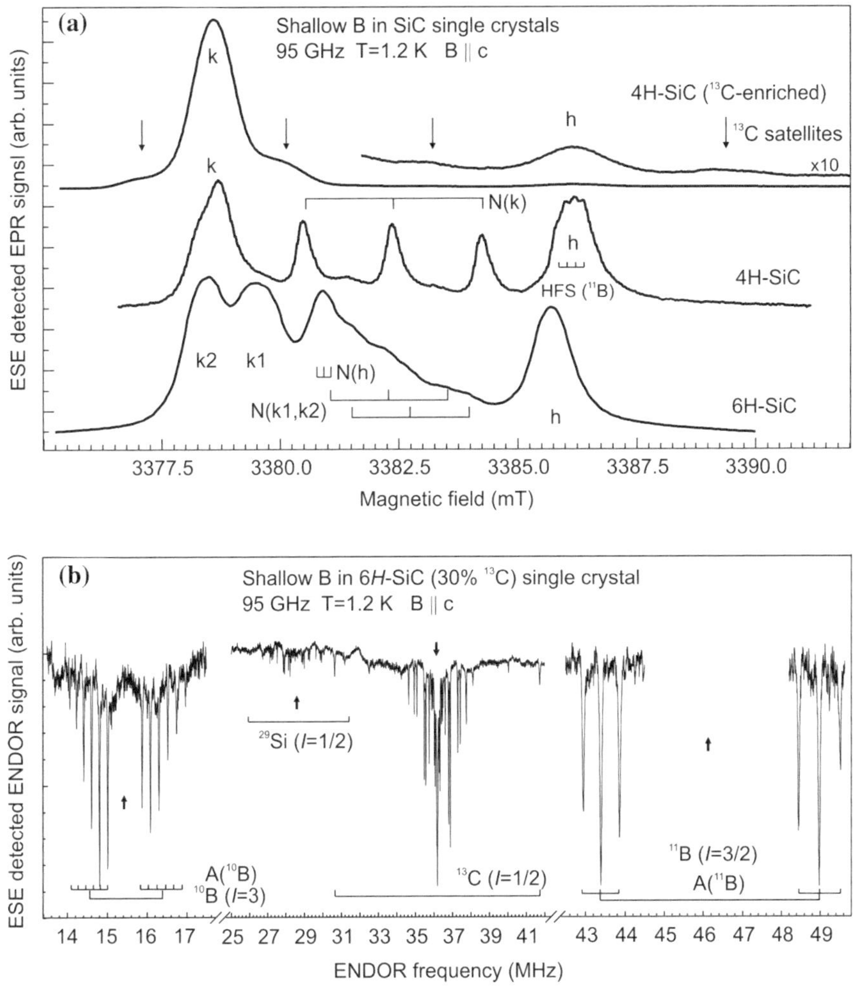

Fig. 4.15 **a** The EPR spectrum at 95 GHz with B ‖ c for the sB acceptor in 4H-SiC single crystal (^{13}C enriched crystal). The spectrum consists of two lines; one is related to a B atom on the quasi-cubic (k) site, and one to a B atom on the hexagonal (h) site. The shoulders are attributed to the HF interaction with the nearest-neighbor ^{13}C nuclei. The middle EPR spectrum is obtained on the 4H-SiC crystal with ^{13}C natural abundance. In addition to the sB lines, three lines are observed of the shallow N donor on the k site in 4H-SiC. This triplet is attributed to the HF interaction with the 14 N nuclear spin (I = 1). The lowest spectrum is obtained for the sB acceptor in 6H-SiC with B ‖ c. In addition to the EPR lines of the sB acceptor on the k1, k2, and h positions, a signal is observed that is attributed to the shallow N donor. All spectra are recorded at 1.2 K. **b** The ENDOR spectrum as measured on the EPR line of the k site in 4H-SiC for B ‖ c at 1.2 K. It shows the clear separation of the ENDOR lines belonging to the different isotopes and nuclei. As the sample is ^{13}C enriched, the ENDOR lines for this isotope are very intense. Only for the outer 13C ENDOR lines there is some overlap with the ^{29}Si and ^{11}B lines. *Arrows* indicate the Larmor frequency of the relevant isotopes in the fields in which the EPR signals were measured

Figure 4.16a shows part of the ^{13}C ENDOR spectra of the sB k1 and k2 site in 6H-Si^{13}C:^{11}B crystal and the k site in 4H-Si^{13}C crystal. The first thing to note is that the k1 spectrum in 6H-SiC is the same, except intensities, as that of the k site in 4H-SiC. This is the case for all measured orientations, which leads us to the conclusion that these two centres have the same position and thus the same surrounding in the crystal. We thus assign the corresponding EPR line to the k1 site, as this site is situated at a similar crystal position as the k site in 4H-SiC. Moreover, these two sites have the same environment in the two layers perpendicular to the c axis above and below the B centre, which we used in the assignment of the ENDOR lines. The EPR line with the smallest g-tensor anisotropy was assigned to the site closest to the h site, as the g tensor of the hexagonal site is axial. As it will turn out, the spin density distribution is not isotropic. The spectrum of k2 contains less lines around the carbon nuclear Zeeman frequency than that of k1 which, combined with the fact that we found one more line at 45 MHz for k2, suggests that for k2 the spin density is more localized than for k1.

Above the k2 ENDOR spectrum, a rescaled part of the k1 spectrum is shown for comparison, from which it can seen that the ENDOR lines belonging to the k1 site are also weakly present in the k2 ENDOR spectrum. Very weakly, the strongest k2 lines can also be seen in the k1 spectrum. This "crosstalk" is the result of the small overlap of the EPR lines of the k1 and k2 sites in 6H-SiC.

To illustrate the complexity of the assignment of the ENDOR lines, we show in Fig. 4.16b the orientational dependence of the inner part of the ^{13}C ENDOR spectrum for the k1 site of sB in 6H-SiC close to the Zeeman frequency of ^{13}C, where B is varied in the (11–20) plane from B // c ($\theta = 0°$, $\phi = 0$) to B $\perp$ c ($\theta = 90°$, $\phi = 0$).

The most remarkable result of the ENDOR study is the assignment of lines in the ^{13}C ENDOR spectra to specific C atoms up to eleven (11) bond lengths away from the B atom. This allows for a determination of the spatial distribution of the unpaired spin density of the sB acceptor, and in particular of an investigation of the difference in the electronic properties of the quasi-cubic and hexagonal sites in 4H-SiC and 6H-SiC. To visualize the delocalization of the electronic wave function, for example, connected to the k1 site, we show in Fig. 4.17 in the form of a contour plot its spin density at C atoms in the main (11–20) plane, that is, the plane containing the c axis. The crystal can be regarded as a collection of staircases along the c axis. The staircase on which the B and the main C are situated is indicated as a full thick line in. The next staircases are located in the one step up and one step down (11–20) planes, are projected onto the plane on which the B and the main C are situated, and are indicated by a dotted line. It is seen that the electron density is distributed in an ellipsoidal shape with the main symmetry axis making an angle of 70° with the c axis, i.e., along the direction of the B-main C line.

All three sB centres in 6H-SiC were shown to have different localizations of the spin density and thus different ionization energies. The question arises whether the anisotropic distribution of the electronic wave function on the remote ^{13}C nuclei as revealed by the ^{13}C ENDOR data can be rationalized on the basis of effective-mass theory (EMT). Shallow gallium (sGa) and shallow aluminum (sAl), as will be

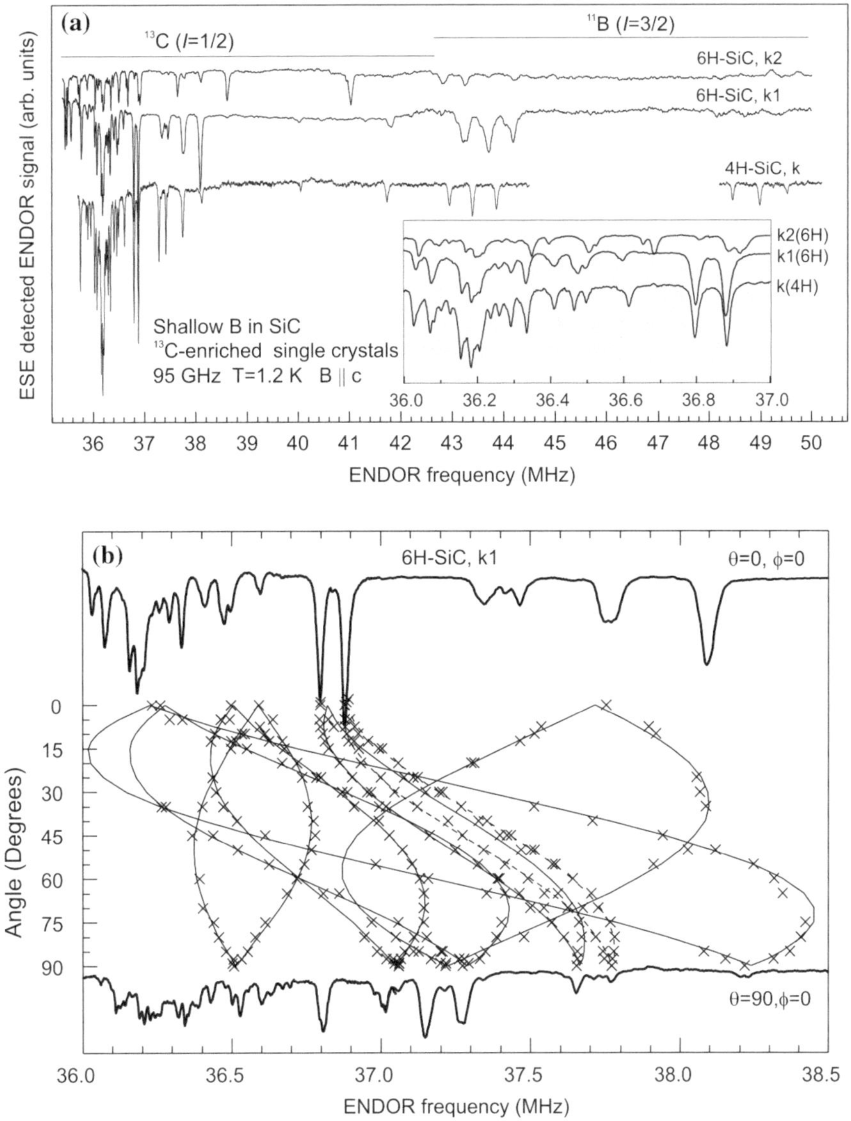

Fig. 4.16 **a** The ^{13}C ENDOR lines with B ∥ c and at 1.2 K for the different sites in 6H-SiC (k1 and k2) and 4H-SiC (k). As can be seen, the spectrum of k is very similar to that of k1. A rescaled part of the spectrum of k1 is shown above the spectrum of k2 in the region between 37 and 38.3 MHz. From the intensity comparison it is clear that the lines that show up in both spectra belong to k1. The two lines at 38.7 and 37.7 MHz belong to k2. These two lines can also be seen weakly in the spectrum of k1. In the inset a blowup of the spectrum around the ^{13}C nuclear Zeeman frequency is shown. **b** The orientational dependence of the HF interaction of the ^{13}C nuclei for the k1 site close to the Zeeman frequency of ^{13}C

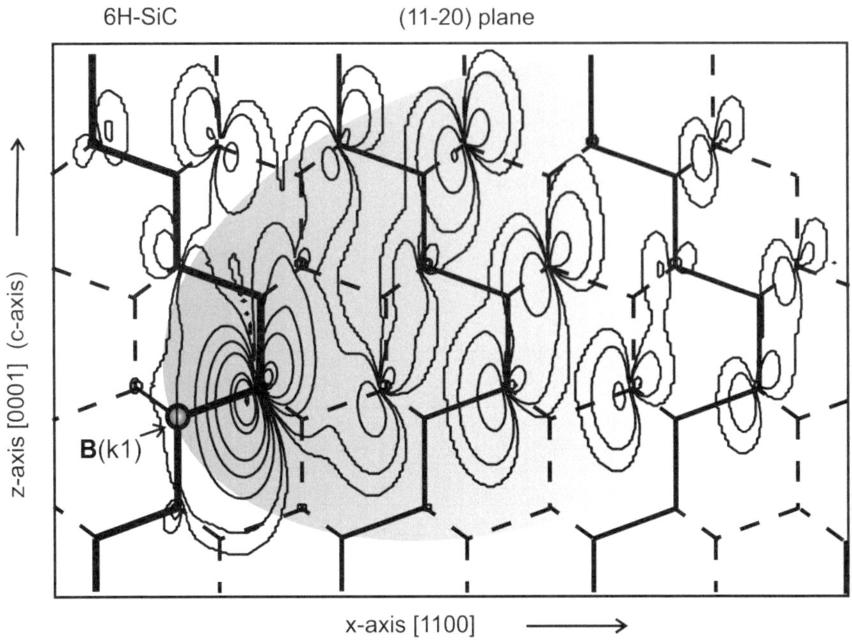

Fig. 4.17 The value of the square of the wave function of the 6H-SiC k1 site in the (11–20) plane, containing the B atom in Si position and the main C atom with the largest spin density (the main plane) is shown as a function of the position in the xz plane using a contour plot. Each line indicates a certain height and the larger the amount of lines the higher is the peak. The B nuclear position is indicated. The position of the main C atom can be recognized from the large number of contour lines. The *full lines* indicate the c axes located in the (11–20) plane, containing the B atom and the main C atom, and the *dotted lines* indicate the projection of the axes located in the ±1 planes into the main plane. It is clear that the spin density is very anisotropic. The distribution points away from the B atom into one direction only along the B-C connection line, making an angle of 70° to the c axis. The nuclei on the other side of the B atom carry an almost negligible amount of spin density

shown later, in contrast to sB, show a highly anisotropic **g**-tensor in agreement with the calculated g anisotropy of the valence-band hole. In the case of the sB acceptor the orbital angular momentum is quenched, and one would not expect to see band-structure properties reflected in the properties of the hole attached to the sB acceptor. Nevertheless we observe that the electronic wave function of the quasi-cubic sB acceptor perpendicular to the c axis extends considerably further than parallel to the c axis, in agreement with the anisotropy of the hole mass: $m_{h\|} = 1.67m_e$ and $m_{h\perp} = 0.62m_e$. Here we remark that the anisotropy in the perpendicular plane is averaged out at higher temperatures, where it is known that the B-main C bond starts to jump between the three possible directions of this bond. The cylindrically symmetric spin density so obtained spreads out further in the perpendicular plane than along the c axis.

Shallow Al. Shallow Al acceptors were detected by ODMR [98–101] and by conventional EPR [91, 96] in 6H- and 4H-SiC crystals. ODMR can be recorded due to existence of spin selection rules for the recombination of electrons and holes (donor-acceptor recombination). The results of ODMR by monitoring of the intensity of donor-acceptor recombination can be understood with the spin Hamiltonian for a weakly coupled donor-acceptor pair:

$$H = H_{\mathrm{D}} + H_{\mathrm{A}} + S_{\mathrm{D}} J S_{\mathrm{A}},$$

where the first and the second terms describe spin Hamiltonian for free donor and free acceptor states, respectively. The third term describes the exchange coupling between the donor and acceptor. For distant donor-acceptor pairs this coupling can be much smaller than Zeeman interaction for either the donor or acceptor and as a result the ODMR spectra of donors and acceptors can be broaden beyond the EPR spectra observed for isolated impurities (for example, by conventional EPR).

Figure 4.18a shows the part of ODMR spectrum for shallow Al acceptors observed at 1.5 K for the orientation between the applied magnetic field and the crystal c-axis at $\theta = 18°$. The ODMR lines correspond to an increase of the luminescence intensity. The spectrum consists of N signal and a group of three anisotropic lines. The group of anisotropic lines was identified as due to the Al shallow acceptor at three nonequivalent silicon lattice sites. Figure 4.14b shows an angular dependence of the ODMR spectra of shallow Al acceptors under a rotation of the crystal from $B \parallel c$ to $B \perp c$. No HF structure has been observed in these spectra. The EPR spectra can be fit to the spin-Hamiltonian with $S = 1/2$ and $g_{\parallel} = 2.412$, 2.400 and 2.325; $g_{\perp} \cong 0$. The g-factor follows a low of the type $g = g_{\parallel}\cos\theta$, where θ is the angle between the magnetic field and crystalline c-axis. Such behaviour indicates an effective-mass-like character for the acceptors, but with reduced orbital contribution due to the localization.

Shallow Ga. In Ga doped 6H-SiC crystals strongly anisotropic signal of shallow Ga acceptors were found in [99–101]. The ODMR spectrum of shallow Ga is shown in Fig. 4.18a. This signal corresponds to the increase of the luminescence intensity. In contrast with shallow Al, the shallow Ga signal have a resolved HF structure. Ga has two stable isotopes, ^{69}Ga and ^{71}Ga, with a natural abundance 60.1 and 39.9%, respectively. Both have a nuclear spin $I = 3/2$. The nuclear g-factors ratio $g_I(^{71}\mathrm{Ga})/g_I(^{69}\mathrm{Ga}) = 1.27$, where $g_I(^{69}\mathrm{Ga})$ and $g_I(^{71}\mathrm{Ga})$ are the nuclear g-factors of ^{69}Ga and ^{71}Ga isotopes, respectively. Since ^{69}Ga is about 60% abundant, one normally expects a spectrum consisting of one or more sets of four equally intense lines. The ^{71}Ga will produce four weaker lines whose spacing is 1.27 times larger (according to the ratio of nuclear g-factors of ^{71}Ga to ^{69}Ga) as compared with those for the ^{69}Ga isotope.

The resolved structure observed in the ODMR spectra was in full agreement with that expected for the HF interaction of unpaired electron with the Ga nuclei. In order to demonstrate that the resolved structure is due to HF interaction with ^{69}Ga and ^{71}Ga nuclei, the simulation of the Ga hyperfine structure were performed. Figure 4.18a shows calculated (bottom) Ga HF structure. This is clear evidence that

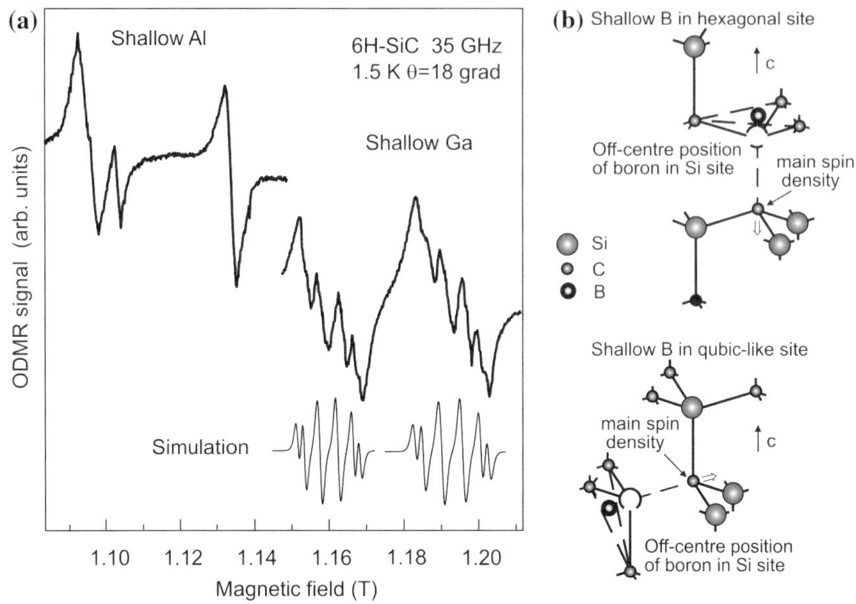

Fig. 4.18 **a** ODMR spectra at 1.6 K and 35.2 GHz in 6H-SiC:Ga and 6H-SiC:Al with magnetic field in the {11–20} plane directed at 18° to the c-axis. **b** A model of shallow B. Direction of off-center shift of B atom corresponds to local z-axis. *Double arrow* shows off-center shift of C with main spin density

the anisotropic ODMR signal is due to the Ga. Two Ga signals seem to belong to different position of Ga (h, k1, k2). Two positions (may be k1, k2) seem to be not resolved.

The angular dependencies of ODMR in 6H-SiC:Ga are presented in Fig. 4.14b. The spectrum can be described by the spin Hamiltonian, where S = 1/2 and I = 3/2. The parameters for the ODMR signals of shallow Ga acceptors are $g_{\|} = 2.27$ and 2.21. For both signals $g_{\perp} \cong 0.6$. The HF interaction constant were measured to be: 1) $A_{\|}(^{69}Ga) \cong 34 \times 10^{-4}$ cm^{-1} and $A_{\|}(^{71}Ga) \cong 43 \times 10^{-4}$ cm^{-1} for the signal with $g_{\|} = 2.27$; 2) $A_{\|}(^{69}Ga) \cong 40 \times 10^{-4}$ cm^{-1} and $A_{\|}(^{71}Ga) \cong 50 \times 10^{-4}$ cm^{-1} for the signal with $g_{\|} = 2.21$. The anisotropic part of the HF interaction constant seems to be small and could not be determined experimentally from ODMR results, thus $A_{\|} \cong a$, where a is isotropic part of HF interaction constant. A decrease of anisotropy for the shallow Ga acceptors as compared to the shallow Al acceptors seems to be due to a larger depth of the Ga level (Fig. 4.14a). The behaviour of Ga shallow acceptors is in agreement with their effective-mass-like character but with reduced orbital contribution due to the localization, which is greater then for shallow Al.

From the values of HF interactions with Ga atom an information can be obtained concerning the degree of localization of the unpaired electron wave function. Unfortunately, it has not been possible to observe the HF structure anisotropy with sufficient reliability to find the anisotropic term b. We could only estimate anisotropic HF interaction constant. One can suggest that sp^3 hybrid orbital is realized and in this case the spin density on the Ga orbital is estimated to be ~5% for the signal with $g_{||} = 2.27$ and ~6% for the signal with $g_{||} = 2.21$. In this case anisotropic term $b \cong 0.05$ mT and this value is within the limits of experimental error of HF structure constant measurements. Most of the residual spin density is probably spread out over sites around the Ga atom. The signal with $g_{||} = 2.27$ seems to correspond to Ga acceptor shallower than that responsible for the signal with $g_{||} = 2.21$. Thus, it follows from the energy levels picture that the signal with $g_{||} = 2.27$ can be attributed to the h-position of the Ga shallow acceptor and the signal with $g_{||} = 2.21$ can be attributed to the cubic-like positions.

Using the results for shallow Ga acceptors we can estimate from EPR linewidth the HF interaction constants of shallow Al acceptors for which the EPR spectra are not resolved. In the case of the low- and high-field lines the HF interaction constants amount to ~0.6 and ~0.7 mT, respectively. The corresponding spin densities on the Al orbital are estimated to be ~2.5% for signal with $g_{||} = 2.412$ and ~3% for signal with $g_{||} = 2.325$. The signal with $g_{||} = 2.412$ seems to correspond to an Al acceptor shallower than that responsible for the signal with $g_{||} = 2.325$. Thus, it follows from the energy levels picture (Fig. 4.14a) that the signal with $g_{||} = 2.412$ can be attributed to the h-position of the shallow Al acceptor and the signal with $g_{||} = 2.325$ can be attributed to the cubic-like positions.

Models of shallow centres. The results for shallow B are strikingly different from those observed for the other group-III shallow impurities, Al and Ga (Fig. 4.14b). In contrast with Al and Ga, no effective-mass-like behaviour was observed for shallow B acceptors (and for Be acceptors [97, 105]). It is the main problem to understand why the Al and Ga shallow acceptors show effective-mass-like properties, while the B and Be shallow acceptors do not. One reason could be the position of energy levels. And the best agreement with g-factors for free holes would be expected for acceptors with the smallest binding energy, a fact which is confirmed experimentally for Al and Ga. But this is not the case for shallow B (and Be) in spite of the fact that the energy positions of the acceptor states of shallow B in the band gap are between of the energy positions of shallow Al and of shallow Ga (Fig. 4.14a).

The structure of the shallow B acceptor has been reinterpreted several times and several contradictory models have been proposed. After the first publication [43], it was believed for a long time that B substitutes for C, as only SHF interaction with ^{29}Si was observed. The results of EPR experiments on a ^{13}C-enriched 6H-SiC crystal [90, 103] and reported a large anisotropic SHF interaction with one ^{13}C nucleus proved that B occupies a Si position, with the main spin density located on the dangling bond of a C atom along the C-B connection line. The observed SHF interaction with ^{29}Si was attributed to the three Si atoms near the C with the maximum spin density.

As it was mentioned the remarkable aspect of the B acceptor is that its g-tensor hardly shows an anisotropy (the average $g \cong 2$, Fig. 4.14b) and this is in stark contrast to the Al and Ga shallow acceptors. This difference in behaviour was explained in [96, 97] by considering the differences in the atomic radii. It was proposed that B, with atomic radii smaller than Si, occupy off-centre positions due to chemical rebounding i.e. they relax away from the neighbouring C along the C-B bond. In this model B is neutral. A similar displacement, but as a manifestation of the Jahn-Teller effect, for B in the h-site in 6H-SiC was proposed in [106] and in 3C-SiC in [92]. The cluster calculation of shallow B centre in 3C-SiC using the self-consistent unrestricted Hartree-Fock-Roothan method was carried out in [107]. This calculation confirmed off-centre position of B in cubic SiC. In [108] the alternative model in which the valence electron of a neighbouring C is donated to B, thus, forming a B-Si-C+ bond where B has negative charge. The unpaired electron of C^+ is uniformly distributed among the three remaining bonding orbitals and B and C relax towards each other.

The presence of two alternative models was a motivation for the authors of [4, 112] to perform an W-band (95 GHz) EPR and ENDOR study of the shallow B acceptor in a ^{13}C-enriched 6H-SiC crystal. The results of these experiments confirmed the model of shallow B which was suggested in [40, 90, 96, 97, 103, 107] that is why in this review only this model will be discussed.

Shallow B substitutes for Si (B_{Si}) but up to 40% of the unpaired spin density is located on one of the nearest C atoms. Al and Ga with shallow levels occupy also substitutional silicon sites in SiC. The difference in ionic radii between the replacing acceptors and Si $\Delta R = (R[A^{3+}] - R[Si^{4+}])$ is −0.19; +0.09 and +0.20 Å for A = B, Al and Ga, respectively. The ionic radii were based on the simplified assumption that a SiC crystal consists of residual Si-ions in 4+ oxidation state and paired binding electrons. In this picture the acceptor has an oxidation state of 3+. The ionic radii of Si^{4+} and A^{3+}, with A = B, Al and Ga, were taken from "Bruker" tables.

The physical behaviour of shallow acceptor in SiC seems to depend on whether the acceptor is larger (Al, Ga) or smaller (B, Be) than Si. There exist mechanism in semiconductors that drives some impurities, e.g., oxygen in silicon or N in diamond and also in silicon, off the nominal substitutional site. In [67, 113] the authors suggested that strong chemical rebounding may be responsible for the off-centre shift of the impurity. Let us place the B at the centre of a regular tetrahedron of carbon hybrids. The total energy can be written as the sum of a bonding energy and repulsion energy. For B in Si substitutional site the bonding matrix element and the repulsive force laws seem to be such that definitely drive the B off centre. The instability that leads to the off-centre shift is purely chemical rebounding. It is energetically favourable for the central B atom to move off centre and form a strong bond with three carbon atoms rather than stay at the centre and bond weakly to four atoms. If we leave the four carbon atoms at their perfect-crystal positions, the C-B distance is ~1.89 Å, which is 1.21 times larger than the normal C-B distance in BC_3 (e.g., in $B(CH_3)_3$ molecule [114]) which is 1.56 Å. We consider the displacement of B atom from a centre of Si site along a C-Si bond direction toward the midpoint between three C atoms which seems to be the energy potential minimum. Figure 4.18b presents a model of the shallow B centre showing the directions of the off-centre shifts of B atom and the disposition of the C atom having the main spin density in axial (h-site) and the nonaxial (k-sites) configuration. The plain of the figure is the {11–20} plane. The local axis z and x lie in this plane, being principal directions of the g-tensor and A-tensor.

In view of this model, the essential point for us is the change in hybridization of this B atom. The bonds between the B and four carbon neighbours could depart from the usual sp^3 tetrahedral bond admixture. If the B moves toward the midpoint between three carbon atoms the molecule BC_3 consisting of the B atom and the three carbons does not keep its sp^3 hybridization and relaxes toward a sp^2 configuration. The $\sim sp^2$ hybrids (natural for B) directed toward the B in a planar configuration will be realized. In this case an instability leads to motion of the B atom either in a c-direction for h-position or B can move in one of the six Si-C directions (except c-axis) for k1 and k2-positions in 6H-SiC. In 3C-SiC B moves in one of four $\langle 111 \rangle$ directions. Around 40% of the spin density is localized in the p_z orbital of a C which moves into the opposite B direction. From this model, one can obtain a qualitative explanation of the properties of the g-tensor. We will assume that the unpaired electron occupies a p_z orbital directed towards the B. The **g**-tensor and **A**-tensor then have cylindrical symmetry around the axis of the p_z-orbital, which coincides with c-axis for axial centre (h-site) and makes angle 110° or 70° to the c-axis for nonaxial centres (k1 or k2). The principal values of g-factor are $g_{\parallel} = g_e$, $g_{\perp} = g_e(1 - \lambda/\Delta E)$, where g_e is the g-factor of the free electron (2.0023), λ is the spin-orbit coupling constant for the C (which is negative), and ΔE is the (positive) energy difference between the ground and excited states. The predictions of these calculations, in particular that $\Delta g_{\perp} > 0$, are in qualitative agreement with the experimental results. Of course, in the nonaxial defect **g**-tensor should be orthorhombic, but the deviation from cylindrical symmetry is a very small.

It was concluded that the C-B bond is a dangling bond, that B is neutral and that there is no direct spin density on B. The finding that the deviation of g_z from ge is zero supports the model in which the unpaired electron is localized in the p_z-orbital because the expectation value of the orbital angular momentum along this direction is zero. From the experimental results obtained, it appears that the relative geometric arrangement of the local axes of the **g**-tensor and **A**-tensor are practically identical in the axial and nonaxial shallow B centre. This confirms the similarity between the two configurations of shallow B.

The model can also explain the opposite signs of the B HF structure constant $A_{\parallel}$ and $A_{\perp}$ (*a* and *b*) that were observed for h-site in [43, 106] and k-site in [92, 108]. There is no direct spin-density on B and consequently, the anisotropic part of the HF interaction is determined by the dipole-dipole interaction with the unpaired spin in the p_z orbital of C. This dipole-dipole interaction can explain the sign and the order of magnitude of the experimentally observed HF parameters [4, 112]. First of all, the sign of the HF constant *b* is predicted to be positive and the HF tensor is axially symmetric with the symmetry axis along the dangling bond, in agreement with the experimental data. To estimate the value of *b*, it was used a point-dipole approximation. The estimate gives $b = +2.7$ MHz, in good agreement with the observed value of about +3 MHz.

The isotropic part of the B HF interaction is very small and negative. There are three mechanisms which can produce an isotropic hyperfine interaction with a nucleus situated in the neighbourhood of a paramagnetic defect. They are the overlap of orbitals on different centres, covalence, and exchange polarization of the core electrons (core polarization). The former two produce a positive spin density and the latter mechanism a negative spin density at the nucleus. The final result in the shallow B centre is a negative spin density at the B nucleus, owing to the dominant contribution of the direct exchange polarization. The observed negative spin density for shallow B comes from the polarization of the s core of the B by direct interaction with p_z-orbital of the C ion. The core polarization represents the difference between the density of spin-up (+) (parallel to unpaired spin) and spin-down (−) electrons at the nucleus. One has to realize that only 0.1% of spin-density in the s-orbital of B gives $a \sim 3$ MHz. This indicates that there is a subtle balance between different types of polarization effects which are responsible for the value of a. This is consistent with the model that there is no direct spin-density on B. In fact, the value of a is the only parameter which differs considerably between the hexagonal and cubic-like sites, as well as between a for shallow B in different polytype of SiC. Qualitatively, polarization effects might be understood on a semi-empirical level for simple organic radicals, but for crystals, the situation is much more complicated.

In contrast to the results for shallow B acceptor Al and Ga atoms occupy the Si-substitutional on-centre position and as a consequence a mass-like behaviour for Al and Ga shallow acceptor was observed. The difference in ionic radii between the replacing Be and silicon $\Delta R = -0.07$ Å, so it is close to ΔR for B and Be seems also to be a good candidate to have off-centre position.

4.2.2.2 Group-III Acceptors with Deep Levels

In this section the EPR, ENDOR and ODMR results on deep B, Al and Ga acceptors in SiC will be presented.

Deep B. The first magnetic resonance observation of deep B acceptors (dB) in SiC was made by using ODMR techniques [100, 101, 115, 116]. A number of anisotropic signal ascribed to deep B acceptors have been observed at 1.5 K in 6H- and 4H-SiC. ODMR spectra were recorded by monitoring of the intensity of the yellow luminescence band in 6H-SiC and its analogue in 4H-SiC. Anisotropic ODMR spectra were shown to belong to the several types of centres since the relative intensities of the signals vary with the emission wavelength within the luminescence band. The ODMR lines of dB showed nearly axial symmetry around hexagonal axis (c-axis) and had g-factors $g_{\parallel} \cong 2.02$–2.03 and $g_{\perp} \cong 2.0$. The low resolution of ODMR method did not allow to observe HF structure from B and unambiguously to connect these spectra with B impurity.

The results of the EPR study of deep B centres were presented in [93–97, 104]. The ENDOR measurements were made in [104, 117, 118]. The X-band EPR spectra observed in a ^{11}B doped 6H-SiC crystal at 4.5 K for magnetic field B at different angles with respect to the c-axis are shown in Fig. 4.19a. The high-field part of the EPR spectra (for the orientations close to B ‖ c) in the region from 326.5

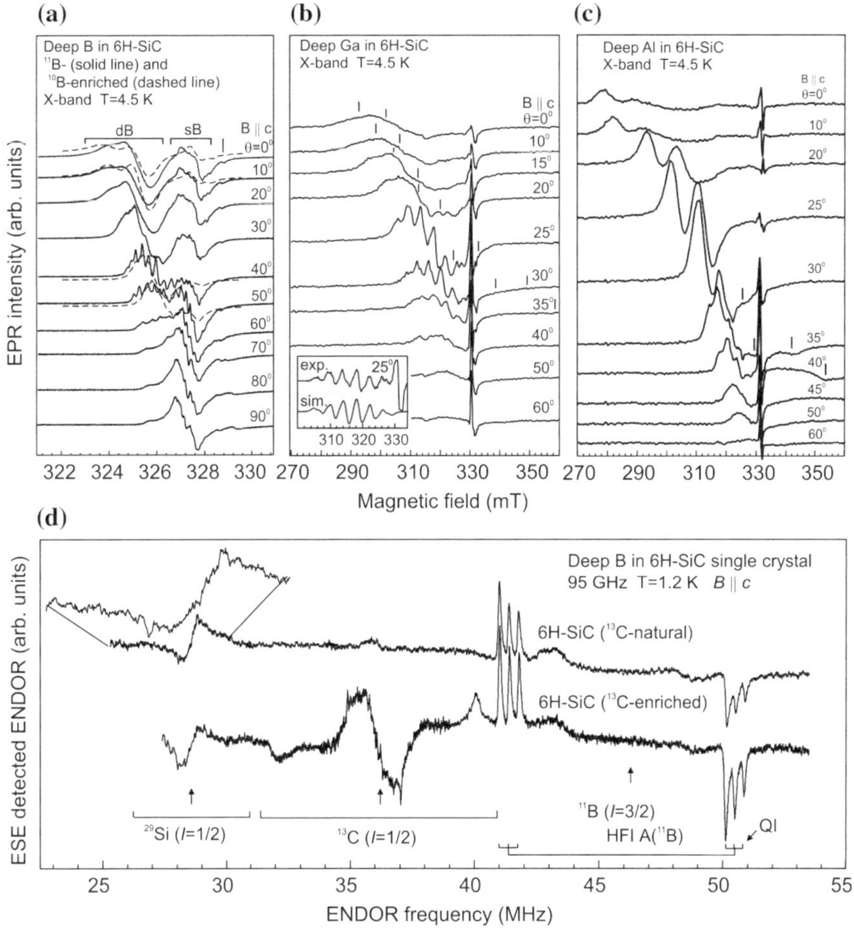

Fig. 4.19 **a** Angular dependence of X-band EPR spectra in 6H-SiC doped with ^{11}B. *Dashed lines* show the EPR spectra in 6H-SiC doped with ^{10}B. **b** Angular dependence of X-band EPR spectra in 6H-SiC:Ga. Calculated dependence for shallow Ga is drawn as *vertical marks*. *Inset* shows Experimental with baseline corrected (*top*) and calculated (*bottom*) Ga HF structure for magnetic field at angle of 25° with respect to c-axis. **c** Angular dependence of X-band EPR spectra in 6H-SiC:Al crystal of p-type. Calculated dependence for shallow Al is drawn as *vertical marks*. **d** ESE-detected ENDOR spectra of deep B in 6H-SiC single crystals of the non-enriched and ^{13}C-samples at 95 GHz and 1.2 K. The lines belonging to ^{13}C are visible in the spectrum of the enriched sample. The spectrum around the ^{29}Si Zeeman frequency is magnified in an *inset* in this figure

to 328.5 mT and low-field part in the region from 323.0 to 326.0 mT belong to different centres as the ratio of the intensities of these signals is sample sensitive. The high-field signal labelled as a sB and marked by a horizontal line belongs to shallow B, since this spectrum can be fitted to a spin Hamiltonian with g-factors and HF structure constants of shallow B at two cubic-like and the hexagonal lattice sites. One can see in the low-field parts of the spectra for the angles $\theta = 0°$, at least, two overlapped broad unresolved lines, which are labelled in Fig. 4.19a as a dB. The intensity of these lines correlates with the characteristic bright-yellow photoluminescence and DLTS spectra of deep B centres. High-temperature heat treatment has a substantial effect on the low-field EPR signal. 5-min annealing at 2350 °C of the samples causes a significant decrease of the low-field EPR signal and a simultaneous decrease in the intensity of the yellow luminescence and DLTS signal of deep B centres. After annealing, the samples exhibit only the intense EPR spectrum of shallow B. The weak single resonance at 3285 G is due to the high field HF component of N.

The unresolved broad lines in Fig. 4.19a were interpreted to arise from deep B centres at inequivalent lattice sites. The spectrum which is ascribed to deep B centres is anisotropic and as magnetic field is rotated away from the c-axis for the angles exceeding 20° the resolved structure appears and the resolution of the spectra increases with increasing angle. This structure is best resolved for $\theta \cong 40°$. For $\theta > 60°$ the resolved structure disappears again and the intensity of deep B EPR spectrum decreases. A direct identification of the chemical species involved in a defect centre can be established by the presence of HF structure in its spectrum. In order to unambiguously demonstrate that the resolved structure is due to HF interaction with ^{11}B nuclei, EPR measurements of 6H-SiC crystals doped ^{10}B isotope were performed. These measurements unambiguously proved that the resolved structure in Fig. 4.19a arises from the HF interaction of the unpaired electron spin with a ^{11}B nucleus and the EPR signals do belong to the B impurity. Identification of the B involved in a defect centre was also established by ENDOR [104, 117].

The g-tensor for the EPR spectra of deep B centres is nearly axial around c -axis and the positions of the groups of lines (which belong to a hexagonal-like and two cubic-like crystallographic inequivalent lattice sites) as well as the structure on each group in principle can be fitted to the axial form of the standard spin Hamiltonian, where z coincides with the c-axis of the crystal, $S = 1/2$, $I = 3/2$ (^{11}B).

The $g_{\parallel}$ value was found to depend on temperature. Upon warming the crystal above 5 K, these lines move to higher magnetic field (g-factors decrease) and start to decay at about 10 K. The resolution of the spectra for ^{11}B that was observed for the angles θ between 20° and 60° depends on temperature and as the temperature was raised above 8 K the EPR lines start to broaden and the resolved HF structure began to disappear. This broadening seems to be due to some thermally activated motion of the holes, connected with deep B.

One notes that a defect located at the h-site would experience a stronger local axial field than the one at the k-site. It seems reasonable that the largest g-factor shift for deep B centres, when the applied magnetic field is parallel to the crystal c-axis occurs for a B occupying the h-lattice site. The more intense line with larger linewidth in Fig. 4.19a (approximately of double intensity), which is shifted to higher magnetic fields seems to belong to two unresolved cubic-like sites k1 and k2. At $T = 4.5$ K $g_{\parallel} \cong 2.03$ and $g_{\perp} \cong 2.01$ for h-site; $g_{\parallel} \cong 2.02$, 2.023 and $g_{\perp} \cong 2.0$, 1.99 for cubic-like sites. The isotropic part of the HF interaction with ^{11}B is estimated to be ~0.2 mT for the h-site and ~0.3 mT for the k-sites. On the other hand, the anisotropic part is small and could not be determined with sufficient reliability from experimental EPR spectra. A results of the X-band ENDOR investigation of deep B [117] gives HF structure constants for deep ^{11}B: a = 5.8 MHz, b < 0.3 MHz (h-site); a = 6.1 MHz, b ≅1.5 MHz and a = 6.2 MHz, b ≅1.5 MHz (cubic-like sites). The quadrupole interaction constant Q with ^{11}B nucleus was found to be equal 0.2 MHz for the h-site and 0.19 MHz for the k-site of B. It should be noted that for deep B the constant Q is approximately four times smaller than the value of Q for shallow B which equals ~0.75 MHz [106]. Moreover, the HF structure constant for shallow B is sharply anisotropic but this is not the case for deep B. The anisotropic part of the HF structure constant for deep B is suppressed either virtually completely for the h-site or partially for the k-site. EPR spectra of deep B were also observed in 4H- and 3C-SiC crystals [118]. A preliminary results of the investigation of EPR spectra of deep B in 3C-SiC show that the these centres have ⟨111⟩ main symmetry axis.

W-band (95 GHz) ENDOR measurements were performed on non-enriched and 30% ^{13}C-enriched crystals. In Fig. 4.19d the spectra for B ǁ c are shown for both crystals, which show ENDOR transitions of ^{29}Si (I = 1/2), ^{13}C (I = 1/2) and ^{11}B (I = 3/2) nuclei. For clarity the spectrum around the silicon nuclear Zeeman frequency is magnified in an inset.

The ^{11}B ENDOR spectrum in Fig. 4.19d consists of three sharp lines and a broad hump, symmetrically located around the ^{11}B nuclear Zeeman frequency. As ^{11}B has a nuclear spin I = 3/2 we expect three lines in each manifold for each site, due to the quadrupole interaction. The ^{11}B hyperfine and quadrupole parameters derived from the HF ENDOR spectra are $A_{\|}$(k1) = 8.8 MHz, $A_{\|}$(h) = 5.4 MHz and P(k1, h) = 190 kHz, in agreement with the values reported for X-band ENDOR in [117].

The ^{13}C ENDOR spectrum with B ǁ c consists of a broad line around the ^{13}C nuclear Zeeman frequency and two separate lines at 32.2 MHz and 40.0 MHz symmetrically positioned around the Zeeman frequency. The latter are probably due to the ^{13}C atoms with the highest spin density. The hyperfine constant $A_{\|}$ = 7.8 MHz. In the broad line surrounding the ^{13}C Zeeman frequency several lines can be recognized with a maximum hyperfine value $A_{\|}$ = 2.2 MHz. The ^{29}Si ENDOR spectrum consists of a broad line at the ^{29}Si nuclear Zeeman frequency. Two lines can be distinguished with a hyperfine splitting $A_{\|}$ = 0.76 MHz.

Deep Ga. A new EPR spectrum was observed in a Ga doped 6H-SiC free-standing epitaxial layer [96, 97]. The spectrum, which depends strongly on the angle between the crystal c-axis and the magnetic field, is shown for the different orientations of the magnetic field in Fig. 4.19b. The striking result is that the great qualitative parallels between EPR spectra of deep B centres (Fig. 4.19a) and spectra which were observed in a Ga doped 6H-SiC (Fig. 4.19b) were found. Quantitative differences have only been observed in the spin Hamiltonian parameters of EPR spectra in the Ga and B doped crystals. It is clearly seen in Fig. 5.19a, b that only g-factor shift and the separations between the hyperfine lines are larger for the EPR spectra of Ga doped crystals compared with those of deep B. The spectrum which is ascribed to Ga centres is anisotropic and as magnetic field is rotated away from the c-axis the resolution of the spectra increases with increasing angle. This structure is best resolved for $\theta \sim 25°$. In addition, the signal with a g-factor of about 2.0 seems to arise from B shallow acceptors (B is trace impurity in silicon carbide). The involvement of one Ga atom in the defect was suggested from the resolved HF structure. In order to demonstrate that the resolved HF structure for the angles exceeding 20° is due to HF interaction with ^{69}Ga and ^{71}Ga nuclei, the simulation of the Ga hyperfine structure were performed.

Inset in Fig. 4.19b shows experimental (top) and calculated (bottom) Ga hyperfine structure for magnetic field at angles of 25° with respect to the c-axis. The simulation is based on the point that experimental spectrum can be resolved into three sets of ^{69}Ga and ^{71}Ga hyperfine lines which seem to be due to the three crystallographic inequivalent lattice sites. Using this information and slightly different g-factors (slightly shifted the centres of gravity of the HF structure) of Ga centres for three lattice sites ($g_1 = 2.121$, $g_2 = 2.096$ and $g_3 = 2.070$), it has been possible to successfully simulate the experimental spectra. As expected, the reasonable agreement between experiments and simulations is obtained using the ratio between hyperfine parameters for the two isotopes which is proportional to the ratio of their nuclear g-factors. We used the same HF interaction constant (4 mT for ^{69}Ga) for three lattice sites, that, probably, is rough approximation. This analysis of the Ga hyperfine interactions confirms unambiguously that one Ga atom is involved in the observed defect. One may, therefore, to call a new Ga centre as a "deep Ga" analogously to a deep B. The ratio of HF structure constant of deep Ga to deep B corresponds to the ratio of the theoretical HF structure constant of Ga atom to B atom [61].

The g-factor tensor for the EPR spectra of deep Ga centres like for deep B is nearly axial around c-axis and the positions of the groups of lines, in principle, can be fitted to the same standard spin Hamiltonian. The principal values of the g-tensor at T = 4.5 K was estimated to be approximately $g_{\parallel} \cong 2.16–2.19$ and $g_{\perp} \cong 2.0$, i.e. no effective-mass-like behaviour was observed for deep Ga acceptors. It is interesting to compare the physical behaviour for the known Ga shallow acceptors [99–101] and new Ga centres. The calculated angular dependencies of the EPR transitions for shallow Ga acceptors with using ODMR data [99] (here the HF structure is not included, i.e. the line position is representing the centre of gravity of the HF structure) are drawn as bars. The difference in the line positions for the Ga shallow and deep acceptors is clearly observed for the angles exceeding 30°. It is seen that for the orientations close to $\theta = 0°$ the EPR spectra of Ga shallow and deep acceptors will be overlapped, however, the intensity of the EPR signal of shallow Ga centres was too low in our experiments and did not mask the signal of deep Ga centres.

Like in the case of deep B centre the $g_{\parallel}$ value for deep Ga centre was found to depend on temperature. Upon warming the crystal ($\theta = 0°$) above 5 K, lines move to higher magnetic field (g-factor decreases) and start to decay at about 8 K. The resolution of the spectra also depends on temperature and as the temperature was raised above 6 K the resolved HF structure ($\theta = 30°$) began to disappear.

Deep Al. The next step was to find deep Al centres in silicon carbide that is of a great interest for silicon carbide applications. It was small probability to find HF structure of Al because such a structure did not observe for shallow Al acceptors [91, 98–101]. Therefore it was the purpose to find EPR spectra in Al doped SiC with the angular and temperature dependencies that are similar to those which was shown observed for B and Ga. The authors of [96, 97] succeeded in the observation of EPR spectra of deep Al centres in SiC.

Figure 4.19c shows angular dependence of the EPR spectra (X-band) that have been observed in Al doped 6H-SiC crystals at 4.5 K. The magnetic field applied was rotated in the {11–20} plane. The spectra depend strongly on the angle between the crystal c-axis and the magnetic field. The spectra consist of several group of lines. We identify one of the group of anisotropic lines as due to the Al shallow acceptor at inequivalent lattice sites. One can see in the low-field parts of the spectra in Fig. 4.19c two broad lines (for the orientations close to $B \parallel c$). The $\cos\theta$ angular dependencies of the EPR transitions for shallow Al acceptors are drawn as solid bars. This indicates an effective-mass-like character for the shallow acceptor. Some disagreement of the positions of the arrows with experimental lines seems to be due to the not precise orientation of the crystal. Upon warming the crystal the EPR spectrum of the Al shallow acceptor decays above 5 K. Another group of anisotropic lines has the angular and temperature dependencies that are similar to those which were observed for deep B and deep Ga. Therefore we will call a new Al centre as a "deep Al" analogously to a deep B and deep Ga. One may distinguish shallow Al EPR signals from those for deep Al because the deep Al signal remain strong up to 7 K while the shallow Al signal are visible only below 5 K. The difference in the line positions for the Al shallow and deep acceptors is clearly observed for the angles exceeding 30° in the same manner as for Ga. For the orientations close to $\theta = 0°$ the EPR spectra of Al shallow and deep acceptors will be overlapped.

Al has one stable isotope ^{27}Al with a natural abundance 100% and a nuclear spin $I = 5/2$. One normally expects a spectrum consisting of one or more sets of six equally intense lines, whose spacing is about 4 times smaller (according to the ratio of calculated HF structure constants of ^{69}Ga to ^{27}Al atoms [61]) as compared with those for the ^{69}Ga isotope.

The principal values of the g-tensor at T = 4.5 K for deep Al in 6H-SiC was estimated to be approximately $g_{\parallel} \cong 2.1$–2.19 and $g_{\perp} \cong 2.0$, i.e. no effective-mass-like behaviour was observed for deep Al acceptors.

In the temperature region ~4–10 K the deep Al centre spectrum in 6H-SiC is observed to change drastically. Upon warming the crystal ($\theta = 0°$) above 5 K shallow acceptor signal disappears and deep Al signal moves to higher magnetic field and decays at about 10 K. These results show that new Al EPR spectrum observed is attributed to the same kind of defect as for deep B and deep Ga.

Some control experiments were made where Al doped p-type 6H-SiC crystals grown with the Cree Corp were used. In these samples a strong signals of deep Al acceptors was observed with the concentration comparable to that for shallow Al acceptors. This is the direct evidence which indicates the existing of deep Al levels in the p-type SiC wafers used for applications.

Aluminum related deep defects in 4H-SiC were identified using X-band EPR and ENDOR in [118]. Figure 4.20a shows the EPR spectra recorded at 4.5 K for several orientations in p-type 4H-SiC crystals heavily doped with Al ($N_{Al} = 10^{20}$ cm^{-3}).

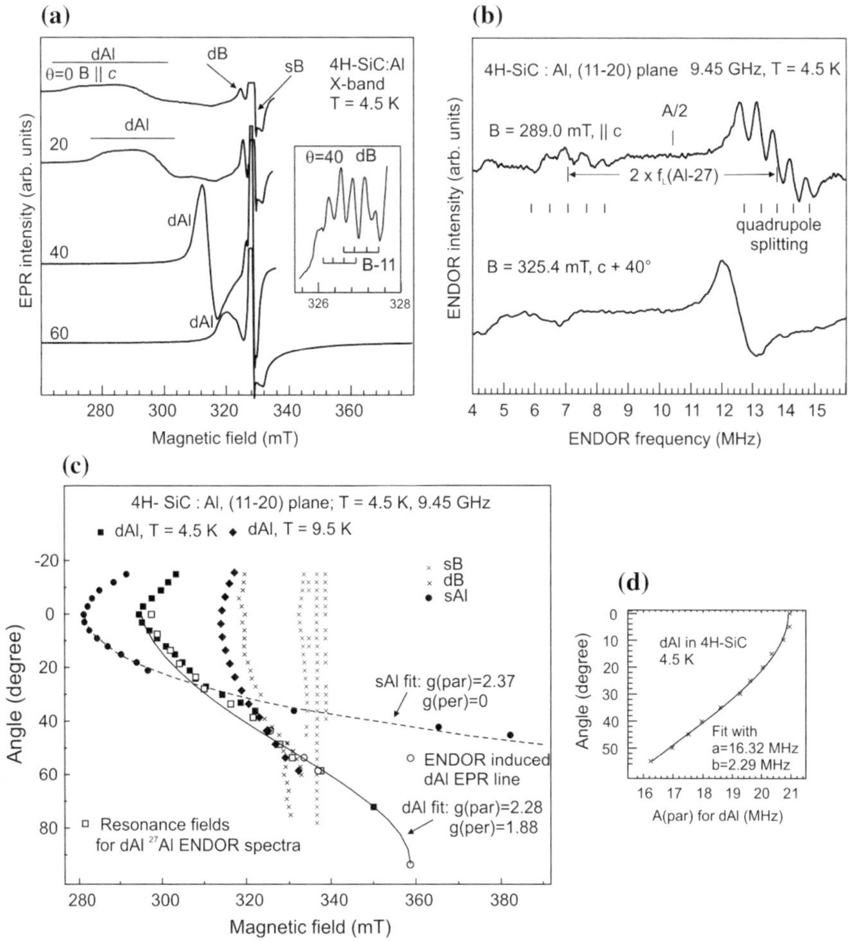

Fig. 4.20 **a** X-band EPR spectra of Al doped 4H-SiC single crystal at 4.5 K in different orientations of a crystal. The central lines originate from boron which induces a shallow and deep acceptor levels and is an omnipresent residual impurity in SiC. *Inset* shows EPR signal of deep B for θ = 40°. **b** ENDOR signal detected on the deep Al acceptor, the two groups of lines are separated by the Larmor frequency of the ^{27}Al nucleus. The five-lines structure arises from quadrupole interaction. **c** X-band (9.45 GHz) EPR angular dependencies in 4H-SiC:Al; rotation in (11–20) plane, T = 4.5 K. The *circles* refer to the shallow Al acceptor, the *squares* to the deep Al acceptor. **d** The angular dependence of hyperfine splitting $A_{\parallel}$ which was found from ENDOR data and fit (*solid line*) with isotropic HF part (**a**) and anisotropic HF part (**b**)

The crystal was Al-doped during growth and contains also trace boron impurity with concentration 10^{17} cm^{-3}. In the orientation B ∥ c (θ = 0°) one can see three distinct EPR signals. One (consisting of two broad lines) is marked as dAl (deep Al). It has angular dependence very close to that observed for deep Al centres in 6H-SIC. Symmetry of this centre is almost axial along the c-axis and the signal can be characterized with electron spin S = 1/2 and the following g-factors: $g_{\|} = 2.35$; 2.23. Two dAl lines appear due to the presence of Al impurity in two non-equivalent h and k lattice sites. As in the case of deep Al and deep B in 6H-SiC we believe that the low-field line in dAl signal belongs to Al in h-site in the lattice. Contrary to previous observation of deep Al in 6H-SiC, in this sample the deep Al EPR signal is not masked by a signal from shallow Al acceptors. EPR intensity of shallow Al signal is at least 200 times of magnitude smaller than that of deep Al and could be hardly observed.

The line marked as sB in 5.20a arises due to shallow B centres. Boron is a common trace impurity in p-type SiC. The signal named in Fig. 4.20a as dB (deep B) has angular and temperature dependence completely the same as that of deep B signals that was found earlier in 4H-SIC after B diffusion. The inset in the figure shows deep B signal for orientation θ = 40° in which the characteristic hyperfine structure due to ^{11}B nuclear spin (I = 3/2) is well resolved for two positions of B atom in the lattice. This signal belongs to deep B centres. It is interesting since up to now EPR of deep B could be observed only in the crystals that were doped by diffusion while this sample was doped during growth and boron is a trace impurity.

In order to determine the g-anisotropy we performed angular dependent EPR measurements and rotated the sample from the magnetic field orientation parallel to the c-axis to perpendicular to the c-axis of the crystal. Figure 4.20c shows X-band (9.45 GHz) EPR angular dependencies in 4H-SiC:Al; rotation in (11–20) plane, T = 4.5 K. The circles refer to the shallow Al acceptor, the squares to the deep Al acceptor. One notes that for approximately 60° off the c-axis this EPR signal merges with the shallow B resonance and thus for the time being limits a precise determination of the g-value for the perpendicular orientation (see below).

To gain further insight into the defect structure ENDOR experiments were performed. Nuclear magnetic resonance transitions are induced between the magnetic field split nuclear spin states of lattice nuclei surrounding the paramagnetic centre, Si and C both with I = 1/2, and of course also of an impurity atom involved in the defect structure. The identification is based on the nuclear Zeeman term which includes the Larmor frequency of the respective nuclei. For a spin S = 1/2 in the presence of hyperfine interactions in ENDOR one observes a pair of lines (see Fig. 4.20b) separated by two times the Larmor frequency of the nucleus involved centered approximately around the frequency position given by half of the hyperfine interaction constant *A*. The characteristic five line spectrum as well as its intensity ratio is caused by the nuclear quadrupole interaction with a single Al nucleus which has a nuclear spin I = 5/2. This together with the Larmor frequency unambiguously identifies Al as one central part of the defect. Figure 4.20d shows the angular dependence of hyperfine splitting $A_{\|}$ which was found from ENDOR data and fit (solid line) with isotropic HF part (a) and anisotropic HF part (b).

This pattern has been measured for various orientations of the crystal with respect to the static magnetic field (ENDOR spectrum for $\theta = 40°$ is also shown in Fig. 4.20b) and helped to establish the g-anisotropy of the EPR line. The high frequency line pattern in Fig. 4.20b can be used to measure a sort of excitation spectrum of the ENDOR, since it is induced only within the EPR line of the defect. It is commonly called ENDOR induced EPR. Since the EPR of deep Al centre is angular dependent and hence occurs at different magnetic field positions the nuclear Larmor frequency has to be adjusted correspondingly. The open circles in Fig. 4.20c are the precise EPR field positions of the Al defect as obtained by above mentioned method: for parallel to the c-axis $g_{\parallel} = 2.28$ and for perpendicular to the c-axis $g_{\perp} = 1.88$ at $T = 4.5$ K. It should be noted that for the shallow Al acceptor one has $g_{\parallel} = 2.37$ and $g_{\perp} = 0$.

Deep Al EPR signal has strong temperature dependence. The anisotropy of the EPR signals decreases when the temperature increases and a signal at high temperatures is nearly isotropic with $g = 2.0$.

Figure 4.21 shows the orientational dependence of the deep B acceptors (a) and deep Al acceptors (b) at 95 GHz and 1.5 K in 6H-SiC, with the magnetic field rotating in the (11–20) plane. The angles θ and ϕ describe the orientation of the magnetic field with respect to the c axis and the orientation of the magnetic field in a plane perpendicular to the c axis, respectively. From the figure it follows that the g tensor principal z axes of all the centres align with the c axis. The slight nonaxiality of the g tensor is clearly visible. Insets show ENDOR spectra fragments which are fingerprints of boron and aluminum and their quadrupole splitting. It was proposed that the deep boron acceptor or deep aluminum acceptor consists of a boron or aluminum on a silicon position with an adjacent carbon vacancy. Apparently, for energetic reasons, this carbon vacancy combines always with a boron or aluminum along the c axis.

Models of deep centres. The experimentally observed features of the EPR spectra of deep Al and deep Ga are identical in most details to those exhibited by deep B.

1. The orientation dependencies of the EPR spectra indicate that the deep centres of B, Al and Ga have the same symmetry nearly axial around hexagonal axis of the crystal. No effective-mass-like behaviour was observed for the group-III impurities deep acceptors.
2. For the angles θ between 20° and 60° the resolved HF structure was observed.
3. The temperature dependencies of the EPR spectra indicate that the deep centres of B, Al and Ga have the same temperature behaviour. In the temperature region ~5–15 K, the deep centres spectra are observed to change drastically. For the orientation close to B ∥ c g-factor appear to decrease and lines to broaden and disappear as the temperature is raised. The resolution of the HF structure (for B and Ga) also depends on temperature and as the temperature was raised above 8 K for B and above 6 K for Ga the resolved structure began to disappear.

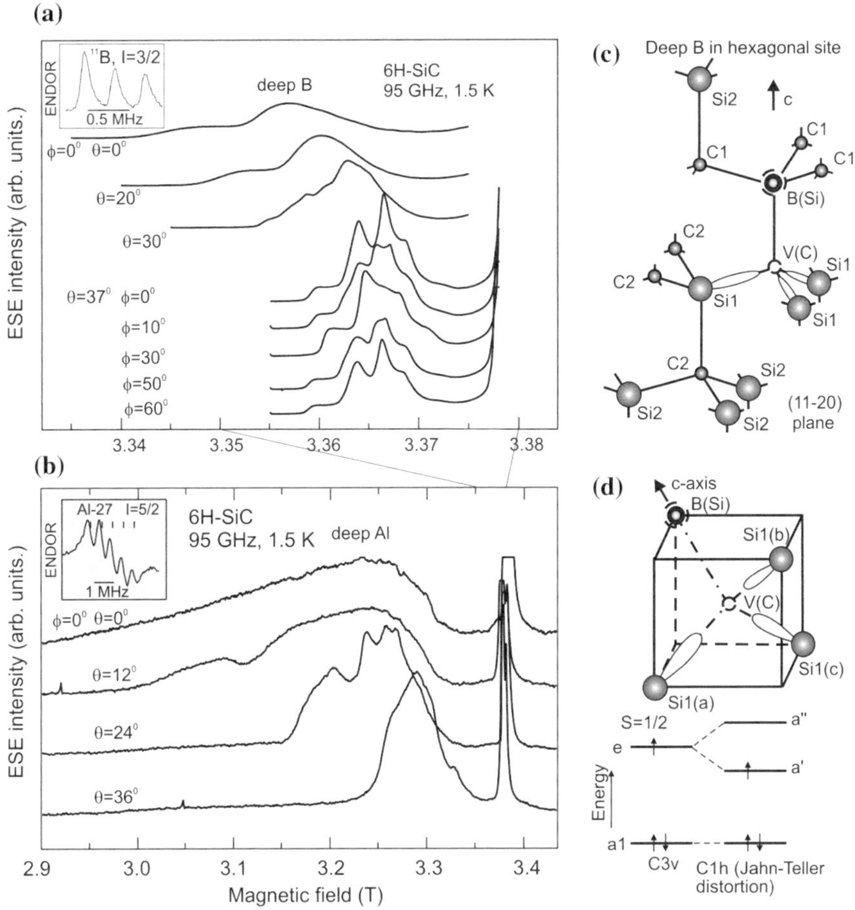

Fig. 4.21 Angular dependences of the high-frequency EPR spectra (95 GHz) of the deep B acceptors (**a**) and deep Al acceptors (**b**) at 1.5 K in 6H-SiC. Magnetic field rotating in the (11–20) plane, where θ indicates the angle between the direction of B and the c axis. Dependence on ϕ, with the magnetic field describing a cone with fixed θ. The slight nonaxiality of the g tensor is clearly visible. *Insets* show ENDOR spectra fragments which are fingerprints of boron and aluminum. **c** A model of deep boron acceptor. The spin density distribution in the deep boron (hexagonal site). B(Si) indicates the position of the boron impurity substituting for silicon, V(C) indicates the position of the vacancy, substituting for carbon. On the ground of ENDOR and HYSCORE data [40] it was suggested the following spin-density distribution. On the boron nucleus the spin density is negligible. About 25–30% resides on each Si1 and on each C1 there is ∼0.2% of spin density. On each C2 ∼ 0.8%, ∼0.2% on each Si2 around C1 and ∼0.1% on each Si2 around C2. The other Si and C have a negligible spin density. **d** Simple LCAO one-electron MO model for the electronic structure of the deep boron acceptor consisting of a boron substituting for a silicon and a neighboring carbon vacancy. The B-V bond has the direction of the c axis ([111] direction in the cube). a, b, and c are the three broken bonds of the Si atoms surrounding the vacancy. Without the Jahn-Teller distortion the point-group symmetry is C3v, with the Jahn-Teller distortion the symmetry is C1h

4. For the orientation close to $\theta = 0$ there exist some distribution in values of the g-factors.
5. The coexisting of shallow and deep group-III acceptors was usually observed. The intensity ratio of the deep acceptor EPR signal to the shallow acceptor signal was sample sensitive.
6. The HF structure constants of the deep acceptors are close to those for shallow acceptors.
7. Quantitative differences have only been observed in the spin Hamiltonian parameters of EPR spectra in the B, Al and Ga doped crystals and in decay temperatures of the different EPR signals.

It is interesting to note that regardless of whether the deep centre contains atom which are larger (Al, Ga) or smaller (B) than silicon, it leads to the same physical behaviour and the microscopic model. We indicate that the model of deep B, deep Al and deep Ga in acceptor state seems to be group-III element-vacancy pair $A_{Si} - V_C$ with A = B, Al and Ga. A_{Si} is B (Al, Ga) substituting for Si, V_C is a C vacancy. Arguments have been put forward that the centre has nearly axial symmetry around c-axis with a strong anisotropy and the HF structure interaction with the B (Al, Ga) nucleus is of the same order as for shallow B, shallow Al and shallow Ga centres, which were established to be B_{Si}, Al_{Si} and Ga_{Si} centres.

As an example, consider the deep B acceptor structure [104] and the spin density distribution with the help of Fig. 4.21c. The proposed model for the deep boron would predict a large HF splitting of the ^{29}Si nuclei surrounding the C vacancy, whereas in the ENDOR spectrum of Fig. 4.19d only ENDOR lines closely around ^{29}Si Zeeman frequency are observed. A rough idea about the HF splitting expected for the ^{29}Si atoms surrounding the C vacancy can be found from the splitting seen in the case of the sB acceptor for the carbon atom carrying the main spin density. This carbon carries 38% of the spin density and its hyperfine interaction is given by isotropic contribution $a_C = 4.1$ mT and anisotropic one $b_C = 1.1$ mT. Assuming a spin density of 25–30% on each of the three silicons, we expect their hyperfine interaction to be of the same order of magnitude. The satellite lines in the dB EPR spectra with HF splitting about 4 mT were ascribed [104] to the ^{29}Si surrounding the vacancy, Si1 in the Fig. 4.21c. These lines exhibit the same temperature and orientational dependence as the main deep boron signal.

The Si nuclei Si1 are calculated to each carry a spin density of 25–30% with an estimated isotropic HF interaction of 3–4 mT. In Fig. 4.21c B(Si) indicates the position of the boron impurity substituting for silicon, V(C) indicates the position of the vacancy, substituting for carbon. On the ground of ENDOR and HYSCORE data [40] it was suggested the following spin-density distribution. On the boron nucleus the spin density is negligible. About 25–30% resides on each Si1 and on each C1 there is ~0.2% of spin density (2.2 MHz). On each C2 ~ 0.8% (7.8 MHz), ~0.2% on each Si2 around C1 (2.6 MHz) and ~0.1% on each Si2 around C2. The other Si and C have a negligible spin density. Similar model is proposed for deep aluminum and deep gallium acceptors.

It was shown that in contrast to the shallow boron acceptor the g tensors for deep B (Al, Ga) are almost axial with g_z parallel to the c axis for the hexagonal as well as for the quasicubic sites, but a small nonaxiality was observed in EPR spectra (see Fig. 4.21). In addition, the EPR linewidth at 95 GHz is about ten times larger than at 9.3 GHz indicating a dominant g strain broadening that is most pronounced with B ∥ c. Figure 4.21d shows a simple one-electron linear combination of atomic orbitals (LCAO) molecular-orbital (MO) description of a carbon vacancy next to a boron substituting for a silicon atom [119]. The atomic orbitals are the dangling bonds of the three silicon atoms a, b, and c surrounding the vacancy. The point-group symmetry is C3v and the totally symmetric orbital (a_1) is expected to be lowest with the degenerate orbital (e) lying above it. By populating these one-electron orbitals with the unpaired electron of the three silicons surrounding the vacancy, we see that one electron occupies the (e) orbital, thus explaining the S = 1/2 character of the defect. Because of the orbital degeneracy a static Jahn-Teller distortion will take place, making one silicon atom inequivalent with respect to the other two and lowering the symmetry to C1h. As a result the (e) orbital will split in an a′ and an a″ component according to the two irreducible representations of C1h. The observed directions of the principal axes and principal values of the g tensor are in agreement with this model. The g_z axis is found parallel to the vacancy-boron axis and the g_x axis is found in the (11–20) plane.

For deep Al one obtains $g_{\|} = 2.28$ and $g_{\perp} = 1.88$ at 4 K. There is an obvious deviation from $g_{\perp} = 0$, which was observed for the shallow Al centres. To explain g-factors anisotropy for deep group-III-related (Al, Ga) acceptors in SiC we will apply the theoretical consideration which was developed for O-centres in oxide perovskites in [120]. It was shown that rather small orthorhombic influences lead to a big deviation from $g_{\perp} = 0$ and $g_{\perp}$ is shifted toward $g_S = 2$, indicating an increased quenching of orbital angular momentum. The microscopic origin of the small orthorhombicities needed to reproduce the measured g parameters was discussed above can be caused by Jahn-Teller effect.

Thus B, Al and Ga do not only form a shallow acceptor level but also deep B, Al and Ga related deep level centre. The possibility that under high doping conditions using B, Al and Ga compensating centres are formed should be taking into consideration.

4.2.2.3 Beryllium and Scandium Acceptors

Beryllium. Beryllium is a double acceptor in SiC. The Be acceptors were studied by EPR and ENDOR [97, 105, 121]. The shape of beryllium EPR spectra and their orientational behavior depend on measurement temperature and undergo changes within the temperature intervals above 10 and 50 K. Below 10 K, one observed EPR spectra of three types of centers with a close-to-axial symmetry, which relate to two quasicubic and one hexagonal beryllium sites in 6H-SiC. Above 50 K, the symmetry of the two beryllium centers assigned by us to quasicubic beryllium sites is lower than axial, and their z axis is aligned with one of the Be-C bond which does not coincide with the c axis. It is believed that because beryllium is smaller in radius than silicon, it occupies, similar to the shallow boron, an off-center position at the silicon site, i.e., that it is shifted from the tetrahedron center toward the center of the plane containing three carbon atoms and away from the fourth carbon with the highest spin density. Unlike the shallow boron which is neutral, beryllium must be negatively charged.

High-frequency (95 GHz) pulsed EPR and ENDOR measurements on the Be acceptors in 6H-SiC were presented in [105]. The different shallow and deep Be acceptor centers that are formed when 6H-SiC is doped with Be and compare as grown Be-doped sample with Be diffusion samples were observed. It is found that in Be diffusion samples three shallow acceptor centers and five deep acceptor centers are formed, whereas in the as grown samples the deep centers are hardly present. It is also found that the Be-related centers can be described using the same kind of model as for the B-related ones. The sBe(h) center and sBe(k1, k2) centers that are found resemble the sB(h) and sB(k1, k2) centers in 6H-SiC and the same model can be used to describe them. It is found the Be substitutes for a Si and that the main part of the spin density, 30%, resides in a dangling pz orbital on the neighbouring C. It should be noted the Be is negatively charged whereas B is neutral. Three of the deep centers found, dBe(h) and dBe(k1, k2), resemble the dB centers found in 6H-SiC. The main part of the spin density is located on the three Si atoms on the other side of the C vacancy that accompanies the Be impurity. Be-C-vacancy direction is along the c-axis for all three sites. As in B-doped 6H-SiC the deep centers are thus characterised by a $g_z > g_x, g_y$, for shallow centers $g_z < g_x, g_y$, and a relatively small quadrupole interaction. These characteristics are shared by the two other deep centers that were found, dBe(k1) and dBe(k2). Only for these centers the Be-C-vacancy axis makes an angle of $\sim 70°$ with the c-axis. From these measurements it can be concluded that the two energy levels, at $E_v + 0.42$ eV and $E_v + 0.60$ eV, found by both electrical and diffusion measurements can be assigned to respectively, the three shallow Be centers and the five deep Be centers.

Scandium. Scandium may be considered as a kind of a bridge between acceptors and transition elements; indeed, on the one hand, it is a Group-III element while, on the other, it is the first element in the transition-metal group. This is seen clearly from Table 4.1, where Sc in neutral state (A^0) occupying the Si site (the four valence electrons bond the four C atoms) acts as acceptor, while when in the A^{2-} state it has one unpaired 3d electron. EPR spectra of at least three Sc acceptor types have been observed [97, 122]. Similar to the shallow B and Be acceptors, these spectra are noticeably temperature dependent. Above 30 K they exhibit axial symmetry, but it decreases at lower temperatures. Besides the spectra of the Sc acceptors, one observed EPR signals that was assigned to Sc^{2+} ions (see transition elements section).

4.3 Deep Level Colour Centres and Shallow Donors in Bulk AlN Crystals: EPR, ENDOR, ODMR and Optical Studies

4.3.1 Introduction

Aluminum nitride (AlN) is a direct-bandgap semiconductor with an energy gap of about 6.0 eV at room temperature. AlN has a considerable potential for optoelectronic devices operating in ultraviolet (UV) spectral region [123]. Nominally undoped AlN crystals have the n-type conductivity and high resistance due to the presence of deep-level defects. N-type conductivity of semiconductors can be affected by a transition of the shallow donors (SDs) to a deep state. Experimental and theoretical studies of semiconductor's electronic properties have demonstrated that the shallow donor could give rise to the two types of electronic states, either a shallow state with a delocalized effective-mass-like wavefunction or a deep state with a localized wavefunction. The latter (usually called the DX center) arises due to the lattice distortions at or near the donor site and exhibits a negative correlation energy U for the electrons trapped at this site [124–126]. Electron paramagnetic resonance (EPR), electron-nuclear double resonance (ENDOR) and optically detected magnetic resonance (ODMR) are the most informative methods for identication of defects in semiconductors. EPR, ENDOR and ODMR studies of SDs in AlN and GaN were presented in [127–132], and deep-level defects in [130, 133, 134]. Results, obtained by EPR and EPR-related methods can be correlated with those, obtained by other experimental techniques such as photoluminescence (PL), optical absorption (OA) and thermoluminescence (TL). As a result, deep-level defects, which are responsible for the optical characteristics of the crystals, can be identified.

High-frequency EPR and ENDOR were demonstrated to be the methods of choice for identication of the effective-mass-like shallow donors and deep-level defects in semiconductors [129, 135–137]. In this paper we report the results of EPR, ENDOR and ODMR experiments on as-grown bulk AlN crystals that prove the presence of deep-level color centers with a localized electronic wavefunction and shallow donors with a strongly delocalized wavefunction. The properties of the SDs in the nitrides remain contradictory. Some theoretical works predicted that the DX state is a stable conguration of a silicon impurity in AlN [138, 139] and, on the contrary, it was argued that oxygen forms the DX center [140]. DX-relaxation of the SDs to a deep state makes these SDs unsuitable for creation of the n-type conductivity in AlN crystals at room temperature. The search of the SDs with a stable state at room temperature is one of the major challenges in materials based on AlN. An important applied problem is to find the impurity, which provide a reliable n-type conductivity at room temperature not affected by the DX-relaxation, on the one hand, and to identify the deep level centers that lead to the ionization of the donor impurities, as well as the appearance of the coloration of the AlN crystals, preventing their application for optoelectronics, on the other. These challenges are addressed in present chapter.

4.3.2 Experimental

Five different bulk AlN crystals grown by sublimation sandwich method in a resistively heated furnace with axially symmetric graphite heating elements have been studied in this chapter. A key feature of this growth technique is the use of a precarbonized tantalum growth crucible [141]. Commercial AlN source powder was puried by annealing in vacuum and then in nitrogen atmosphere at 2200 °C leading to reduction of the oxygen content in a source powder from a few percent to 10^{19} atoms/cm3. Following growth of the AlN crystals was carried out on the Si-faced (0001) 6H-SiC 0.3–0.5 mm thick plate in closed tantalum carbide containers. The source and the substrates temperature was kept at the temperature range of 2050–2150 °C. Then, 0.3–0.5 mm thick Al-faced (0001) AlN layers grown on 6H-SiC seeds were cut out and used as a seeds for the further growth of bulk AlN crystals. The latter were cut into plates and used in our measurements. We investigated four different AlN crystals. Samples #1 and #2 were two types of single crystalline AlN substrates, 16 mm in diameter and 0.5 mm thick cut from the AlN ingots. Figure 4.22a shows the images of the AlN ingot (a) and a single crystalline AlN substrate, which was both-side polished and labelled as sample #1 (b). The sample had yellow-brown color. The sample #2 had a similar size and form but on contrast to sample #1 was almost colorless due to the different growth conditions. Lighter crystal (#2) was grown at a relatively low growth temperature of 2120 C and the temperature difference between the source and the seed was kept at 20 °C. The growth temperature of the darker crystal (#1) was about 2200 °C, a temperature drop was less than 10 s. As the source material for sample #3 we used commercially available AlN powder containing iron impurities. Samples #4 and #5 were grown by sublimation of the AlN charge placed in the hot zone of a crucible with following condensation in a cooler region [129].

Continuous wave (cw) EPR at the X-band (9.4 GHz) was used in the studies. For light illumination inside the cavity a light of a xenon lamp was used. The high-energy end of the lamp spectrum was cut-off with the edge filters. For the X-ray irradiation a tube having a molybdenum anode was used. The irradiation time was 5–10 min at a tube current of 15 mA and voltage of 55 kV.

High-frequency EPR and ENDOR experiments were performed at 10–300 K on Bruker Elexsys 680 spectrometer operating at 94.9 GHz (W-band). The spectra were recorded by monitoring the electron-spin echo (ESE) signal.

Optically detected magnetic resonance (ODMR) at 35 GHz (Q-band) was measured by monitoring luminescence excited by the ultraviolet (UV) light of a deuterium arc lamp with appropriate glass filters and analyzed with a grating monochromator. The ODMR spectra were recorded at 2 K in a cryostat with an

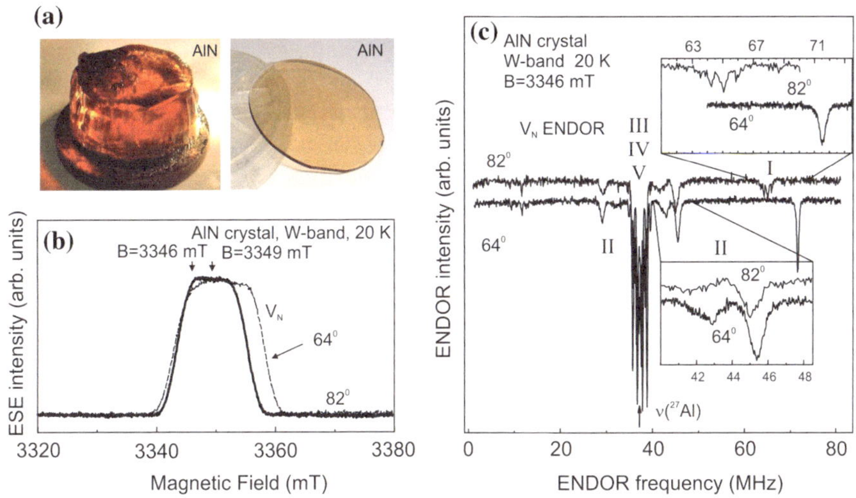

Fig. 4.22 **a** The images of AlN ingot (*top*) and both-side polished single crystalline AlN substrate (*bottom*), sample #1. **b** ESE detected EPR spectra of V_N centers in sample #1 for θ = 82° (B⊥c − 8°) and θ = 64° at 20 K; **c** ESE detected ENDOR signal for V_N centers in sample #1; (*insets*) An expanded view of high-frequency parts of the ENDOR spectra for θ = 82° (*top*) and θ = 64° (*bottom*). The nuclear Zeeman frequency of ^{27}Al (ν(^{27}Al)) is indicated by an *arrow*. B was set at 3346 mT

optical access. The microwave power in the cavity of the ODMR spectrometer was on-off modulated at a sound frequency, and the microwave-induced changes in the luminescence intensity were detected with a lock-in detector. Optical absorption in the 200–900 nm range was measured at 80 and 300 K using a Hitachi spectrophotometer.

4.3.3 *Colour Centres in AlN*

4.3.3.1 High-Frequency EPR and ENDOR

In this section we report the results of high-frequency EPR and ENDOR experiments that aimed in better understanding on main deep level intrinsic defect in AlN —neutral nitrogen vacancy. W-band (94 GHz) ESE detected EPR spectra registered in sample #1 at 20 K for two orientations of the magnetic field with respect to the crystal c axis are shown in Fig. 4.22b. Previous results from the EPR, ODMR and ENDOR studies have shown that these spectra originate form anisotropic hyperfine (HF) interaction with one ^{27}Al nucleus (I = 5/2, 100% abundance). The values of the HF structure of this interaction with one ^{27}Al are $A_{||}$ = 4.0 mT, $A_{\perp}$ = 1.9 mT, and perfectly explain the flat-topped line shape of the spectra. Several models were recently proposed for this defect: (i) interstitial Al atom [133],

(ii) nitrogen vacancy [133, 134] or (iii) neutral oxygen substituting for nitrogen [134]. Finally in [130] the model of the neutral nitrogen vacancy (V_N) was confirmed to be right.

To improve the better understanding of the neutral nitrogen vacancy and provide the information about the distribution of the unpaired electron of the vacancy and electric field gradient within several nearest shells we performed the ENDOR studies. Figure 4.22c shows ESE detected W-band ^{27}Al ENDOR spectra for the nitrogen vacancy centre measured at 20 K for two orientations of the magnetic field with respect to the crystal hexagonal c axis. Magnetic field B was set at 3346 mT. Groups of lines symmetrically located around the ^{27}Al nuclear Zeeman frequency $\nu(^{27}Al)$ are due to the HF and quadrupole interactions with the aluminium nuclei. We labelled these groups by I, II, III, IV and V; the group number increases as we approach the $\nu(^{27}Al)$, indicating the decrease of the HF interaction with distant Al shells. High-frequency parts of groups I and II of the ENDOR spectra are shown on the expanded scale in the insets in Fig. 4.22c. The most distant group of lines labelled by I in the frequency range above 63 MHz corresponds to the largest HF interaction with one aluminium atom. This interaction was studied in details by means of the X-band ENDOR [134], where the HF constants were determined to be $A_{\|} = 111.30$ MHz, $A_{\perp} = 54.19$ MHz. We use these data to determine the crystal orientation as it was difficult to properly align the crystal orientation during the mounting procedure, and a small misalignment was expected. The effective values of the HF interaction are $A_{eff} = 68.942$ MHz for the upper spectrum and $A_{eff} = 54.19$ MHz for the lower spectrum. Knowing that $A_{eff}^2 = A_{\|}^2 \cos^2\theta + A_{\perp}^2 \sin^2\theta$, where θ is an angle between the defect symmetry axis and the direction of the magnetic field, we can determine orientation of the crystal.

In case of the axial symmetry, the symmetry axis is the hexagonal c axis of the crystal. Thus, the angles θ for which the ESE and ENDOR spectra were observed in our experiments are $\theta = 82°$ and $\theta = 64°$ (Fig. 5.22b, c). To analyze the V_N EPR and ENDOR spectra we use the spin Hamiltonian

$$\hat{H} = \mu_B \vec{B} \cdot \hat{g} \cdot \hat{\vec{S}} + \sum_i (\hat{\vec{S}} \cdot \hat{A}_i \cdot \hat{\vec{I}}_i + \hat{\vec{I}}_i \cdot \hat{P}_i \cdot \hat{\vec{I}}_i - g_N \mu_N \vec{B} \cdot \hat{\vec{I}}_i) \tag{4.11}$$

Here, $\hat{\vec{S}}$ is the electron spin operator with S = 1/2, $\hat{\vec{I}}_i$ are the nuclear spin operators. $\hat{\vec{I}}_i$ terms are summed over the Al and N nuclei that interact with the electron. $\hat{g}$ tensor reflects an axial symmetry around the crystal c axis and its principal values are $g_{\|} = 2.002$ and $g_{\perp} = 2.006$. The second, third, and fourth terms in (4.11) reflect the hyperfine, quadrupole, and nuclear Zeeman interactions, respectively.

The HF interaction parameters can be described in terms of the isotropic part a, and anisotropic parts b and b', which are related to the principal HF tensor coordinates (xyz): $A_{xx} = a - b + b'$, $A_{yy} = a - b - b'$, $A_{zz} = a + 2b$. Diagonal matrix elements of the nuclear quadrupole interaction in the principal coordinates are $P_{xx} = -q + q'$, $P_{yy} = -q - q'$, $P_{zz} = 2q$. Here, b' and q' denote the deviation from the axial symmetry and equal to zero in our case. In case of the axial symmetry the third term in (4.11), which reflects the quadrupole interaction of the Al (N) nuclei, can be written as $H_Q = P[I_z^2 - 1/3I(I+1)]$, here $P = 3/2P_{zz} = 3q$.

Isotropic component of the HF interaction $a_i = (8\pi/3)g_e\mu_B g_{ni}\mu_N|\psi(r_i)|^2$ reflects the spin density of the electron wavefunction (ψ) at the site of the nucleus (r_i) and anisotropic component $b_i = (2/5)g_e\mu_B g_{ni}\mu_N \langle r_{3p}^{-3}\rangle$ reflects the axial symmetry around the 3p function axis of the Al atom (or 2p for the N atom). Here, g_e is the electronic g-factor, g_{ni} is the g-factor of nucleus i, and μ_N is the nuclear magneton. The related ENDOR transitions frequencies for a paramagnetic center with S = 1/2 are $\nu_{ENDORi} = 1/h|g_{ni}\mu_N B_0 \pm ½[a_i + b_i(3\cos^2\theta - 1)]|$, where B_0 is the magnetic field corresponding the EPR resonance conditions. Each i nucleus gives rise to two ENDOR transitions symmetrically placed around its nuclear Zeeman frequency $g_{ni}\mu_N B_0/h$ when the quadrupole interaction is neglected and $½[a_i + b_i(3\cos^2\theta - 1)]| < g_{ni}\mu_N B_0$. The "+" and "−" signs in the equation denote ENDOR lines for $M_S = +1/2$ and $M_S = -1/2$, respectively.

For the nuclear spin I > 1/2 the quadrupole interaction of the nucleus with an electric field gradient must be taken into account by $q_i = (eQ_0)/[4I(2I-1)]V_{zz}(r_i)$ in case of the axial symmetry of the center. Here Q_0 is the electric quadrupole moment in multiples of $|e| \times 10^{-24}$ cm^2 and $V_{zz}(r_i)$ is the electrical field gradient. Additional term $1/h[3m_q q_i(3\cos^2\theta - 1)]$ should be added to calculate ν_{ENDOR}, where m_q is the average value of the nuclear quantum states m_I and $m_{I'}$ between which the nuclear transition takes place.

For ^{27}Al nuclear spin is I = 5/2, giving five m_q-values: ±2, ±1 and 0. Thus, the quintet character of the lines in group I is due to the quadrupole interaction of the ^{27}Al. Low-frequency parts of the ^{27}Al ENDOR spectra that should be observed in group I are not visible in Fig. 4.22c because of the low intensity of the signals in this frequency range. The angular dependence observed for the quadrupole splitting is typical for the axial symmetry of the quadrupole interaction of the nuclear spin $I = 5/2$ along the c axis. The lines are degenerate for θ = 54.70°.

The HF structure parameters for group I are $A_\parallel = 111.30$ MHz, $A_\perp = 54.2$ MHz. Thus the isotropic component a_I is equal to 73.2 MHz and anisotropic component $b_I = 19.0$ MHz. The quintet character of the lines included in group I of the ^{27}Al ENDOR spectra for the nitrogen vacancy center is due to the quadrupole interaction of the ^{27}Al nucleus with an electric field gradient. For θ = 82° the quadrupole splitting is of 0.807 MHz and for θ = 64 it is about 0.366 MHz. Thus, quadrupole parameter is P = 0:864 MHz and almost coincides with the parameters derived from the ENDOR studies in the X-band [134].

The HF and quadrupole interactions with other Al and N nuclei will be discussed in this chapter. The surrounding ions of the V_N center, for which the HF and quadrupole interactions were resolved by EPR and ENDOR, are presented in Fig. 4.23a. The unpaired electron is well localized inside the shells formed by Al and N atoms, labelled by I, II, III, IV, V and N. The unpaired electron density outside these shells must be very small. We now consider the ENDOR lines labelled by II (Fig. 4.22c). The quadrupole splitting is not observed for this group indicating that the symmetry axis of this interaction is directed along one of three

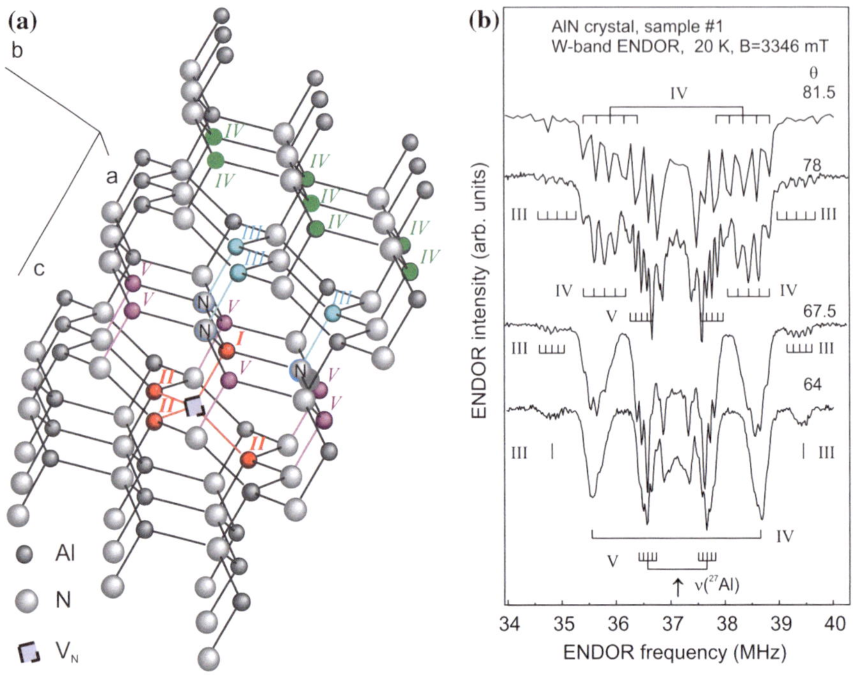

Fig. 4.23 **a** AlN crystal lattice with V_N. Groups of Al and N nuclei indicated by I, II, III, IV, V and N correspond to groups of lines observed in the ^{27}Al and ^{14}N ENDOR spectra of V_N. Unpaired electron density outside these shells is very small. **b** Central part of ^{27}Al ENDOR signal for V_N centers in the sample #1 measured for θ = 81.5° (90°−8.5°), 78°, 67.5°, and 64° at 20 K for the magnetic field of 3346 mT. ν(^{27}Al) is indicated by an *arrow*

bonds that do not coincide with the c axis. This suggests that this signal most likely reflects the interaction with three aluminum atoms located in the basal plane (Fig. 4.23a). Two pairs of lines observed in this group are due to anisotropy of the HF interaction and nonequivalence of these three axes with respect to the direction of magnetic field. The HF parameters are 15.78 MHz (θ = 82°) and 16.674 MHz (θ = 64°) for the group with the larger HF splitting, and 8.872 MHz (θ = 82°) and 11.516 MHz (θ = 64°) for the group with the smaller HF splitting. The absence of the ENDOR signal that refer to the Al nucleus directed along the c axis in group II lines eliminates the antisite position of aluminum. Should such interaction with the Al nucleus occur, additional resolved quadrupole structure, similar to the structure observed for the group I would be observable in the ENDOR spectra.

The second shell is the nitrogen shell and the ^{14}N ENDOR signal for V_N consists of two almost isotropic lines with a maximum splitting of 2.4 MHz. We refer them to the interaction with three nitrogen atoms, labelled by N, that bound to the aluminum nucleus labelled by I (see Fig. 4.23a).

The ^{14}N ENDOR for the shallow donors in bulk AlN crystal shows the broad unresolved ENDOR line at ν(^{14}N) which is caused by superposition of multiple

ENDOR lines with a very small HF parameter, since the donor wavefunction spreads considerably in space [129]. In opposite, in case of deep V_N donors only several narrow lines near $\nu(^{14}N)$ are observed. This indicates that the spin density of unpaired electron that belong to the nitrogen vacancy is strongly localized and distributed over a smaller volume as one can expect to be in case of the deep donor center.

Figure 4.23b shows the central part of the W-band ^{27}Al ENDOR signal for V_N centers at 20 K with the angles of $\theta = 81.5°$ $(90° - 8.5°)$, 78°, 67.5° and 64° between the direction of the magnetic field and the crystal c axis. Spectra were measured with B set at 3346 mT. $\nu(^{27}Al)$ is indicated by an arrow. Several groups of lines arranged symmetrically with respect to the $\nu(^{27}Al)$ relate to the hyperfine and quadrupole interactions with following aluminum shells (III, IV, and V in Fig. 4.23a). The sequence of Al shells corresponds to the reduction of the hyperfine splitting with increase of the distance from the vacancy. As we will show now this assignment is also confirmed by the angular dependencies of the hyperfine and quadrupole splittings.

Groups III and IV include five lines due to the quadrupole splitting. The angular dependencies of these splittings are similar to the angular dependence of the quadrupole splitting caused by the interaction with one aluminum nucleus (labelled as I, Fig. 4.23a). This suggests that the symmetry of these interactions should be similar and we can make the assumption about the direction of the bonds between the vacancy and the aluminum atoms.

The HF parameters of interaction with Al nuclei in group III are $A_{\parallel} = 5.82$ MHz and $A_{\perp} = 4.33$ MHz, thus the isotropic component of this interaction is $a_{III} = 4.83$ MHz and anisotropic component is $b_{III} = 0.50$ MHz (if $A_{\parallel}$ and $A_{\perp}$ are of the same sign). The quintet character of the ENDOR spectrum for group III lines in Fig. 4.23b for $\theta = 67.5°$ and $\theta = 78°$ is due to the nuclear electric quadrupole interaction. These data give P = 0.19 MHz. For $\theta = 64°$ quadrupole splitting is not resolved and for orientation of $\theta = 81.5°$ the spectrum is written with low resolution. Al atoms included in group III are shown in Fig. 4.23a.

HF structure constants for group IV are $A_{\parallel} = 5.12$ MHz and $A_{\perp} = 2.39$ MHz, isotropic component $a_{IV} = 3.3$ MHz and anisotropic component $b_{IV} = 0.91$ MHz. The quintet character of the ENDOR spectrum for group IV in Fig. 4.23b is also caused by the nuclear electric quadrupole interaction. These data give the quadrupole parameter of P = 0.21 MHz. Al atoms included in group IV are shown in Fig. 4.23a. It should be noted that the ratio of the ENDOR signal intensities for groups III and IV is in the qualitative agreement with the number of aluminum atoms in these groups.

The value and anisotropy of the HF interactions allow determination of the s and p spin densities on the Al atoms in groups I, III and IV. To do so we approximate the wave function for the unpaired electron by a linear combination of atomic orbitals centered on the Al atoms near the V_N: $\Psi = \Sigma\eta_i\psi_i$. At each j site we approximate ψ_i as a hybrid 3s-3p orbital: $\psi_i = \alpha_i(\psi_{3s})_i + \beta_i(\psi_{3p})_i$, $\alpha_i^2 + \beta_i^2 = 1$.

Using the values of hyperfine parameters a_i and b_i, determined for each group of Al atoms (I, III and IV), we determine the corresponding molecular wave-function coefficients (α_i^2, β_i^2 and η_i^2).

Group I: 1 Al atom, $a_i = 73.2$ MHz, $b_i = 19$ MHz, $\alpha_i^2 = 0.08$, $\beta_i^2 = 0.92$, $\eta_i^2 = 0.25$;

Group III: 3 Al atoms, $a_i = 4.83$ MHz, $b_i = 0.5$ MHz, $\alpha_i^2 = 0.16$ $\beta_i^2 = 0.84$ $\eta_i^2 = 0.022$;

Group IV: 7 Al atoms, $a_i = 3.3$ MHz, $b_i = 0.91$ MHz, $\alpha_i^2 = 0.7$, $\beta_i^2 = 0.93$, $\eta_i^2 = 0.09$.

From these coefficients we can conclude that the highest spin density (~25%) is localized on one Al atom directed along the *c*-axis of the crystal (labelled as I in Fig. 4.23a). Sharp deviation from sp^3 hybrid orbitals is observed for all interactions under consideration.

Resolved nuclear quadrupole splitting allows direct determination of the electric field gradient at the nuclear position. Usually two sources of the electric gradient $V_{zz}(r_i)$ are considered: intrinsic electric-field gradients due to the hexagonal crystal structure of AlN, and the unpaired charge density in the 3p orbital of the ion itself. The magnitude of the latter can be estimated from the anisotropic part of the HF interaction *b*.

It was discussed before that $P_i = 3q_i = 3(eQ_0)/[4I(2I - 1)]V_{zz}(r_i)$. For ^{27}Al nuclei $I = 5/2$ and $Q_0(^{27}Al) = 0.150$. Knowing the values of quadrupole interactions for groups I, III and IV, determined from our experiments, and the value of quadrupole interaction for SDs in AlN [129] (the quadrupole interaction for SDs of P = 0.135 MHz can be considered as minimal because it can be associated with the interaction with the remote Al shells, for which the hyperfine splitting is close to zero), we can calculate the electric field gradients and estimate the contribution caused by the influence of AlN crystal field. Quadrupole splittings, caused by Al nuclei surrounding VN, which were included in groups I, III and IV, and the values of the electric field gradient are the next:

Group I: 1 Al atom, P = 0.864 MHz, $V_{zz}(r_i) = 31.76 \times 10^{20}$ V/m^2
Group III: 3 Al atom, P = 0.19 MHz, $V_{zz}(r_i) = 6.98 \times 10^{20}$ V/m^2
Group IV: 7 Al atom, P = 0.21 MHz, $V_{zz}(r_i) = 7.71 \times 10^{20}$ V/m^2

For comparison, the quadrupole splitting for SDs in AlN and the values of the electric field gradient are given:

AlN crystal: P = 0.135 MHz, $V_{zz}(r_i) = 4.9 \times 10^{20}$ V/m^2.

Interpretation of the remaining lines in the ENDOR spectra requires further research. We can assume that the signals indicated as V in Fig. 4.23b correspond to the hyperfine interaction with aluminum atoms, indicated by V on the AlN model shown in Fig. 4.23a. The quadrupole splitting is observable for $\theta = 64°$ with the hyperfine parameter of A = 1.092 MHz and quadrupole splitting parameter of P = 0.075 MHz.

4.3.3.2 Optical Absorption and Thermo-luminescence

Crystal coloration depends on the content of intrinsic and impurity defects. In many wide-band-gap semiconductors and ionic crystals the coloration is due to the so-called color centers, frequently different vacancy-comprising defects, for example, anion vacancy (F-center) in ionic crystals, or the nitrogen-vacancy defect in diamond. The nature of the color centers, as a rule, is studied by optical and EPR techniques with their subsequent correlation [125].

Apart from the fundamental and excitonic absorption near the bandgap energy, bulk AlN crystals generally show absorption bands in the visible and UV range [142–146]. The color of the as-grown samples strongly depends on the growth conditions. Determination of the nature of the color centers is crucially important for further applications.

Here we are interested in the optical transitions that cause the absorption in the blue spectral region. Two samples under investigation have markedly different color: sample #2 was almost transparent and on-contrast sample #1 had yellow-brown color due to the absorption in the blue spectral range. To identify the optical transitions that cause this absorption both samples was subjected to the X-ray irradiation.

Optical absorption spectra measured in AlN sample #2 before (dashed curve) and after (bold curve) 15-min X-ray irradiation are shown in Fig. 4.24. Broad absorption band at 450 nm appeared after X-ray irradiation was assigned in the previous literature to the neutral nitrogen vacancy center (see [20]). However, the lack of the correlation with the EPR data, leaded to the further assignment of this band to the oxygen-related center. We performed the EPR measurements and the spectra taken at B⊥c orientation of the magnetic field in the same sample #2 before and after X-ray irradiation are shown in the inset. EPR signals appear only after X-ray irradiation at room temperature and originate from the neutral nitrogen vacancy V_N. Correlation between the EPR spectra and the optical absorption unambiguously evidence that the nitrogen vacancy in the neutral charge state is a color center in AlN crystals. Initially V_N donors in sample #2 were in the non-paramagnetic singly ionized charge state (V_N^+). Short X-ray irradiation at room temperature converts ionized donors to the paramagnetic neutral state by trapping a radiation-induced electron, while the corresponding radiation-induced hole is trapped elsewhere in the crystal on the unidentified traps.

In sample #1 the EPR spectra of V_N centers and the optical absorption band at 450 nm were observed even before the X-ray irradiation. X-ray irradiation of the sample at room temperature leaded to significant increase of the intensity of the V_N EPR signal and the optical absorption at 450 nm. Thus, in the sample #1 V_N centers are partially in the neutral charge state even before the X-ray irradiation, so accounting for the yellow-brown color of the sample, and consequently, the V_N EPR signal and 450 nm optical absorption observed in the as-grown crystals.

Thus, the presence of the neutral nitrogen vacancies in AlN can be revealed by monitoring the intensity of the optical absorption and the concentration of the nitrogen vacancies can be estimated.

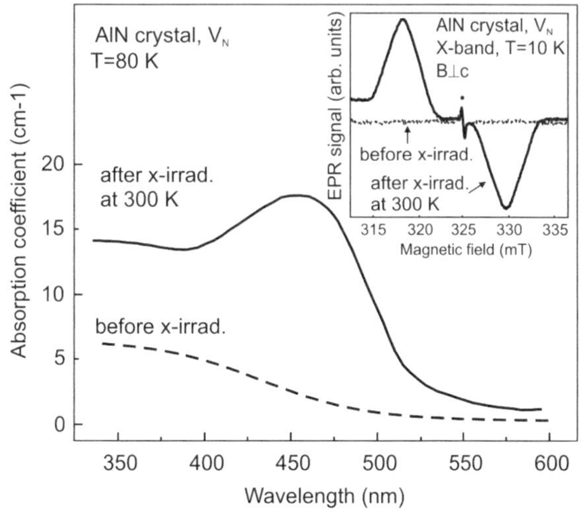

Fig. 4.24 Optical absorption of AlN sample #2 before (*dashed curve*) and after (*bold curve*) X-ray irradiation; (*inset*) EPR spectra measured in the same AlN sample before and after X-ray at B⊥c, the reference signal from quartz is marked with an *asterisk*

High-temperature annealing of the X-irradiated AlN samples at 300–500 K resulted in the thermo-luminescence (TL) due to recombination of the deep-level centers. The spectral dependence of the TL corresponds to the UV range with a maximum at ~360 nm. On the basis of these measurements the level depth was estimated: using the first-order kinetic the E_a was shown to be in the range between 0.5 and 0.9 eV. The activation energy of the V_N is believed to be about 0.75 eV because this correlates well with the data obtained by EPR. The intensity of the EPR signal decreases to its primary magnitude with the increase of the annealing temperature.

These experiments demonstrate that the V_N center is a deep donor center with a spin density mainly localized on one Al nuclei located in the first coordination shell of the vacancy. The direction of the bond between the vacancy and the Al nucleus coincide with the crystal c axis. Optical absorption studies show that the neutral nitrogen vacancy V_N is the color center. The optical absorption band of the vacancy is in the visible region with a maximum at ~450 nm. The donor energy level of ~0.75 eV relative to the conduction band is determined by means of thermo-luminescence. These results are in line with the theoretical studies [147] where the nitrogen vacancy in the neutral charge state was shown to act as a deep donor in AlN.

4.3.3.3 Optically Detected Magnetic Resonance

Bulk AlN crystals are very rich in their optically detected magnetic resonance spectra11. Samples #1 and #2 exhibit intense PL, extending from the visible to the near infrared spectral regions, with a broad peak shifting towards the lower energy for the lower excitation wave-lengths. Using different photoexcitation energies, it was possible to extract a number of ODMR signals associated with the different centers contributing to the PL.

Figure 4.25a shows the 35 GHz ODMR spectra recorded in AlN crystal (sample #1) at different orientations of the magnetic field relative to the c axis of the crystal in the (11–20) plane. The chemical identification of the impurity involved is quite complicated because the HF structure is not observed. However, the value of the HF interactions differ for N and Al positions of the defect and the position of the binding core can be obtained from the EPR linewidth. The larger gyromagnetic ratio and spin of ^{27}Al nucleus indicates that the HF interaction with the Al nuclei should dominate and the effects of the N nuclei could be neglected in AlN. Since the ODMR lines are broad, the main contribution in the linewidth is due to the interaction with ^{27}Al.

The angular dependencies can be well-fitted by standart spin Hamiltonian with $S = 1$, $D = 940 \times 10^{-4}$ cm^{-1} and $\alpha = 55°$ (Fig. 4.25a). Here α correspond to the angle between the principal z axis of the centers, which is in the (11–20) plane and the c axis. In this direction a pair of vacancies with the shortest relative distance occupies two neighboring (0001) planes with the lattice separation of 4.4 Å. As

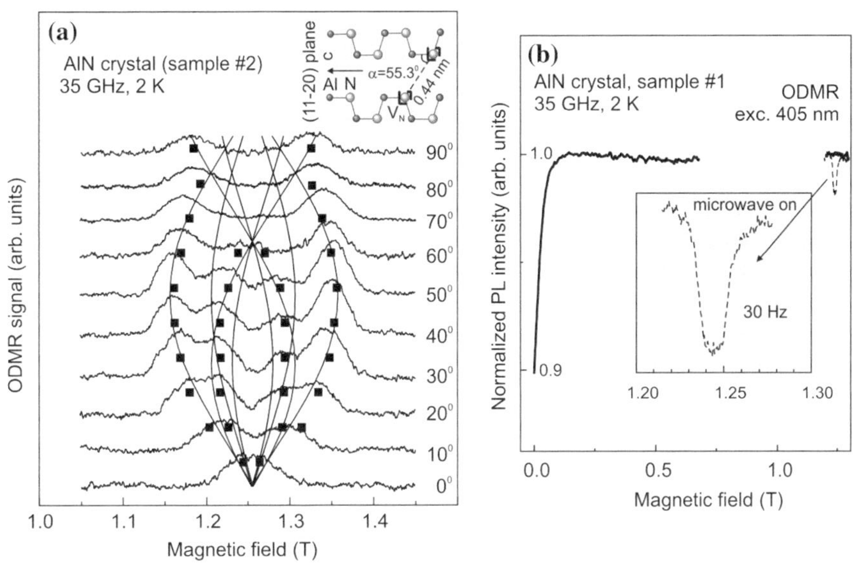

Fig. 4.25 **a** Angular dependencies of the ODMR spectra in AlN crystals (sample #2) for B rotated in (11–20) plane. 0° corresponds to the angle $\theta = 7°$ between the *c* axis and the direction of the magnetic field. Calculated angular dependencies of the resonance transitions for $S = 1$ and $D = 940 \times 10^{-4}$ cm^{-1}. The experimental positions of the ODMR lines are shown by *black squares*. (*Inset*) Model for pair of the nitrogen vacancies in AlN. **b** Dependence of the luminescence intensity on the magnetic field in AlN crystal (sample#1) without (*bold curve*) and with application of 35 GHz microwaves (*dashed*). *Inset* shows the ODMR signal recorded at $\theta = 50°$ with microwaves on-off modulated at 30 Hz at T = 2 K on the extended scale

soon as it is not possible to follow all lines throughout the angular dependence because of the variations in the intensities of individual components and superposition of the spectral lines, which are very broad, we cannot rule out a slight disorientation of about 5°. In addition, a small deviation from the axial symmetry (E parameter in the spin Hamiltonian) is observed. We suggest that this defect is the exchange-coupled pair of the nitrogen vacancies V_N-V_N in AlN. The model of this defect is shown in Fig. 4.25a (inset). There are six crystal directions, and therefore divacancies are distributed equally along all of them. The carbon divacancy with a similar structure was observed in hexagonal SiC crystals [148–150].

ODMR was recorded as the microwave-induced variations of the intensity of the luminescence excited with a deuterium arc lamp (250–400 nm) and detected in the range of 550–600 nm at 2 K. The microwave frequency was 35.1 GHz, the microwave power 50 mW, and the modulation frequency 85 Hz.

AlN crystals containing V_N color centers (deep donors) exhibit ODMR signals of the V_N defects, corresponding to a resonance reduction of the luminescence intensity, as well as a reduction of the luminescence intensity in zero magnetic field even without application of the microwaves. This is what is observed in Fig. 4.25b, where the dependence of the luminescence intensity in AlN crystals (sample #1) on the magnetic field is shown without (bold curve) and with application of the microwaves (dashed curve). ODMR was recorded using 35 GHz microwaves on-off modulated at 30 Hz. A strong increase (about 10%) in the luminescence intensity is observed in sufficiently weak magnetic fields (up to 10 mT), irrespective of the microwave field. In the region of $g = 2.00$ a negative ODMR signal with anisotropic linewidth can be seen. The flat-topped line shape, the value, and the anisotropy of the linewidth are in a good agreement with that for the V_N color centers, therefore, the ODMR signal seems to belong to the V_N donors.

The magnitude of the ODMR signal is temperature independent at least in the range of 2–10 K. This indicates that the observed effects are not due to the thermalization, but due to the spin dependence of the transition probabilities, and can be explained by assuming the existence of spin-dependent non-radiative process that shunts the luminescence channel. When the magnetic field exceeds the internal magnetic field, comparable with the width of the magnetic resonance line, spin selection takes place due to (for example) hyperfine interaction. As a result the efficiency of the non-radiative process decreases and the luminescence intensity increases. This non-radiative channel can be activated at the moment of the magnetic resonance, which alters the orientation of the electron spins of one of the partners of the investigated process. Such a resonance corresponding to a reduction of the luminescence intensity of the V_N deep donors provides a direct proof of participation of these donors in the non-radiative process. Similar effects were observed for defects in SiC crystals [151] and for color centers (F-centers) in alkali halide crystals [152]. F-centers are among the simplest defects in ionic crystals: an unpaired electron located in an anion vacancy. Thus, a nitrogen vacancy with an unpaired electron in AlN is almost the analogue of the color center in ionic crystals.

4.3.4 *Shallow Donors in AlN*

The behavior of the *n*-type conductivity in the nitride semiconductors is a subject of intense investigation. The microscopic identity of the shallow donors is of great importance. An incorporation of impurities on both cationic and anionic sublattices was considered: oxygen on N position and group IV impurities (C, Si) on Al position along with nitrogen vacancy have been discussed.

4.3.4.1 Shallow Donors with DX Behavior

W-band (95 GHz) pulse experiments: observation of the triplet meta-stable state of shallow donor pairs in AlN crystals with a negative U behavior. As mentioned in the introduction, the important factor that may affect the doping is a transition of donors to a *DX*-like state. The formation of so called *DX* centers leads to self-compensation of the shallow donor, according to the reaction [126]:

$$2d^0 = d^+ + DX^- + U$$

Here d denotes a substitutional shallow impurity and *DX* the displaced deep state. The superscripts specify the charge states and U stands for the negative correlation energy. Shallow donor can lower its energy by the capture of a second electron following by a large lattice relaxation of the donor impurity off the substitutional site. The energy gain associated with electron pairing in the dangling bonds of a defect, and coupled with a large lattice relaxation, was suggested by Anderson may overcome the Coulombic repulsion of the two electrons, supplying a net effective attractive interaction between the electrons at one site (Anderson negative-U system). Electrons would therefore be trapped by pairs at the defect, providing no paramagnetism. The theoretical suggestions and the experimental evidence of the negative-U properties for point defects in semiconductors were first published in [124, 125]. High-frequency EPR and ENDOR experiments on as-grown bulk AlN single crystals unambiguously reveal the presence of two types of shallow effective-mass-like donors. The shallow character of the wave function is evidenced by the multitude of ^{27}Al ENDOR lines. The light-induced shallow donors were discovered to create coupled pairs with exchange interaction of about 20 cm^{-1} and with lowest triplet meta-stable state which shows a negative U behaviour. Since the *DX* formation reaction leads to a self-compensation, it is of great fundamental as well as practical interest to check whether a similar reaction occurs in AlN.

There were two types of EPR studies of donors in crystals: the high-frequency (W-band) electron spin echo (ESE), ENDOR [7] and the CW experiments at X-band. Figure 4.26a (right inset) shows the ESE-detected EPR spectra measured at 94.9 GHz in as-grown undoped AlN single crystal (sample #5) at 1.8 K after cooling down from room temperature to 1.8 K in the dark (dashed line) and after

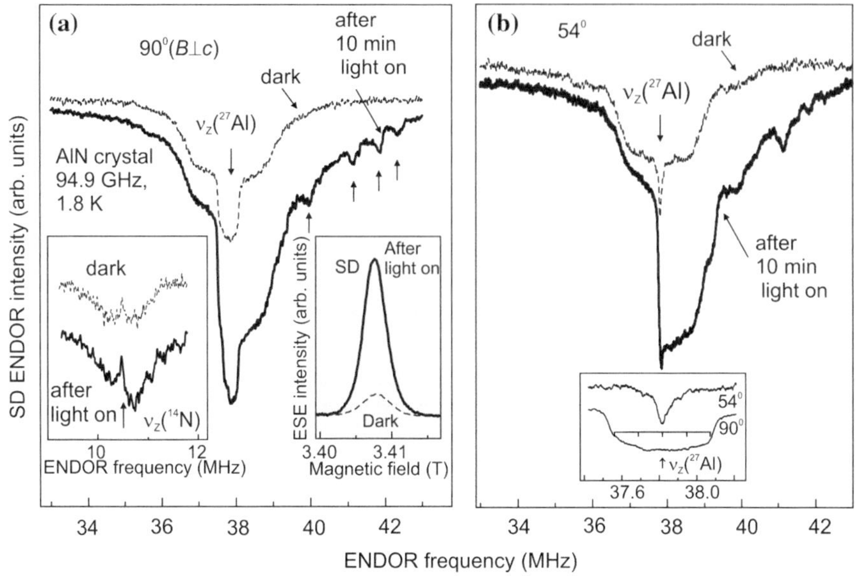

Fig. 4.26 **a** Shallow donor (*SD*) ESE detected ENDOR signal in as-grown undoped AlN single crystal at 1.8 K after cooling down from room temperature to 1.8 K in the dark (*dashed line*) and after 10 min light illumination (*bold line*); **a** ^{27}Al ENDOR signals measured in orientation θ = 90°, B⊥c, (*left inset*) ^{14}N ENDOR signals measured in orientation θ = 90°; (*right inset*) ESE-detected EPR spectra of SD at 94.9 GHz at 1.8 K in the orientation B⊥c after cooling down from room temperature to 1.8 K in the dark (*dashed line*) and after 10 min light illumination (*bold line*). **b** ^{27}Al ENDOR signals measured in orientation θ = 54°; (*inset*) the central part of the SD ENDOR signal of ^{27}Al nuclei measured after cooling down from room temperature to 1.8 K in the dark in two orientations: θ = 54° and θ = 90° which reveals the quadrupole interaction

10 min light illumination (bold line). The EPR signals were detected with the magnetic field perpendicular to the crystal *c* axis (*B*⊥*c*). The observed high field EPR signal is characterized by slightly anisotropic *g* factor of $g_{//} = 1.9900$ and $g_{\perp} = 1.9894$. This g factor is somewhat smaller then g = 2.00 as expected for shallow donors in a wide-band-gap semiconductor such as AlN. The anisotropy is consistent with the hexagonal symmetry of the AlN crystal. These factors support the association of the single resonance in Fig. 4.26a (right inset) with shallow donors (SD). After cooling in the dark, only weak EPR signal of SD is observed. After illumination (light with wave length shorter then 700 nm), a strong EPR signal of SD appears. This EPR signal, once excited at low temperature, persists at low temperature after switching off the light. The light-induced EPR signal of SD disappears after heating above 200 K.

The EPR line of SDs does not provide information on the chemical nature of the donor species present and the nearest neighbors since no resolved hyperfine structure is observed. Therefore ENDOR spectroscopy has been performed at 95 GHz and at 1.5–2 K using a method which is based on the ESE detected ENDOR. Figure 4.26 shows the shallow donor ESE detected ENDOR signal of

^{27}Al nuclei (I = 5/2, abundance 100%) at 1.8 K after cooling down from room temperature to 1.8 K in the dark (dashed) and after 10 min light illumination (bold). The ENDOR signals were measured in two orientations: $\theta = 90°$, $B \perp c$ (a) and $\theta = 54°$ (b). The nuclear Zeeman frequency of ^{27}Al is shown by arrow. For the understanding of the ENDOR results one can consider the isotropic HF interaction a_i which reflects the spin density of the shallow donor electron wave function (Ψ) at the site of the nucleus (r_i)

$$a_{\mathrm{i}} = (8\pi/3) g_e \mu_B g_{ni} \mu_N |\Psi(r_i)|^2,$$

where g_e is the electronic g factor, μ_B is the electronic Bohr magneton, g_{ni} is the g factor of nucleus i, and μ_N is the nuclear magneton. The related ENDOR transition frequencies for shallow donor are

$$\nu_{\mathrm{ENDOR}i} = h^{-1} |-g_{ni} \mu_N B_0 + M_S a_i|,$$

where for S = 1/2 each nucleus i gives rise to two ENDOR transitions symmetrically placed around its nuclear Zeeman frequency $g_{ni}\mu_N B_0/h$ when the quadrupole interaction is neglected and when $a_i < g_{ni}\mu_N B_0$, which is the case in the spectra in Fig. 4.26. This symmetrical behavior is indeed observed for "dark" SD signal although the HF lines are not resolved as in the case of SDs in AgCl, SiC or ZnO. In contrary, considerable difference in intensity is observed for ENDOR signals of light-induced SDs that are positioned in Fig. 4.26 above the nuclear Zeeman frequency of ^{27}Al and below this frequency. This difference in the intensities strongly depends on the temperature and increases dramatically when the temperature reduces. The ENDOR spectra consist of a multitude of lines which proved to be isotropic apart from a few lines in the Al spectrum which seem to exhibit a small orientation dependence of linewidth, and which are indicated by arrows. These lines correspond to the HF interaction constants a_i of 9.04, 7.99, 6.56, and 4.19 MHz.

Based on the expressions for the ENDOR frequencies, we can now qualitatively understand the ENDOR spectrum of light-induced SD shown in Fig. 4.26 if to suggest that we are dealing with a triplet ground state with a ferromagnetic ordering of the two shallow donor spins. The more intensive signals covering the frequency region in Fig. 4.26 above the nuclear Zeeman frequency is related to ENDOR transitions in the lower $M_S = -1$ sublevel, whereas the less intensive signals covering the frequency region below the nuclear Zeeman frequency corresponds to the transition in the M_S = 1 sublevel. One can see that in the ENDOR spectra at 1.8 K the intensities of the fine-structure components differ strongly due to the extreme difference in the populations of the triplet sublevels at this low temperature and the large Zeeman splitting. According to the equation for $\nu_{\mathrm{ENDOR}i}$ the ENDOR transition in the $M_S = -1$ sublevel will lie at a frequency $h^{-1}(1/2a_i)$ above or below the nuclear Zeeman frequency $g_{ni}\mu_N B_0$, depending on the sign of the HF constant a_i. From the observation that the intensive group of lines which correspond to the transitions in the $M_S = -1$ level is positioned above the ^{27}Al Zeeman frequencies and the fact that for ^{27}Al g_N has a positive sign, we can derive that HF interaction for these lines a_i has a positive sign.

If the triplet state is a case the similar asymmetric line should be observed for nitrogen ENDOR. Indeed, strong temperature dependent asymmetry of ENDOR line from N similar to that for Al is observed. Figure 4.26a (left inset) shows ENDOR signal of ^{14}N nuclei at 1.8 K after cooling down from room temperature to 1.8 K in the dark (dashed) and after 10 min light illumination (bold).

The central part of the shallow donor ENDOR signal of ^{27}Al nuclei measured in the sample 1 after cooling down from room temperature to 1.8 K in the dark in two orientations: $\theta = 54°$ (1) and $\theta = 90°$ (2) is presented in Fig. 4.26b (inset). One can see that linewidth of the central part changes from 0.13 MHz for $\theta = 54°$ to 0.57 MHz for $\theta = 90°$ which reveals the quadrupole interaction (QI). AlN crystallizes in the hexagonal wurtzite structure with an accompanying intrinsic electric-field gradients at the nuclear sites. To account for the quadrupole interaction, for axial symmetry the term

$$h^{-1} m_q 3 q_i (3\cos^2\theta - 1)$$

must be added to the equation of ν_{ENDOR} where m_q is the average value of the two nuclear quantum states m_I, $m_{I'}$, between which the nuclear transition takes place. For axial symmetry one has

$$q = (eQ_0)/[4I(2I-1)]V_{ZZ}(r_i),$$

where Q_0 is electric quadrupole moment in multiples of $|e| \times 10^{-24}$ cm^2 and $V_{ZZ}(r_i)$ is electrical field gradient. For ^{27}Al nuclei nuclear spin $I = 5/2$ and $Q_0(^{27}\mathrm{Al}) = 0.150$. For $I = 5/2$ there are five m_q-values: $m_q = \pm 2, \pm 1$ and 0. Thus, the quintet character of the lines for AlN single crystal (Fig. 4.26b) comes from quadrupole interaction for remote Al shells. The angular dependence observed for quadrupole splitting which is of 0.135 MHz for the orientation $\theta = 90°$ is typical for axial symmetry along c axis for nuclear spin $I = 5/2$. The intrinsic electric-field gradient at the Al nuclear sites was shown to be about 1.5 times smaller compared to that for Zn nuclei in ZnO.

A strong intensity of the ENDOR line on ^{27}Al Zeemann frequency from large number of the remote ^{27}Al nuclear (about 100% decrease of the ESE signal) could be due to dynamical nuclear polarization similar for the case of the shallow donors in ZnO. This effect supports a suggestion that a donor wave-function spread out considerably in space. Therefore one can discuss these results in a model of a shallow donor in the Effective-Mass-Theory (EMT). The wavefunction of a shallow donor in EMT is hydrogen 1s like $\Phi(r) \sim \exp(-r/r_D)$ where r_D is effective Bohr radius of the shallow donor. The effective Bohr radius $a_{eff} = a_0 \cdot \varepsilon/m^*$. In AlN the experimental values of the effective mass m^* is $0.33m_0$, the dielectric constant $\varepsilon = 9.14$. Thus from EMT one expect $a_{eff} \sim 1.5$ nm. This is almost the same value as for the shallow EMT like donor in ZnO.

The HF structure is not resolved and determines the EPR line widths. The line width of an unresolved EPR line ΔB for the case where this width is dominated by HF interactions is given by

$$\Delta B = \frac{2}{g\mu_B}[\sum_{i}^{R} f(^N X) n_i(X) \frac{a_i^2(^N X) I(I+1)}{3}]^{\frac{1}{2}},$$

where f is the relative concentration of the given isotope; $f(^{27}Al) = 1.0$ (natural abundance 100%), $f(^{14}N) = 0.996$, ^{N}X stands for ^{27}Al or ^{14}N, n_i is the number of equivalent sites for X atoms in the i-th coordination shell, a_i is the isotropic HF interaction constant for ^{N}X atoms occupying equivalent sites in the i-th coordination shell that is proportional to the wave-function density at the nucleus site $|\psi(\vec{r}_i)|^2$, g is the electronic g factor, μ_B is the Bohr magneton, I is the nuclear angular momentum of the ^{N}X isotope ($I = 5/2$ for ^{27}Al and $I = 1$ for ^{14}N isotopes). As follows from equation for linewidth, the concentration of isotopes with nonzero nuclear spins has a considerable effect on the linewidth if the isotropic HF interaction constant for these isotopes is large enough. The major contribution to the line width is due to several coordination shells closest to the center of the shallow donor. The larger gyromagnetic ratio and spin of ^{27}Al nucleus indicates that the HF interaction with the Al nuclei should dominate and the contribution to the linewidth of the N nuclei may be neglected.

It is of interest to compare the linewidths for the shallow donors in AlN and ZnO, since the Bohr radii for shallow donors in these crystals have close magnitudes. The linewidth, which is 0.5 mT for hydrogen shallow donors in the ZnO, narrows strongly as compared to that measured in the crystal AlN, 5.0 mT. As follows from the equation for linewidth, this narrowing is due to low concentration of ^{67}Zn isotope having nuclear magnetic moment ($I = 5/2$, 4.1%) compare with ^{27}Al (I = 5/2, 100%) and, in addition, HF structure constant for the free Zn atom according to [30] A = 2087 MHz is much smaller compare to the free Al atom A = 3911 MHz. This narrowing corresponds to ΔB(AlN)/ΔB(ZnO) = $(1.0/0.041)^{1/2}$ A(^{27}Al)/A(^{67}Zn) = 4.94 × 3911/2087 = 9.26. Expected ΔB(AlN) = 0.5 × 9.26 = 4.6 mT is close to the experimental linewidth.

The microscopic nature of the shallow donor could be identified, if the HF interaction of the central nucleus could be determined experimentally. In the present experiments, a search for Si, C or O was unsuccessful, probably due to the low natural abundance of the isotopes with nuclear magnetic moments. Successful experiments would demand samples intentionally doped with, e.g. isotopically enriched ^{29}Si, ^{13}C or ^{17}O. The character of the HF interaction with Al shells is different for dark and light-induced shallow donors, which could give an information about the position of the shallow donor in cationic (Al) or anionic (N) sublattice. The strong HF interaction for light-induced SD supports the assignment to the impurity in anionic sublattice (e.g. oxygen in N position).

It was shown [153, 154] that, by orthogonalizing a suitable envelope function $\Phi(r)$ to the cores of the lattice ions in order to allow for the Pauli principle, the spin density on nucleus i may be written as an "amplification factor" K_i times the density of the envelope function Φ on nucleus i. If the envelope function remains approximately constant within each ion core (which is the case for a strongly delocalized envelope function Φ), the value of K_i will depend only on the species of ion i, and not on its position in the lattice. Thus $|\Psi(r_i)|^2 = K_i\ |\Phi(r_i)|^2$, for crude estimation $K_i = [|\langle\Phi|\Psi_{ns}\rangle|^2 \cdot |\Psi_{ns}\ (r_i)|^2]/|\Phi(r_i)|^2$, here $\langle\Phi\ |\Psi_{ns}\rangle$ is overlap integral, Ψ_{ns} is wave function of outer-shell ns electron of the lattice or donor impurity ion.

For outer-shell 3s atomic orbitals of Al atom $|\Psi_{3s}(0)|^2 = 3.356$ a.e. [155], the overlap integral $|\langle\Phi|3\ s(\mathrm{Al}^{\circ})\rangle|^2 = 0.004313$ were calculated for the nearest-neighbor position of Al with a Coulombic center on N site. The theoretical amplification factors for Al is $K_{\mathrm{Al}}(\mathrm{t}) = 1590$. The largest experimental HF interaction constant of light-induced shallow donors is 9.044 MHz which is suggested to reflect the HF interaction with the nearest-neighbor position of Al for a Coulombic center on N site. The amplification factors for Al, obtained with using the experimental values of 9.044 MHz is $K_{\mathrm{Al}}(\mathrm{e}) \cong 1000$. We used for the calculation of $K_{\mathrm{Al}}(\mathrm{ex})$ the same $|\Phi(0)|^2$ values as for the calculation of K(theor) thus the K(ex) values are only partly based on experimental findings.

The formation of pairs of the shallow donors is caused by the relatively strong isotropic exchange interaction arising between neighboring donors with overlapping wave functions. The Hamiltonian of an exchange-coupled pair with spins $S_1 = S_2 = 1/2$ and the Zeeman interaction has the form

$$\hat{H} = J\vec{S}_1 \cdot \vec{S}_2 + \mu_B\vec{B} \cdot \overleftrightarrow{g} \cdot (\vec{S}_1 + \vec{S}_2),$$

where J is the isotropic-exchange constant and $\overleftrightarrow{g}$ is the g tensor of the dimer, which is practically equal to the g tensor of an isolated donor. If $J \gg g\mu_B B$ (which is the case for pair centers in our experiments), the system can be conveniently described in terms of the total spin, equal to 0 or 1. For a system with spin $S = 1$ the HF constants become twice as small compared to those of isolated shallow donors.

The exchange interaction energy increases exponentially with decreasing separation between isolated donors. This energy can be estimated from the formula derived for the exchange interaction between two atoms in a hydrogen molecule which was modified in semiconductor crystals with regard to the m^* and ε [156]

$$J = \left(\frac{m*}{\varepsilon^2}\right) A_H \exp[-B_H \left(\frac{m*}{\varepsilon}\right)^3 V],$$

where for AlN, $m^* = 0.33m_e$ is the average effective mass of a donors, $\varepsilon = 9.14$ is the dielectric constant, $A_H = 9.66$ eV and $B_H = 7.84 \times 10^{22}$ cm^{-3} are coefficients calculated for the hydrogen molecule, and $V = 4/3\ \pi r^3$ with r being the separation between the interacting donor atoms. For AlN, we obtain $J = 0.0382\exp(-3.674 \cdot 10^{18} \cdot V)$.

To find J, the spin-lattice relaxation rate was measured for different temperatures. Experimental results for the spin-lattice relaxation rate $1/T_1$ for ESE signal of shallow donors in AlN measured at 95 GHz after cooling down in the dark (filled circles) and after 10 min light illumination (open circles) with B perpendicular to the *c* axis in the temperature range from 1.5 to 5 K are presented in Fig. 4.27a. The data are fitted accurately by the formula $1/T_1$ [ms^{-1}] = 1.1 for dark signal (solid line) and $1/T_1$ [ms^{-1}] = $0.12 + 4 * 10^6 * \exp(-34/T_1)$ for light-induced ESE signal (dashed line), showing that at higher temperature relaxation is dominated by an Orbach process [157] for light-induced ESE signal and at lower temperatures by the direct process in the both cases. At the lowest temperatures the relaxation rate is chiefly due to the spontaneous emission of phonons.

Suppose the exchange coupled SD pair has a set of energy levels such as that shown in Fig. 4.27b, where there are three low-laying spin sublevels of a triplet state and an excited singlet state, whose energy less than the maximum phonon energy. It is then possible for a center, say, higher triplet sublevels to absorb a phonon of the appropriate frequency by a direct process, and be excited to the excited singlet state. In this state it emits a second phonon by spontaneous or induced emission and falls down to the lowest triplet sublevel. This gives an indirect transfer of center from the upper sublevels of the triplet state to the lowest sublevel, and constitutes a relaxation process that may be faster than the direct transfer between these sublevels because of much higher density of phonons of

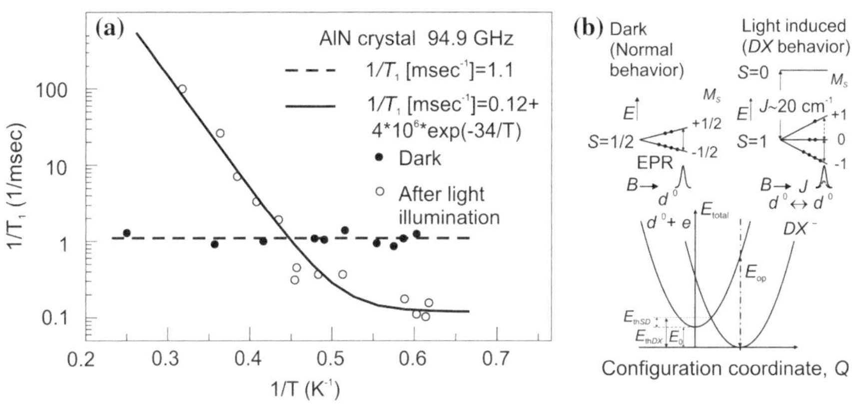

Fig. 4.27 **a** The temperature dependence of spin-lattice relaxation time T_1 for ESE signal of shallow donors in AlN measured at 94.9 GHz after cooling down in the dark (*filled circles, black*) and after 10 min light illumination (*open circles, red*). The *dashed* and *bold lines* are a fit of the temperature dependences of spin-lattice relaxation time T_1 for dark and light-induced ESE signals, respectively. **b** (*top*) Energy levels diagrams for normal shallow donors in AlN measured after cooling down in the dark (S = 1/2) and the light-induced shallow donor pairs coupled by exchange interaction J ≈ 20 cm^{-1} with DX behavior. A schematic diagram of the positions of the singlet and triplet states of the shallow donor pair as a function of the magnetic field is presented. The *solid arrow* and *solid EPR line* indicate the observed transition and the signal for the lowest populated energy levels $M_S = -1 \leftrightarrow M_S = 0$, the *broken arrow* and the *broken EPR line* represent the transition and the signal for upper less populated levels $M_S = 0 \leftrightarrow M_S = +1$. The Boltzmann distribution of the populations of the levels is symbolically indicated by different numbers of *filled circles*. (*Bottom*) Configuration-coordinate diagram for DX^- centers and shallow donor pairs d^0 coupled by the exchange interaction in AlN

energy between the triplet and the singlet states. Taking the exchange energy J as a fitting parameter, satisfactory agreement between simulation and experimental data can be achieved for $J \cong 20\ \mathrm{cm}^{-1}$. For this value the corresponding separation between the interaction donors was found to be $\sim$5.5 nm.

In Fig. 4.27b (top), the energy levels diagrams for normal shallow donors measured after cooling down in the dark ($S = 1/2$) and the light induced shallow donors with DX behavior are shown. A schematic diagram of the positions of the singlet and triplet states of the shallow donor pair coupled by exchange interaction as a function of the magnetic field are presented in the central part of Fig. 4.27b. The solid arrow and solid EPR line indicate the observed transition and the signal for the lowest populated energy level, the broken arrow and EPR line represent the transition and the signal for upper less populated levels. The Boltzmann distribution of the populations of the levels is symbolically indicated by different numbers of filled circles. There is no chance to separate the both transitions in EPR, however, they could be easily separated in the high-frequency ENDOR spectra.

Figure 4.27b (bottom) shows a configuration-coordinate diagram for DX^- centers and shallow donor pairs d^0 coupled by the exchange interaction in AlN. The low parabola represents the DX^- state. DX^- center which is occupied by two electrons is more stable then the shallow donor pairs that are in a meta-stable state with energy E_0. Due to strong coupling between the electronic and vibrational systems the donor level with two electrons drops deep into the gap forming DX^- center. Large Stokes shift between its optical ionization energy (E_{optic}) and thermal ionization energy ($E_{\mathrm{therm}DX}$) is observed for DX^- centers. The illumination at low temperature transforms the stable d^+ and DX^- states into meta-stable state of two shallow donors d^0 coupled by the exchange interaction due capturing a free electron by ionized donor (d^+ state). These d^0 states which represent the upper parabola generate the observed EPR signal of shallow donors with $S = 1$. The light induced EPR signal of SD vanishes above 200 K. The metastable state is separated from the d^+ and DX^- state by an energy barrier E_{thermSD} which prevents the DX^--formation reaction to return back to the stable state. When the thermal energy is high enough to overcome the barrier E_{thermSD} (200 K) the EPR signal disappears.

X-band (9.5 GHz) continious wave (CW) experiments. Two shallow donors (presumably oxygen located on the nitrogen site and carbon located on the aluminum site) are suggested to exhibit the DX-relaxation. Third shallow donor (presumably silicon on the Al site) shows the shallow donor behavior up to the room temperature and can be observed without light excitation at temperatures above 200 K. The values of the Bohr radius of the shallow donors are estimated to be $\sim$1.5 nm [158]. X-band EPR spectra recorded at 15 K after light illumination in samples #1 and #3 for perpendicular and parallel orientations of magnetic field with respect to the hexagonal c-axis (B⊥c) are shown in Fig. 4.28a. Without light illumination the EPR signals in samples #1 and #3 were not observed. Strong EPR signals appeared after light illumination with a light at the wavelengths shorter than 650 nm in samples #1 and #3 with identical lightly anisotropic g-factors typical for the shallow donors in AlN and strongly different anisotropic EPR linewidth ΔB.

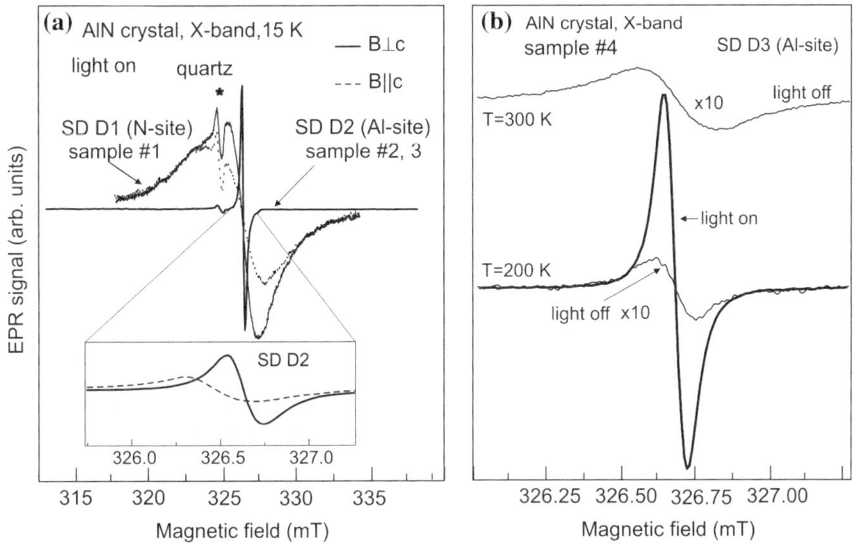

Fig. 4.28 **a** X-band EPR spectra recorded in samples #1 and #3 (#2) at 15 K after light illumination for perpendicular orientation of magnetic field with respect to the c-axis, B⊥c (*solid line*) and parallel orientation, B||c (*dashed line*). Note, D2 shallow donors were observed in sample #2 together with V_N color centers. The reference signal from quartz is marked with an *asterisk*; (*inset*) Expanded scale EPR spectra of D2 shallow donors recorded in sample #3 for B⊥c and B||c. b X-band EPR spectra of D3 SD in sample #4 for B||c orientation. *Upper spectrum* was recorded at room temperature without light illumination and is depicted with x10 magnification. *Bottom spectra* were recorded at T = 200 K without light illumination, and under light illumination

Expanded scale EPR spectra of D2 shallow donors recorded in sample #3 for B⊥c and B||c are shown in Fig. 4.28a (inset).

Observed EPR spectra of SDs are characterized by different EPR linewidths. Since the EPR linewidth is mainly determined by unresolved hyperfine interactions with ligands (Al or N), it is possible to determine the position of the impurity in the crystal lattice. The larger gyromagnetic ratio and spin of ^{27}Al nucleus indicate that the HF interaction with the Al nuclei should dominate and the effects caused by the N nuclei could be neglected. Thus, the main contribution to the linewidth is due to the interaction with ^{27}Al. The EPR linewidth of the D1 SDs is much larger compared with that of D2 SDs. For this reason one believes that the Coulombic center of the light-induced D1 SDs is located on the N site and; on the contrary, the Coulombic center of the light-induced D2 SDs is located on the Al site.

Observed EPR spectra can be described with the conventional spin Hamiltonian for the Zeeman interaction of the SD (S = ½, anisotropic g-factor for axial symmetry along the c-axis of the crystal). The HF interactions with the ligand nuclear spins are not resolved and contribute to the EPR linewidth ΔB: Parameters of the EPR spectra of D1 and D2 SDs at 15 K are shown below:

D1: $g_{||} = 1.986$, $g_{\perp} = 1.985$, $\Delta B_{||} = 4.1$ mT, $\Delta B_{\perp} = 2.3$ mT;
D2: $g_{||} = 1.992$, $g_{\perp} = 1.991$, $\Delta B_{||} = 0.7$ mT, $\Delta B_{\perp} = 0.3$ mT.
The linewidth of the EPR spectra depicted in Fig. 4.28a changes along with the rotation of the magnetic field from the c-axis to the basal plane.

These light-induced EPR signals, once excited at low temperature, persisted after switching off the light and disappeared only after sample heating. Using different edge filters to cut off the high-energy end of the lamp spectrum, the dependencies of the EPR signal intensity on the photon energy were measured for the SDs in sample #1 and in sample #3 (#2). The intensity of the EPR signal of the SDs remains almost constant up to 30 K and then quickly drops below the detection limit. Thus, it is very possible that these SDs are forming the DX center and showing a negative correlation energy U: That is initially these SDs are neutral and paramagnetic. Due to instability of the system it becomes energetically more favorable to move the electron from one donor (leaving it in a non-paramagnetic d^+ charge state) to another donor; the latter transforms into non-paramagnetic deep-level $DX^-(d^-)$ center with two electrons. Due to the strong coupling between the electronic and vibrational systems the energy level of the DX^- center drops deep into the bandgap, forming a stable non-paramagnetic deep donor state.

4.3.4.2 Shallow Donors with Normal Behavior

DX relaxation of shallow donors to a deep state makes these shallow donors unsuitable for creation of the n-type conductivity in AlN crystals at room temperature. Naturally, the search for shallow donors with a stable state at room temperature is a major challenge in materials based on AlN. The impurities responsible for the creation of D1 and D2 SDs apparently are not good candidates for applications. A number of works were devoted to elucidating the suitability of the silicon impurity to create a shallow donor in AlN. As mentioned earlier, contradictory arguments were presented in literature in relation to the DX-relaxation of the Si shallow donors. It was assumed that the EPR spectra characterized by a very narrow line ($\sim$0.1 mT) with isotropic g-factor and temperature-dependent linewidth belong to the silicon impurity [127, 128, 132]. These EPR spectra were observed only after optical excitation at low temperature and disappeared at temperatures of about 50 K. Thus, in order to explain these results, two models were proposed: (1) DX-relaxation of the Si center [127] and (2) the compensation of the donors by acceptors in the lower half of the band-gap, without the assumption of Si DX center formation [132]. In this paper we present results that confirm that the EPR spectra of these centers can be observed at room temperature (RT) without optical excitation, so that the DX relaxation model for Si-related SDs can be ruled out from consideration.

Figure 4.28b shows X-band EPR spectra of SDs (labelled as D3) in sample #4 for B||c. Upper spectrum was recorded at room temperature without light illumination and is depicted with $\times$10 magnification. Bottom spectra were recorded at T = 200 K without light illumination, and under light illumination. Decay time of the light-induced EPR signal intensity at 200 K to its original value after the illumination was switched off is about of 20 s.

The EPR linewidth of D3 SDs depends on the temperature and the intensity of the optical excitation. The EPR linewidth decreases with temperature increase in the range of 4–30 K from 0.2 mT at 4 K to a value below 0.1 mT at 30 K. The EPR signal is identical to that reported in [127, 128] for Si-doped AlN films, grown by plasma-induced molecular beam epitaxy and interpreted as arising from a silicon donor band. A decrease in linewidth with the temperature increase has been observed for donors in other semiconductors [159] and this effect was attributed to an averaging of the ligand hyperfine interactions through a motion of the electron from one donor site to another or to the enhancement of the exchange interaction between donors.

The identification of the binding core of the SDs labelled as D1, D2 and D3 from the EPR measurements was unsuccessful since the HF structure was not resolved. As was mentioned before, one believes that the Coulombic center of the light-induced D1 SD is located on the N site and a possible candidate then might be the oxygen (O_N). In contrast, the Coulombic centers of the D2 and D3 SDs are located on the Al site and are proposed to be the carbon and silicon in Al position (C_{Al} and Si_{Al}), respectively, as possible candidates.

4.4 Transition and Rare-Earth Elements Impurities in SiC, GaN and AlN Crystals

4.4.1 Transition-Metal Impurities in SiC

Transition-metal elements ($3d^n$, $4d^n$, $5d^n$) can exist in SiC as residual impurities and create deep levels in the band gap. As a rule, each impurity enters the crystal in multiple charge states, underlining their role as deep level defects in SiC, and affects substantially the electrical and optic characteristics of the material. Their electrical activity can be of importance already at very low defect concentrations. Sensitive methods are therefore required to identify and characterize a given transition metal in semiconductor host. Since transition metals have an unfilled d-shell and unpaired electronic spin, magnetic resonance techniques, as electron paramagnetic resonance (EPR) and optically detected magnetic resonance (ODMR), are ideally suited for this purpose. Controlled incorporation of transition-metal impurities appears very promising for development of semi-insulating substrates in device fabrication. An encyclopedic knowledge about incorporation of various transition elements in the silicon lattice has been elaborated by EPR and other techniques [160, 161]. Most of the available information about the transition metals arises from EPR experiments. Ludwig and Woodbury initiated a systematic study of the 3d transition-element impurities in silicon, using EPR and ENDOR, several charge states were observed and a simple physical picture of their properties emerged [162]. Starting from the free ion $3d^n4s^m$ configuration for a particular charge state, all $n + m$ electrons go into these orbitals for the non-bonding interstitial case. Substitutional impurities require four electrons to complete their bonds to the four silicon neighbors, $n + m - 4$ remain to go into the 3d orbitals. In both cases, the levels are filled according to Hund's rule, electrons paired with maximum spin S, first filling the lower level, spin-up, then the upper, spin-up, before filling, spin-down, in the lower, etc. The repulsive electron–electron interactions between the localized 3d orbitals, which force maximum spin, therefore dominate over the crystal-field energy.

Research on transition-element impurities in SiC before 2000 have been published in several reviews: 3d-group (Sc, Ti, V, Cr) and 4d-group (Mo) transition ions were investigated in SiC using EPR and ODMR [163, 164]. Table 4.8 lists transition-metal elements in various charge states which have been investigated in SiC [165–169].

Here it should be pointed out that all available data are consistent with the assumption that impurity ions of transition elements substitute for silicon in SiC. The opinions relating to SiC are rather contradictory, and they draw primarily on the fact that transition elements occupy in silicon preferably interstitial sites.

Chromium. Studies of 6H-SiC:Cr crystals reveal EPR signals due to chromium ions in the charge states Cr^{3+} ($3d^3$, S = 3/2), and Cr^{2+} ($3d^4$, S = 2). The chromium was identified from the hyperfine structure of the ^{53}Cr nuclei (natural abundance 9.5%, nuclear spin I = 3/2). The 6H-SiC lattice has three inequivalent sites, namely, a hexagonal (h) and two quasicubic ones (k1 and k2). The Cr^{3+} EPR signals (Fig. 4.29a) observed originated from two inequivalent lattice sites (apparently k1 and k2). The HF structure was reliably identificated only in the B ∥ c orientation and was found to be: for the k1 site $A_{\parallel} = 8.67 \cdot 10^{-4}\ cm^{-1}$, and for k2 site $A_{\parallel} = 9.11 \cdot 10^{-4}\ cm^{-1}$.

Anisotropy of the EPR spectra is characteristic for an S = 3/2 system in a strong axial crystalline field. In terms of real spin S = 3/2 EPR spectrum can be described by a spin Hamiltonian of the form

Table 4.8 Transition-metal impurities in different charge states studied in SiC crystals (the charge states with zero spin were studied by indirect methods by monitoring a change of the EPR spectra in the process of an optical illumination) [5]

Free atom	Sc ($3d^14\ s^2$)	Ti ($3d^24s^2$)	V ($3d^34s^2$)	Cr ($3d^54\ s^1$)	Mo ($4d^55s^1$)	Ta ($5d^36s^2$)
A^{2-}	$Sc^{2+}(3d^1)$ S = 1/2			$Cr^{2+}(3d^4)$ S = 2		
A^-	$Sc^{3+}(3d^0)$ S = 0	$Ti^{3+}(3d^1)$ S = 1/2	$V^{3+}(3d^2)$ S = 1	$Cr^{3+}(3d^3)$ off-center S = 3/2	$Mo^{3+}(4d^3)$ S = 3/2	$Ta^{3+}(5d^2)$? S = 1
A^0	Sc acceptor S = 1/2	$Ti^{4+}(3d^0)$ S = 0	$V^{4+}(3d^1)$ S = 1/2	$Cr^{4+}(3d^2)$ S = 1	$Mo^{4+}(4d^2)$ S = 1	
A^+			$V^{5+}(3d^0)$ S = 0	$Cr^{5+}(3d^1)$ S = 1/2	$Mo^{5+}(4d^1)$ S = 1/2	

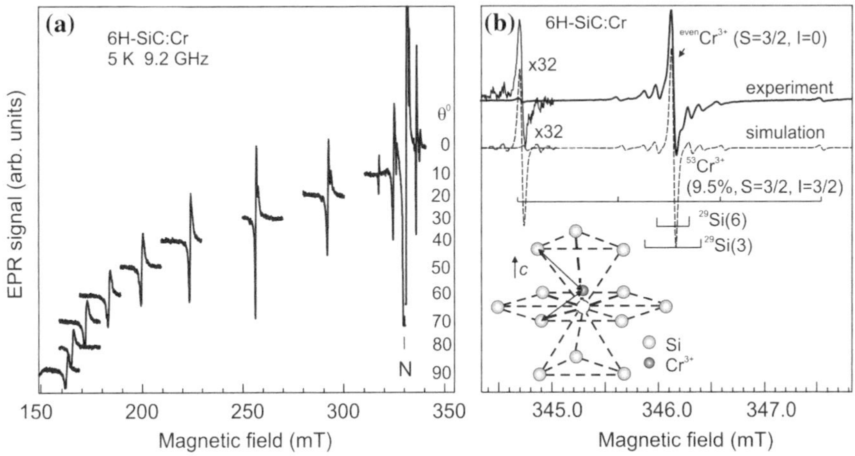

Fig. 4.29 **a** Angular dependence of the X-band EPR spectra of Cr^{3+} in 6H-SiC at 5 K. **b** EPR spectrum for one of the quasicubic sites of Cr^{3+} ions in 6H-SiC obtained at 9.5 GHz and 4.5 K with the magnetic field parallel to the hexagonal axis of the crystal. Shown below is a simulated spectrum calculated with the following parameters: HF structure constant for ^{53}Cr of 26.5 MHz, SHF structure constant with ^{29}Si for six equivalent Si atoms in the second coordination sphere of 8.4 MHz, and SHF structure constant with ^{29}Si for three equivalent Si atoms in the second coordination sphere of 14.56 MHz. (*Inset*) Model for the Cr^{3+} off-center position

$$H = g_{\|}\mu_B B_z S_z + g_{\perp}\mu_B(B_x S_x + B_y S_y) + D\left[S_z^2 - 1/3S(S+1)\right] + \boldsymbol{S}\cdot\boldsymbol{A}_{\mathrm{TM}}\cdot\boldsymbol{I}_{\mathrm{TM}} + \Sigma\boldsymbol{S}\cdot\boldsymbol{A}_{\mathrm{i}}\cdot\boldsymbol{I}_{\mathrm{i}} \tag{4.12}$$

with $S = 3/2$, z denotes the c-axis of the crystal (principal axis of the centre); D is the axial fine structure parameter, $\boldsymbol{A}_{\mathrm{TM}}$ is tensor for the HF interaction with odd isotopes of a transition metal (TM), $\boldsymbol{I}_{\mathrm{TM}}$—nuclear spin momentum for odd isotopes of a TM, $\boldsymbol{A}_{\mathrm{i}}$—is tensor for the HF interaction with nuclear spin momentum $\boldsymbol{I}_{\mathrm{i}}$ of i-th ligand atom: ^{29}Si (4.67%, I = 1/2) or ^{13}C (1.11%, I = 1/2).

It is possible to use effective spin approximation since the magnitude of the zero-field splitting 2D is much larger than the microwave energy at the X-band (the strong zero-field limit): $2D \gg g\mu_B B$. Thus only transitions within the $M_S = \pm 1/2$ Kramers doublet can be detected, which is described by an effective spin S′ = ½ and the spin Hamiltonian with effective g′ factor

$$H = g'_{\|}\mu_B B_z S'_z + g'_{\perp}\mu_B(B_x S'_x + B_y S'_y), \tag{4.13}$$

where g' is the effective g-factor and S' is an effective spin $S' = 1/2$. From experimental data the following values of effective g factor have been found: $g_{\|}' = 4.00$, $g_{\perp}' = 1.97$ for the k1 site and $g_{\|}' = 4.02$, $g_{\perp}' = 1.96$ for the k2 site. The experimental values g′ are related to the true values of the g factor for S = 3/2 in the following way: $g_{\parallel}' = g_{\parallel}$, $g_{\perp}' = 2\, g_{\perp}[1 - (3/16)(h\nu/2D)^2]$. It was estimated that D > 40 GHz.

For illustration of the HF interactions consider the Cr^{3+} ion whose EPR spectrum [165] is well accounted for by the off-center position of chromium at the silicon site (Fig. 4.29b). The observed superhyperfine (SHF) structure (hyperfine structure with ligand atoms in environments) can be explained as due to interaction with six equivalent Si atoms and three equivalent Si atoms in the second coordination sphere (Fig. 4.29b, inset). This arrangement appears when the chromium ion is shifted along the c axis. No deviation from the central position was found for Cr^{2+} ions (which were observed only in 6H-SiC crystals grown on the C side) with an HF constant of 15.8 G, which is ~1.7 times larger than that for Cr^{3+} ion. No deviation from the central position was observed for other transition-metal ions either. In particular, in the case of Mo^{4+} ions one can isolate only one constant of superhyperfine interaction with 12 equivalent Si atoms in the second coordination sphere, which was found to be 8.2 G (23 MHz).

Besides chromium ions in SiC in a regular environment, EPR spectra of complexes with a spin $S \nabla 3/2$, which apparently contain chromium with local axes along the Si–C bonds were observed [170]. In addition, EPR spectra that can be assigned to Ta^{3+} ions with an HF constant of the order of 150 G, and they were found to correlate with five IR luminescence lines with wavelengths at 5 K of 1.074, 1.049, 1.031, 1.011, and 0.999 eV.

Iron. The first observation of iron impurity EPR in SiC:Fe crystals were published in [171, 172]. Iron exists in Fe^{3+} ($3d^5$) charge state with electron spin S = 5/2 and seems to occupy silicon sites in the 6H-SiC lattice. The EPR spectrum of Fe^{3+} in SiC had the characteristic anisotropy of an S = 5/2 system in a strong axial crystalline field with fine structure parameter D = 0.25 cm^{-1}. The g-factor was found to be nearly isotropic with g = 1.99. Possibility of using iron doping to obtain semi-insulating SiC crystals was discussed.

Iron-doped 6H-SiC bulk crystals were grown by the sublimation sandwich method [14, 15]. The growth process was carried out in Ar atmosphere at temperature 2500 °C. Growth rate was about 0.2 mm/hour. Doping with iron was performed during growth process. Crystals were of n-type conductivity due to presence of N-donors with concentration $N_D = 10^{17} cm^{-3}$. The samples have the shape of 4 × 8 mm platelet and were oriented for rotation in the {11–20} plane.

Figure 4.30a shows the EPR spectra of 6H-SiC crystal doped with iron, recorded in several orientations of crystal under rotation in {11–20} plane at 65 K. Angles θ represent angles between hexagonal c-axis and direction of magnetic field B, θ = 0 means B ∥ c.

One can see that spectra consist of a great number of EPR lines with different angular dependencies. The most interesting feature in these spectra is a line marked with an arrow in orientation B⊥c. It have characteristic anisotropy of $g_{\perp} = 6$, $g_{\parallel} = 2$ and may arise due to a paramagnetic center with S = 5/2 in strong axial crystalline field. Vanishing of this signal at angles smaller than 20° is also consistent with this assumption. As will be shown later other signals connected with

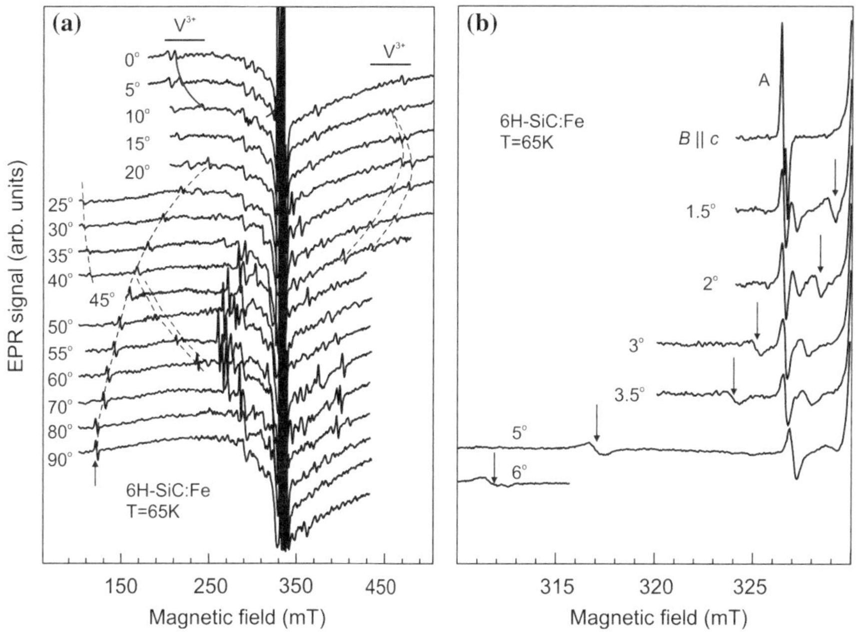

Fig. 4.30 **a** EPR spectra in different orientations of the 6H-SiC:Fe crystal at 65 K. Lines connect the signals that arise due to different transitions within a 3d-shell of Fe3+ ion. Positions of the two groups of lines of the V^{3+} impurity in one lattice site of 6H-SIC are marked for B ‖ c. **b** EPR spectra on a large scale at the angles close to B ‖ c. *Arrows* mark a signal that belong to Fe^{3+}. A strong line at 330mT belongs to nitrogen donors

lines in Fig. 4.30a belong to other transitions of the same center. Since crystal which we have studied was doped with iron it is natural to ascribe these signals to Fe^{3+} ($^6S_{5/2}$, $3d^5$) impurity center with S = 5/2. Iron has one stable odd isotope ^{57}Fe with nuclear spin I = 1/2. The ^{57}Fe isotope, however, have low natural abundance of 2.15% and this value is too small to find hyperfine components due to interaction with nuclear spin of this isotope.

The EPR spectra on a large scale at the angles close to B ‖ c are shown in Fig. 4.30b. One can see a number of EPR lines which could not be seen in Fig. 4.30a, because they were masked by the signal of nitrogen (a strong line at 330mT belongs to nitrogen donors). One of these lines (marked with arrows) have very strong anisotropy and it's intensity lowers with increasing an angle this line is almost invisible at θ = 7°. Such behaviour is consistent with S = 5/2 center in strong axial crystalline field. We believe that this signal also belong to Fe^{3+} impurity center. Some signals in Fig. 4.30a arise due to the presence of V^{3+} (S = 1) impurity (see Table 4.8). Positions of the two groups of lines which belong to V^{3+} impurity in one quasi-cubic lattice site are marked for B ‖ c orientation in

Fig. 4.30a. Hyperfine splitting within groups caused by ^{51}V (nuclear spin I = 7/2) is difficult to observe on such small-scale figure.

Besides lines which belong to iron impurity in Fig. 4.30a one can see a number of EPR signals, that appear at angles greater than 15° on both sides of nitrogen signal. In Fig. 4.30b some additional EPR lines (marked as A) could also be observed. These signals will be discussed later. A strong wide isotropic line on which nitrogen signal is overlapped does not belong to the sample.

Experimentally measured angular dependence of the signals, which were ascribed to Fe^{3+}, is shown with circles in figure Fig. 4.31a. The magnetic field applied was rotated in the {11–20} plane. Angle θ is an angle between c-axis of 6H-SiC crystal and applied magnetic field.

This dependence of Fe^{3+} ($3d^5$) ion in axial (C3v) crystalline field can be described by the spin Hamiltonian of the form [173]:

$$\hat{H} = g\mu_B\vec{B} \cdot \hat{\vec{S}} + B_2^0O_2^0 + B_4^0O_4^0 + B_4^3O_4^3, \tag{4.14}$$

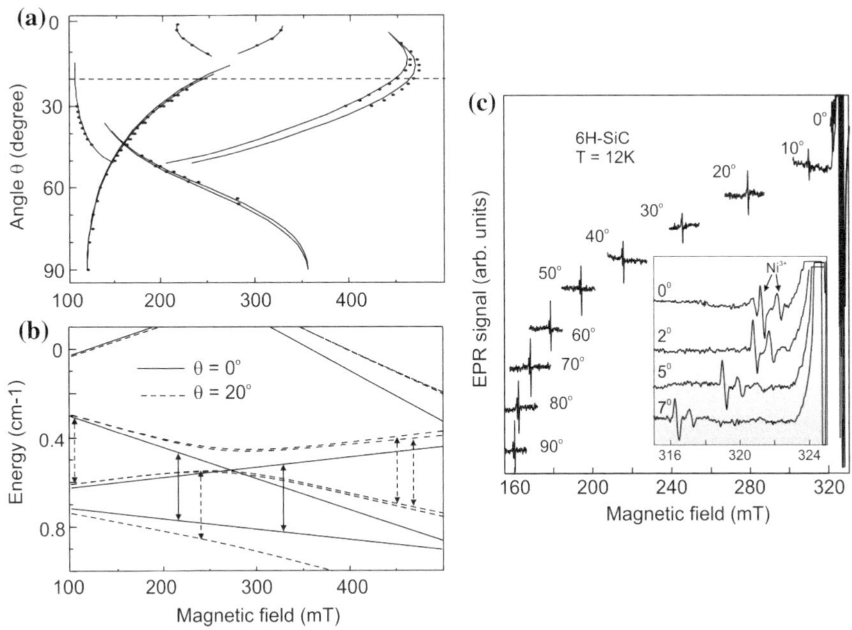

Fig. 4.31 **a** Experimental (*circles*) and theoretical (*lines*) angular dependence for Fe^{3+} EPR signals in 6H-SIC. Calculations were made using spin Hamiltonian (3) with parameters listed in Table 1. **b** Part of energy levels of the S = 5/2 center under combined action of the weak cubic and strong axial crystalline fields. Parameter D is supposed to be positive. *Solid and dashed lines* give energy levels and allowed transitions for θ = 0° and 20°, respectively. **c** Angular dependence of the EPR spectra of Ni^{3+} in 6H-SiC at 12 K after excitation by near-IR light. *Inset* shows part of the dependence on the expanded scale

where equivalent operators are given by:

$$\begin{aligned} O_2^0 &= 3S_z^2 - S(S+1) \\ O_4^0 &= 35S_z^4 - 30S(S+1)S_z^2 + 25S_z^2 - 6S(S+1) + 3S^2(S+1)^2 \\ O_3^4 &= 1/4\big(S_z(S_+^3 + S_-^3) + (S_+^3 + S_-^3)S_z\big). \end{aligned} \tag{4.15}$$

Here μ_B is the Bohr magneton, electron spin S = 5/2 and the electron g-factor is assumed to be isotropic. The z-axis is directed along hexagonal axis of the 6H-SiC crystal. This spin Hamiltonian can be rewritten for S = 5/2 in another form:

$$\hat{H} = g\mu_B\vec{B}\cdot\hat{\vec{S}} + \frac{1}{6}a(\hat{S}_\xi^4 + \hat{S}_\eta^4 + \hat{S}_\zeta^4 - \frac{707}{16}) + D(\hat{S}_z^2 - \frac{35}{12}) + \frac{7F}{36}(\hat{S}_z^4 - \frac{95}{14}\hat{S}_z^2 + \frac{81}{16}). \tag{4.16}$$

A system of coordinates, ξηζ, arises from three perpendicular cubic crystal-field axes of the fourth order with the center at the Si site of the SiC crystal. The z-axis is parallel to the hexagonal c-axis of the crystal (the same as the $\langle 111 \rangle$ axis of the ξηζ system) while x-axis lies in one of the {11–20} planes and y-axis is perpendicular to both x- and z-axes. Parameters D, F and a characterize the axial and cubic crystal fields and are related to parameters B_k^q as follows:

$$\begin{aligned} 3B_2^0 &= D \\ 60B_4^0 &= \frac{F-a}{3} \\ 60B_4^3 &= \frac{20}{3}\sqrt{2}a \end{aligned}$$

It should be noted that spin-Hamiltonian (4.16) has been used before to analyze the EPR spectra of Fe^{3+} in ZnO [174] and GaN [175, 176] which crystallize in the hexagonal (wurtzite) structure. Parameters of the spin Hamiltonian (4.16) for Fe^{3+} ion in 6H-SiC and GaN (GaN data presented for comparison) are given in Table 4.9.

In principle spin Hamiltonians (4.16) for iron should include a term which describe hyperfine interaction of unpaired electron with nuclear spin I = 1/2 of ^{57}Fe isotope. However we have not observed hyperfine splitting of Fe^{3+} lines because ^{57}Fe isotope have low natural abundance of 2.15% and hyperfine structure was not found in non-enriched samples.

Table 4.9 Parameters of the spin Hamiltonian (4.16) for Fe^{3+} ion in 6H-SiC and GaN (GaN data presented for comparison)

	6H-SiC	GaN
$g_\parallel$	1.99	1.990
$g_\perp$	1.99	1.997
$\lvert D\rvert$, 10^{-4} cm^{-1}	2500	713
$\lvert a - F\rvert$, 10^{-4} cm^{-1}	27	52
$\lvert a\rvert$, 10^{-4} cm^{-1}	164	48

In the absence of crystalline field the ground state of Fe^{3+} ion $^6S_{5/2}$ is sixfold degenerated according to electron spin S. Cubic field partially lift the degeneracy giving one doubly degenerated and one fourfold degenerated level. In strong axial field degeneracy is lifted, leaving three Kramers' doublets which correspond to the values of $M_S \pm 1/2$, $\pm 3/2$ and $\pm 5/2$.

Figure 4.31a shows experimental (circles) and theoretical (lines) angular dependence for Fe^{3+} EPR signals in 6H-SIC. Calculations were made using spin Hamiltonian (4.16) with parameters listed in Table 4.10. In Fig. 4.31b a part of the energy levels of the center with S = 5/2 under combined action of the weak cubic and strong axial crystalline fields is shown. Energies are given with respect to position of Fe^{3+} ($^6S_{5/2}$) free ion ground state level. Parameter D is supposed to be positive. Changing the sign of D will only inverse the order of energy levels and will not affect the EPR spectra. Solid lines in Fig. 4.31b give energy levels for $\theta = 0°$, while dashed lines represent levels for $\theta = 20°$. Solid and dashed vertical lines show allowed transitions for these orientations of the crystal and correspond to experimental points for $\theta = 0°$ and $\theta = 20°$ in Fig. 4.31b.

Table 4.10 Spin-Hamilton parameters of TM impurities in 4H-SiC, 6H-SiC and 15R-SiC ([163] and references therein)

Polytype/ion	Site	$g_{\parallel}$	$g_{\perp}$	D (10^{-4}) cm^{-1}	$A_{\parallel}$ (10^{-4}) cm^{-1}	$A_{\perp}$ (10^{-4}) cm^{-1}
4H-SiC						
$^{48}Ti^{3+}(3d^1)$	h	1.705	0			
4H-SiC						
$V^{3+}(3d^2)$	k	1.963	1.976	884	57.7	59.0
	h	1.962	1.958	3459	57.7	63.0
$V^{4+}(3d^1)$	h	1.748	0		78.7	
6H-SiC						
$V^{3+}(3d^2)$	k1	1.963	1.980	244	81.7	59.0
	k2	1.961	1.960	991	58.4	58.7
	h	1.976	1.961	3560	54.0	63.0
$V^{4+}(3d^1)$	k1	1.967	1.937	–	64.0	86.4
	k2	1.946	1.937	–	63.4	86.4
	h	1.749	0	–	77.4	–
6H-SiC						
$Cr^{2+}(3d^4)$	k(?)	1.987	1.942	12850		
$Cr^{3+}(3d^4)$	k1 (?)	1.973	$g_{\perp}$*= 3.95	15230	8.84	
	k2 (?)	1.961	$g_{\perp}$* = 3.99	>13,300	9.11	

(continued)

Table 4.10 (continued)

Polytype/ion	Site	$g_{\parallel}$	$g_{\perp}$	D (10^{-4}) cm^{-1}	$A_{\parallel}$ (10^{-4}) cm^{-1}	$A_{\perp}$ (10^{-4}) cm^{-1}
4H-SiC						
$Mo^{4+}(4d^2)$	k	1.982	1.977	1098		
6H-SiC						
$Mo^{3+}(4d^3)$	k	1.945	$g_{\perp}^* = 3.94$	>20,000	22.0 ($\theta = 15°$)	
$Mo^{4+}(4d^2)$	k1	1.977	1.976	1018	30.5	30.4
	k2	1.975	1.977	1108	30.4	30.5
$^{ev}Mo^{5+}(4d^1)$		1.968	1.975	–	–	–
$^{95}Mo^{5+}(4d^1)$		1.968	1.975		23.0	5
$^{97}Mo^{5+}(4d^1)$		1.968	1.975		23.5	4
15R-SiC						
$Mo^{3+}(4d^3)$	k	1.945	$g_{\perp}^* = 3.94$	>20,000	26.3 ($\theta = 20°$)	
$Mo^{4+}(4d^2)$	k	1.978	1.983	902	29.6	29.6
	k′	1.978	1.980	1126	30.5	30.5
4H-SiC						
TiN-pair	k	$g_1\langle 0001\rangle = 1.941$ $g_2\langle 1010\rangle = 1.963$ $g_3\langle 1120\rangle = 1.939$			A (Ti) = 8.67 (B‖c)	
	h	$g_1\langle 0001\rangle = 1.706$ $g_2\langle 1010\rangle = 0$ $g_3\langle 1120\rangle = 0$			A (N) = 0.90 (B⊥c)	
6H-SiC						
TiN-pair	k1	$g_1\langle 0001\rangle = 1.943$ $g_2\langle 1010\rangle = 1.964$ $g_3\langle 1120\rangle = 1.940$				
	k2	$g_1\langle 0001\rangle = 1.930$ $g_2\langle 1010\rangle = 1.894$ $g_3\langle 1120\rangle = 1.879$				
	h	$g_1\langle 0001\rangle = 1.854$ $g_2\langle 1010\rangle = 1.856$ $g_3\langle 1120\rangle = 1.904$			A (Ti) = 9.07 (B‖c)	

An investigation of the angular dependence of the Fe^{3+} EPR spectrum allowed to find the best-fit parameters for spin Hamiltonian of (4.16). Results are listed in Table 4.10. The calculated angular dependence of fine-structure line positions of Fe^{3+} ions obtained at 9.25 GHz are plotted as lines in Fig. 4.31b. It is evident that calculated angular dependence is in good agreement with experimental points, thus proving that these signals are caused by S = 5/2 center.

One can see in Fig. 4.31a that fine structure lines of Fe^{3+} are split in arbitrary orientations of the crystal in respect to the magnetic field. Corresponding splitting of energy levels for $\theta = 20°$ can be seen in Fig. 4.31b. The doublet structure results from the difference in the crystalline electric fields at the two Si positions. These two positions could not be distinguished crystalographically but are magnetically non-equivalent, having different sets of cubic field axes $\xi\eta\zeta$. Thus the lines split into doublets because of the influence of the term *a* in the spin Hamiltonian of (4.16) on two Fe^{3+} ions with different cubic field axes. In orientations B ∥ c and B ⊥ c these sites are wholly equivalent and no splitting is observed. The EPR signals of Fe^{3+} could be observed at temperatures from 4 K up to room temperature with maximum intensity at about 100 K.

Here we summarize the main features of Fe^{3+} in 6H-SiC. In Table 4.10 we list parameters of the spin Hamiltonian (4.16) for 6H-SiC and GaN. Gallium nitride is a wide-bandgap semiconductor with hexagonal lattice structure similar to that of 6H-SiC.

Iron exists in Fe^{3+} charge state, electronic configuration $3d^5$, $^6S_{5/2}$. The g-factor is nearly isotropic and it's value is close to 2 in accordance with $^6S_{5/2}$ ground state. It is known that in crystals with strong covalence (as SiC and GaN) the g-factor is usually a little bit smaller than free electron g-factor.

The parameter of the axial crystalline field D have been found to be strong and much greater than cubic field parameter *a*. Small cubic crystalline field produce only weak influence on the position of the signals in the EPR spectrum. In particular it causes splitting of lines in angular dependence of the EPR spectra. The same splitting have been observed in [174] for orientation dependence of Fe^{3+} EPR signals in GaN. Considerable difference was observed in the values of axial crystal field parameter D: in SiC it is 3.5 times stronger than in GaN.

From EPR spectra observed in X-band at 65 K it is only possible to find relative signs of crystal field parameters. We have found that D, *a*-F and *a* have the same sign. As for iron in GaN, authors of [175, 176] managed to find signs from optically detected magnetic resonance (ODMR) measurements at 24 GHz and 1.5 K. At these conditions thermal depopulation of electronic Zeeman levels is prominent and affects intensities of ODMR signals. They have found that parameter D is negative while *a*-F and *a* are positive.

Since odd ^{57}Fe isotope has small natural abundance (2.15%) a ratio between the central and hyperfine components should be ~1/100. Thus observed intensities of the central components are too small to observe hyperfine lines due to ^{57}Fe. Expected hyperfine splitting is also expected to be small, about 8mT for a crystal with strong covalence bonding like SiC. This value is comparable with observed linewidths for iron signals and hyperfine splitting may not be resolved. In GaN [174, 175] hyperfine structure also was not observed since linewidth of the signals is even larger than in SiC.

One of the most important points is simultaneous presence of Fe^{3+} and V^{3+} signals in the EPR spectra of 6H-SiC. This implies that positions of energy levels of these ions in 6H-SiC bandgap are rather close. It is known that energy level position of V^{3+} in 6H-SiC bandgap is 0.6 eV below the conduction band [163]. So one may expect that Fe^{3+} produces energy level which also lies deep in the 6H-SiC bandgap. Furthermore, energy level of vanadium in V^{4+} charge state lies near the middle of the bandgap [163] and V^{4+} is used to produce semi-insulating SiC layers. Thus we can expect that if Fe^{4+} charge state is stable then it can also produce energy level in the middle of the 6H-SiC bandgap. This will allow to use Fe^{4+} impurity to obtain semi-insulating 6H-SiC layers instead of vanadium.

In 6H-SiC there are three non-equivalent lattice sites: two quasi-cubic and one hexagonal one. However EPR of Fe^{3+} is observed for only one lattice site. This may imply that iron energy levels within the bandgap could depend significantly on the lattice site occupied. Similar situation was observed for Mo^{3+} and Mo^{4+} ions in 6H-SIC [172]. For the reasons mentioned in the above paragraph it is important to find EPR of iron in another charge state Fe^{4+} with $S = 2$. Up to now we have not found Fe^{4+} perhaps because position of the Fermi level prevent an EPR observation of iron in this charge state. Fe^{3+} ions were supposed to occupy silicon sites in 6H-SiC lattice.

In both semiconductors SiC and GaN EPR of Fe^{3+} can be observed up to room temperature, proofing that electronic configuration is $^6S_{5/2}$. This possibility is caused by very small spin-orbit interaction with excited states which lie very high above ground level. From intensities of iron EPR signals in SiC the concentration of iron impurity in the sample has been estimated to be approximately $10^{17}cm^{-3}$.

We will not give here detailed discussion of the nature of other EPR signals that were observed in the sample (see Fig. 4.30a, b). Further investigation is needed to clear this problem. It seems, however, that symmetry axes of the centers that cause these signals do not coincide with c-axis. For signals in Fig. 4.30a it is evident from the fact that angular dependencies of these lines have extreme positions not at $B \parallel c$ or $B \perp c$ but at approximately $\theta = 55°$. Thus we can assume that these signals are caused by centers oriented at about 55° to hexagonal axis of the crystal. Other signals marked as A in Fig. 4.30b again seem to have symmetry axis non-coincident with the c-axis. The fact that at $B \parallel c$ we have two nearly overlapping lines which split with increasing angle θ leads to the conclusion that we observe EPR signals from a center that has a number of orientations in the lattice which become equivalent when magnetic field is parallel to c-axis. Splitting at $B \parallel c$ may be caused by small misalignment of the crystal.

Important point to note is that all these low-symmetry signals have almost the same temperature dependence as Fe^{3+}. Thus we can assume that they contain iron impurity. We have found that linewidths of iron signals and of the lines that are visible in Fig. 4.30a at angles greater than 15° are of the same order (approximately 1mT). This is another evidence that these signals arise due to the centers that contain iron. On the other hand, lines marked as "A" in Fig. 4.30b have much smaller linewidth (≈0.05 mT) and thus belong to another center than the signals in Fig. 4.30a. Direct evidence could have been obtained from EPR measurements on ^{57}Fe-enriched samples in which hyperfine structure due to interaction with iron nuclear spin $I = 1/2$ would allow unambiguous identification of impurity.

Nickel. The first observation of nickel impurity EPR in SiC crystals were published in [172]. In the sample, where EPR signal of Fe^{3+} were found, other signals could be observed if the sample is illuminated with visible or infrared light. In Fig. 4.31c the EPR spectra observed in 6H-SiC:Fe crystal at 12 K in different orientations of the crystal with respect to external magnetic field under rotation in {11–20} plane are shown. The two strongly anisotropic EPR lines could be seen. The inset in Fig. 4.31c shows EPR spectra on expanded scale at angles close to B || c. Two lines is clearly seen in such large-scale figure.

The observed EPR spectra could be described in terms of effective spin $S' = 1/2$ with a spin Hamiltonian (4.13). From experimental data the following values of effective g-factors have been obtained: $g_{\|}'(1) = 2.032$, $g_{\|}'(2) = 2.026$ $g_{\perp}'(1) = 4.10$, $g_{\perp}'(2) = 4.08$.

Anisotropy of the EPR spectra is characteristic for an S = 3/2 system in a strong axial crystalline field and positive g shift is consistent with the electron configuration d^7. It is possible to use effective spin approximation since the magnitude of the zero-field splitting is much larger than the microwave energy at the X-band and as a result only transitions within the $M_S = \pm 1/2$ Kramers doublet can be detected. In terms of real spin S = 3/2 EPR spectrum can be described by a spin Hamiltonian of the form (4.12) with $S = 3/2$.

Investigation of angular dependence of new signal allowed to find a best-fit parameters for spin Hamiltonian (4.12): $g_{\|}(1) = 2.032$, $g_{\|}(2) = 2.026$, $g_{\perp}(1) = 2.05$; $g_{\perp}(2) = 2.06$, $|D| > 1.5\ \text{cm}^{-1}$.

The signal could be observed at the temperatures from 4 K up to 55 K with maximum intensity at 12–20 K.

The question arises about the nature of the center. It should be an ion in d^7 electronic configuration with small natural abundance of odd isotopes, since no hyperfine structure was observed. Two candidates are possible: Fe^+ and Ni^{3+}. On the one hand, the crystal studied was doped with iron and Fe^{3+} EPR signal have been found in it. Incident light may cause recharging of iron ions. On the other hand, no change in Fe^{3+} EPR intensity was observed when the light is turned on. Moreover, Fe^+ charge state is not very probable for substitutional Fe ion, rather for interstitial one. No direct EPR evidence about interstitial transition ions in SiC is available up to now. At last, we have found this signal in the 4H-SiC crystal, heavily doped with Al where no trace of Fe^{3+} signal could be found. Thus we believe that this new signal belongs to Ni^{3+} ions. Nickel has one stable odd isotope ^{61}Ni (I = 3/2) with low natural abundance of 1.1%. So there is no surprise that no hyperfine interaction have been observed. We believe that two lines that were observed belong to Ni^{3+} impurity in two crystallographicaly non-equivalent lattice sites.

The most interesting feature of Ni^{3+} EPR signal is its photosensitivity. It has very low (nearly zero) intensity before illumination. Only irradiation with visible or near-IR light make it possible to detect EPR. After the signal appeared, its intensity does not change after the light is turned off at low temperatures. Heating of the sample up to the temperatures >100 K is needed to destroy the EPR signal. Investigation of dependence of EPR spectra intensity on wavelength of the light could give information about position of the energy level of Ni^{3+} in 6H-SIC bandgap. Now it is only possible to state that the Ni^{3+} signal have maximum intensity when exciting with 1000–1500 nm light.

Molybdenum. The EPR signals of molybdenum in SiC crystal have been observed in three charge states, Mo^{3+}, Mo^{4+} and Mo^{5+} [163]. The HF structure from interaction with the nuclear spins of odd isotopes of Mo was observed. Two Mo isotopes with nuclear spins: ^{95}Mo (15.9%, I = 5/2) and ^{97}Mo (9.6%, I = 5/2) permitted unambiguous chemical identification of the impurity; the electron spin value allowed to find the charge state of the Mo ion in SiC. Weakly n-type crystals exhibit signals due to Mo^{4+} ($4d^2$, S = 1) occupying the k1 and k2 sites, and in strongly n-type crystal only k2-site EPR signal can be detected, which shows that the Mo^{4+} impurity sitting at different sites produces levels with strongly different energies in the band-gap. The line positions can be described by spin Hamiltonian (4.12) for S = 1. Its parameters are for k1 site: $g\|$ = 1.977, $g_\perp$ = 1.976, |D| = 1018 · 10^{-4} cm^{-1}; for k2 site: $g\|$ = 1.975, $g_\perp$ = 1.977, |D| = 1108 · 10^{-4} cm^{-1}.

The EPR signals of Mo^{3+} ion ($4d^3$, S = 3/2) were observed in n-type crystals. The line positions can be described by spin Hamiltonian (4.12) for S = 3/2. Its parameters are for one site: $g\|$ = 1.945, $g_\perp$ = 1.969 and |D| > 2 cm^{-1}.

The EPR signals for Mo^{5+} ion ($4d^1$, S = 1/2) were also detected and can be described by the conventional spin Hamiltonian for S = ½ with the parameters $g\|$ = 1.9679, $g_\perp$ = 1.9747. In p-type SiC crystals Mo signals were not observed, which suggests that the nonparamagnetic Mo^{6+} state ($4d^0$, S = 0) is here in equilibrium.

Scandium. The two types of the EPR spectra which were assigned to scandium acceptors and Sc^{2+} ions observed in 6H-SiC crystals. The EPR spectra of scandium acceptors are characterized by a relatively small hyperfine interaction constants corresponding to their values of hyperfine coupling constants for other elements of Group III in the SiC—acceptors, boron, aluminum and gallium. The EPR spectra of scandium acceptors undergo significant changes in the temperature range 20–30 K. In the low-temperature (LT) phase of the EPR spectra are characterized by orthorhombic symmetry, while the high-temperature (HT) phase has a higher axial symmetry. EPR spectra appearing at temperatures above 40 K, and assigned to ions Sc^{2+} ($3d^1$, S = ½) or A^{2-} (see Table 4.8), have a substantially large hyperfine structure constants and narrower line compared with the EPR spectra of scandium acceptors. The parameters of the EPR spectra close to those of Sc^{2+} ($3d^1$) in ionic crystals and ZnS, then the parameters of the EPR spectra of scandium acceptors longer match the holes localized on the atoms of the third group. It is concluded that the scandium atoms in all the centers occupy the position of silicon.

At temperatures close to 40 K, one observed Sc acceptor signals Sca(HT) where HT stands here for high temperature and signals due to Sc^{2+} ions. Both centers have axial symmetry relative to the c axis and can be described by a spin Hamiltonian

$$H = \mu_B \boldsymbol{B} \cdot \boldsymbol{g} \cdot \boldsymbol{S} + \boldsymbol{S} \cdot \boldsymbol{A} \cdot \boldsymbol{I},$$

where S = ½ and $\boldsymbol{A}$ is the HF structure tensor for HF interaction with nuclear spin of ^{45}Sc isotope (100%, I = 7/2). The z axis is parallel to the c-axis of the 6H-SiC crystal. The Sc impurity was identified from the HF structure of the spectra caused by interaction with ^{45}Sc isotope. The spin Hamiltonian parameters were found for Sc-related EPR spectra.

The Sc_a(HT) acceptor has the following parameters: $g_{\|} = 2.0016$, $g_{\perp} = 2.0011$, $A_{\|} = 10.1 \cdot 10^{-4}$ cm^{-1}, $A_{\perp} = 22.6 \cdot 10^{-4}$ cm^{-1}. For Sc^{2+} the parameters are: $g_{\|} = 2.0047$, $g_{\perp} = 2.0002$, $A_{\|} = 44.6 \cdot 10^{-4}$ cm^{-1}, $A_{\perp} = 8.4 \cdot 10^{-4}$ cm^{-1}.

As the temperature is lowered to 4 K, these spectra disappear to be replaced by at least two different EPR signals due to lower-symmetry Sc acceptors, denoted by Sc_a(LT) and Sc_{a1}(LT) with LT standing for low temperature. The parameters of Sc_a(LT) are: $g_x = 2.001$, $g_y = 2.016$, $g_z = 2.008$, $A_x = 15.0 \cdot 10^{-4}$ cm^{-1}, $A_y = 3.0 \cdot 10^{-4}$ cm^{-1}, $A_z = 5.0 \cdot 10^{-4}$ cm^{-1}. The Sc_a(LT) spectra were not studied thoroughly, but qualitatively the local axes of the corresponding centers can be identified as directed along the Si–C bonds which do not coincide with the c-axis. The low symmetry of the low-temperature Sc signals may be due to spin-density redistribution and, possibly, to off-center position of Sc atom in Si position.

Niobium. Niobium-related impurity centers have been identified in [177]. An analysis of the HF interaction with ^{93}Nb, ^{29}Si and g-factors for Nb-related EPR spectra in 4H- and 6H-SiC was presented. Comparing HF data obtained by EPR and by HF calculations for the neutral isolated substitutional Nb, Nb^0_{Si}, and the Nb-vacancy complex provides support for the identification of the defect as $Nb_{Si}V^0_C$.

4.4.2 Transition-Metal Impurities in AlN and GaN

Aluminum Nitride. EPR spectra in different orientations of the magnetic field with respect to the c-axis of the AlN crystal were recorded at X-band (9.3 GHz) [178]. The angular dependences upon rotating the magnetic field in a (11–20) plane is shown in Fig. 4.32a. A distinct feature of these spectra is a strong line in the B⊥c (90°) orientation that shifts to lower magnetic fields with the rotation of the crystal toward B ∥ c (0°) orientation. This line is marked as L1 for θ = 90°. The positions of this signal in different orientations of the sample are connected in Fig. 4.32a with dashed line. Along with the movement, the intensity of this line decreases and it could not be observed at the angles smaller than 10° with respect to the c-axis. The line could be observed at the temperatures from 5 K up to the room temperature with maximum intensity at 40 K. The linewidth decreases from 4.5 mT in B⊥c orientation to 2.8 mT at θ = 25°. In the spectra (Fig. 4.32a) there are also three lines marked as L2, L3, L4 for θ = 95°. The nature of three additional lines in spectra (Fig. 4.32a) marked as L2, L3, L4 θ = 95° will be discussed later on.

The same crystal has been studied at the Q-band ‖35 GHz|. As shown in Fig. 4.32b the spectra have been recorded in several orientations of the sample with respect to the magnetic field. Two strong anisotropic lines dominate the spectra. The first ‖narrow| line shows the same type of the angular dependence as L1 signal in Fig. 4.32a and can be observed even at the room temperature. Its effective *g* factor is found to be equal at X-band and Q-band.

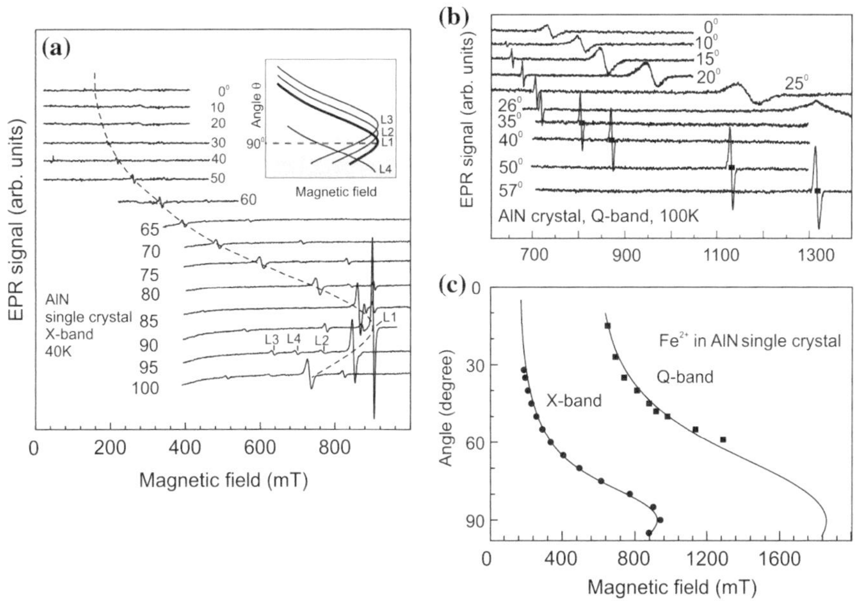

Fig. 4.32 **a** Experimental X-band EPR spectra in AlN single crystal at T = 40 K recorded in different orientations of the magnetic field with respect to the c-axis and simulated angle dependence for L1 line (*dashed line*), rotation in (11–20) plane. A rough sketch of the angular dependencies of the L1-L4 lines is shown in the *inset*. **b** Experimental angular dependence of EPR spectra in Q-band in AlN single crystal at T = 100 K. **c** Experimental data (*circles* and *squares*) and theoretical calculations (*lines*) of the angular dependences of L1 EPR signals arising due to the Fe^{2+} center detected in X-band and Q-band

The second (broad) line in Fig. 4.32b exhibits even stronger angular dependence. The range of our magnet allowed us to trace the angular dependence for angles smaller than 25°.

The L1 line angular dependences at X-band and Q-band shown in Figs. 5.32a, b are typical for S = 2 systems in an axial (C_{3v}) crystalline field and can be roughly described by the conventional spin Hamiltonian of the form

$$H = g_{||}\mu_B B_z S_z + g_{\perp}\mu_B (B_x S_x + B_y S_y) + D\left[S_z^2 - 1/3S(S+1)\right], \qquad (4.12')$$

where z is directed along the hexagonal c-axis of the crystal, parameter D characterizes the axial crystal field. The large D value leads within the S = 2 manifold to a splitting into the spin states $M_S = 0, \pm1, \pm2$. The EPR spectra consist of one line which corresponds to the $M_S = -1 \leftrightarrow M_S = 1$ transition within the ground state. The investigation of the angular dependence of the spectrum allowed us to find the best-fit parameters for spin Hamiltonian: $g_{||} = 2.003$, $g_{\perp} = 2.12$, $D = 7.93\ \text{cm}^{-1}$. In Fig. 4.32c measured (circles for X-band and squares for Q-band) and calculated (lines) angular dependencies of the L1 signal are plotted, showing a good agreement between experimental and calculated dependencies proving accuracy of found spin Hamiltonian parameters.

Discussing the origin of the L1 signal one should consider transition metal ions with an unfilled *d*-shell. The $S = 2$ value and positive g shift is consistent with the electron configuration d^6. Taking into account that iron is a common trace impurity in GaN crystals, one can assume that the observed signal belongs to the non-Kramers ion $Fe^{2+}(3d^6)$ with $S = 2$. Iron has one stable odd isotope ^{57}Fe with nuclear spin $I = 1/2$. The ^{57}Fe isotope, however, has a low natural abundance of 2.15% and thus the observation of hyperfine ||HF) components arising due to the interaction with the nuclear spin of this isotope becomes complicated. The expected HF splitting is smaller than the observed linewidths for the iron signal and the HF splitting may not be resolved.

In an axially distorted tetrahedral environment the free ion ground state ^5D of $Fe^{2+}(3d^6)$ is split by the action of the cubic field, the axial field and spin-orbit interaction as shown in Fig. 4.33a. At a site of tetrahedral symmetry (Td), the lowest free ion term, ^5D, of Fe^{2+} is split by the crystalline field into an orbital doublet ^5E and triplet $^5T_2$, crystal-field theory predicts that ^5E is the ground state. The axial field splits ^5E state leaving the orbital singlet state ^5A as the lowest. The further splitting is caused by the spin-orbit and the Zeemann interactions. The energy level diagram for the angle of 10° and the EPR transitions at X-band and Q-band are presented in the right panel of Fig. 4.33a.

The values of the g factor and zero-field splitting are in good agreement with those found in two II–IV–V compounds: $CdSiP_2$ and $ZnGeP_2$ [180, 181]. Moreover, Fe^{2+} signal could be observed at room temperature in AlN as well as in $CdSiP_2$ crystals [180].

Other lines of smaller intensity are clearly seen in the spectra in Fig. 4.32a, marked as L2, L3, and L4 for $\theta = 95°$. The angular dependencies of their positions, intensities and linewidths, their temperature dependencies are absolutely the same as for the L1 signal. A rough sketch of the angular dependencies of the L1–L4 lines is shown in the inset in Fig. 4.32a. These three lines seem to belong to the same

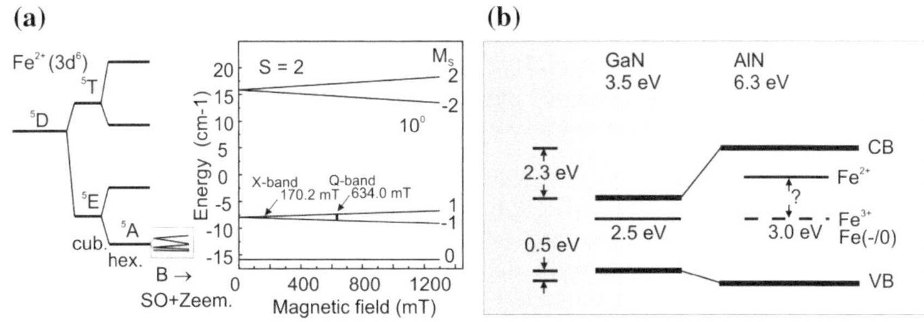

Fig. 4.33 **a** The energy level diagram for Fe^{2+} in the crystal field of AlN and Zeemann levels for angle of 10° and the EPR transitions at X-band and Q-band (the *right panel*). **b** Experimental Band offsets $\Delta E_V = 0.5 \pm 0.1$ eV and $\Delta E_C = 2.3 \pm 0.1$ eV of the GaN/AlN system, as predicted from the common (-/0) acceptor reference level of iron [176, 179]

center (like L1), localized in different parts of the sample. We believe that our sample consists of at least four parts (domains) that are misaligned with respect to the c-axis. Thus, the same center in different parts (domains) of the sample will give rise to the signals, which angular dependencies are shifted with respect to each other. The misalignments of the three parts are −5°, −10°, and +15° with respect to the c-axis of the largest part (domain) that gives rise to the most intense signal (L1).

The striking result is that the iron in investigated AlN crystal has the Fe^{2+} charge state in opposite to GaN compound where Fe^{3+} was observed [176, 179]. The non-Kramers ions Fe^{2+} is usually difficult to detect by EPR. It is also remarkable that their EPR spectrum in AlN crystals is still observable at the room temperature with relatively narrow linewidths. Figure 4.33b shows the valence-band (VB) off-set as $\Delta E_V = 0.5 \pm 0.1$ eV and conductive-band (CB) off-set as $\Delta E_C = 2.3 \pm 0.1$ eV of the GaN/AlN system, as predicted from the common Fe^{3+} (-/0) acceptor reference level of iron according to the Langer–Heinrich model, which is suggested to give a good prediction of the band offsets at an actual AlN/GaN interface [182]. In addition Heitz et al. [183] reported that the deep Fe 3+/2+ acceptor level is 3.17 eV above the VB maximum in GaN. Assuming band offsets in nitrides following the internal reference rule, the Fe 3+/2+ acceptor level is expected to be about 3.96 eV above the VB of AlN, where the VB offset for GaN/AlN is 0.79 eV [184]. The Fe^{2+} level in AlN could be located in the band gap in contrast to GaN where this level seems to be located in the CB (Fig. 4.33b).

At least three more anisotropic EPR signals have been observed in AlN crystal (sample #1) at low temperatures. Figure 4.34a shows an angular dependence of two of these signals in the X-band at 15 K. These EPR signals with linewidth of about 2 mT can be observed in narrow temperature range 5–15 K.

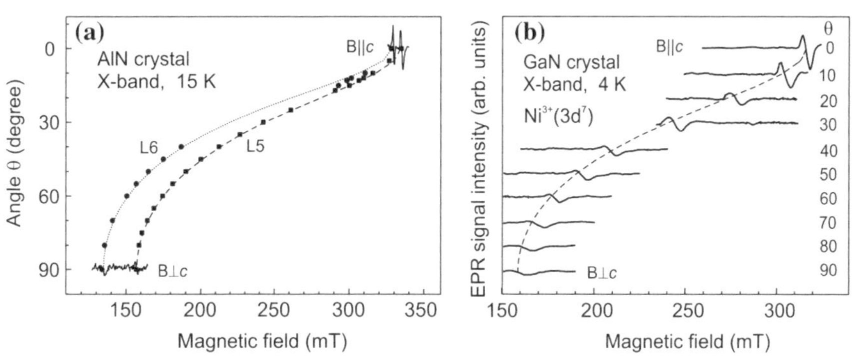

Fig. 4.34 **a** Angular dependences of L5 and L6 EPR signals in AlN single crystal, X-band, T = 15 K; theoretical angular dependences of L5, L6 lines, calculated using spin Hamiltonian of (4.13). For L5 center additional calculations were made calculated using spin Hamiltonian of (4.12′) with following parameters: $S = 3/2$, $g_{\|} = 1.975$, $g_{\perp} = 2.10$, $D = 2\ cm^{-1}$. **b** Angular dependence of Ni^{3+} EPR line for rotation of the GaN crystal in {11–20} plane. *Dashed line* was calculated using spin Hamiltonian parameters, listed in the text

These EPR spectra seem to have the characteristic anisotropy of S = 3/2 system in a strong axial crystalline field. In AlN a combination of trigonal and tetrahedral crystalline fields yields to the splitting of 4F ground term of the free ion by a cubic tetrahedral part of crystalline field into two orbital triplets $^4T_2$, $^4T_1$ and a ground-state orbital singlet $^4A_2$ with following split of $^4A_2$ ground state into two Kramers doublets by the combined action of the trigonal field and the spin-orbit interaction. Omitting the HF interactions, the EPR spectrum can be described by a spin Hamiltonian of the form (4.12′) with S = 3/2. Since the magnitude of the zero-field splitting 2D is much larger than the microwave energy at the X-band (the strong zero-field limit), only the transitions within the $M_S = \pm 1/2$ Kramers doublet can be detected. Practically, it is impossible to determine all parameters of the spin Hamiltonian of (4.12′) for a real magnetic field in the strong zero-field limit at the X-band and our purpose was only to estimate the zero-field splitting parameter. In the magnetic field range available in our EPR experiments (0–1.4 T), only the $M_S = ½ \leftrightarrow M_S = -1/2$ transitions were observed. Our analysis has shown that $D \geq 1.5\ cm^{-1}$.

The spin Hamiltonian of (4.12′) can be transformed in the strong zero-field limit as (4.13) with the effective g factor g′ and an effective spin S′ = ½. The effective g′ values are $g_{\|}' = 1.975$, $g_{\perp}' = 4.2$ for the L5 signal. For the L6 signal only $g_{\perp}'$ could be found with reasonable accuracy: 4.9 and 4.7, respectively. The $g_{\|}'$ values for L6 signal could only be estimated to be equal to 2.00 ± 0.02. Using the perturbation theory up to the third order, it was shown in [179] that experimental g′ values of transition within the $M_S = \pm 1/2$ Kramers doublet of an orbital singlet system with S = 3/2 give $g_{\|}' = g_{\|}$, $g_{\perp}' = 2\, g_{\perp}[1 - 3/16(h\nu/2D)^2]$. Dividing $g_{\perp}'$ by two gives a rough approximation of $g_{\perp}$. Figure 4.34a shows the theoretical angular dependence of the EPR spectra, calculated using spin Hamiltonian of (4.12′) with following parameters: S = 3/ 2, $g_{\|} = 1.975$ $g_{\perp} = 2.10$, $D = 2\ cm^{-1}$ (for L5 signal). The g shift, $\Delta g = g - g_e$ (g_e is the free-electron g factor), is given approximately by $\Delta g = -\ 8\lambda/10Dq$, where λ is the spin-orbit coupling constant (for Ni^{3+} free-ion λ is approximately $-238\ cm^{-1}$) and 10Dq is the energy splitting between $^4A_2$ and $^4T_2$. For tetrahedral symmetry additional terms, cubic in spin operators and linear in the magnetic field, should appear in the S = 3/2 spin Hamiltonian of (4.12′). However, these terms should be small and, therefore, will not modify the angular dependence predicted by the spin Hamiltonian of (4.13).

Attribution of the observed S = 3/2 signals is a matter of discussion. The HF structure of the signal could not be observed, perhaps due to the low signal intensity, so the most possible candidates should be transition metal ions with small abundance of odd isotopes with nonzero nuclear spin, i.e., $Cr^{3+}(3d^3)$ or $Ni^{3+}(3d^7)$. As chromium has only one stable isotope with nonzero nuclear spin, ^{53}Cr (I = 3/2, natural abundance 9.1%), the intensity of one line of HF structure should be approximately 2.5% of the central line that is caused by Cr isotopes with zero nuclear spin and taking into account the small intensity of the signal, it should be difficult to observe the HF structure arising from interaction with the ^{53}Cr nuclei. Considering nickel which also has only one stable odd isotope ^{61}Ni (I = 3/2, natural abundance 1.13%) the intensity of the HF components, as a fraction of the main line, is expected to be about 0.25% only and the HF structure could not be observed in case of weak central line.

To summarize, EPR measurements at 9.4 and 35 GHz ranges have been used to study defects in AlN single crystals grown by a sublimation sandwich method. Due to different source material two types of the crystals were grown: reddish colored sample and slightly amber colored sample. These studies reveal the presence of iron impurity in the reddish sample. The spectra of substitutional Fe^{2+} are highly anisotropic and could be observed up to the room temperature. The EPR spectra can be described by the conventional axial symmetry spin Hamiltonian with $S = 2$, $g_{\parallel} = 2.003$, $g_{\perp} = 2.12$, $D = 7.93\ cm^{-1}$. In addition to Fe^{2+} at least three anisotropic EPR signals have been observed, which have the characteristic anisotropy of an $S = 3/2$ system in a strong axial crystalline field and possible candidates are Cr^{3+} or Ni^{3+}.

Gallium Nitride. Gallium Nitride (GaN) is considered to be one of the most promising semiconductor materials for the construction of short-wavelength emitting devices, such as blue diodes and lasers [185]. In spite of impressive technological achievements of the last years, there are still substantial gaps in our knowledge of the basic physical properties of GaN. One of them concerns the identification of transition metal impurities and related complexes in GaN. The EPR proved to be a powerful method for the identification of transition metal impurities in semiconductors, as an example, Fe^{3+} ions were investigated in GaN using EPR [175, 176].

The sublimation sandwich-method was used for growing GaN crystals as thick as 0.1 mm [186, 187]. It was reported [188, 189], that these crystals have good characteristics, no worse than in thin GaN layers grown by other techniques. These crystals have shown at least three zero-phonon lines of photoluminescence in infrared spectrum range: 1.3, 1.19, 1.047 eV [188, 189]. It was supposed, that these lines belong to transitions within 3d levels of trace impurities of transition metal ions. The investigations of optically detected magnetic resonance have shown the correspondence between the EPR spectra, attributed to Fe^{3+}, and 1.3 eV luminescence line. The nature of other lines was widely discussed. The experimental results on the 1.047 eV emission [190] fit to a $^4T_2(F)$-$^4A_2(F)$ internal electronic transition of a transition metal with a $3d^7$ electronic configuration. It was suggested [191] that the best candidate was Co^{2+}, but the author could not completely exclude the Ni^{3+}. Thus, only correlation with the EPR seems can clarify this problem. It should be noted that the emission at 1.047 eV has not been observed as natural contaminant in GaN samples grown by metal organic vapour phase epitaxy and vapour phase epitaxy.

We report here EPR results for Fe^{3+}, Mn^{2+}, Ni^{3+} and EPR spectra of two new axial centres in GaN grown by the sublimation sandwich method. Preliminary reports have been published elsewhere [191].

GaN crystals grown on 6H-SiC substrates by the sublimation sandwich method [186, 187] were investigated. The thickness of GaN epitaxial layers was 0.1–0.2 mm and one can say with reasonable confidence that GaN was a bulk material in the experiments. Epitaxial layers on substrates and free-standing layers were used. No intentional doping of samples was performed. The samples were oriented for rotation in the {11–20} plane. The EPR spectra were studied on X-band (9.25 GHz) EPR spectrometer in the temperature range 4–150 K.

The EPR spectra of Mn^{2+} and Fe^{3+} ions in GaN measured in two orientations B ∥ c and B ⊥ c of the crystals are shown in Fig. 4.35a. The upper pattern of Mn^{2+} EPR spectrum was recorded in the iron free sample in orientation B ∥ c. The separations between the lines are about 7 mT. We could observe other four groups with smaller intensities on both sides of central one, which proves that this spectrum belongs to an ion with electron spin S = 5/2. All the groups have the same HF structure as the central one. Among the transition metal elements only manganese has a 100% abundant isotope with nuclear spin I = 5/2 and the observed splitting corresponds to that of Mn^{2+} in $3d^5$ ($^6S_{5/2}$) state. The group in Fig. 4.35a belongs to the Mn^{2+} central fine structure transition ($M_S = 1/2 \leftrightarrow M_S = -1/2$), split by the HF interaction with nuclear spin I = 5/2. This spectrum can be described by the spin Hamiltonian (4.16) with the addition of the HF interaction term **S·A·I** [192], where S = 5/2 and I = 5/2 are the electron and nuclear spins, **A** is tensor, describing the HF interaction with the impurity nucleus.

An investigation of the angular dependence of the spectrum of Mn^{2+} allowed to find the best-fit parameters for spin Hamiltonian of (4.16). The results are listed in Table 4.11 (the small parameters $a - F \cong 4 \cdot 10^{-4}$ cm^{-1} and $a \cong 5 \cdot 10^{-4}$ cm^{-1} are not pointed out in the table). The measured and calculated angular dependencies of fine-structure line positions of Mn^{2+} ions obtained at 9.25 GHz are plotted as open

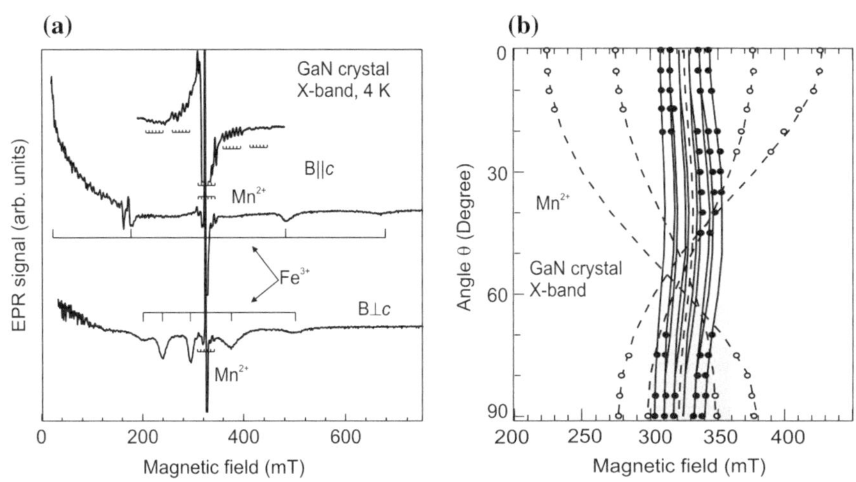

Fig. 4.35 **a** EPR spectra of Mn^{2+} and Fe^{3+} ions in GaN measured in two orientations B ∥ c and B ⊥ c of the crystals. The upper pattern of Mn^{2+} EPR spectrum was recorded in the iron free sample in orientation B ∥ c. **b** The measured (*open circles*) and calculated (*dashed lines*) angular dependencies of fine-structure line positions of Mn^{2+} ions in GaN obtained at 9.2 GHz. The *solid circles* and *solid lines* represent the measured and calculated the HF structure positions, respectively, which are plotted only for central transition $M_S = 1/2 \leftrightarrow M_S = -1/2$

Table 4.11 Parameters of the spin Hamiltonians for Fe^{3+}, Mn^{2+} and Ni^{3+} ions in GaN and ZnO (ZnO data presented for comparison)

	GaN				ZnO			
	$g_{\parallel}$	$g_{\perp}$	$\|D\|$ 10^{-4} cm^{-1}	A 10^{-4} cm^{-1}	$g_{\parallel}$	$g_{\perp}$	D 10^{-4} cm^{-1}	A 10^{-4} cm^{-1}
Fe^{3+} $(3d^5)$	1.990	1.997	713		2.006	2.006	−595	
Mn^{2+} $(3d^5)$	1.999	1.999	240	70	2.001	2.001	−236	74.1
Ni^{3+} $(3d^7)$	S′ = 1/2 2.10 S = 3/2 2.10	≅4.2 ≅2.1	$\geq 1.5 \cdot 10^4$		S′ = 1/2 2.142 S = 3/2 2.142	4.318 ≅2.16	$2 \cdot 10^4$	

circles and dashed lines, respectively, in Fig. 4.35b. The fine-structure positions have been estimated as the centre of gravity of the measured HF structure transitions. The solid circles and solid lines represent the measured and calculated HF structure positions, which are plotted only for central transition ($M_S = 1/2 \leftrightarrow M_S = -1/2$).

With increasing temperature EPR signals of Mn^{2+} and Fe^{3+} decrease in a similar manner. Fe^{3+} and Mn^{2+} EPR signals are detectable up to ~100 K.

An intense anisotropic EPR line was observed in some GaN crystals. Figure 4.34b shows an angular dependence of these lines in GaN crystal at the X-band. The magnetic field applied was rotated in the {11–20} plane. This EPR spectrum has the characteristic anisotropy of an S = 3/2 system in a strong axial crystalline field and positive g shift which is consistent with the electron configuration d^7. The EPR line can be observed up to ~150 K and the linewidth is very sensitive to the orientation of the crystal in magnetic field. The intensities of the EPR signals for $\theta \neq 0$ at 77 K are lower and depend more on the angle between magnetic field and c-axis then those at 4 K. Therefore it has not been possible to detect the signals at all angles at 77 K.

We attribute this spectrum to the trace impurity of nickel in the charge state Ni^{3+}. Isolated substitutional Ni^{3+} ion has a $3d^7$ electronic configuration. In GaN there is a combination of trigonal and tetrahedral crystalline fields. The ^4F ground term of the free ion is split by a cubic tetrahedral part of crystalline field into two orbital triplets, $^4T_2$ and $^4T_1$, and a ground-state orbital singlet, $^4A_2$. The $^4A_2$ ground state is split into two Kramers doublets by the combined action of the trigonal field and the spin-orbit interaction. Omitting the hyperfine interactions, the EPR spectrum can be described by a spin Hamiltonian of the form (4.12′) and (4.13). The g values are listed in Table 4.11.

In [180], an analogy was revealed between the parameters of Fe^{3+} ions in GaN and ZnO crystals. It was taken into account that both materials have the same hexagonal wurzite structure and close by similar physical parameters. We will follow this analogy in the case of Mn^{2+} and Ni^{3+} and the Table 4.11 lists corresponding values previously obtained for ZnO: Mn^{2+} [193] and ZnO: Ni^{3+} [194] (the parameters for Mn^{2+} in ZnO, $a - F = 5.2 \cdot 10^{-4}$ cm^{-1} and $a = 6.2 \cdot 10^{-4}$ cm^{-1}, are not pointed out in the Table 4.11). As illustrated in Table 4.4, a good correspondence is observed between the parameters of Fe^{3+}, Mn^{2+} and Ni^{3+} in GaN and ZnO. The small quantitative variation in the EPR parameters seems to reflect the changing degree of covalency. By analogy with Fe^{3+} in GaN we suppose that manganese and nickel occupy gallium sites in the GaN lattice. Since no HF structure for the line which we attributed to Ni^{3+} was observed we could not completely exclude some impurities which are isoelectronic to Ni^{3+} and have small concentration of odd isotopes or a very small value of nuclear magnetic moment, e.g., Fe^{+} or ions with $4d^7$ and $5d^7$ configurations. 1.047 eV belongs to transition $^4T_2(F)$-$^4A_2(F)$ within 3d levels of Ni^{3+} ion with a $3d^7$ electronic configuration. This luminescence line has been observed only in GaN samples grown by the sandwich technique and seems to correlate with EPR spectra of Ni^{3+}. This assumption is consistent with the experiments of [190] in which photoluminescence results on the 1.047 eV emission fit to $^4T_2(F)$-$^4A_2(F)$ internal electronic transition of a transition metal with a $3d^7$ electronic configuration. It should be noted that Fe^{3+} (the same charge state as Ni^{3+}) has the stable configuration in n-type GaN material and Fe^{3+} EPR spectra have been observed in the same samples in which the EPR spectra of Ni^{3+} were recorded.

The EPR signal of Mn^{2+} was observed and analyzed in Mn-doped MBE-grown GaN and AlN films [195]. All observed transitions are consistent with the spin Hamiltonian for isolated substitutional Mn^{2+} centers with parameters for GaN: $g = 1.9994$, fine structure parameter $D = -218$ to -236 G, HF structure constant $A = 6.9$ mT; and for AlN: $g = 2.0004$, $D = -64.8$ mT; $A = 6.9$ mT. The differences of the axial crystal field parameters D are correlated with the macroscopic strain in GaN : Mn films and agree well with predictions of the superposition theory of crystal fields. The paramagnetic Mn^{2+} impurities were concluded to occupy almost substitutional sites in GaN and AlN. At the investigated Mn concentration of 10^{20} cm^{-3}, they are present as isolated paramagnetic centers, and exchange effects are negligible.

4.4.3 Rare-Earth Element Impurities in SiC

4.4.3.1 Introduction

Rare-earth doped semiconductors attract an increasing interest due to their possible applications in light emitting diodes or diode lasers. The most promising property of these materials is the possibility of excitation of narrow temperature-independent luminescence. Since the luminescence is due to an intra-4f shell transition, the influence of the crystal field of host lattice is weak and this luminescence was found to be fairly independent on the host materials. Erbium doped semiconductors have attracted particular attention because the Er^{3+} intra-4f-shell transition $^4I_{13/2} \rightarrow {}^4I_{15/2}$

at 1.54 μm matches the minimum in the absorption of silica-based fiber-optic communication systems. A number of papers have been published on the photoluminescence properties of rare-earth elements in III–V compounds [196] and silicon [197]. The techniques used to incorporate these elements were ion implantation, liquid-phase epitaxy, molecular-beam epitaxy.

At present, the main obstacle for applications is the low luminescence yield, particularly at room temperature. The quenching of the emission intensity decreases with increasing energy gap of the semiconductor host. This trend is particularly important for the devices' aspects of Si : Er, since it seems to imply a basic principle suggesting that Si:Er will not give a reasonable yield at room temperature [198]. In [199] intense erbium—1.54 μm photoluminescence was observed in ion-implanted SiC crystal which is wide-band gap semiconductor. The maximum penetration of the erbium was about 0.3 μm.

Electron paramagnetic resonance (EPR) has proven to be an extremely powerful tool for the study of defects in semiconductors. The reason for this is that the EPR spectra usually contain highly detailed microscopic information about the structure of the defects, details that often cannot be obtained in any other way. However, ionic implantation produces a very thin layer and EPR measuring of such a layer has always been a big problem. The second problem is that the defects in the layer produced by ionic implantation, on the one hand and in as-grown bulk material, on the other, as a rule, differ structurally. What is more, the impurities which could be introduced by ionic implantation could often not be doped in the process of the crystal growth. In recent paper [200] EPR spectrum from erbium/oxygen complex (Er^{3+} surrounded by six O atoms) with a structure similar to that of the orthorhombic site of Er_2O_3 embedded with well-defined orientations within the Si lattice. The hyperfine structure could not be observed above the noise level and the possibility of Er-pairs must not be ruled out.

SiC was believed can be used to advantage, because, on the one hand, this material has a wide band gap, which is essential for efficient Er^{3+} luminescence and, on the other, SiC : Er microelectronics can apparently be directly matched to silicon-based devices. The problem of incorporating Er^{3+} ions in bulk SiC crystals in the course of growth has been solved [201]. Strong EPR signals of several types of Er^{3+} ions have been detected for the first time in these crystals. More details on studies of rare-earth elements in SiC can be found in the report [169, 170, 202, 203].

4.4.3.2 EPR of Er^{3+} Ions in 6H-SiC

The samples were erbium doped bulk 6H-SiC crystals which were grown by the sublimation sandwich-method [47] in vacuum at temperatures 1850–1900 °C. The source of the impurity was metallic erbium which was directly placed inside the tantalum container. As-grown crystals were of n-type, owing to nitrogen donors. The using of the container from the tantalum material allowed to decrease the concentration of nitrogen donors up to $\sim 10^{16}$ cm^{-3} level.

The EPR spectra were studied in the temperature range 4.5–300 K using an X-band (9.3 GHz) spectrometer. The samples were oriented for rotation in the (11–20) plane. The samples had the shape of a platelet (3 × 4 mm^2) with the face perpendicular to the hexagonal axis (c-axis) and the thickness was about 1 mm. The observed EPR spectra in Er doped 6H-SiC crystal were composed of many lines whose positions varied with orientation in the region of magnetic field from 50 mT to 800 mT. The EPR spectra of, at least, three types of centres having qualitatively different orientation dependencies have been observed.

Low symmetry Er^{3+} centres. Figure 4.36a shows the low-field part of X-band EPR spectra observed in a Er doped 6H-SiC crystal at 12 K for magnetic field at different angles θ with respect to the c-axis. In principal, it can be resolved into three sets of sharp lines, designated LS_1, LS_2 and LS_3 (LS = low symmetry) in Fig. 4.36a. The peak-to-peak linewidths of the lines are ~0.1 mT. The signals labelled by LS_1, LS_2 and LS_3 belong to different centres as the ratio of the intensities of these signals is sample and temperature sensitive. These spectra were believed can be attributed to three different lattice positions. 6H-SiC has over-all hexagonal symmetry and for 6H-SiC there are a hexagonal-like (h) and two cubic-like (k1, k2) crystallographic inequivalent lattice sites, when the first- and second nearest neighbours are considered. The simplification of the spectrum occurred with B ∥ c. The spectrum of every lattice position shows the typical orthorhombic symmetry. The similarity of the EPR spectra's orientation dependencies implies that all centres have the same symmetries. Quantitative differences have only been observed in the spin Hamiltonian parameters and in decay temperatures of the different LS_1, LS_2 and LS_3 EPR signals. There are six sites that are magnetically inequivalent; these coalesce into four in the {11–20) plane. The spectra of these sites share a common point when the magnetic field is along the c-axis. As the field direction changes in the (11–20) plane, the lines split as one can see in Fig. 4.36a.

A direct identification of erbium ions has been established by the presence of hyperfine interaction with ^{167}Er nuclei [see Fig. 4.36a (insets)]. Erbium has only one stable odd isotope, ^{167}Er (natural abundance 22.8%), having nuclear spin I = 7/2 and one normally expects a spectrum consisting of one intense main line and eight equally intense weak satellite lines. Their intensity, as a fraction of the main line, is expected to be about 3% only and the hypertine structure could not be observed for all orientations. Figure 4.36a (insets) shows hyperfine structure for $^{167}Er^{3+}$ ions in 6H-SiC : Er crystal in three orientations of crystal under rotation in plane near to the (11–20) plane. In orientations B ∥ c and B ⊥ c the positions of low symmetry (LS) and axial symmetry (Ax) centres are indicated. The EPR spectrum recorded at a small deviation from the parallel orientation (~3°) is presented to demonstrate the same nature of the central lines and the hyperfine components. The separations between the lines are about 8 mT for B ∥ c. Each of the spectra LS_1, LS_2 and LS_3 can be fitted to the conventional spin Hamiltonian

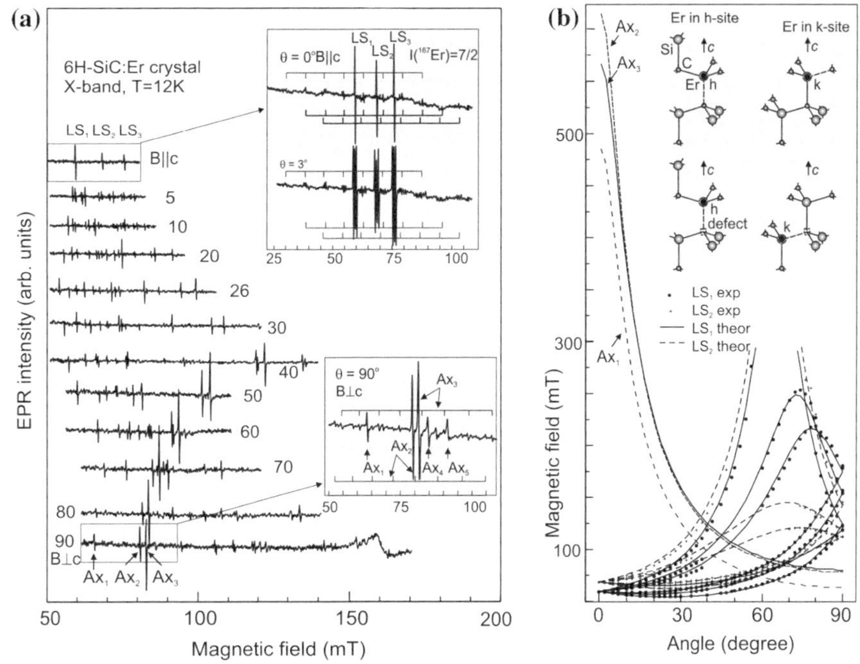

Fig. 4.36 **a** The X-band EPR spectrum of Er^{3+} ions in 6H-SiC : Er crystal in some orientations of crystal under rotation in plane near to the (11–20) plane. In orientations B ∥ c and B ⊥ c the positions of low symmetry (LS) and axial symmetry (Ax) centres are indicated; (*insets*) Hyperfine structure for $^{167}Er^{3+}$ ions in 6H-SiC: Er crystal in three orientations of crystal under rotation in plane near to the (11–20) plane. In orientations B ∥ c and B ⊥ c the positions of low symmetry (LS) and axial symmetry (Ax) centres are indicated. The EPR spectrum recorded at a small deviation from the parallel orientation ($\sim 3°$) is presented to demonstrate the same nature of the central lines and the hyperfine components. **b** The angular dependencies of line positions of Er^{3+} ions for LS_1 and LS_2 centres and axial centres Ax_1, Ax_2 and Ax_3 in 6H-SiC obtained at 9.25 GHz with the field rotated in the (11–20) plane; (*inset*) Possible structures of different Er^{3+} centers according to the EPR data of axial centers and orthorhombic centers

$$H = \mu_B \mathbf{B} \cdot \mathbf{g} \cdot \mathbf{S} + \mathbf{S} \cdot \mathbf{A} \cdot \mathbf{I},$$

with S = l/2 and I = 7/2 which is the spin of the ^{167}Er nucleus. The first term is the interaction of the electron spin with the external magnetic field, the second term describes the hyperfine interaction of the electron spin with ^{167}Er nuclei and **A** is the corresponding tensor (this term is zero for even Er isotopes). The local z-axis coincides with one of the six directions SiC, which is not coincide with c-axis, x and y axes are in a plane perpendicular to the z axis and x axis lies in one of the {11–20} planes. The principal g values for LS_1 spectra are $g_{lx} = 12.2$, $g_{ly} = 3.35$, $g_{lz} = 1.5$; for LS_2: g values: for LS_2 they are $g_{2x} = 10.6$, $g_{2y} = = 6.16$, $g_{2z} = 1.26$. For LS_3: $g_{3x} = 9.6$, $g_{3y} = 7.52$, $g_{3z} = 1.45$.

The measured and calculated angular dependencies of EPR line positions of Er^{3+} ions (only for LS_1 and LS_2) obtained at 9.25 GHz are plotted as open circles and solid lines (LS_1), crosses and dashed lines (LS_2), in Fig. 4.36b. It should be noted that a slight misalignment ($\sim 7°$) of the sample occurred so that the magnetic field was not wholly in the (11–20] plane. This small misalignment has been taken into account in the analysis and did not affect the conclusions. The defect coordinate system for orthorhombic centres is tilted by small angle of $\sim 5°$ around y axis.

The intensities of the EPR lines were strongly temperature dependent with maximum intensity at about 15 K and the spectra were unobservable above 30 K.

Axial Er^{3+} centres. In addition to the EPR spectrum of the orthorhombic Er^{3+} centres, EPR lines in the region from 100 to 600 mT have been observed which clearly show axial symmetry with crystalline c-axis being the axis of the g tensor. The part of these spectra is shown in Fig. 5.36a, b (inset). One can see three lines labelled as Ax_1, Ax_2 and Ax_3. The EPR spectra can be described by the spin conventional spin Hamiltonian with axial symmetry. The principal g values for Ax_1 centre are $g_{1\parallel} = 1.359$, $g_{1\perp} = 10.251$; for Ax_2 centre $g_{2\parallel} = 1.073$, $g_{2\perp} = 8.284$ and for Ax_3 centre $g_{3\parallel} = 1.164$, $g_{3\perp} = 8.071$ where the parallel axis coincides with the c-axis of the crystal. The relative intensities of all the EPR spectra depend on the temperature and again the maximum signal intensity is observed at 15 K and the spectra were unobservable above 30 K. The angular dependencies of EPR line positions of Er^{3+} ions obtained at 9.25 GHz are plotted as lines in Fig. 4.36b. The hyperfine structures from ^{167}Er have also been observed for Ax_2 and Ax_3 axial centres. The separations between the lines were about 10 mT for $\theta = 45°$.

In addition to the spectra above, an EPR line in the region 800 mT which seems to show axial symmetry with $g_{\parallel} = 0.776$ was also observed. A striking feature in this observation is a large variation of the signal intensity with the orientation of the applied magnetic field. This seems to be a result of the highly anisotropic g-values and it led to difficulties in following the line over the angles larger than 10°. This spectrum seems also to belong to Er^{3+} although the hyperfine structure associated with the ^{167}Er isotope could not be clearly observed above the noise level. The new EPR lines can probably be attributed to the first excited state of the Er^{3+} ions as it was observed in ionic crystal. By assuming that this dependence is exponential, one obtains the energy of the first excited state of erbium in 6H-SiC amounting to $\sim 15\ cm^{-1}$ (one reasoned that the energy of the ground state must be zero).

The nature of Ax_1 centre is not clear up to now, however there is some evidence which indicates that this spectrum belongs to Dy^{3+} ions. Several other resonance lines have been noted in every Er-doped Sic sample so far examined, but these lines have not yet been identified. These resonance lines (noted as Ax_4 and Ax_5) exhibit axial symmetry with following g-factors: for Ax_4: $g_{\parallel} = 2.92$, $g_{\perp} = 7.69$; for Ax_5: $g_{\parallel} = 4.32$, $g_{\perp} = 7.22$. The trace amounts of the other rare-earth elements had probably been present as impurities in the erbium starting material. Analogous result was observed in $Y_3A1_5O_{12}$: Er crystal in which in addition to the EPR spectra of the Er^{3+} centres the signals of the trace impurities of other rare-earth elements as Nd^{3+} and Yb^{3+} were observed [204].

Discussion. The resonances are attributed to Er^{3+} as erbium is not expected to attain any other valence under normal conditions. The electronic configuration of Er^{3+} is $4f^{11}$ with a free-ion ground state of $^4I_{15/2}$. The 16-fold ground-state degeneracy will be split by crystal field into a number of quartet and doublet Stark levels, for example, an electric field of Td symmetry splits this state into three Γ_8 quartets and two doublets, Γ_6 and Γ_7. According to the [205], the ground state is either Γ_6 or Γ_7 and it depends on the ratio of the fourth to sixth order crystal field terms. The isotropic g values for Er^{3+} ion was found to be $g(\Gamma_6) = 6.772$ and $g(\Gamma_7) = 5.975$. The average g values $\langle g \rangle$ defined by $\langle g \rangle = (g\| + 2\, g_{\perp})/3$ for Ax_2 and Ax_3 axial centres will be 5.88 and 5.77, respectively; for orthorhombic centres $\langle g \rangle = (g_x + g_y + g_z)/3$ will be 5.68 (LS_1), 6.0 (LS_2) and 6.17 (LS_3). The average g values can be compared with the isotropic cubic g factor but one should assume that the axial (and orthorhombic) field is small compared to the cubic field. This suggests that the parent cubic ground state may have Γ_7 representation for axial and orthorhombic Er^{3+} centres. The Γ_7 g value should be corrected for covalent bonding and the reduction may be due to covalency. The average g value for Ax1 axial center is $\langle g \rangle = 7.29$ and it fit very well for Γ_6 ground state of Dy^{3+} ion which is $g(\Gamma_6) = 7.56$ [206]. The average g-values for Ax_4 and Ax_5 are 6.1 and 6.25 respectively. Since these average g-values can fit for Dy^{3+} as well as Er^{3+}, thus, these centres are not identified yet.

Comparison of the ionic radii of rare-earth ions with the silicon and carbon radii leads one to expect that rare-earth impurities would occur substitutionally for silicon in SiC. Presumably, erbium substitutes for silicon and lies in a regular environment for axial centre. The orthorhombic Er^{3+} has a more complicated structure and it seems to include another defect at carbon position along with Er^{3+} ion. Er^{3+} ion is substitutionally incorporated probably in association with nearest oxygen atom or carbon vacancy such that the lines joining them to the Er^{3+} ion are one of the Si-C directions which do not coincide with c-axis for hexagonal and qubic-like sites. In addition, the EPR spectrum of the excited state of Er^{3+} in 6H-SiC seems to be observed at higher temperature.

Figure 4.36b (inset) demonstrates the models proposed for the erbium centers in 6H-SiC on the basis of the EPR data. In the axial centers, erbium replaces silicon and occupies the hexagonal or cubic-like site in the regular lattice. The differences in the EPR parameters of the three centers are due to the presence of three possible erbium sites in the 6H-SiC crystals, namely, the hexagonal and two cubic-like sites. The orthorhombic Er^{3+} centers have a more complex structure displayed for the hexagonal and quasi-cubic sites. It seems likely that these centers involve another defect in the carbon site near erbium. The presence of three types of these centers is also explained by the hexagonal and two quasi-cubic erbium sites in the lattice. It is quite possible that, in the orthorhombic centers, the Er^{3+} ion forms the complex with one oxygen atom or carbon vacancy in such a way that the line connecting them coincides with one of the Si–C bonds, making an angle of $\sim 70°$ with the hexagonal axis c. A part of the axial centers can also be associated with the complexes whose composition involves the erbium ions; in this case, the defect (for example, the oxygen atom or carbon vacancy) is located so that the line connecting the erbium ion and the defect coincides with the Si–C bond aligned along the c-axis.

4.4.3.3 Erbium Luminescence in SiC Bulk Crystals

The luminescence at a wavelength of about 1.54 μm was revealed in the 6H-SiC:Er crystals, which showed the EPR spectra for several types of Er^{3+} centers with different local symmetry. The EPR signals for at least seven Er^{3+} centers (designated as Ax_1–Ax_7) with an axial symmetry relative to the hexagonal crystal c-axis were recorded in these crystals. In addition to the axial centers, the EPR spectra displayed signals of the Er3 + ions (denoted as LS_1, LS_2, and LS_3) [169, 170, 202, 203] characterized by an orthorhombic symmetry. For the centers with the orthorhombic symmetry, the local z-axis coincides with one of six directions of the Si–C bonds forming an angle of $\sim 70°$ with the c-axis. For the orthorhombic centers and a number of axial centers, the EPR spectra showed a hyperfine structure owing to the interaction with the ^{167}Er nucleus, which made it possible to directly identify the erbium ions and, furthermore, to establish that the structure of each center involves only one erbium ion.

Figure 4.37a displays the photoluminescence spectra of the 6H-SiC : Er crystal in the wavelength range 1.1–1.7 μm at three temperatures (77, 210, and 300 K). It is seen that the photoluminescence spectra exhibit the EPR signals of the Er^{3+} ions. The photoluminescence was excited by the visible light (400–650 nm) of a mercury lamp. The spectrum in the range of 1.54 μm is typical of the $^4I_{13/2} \rightarrow {}^4I_{15/2}$ transition within the 4f shell of the Er^{3+} ions. The photoluminescence spectrum in the range of 1.54 μm at a temperature of 300 K is depicted on an enlarged scale in inset of Fig. 4.37a. About 20 photoluminescence bands can be separated in this spectrum, and the most intense bands are located at wavelengths of about 1.53 and 1.56 μm. The erbium photoluminescence can be excited over a wide range of wavelengths from 320 to 600 nm; in this case, no substantial changes in the relative intensities of particular photoluminescence bands of the Er^{3+} ions were observed in the range of 1.54 μm.

In the range 1.3–1.4 μm, three luminescence bands correspond in location to the luminescence bands of the vanadium ions V^{4+} in the 6H-SiC crystal [207]. Vanadium is a typical uncontrollable impurity in silicon carbide.

The main feature of the observed photoluminescence of erbium is the unusual temperature dependence of its intensity. This dependence for the photoluminescence band of erbium at a wavelength of 1.531 μm is depicted by dark circles in Fig. 4.37b. As the temperature increases beginning with 77 K, the intensity of erbium photoluminescence rapidly increases and reaches a maximum value at approximately 240 K. With a further increase in the temperature, the intensity of erbium photoluminescence decreases and becomes equal to about 50% of the maximum value at a temperature of 300 K. The photoluminescence is reliably

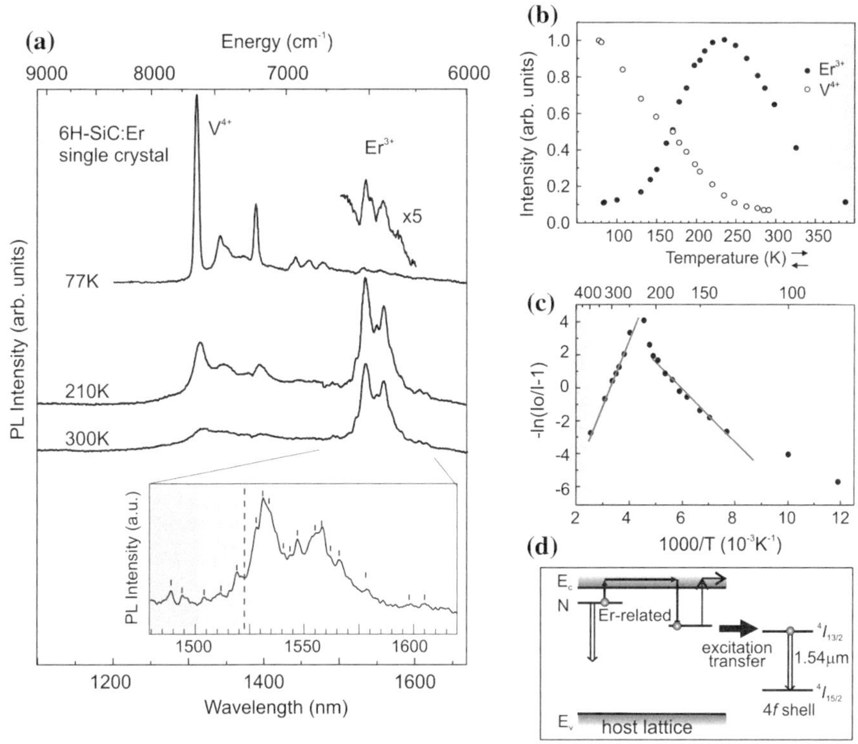

Fig. 4.37 **a** Photoluminescence spectra of the Er^{3+} ions in 6H-SiC:Er crystal in the range 1100–1700 nm at temperatures of 77, 210, and 300 K. Photoluminescence was excited by the visible light of a mercury lamp (400–650 nm). (*Inset*) Photoluminescence spectrum of Er^{3+} ions presented on an enlarged scale. *Vertical marks* indicate selected bands. The *vertical dashed line* conventionally separates the transitions from the low-lying Stark level of the excited subterm $^4I_{13/2}$. **b** Temperature dependences of the intensity of photoluminescence for the 1531-nm band of erbium and the high-energy band of vanadium in the 6H-SiC:Er crystals. **c** Dependences shown in Fig. 4.37b for erbium but constructed on the $\ln(I_0/I - 1) - 1/T$ coordinates for two temperature ranges corresponding to the flare-up and quenching of the photoluminescence. **d** Possible scheme of the energy levels and mechanisms of energy transfer in the 6H-SiC: Er crystals. *Single arrows* show thermal release of electrons from the donor level, retrapping in the erbium-related level, and thermal detrapping of electrons from the erbium level. *Double arrows* demonstrate radiative transitions, and the *heavy arrow* indicates the excitation transfer from the erbium-related level to the 4f shell of Er^{3+} ion

recorded up to ~400 K. Figure 4.37b also demonstrates the temperature dependence of the intensity for the high-energy band of the luminescence of vanadium (open circles). In this case, an increase in the temperature starting with 77 K is accompanied with a drastic decrease in the luminescence intensity. As can be seen from Fig. 4.37b, an increase in the intensity of the erbium photoluminescence and a decrease in the intensity of the vanadium luminescence are observed in about the same range of temperatures. Moreover, it should be mentioned that the temperature dependence of the intensity of the donor–acceptor photoluminescence observed in the visible range for the 6H-SiC : Er crystals has about the same shape as the corresponding dependence for the IR photoluminescence of vanadium.

The temperature flare-up and the temperature quenching of the photoluminescence of erbium can be described by the known formula

$$I = \frac{I_0}{1 + A\exp\left(-\frac{E_A}{kT}\right)}, \tag{4.17}$$

where E_A is the activation energy of the process, I_0 is the intensity of luminescence without quenching (at ~240 K), and A is the constant depending on the probability of radiative recombination and the constant in the Boltzmann relation. The formula (4.17) enables one to determine the activation energies for the flare-up and quenching of the luminescence from the slope of the straight line on the $\ln(I_0/I) - 1/T$ coordinates. The temperature dependences of the photoluminescence intensity on these coordinates for two temperature ranges corresponding to the flare-up and quenching of the erbium photoluminescence are displayed in Fig. 4.37c. The activation energies E_A for the flare-up and quenching of the Er^{3+} luminescence are evaluated to be equal to ~130 ± 20 and ~350 ± 20 meV, respectively. The theoretical dependences calculated according to the formula (4.17) with the use of the activation energies given above are depicted by solid lines in Fig. 4.37c. In the description of the above processes, we proceeded from rough estimates and approximated each process by only one exponent, even though, in principle, it is not improbable that the process is more complex and should be described by the sum of exponents with several activation energies; in the latter case, the activation energies given above correspond to the averaged energies.

The experimental results obtained in the present work do not permit us to directly assign the photoluminescence bands to the specific (axial or orthorhombic) erbium centers, which manifest themselves in the EPR spectra. It should only be remarked that the photoluminescence was observed solely in the crystals that showed the EPR spectra of the Er^{3+} centers. The symmetry of all the Er^{3+} centers in the 6H-SiC crystal is lower than the cubic symmetry. Therefore, all these centers can contribute to the luminescence (in the octahedral complex, the intracenter luminescence of the Er^{3+} ions cannot be observed, because the f–f transitions are parity-forbidden). The ground state $^4I_{15/2}$ in the crystal field of this symmetry is split into the eight doubly degenerate Stark levels (Kramers doublets), and, hence, at least eight luminescence bands should be observed for each erbium center. Taking into account the fact that, according to the EPR data, there are about ten different Er^{3+} centers, a large number of luminescence bands should be observed for the $^4I_{13/2} \rightarrow {}^4I_{15/2}$ transition in the Er^{3+} ions. Since all the luminescence bands are located in a narrow spectral range, their large number apparently results in considerable widths of the photoluminescence bands at temperatures above 77 K and the spectral resolution provided by the instrument employed. For the symmetry lower than the cubic symmetry, the first excited subterm $^4I_{13/2}$ is split into seven doubly degenerate energy levels. At sufficiently low temperatures (when the thermal energy is less that the splitting of energy levels by the crystal field), only the transitions from the lowest-lying state of the $^4I_{13/2}$ excited subterm can occur, and eight photoluminescence bands should be observed for each erbium center. The higher-lying states of the $^4I_{13/2}$ subterm split by the crystal field are not involved in the luminescence process at low temperatures. The higher-lying states of the luminescent levels can be filled with an increase in the sample temperature. The filling of these levels should lead to the appearance of additional photoluminescence bands on the high-energy side of each band observed at low temperatures [208]. A number of these bands can be seen on the high-energy side of the most intense bands corresponding to the transitions in the Er^{3+} ions, because a temperature of 77 K is high enough for the Stark levels of the $^4I_{13/2}$ subterm to be filled. These bands are conventionally separated by the vertical dashed line in Fig. 4.37a (inset).

Let us now discuss the possible mechanisms of excitation, temperature flare-up, and temperature quenching of the photoluminescence of erbium ions in SiC. According to our experimental data, the temperature behavior of the erbium photoluminescence virtually does not depend on the excitation quantum energy, so that, apparently, the direct optical excitation of the 4f shell, followed by the radiative relaxation from the excited state to the ground state is of little importance.

Another mechanism explaining the excitation of the luminescence of a rare-earth ion involves the energy transfer from the electron–hole pair (excited, for example, upon interband absorption of the light) to the ion. This mechanism can be efficient when an energy level (or several energy levels), which is related to the rare-earth ion, occurs in the forbidden gap of a semiconductor. In the case when the rare-earth ion gives rise to the energy level in the forbidden gap of the semiconductor, the excitation of carriers from the valence band (conduction band) to this level can produce a bound electron–hole pair or a bound exciton with the subsequent transfer of the recombination energy to the 4f shell of the rare-earth ion. The most probable mechanism of this energy transfer is the so-called impurity Auger recombination. The intracenter luminescence corresponds to the transitions between the 4f levels lying outside the forbidden gap. The 4f electrons are tightly bound to the ion, and their states can be treated as the internal states independent of the band structure of the matrix. In the absence of tight binding, the energy transfer between electronic states of the crystal and strongly localized states of the 4f electrons is determined by the Coulomb interaction. The theoretical treatment of the excitation suggests that the intracenter Auger process provides the energy transfer to the 4f electrons through the dipole and exchange interactions. The efficiency of this process drastically increases if the state (related to the rare-earth center) in the forbidden gap allows the localization of energy in the form of bound excitons (or bound electron–hole pairs).

In order to elucidate the nature of these levels, we perform a qualitative analysis of the model according to which a rare-earth ion replaces the silicon atom in the SiC crystal. Let us consider the so-called "vacancy" model, which was elaborated for transition metal ions in semiconductors [170].

According to this model, the elimination of a host crystal atom (for example, a silicon atom in the Si or SiC crystal) from a semiconductor lattice leads to the formation of a vacancy with a certain set of energy levels and wave functions. When an impurity atom occupies this vacancy, the set of energy levels and wave functions, as a rule, are changed but not too strongly to lose their identity. The impurity atom gives rise to its intrinsic levels; however, these levels are usually located below the top of the valence band. A similar model for rare-earth elements in the crystals A_3B_5 and silicon was advanced in [209, 210]. As a result of the removal of an atom from the host semiconductor lattice, the vacancy states characterized by the a_1 and t_2 symmetries are formed in the forbidden gap of the crystal. The next stage involves the formation of the substituting rare-earth element center, which can be treated as the interaction between the rare-earth atom and the vacancy. The rare-earth atom shows the $4f^n6s^2$ electronic structure. The ground state of the 4f electrons can have three groups of energy levels transformed according to the a_2, t_1, and t_2 irreducible representations of the T_d group, and the 6s shell is characterized by the a_1 symmetry. Therefore, the interaction between the 6s shell of the rare-earth atom and the a_1 state of the vacancy leads to the formation of bonding and antibonding states. Note that the lower-lying bounding state is occupied and most likely occurs in the valence band, whereas the higher-lying antibonding state (related to the rare-earth atom) is unoccupied, can lie near the conduction band, and can serve as an electron trap. It is assumed that, for erbium in SiC, this state is almost completely formed by the 6s shell of erbium and the 2s and 2p valence electrons of carbon. For simplicity, let us consider the set of erbium energy levels in cubic SiC. The state of the Er^{3+} ion in SiC corresponds to the A-state; i.e., it is negatively charged. Hence, it is necessary to consider the negatively charged silicon vacancy with five electrons, whereas six electrons are required for the filling of the t_2 state. Reasoning from the experimentally found charge state of Er^{3+} in SiC, we should assume that one of the 4f electrons of erbium transfers to the t_2 level. As a result, we have the completely occupied level, which is located near the bottom of the valence band and, apparently, can serve as a specific hole trap.

The temperature dependence of the photoluminescence intensity in the 6H-SiC : Er crystal is characterized by two main ranges (Fig. 5.37b, c). In the low-temperature range (77–240 K), the efficiency of excitation of the erbium luminescence increases with an increase in the temperature. In the high-temperature range (240–400 K), an increase in the temperature is accompanied by the quenching of the erbium luminescence. Similar dependences were also observed for erbium in silicon but at substantially lower temperatures [211].

The flare-up of the erbium luminescence can be roughly described by the above formula. The experimentally observed activation energy E_A of luminescence in this process is equal to $\sim 130 \pm 20$ meV. Since this energy approximately corresponds to the location of donor nitrogen levels with respect to the conduction band, it is reasonable to suppose that an increase in the erbium luminescence is associated with the thermal ionization of nitrogen donors with the subsequent trapping of electrons in the deeper erbium-related levels.

A possible scheme of energy levels in the 6H-SiC : Er crystal is depicted in Fig. 4.37d. We believe that, at low temperatures, the carriers are more efficiently trapped in the usual donor levels (most likely, nitrogen donors in our experiments), followed by the recombination. This is corroborated by the intense donor–acceptor luminescence at low temperatures. This luminescence is quenched in about the same temperature range in which the flare-up of the erbium luminescence takes place. As the temperature increases, the electrons undergo a thermal release from the donor levels and are retrapped in the deeper erbium-related levels with the subsequent transfer of energy to the 4f shell of the Er^{3+} ion (see scheme in Fig. 4.37d).

Now, let us dwell briefly on the quenching of the erbium luminescence at high temperatures. First and foremost, we should note that our results obtained for the temperature quenching of photoluminescence in the SiC bulk crystals at high temperatures are in qualitative agreement with the experimental data on the luminescence in ion-implanted layers [199]. This indicates that the processes of photoluminescence quenching occur through similar mechanisms. The quenching of photoluminescence is caused by the fact that the excitation of luminescence is accompanied by competing processes: the release of carriers (most likely, electrons) from the erbium-related levels followed by the nonradiative recombination. The experimentally observed activation energy for the luminescence quenching $E_A \sim 350$ meV apparently corresponds to the energy of the erbium-related level. The mechanism of the high-temperature quenching is not conclusively elucidated. It can be assumed that, after the ionization of the erbium-related level, the nonradiative recombination at high temperatures occurs at the expense of the Auger recombination with free carriers. As was noted by a number of researchers, this process represents the predominant nonradiative recombination channel for rare-earth impurities in semiconductors. The merits of wide-gap semiconductors are evident owing to the presence of the deeper erbium-related levels and, hence, the higher temperatures of the luminescence quenching. A similar approach can also be applied in describing the quenching of the vanadium luminescence and the donor–acceptor recombination luminescence in the temperature range 77–240 K with the activation energy $E_A \sim 130$ meV. It should be emphasized that, in silicon crystals, the high quantum yield of luminescence was observed only in the case when the Er^{3+} ion was in a strong negative electric field of ligands (for example, oxygen or fluorine). Moreover, it was noted that this field plays an essential role in an increase in the optical activity of Si : Er [212]. It is reasonable that, in the case of SiC, this role can be played by carbon, because the SiC crystal possesses a considerable degree of ionicity; i.e., in actual fact, we are dealing with the $Si^{+}C^{-}$ crystal. Therefore, the advantages of SiC also reside in the fact that the Er^{3+} ion should be surrounded by a negative ligand field in a natural way. This implies that, unlike the silicon crystal, there is no need for an additional doping with oxygen.

References

1. Landau, L.: Electron motion in crystal lattices. Phys. Z. Sowj. Un. **3**, 664 (1933)
2. Kabler, M.N.: In: Crawford, J.H., Slifkin, L.M. (eds.) Point Defects in Solids. Plenum Press, New York (1972)
3. Song, K.S., Williams, R.T.: Self-Trapped Excitons. Springer, Berlin, Heidelberg (1993)
4. Känzig, W.: Electron spin resonance of *V*1-centers. Phys. Rev. **99**, 1890–1891 (1955)
5. Castner, T.G., Känzig, W.: The electronic structure of *V*-centers. J. Phys. Chem. Solids **3**, 178–195 (1957)
6. Gazzinelli, R., Mieher, R.L.: Electron-nuclear double resonance of the self-trapped hole in LiF. Phys. Rev. **175**, 395–411 (1968)
7. Höhne, M., Stasiw, M.: Detection of Ag^{2+} in doped AgCl crystals by ESR. Phys. Status Solidi **25**, k55–k57 (1968)
8. Höhne, M., Stasiw, M.: ESR detection of self-trapped holes in AgCl. Phys. Status Solidi **28**, 247–253 (1968)
9. Fukui, M., Hayashi, Y., Yoshioka, H.: The ESR study of growth and decay processes of self-trapped holes in AgCl crystals doped with Cu. J. Phys. Soc. Jpn. **34**, 1226–1233 (1973)
10. Kanzaki, H.: Photographic Sci. Eng. **24**, 219 (1980)
11. Hayes, W., Owen, I.B.: EPR of the self-trapped exciton in AgCl. J. Phys. C **9**, L69–L71 (1976)
12. Murayama, K., Morigaki, K., Sakuragi, S., Kanzaki, H.: Optically detected ESR of the excited states in silver halides. J. Phys. Soc. Jpn. **41**, 1617–1624 (1976)
13. Hayes, W., Owen, I.B., Walker, P.J.: Optically-detected EPR of excitons in AgCl, AgCl: Br and AgBr. J. Phys. C **10**, 1751–1759 (1977)
14. Hayes, W., Owen, I.B.: Comparison of effects of X-rays and ultraviolet light in the creation of excitons in AgCl. J. Phys. C **11**, L607–L610 (1978)
15. Marchetti, A.P., Tinti, D.S.: Low-temperature photophysics of crystalline AgCl. Phys. Rev. B **24**, 7361–7370 (1981)
16. Yoshioka, H., Sugimoto, N., Yamaga, M.: ODMR of self-trapped excitons in AgCl crystals: analysis of polarization and intensity of ODMR. J. Phys. Soc. Jpn. **54**, 3990–4004 (1985)
17. Sugimoto, N., Yoshioka, H., Yamaga, M.: Optically detected magnetic resonance of trapped excitons $(AgCl_6)^{5-}$ and $(AgBrCl_5)^{5-}$ in AgCl crystals. J. Phys. Soc. Jpn. **54**, 4331–4339 (1985)
18. Yamaga, M., Sugimoto, N., Yoshioka, H.J.: Dynamic Jahn-Teller effect in the ODMR spectrum of the self-trapped exciton in AgCl crystals. Phys. Soc. Jpn. **54**, 4340–4344 (1985)
19. Spoonhower, J.P., Ahlers, F.J., Eachus, R.S., McDugle, W.G.: Recombination photophysics in AgCl. J. Phys.: Condens. Matter **2**, 3021–3030 (1990)
20. Donckers, M.C.J.M., Poluektov, O.G., Schmidt, J., Baranov, P.G.: Exchange splitting of self-trapped excitons in AgCl from optically detected EPR at 95 GHz. Phys. Rev. B **45**, 13061–13063 (1992)
21. Poluektov, O.G., Donckers, M.C.J.M., Baranov, P.G., Schmidt, J.: Dynamical properties of the self-trapped exciton in AgCl as studied by time-resolved EPR at 95 GHz. Phys. Rev. B **47**, 10226–10234 (1993)
22. Romanov, N.G., Baranov, P.G.: Multiquantum ODMR spectroscopy of semiconductors and silver chloride. Semicond. Sci. Technol. **9**, 1080–1085 (1994)
23. Baranov, P.G., Romanov, N.G.: Magnetic resonance in micro- and nanostructures. Appl. Magn. Reson. **21**, 165–193 (2001)
24. Bennebroek, M.T., Arnold, A., Poluektov, O.G., Baranov, P.G., Schmidt, J.: Spatial distribution of the wave function of the self-trapped exciton in AgCl. Phys. Rev. B **53**, 15607–15616 (1996)
25. Stoneham, A.M.: Theory of Defects in Solids. Clarendon, Oxford (1985)

26. Bennebroek, M.T., Poluektov, O.G., Zakrzewsky, A.J., Baranov, P.G., Schmidt, J.: Structure of the intrinsic shallow electron center in AgCl studied by pulsed electron nuclear double resonance spectroscopy at 95 GHz. Phys. Rev. Lett. **74**, 442–445 (1995)
27. Bennebroek, M.T., Arnold, A., Poluektov, O.G., Baranov, P.G., Schmidt, J.: Shallow electron centers in silver halides. Phys. Rev. B **54**, 11276–11289 (1996)
28. Baetzold, R.C., Eachus, R.S.: The possibility of a split interstitial silver ion in AgCl. J. Phys.: Condens. Matter **7**, 3991–3999 (1995)
29. Gourary, B.S., Adrian, F.J.: Approximate wave functions for the *F* center, and their application to the electron spin resonance problem. Phys. Rev. **105**, 1180–1192 (1957)
30. Zhitnikov, R.A., Baranov, P.G., Melnikov, N.I.: Ag_2^+ molecular ions in a KCl crystal. Phys. Status Solidi (b) **59**, K111–K114 (1973)
31. Eachus, R.S., Graves, R.E., Olm, M.T.: Observation by ESR of electrons localized at intrinsic shallow traps in silver halide systems. Phys. Status Solidi (b) **152**, 583–592 (1989)
32. Sakuragi, S., Kanzaki, H.: Identification of shallow electron centers in silver halides. Phys. Rev. Lett. **38**, 1302–1305 (1977)
33. Bennebroek, M.T., Duijn-Arnold, Av, Schmidt, J., Poluektov, O.G., Baranov, P.G.: Self-trapped hole in silver chloride crystals: a pulsed EPR/ENDOR study at 95 GHz. Phys. Rev. B **66**, 054305 (2002)
34. Grachev, V.G.: Correct expression for the generalized spin Hamiltonian for a noncubic paramagnetic center. Sov. Phys. JETP **65**, 1029–1035 (1987)
35. Moreno, M.: Microscopic characterization of impurities in insulators through EPR. J. Phys. Chem. Solids **51**, 835–859 (1990)
36. Valiente, R., Aramburu, J.A., Barriuso, M.T., Moreno, M.: Electronic structure of Ag^{2+} impurities in halide lattices. J. Phys. C **6**, 4515–4525 (1994)
37. Feher, G.: Electron spin resonance experiments on donors in silicon. I. Electronic structure of donors by the electron nuclear double resonance technique. Phys. Rev. **114**, 1219–1244 (1959)
38. Fletcher, R.C., Yager, W.A., Pearson, G.L., Merritt, F.R.: Hyperfine splitting in spin resonance of group V donors in silicon. Phys. Rev. **95**, 844–845 (1954)
39. Ivey, J.L., Mieher, R.L.: Ground-state wave function for shallow-donor electrons in silicon. II. Anisotropic electron-nuclear-double-resonance hyperfine interactions. Phys. Rev. B **11**, 849–857 (1975)
40. van Duijn-Arnold, A., Mol, J., Verberk, R., Schmidt, J., Mokhov, E.N., Baranov, P.G.: Spatial distribution of the electronic wave function of the shallow boron acceptor in 4H- and 6H-SiC. Phys. Rev. B **60**, 15829–15847 (1999)
41. Choyke, W.J., Patrick, L.: Optical properties of cubic SiC: luminescence of nitrogen-exciton complexes, and interband absorption. Phys. Rev. **127**, 1868–1877 (1962); Choyke, W.J., Hamilton, D.R., Patrick, L.: Optical properties of cubic SiC: luminescence of nitrogen-exciton complexes, and interband absorption. Phys. Rev. **133**, A1163–A1166 (1964)
42. Colwell, P.J., Klein, M.V.: Raman scattering from electronic excitations in n-Type silicon carbide. Phys. Rev. B **6**, 498–515 (1972)
43. Woodbury, H.H., Ludwig, G.W.: Electron spin resonance studies in SiC. Phys. Rev. **124**, 1083–1089 (1961)
44. Hardeman, G.E.G., Gerritsen, G.B.: Displacement phenomena of boron acceptors in 6H SiC. Phys. Lett. **20**, 623–624 (1966)
45. Patrick, L.: Kohn-Luttinger interference effect and location of the conduction-band Minima in 6H SiC. Phys. Rev. B **5**, 2198–2206 (1972)
46. Kalabukhova, E.N., Kabdin, N.N., Lukin, S.N.: Sov. Phys. Solid State **29**, 1461–1462 (1987)
47. Greulich-Weber, S., Feege, M., Spaeth, J.-M., Kalabukhova, E.N., Lukin, S.N., Mokhov, E. N.: On the microscopic structures of shallow donors in 6H SiC: studies with EPR and ENDOR. Solid State Commun. **93**, 393–397 (1995)

48. Greulich-Weber, S.: EPR and ENDOR investigations of shallow impurities in SiC polytypes. Phys. Status Solidi A **162**, 95–151 (1997)
49. Kohn, W., Luttinger, J.M.: Theory of donor levels in silicon. Phys. Rev. **97**, 1721 (1955)
50. Kohn, W.: Shallow impurity states in silicon and germanium. In: Seitz, F., Turnbull, D. (eds.) Solid State Physics, vol. 5, pp. 257–320. Academic Press Inc., New York (1957)
51. van Duijn-Arnold, A., Zondervan, R., Schmidt, J., Baranov, P.G., Mokhov, E.N.: Electronic structure of the N donor centre in *4H*-SiC and *6H*-SiC. Phys. Rev. B **64**, 085206 (2001)
52. Szász, K., Trinh, X.T., Son, N.T., Janzén, E., Gali, A.: Theoretical and electron paramagnetic resonance studies of hyperfine interaction in nitrogen doped 4H and 6H SiC. J. Appl. Phys. **115**, 073705 (2014)
53. Son, N.T., Janzén, E., Isoya, J., Yamasaki, S.: Hyperfine interaction of the nitrogen donor in *4H*-SiC. Phys. Rev. B **70**, 193207 (2004)
54. Savchenko, D.V., Kalabukhova, E.N., Kiselev, V.S., Hoentsch, J., Poppl, A.: Spin-coupling and hyperfine interaction of the nitrogen donors in 6H-SiC. Phys. Status Solidi B **246**, 1908–1914 (2009)
55. Savchenko, D.V., Kalabukhova, E.N., Poppl, A., Shanina, B.D.: Electronic structure of the nitrogen donors in 6H SiC as studied by pulsed ENDOR and TRIPLE ENDOR spectroscopy. Phys. Status Solidi B **249**, 2167–2178 (2012)
56. Baranov, P.G., Ber, B.Ya., Godisov, O.N., Il'in, I.V., Ionov, A.N., Mokhov, E.N., Muzafarova, M.V., Kaliteevskii, A.K., Kaliteevskii, M.A., Kop'ev, P.S.: Probing of the shallow donor and acceptor wave functions in silicon carbide and silicon through an EPR study of crystals with a modified isotopic composition. Phys. Solid State **47**, 2219–2232 (2005)
57. Son, N.T., Isoya, J., Umeda, T., Ivanov, I.G., Henry, A., Ohshima, T., Janzén, E.: EPR and ENDOR studies of shallow donors in SiC. Appl. Magn. Reson. **39**, 49–85 (2010)
58. Spaeth, J.-M., Niklas, J.R., Bartram, R.H.: Structural Analysis of Point Defects in Solids. An Introduction to Multiple Magnetic Resonance Spectroscopy. Springer, Berlin, Heidelberg (1992)
59. Kalabukhova, E.N., Lukin, S.N., Mitchel, W.C.: Electrical and multifrequency EPR study of nitrogen and carbon antisite-related native defect in n-type as-grown 4H-SiC. Mater. Sci. Forum **433–436**, 499–502 (2003)
60. Son, N.T., Isoya, J., Yamasaki, S., Janzen, E.: Hyperfine interaction of nitrogen donor in 4H-SiC studied by pulsed-ENDOR. In: Nipoti, R., Poggi, A., Scorzoni, A. (eds.) Book of Abstracts of the 5th European Conference on Silicon Carbide and Related Materials, Bolonga, Italy, 2004, pp. 351–354. CNR–IMM, Area Della Ricerca, Bologna (2004)
61. Morton, J.R., Preston, K.F.: Atomic parameters for paramagnetic resonance data. J. Magn. Reson. **30**, 577–582 (1978)
62. Smith, W.V., Sorokin, P.P., Gelles, I.L., Lasher, G.J.: Electron-spin resonance of nitrogen donors in diamond. Phys. Rev. **115**, 1546–1553 (1959)
63. Keith, L.: Brower: Jahn-Teller-distorted nitrogen donor in laser-annealed silicon. Phys. Rev. Lett. **44**, 1627–1629 (1980)
64. Murakami, K., Kuribayashi, H., Masuda, K.: Motional effects between on-center and off-center substitutional nitrogen in silicon. Phys. Rev. B **38**, 1589–1592 (1988)
65. Messmer, R.P., Watkins, G.D.: Molecular-orbital treatment for deep levels in semiconductors: substitutional nitrogen and the lattice vacancy in diamond. Phys. Rev. B **7**, 2568–2590 (1973)
66. DeLeo, G.G., Fowler, W.B., Watkins, G.D.: Theory of off-center impurities in silicon: substitutional nitrogen and oxygen. Phys. Rev. B **29**, 3193–3207 (1984)
67. Pantelides, S.T., Harrison, W.A., Yndurain, F.: Theory of off-center impurities in semiconductors. Phys. Rev. B **34**, 6038–6040 (1986)
68. Anderson, F.G.: Pseudo-Jahn-Teller effect for distortions involving oxygen and nitrogen impurities in silicon. Phys. Rev. B **39**, 5392–5396 (1989)
69. Veinger, A.I., Zabrodski, A.G., Lomakina, G.A., Mokhov, E.N.: Sov. Phys. Solid State **28**, 917 (1986)

70. Kalabukhova, E.N., Lukin, S.N., Mokhov, E.N.: Electron spin resonance in 2 mm range of transmutation phosphorus impurity in 6H-SiC. Sov. Phys. Solid State **35**, 361–363 (1993)
71. Heissenstein, H., Peppermueller, C., Helbig, R.: Characterization of phosphorus doped *n*-type 6H-silicon carbide epitaxial layers produced by nuclear transmutation doping. J. Appl. Phys. **83**, 7542–7546 (1998)
72. Troffer, T., Peppermueller, C., Pensl, G., Rottner, K., Schoerner, A.: Phosphorus-related donors in 6H-SiC generated by ion implantation. J. Appl. Phys. **80**, 3739–3743 (1996)
73. Heissenstein, H., Helbig, R.: Characterization of SiC: P prepared by nuclear transmutation due to neutrons. Mater. Sci. Forum **353–356**, 369–372 (2001)
74. Baranov, P.G., Ilyin, I.V., Mokhov, E.N., von Bardeleben, H.J., Cantin, J.L.: EPR study of shallow and deep phosphorous centers in *6H*-SiC. Phys. Rev. B **66**, 165206 (2002)
75. Gali, A., Deak, P., Briddon, P.R., Devaty, R.P., Choyke, W.J.: Phosphorus-related deep donor in SiC. Phys. Rev. B **61**, 12602–12604 (2000)
76. Isoya, J., Ohshima, T., Morishita, N., Kamiya, T., Itoh, H., Yamasaki, S.: Pulsed EPR studies of shallow donor impurities in SiC. Phys. B **340–342**, 903–907 (2003)
77. Son, N.T., Henry, A., Isoya, J., Katagiri, M., Umeda, T., Gali, A., Janzen, E.: Electron paramagnetic resonance and theoretical studies of shallow phosphorous centers in 3*C*-, 4*H*-, and 6*H*-SiC. Phys. Rev. B **73**, 075201 (2006)
78. Rauls, E., Pinheiro, M.V.B., Greulich-Weber, S., Gerstmann, U.: Reassignment of phosphorus-related donors in SiC. Phys. Rev. B **70**, 085202 (2004)
79. Baranov, P.G., Ber, B.Ya., Ilyin, I.V., Ionov, A.N., Mokhov, E.N., Muzafarova, M.V., Kaliteevskii, M.A., Kop'ev, P.S., Kaliteevskii, A.K., Godisov, O.N., Lazebnik, I.M.: Peculiarities of neutron-transmutation phosphorous doping of 30Si enriched SiC crystals: electron paramagnetic resonance study. J. Appl. Phys. **102**, 063713 (2007)
80. Pensl, G., Helbig, R.: Silicon carbide (SiC)—recent results in physics and in technology. In: Roessler, U. (ed.) Festkoerperprobleme: Advances in Solid State Physics, vol. 30, pp. 133–156. Vieweg, Braunschweig (1990)
81. Lomakina, G.A.: Electrical properties of hexagonal SiC with N and B impurities. Sov. Phys. Solid State **7**, 475–479 (1965)
82. Anikin, M.M., Lebedev, A.A., Syrkin, A.L., Suvorov, A.V.: Investigation of deep levels in SiC by capacitance spectroscopy methods. Sov. Phys. Semicond. **19**, 69 (1985)
83. Kuwabara, H., Yamada, S.: Free-to-bound transition in β-SiC doped with boron. Phys. Stat. Sol. (a) **30**, 739–746 (1975)
84. Ikeda, M., Matsunami, H., Tanaka, T.: Site effect on the impurity levels in 4H, 6H, and 15R SiC. Phys. Rev. B **22**, 2842–2854 (1980)
85. Suttrop, W., Pensl, G., Laning, P.: Boron-related deep centers in 6H-SiC. Appl. Phys. A **51**, 231–237 (1990)
86. Ballandovich, V.S., Mokhov, E.N.: Semiconductors **29**, 187 (1995)
87. Ikeda, M., Matsunami, H., Tanaka, T.: Site-dependent donor and acceptor levels in 6H-SiC. J. Luminesc. **20**, 111–129 (1979)
88. Ludwig, G.W., Woodbury, H.H.: In Ehrenreich, H., Seitz, F., Turnbull, D. (eds.) Solid State Physics, vol. 13, pp. 223–304. Academic, New York (1962)
89. Feher, G., Hensel, J.C., Gere, E.A.: Paramagnetic resonance absorption from acceptors in silicon. Phys. Rev. Lett. **5**, 309–312 (1960)
90. Zubatov, A.G., Zaritskii, I.M., Lukin, S.N., Mokhov, E.N., Stepanov, V.G.: Sov. Phys. Solid State **27**, 197 (1985)
91. Maier, K., Schneider, J., Wilkening, W., Leibenzeder, S., Stein, R.: Electron spin resonance studies of transition metal deep level impurities in SiC. Mater. Sci. Eng. B **11**, 27–30 (1992)
92. Baran, N.P., Bratus', V.Ya., Bugai, A.A., Vikhnin, V.S., Klimov, A.A., Maksimenko, V.M., Petrenko T.L., Romanenko, V.V.: Phys. Solid State **35**, 1544 (1993)
93. Baranov, P.G., Khramtsov, V.A., Mokhov, E.N.: Chromium in silicon carbide: electron paramagnetic resonance studies. Semicond. Sci. Technol. **9**, 1340–1345 (1994)
94. Baranov, P.G., Mokhov, E.N.: Inst. Phys. Conf. Ser., No 142, Chapter 2, pp. 293–296. IOP Publishing Ltd. (1996)

95. Baranov, P.G., Mokhov, E.N.: Electron paramagnetic resonance of deep boron in silicon carbide. Semicond. Sci. Technol. **11**, 489–494 (1996)
96. Baranov, P.G., Ilyin, I.V., Mokhov, E.N.: Electron paramagnetic resonance of the group-III deep acceptor impurities in SiC. Solid State Commun. **100**, 371–376 (1996)
97. Baranov, P.G., Mokhov, E.N.: Electron paramagnetic resonance of beryllium, deep boron, and scandium acceptors in silicon carbide. Phys. Solid State **38**, 798–807 (1996)
98. Lee, K.M., Dang, Le Si, Watkins, G.D., Choyke, W.J.: Optically detected magnetic resonance study of SiC:Ti. Phys. Rev. B **32**, 2273–2284 (1985)
99. Baranov, P.G., Vetrov, V.A., Romanov, N.G., Sokolov, V.I.: Sov. Phys. Solid State **27**, 2085 (1985)
100. Baranov, P.G., Romanov, N.G.: ODMR study of recombination processes in ionic crystals and silicon carbide. Appl. Magn. Res. **2**, 361–378 (1991)
101. Baranov, P.G., Romanov, N.G.: Acceptors in silicon carbide: ODMR data. Mater. Sci. Forum **83–87**, 1207–1212 (1992)
102. Baranov, P.G.: Acceptor Impurities in Silicon Carbide: Electron Paramagnetic Resonance and Optically Detected Magnetic Resonance Studies, Defect and Diffusion Forum, vols. 148–149, pp. 129–160. Scitec Publications, Switzerland (1997)
103. Matsumoto, T., Poluektov, O.G., Schmidt, J., Mokhov, E.N., Baranov, P.G.: Electronic structure of the shallow boron acceptor in 6H-SiC:mA pulsed EPR/ENDOR study at 95 GHz. Phys. Rev. B **55**, 2219–2229 (1997)
104. van Duijn-Arnold, A., Ikoma, T., Poluektov, O.G., Baranov, P.G., Mokhov, E.N., Schmidt, J.: Electronic structure of the deep boron acceptor in boron-doped 6H-SiC. Phys. Rev. B **57**, 1607–1619 (1998)
105. van Duijn-Arnold, A., Schmidt, J., Poluektov, O.G., Baranov, P.G., Mokhov, E.N.: Electronic structure of the Be acceptor centres in 6H-SiC. Phys. Rev. B **60**, 15799–15809 (1999)
106. Petrenko, T.L., Teslenko, V.V., Mokhov, E.N.: Electron nuclear double resonance and electron structure of boron impurity centers in 6H-SiC. Sov. Phys. Semicond. **26**, 874 (1992)
107. Petrenko, T.L., Bugai, A.A., Baryakhtar, V.G., Teslenko, V.V., Khavryutchenko, V.D.: Cluster calculation of the boron centre in SiC. Semicond. Sci. Technol. **9**, 1849–1852 (1994)
108. Muller, R., Feege, M., Greulich-Weber, S., Spaeth, J.-M.: ENDOR investigation of the microscopic structure of the boron acceptor in 6H-SiC. Semicond. Sci. Technol. **8**, 1377–1384 (1993)
109. Reinke, J., Muller, R., Feege, M., Greulich-Weber, S., Spaeth, J.-M.: Endor and ODEPR investigation of the microscopic structure of the boron acceptor in 6H-SiC Mat. Sci. Forum **143–147**, 63–68 (1994)
110. Reinke, J., Greulich-Weber, S., Spaeth, J.-M., Kalabukhova, E.N., Lukin, S.N., Mokhov, E. N.: Inst. Phys. Conf. Ser. (UK) **137**, 211 (1994)
111. Adrian, F.J., Greulich-Weber, S., Spaeth, J.-M.: Further evidence for the B_{SI}^{-}-C^{+} model of the boron acceptor in 6H silicon carbide from a theoretical analysis of the hyperfine interactions. Solid State Comm. **94**, 41–44 (1995)
112. Ionov, A.N., Baranov, P.G., Ber, B.Y., Bulanov, A.D., Godisov, O.N., Gusev, A.V., Davydov, V.Yu., Ilyin, I.V., Kaliteevskii, A.K., Kaliteevski, M.A., Safronov, A.Yu., Lazebnik, I.M., Pohl, H.-J., Riemann, H., Abrosimov, N.V., Kop'ev, P.S.: Neutron transmutation doping of silicon ^{30}Si monoisotope with phosphorus. Tech. Phys. Lett. **32**, 550–553 (2006)
113. Bachelet, G., Baraff, G.A., Schulter, M.: Defects in diamond: the unrelaxed vacancy and substitutional nitrogen. Phys. Rev. B **24**, 4736–4744 (1981)
114. Gray, H.B.: In: Benjamin, W.A. (ed.) Electrons and Chemical Bonding. New York, Amsterdam (1965)
115. Romanov, N.G., Vetrov, V.A., Baranov, P.G., Mokhov, E.N., Oding, V.G.: Sov. Tech. Phys. Lett. **11**, 483 (1985)

116. Baranov, P.G., Romanov, N.G., Vetrov, V.A., Oding, V.G.: In: Anastassakis, E.M., Joannopoulos, J.D. (eds.) Proceedings of the 20th International Conference on the Physics of Semiconductors, vol. 3, p. 1855. World Scientific, Singapore (1990)
117. Baranov, P.G., Mokhov, E.N., Hofstaetter, A., Scharmann, A.: Electron-nuclear double resonance of deep-boron acceptors in silicon carbide. JETP Lett. **63**, 848–854 (1996)
118. Meyer, B.K., Hofstaetter, A., Baranov, P.G.: On the identification of Al related deep centre in 4H-SiC—self-compensation in SiC? Mater. Sci. Forum **264–268**, 591–594 (1998)
119. Watkins, G.D.: Defects in Irradiated Silicon: electron paramagnetic resonance and electron-nuclear double resonance of the aluminum-vacancy pair. Phys. Rev. **155**, 802–815 (1967)
120. Maiwald, M., Schirmer, O.F.: O^- dynamic Jahn-Teller polarons in $KTaO_3$. Europhys. Lett. **64**, 776–782 (2003)
121. Hofstaetter, A., Meyer, B.K., Scharmann, A., Baranov, P.G., Ilyin, I.V., Mokhov, E.N.: X-Band ENDOR of boron and beryllium acceptors in silicon carbide Mater. Sci. Forum **264–268**, 595–598 (1998)
122. Baranov, P.G., Ilyin, I.V., Mokhov, E.N., Roenkov, A.D., Khramtsov, V.A.: Electron paramagnetic resonance of scandium in silicon carbide. Phys. Solid State **39**, 44–48 (1997)
123. Monemar, B., Paskov, P.P., Bergman, J.P., Toropov, A.A., Shubina, T.V.: Recent developments in the III-nitride materials. Phys. Status Solidi B **244**, 1759–1768 (2007)
124. Baraff, G.A., Kane, E.O., Schluter, M.: Silicon vacancy: a possible "Anderson negative-*U*" system. Phys. Rev. Lett. **43**, 956–959 (1979)
125. Watkins, G.D., Troxell, J.R.: Negative-*U* properties for point defects in silicon. Phys. Rev. Lett. **44**, 593–596 (1980)
126. Chadi, D.J., Chang, K.J.: Theory of the atomic and electronic structure of *DX* centers in GaAs and $Al_xGa_{1-x}As$ alloys. Phys. Rev. Lett. **61**, 873–876 (1988)
127. Zeisel, R., Bayerl, M.W., Goennenwein, S.T.B., Dimitrov, R., Ambacher, O., Brandt, M.S., Stutzmann, M.: *DX*-behavior of Si in AlN. Phys. Rev. B **61**, R16283–R16286 (2000)
128. Bayerl, M.W., Brandt, M.S., Graf, T., Ambacher, O., Majewski, J.A., Stutzmann, M., As, D. J., Lischka, K.: *g* values of effective mass donors in $Al_xGa_{1-x}N$ alloys. Phys. Rev. B **63**, 165204 (2001)
129. Orlinskii, S.B., Schmidt, J., Baranov, P.G., Bickermann, M., Epelbaum, B.M., Winnacker, A.: Observation of the triplet metastable state of shallow donor pairs in AlN crystals with a negative-*U* behavior: a high-frequency EPR and ENDOR study. Phys. Rev. Lett. **100**, 256404 (2008)
130. Soltamov, V.A., Ilyin, I.V., Soltamova, A.A., Mokhov, E.N., Baranov, P.G.: Identification of the deep level defects in AlN single crystals by electron paramagnetic resonance. J. Appl. Phys. **107**, 113515 (2010)
131. Glaser, E.R., Kennedy, T.A., Carlos, W.E., Freitas Jr., J.A., Wickenden, A.E., Koleske, D. D.: Optically detected electron-nuclear double resonance of epitaxial GaN. Phys. Rev. B **57**, 8957–8965 (1998)
132. Son, N.T., Bickermann, M., Janzen, E.: Shallow donor and *DX* states of Si in AlN. Appl. Phys. Lett. **98**, 092104 (2011)
133. Mason, P.M., Przybylinska, H., Watkins, G.D., Choyke, W.J., Slack, G.A.: Optically detected electron paramagnetic resonance of AlN single crystals. Phys. Rev. B **59**, 1937–1947 (1999)
134. Evans, S.M., Giles, N.G., Halliburton, L.E., Slack, G.A., Schujman, S.B., Schowalter, L.J.: Electron paramagnetic resonance of a donor in aluminum nitride crystals. Appl. Phys. Lett. **88**, 062112 (2006)
135. Hofmann, D.M., Hofstaetter, A., Leiter, F., Zhou, H., Henecker, F., Meyer, B.K., Orlinskii, S.B., Schmidt, J., Baranov, P.G.: Hydrogen: a relevant shallow donor in zinc oxide. Phys. Rev. Lett. **88**, 045504 (2002)
136. Orlinskii, S.B., Schmidt, J., Baranov, P.G., Dyakonov, V.: Identification of shallow Al donors in Al-doped ZnO nanocrystals: EPR and ENDOR spectroscopy. Phys. Rev. B **77**, 115334 (2008)

137. Duijn-Arnold, A.v., Ikoma, T., Poluektov, O.G., Baranov, P.G., Mokhov, E.N., Schmidt, J.: Electronic structure of the deep boron acceptor in boron-doped 6*H*-SiC. Phys. Rev. B **57**, 1607–1619 (1998)
138. Park, C.H., Chadi, D.J.: Stability of deep donor and acceptor centers in GaN, AlN, and BN Phys. Rev. B **55**, 12995–13001 (1997)
139. Boguslawski, P., Bernholc, J.: Doping properties of C, Si, and Ge impurities in GaN and AlN. Phys. Rev. B **56**, 9496–9505 (1997)
140. Van de Walle, C.G.: DX-center formation in wurtzite and zinc-blende $Al_xGa_{1-x}N$. Phys. Rev. B **57**, R2033–R2036 (1998)
141. Mokhov, E.N., Avdeev, O.V., Barash, I.S., Chemekova, T.Y., Roenkov, A.D., Segal, A.S., Wolfson, A.A., Makarov, Yu.N., Ramm, M.G., Helava, H.: Sublimation growth of AlN bulk crystals in Ta crucibles. J. Cryst. Growth **281**, 93–100 (2005)
142. Slack, G.A., Schowalter, L.J., Morelli, D., Freitas, J.A.: Some effects of oxygen impurities on AlN and GaN. J. Cryst. Growth **246**, 287–298 (2002)
143. Makarov, Yu.N., Avdeev, O.V., Barash, I.S., Bazarevskiy, D.S., Chemekova, T.Yu., Mokhov, E.N., Nagalyuk, S.S., Roenkov, A.D., Segal, A.S., Vodakov, Yu.A., Ramm, M.G., Davis, S., Huminic, G., Helava, H.: Experimental and theoretical analysis of sublimation growth of AlN bulk crystals. J. Cryst. Growth **310**, 881–886 (2008)
144. Lu, P., Collazo, R., Dalmau, R.F., Durkaya, G., Dietz, N., Sitar, Z.: Different optical absorption edges in AlN bulk crystals grown in *m*- and *c*-orientations. Appl. Phys. Lett. **93**, 131922 (2008)
145. Yang, S., Evans, S.M., Halliburton, L.E., Slack, G.A., Shujman, S.B., Morgan, K.E., Bondokov, R.T., Mueller, S.G.: Electron paramagnetic resonance of Er^{3+} ions in aluminum nitride. J. Appl. Phys. **105**, 023714 (2009)
146. Bickermann, M., Epelbaum, B.M., Filip, O., Heimann, P., Nagata, S., Winnacker, A.: UV transparent single-crystalline bulk AlN substrates. Phys. Status Sol. C **7**, 21 (2010)
147. Stamp, C., Van de Walle, C.G.: Theoretical investigation of native defects, impurities, and complexes in aluminum nitride. Phys. Rev. B **65**, 155212 (2002)
148. Vainer, V.S., Veinger, V.I., Il'in, V.A., Tsvetkov, V.F.: Electron spin resonance associated with the triplet state of secondary thermal defects in 6H-SiC. Sov. Phys. Solid State **22**, 2011–2013 (1980) [Fiz. Tverd. Tela. (Leningrad) **22**, 3436–3439 (1980)]
149. Vainer, V.S., Il'in, V.A.: Electron spin resonance of exchange-coupled vacancy pairs in hexagonal silicon carbide. Sov. Phys. Solid State **23**, 2126–2133 (1981) [Fiz. Tverd. Tela. (Leningrad) **23**, 3659–3671 (1981)]
150. Pavlov, N.M., Iglitsyn, M.I., Kosaganova, M.G., Solomatin, V.N.: Spin-1 centers in silicon carbide irradiated with neutrons and α-particles. Sov. Phys. Semicond. **9**, 845–849 (1975) [Fiz. Tekn. Poluprovodn. **9**, 1279–1285 (1975)]
151. Romanov, N.G., Vetrov, V.A., Baranov, P.G.: Optically detected magnetic resonance in silicon carbide containing radiation and thermal defects. Sov. Phys. Semicond. **20**, 96 (1986)
152. Jaccard, C., Ruedin, Y., Aegerter, M., Schnegg, P.-A.: Weak magnetic field enhancement of the luminescence from F centre pairs in alkali halides. Phys. Stat. Sol. B **50**, 187–198 (1972)
153. Gourary, B.S., Adrian, F.S.: Wave functions for electron-excess color centers in alkali halide crystals. Solid State Phys. **10**, 127–247 (1960)
154. Bennebroek, M.T., Arnold, A., Poluektov, O.G., Baranov, P.G., Schmidt, J.: Shallow electron centers in silver halides. Phys. Rev. B **54**, 11276–11289 (1996). and references therein
155. Clementi, E., Roetti, C.: Roothaan-Hartree-Fock atomic wavefunctions: basis functions and their coefficients for ground and certain excited states of neutral and ionized atoms, $Z \leq 54$. At. Data Nucl. Data Tables **14**, 177–478 (1974)
156. Sonder, E., Schweinler, H.C.: Magnetism of interacting donors. Phys. Rev. **117**, 1216–1221 (1960)
157. Abragam, A., Bleaney, B.: Electron Paramagnetic Resonance of Transition Ions. Oxford University Press, Oxford (1970)

158. Yavkin, B.V., Soltamov, V.A., Babunts, R.A., Anisimov, A.N., Baranov, P.G., Shakhov, F. M., Kidalov, S.V., Vul', A.Ya., Mamin, G.V., Orlinskii, S.B.: Defects in nanodiamonds: application of high-frequency cw and pulse EPR, ODMR. Appl. Magn. Reson. **45**, 1035–1049 (2014)
159. Carlos, W.E., Freitas, J.A., Asif Khan, M., Olson, D.T., Kuznia, J.N.: Electron-spin-resonance studies of donors in wurtzite GaN. Phys. Rev. B **48**, 17878–17884 (1993)
160. Watkins, G.D.: EPR of defects in semiconductors: past, present, future. Phys. Solid State **41**, 746–750 (1999)
161. Weber, E.R.: Transition metals in silicon. Appl. Phys. A **30**, 1–22 (1983)
162. Ludwig, G.W., Woodbury, H.H.: Electron Spin Resonance in Semiconductors. In: Seitz, F., Turnbull, D. (eds.) Solid State Physics, vol. 13, pp. 223–304. Academic Press, N.Y. (1962)
163. Baur, J., Kunzer, M., Schneider, J.: Transition metals in SiC polytypes, as studied by magnetic resonance techniques. Phys. Stat. Sol. (a) **162**, 153–172 (1997)
164. Baranov, P.G.: Radiospectroscopy of wide-gap semiconductors: SiC and GaN. Phys. Solid State **41**, 712–715 (1999)
165. Baranov, P.G., Khramtsov, V.A., Mokhov, E.N.: Chromium in silicon carbide: electron paramagnetic resonance studies. Semicond. Sci. Technol. **9**, 1340–1345 (1994)
166. Dombrowski, K.F., Kunzer, M., Kaufmann, U., Schneider, J., Baranov, P.G., Mokhov, E.N.: Identification of molybdenum in 6H-SiC by magnetic resonance techniques. Phys. Rev. B **54**, 7323–7327 (1996)
167. Baur, J., Kunzer, M., Dombrowski, K.F., Kaufmann, U., Schneider, J., Baranov, P.G., Mokhov, E.N.: Identification of multivalent molybdenum impurities in SiC crystals. Inst. Phys. Conf. Ser. **155**, 933–936 (1997)
168. Baranov, P.G., Ilyin, I.V., Mokhov, E.N.: Identification of iron transition group trace impurities in GaN bulk crystals by electron paramagnetic resonance. Solid State Commun. **101**, 611–615 (1997)
169. Baranov, P.G., Ilyin, I.V., Mokhov, E.N.: Electron paramagnetic resonance of erbium in bulk silicon carbide crystals. Solid State Commun. **103**, 291–295 (1997)
170. Baranov, P.G., Ilyin, I.V., Mokhov, E.N., Roenkov, A.D., Khramtsov, V.A.: Transition and rare-earth elements in the SiC and GaN wide-gap semiconductors: recent EPR studies. Phys. Solid State **41**, 783–785 (1999)
171. Baranov, P.G., Ilyin, I.V., Mokhov, E.N., Khramtsov, V.A.: Identification of iron in 6H–SiC crystals by electron paramagnetic resonance. Semicond. Sci. Technol. **16**, 39–43 (2001)
172. Baranov, P.G., Ilyin, I.V., Mokhov, E.N., Khramtsov, V.A.: Identification of iron and nickel in 6H-SiC by electron paramagnetic resonance. Mater. Sci. Forum **353–356**, 529–532 (2001)
173. Abragam, A., Bleaney, B.: Electron Paramagnetic Resonance of Transition Ions. Clarendon Press, Oxford (1970)
174. Schneider, J.: Notizen: Paramagnetische resonanz von Fe^{+++}-Ionen in synthetischen ZnO-kristallen. Z. Naturforsch A **17**, 189–190 (1962)
175. Maier, K., Kunzer, M., Kaufmann, U., Schneider, J., Monemar, B., Akasaki, I., Amano, H.: Iron acceptors in gallium nitride (GaN). Mat. Sci. Forum **143–147**, 93–98 (1994)
176. Baur, J., Maier, K., Kunzer, M., Kaufmann, U., Schneider, J.: Determination of the GaN/AlN band offset via the (-/0) acceptor level of iron. Appl. Phys. Lett. **65**, 2211–2213 (1994)
177. Son, N.T., Trinh, X.T., Gällström, A., Leone, S., Kordina, O., Janzén, E., Szász, K., Ivády, V., Gali, A.: Electron paramagnetic resonance and theoretical studies of Nb in 4H- and 6H-SiC. J. Appl. Phys. **112**, 083711 (2012)
178. Soltamov, V.A., Ilyin, I.V., Soltamova, A.A., Mokhov, E.N., Baranov, P.G.: Identification of the deep level defects in AlN single crystals by electron paramagnetic resonance. J. Appl. Phys. **107**, 113515 (2010)
179. Baur, J., Kunzer, M., Maier, K., Kaufmann, U., Schneider, J.: Determination of the GaN/A1 N band discontinuities via the $(^-/_0)$ acceptor level of iron. Mater. Sci. Eng., B **29**, 61–64 (1995)

180. Kaufmann, U.: EPR and optical absorption of Fe^{+}, Fe^{2+}, Fe^{3+}, and Fe^{4+} on tetragonal sites in $CdSiP_2$. Phys. Rev. B **14**, 1848–1857 (1976)
181. Gehlhoff, W., Azamat, D., Hoffmann, A., Dietz, N., Voevodina, O.V.: Transition metals in $ZnGeP_2$ and other II–IV–V_2 compounds. Phys. B **376–377**, 790–794 (2006)
182. Langer, J.M., Heinrich, H.: Deep-level impurities: a possible guide to prediction of band-edge discontinuities in semiconductor heterojunctions. Phys. Rev. Lett. **55**, 1414–1417 (1985)
183. Heitz, R., Maxim, P., Eckey, L., Thurian, P., Hoffmann, A., Broser, I., Pressel, K., Meyer, B. K.: Excited states of Fe^{3+} in GaN. Phys. Rev. B **55**, 4382–4387 (1997)
184. Nepal, N., Nakarmi, M.L., Ang, H.U., Lin, J.Y., Jiang, H.X.: Growth and photoluminescence studies of Zn-doped AlN epilayers. Appl. Phys. Lett. **89**, 192111 (2006)
185. Morkoc, H., Strite, S., Gao, G.B., Lin, M.E., Sverdlov, B., Burns, M.: Large-band-gap SiC, III-V nitride, and II-VI ZnSe-based semiconductor device technologies. J. Appl. Phys. **76**, 1363–1398 (1994)
186. Vodakov, YuA, Mokhov, E.N., Roenkov, A.D., Saidbekov, D.T.: Effect of crystallographic orientation on the polytype stabilization and transformation of silicon carbide. Phys. Stat. Sol. (a) **51**, 209–215 (1979)
187. Vodakov, YuA, Karklina, M.I., Mokhov, E.N., Roenkov, A.D.: Inorganic Mater. **17**, 537 (1980)
188. Wetzel, C., Volm, D., Meyer, B.K., Pressel, K., Nilsson, S., Mokhov, E.N., Baranov, P.G.: GaN on 6H-SiC—structural and optical properties. Mat. Res. Soc. Symp. Proc. **339**, 453–458 (1994)
189. Wetzel, C., Volm, D., Meyer, B.K., Pressel, K., Nilsson, S., Mokhov, E.N., Baranov, P.G.: GaN epitaxial layers grown on 6H-SiC by the sublimation sandwich technique. Appl. Phys. Lett. **65**, 1033–1035 (1994)
190. Pressel, K., Nilsson, S., Heitz, R., Hoffmann, A., Meyer, B.K.: Photoluminescence study of the 1.047 eV emission in GaN. J. Appl. Phys. **79**, 3214–3218 (1996)
191. Baranov, P.G., Ilyin, I.V., Mokhov, E.N.: Identification of iron transition group trace impurities in GaN bulk crystals by electron paramagnetic resonance. Solid State Commun. **101**, 611–615 (1997)
192. Abragam, Bleaney, B.: Electron Paramagnetic Resonance of Transition Ions, vol. 1, ch. 7. Clarendon Press, Oxford (1970)
193. Hausmann, A.: The cubic field parameter of $^6S52$ ions in zinc oxide crystals. Sol. St. Comm. **6**, 457–459 (1968)
194. Holton, W.C., Schneider, J., Estle, T.L.: Electron paramagnetic resonance of photosensitive iron transition group impurities in ZnS and ZnO. Phys. Rev. **133**, A1638–A1641 (1964)
195. Graf, T., Gjukic, M., Hermann, M., Brandt, M.S., Stutzmann, M., Ambacher, O.: Spin resonance investigations of Mn^{2+} in wurtzite GaN and AlN films. Phys. Rev. B **67**, 165215 (2003)
196. Masterov, V.F.: Electronic structure of the rare earth impurities in III–V compounds. Fiz. Tekh. Poluprovodn. **27**, 1435–1452 (1993)
197. Michel, J., Benton, J.L., Ferrante, R.F., Jacobson, D.C., Eaglesham, D.J., Fitzgerald, E.A., Xie, Y.-H., Poate, J.M., Kimerling, J.: Impurity enhancement of the 1.54-μm Er3+ luminescence in silicon. Appl. Phys. **70**, 2672–2678 (1991)
198. Jantsch, W., Przybylinska, H.: In: Schefler, M., Zimmermann, R. (eds.) Proceedings of 23rd International Conference on Physics Semiconductor (Berlin, July 21–26, 1996), pp. 3025–3032. World Scientific, Singapore-New Jersey-London-Hong-Kong (1996)
199. Choyke, W.J., Devaty, R.P., Clemen, L.L., Yoganathan, M., Pensl, G., Haessier, Ch.: Intense erbium-1.54-μm photoluminescence from 2 to 525 K in ion-implanted 4H, 6H, 15R, and 3C SiC. Appl. Phys. Lett. **65**, 1668–1670 (1994)
200. Carey, J.D., Donegan, J.F., Barklie, R.C., Priolo, F., Franzo, G., Coffa, S.: Electron paramagnetic resonance of erbium doped silicon. Appl. Phys. Lett. **69**, 3854–3856 (1996)
201. Vodakov, YuA, Mokhov, E.N., Ramm, M.G., Roenkov, A.D.: Epitaxial growth of SiC layers by sublimation sandwich method. Krist. und Techn. **14**, 729–740 (1979)

202. Baranov, P.G., Il'in, I.V., Mokhov, E.N., Pevtsov, A.B., Khramtsov, V.A.: Erbium in silicon carbide crystals: EPR and high-temperature luminescence. Phys. Solid State **41**, 32–34 (1999)
203. Babunts, R.A., Vetrov, V.A., Il'in, I.V., Mokhov, E.N., Romanov, N.G., Khramtsov, V.A., Baranov, P.G.: Properties of erbium luminescence in bulk crystals of silicon carbide. Phys. Solid State **42**, 829–835 (2000)
204. Baranov, P.G., Zhekov, V.I., Murina, T.M., Prokhorov, A.M. Khramtsov, V.A.: Sov. Phys. Solid State **29**, 723 (1987) [Fiz. Tverd. Tela **29**, 1261 (1987)]
205. Lea, K.R., Leask, M.J.M., Wolf, W.P.: The raising of angular momentum degeneracy of *f*-Electron terms by cubic crystal fields. J. Phys. Chem. Solids **23**, 1381–1405 (1962)
206. Watts, R.K., Holton, W.C.: Paramagnetic-resonance studies of rare-earth impurities in II–VI compounds. Phys. Rev. **173**, 417–426 (1968)
207. Kunzer, M., Müller, H.D., Kaufmann, U.: Magnetic circular dichroism and site-selective optically detected magnetic resonance of the deep amphoteric vanadium impurity in 6*H*-SiC. Phys. Rev. B **48**, 10846–10854 (1993)
208. Choyke, W.I., Devaty, P.R., Yoganathan, M., Pensl, G., Edmond, J.A.: In: Ammerlaan, C.A. J., Pajot, B. (eds.) Proceedings of International Conference on Shallow-Level Centers in Semiconductors (Amsterdam, 17–19 July, 1996), pp. 297–302. World Scientific Publishing Company (1997)
209. Il'in, N.P., Masterov, V.F.: A calculation of energy spectra of binary semiconductors doped with rare-earth elements. Fiz. Tekh. Poluprovodn. **29**, 1591–1602 (1995)
210. Il'in, N.P., Masterov, V.F.: Electronic structure of the $Er\text{-}O_6$ complex in silicon. Semicond. **31**, 886–892 (1997)
211. Przybylinska, H., Jantsch, W., Suprun-Belevitch, Yu., Stepikhova, M., Palmetshofer, L., Hendorfer, G., Kozanecki, A., Wilson, R.J., Sealy, B.J.: Optically active erbium centers in silicon. Phys. Rev. B **54**, 2532–2547 (1996)
212. Kimerling, L.S., Kolenbrander, K.D., Michel, J., Palm, J.: Light emission from silicon. Solid State Phys. **50**, 333–381 (1996)

Chapter 5
Magnetic Resonance in Semiconductor Micro- and Nanostructures

Nowadays semiconductor and solid-state physics appears to be the physics of systems with reduced dimensionality. Fabrication of single or periodic potential wells by simply combining two materials with different energy gaps and spatial dimensions confining the motion of electrons and holes results in exciting new effects that originate in the size dependence of quantum phenomena. The role of bandgap engineering, as it affects charge distribution and confinement, electronic wave-function mixing, and carrier transport behavior in semiconductor heterostructures, is critically important to the realization of next-generation device technology.

5.1 High-Frequency EPR and ENDOR Spectroscopy on Semiconductor Quantum Dots

High-frequency electron paramagnetic resonance (EPR) and electron-nuclear double resonance (ENDOR) spectroscopy is shown to be excellent tools for the investigation of the electronic properties of semiconductor quantum dots. The great attractions of these techniques are that, in contrast to optical methods, they allow the identification of the dopants and provide information about the spatial distribution of the electronic wave function. This latter aspect is particularly attractive because it allows for a quantitative measurement of the effect of confinement on the shape and properties of the wave function.

EPR and ENDOR results are presented on doped ZnO quantum dots. Shallow donors (SDs), related to interstitial Li and Na atoms and substitutional Al atoms, have been identified in this material by pulsed high-frequency EPR and ENDOR spectroscopy. The shallow character of the wave function of the donors is evidenced by the multitude of ENDOR transitions of the ^{67}Zn nuclear spins and by the hyperfine (HF) interaction of the ^{7}Li, ^{23}Na and ^{27}Al nuclear spins that are much smaller than for atomic lithium, sodium and aluminum free atoms. The EPR signal

P.G. Baranov et al., *Magnetic Resonance of Semiconductors and Their Nanostructures*, Springer Series in Materials Science 253,
DOI 10.1007/978-3-7091-1157-4_5

of an exchange-coupled pair consisting of a shallow donor and a deep acceptor has been identified in ZnO nanocrystals with radii smaller than 1.5 nm. From ENDOR experiments it is concluded that the deep Na-related acceptor is located at the interface of the ZnO core and the $Zn(OH)_2$ capping layer, while the shallow donor is in the ZnO core.

The spatial distribution of the electronic wave function of a shallow donor in ZnO semiconductor QDs has been determined in the regime of quantum confinement by using the nuclear spins as probes. Hyperfine interactions as monitored by ENDOR spectroscopy quantitatively reveal the transition from semiconductor to molecular properties upon reduction of the size of the nanoparticles. In addition, the effect of confinement on the g-factor of SDs in ZnO as well as in CdS QDs is observed.

It is shown that an almost complete dynamic nuclear polarization (DNP) of the ^{67}Zn nuclear spins in the core of ZnO quantum dots and of the ^{1}H nuclear spins in the $Zn(OH)_2$ capping layer can been obtained. This DNP is achieved by saturating the EPR transition of the shallow donors present in the QDs with resonant high-frequency microwaves at low temperatures. This nuclear polarization manifests itself as a hole and an antihole in the EPR absorption line of the SD in the QDs and a shift of the hole (antihole). The enhancement of the nuclear polarization opens the possibility to study semiconductor nanostructures with nuclear magnetic resonance (NMR) techniques.

Co- and Mn-doped ZnO quantum dots with $ZnO/Zn(OH)_2$ core-shell structure were studied using high-frequency EPR, ESE and ENDOR at low temperature. The shape of the EPR spectrum of cobalt ions was observed to change as a result of Co^{2+} coupling with optically created shallow donors. This, along with a shift of SDs line, is a direct demonstration of interaction between the magnetic ion and donor electron in confined system of ZnO QD. ENDOR resonance of the ^{1}H hydrogen nuclei detected by the EPR signal of Co^{2+} and Mn^{2+} evidence the hyperfine coupling between these ions, located in the ZnO core, and the protons outside the quantum dot core in the shell.

By means of ENDOR, it is shown that the Al impurity, which acts as a shallow donor in ZnO, leads to a significant reduction of the electric field gradient in ZnO single crystals. In ZnO quantum dots, however, the gradient on the Al sites remains virtually unchanged. When the Zn^{2+} ion is substituted by Mn^{2+} in a ZnO single crystal, the electric field gradient slightly increases (by about 20%). Therefore, the Mn^{2+} ions can be used as probes to monitor the electric field gradients in ZnO crystals.

The intentional introduction of impurities is fundamental for the control of the electronic and optical properties of bulk semiconductors, and has led to a myriad of technological applications. These successes have stimulated similar efforts to dope colloidal semiconductor nanocrystals. The remarkable and attractive feature of colloidal semiconductor nanocrystals is that, owing to their nanoscale dimensions, size effects can be fully exploited to tailor the material properties [1]. Quantum confinement effects become increasingly important as the dimensions of the nanocrystal decrease below a certain critical limit (viz., the spatial extension of the electron wavefunction in the material), leading to size- and shape-dependent electronic structure. Further, as the size of a nanocrystal decreases, the surface to volume ratio increases dramatically. This has important consequences, one of them being that the nanocrystal becomes easily dispersible in solvents (i.e., stable colloidal suspensions can be obtained), making fabrication and processing in solution possible, which is an essential advantage of colloidal nanocrystals over nanomaterials prepared by other techniques. Besides, colloidal chemistry methods are cheaper and easier to upscale, and are also highly versatile in terms of composition, size, shape and surface control. The potential of doped colloidal semiconductor nanocrystals has spurred an intense research activity over the past decades. Unfortunately, the efforts to dope semiconductor nanocrystals in a controlled way have met with limited success. One of the difficulties in the field is that the optical spectroscopic techniques, that have been mainly used to study doped nanocrystals, fail to identify the chemical nature of the dopants and their location in the nanocrystal. EPR and ENDOR techniques are methods of choice to identify dopants in these nanocrystals and to obtain information about their electronic properties that remain hidden for optical spectroscopic techniques [2].

ZnO with a direct band gap of 3.3–3.4 eV, attracts considerable attention [3–12]. The attraction of ZnO quantum dots is that the confinement of the electronic wave function allows the tuning of the optical and electronic properties. The effect of confinement on the electronic energy levels can easily be made visible by the change in the optical absorption spectra. The high free-exciton binding energy of 60 meV makes excitons stable at room temperature. This binding energy increases further in confined systems resulting in a luminescence efficiency of ZnO nanocrystals that is much higher than that of bulk ZnO crystals. Doped ZnO nanocrystals, which can be easily processed at temperatures much lower than those for bulk ZnO crystals, are of particular interest because of their potential for use in light-emitting devices.

The fabrication of nanocomposites by combining ZnO nanocrystals and conjugated polymers is an attractive field in organic optoelectronics because the efficient luminescence from both materials is combined. The expectation is that these materials can be used in forward flat panel displays and in lighting applications and may allow the realization of photo-induced charge transfer in organic-inorganic hybrids for photovoltaic applications [13].

The group III elements (Al, Ga and In) are expected to form shallow donors in single crystals of ZnO by replacing Zn atoms. Indeed, by using EPR spectroscopy, In and Ga shallow donors were identified on the basis of the resolved hyperfine (HF) structure in the EPR lines [10, 14–16]. However, since such a hyperfine splitting is absent in the EPR signals of Al-doped ZnO single crystals, the identity of Al as the core of the shallow donor has not been unambiguously ascertained.

As yet, Al, Ga and In have not been identified as shallow donors in ZnO nanocrystals. Stimulated by the results of high-frequency EPR and ENDOR investigations, that allowed the identification of shallow donors in bulk wide-band-gap semiconductors, such as AgCl, ZnO and AlN, and the spatial distribution of their wave functions [17–20], it have been decided to apply these techniques to doped semiconductor nanocrystals. In this chapter, the results obtained on ZnO nanocrystals are reported. The reason to choose this material is motivated to a large extent by the ease with which ZnO nanocrystals can be produced by colloidal chemistry methods.

The first, rather unexpected, result of the EPR and ENDOR investigations is that interstitial Li and Na form shallow donors in ZnO nanocrystals [2]. The identification was based on the observation of ENDOR signals of the ^{7}Li ($I = 3/2$) and ^{23}Na ($I = 3/2$) spins in Li-doped and Na-doped ZnO nanocrystals. These observations demonstrate the attraction of EPR and ENDOR spectroscopy for studying semiconductor nanostructures, since here traditional methods of measuring n- and p-type conductivity as applied to bulk semiconductors cannot be used.

The identification of Li and Na as interstitial shallow donors in ZnO nanocrystals is of fundamental interest, but is probably not of great importance for practical applications in devices where one needs to control the concentration of donor impurities. For this purpose one would prefer to dope the particles with substitutional impurities like Al, Ga or In. By applying the EPR and ENDOR techniques to ZnO nanoparticles loaded with Al one was able to show that Al forms a shallow substitutional donor in ZnO nanoparticles. This identification was ascertained by the EPR signal of this donor, the ENDOR signals of the ^{27}Al nuclear spin and of the ENDOR signal of a multitude of ^{67}Zn nuclear spins.

A prerequisite for the observation of the EPR signal of the unpaired spin of shallow donors at liquid-helium temperatures is that the ZnO QDs are first illuminated with above-band-gap light. This observation shows that there must be deep acceptors present in the nanocrystals that capture the thermally excited donor electrons at room temperature. Apparently these electrons remain frozen at the acceptor when the material is cooled in the dark to low temperature. The above-band-gap light transfers the electron from the acceptor to the donor and makes both sites paramagnetic.

The combination of EPR and optical experiments allowed to demonstrate [7] that donor-acceptor pairs are formed in the confined structure of ZnO nanoparticles between the shallow, interstitial Li donor and a deep Na-related acceptor. From ENDOR experiments it is concluded that these deep acceptors are located at the ZnO/Zn$(OH)_2$ interface.

The saturation of the EPR transition of the shallow donors in ZnO single crystals at high frequency and low temperature leads to an almost complete polarization of the ^{67}Zn ($I = 5/2$) nuclear spins [20]. During the EPR experiments on the SDs in ZnO quantum dots it was observed that prolonged irradiation of the EPR transition of this donor produces a hole in the EPR resonance line. It was shown [9] that this hole burning is caused by an almost complete polarization of the ^{67}Zn nuclear spins in ZnO core and of ^{1}H ($I = 1/2$) spins of the Zn$(OH)_2$ capping layer.

The preparation of the samples of free-standing hydroxyl-capped ZnO nanocrystals in the form of dry powders was achieved using a modified version of methods reported in the literature [21–24]. The method is based on the hydrolysis of Zn^{2+} ions in absolute alcohols (ethanol or 1-butanol), using either LiOH $\cdot$ H_2O for the Li-doped nanocrystals [3, 5] or NaOH for the Na-doped nanocrystals [7]. LiOH $\cdot$ H_2O in ethanol is quickly injected into a Zn(Acetate)$_2$ · 2H$_2$O ethanolic solution under vigorous stirring at room temperature. As a result suspension of negatively charged ZnO nanocrystals is formed. Free-standing hydroxyl-capped ZnO:Li nanocrystals ([Li] ZnO $\leq$ 0.4 at%) in the form of dry powders are obtained by adding heptane, isolating the precipitate by centrifugation, and drying under vacuum. Nanocrystal diameter: (2–6 nm ± 10%), was controlled by growth duration (5 min to 2 days). Size determined by TEM and powder X-ray diffraction, based on the peak broadening due to the finite crystallite sizes (Scherrer's equation), in addition, the optical properties of ZnO nanocrystals were controlled by ultraviolet (UV)-visible absorption spectroscopy, based on the size dependence of the band gap owing to quantum confinement effects and using a calibration curve [21, 22].

Free-standing ZnO nanocrystals with different diameters of 2.2, 2.34, 2.8, 3.4, 4.0, 4.2, 5.6, and 6 nm were prepared. The surface of the as-prepared ZnO dots is capped by a thin layer (about one monolayer) of Zn(OH)$_2$ and thus the dots consist of a ZnO/Zn(OH)$_2$ core-shell structure. Co- or Mn-doped ZnO nanocrystals are obtained by partially replacing Zn(Acetate)$_2$ · 2H$_2$O by Co(Acetate)$_2$ · 4H$_2$O or Mn(Acetate)$_2$ · 4H$_2$O.

Free-standing polyphosphate capped CdS nanocrystals with diameters of 2 and 3.3 nm were prepared using a modified version of the method described in [25] by replacing the Zn salts by Cd salts. The nanocrystals were illuminated with UV above-band-gap light to create paramagnetic electron and hole centres.

Figure 5.1a shows absorption and photoluminescence (PL) spectra of ZnO nanocrystals with an average diameter of 2.9 nm. Shift to higher energies with decreasing nanocrystal size could be observed. The exciton PL (narrow line) and

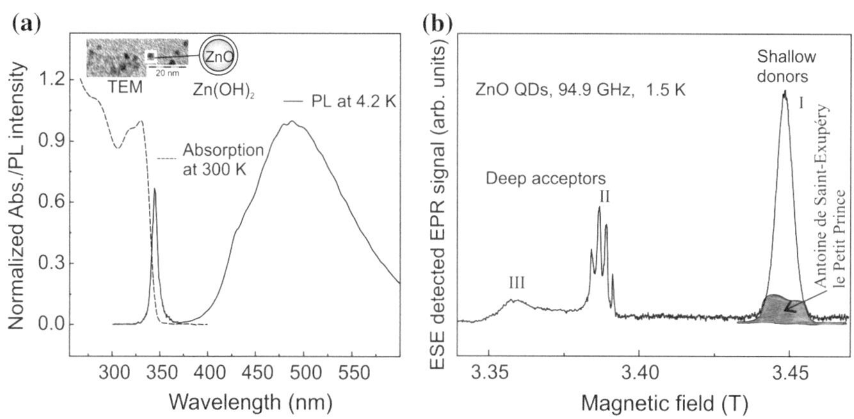

Fig. 5.1 **a** Absorption and photoluminescence of ZnO QDs with an average diameter of 2.9 nm; (*inset*) transmission electron microscopy (TEM) image of ZnO QDs and a model of the ZnO/Zn(OH)$_2$ core-shell structure. **b** The ESE-detected EPR spectrum at 94.9 GHz and T = 1.5 K of a dry powder sample of Li-doped ZnO quantum dots with an average diameter of 3.4 nm after 30 min UV irradiation; hat from The Little Prince by Antoine de Saint-Exupéry, symbolizing the hidden nature of the unresolved single EPR line of the shallow donor (SD)

defect related PL (broad band) dominate; TEM image of ZnO nanoparticles and a model of the ZnO/Zn(OH)$_2$ core-shell structure are presented in the inset.

The EPR and ENDOR experiments were performed at temperatures ranging from 1.2 to 10 K using pulsed EPR spectrometers operating at 95 GHz [26] and 275 GHz [27]. The great advantage of working at this high microwave frequency is that a high spectral resolution is obtained in the EPR as well as in the ENDOR spectra. In addition the use of pulsed microwave techniques facilitates considerably the observation of the ENDOR spectra. The EPR spectra were recorded by monitoring the electron spin echo (ESE) signal following a microwave $\pi/2$- and a π-pulse sequence as a function of the magnetic field. The ENDOR spectra were obtained by monitoring the intensity of the stimulated echo, following three microwave $\pi/2$-pulses, as a function of the frequency of a radio-frequency pulse applied between the second and third microwave pulses (Mims' type the stimulated-echo pulse sequence) [28].

5.1.1 The Identification of the Binding Core of Shallow Donors in ZnO Quantum Dots

Figure 5.1b shows the ESE-detected EPR spectrum of a dry powder sample of Li-doped ZnO QDs with an average diameter of 3.4 nm [2, 3]. The EPR spectrum appears after illumination with above-band-gap light during 30 min at 1.6 K and persists at low temperature after switching off the light. The spectrum disappears when the temperature is increased above 200 K. The signal labeled (I) at 3.4600 T with a line width of 6.0 mT is assigned to the shallow donor. Its average g-value g_{av} = 1.9666 differs somewhat from the $g_{\|}$ = 1.9569 and $g_{\perp}$ = 1.9552 values obtained for the interstitial-hydrogen donor in a single crystal of ZnO [9]. The line width however corresponds very well with the value $g_{\|} - g_{\perp}$ = 0.0017 obtained for the shallow donor in a ZnO single crystal and taking into account the random character of the powder sample. The result of averaging of the signal, assuming that the ZnO nanocrystals are randomly oriented, gives rise to a line width of about 3 mT which is slightly smaller then that of the shallow donor signal observed in ZnO nanocrystals. The size dispersion of the ZnO nanocrystals of about 5% is probably responsible for the additional broadening of the line. The EPR signals labeled (II) and (III) are assigned to deep acceptors [2, 3].

First, we will concentrate on the EPR signal (I) assigned to the shallow donor. The EPR signal (I) in Fig. 5.1b is assigned to a donor because g_{av} is smaller than the g-value of a free electron. The shallow character becomes clear from the dependence of g_{av} on the size of the QDs as we will discuss in the next section. This shift towards the free-electron g_e-value with decrease of QD size is caused by the confinement of the hydrogen-like 1s-type wave function of shallow donors when the Bohr radius becomes comparable to the size of the QD.

The EPR spectra in Fig. 5.1b do not provide information on the chemical nature of the donor since no resolved hyperfine structure is observed, as is the case for the In and Ga shallow donors in ZnO bulk crystals. A hat from The Little Prince by Antoine de Saint-Exupéry (Fig. 5.1b) symbolizes the hidden nature of the unresolved single EPR line of the SDs.

To identify the binding core, ENDOR experiments were performed. In Fig. 5.2 the ENDOR signals are presented as obtained on the EPR signal (I) of the shallow donor. To understand these results we consider the isotropic hyperfine interaction or Fermi contact term a_i which reflects the spin density of the donor electron wave function (Ψ) at the site of the nucleus (r_i)

$$a_i = (8\pi/3) g_e \mu_B g_{ni} \mu_N |\Psi(r_i)|^2,$$

where g_e is the electronic g factor, μ_B is the Bohr magneton, g_{ni} is the g factor of nucleus i, and μ_N is the nuclear magneton. The related ENDOR transition frequencies are (see also Chaps. 2 and 4)

$$\nu_{\mathrm{ENDOR}i} = h^{-1} |g_{ni} \mu_B B_0 \pm a_i/2|.$$

As seen in Fig. 5.2a, symmetrically around the Zeeman frequency of ^{7}Li (I = 3/2, abundance 92.5%) at 57.1 MHz two ENDOR lines separated by 90 kHz are present, which are assigned to ^{7}Li. The observation of the ENDOR transitions of the ^{7}Li nuclear spins in the EPR signal of the shallow donor gives an unambiguous identification of the shallow donor as a Li-related centre [2, 3]. The transparent hat from The Little Prince by Antoine de Saint-Exupéry symbolizes the effectiveness of the ENDOR method in determining the hyperfine interactions, hidden under the unresolved line. Moreover, the observed HF splitting gives direct information about the density of unpaired electron spin of the SD at the Li nucleus since the isotropic HF splitting is proportional to the wave-function density. This finding confirms the results of Park et al. [29], who predicted on the basis of theoretical calculations that Li and Na at interstitial sites in ZnO behave as shallow donors. Apparently, Li^+ forms an interstitial core for the shallow donor electron in the ZnO nanocrystal, similar to hydrogen in ZnO single crystals [18]. Further, it is seen in Fig. 5.2c that symmetrically around the Zeeman frequency of ^{67}Zn (I = 5/2, abundance 4.1%) at 9.2 MHz a broad, unresolved set of ENDOR lines of ^{67}Zn spins is present. From the multitude of lines it is clear that we are indeed dealing with a delocalized electron of a shallow donor that interacts with a large number (tens) of ^{67}Zn nuclei [2, 3].

Figure 5.2d shows an ENDOR line with a width $\Delta\nu$ = 60 kHz present exactly at the Zeeman frequency of ^{1}H. From the width one can deduce a ^{1}H HF interaction smaller than 60 kHz. This should be compared to the previous observation on the hydrogen-related shallow donor in a bulk crystal of ZnO where two ENDOR lines were found with a hyperfine splitting of 1.4 MHz [18]. The observed ENDOR lines were concluded to originate in the hydrogen atoms present in the $Zn(OH)_2$ capping layer where the density of the electronic wave function is very small [2, 3].

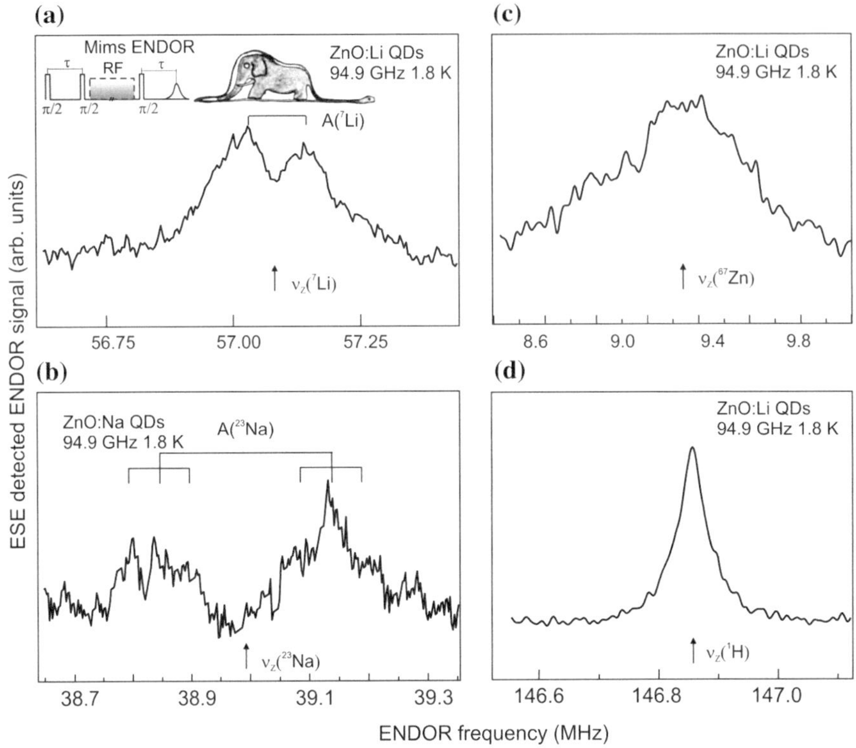

Fig. 5.2 **a** The ESE-detected ENDOR transitions of the ^{7}Li spins as observed in the EPR signal I (see Fig. 5.1b) of the shallow donors in Li-doped ZnO quantum dots with an average size of 3.4 nm recorded at 94.9 GHz and T = 1.8 K. The two ENDOR transitions are symmetrically placed around the Zeeman frequency of ^{7}Li (marked by an *arrow*); (*top*) illustration of ESE-detected ENDOR experiment. The transparent hat from The Little Prince by Antoine de Saint-Exupéry symbolizes the effectiveness of the ENDOR method in determining the hyperfine interactions, hidden under the unresolved line. **b** The ESE-detected ENDOR transitions of the ^{23}Na nuclear spins as observed in the EPR signal of the shallow donor in Na-doped ZnO quantum dots with an average size of 3.0 nm recorded at 94.9 GHz and T = 1.8 K. The two ENDOR transitions are symmetrically placed around the Zeeman frequency of ^{23}Na (marked by an *arrow*). **c** and **d** The ESE-detected ENDOR transitions of the ^{67}Zn (c) and ^{1}H (d) nuclear spins as observed in the EPR signal I

To check whether interstitial Na can also act as a shallow donor in ZnO it has performed similar EPR and ENDOR experiments on ZnO QDs that were prepared using NaOH instead of LiOH as catalyst. In such ZnO QDs with a diameter of 3.0 nm again three EPR signals in analogy to the Li-doped sample were observed. The EPR signal similar to (I) in Fig. 5.1b, with a $g_{av} = 1.9592$, is assigned to a shallow Na-related donor. Figure 5.2b shows the result of an ENDOR experiment on this signal, that reveals two transitions with a splitting of 300 kHz symmetrically placed around the Zeeman frequency of ^{23}Na at 38.97 MHz. This is considered as a

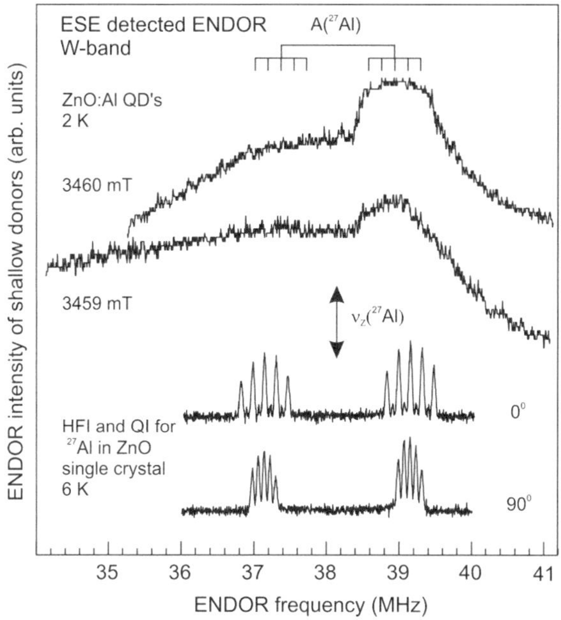

Fig. 5.3 (*Top two spectra*) The ESE-detected ENDOR signal of the ^{27}Al nucleus observed in the EPR signal of the shallow donor in ZnO:Al quantum dots with a radius of about 2.8 nm recorded at 94.9 GHz and T = 1.8 K for two values of the magnetic field: B_0 = 3459 mT and B_0 = 3460 mT within the shallow donor EPR line. (*Bottom two spectra*) ESE-detected ENDOR spectra of the ^{27}Al nuclei in a ZnO single crystal containing the Al impurity, measured at W-band at 6 K in two orientations of the magnetic field: parallel (0°) and perpendicular (90°) to the c axis (are given for comparison)

proof of the presence of a shallow donor related to interstitial Na in the ZnO nanocrystal.

Figure 5.3 shows the ENDOR spectrum of ^{27}Al nuclei as observed in the EPR signal of the shallow donor in ZnO:Al nanocrystals with a radius of about 2.8 nm for two values of the magnetic field: B_0 = 3459 mT and B_0 = 3460 mT. The shape of the ENDOR spectrum of the ^{27}Al nuclear spins as observed in the ZnO nanocrystals is caused by the distribution of quadrupole and HF interactions. The ENDOR spectrum is a result of averaging of these interactions in the randomly oriented ZnO nanocrystals. It is seen that symmetrically around the nuclear Zeeman frequency of ^{27}Al (I = 5/2, abundance 100%) at 38.4 MHz two broad ENDOR lines separated by 1.45 MHz are present, which are assigned to HF interaction with the ^{27}Al nucleus. This splitting corresponds to a ^{27}Al hyperfine interaction constant $A(^{27}\text{Al})$ = 1.45 MHz. For B_0 = 3460 mT a box-like distribution of ENDOR lines is observed. Since the EPR spectrum of the shallow donors is anisotropic, this field selects mainly a set of nanocrystals with their c-axes perpendicular to the magnetic field. The box-like form of the ENDOR spectrum is explained by the quadrupole interaction of the ^{27}Al nuclei that gives rise to five unresolved ENDOR lines.

To account for the quadrupole interaction in the case of axial symmetry the term $h^{-1}m_q 3q_i(3\cos^2\theta - 1)$ must be added to the equation of $\nu_{\rm ENDOR}$, where m_q is the average value of the two nuclear quantum states m_I, m_I', between which the nuclear transition takes place. For axial symmetry one has

$$q = (eQ_0)/[4I(2I - 1)]V_{ZZ}(r_i),$$

where Q_0 is the electric quadrupole moment in multiples of $|e| \times 10^{-24}$ cm^2 and $V_{ZZ}(r_i)$ is the electric field gradient. For ^{27}Al nuclei, the nuclear spin $I = 5/2$ and Q_0 (^{27}Al) = 0.150. For $I = 5/2$, there are five m_q-values: $m_q = \pm 2, \pm 1$ and 0. Thus, the quintet character of the ENDOR spectrum should be observed for Al nucleus, but the quadrupole splitting is not resolved in Fig. 5.3 for QDs. Figure 5.3 (bottom two spectra) shows ESE-detected ENDOR spectra of the ^{27}Al nuclei in a ZnO single crystal containing the Al impurity, measured at W-band at 6 K in two orientations of the magnetic field: parallel (0°) and perpendicular (90°) to the c axis which are given for comparison [30].

It is important to note that the value of the quadrupole interaction of ^{27}Al is almost equal to that of the ^{67}Zn nuclear spins in ZnO. First, the values of the quadrupole moments and, second, the nuclear spins of ^{27}Al and ^{67}Zn are the same. It is shown that the intrinsic electric-field gradients at the Zn nuclear sites and Al site are virtually the same. This finding is taken as a proof that Al enters substitutionally in the ZnO nanocrystals and it is centrally located at a Zn position and forms a core for the shallow donor electron in the ZnO nanoparticle. This is not obvious because the small radius of Al^{3+} (0.51 Å compared to 0.74 Å for Zn^{2+}) could drive the impurity into an off-center position.

In general, quadrupole interactions should also be observed in the ENDOR spectra presented in Fig. 6.2a, b for Li- and Na related SDs, since ^{7}Li and ^{23}Na have nuclear spins $I = 3/2$ and quadrupole moments $Q_0(^7\text{Li}) = -0.040$ and $Q_0(^{23}\text{Na}) = 0.$ 103. However, the quadrupole splitting is not resolved and could only be estimated from the line width. The pattern with the assumed quadrupole splitting for the Na-related shallow donors is presented in Fig. 5.2b. The estimations, which are more reliable for Na-related SDs (larger quadrupole moment), show that the electric-field gradients at the Li and Na nuclear sites are about two times smaller than the intrinsic electric-field gradient at the Zn site. This is in line with the suggestion that Li and Na occupy an interstitial position to create the shallow donor.

5.1.2 Probing the Wave Function of Shallow Donors and Confinement Effects in ZnO and ZnSe Quantum Dots

The ENDOR studies allow us to probe the effects of confinement on the spatially extended wave function of the shallow donor by measuring the isotropic HF interaction, which reflects the spin density at the site of the nucleus, and by varying

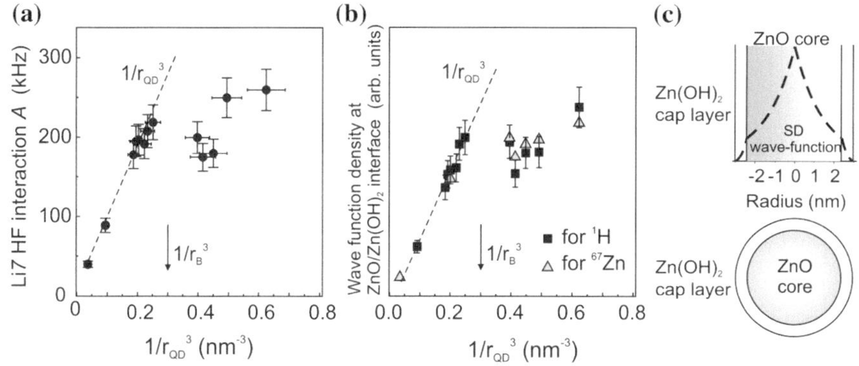

Fig. 5.4 **a** Isotropic hyperfine interaction A of the ^{7}Li nuclear spin of the shallow Li-related donors in ZnO quantum dots with radii between 3.0 and 1.1 nm. The *black circles* indicate the hyperfine splitting as observed in the ENDOR spectra at T = 1.2 K. The error bar in the values of A_{iso} is estimated from the noise in the ENDOR spectra. The variation in the size of the particles is derived from TEM and XRD measurements. The *dashed line* is a fit to the measured values of A for the QDs with radii between 3.0 and 1.5 nm using the function $\sim r_{QD}^{-3}$. **b** Variation of the wave function density at the interface of the ZnO core and the $Zn(OH)^2$ capping layer for the QDs as calculated from the dip in ENDOR of the ^{7}Li nuclear spin (core side, *black square*) and from the width of the ENDOR line of the ^{1}H nuclear spins in the $Zn(OH)_2$ capping layer (*open triangle*). The *dashed line* is a fit to the measured values for the QDs with radii between 3.0 and 1.5 nm using the function $\sim r_{QD}^{-3}$. **c** Sketch of the wave function of the shallow donor in quantum dot and a model of the $ZnO/Zn(OH)_2$ core–shell structure

the particle size in the quantum-size regime. The ^{7}Li ENDOR signals are excellently suited for this purpose and thus it has been measured the dependence of the splitting of the two ^{7}Li hyperfine components on the radius R of the ZnO core of the QDs. Figure 5.4a shows the values of the isotropic HF interaction A of the ^{7}Li nuclear spin of the shallow Li-related donor in ZnO QDs with radii between 3.0 and 1.1 nm as observed in the ENDOR spectra at T = 1.2 K. The error bar in the values of A is estimated from the noise in the ENDOR spectra. The variation in the size of the particles is derived from TEM and X-ray diffraction (XRD) measurements. The dashed line in Fig. 5.4a is a fit to the measured values of A for QDs with radii between 3.0 and 1.5 nm using the function $\sim r_{QD}^{-3}$. Down to r = 1.5 nm, a value which is equal to the Bohr radius of the shallow donor r_B, the experimental results follow quite closely this dependence, while for smaller radii there is a considerable deviation [5].

In addition to the hyperfine interactions with the ^{7}Li nucleus, information can be also obtained from the hyperfine interactions with the ^{67}Zn nuclei. In Fig. 5.5 the ENDOR spectra of the ^{67}Zn nuclei are presented (symmetrically placed around the ^{67}Zn Zeeman frequency at about 9.2 MHz) for ZnO QDs of various radii: 3.0, 1.6 and 1.17 nm. The spectra consist of a multitude of ENDOR lines that are related to various Zn shells. The shape of the ENDOR spectrum of the ^{67}Zn nuclear spins as observed in the ZnO QDs is caused by the distribution of quadrupole and HF

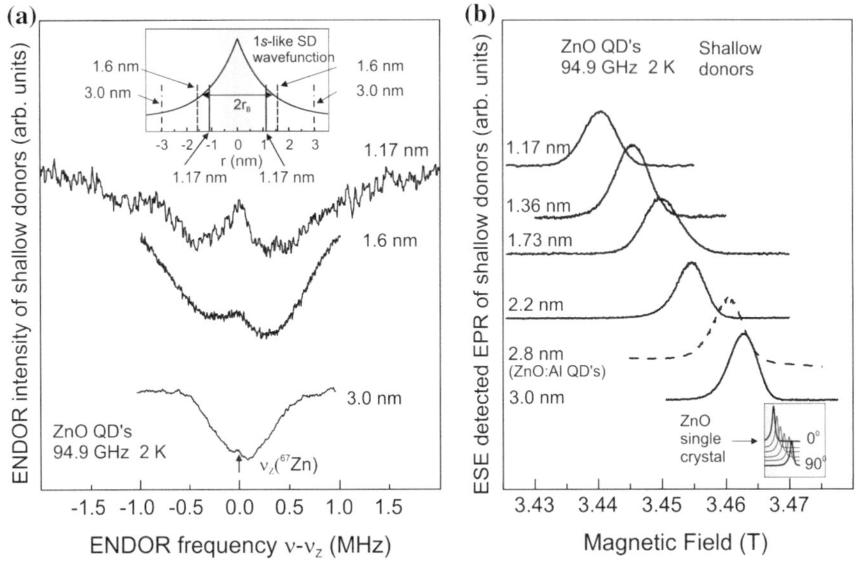

Fig. 5.5 **a** ENDOR spectra of the ^{67}Zn nuclear spins in the ZnO quantum dots with radii of 1.17, 1.6 and 3.0 nm recorded at 94.9 GHz and T = 2 K. Each spectrum consists of many unresolved lines placed symmetrically around the Zeeman frequency (marked by *arrow*) of the ^{67}Zn nuclear spins. *Inset* shows a sketch of the 1-*s* like wave function of the shallow donor in quantum dot with Bohr radius r_B, *vertical marks* indicate the limits of the QD ZnO core. **b** ESE-detected EPR spectra of shallow donors in ZnO QDs with radii of 1.17, 1.36, 1.73, 2.2, 2.8 and 3 nm recorded at 94.9 GHz and T = 2 K. The *inset* shows the angular dependence of the ESE-detected EPR spectra of H-related shallow donors in ZnO single crystal presented in the same magnetic-field scale. The *dashed line* is the ESE-detected EPR spectrum of ZnO:Al QDs

interactions inside the nanocrystal. The broad line is a result of averaging of these interactions in the randomly oriented ZnO nanocrystals.

The remarkable observations in the ENDOR spectra in Fig. 5.5a are that the distribution of ENDOR lines broadens upon reduction of the size of the QDs and that a dip develops around the Zeeman frequency of the ^{67}Zn nuclear spins. The dip becomes more prominent and broader when the radius of the ZnO core is reduced from 3.0 to 1.6–1.17 nm. The broadening of the ENDOR band indicates that the maximum density of the electronic wave function increases when reducing the size of the nanoparticles. The disappearance of the ENDOR signals close to the ^{67}Zn Zeeman frequency shows that remote shells are missing in the QD. The dip in the ENDOR spectrum indicates that for the small QDs, the Zn nuclei at the interface carry a nonzero spin density. This conclusion is in line with the observation that the electronic density at the ZnO/Zn$(OH)_2$ interface, as measured from the line width of the ^{1}H ENDOR signal, increases with decreasing size of the QDs [5]. This width is taken as a measure of the distribution of the hyperfine fields in the capping layer.

The estimates of the electronic density at the ZnO/Zn$(OH)_2$ interface, using either the width of the dip in the ^{67}Zn ENDOR signal or the line width of the ^{1}H

ENDOR signal, and using the amplification factors for Zn of 1500 and for H of 20 [18], give about the same value, in agreement with the wave-function continuity principle. The variation of the density of the wave function at the interface between the ZnO core and the $Zn(OH)_2$ capping layer for the QDs with radii between 3 and 1.1 nm are presented in Fig. 5.4b. For QDs with radii between 3 and 1.5 nm, the ${r_{QD}}^{-3}$ dependence describes the experimental data but for smaller radii, a deviation from this dependence is observed.

To test whether the observed size dependence of the density of the wave function at the ^{7}Li nucleus in the ZnO core and its distribution in the $Zn(OH)_2$ capping layer can be explained with the effective-mass approximation, a trial wave function with appropriate boundary conditions to simulate the envelope function of the shallow donor electron has been introduced [5]. By using a variational procedure in which the total energy is calculated numerically and minimized, it has been derived the density of the wave function at the centre of the ZnO core, where the Li nucleus is supposed to be located, and at the $ZnO/Zn(OH)_2$ interface. This simple analytical EMA-based model gives a good account for the envelope function of the shallow donor electron for large nanocrystals, but it does not give a stable solution in the quantum confinement regime, i.e., in semiconductor nanocrystals with radii of the order of or smaller than the Bohr radius in the bulk material. There are two possible reasons for this failure. First, it is not permitted to use the bulk value for the effective mass of the electron. This parameter reflects the effect of the periodic potential of a (infinite) semiconductor crystal and this approximation breaks down for the nanometer-sized nanocrystal. In other words, the allowed values for the wave vector $\boldsymbol{k}$ become discrete and the energy eigenvalues are those for an electron of mass m_0 in a box. Secondly, the definition of a dielectric constant as a consequence of the Lorentz relation does not apply to the nanocrystal. As demonstrated in [31], the effective screening function in a confined system depends on the size of and on the position in the nanoparticle. It is concluded that EMA-based model does not yield an appropriate description of the electronic wave function when the radius of the QD is reduced below the Bohr radius of the shallow donor. An appropriate description of the electronic wave function is believed may be found by using molecular-cluster-type calculations as carried out in [32] for the electronic wave function of shallow P donors in Si nanoparticles.

The shallow character of the Li-related donor is also visible in the dependence of the g-factor on the size of the ZnO QD as shown in Fig. 5.5b. In this figure one haa also incorporated a recording of the EPR signal of the substitutional Al donor in ZnO nanocrystals. The shift of the g-value towards the free-electron value when reducing the size of the nanoparticles is caused by the confinement of the hydrogen-like 1s-type wave function of shallow donors when the Bohr radius becomes comparable to the size of the nanocrystals. The effect is explained by the reduction of the admixture of valence-band states and higher-lying conduction bands into the lowest conduction band by the increase of the band-gap energy and the energy of higher-lying conduction bands upon the reduction of the size of the nanocrystals [33]. For comparison, the ESE-detected EPR spectra of the shallow hydrogen-related donors in a single crystal of ZnO are presented in the top of

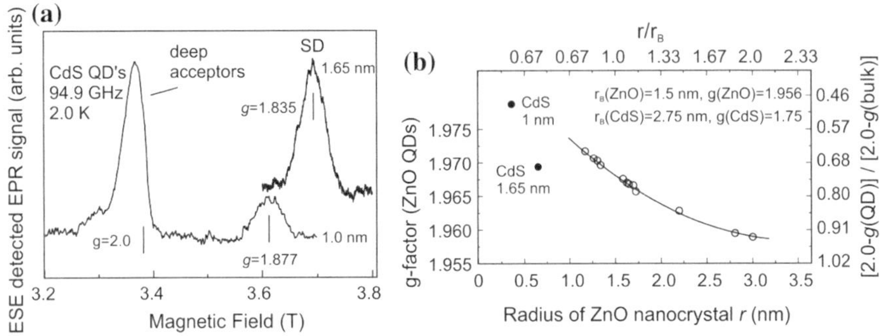

Fig. 5.6 **a** ESE-detected EPR spectra of SDs in CdS QDs with radii of 1.65 and 1.0 nm recorded at 94.9 GHz and T = 2.0 K. **b** Dependence of the *g*-factor of the shallow donor on the size of ZnO QD. The *upper scale* gives a QD radius normalized to the Bohr radius r_B of the SD in the related bulk semiconductor. The *right scale* gives a *g*-factor shift of the SD in QD from g = 2.0 normalized to that for the bulk material. The *filled circle* indicates the *g*-factor shift in CdS QDs on the normalized scales

Fig. 5.5b for several orientations of magnetic field with respect to the *c*-axis. The frequency, the temperature and the magnetic-field scales are the same for ZnO QDs and for ZnO single crystals. Averaging of the signal, on the assuming that the ZnO nanocrystals are randomly oriented, gives a line width of about 3 mT. The size dispersion of the ZnO nanocrystals of about 5% is probably responsible for the additional broadening of the line.

Figure 5.6a shows the ESE-detected EPR spectrum of shallow donors in CdS quantum dots with radii of 1.65 and 1 nm registered after 30-min light excitation at 94.9 GHz and $T = 2.0$ K. The influence of the confinement effect on the *g*-factor of these SDs is visible in Fig. 5.6b. In the same figure the dependence of the *g*-factor of the shallow Li-related donor on the size of ZnO QDs due to confinement effect is also displayed. The upper scale gives a QD radius normalized to the Bohr radius of the shallow donor in the related bulk semiconductor. The left scale gives the *g*-factor shift of the shallow donor in QDs from $g = 2.0$ normalized to that for the bulk material. The filled circle indicates the *g* shift in CdS QDs on a normalized scale. One can see that the data on ZnO and CdS QDs cannot be fitted with the same dependence. It remains for theoreticians to explain these different behaviors.

5.1.3 Dynamic Nuclear Polarization of Nuclear Spins

Figure 5.9a shows shift of the hole in the EPR transition of the shallow Li donor in ZnO QDs with radius of 1.4 nm induced by cw microwave irradiation at 94.9 GHz and T = 2 K. The lowest EPR line (not labeled) is recorded without pre-irradiation. The second recording from the bottom is obtained after cw microwave irradiation during 3 min at the centre of the unperturbed line. It is seen that after the irradiation a hole is burnt in the line and that simultaneously a new peak (an antihole) appears at the low-field side of the hole. The next curves are observed after cw microwave irradiation during 3 min at the maximum of the antihole of the previous recording. Finally, the antihole stabilizes at a position shifted by 7 mT with respect to the original position. The spectrum labeled by (0) represents the difference between two upper curves. Curves (15) and (75) are recorded 15 and 75 min after the pre-irradiation, respectively (the upper unperturbed line was subtracted from each spectrum). It is seen that slowly the hole and "antihole" decrease and EPR spectrum recovers its initial unperturbed form. Before the EPR experiment ultraviolet light illuminated the sample at 1.4 K during 30 min.

The creation of the hole and antihole in the EPR line is caused by dynamic nuclear polarization (DNP) of the ^{67}Zn (I = 5/2) nuclear spins and, as will be shown below, of the ^{1}H (I = 1/2) nuclear spins in the $Zn(OH)_2$ capping layer. The polarized nuclear spins create an internal magnetic field and, as a result, the resonance line of the electron spins, subjected to the microwave irradiation, shifts to a lower external field value resulting in the hole and the antihole in the inhomogeneously broadened EPR line. In the DNP process of shallow H donors in a single crystal of ZnO a similar effect was observed [20]. Since the line width of the Li donor in the random sample of ZnO nanocrystals depends to a large extent on the anisotropy of the g-tensor, the hole in the line corresponds to electron spins of Li donors in ZnO particles with a given orientation of their hexagonal crystal axis with respect to the external magnetic field. The hole decays slowly and disappears in about 75 min. This disappearance is caused by the nuclear spin-lattice relaxation that gradually restores the non-thermal nuclear spin polarization to its equilibrium value.

The striking result is that the intensity of the induced hole depends on the orientation of dry powder sample. The hole was observed to disappear after a rotation by 90° but after the reverse rotation or a rotation of 180°, the hole reappears at the same position [9]. The attractive feature of this experiment is that one can select a particular orientation of nanoparticles in a random sample.

In Fig. 5.7b the shift of the hole and the antihole versus the number of 3-min cycles of microwave pre-irradiation is shown for the shallow donor in ZnO QDs with diameters of 2.8, 3.4, and 4.2 nm [9]. For comparison, a similar curve is shown for the shift of the EPR line of the shallow donor in a single crystal of ZnO [20]. It is seen that for particles with an average diameter of 4.2 nm, the maximum shift is about the same as for the single crystal but that the shift increases considerably when performing the experiments on nanocrystals with diameters of 3.4 and 2.8 nm. In the next paragraphs there will be presented arguments to explain the increase of this shift by DNP of the ^{67}Zn nuclear spins and of the ^{1}H nuclear spins in the $Zn(OH)_2$ capping layer, in combination with the effect of confinement of the electronic wave function of the shallow donor.

To check whether a polarization of the ^{1}H nuclear spins affects the resonance line of the electron spin ENDOR experiments on the protons have been carried out. Curve (1) in Fig. 5.7c represents again the unperturbed EPR line. Curve (2) is the result obtained after 3 min of microwave irradiation at the peak of the line. Curve

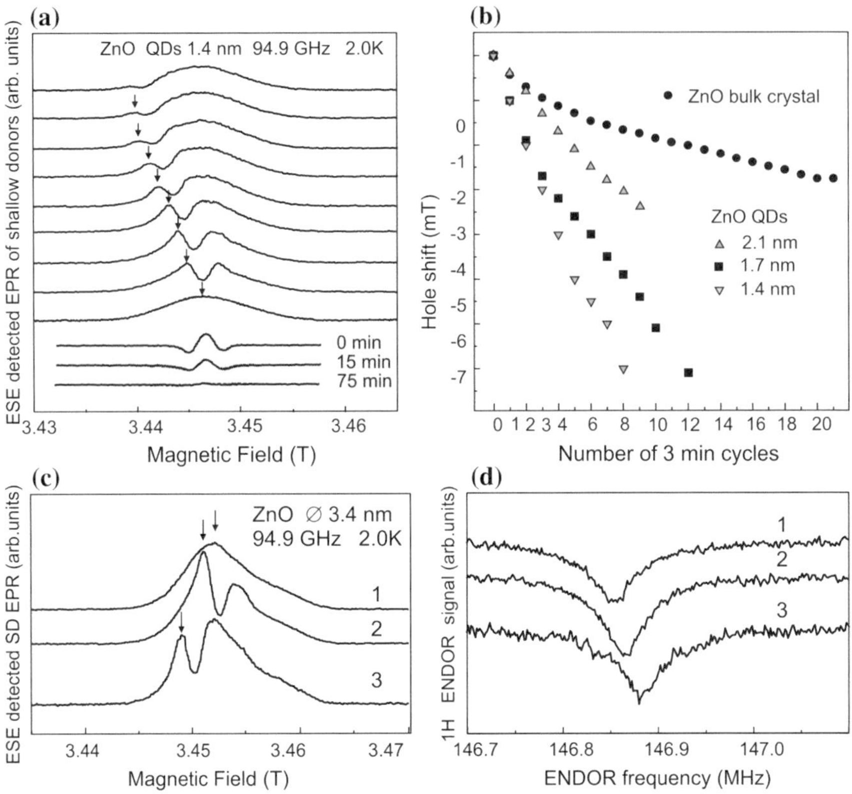

Fig. 5.7 **a** Shift of the hole in the EPR transition of the shallow Li donor in ZnO QDs with radius of 1.4 nm induced by cw microwave irradiation at 94.9 GHz and T = 2 K. The lowest EPR line (not labeled) is recorded without pre-irradiation. The second recording from the bottom is obtained after cw microwave irradiation during 3 min at the centre of the unperturbed line. The next curves are observed after cw microwave irradiation during 3 min at the maximum of the antihole of the previous recording. Finally, the antihole stabilizes at a position shifted by 7 mT with respect to the original position. The spectrum labeled by (0) represents the difference between two upper curves. Curves (15) and (75) are recorded 15 and 75 min after the pre-irradiation, respectively (the upper unperturbed line was subtracted from each spectrum). It is seen that slowly the hole and "antihole" decrease and EPR spectrum recovers its initial unperturbed form. **b** Shift of the hole and the antihole versus the number of 3-min cycles of cw microwave irradiation in the 94.9 GHz EPR line of shallow Li-related donors in ZnO QDs with radii of 1.4, 1.7 and 2.1 nm, and that of the shallow H donors in a ZnO single crystal; T = 2 K. **c** Curve (*1*) represents the unperturbed EPR transition of the shallow donor in ZnO QDs with a diameter of 3.4 nm. Curve (*2*) is obtained after cw microwave irradiation during 3 min at the centre of the unperturbed EPR line. Curve (*3*) is recorded after cw microwave irradiation during 3 min at the peak of the antihole in curve (*2*). The *arrows* indicate the positions at which the ENDOR spectra displayed in (**b**) are taken. ν = 94.9 GHz, T = 2 K. **d** ENDOR spectra of ^{1}H nuclei detected in the EPR transition of the shallow donor in ZnO QDs with a diameter of 3.4 nm at different magnetic-field positions in the EPR line as indicated in (**c**)

(3) is recorded after 3 min of microwave irradiation at the peak of curve (2). In Fig. 5.7d the ENDOR signals of the ^{1}H nuclear spins detected in curves (1), (2) and (3) of Fig. 5.7c at the positions indicated by the arrow are presented. First of all, it is seen that strong ENDOR signals are obtained around the Zeeman frequency of the ^{1}H nuclear spins, indicating that the wave function of the shallow donor extends into the capping layer. Second, the resonance frequency of the ^{1}H nuclear spins shifts to higher frequency when the magnetic field at which the ENDOR experiment is carried out moves to lower field values. This behavior is explained by the polarization of the ^{1}H nuclear spins, which produce an internal field and thus shift their resonance frequency to the higher values.

The experiments presented above demonstrate that the ^{67}Zn as well as the ^{1}H nuclear spins become polarized when saturating the EPR transition of the shallow donor in the ZnO QDs. The dynamic polarization of the ^{67}Zn nuclear spins shows the same behavior as observed for the ^{67}Zn spins in a single crystal of ZnO doped with the shallow donor [20]. In particular, the maximum shift of the hole observed in the ZnO nanoparticles with a diameter of 4.2 nm is the same as the shift of the EPR line of the shallow donor in the single crystal of ZnO. This is reasonable because it was shown that in the case of the ZnO single crystal the maximum shift of 2.8 mT could be simulated by considering the complete polarization of all ^{67}Zn spins in a sphere with a radius of about 2.0 nm [20]. When the diameter of the QD becomes smaller, a considerable increase of the maximum shift the hole is observed (see Fig. 5.7b). In principle, there are two possible explanations for this observation. First, it is known from previous experiments (Sect. 6.12) that, as a result of the confinement, the density of the wave function at the position of the ^{67}Zn spins in the QD increases when the radius of the particles becomes of the order of the Bohr radius or smaller [7]. As a result, the hyperfine interaction increases and thus the shift of the resonance field of the electron spin increases when the nuclear spins become polarized. It have been carried out a numerical calculation of the local field in the ZnO QDs with a diameter of 2.8 nm produced by a complete polarization of the ^{67}Zn nuclear spins. This calculation was performed in the same way as for the shift of the EPR line observed for the shallow donor in a single crystal of ZnO [20]. The only differences are that it was taken into account the effect of confinement of the electronic wave function and the resulting increase of the hyperfine interaction with the ^{67}Zn nuclear spins and that only about 7 shells of Zn atoms have to be considered in the 2.8 nm particles. The resulting shift amounts to 3.0–4.0 mT, depending somewhat on the estimated value of the Bohr radius. The second possibility is that the ^{1}H nuclear spins in the $Zn(OH)_2$ capping layer also become polarized. When assuming a monolayer of $Zn(OH)_2$ with completely polarized ^{1}H spins and a hyperfine interaction of about 100 kHz, as derived from the width of the ENDOR lines in Fig. 5.2d, it was estimated that these polarized ^{1}H nuclear spins produce a maximum hole shift of about 1–2 mT in the 2.8 nm ZnO nanoparticles. The conclusion is that the increase of the shift of the hole when reducing the size of the QD is a result of both the increase of the hyperfine interaction with the polarized ^{67}Zn nuclear spins in the ZnO core and the polarization of the ^{1}H nuclear spins in the $Zn(OH)_2$ capping layer.

The polarization of the ^{67}Zn nuclear spins in the case of the shallow donor in the single crystal of ZnO is caused by an Overhauser effect [20, 34]. Here a cross relaxation process, in which electron spins and nuclear spins undergo flip-flop motions, transfers the electron spin polarization to the nuclear spins upon saturation of the electron spin transition. To make this process efficient, a rapid modulation of the hyperfine interaction is required to induce a fast spin-lattice relaxation rate of the electron spins and a fast cross relaxation rate, i.e., a fast flip-flop motion. In semiconductors this modulation is provided by the rapid motion of conduction electrons or by the exchange interaction of shallow donors that at high concentrations have a sufficient overlap of their wave function [35, 36]. Since the shallow donors in bulk ZnO are at low concentration and do not show any sign of a rapid exchange, it was proposed that the modulation of the hyperfine interaction is caused by the fluctuations of the zero-point vibrations of the phonon system. This seems a reasonable suggestion because at the high magnetic field (10 T) and low temperatures (5 K) at which the experiments were carried out the Boltzmann factor is no less than 20 and the spontaneous-emission processes dominate the one-phonon-type, spin-lattice relaxation of the electron spins.

In the ZnO nanocrystals, only one donor electron spin is present in a particle and the effect of exchange resulting from an overlap of electronic wave functions can be excluded. Since here the EPR experiments are carried out at 95 GHz in a magnetic field of 3.4 T and at temperatures of 1.5–2.0 K spontaneous-emission-type processes are also expected to dominate the spin-lattice relaxation. At 95 GHz the typical wavelength of phonons in ZnO nanocrystals required to induce the one-phonon relaxation process of the electron spin of the shallow Li donor is determined by the velocity of sound in this material. When using the value $v = 3 \times 10^3\ ms^{-1}$ as a reasonable estimate of this velocity one derives that the wavelength of phonons at 95 GHz, the frequency at which the EPR experiments are carried out, is about 30 nm, i.e., about ten times larger than the average size of the nanoparticles. One thus concludes that the phonons required to induce the one-phonon-type spin-lattice and cross relaxation do not fit into the particles. The remarkable observation is that the spin-lattice relaxation rates observed for the shallow H-donor in the bulk ZnO single crystal and for the shallow Li-donor in the ZnO nanocrystal at low temperature are about the same (about $10^3\ s^{-1}$). To explain the relatively fast one-phonon type spin-lattice relaxation of the shallow Li-donor in the ZnO nanocrystals it was proposed that in the dry powder, used in the experiments, the particles are in physical contact with each other and therefore the phonons are not confined to one particle. To check whether this explanation is correct it would be attractive to perform similar experiments on samples of ZnO nanoparticles dissolved in an organic glass. In such a sample the velocity of sound in the ZnO particles is much higher than in the glassy host material and one may expect that phonons will reflect at the interfaces, thus leading to a better confinement of the phonons in the ZnO nanoparticles. A measurement of the spin-lattice relaxation of the shallow donor as a function of the size of the nanoparticles could then be used to confirm the validity of this contention and would allow for a measurement of the distribution of phonon modes in the nanocrystals.

5.1.4 *Donor–Acceptor Pairs in the Confined Structure of ZnO Quantum Dots*

Deep-level acceptors in ZnO quantum dots. The EPR signal (II) in Fig. 5.1b, which is presented on an enlarged scale in Fig. 5.8a, exhibits a nearly isotropic hyperfine splitting that suggests a hyperfine interaction with a nucleus with spin $I = 3/2$ with an almost 100% abundance. This observation favours a Na-related centre and indeed the ENDOR spectrum of this EPR signal reveals two transitions at 4.2 and 72.0 MHz as shown in Fig. 5.8b. These two ENDOR frequencies are given by $\nu_{\text{ENDOR}} = h^{-1}|g_n\mu_B B_0 \pm A/2|$, when the quadrupole interaction is neglected and where $g_n\mu_B B_0/h$ is the Zeeman frequency for sodium. One thus finds for the hyperfine constant $A = 67.8$ MHz, which corresponds to a HF splitting of 2.4 mT as observed in the EPR line shape. This splitting of 2.4 mT is about 7% of the HF constant for free Na^0 [37].

The conclusion that this deep Na-related centre is located close to or at the $ZnO/Zn(OH)_2$ interface is drawn from the observation that not only the ENDOR signals of the ^{23}Na ($I = 3/2$) nuclear spins can be observed in the EPR signal (II) but also the ENDOR signal of ^1H ($I = 1/2$) nuclear spins (see curve (1) in Fig. 5.8c). The line width of 1.0 MHz is about 8 times larger than that of the ^1H ENDOR signals observed in the ESE-detected EPR signal of the shallow donor (see curve (2) in Fig. 5.8c). This shows that the density of the electronic wave function of the Na-related acceptor is relatively large in the $Zn(OH)_2$ capping layer.

Figure 5.11 also shows the EPR signal of an additional deep acceptor with a g-factor that is typical for deep acceptors in ZnO bulk material (signal (III) in Figs. 6.1b and 6.8a). It should be noted that three types of deep acceptors in ZnO have similar structures and g-values. These centres are, respectively, the substitutional Li or Na impurity (Li_{Zn} or Na_{Zn}), or the Zn vacancy (V_{Zn}) [38, 39]. ZnO has a wurtzite structure, with the Zn ions surrounded by distorted tetrahedrons of oxygen ions. The hole in the three centres is located on one of the O^{2-} ligands and one thus has an O^- ligand with the unpaired spin located on one of the four p-bonds. There are three non-axial bonds that do not possess perfect axial symmetry and the g-tensor of a hole on one of these three oxygen atoms is described by three different principal values g_Z, g_X and g_Y. The oxygen in the c-axis direction is nonequivalent to the other three oxygen atoms, and the g-tensor for a hole on this oxygen atom can be described by $g_{\|}$ and $g_{\perp}$. For the Li_{Zn} or Na_{Zn} impurities, the preferred site is the axial oxygen atom due to the energy difference between the axial and non-axial state. For the V_{Zn} acceptor only the non-axial state was observed in bulk ZnO [39].

We have compared the line shape of the EPR signal in Fig. 5.8a with a simulated curve using the known anisotropy of the g-tensor of the deep Li_{Zn} and V_{Zn} acceptors [38, 39] and assuming that the ZnO nanoparticles are randomly oriented. In Fig. 5.8a one presents the simulations for the EPR line shapes of the deep Li_{Zn} acceptors with axial symmetry (dashed line) and for the Zn vacancy (V_{Zn}) deep acceptor (dotted line). The parameters used for Li_{Zn} are $g_{\|} = 2.0028$, $g_{\perp} = 2.0253$ and for V_{Zn} $g_Z = 2.0033$, $g_X = 2.0192$, $g_Y = 2.0188$ [38, 39]. The dashed-dotted line in Fig. 5.8a

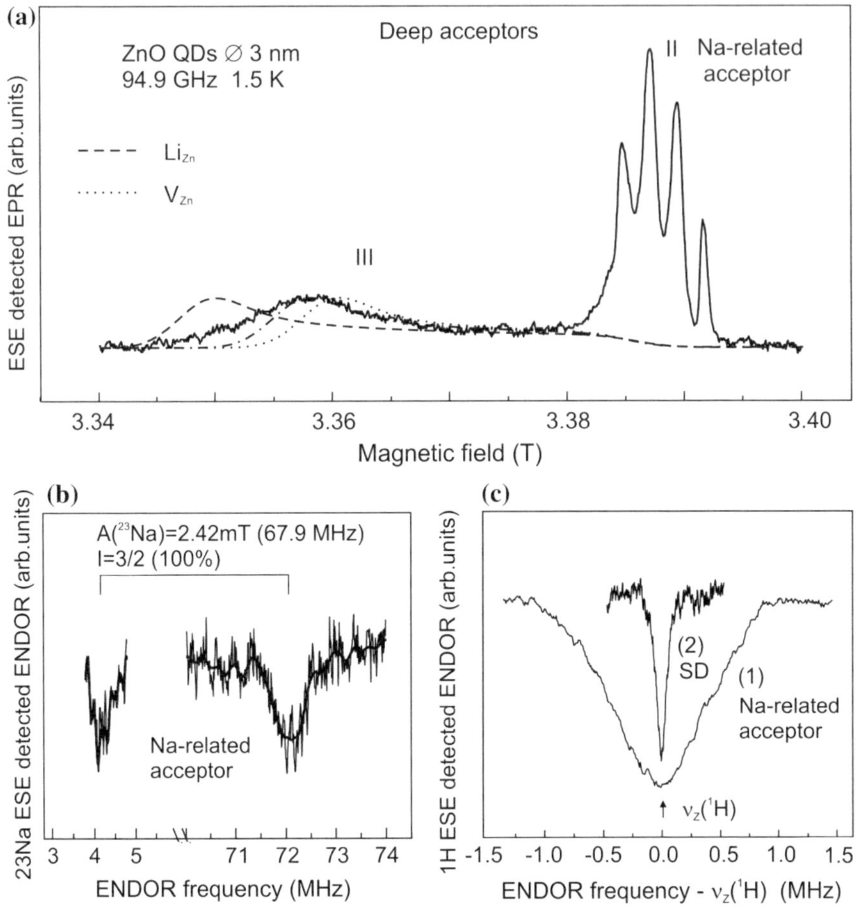

Fig. 5.8 **a** EPR spectrum of deep acceptors in the ZnO QDs with radii of 1.5 nm taken at 94.9 GHz and T = 1.5 K. The simulations for the Li_{Zn} deep acceptors (*dashed line*) and for the V_{Zn} deep acceptors (*dotted line*). The *dashed-dot line* is the result of a fitting using $g|$ = 2.0033 and $g_\perp$ = 2.0205. **b** ESE-detected ENDOR signals of ^{23}Na nuclei as observed in the EPR signal of a deep Na-related acceptor centre in ZnO quantum dots. The sample has been illuminated during 30 min with UV light prior to the EPR and ENDOR recording ν = 94.9 GHz, T = 2 K. **c** ENDOR transition of the ^{1}H nuclear spin observed at 94.9 GHz in the EPR signal (II) of the deep Na-related acceptor (*1*) and in the EPR signal (I) of the shallow Li donor (*2*)

results from a simulation using $g|$ = 2.0033, $g_\perp$ = 2.0205. In all simulations a line width of 5 mT was assumed. From a comparison of the simulated and the experimental curves the EPR line in Fig. 5.8a was concluded to originate either from the deep V_{Zn} or the deep Li_{Zn} acceptor, or a combination of these two centres. The simulated EPR spectrum of the Na_{Zn} acceptor considerably deviates from the experimental spectrum. The shape and position of the signal III in Fig. 5.8a is slightly sample dependent, and the main contribution to the signal was concluded to be coming from the deep Zn vacancy acceptor which might be introduced during the growth of the nanocrystals. This would thus represent the direct observation of a vacancy in semiconductor nanocrystals induced during nanocrystal growth.

Exchange-coupled donor-acceptor (or electron-hole) pairs in semiconductors are observable in the EPR spectrum when the electronic wave functions of the donor and acceptor start to overlap significantly. In intentionally doped semiconductors these pairs are difficult to observe because the concentration of donors and acceptors has to be relatively high. An interesting case is formed by the Frenkel pairs, as observed for instance in ZnSe [39]. These pairs consist of a Zn vacancy and a Zn interstitial (V_{Zn}-Zn_i) which stabilize at such a short mutual distance that the exchange interaction, resulting from the overlap of the wave functions, forms the dominant term in the spin Hamiltonian. This spin Hamiltonian then takes the form

$$\hat{H} = g_D \mu_B \hat{\vec{S}}_D \cdot \vec{B} + g_A \mu_B \hat{\vec{S}}_A \cdot \vec{B} + J \hat{\vec{S}}_D \cdot \hat{\vec{S}}_A. \tag{5.1}$$

Here the first two terms describe the Zeeman energies of the donor and acceptor with electron spins $S_D = S_A = ½$ and isotropic g-factors g_D and g_A. J is the isotropic exchange interaction between the donor and the acceptor. For large values of J such that

$$|J| \gg |g_D - g_A| \mu_B B,$$

where μ_B and B are the Bohr magneton and the magnetic field, respectively, a single EPR signal arises at the mean g-value $g_P = (g_D + g_A)/2$ [40–42].

In nominally undoped ZnO crystals, the EPR signal of interstitial hydrogen shallow donors has recently been observed [18]. The Bohr radius of the hydrogen-like 1s-type wave function of this interstitial-hydrogen donor is a_D = 1.5 nm. As yet, only deep acceptors with a localized wave function have been detected in this material. Bohr radius a_D of the donor is much larger than the Bohr radius a_A of the acceptor, the exchange interaction between the shallow donor and the deep acceptor depends exponentially on the donor-acceptor distance, i.e., $J = J_0 \exp(-2r/a_D)$ [43], where J_0 is the limiting exchange interaction for $r = 0$. The value of J_0 can be estimated from a comparison with the particular case of the self-trapped exciton in AgCl. This self-trapped exciton consists of a hole with a deep level localized on an Ag^+ site (Ag^{2+} centre) and an electron shallowly bound to this centre. The Bohr radius of the hydrogen-like 1s-type wave function of this electron, a_D, is also 1.5 nm and it was found that $J_0 = 5.37\ cm^{-1}$ [44]. When using this value of J_0 it is derived that the distance between shallow donors and deep acceptors in ZnO should be smaller than 2 nm in order to observe the pair signals. For the typical concentrations of 10^{16}–$10^{17}\ cm^{-3}$ of these donors and acceptors in bulk ZnO, their average distance is so large that the probability to find donor-acceptor pairs with an appreciable exchange interaction is negligible.

The probability that donor-acceptor pairs are found in ZnO QDs is considerably higher than in bulk ZnO because it is now known that donors and acceptors can be

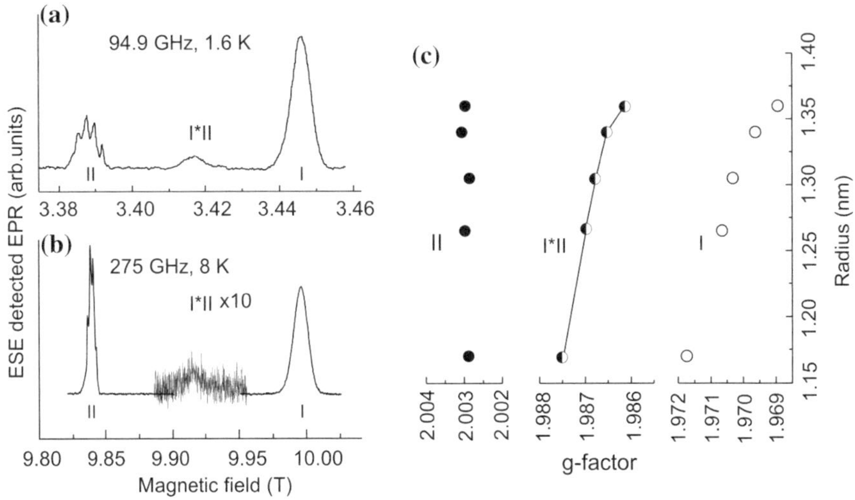

Fig. 5.9 ESE-detected EPR spectra at 94.9 GHz (**a**) and at 275 GHz (**b**) of Li-doped ZnO quantum dots with an average radius of 1.3 nm after UV illumination during 30 min. The signals marked (I) and (II) arise from an isolated shallow Li-related donor and a deep acceptor, respectively. A donor-acceptor pair formed by the shallow donor and the deep acceptor causes the EPR signal marked by (I*II). **c** The variation of the *g*-values of the EPR signals (I), (II) and (I*II) as a function of the radius of the ZnO QD

introduced in these materials. In particular, in nanocrystals of ZnO with radii smaller than 2 nm, that can be routinely produced, one expects that the formation of donor-acceptor pairs will occur. Indeed, evidence for the formation of such pairs have been found as shown in Fig. 5.9a, where the ESE-detected EPR spectrum of Li-doped ZnO nanocrystals with an average radius of 1.3 nm at 94.9 GHz and T = 1.6 K is displayed. In Fig. 5.9b a similarly detected EPR spectrum at 275 GHz and T = 8 K is shown. The spectrum in Fig. 5.13a looks very similar to the one observed at 95 GHz for ZnO QDs with a radius of 1.7 nm (Fig. 5.1b). The signal (I) belongs to the shallow Li-related donors and signal (II) to the deep Na-related centre. The important difference is that a new EPR signal (I*II) is visible halfway between the signal of the shallow donor (I) and the signal of the deep Na-related centre (II). Signal (I*II) is attributed to an exchange-coupled pair formed by the shallow donor and the deep Na-related centre.

The arguments leading to the assignment of signal (I*II) to the exchange-coupled pair of the shallow donor and the deep Na-acceptor are the following. First, its *g*-value g_P is the average of the *g*-values g_D of the shallow Li-donor and the *g*-value g_A of the deep Na-acceptor: $g_P = (g_D + g_A)/2$. This is the *g*-value that one predicts for an exchange-coupled donor-acceptor pair when the exchange coupling $|J| \gg |g_D - g_A|\mu_B B_0$. In Fig. 5.9b, where the EPR spectrum recorded at 275 GHz is presented [7], the pair signal is again exactly halfway between the signals (I) and (II), which are now separated by an interval that is larger by a factor 275/94.9 = 2.90. Second, the pair signal is only visible in ZnO nanoparticles with a radius smaller than 1.5 nm. Apparently, for these particles the exchange interaction, which depends exponentially on the distance, is large enough to create pairs observable in the EPR spectrum.

In Fig. 5.9c the variation of the g-values of signals (I), (II) and (I*II) is presented as a function of the radius of the ZnO QDs. It is seen that the g-value of signal (II) is independent of the size of the nanoparticles, typical for a deep centre with a localized wave function. The variation of the g-value of signal (I) results from the confinement effect on the wave function of the shallow donor [3, 5]. It is seen that the variation of the g-value of signal (I*II) is half that of the variation of the g-value of signal (I), as expected for an exchange-coupled pair formed by the shallow Li-related donor and the deep Na-related acceptor.

Further support for the assignment of signal (II) as arising from a deeply trapped hole is provided by isochronal annealing experiments [7]. The thermally induced reduction in concentration of the paramagnetic donors and acceptors was monitored by the change in intensity of the EPR signals. The reduction in intensity of signal (I) of the shallow donors and the signal (II) of the Na-related centres were shown to be qualitatively similar. The intensity of the two signals starts to reduce substantially at 50 K. After the annealing at 150 K the two signals have completely disappeared. The observations lend support to the idea that the thermally released donor electron is captured by the Na-related centre making both centres non-paramagnetic. The deep Na-related centre was concluded must have acceptor-like properties. The recombination of the donors and acceptors as observed in the EPR-detected annealing experiments is accompanied by an intense thermoluminescence [7].

The comparison of the EPR/ENDOR experiments and the optical experiments show that the emission seems to be at least partly due to a charge-transfer transition of an electron from the shallow donor to the deep Na-related trap, which is located close to the ZnO/$Zn(OH)_2$ interface. This conclusion is drawn from the observation of the ENDOR signal of ^1H nuclear spins in $Zn(OH)_2$ cap layer in the EPR signal of the deep Na-related trap. This shows that the density of the electronic wave function of the Na-related acceptor is relatively large in the $Zn(OH)_2$ capping layer. Further support for this conclusion follows from the effect of saturating the ^1H nuclear-spin transition on the Na-related EPR signal. The whole line shape was shown [7] to undergo a reduction in intensity, demonstrating that the hyperfine interaction with the ^1H nucleus is related to the deep Na acceptor, and not to an EPR signal of another centre that might coincide with the signal of the Na-related acceptor. Before the UV illumination this centre is not paramagnetic. Upon above-band-gap illumination an electron from this centre is transferred to the interstitial shallow Li-donor, making both centres paramagnetic. Since a relatively strong hyperfine interaction of this localized Na-related centre with a proton of the OH^- ligands of Zn in the capping layer is observed, this complex is expected to be located at or near to the ZnO/$Zn(OH)_2$ interface. The hole is then most probably located on Na because about 7% of the spin density is found on this atom.

As mentioned above, the EPR signals of the isolated shallow donor and the Na-related deep acceptor decrease simultaneously and irreversibly when the temperature is increased. Simultaneously, the EPR signal of the exchange-coupled pair starts to decrease irreversibly at lower temperatures. The signal completely disappears after annealing at 20 K. The difference in the intensity ratio of the EPR signal of (I) and (II) at 275 and 94.9 GHz (Fig. 5.9) is due to the difference in the nature of the EPR line broadening. The line width of the shallow donor is caused by g-factor anisotropy, which results in an increase of the line width at 275 GHz compared to 94.9 GHz and in a decrease of the signal intensity. The line width of the Na-related hole is dominated by hyperfine interaction, which does not depend on the microwave frequency. As a result, one observes the same EPR intensity of the Na-related acceptors at 275 and 94.9 GHz.

The magnetic-field positions of the EPR transitions at 94.9 GHz of two exchange-coupled spins $S = 1/2$ of an acceptor with $g_A = 2.0023$ and a donor with $g_D = 1.97$ as a function of the absolute value of J were calculated. Such a coupled pair gives rise to a singlet and a triplet state with four possible EPR transitions. The transitions of a strongly coupled pair appear in the EPR spectrum when $J > 0.7$ cm^{-1} and their position does not change appreciably when J is further increased. When J is varied between 5×10^{-3} and 0.7 cm^{-1}, the positions of the EPR lines change rapidly. As the value of J is decreased further, the EPR spectrum remains virtually unchanged and consists of two lines corresponding to the isolated donors and acceptors. Thus, EPR signals can be observed only in the case of strong ($J > 0.7$ cm^{-1} at 94.9 GHz) and weak exchange coupling. For intermediate values of J, the positions of the EPR lines depend on J, and averaging over all possible values of J results in a broad spectrum with a complicated shape and low intensity that is difficult to detect. The sign of J is further concluded to be negative because at 1.2 K the triplet state is observed.

The most probable position of the shallow interstitial Li-donor, with its large Bohr radius, is near the centre of a ZnO nanocrystal, while the deep Na-related acceptor is localized near the surface of the nanocrystal. Using the expression $J = J_0\exp(-2r/a_D)$ with $J_0 = 5.37$ cm^{-1} and using the radius of the nanoparticle $r = 1.3$ nm one can estimate the exchange interaction J to be about 1 cm^{-1}. This is in a good agreement with the value derived from the simulated EPR line positions.

It should be noted, that the Na-related surface acceptor is perhaps not the only deep acceptor that can capture the electron of the shallow Li-donor because the intensity of the EPR signal of the deep Na-acceptors was shown [7] to decay faster than that of the shallow Li-donor. As it have been shown in [5, 6], the EPR signal of an additional deep acceptor is observable with g-factor that is typical for deep acceptor in ZnO bulk material. This signal is shown in Fig. 5.9, where also the signal of the deep Na-related surface acceptor is visible. From a comparison of the simulated and the experimental curves, the EPR line in Fig. 5.9 was concluded to originate either from the deep V_{Zn} or the deep Li_{Zn} acceptor, or a combination of these two centres.

The concentration of shallow Li-related donors is estimated to be as low as 10^{16}–10^{17} cm^{-3}, as can be deduced from the intensity of the EPR signals, and consequently only one out of 10^3–10^4 particles carries a Li-donor. Based on statistical arguments, it is concluded that the concentration of the deep Na-related surface acceptor must be much higher to have an appreciable probability for pair formation in a nanoparticle. Such a high concentration of Na-related acceptors is not unreasonable because it is a surface defect. These deep surface acceptors are probably introduced during the preparation of the nanocrystals by incorporation of Na impurities from the chemicals, solvents and glassware. Surface adsorption of cations is very likely due to the large surface-to-volume ratio of nanocrystals and would be particularly favored during the precipitation since the ZnO nanocrystals are negatively charged [21, 22], providing a driving force for the nanocrystals to scavenge cations from solution upon addition of a low dielectric solvent such as heptane. Rinsing with heptane and acetone would probably succeed in removing small cations such as Li^+ from the surface, but not large cations such as Na^+. This implies that even small Na concentrations would be effectively incorporated in the nanoparticles, making it difficult to prepare Na^+-free $ZnO/Zn(OH)_2$ nanocrystals. Indeed, even in ZnO nanocrystals synthesized in plasticware using compounds and solvents with the lowest commercially available concentration of Na impurities, the EPR signal of the deep Na-acceptor is still present in comparable strength [7]. Assuming that the incorporation of Na^+ is nearly quantitative, about 35% of the ZnO nanocrystals with an average radius of 1.3 nm would contain a Na^+ ion, even at the lowest Na concentration achievable in our experiments.

The question arises why the EPR signals of the shallow Li-donor and of the Na-related surface acceptor have about equal intensity. One can speculate that the majority of the deep Na-related acceptors are not observable in the EPR spectrum because they can only become paramagnetic upon illumination when they can transfer an electron to the Li-related donor. In this respect, it is low probability that the electrons and holes produced upon illumination are captured in different nanocrystals because then the probability for pair formation would be too low for their observation in the EPR spectrum.

Finally, it is suggested that similar deep surface acceptors might be present in CdS and CdSe nanocrystals, which are known to exhibit blinking behaviour when a single nanocrystal is optically excited [45]. It is currently thought that such deep acceptors at the surface or in the capping layer of CdS and CdSe nanoparticles are essential to explain the lengthening of the "on" periods in the blinking. It would be interesting to study CdS and CdSe nanoparticles with EPR and ENDOR techniques to corroborate this suggestion.

To summarize, EPR and ENDOR experiments on ZnO quantum dots having $ZnO/(ZnOH)_2$ core/shell structures reveal the presence of shallow donors related to interstitial lithium and sodium atoms and substitutional aluminum. The shallow character of the wave function is evidenced by the multitude of ^{67}Zn ENDOR lines and further by the hyperfine interactions with the ^{7}Li, ^{23}Na and ^{27}Al nuclei that are much smaller than for atomic lithium, sodium and aluminum.

The results show that one can monitor the change of the electronic wave function of a shallow donor in a ZnO semiconductor nanoparticle when entering the regime of quantum confinement by using the nuclear spins in the semiconductor nanocrystals as probes. The model based on the effective-mass approximation does not yield an appropriate description of the electronic wave function when the radius of the nanoparticle is reduced below the Bohr radius. It is suggested that molecular, cluster-type calculations should be carried out to describe the observed behavior.

The influence of confinement effects on the value of the *g*-factor of SDs in ZnO and CdS QDs was demonstrated. Dynamic nuclear polarization of nuclear spins in ZnO quantum dots has been observed, where almost complete polarization of ^{67}Zn and of ^{1}H nuclear spins in the ZnO/Zn$(OH)_2$ core-shell structure have been obtained by saturating the EPR transition of the SD present in the ZnO QDs. DNP manifests itself via the creation of a hole and an antihole in the EPR absorption line of the SD in QDs.

In this review, the combination of EPR and optical experiments allowed to demonstrate that donor-acceptor pairs are formed in the confined structure of ZnO nanoparticles between the shallow donor and a deep Na-related acceptor. From ENDOR experiments it is concluded that these deep acceptors are located at the ZnO/Zn$(OH)_2$ interface. Moreover, the arguments are supplied that Zn-vacancy related deep acceptors are incorporated in the ZnO nanocrystals during growth.

5.1.5 Manganese and Cobalt Doped ZnO Quantum Dots

Impurity and intrinsic defects in semiconductors determine their properties. Manganese (Mn) and cobalt (Co) doped ZnO colloidal quantum dots are promising classes of diluted magnetic semiconductors, which consist of a ZnO nanocrystal core and Zn$(OH)_2$ shell [46–49].

A study of Co- and Mn-doped ZnO QDs which consist of a ZnO/Zn$(OH)_2$ core-shell structure using high-frequency EPR, ESE and ENDOR spectroscopy at low temperature is presented here.

The shallow donors in ZnO QDs were subsequently generated by illumination from a mercury arc. The EPR experiments were performed at temperatures ranging from 1.2–50 K using a pulsed EPR spectrometers operating at 94.9 GHz. The EPR spectra were recorded by detecting the ESE signal as a function of the magnetic field. ENDOR spectra were obtained by Mims' type the stimulated-echo pulse sequence.

Co- and Mn-doped ZnO colloidal quantum dots (QDs) consist of a ZnO nanocrystal core and Zn$(OH)_2$ shell. High-frequency ESE detected EPR was used to study isolated Co^{2+} or Mn^{2+} in the Zn sites of ZnO nanocrystal core. The EPR spectra of substitutional Co^{2+} ion in ZnO single crystal can be described by the standard axial spin-Hamiltonian for electron spin $S = 3/2$ ($^4A_2$ ground state) with anisotropic *g*-factor: $g_{\|} = 2.24$, $g_{\perp} = 2.28$, axial fine-structure constant $D = 2.75\ \mathrm{cm}^{-1}$ and anisotropic hyperfine interaction constant ($I = 7/2$) $A_{\|} = 15.9 \cdot 10^{-4}\ \mathrm{cm}^{-1}$ and $A_{\perp} = 2.9 \cdot 10^{-4}\ \mathrm{cm}^{-1}$ [50–52]. The broad ESE-detected EPR signal is observed in Co-doped ZnO quantum dots which is a result of averaging fine-structure and hyperfine interactions for Co^{2+} ions in the randomly oriented ZnO nanocrystals. After illumination

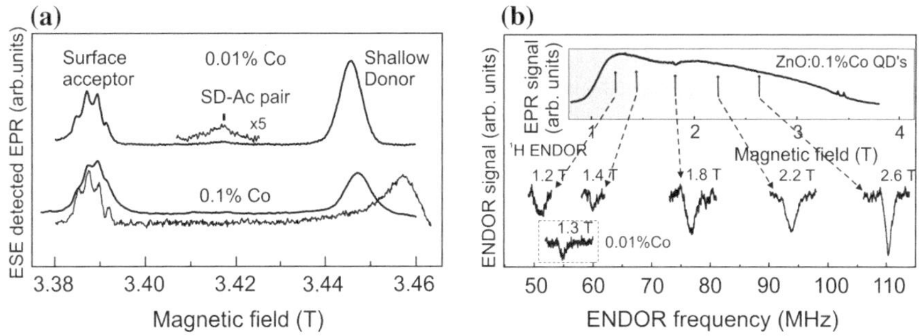

Fig. 5.10 **a** EPR spectra of the light-induced shallow donors and deep Na-related surface acceptors. **b** ESE-detected ENDOR signals of the ^{1}H nuclear spins observed in the EPR signal of Co^{2+} in ZnO:0.1%Co quantum dot shell with a radius of about 2.75 nm for five values of the magnetic field indicated on EPR signal (*inset*) by an *arrows*. In addition ESE-detected ENDOR signals of the ^{1}H nuclear spins in ZnO:0.01%Co quantum dot shell at 1.3 T is presented

with light from a mercury arc, a strong EPR signal of shallow donors appears in ZnO QDs. This EPR signal, once excited at low temperature, persists at low temperature after switching off the light, however disappears after heating above 200 K.

EPR spectra of the light-induced the shallow donors (SDs) and surface deep Na-related acceptors are shown in Fig. 5.10a. A shift of the light induced EPR line of shallow donors caused by the presence of cobalt magnetic impurity was observed. This shift depends on conditions of UV excitation (two different positions of SDs line in ZnO:0.1% Co).

The introduction of shallow donors in a quantum dot with a low concentration of cobalt changes the shape of the EPR spectrum and results in virtually identical EPR spectra of cobalt ions in the quantum dots with higher concentration of cobalt. That is to say that the introduction of SDs raises interaction between cobalt ions, apparently due to an indirect interaction through the shallow donor, the wave function which almost fills the entire space of the quantum dot. In other words, in this case as it were, increases the effective concentration of cobalt in quantum dot. The shape of the EPR spectrum of cobalt ions changed as a result of Co^{2+} coupling with optically created shallow donors. This, along with a shift of SDs line, is a direct demonstration of interaction between the magnetic ion and donor electron in confined system of ZnO QD. The diluted magnetic semiconductors in which a fraction of nonmagnetic elements is substituted by magnetic transition-metal ions seems to be suitable candidates to obtain ferromagnetic semiconductors with high Curie temperatures. By interaction between the localized magnetic moments and carriers (shallow donors), is suggested to control magnetic order by means of carrier concentration [50].

Figure 5.10b shows ESE-detected ENDOR signals of the ^{1}H nuclear spins observed in the EPR signal of Co^{2+} in ZnO:0.1%Co quantum dot shell with a radius of about 2.75 nm for five values of the magnetic field indicated on EPR signal (inset) by arrows. In addition, the ENDOR signal of ^{1}H in ZnO:0.01%Co quantum dot shell for 1.3 T magnetic field is depicted. ENDOR resonances of the ^{1}H hydrogen nuclei detected by the EPR signal of Co^{2+} evidence the hyperfine interaction between unpaired electrons of Co^{2+} ion located in the ZnO nanocrystal core, and the hydrogen nuclei (protons) outside the quantum dot core in the $Zn(OH)_2$ shell.

ESE experiments reveal a long spin-lattice relaxation time (T_1) of about 10 μs and spin coherence time of about 1 μs. Hyperfine interactions between Co electron spins in-side QD core and ^{1}H nuclear spins outside the QD core (in QD shell) are observed via ENDOR, revealing an important contribution to spin decoherence in colloidal QDs. In previous studies (see, e.g., [2]) it was found the interaction of the electron spin of the shallow donors inside the quantum dot with the nuclear magnetic moments of hydrogen in the shell. Dynamic nuclear polarization of nuclear spins in ZnO quantum dots has been observed, where almost complete polarization of ^{67}Zn (core) and of ^{1}H nuclear spins (shell) in the $ZnO/Zn(OH)_2$ core/shell structure have been obtained by saturating the EPR transition of the SD present in the ZnO QDs. The ESE-detected EPR spectrum at 94.9 GHz of a dry powder sample of Mn-doped ZnO quantum dots with an average diameter of 2.8 nm at different temperatures is shown in Fig. 5.11. One can see that at low temperatures, intense broad line on which background the Mn^{2+} hyperfine structure is visible. This structure seems to belong to the averaged spectrum of Mn^{2+} in the ZnO nanocrystal core. For comparison in the top the angular dependence of the ESE-detected EPR spectra of Mn^{2+} in ZnO single crystal presented in the same magnetic-field scale. Mn is substitutional on Zn sites yielding a $3d^5$ electron valence configuration and a $^{6}A_1$ ground state, with a nuclear spin $I = 5/2$ and an electron spin $S = 5/2$. The EPR data is described by standard spin Hamiltonian with parameters: $g = 2.0012$, axial fine-structure constant $D = -236.2 \cdot 10^{-4}\ \mathrm{cm}^{-1}$ and isotropic hyperfine interaction constant $A = 74.1 \cdot 10^{-4}\ \mathrm{cm}^{-1}$ [53]. An increase in temperature leads to the disappearance of the broad line, while only an average hyperfine structure of Mn^{2+} remains, which is practically unchanged as the temperature rises. It was assumed that at low temperatures below 4 K the magnetic ordering in the system of Mn-doped ZnO quantum dots occurs, which leads to superparamagnetism or ferromagnetism. Figure 5.16 (inset) shows the ESE-detected ENDOR signals of the ^{1}H nuclear spins observed in the EPR signal of Mn^{2+} in ZnO:0.1%Mn quantum dot shell with a radius of about 2.8 nm for two values of the magnetic field indicated on EPR signal by an arrows (Fig. 5.11). It can be seen that the signal of ENDOR resonances of the ^{1}H hydrogen nuclei is observed only in a magnetic field, which coincides with the line of the EPR of Mn^{2+} ions in the ZnO nanocrystal core. ENDOR resonances of the ^{1}H hydrogen nuclei detected by the EPR signal of Mn^{2+} evidence the hyperfine interaction between these ions, located in the ZnO nanocrystal core, and the hydrogen nuclei outside the quantum dot core in the shell.

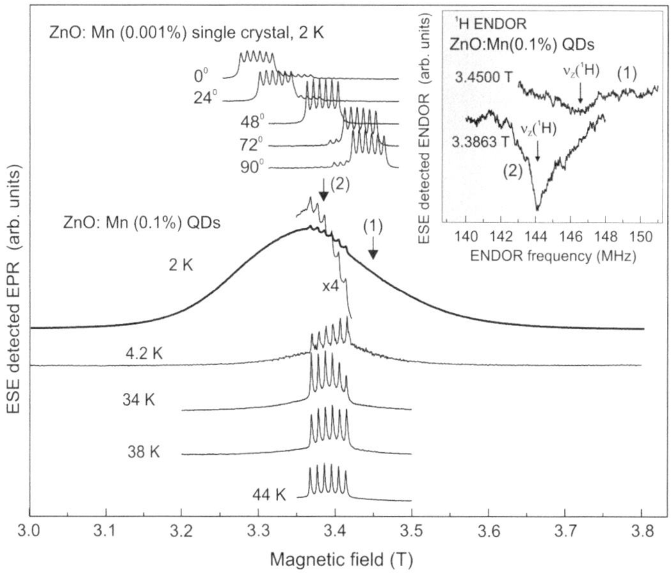

Fig. 5.11 ESE-detected EPR spectrum at 94.9 GHz of a dry powder sample of Mn-doped ZnO quantum dots (0.1% Mn) with an average diameter of 2.8 nm at different temperatures. (*Top*) The angular dependence of the ESE-detected EPR spectra of Mn^{2+} in ZnO single crystal presented in the same magnetic-field scale. *Inset* shows ESE-detected ENDOR signals of the ^1H nuclear spins observed in the EPR signal in ZnO:Mn quantum dot shell for two values of the magnetic field indicated on EPR signal by an *arrows*

To summarize, high-frequency EPR, ESE and ENDOR at 94.9 GHz were applied to study of Co- and Mn-doped ZnO QDs which consist of a ZnO/$Zn(OH)_2$ core-shell structure. The shape of the EPR spectrum of cobalt ions changed as a result of Co^{2+} coupling with optically created shallow donors. This, along with a shift of SDs line, is a direct demonstration of interaction between the magnetic ion and donor electron in confined system of ZnO QD. At low temperatures the magnetic ordering in the system of Mn-doped ZnO quantum dots is assumed to occur.

ENDOR resonance of the ^1H hydrogen nuclei registered by the EPR signal of Co^{2+} and Mn^{2+} evidence the hyperfine coupling between these ions, located in the ZnO core, and the protons outside the quantum dot core in the shell.

5.2 Application of Optically Detected Magnetic Resonance and Level Anticrossing Spectroscopy for the Investigations of Semiconductor Nanostructures

The traditional methods of radiospectroscopy are hardly applicable for low-dimensional systems because of a decreased active volume and the not high enough sensitivity of the methods, but this difficulty can be overcome by employing optically detected magnetic resonance (ODMR). ODMR is now well establishes as a powerful tools in the semiconductor and solids state physics which allows identification of the luminescence and optical absorption features and provides detailed information about the electronic structure of the defects and excitons (see, for example [54, 55]). High sensitivity, extreme resolution and spatial selectivity of ODMR and optically detected cyclotron resonance (ODCR) [56] make these techniques very suitable for a study of defects, carriers and excitons in quantum wells (QWs), superlattices (SLs), and quantum dots (QDs) and nanocrystals. The interest of the ODMR studies of low-dimensional systems lies in the possibility to investigate spatial confinement effects on the physical characteristics of defects, excitons and recombination processes and also in application of this method for local diagnostics of nanostructures. A limitation of the ODMR technique is that the recombination rate should be comparable to or less than the microwave-induced transition rate between the exchange and magnetic-field-split energy levels, i.e., the lifetime has to be at least a few tenths of a microsecond. This limitation does not exist for level anticrossing (LAC) spectroscopy, which can be considered as magnetic resonance at zero frequency.

The electron and hole cyclotron resonance was found in [56] to suppress dramatically the luminescence intensity of excitons and electron-hole drops. Besides the enhanced sensitivity and spatial selectivity, ODCR possesses such important merits as the possibility of studying undoped samples with simultaneous observation of the electron and hole cyclotron resonance, investigation of the dynamics of carriers and the nature of their trapping and recombination, probing the band structure and effects of carrier localization in thin epitaxial layers and nanostructures. Nowadays ODCR has become a powerful tool in semiconductor research [57, 58] which is employed particularly effective in studies of low-dimensional systems [59–62].

A series of pioneering studies of tunneling and photostimulated recombination processes in irradiated ionic crystals by ODMR was made [63, 64]. The effect of impurity spin polarization on recombination luminescence intensity was discovered, and energy transfer processes were investigated in ionic and semiconductor crystals doped with magnetic impurities [55, 65]. Optical pumping and cross relaxation were employed to develop novel techniques for optical detection of electronic and nuclear magnetic resonance without application of microwaves [66]. Numerous studies of donor-acceptor recombination in silicon carbide (SiC) [67, 68] were carried out. The multiquantum ODMR spectroscopy was used to probe semiconductors and silver halides [68, 69].

In this chapter the results of application of ODMR and LAC spectroscopy for a study of semiconductor micro- and nanostructures: (i) ODMR study of GaAs/AlAs and GaAs/AlGaAs quantum wells and superlattices and (ii) silver halide micro and nanocrystals embedded in crystalline alkali halide matrices which are presented in Sects. 6.1 and 6.2.

5.2.1 *ODMR in GaAs/AlAs, InAs/GaAs Quantum Wells, Quantum Dots and Superlattices*

5.2.1.1 Introduction

In GaAs/AlAs superlattices both type II and type I band alignment can be obtained [70]. In type I structures the electron and the hole are both confined in the same GaAs layer. In type II SL the electrons and the holes are spatially separated in the adjacent AlAs in GaAs layers, respectively, and their envelope functions overlap at interfaces. Excitons that are formed by these electrons and holes are localized at the interfaces and are very sensitive to the interface microstrostructure. They can be used as probes to study the properties of the interfaces.

A transmission electron microscopy (TEM) micrograph of the cross-section of a typical MBE grown type II GaAs/AlAs SL is shown in Fig. 5.12a together with the simplified band structure. The SL plane (001) is normal to the plane of the figure. The Γ and X conduction band levels and Γ valence band levels in bulk GaAs and AlAs are shown by narrow lines, the SL states—by thick lines. As shown in the figure, the excitons which have in-plane diameter of the order of 10 nm can be localized both at the normal (AlAs on GaAs) and inverted (GaAs on AlAs) interface. The band structure in Fig. 5.12a is shown for abrupt interfaces, i.e., is idealized.

When a GaAs quantum well thickness is greater than 3.6 nm, the lowest-energy subbands of the conduction and valence bands are Γ states in the QW. Below a GaAs thickness of 3.6 nm and provided that the AlAs thickness is not too small the electron and hole wave functions are derived from X states in AlAs and Γ states in GaAs and the SL is of type II. The quantum confinement makes the X valley in AlAs split into two states, i.e., X_z along the growth direction [001] and X_{xy} perpendicular to it. Owing to the competition between confinement and strain effects, the lowest electron state is X_z for AlAs thickness below 7 nm or X_{xy} otherwise. Optical transitions are direct in type I systems and typical recombination times lie in nanosecond and subnanosecond regions. In type II SL the optical transitions are indirect and therefore forbidden. For a recombination that creates a heavy hole (HH) with X_z electrons the transitions become weakly allowed due to the Γ-X coupling. The border between type II and type I systems depends on the SL

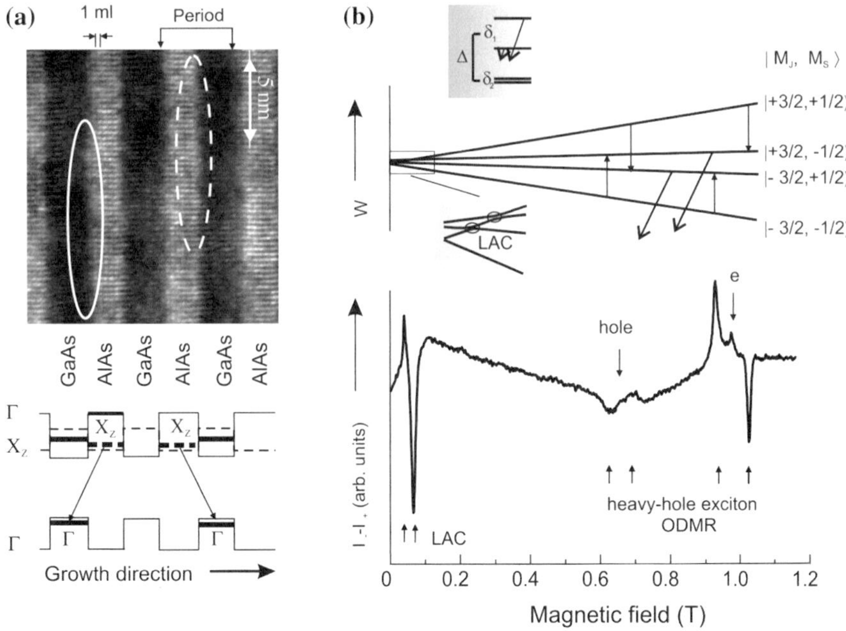

Fig. 5.12 **a** Cross-sectional TEM photograph of a typical MBE-grown GaAs/AlAs SL and the valence and conduction band structure. *Narrow lines* show energy of Γ (*solid lines*) and X (*dashed lines*) states in the bulk materials, thick lines correspond to the superlattice states. The recombination transitions are shown for excitons localized at the opposite interfaces in type II SL. **b** Heavy-hole exciton energy levels in zero magnetic field (*inset*) and in the longitudinal magnetic field with 24 GHz electron paramagnetic resonance (EPR) transitions; (*bottom*) typical LAC and ODMR spectra recorded via circular polarization of emission in type II SL (1.9 nm GaAs/2.1 nm AlAs)

composition. About 20 different samples in different points of type I SL and X_z area of type II SLs have been studied [55].

Type II GaAs/AlAsSLs were widely studied by ODMR which made it possible to measure directly electron and hole g-factors and exciton exchange (fine structure) splittings [71, 72]. The order of the conduction band valleys X_z and X_{xy} could be determined from the anisotropy of the electron *g*-factor [73]. The experimental dependences of the exchange splittings and *g*-factors of localized excitons on the parameters of type II SLs were applied for a local ODMR diagnostics of SLs [74]. Early ODMR studies of the type II to type I transition were mainly focused on the extreme cases of type II and type I recombination [75, 76]. In [75] the first observation of exciton level-anticrossing in type I QW was reported which allowed to measure the exciton exchange splitting and to develop the LAC spectroscopy of type-I QWs and SLs [77]. Application of LAC for the investigations of type I QWs was also reported in [78, 79].

5.2.1.2 Experimental

GaAs/AlAs and GaAs/$Al_xGa_{1-x}As$ (x = 0.3–0.39) QWs and SLs were grown by molecular-beam epitaxy (MBE) on the (001) semi-insulating GaAs substrate which was kept at 580–640 °C. The thickness of the GaAs and AlAs layers varied within 3–30 monolayers (almost the same for GaAs and AlAs, 0.283 nm). The period of SLs and the mean concentration in Al were checked by X-ray diffraction (XRD). The results of the investigation of several MBE grown QWs and SLs are presented. One of the samples [2] was grown without the rotation of the substrate holder, which produced a spatial gradient of composition in the layer plane. The GaAs/AlAs composition determined in this sample from the XRD data was found to vary from (2.08 nm/1.22 nm) to (2.28 nm/1.12 nm) over a length of 23 mm. It is to be noted that the XRD diagnostics gives an average characteristics of SL. Since we deal with excitons localized by the interface roughness, the local parameters of SL at the localization site may be somewhat different.

LAC and ODMR were investigated with the ODMR spectrometer operating at 1.6 K and providing the magnetic field up to 4.5 T. Most of ODMR studies were made at 35 GHz. The samples were placed in the center of a cylindrical H_{011} microwave cavity, which had holes for excitation and emission light. Its unloaded Q-factor was above 3000. The maximum microwave power in the cavity was 900 mW. The sample could be moved along the vertical axis of the microwave cavity allowing spatially resolved measurements of ODMR and LAC at different points on the sample surface. Luminescence was excited far above the band gap with an argon-ion laser and detected with a grating monochromator and a photomultiplier (PM) tube. ODMR was detected by monitoring circular polarization of luminescence. LAC was detected via both circular and linear polarization of light. Some measurements were made with a 24 GHz ODMR spectrometer of the same type.

5.2.1.3 Energy Levels of Excitons, LAC and ODMR in Type II GaAs/AlAs Superlattices

The energy levels of excitons (or pairs of electrons and holes when the electron-hole exchange interaction is involved) which are formed in GaAs/AlAs QW and SL by electrons with S = 1/2 and holes with J = 3/2 can be described by the spin Hamiltonian

$$\hat{H}_{exc} = \hat{H}_e + \hat{H}_h + \hat{\vec{J}}\vec{c}\hat{\vec{S}}, \tag{5.2}$$

where the first two terms are the electron and hole Zeeman interactions and the third term is their exchange interaction. A spin Hamiltonian $\hat{H}_e$ for an electron with S = 1/2 is

$$\hat{H}_e = g_{e//}\mu_B\hat{S}_Z B_Z + g_{e\perp}\mu_B(\hat{S}_X B_X + \hat{S}_Y B_Y), \tag{5.3}$$

where μ_B is the Bohr magneton, $g_{e\|}$ and $g_{e\perp}$ are the components of the g-tensor along the growth axis (z-axis) and perpendicular to it. The valence band in GaAs/AlAs QW and SL is split by spin-orbit coupling into an upper band (J = 3/2) and a lower band (J = 1/2). The upper band is further split into two components with $J_z = \pm 3/2$ (heavy hole) and $J_z = \pm 1/2$ (light hole), separated by more than 10 meV. The light hole states are not populated at low temperature. The heavy hole states $J_z = \pm 3/2$ can be described using the effective spin $S^* = 1/2$:

$$\hat{H}_h = g_{h//}\mu_B\hat{S}^*_Z B_Z + g_{h\perp}\mu_B(\hat{S}^*_X B_X + \hat{S}^*_Y B_Y). \tag{5.4}$$

In GaAs/AlAs QW and SL all four exciton levels are split in zero field as shown in the inset in Fig. 5.12b. For type II SLs this is explained by low local symmetry C_{2v} of interfaces at which excitons are localized [70, 80] whereas for type I QWs and SLs this effect is related to the anisotropic exciton localization potential [81] due to interface islands. The isotropic exchange splitting Δ and the splittings between the exciton radiative (δ_1) and non-radiative (δ_2) levels shown in the upper part of Fig. 5.12b are connected with the components c_x, c_y and c_z of the exchange tensor $\vec{c}$ in (5.2) as follows:

$$\Delta = c_z/2,\ \delta_1 = (c_x + c_y)/2,\ \delta_2 = (c_x - c_y)/2. \tag{5.5}$$

In the longitudinal magnetic field, the exciton levels are as shown in the upper part of Fig. 5.12b. A typical LAC and ODMR spectrum of a type-II GaAs/AlAs SL are presented in Fig. 5.12b. Two types of resonance signals are observed.

1. The resonance signals connected with the microwave-induced electron-spin-flip transitions between the radiative and non-radiative levels of localized excitons, i.e., ODMR. The two pairs of ODMR signals belong to heavy-hole excitons. The arrows marked as "hole" and "e" show the ODMR signals of e-h pairs.
2. The resonance signals in low magnetic field, which do not depend on microwaves and are produced by the state mixing at LAC of optically allowed and forbidden exciton levels.

The positions of LAC and ODMR signals of excitons and their angular variations are described by (5.2), which allows obtaining all parameters of g-factors and exchange splittings from LAC and ODMR spectra and their angular dependences.

5.2.1.4 Dependences of the Exciton g-Factors and Exchange Splittings on the Parameters of Superlattices

The parameters of excitons have been studied in a number of type II SLs with different periods and compositions. LAC and ODMR spectra recorded in SLs with

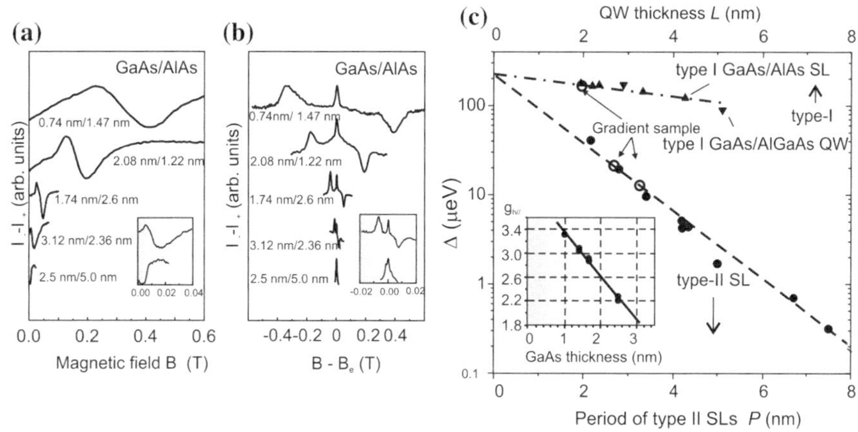

Fig. 5.13 LAC (**a**) and 35 GHz ODMR (**b**) spectra in type II GaAs/AlAs superlattices with different GaAs/AlAs composition as indicated. T = 1.6 K, B || [001], B_e is the resonance field of the electron ODMR signal in distant e-h pairs. **c** Experimental dependences of the isotropic exchange splitting Δ in type II GaAs/AlAs superlattices on the period *P* (*circles*), in type I GaAs/AlAs SLs (*triangles*) and GaAs/AlGaAs QWs (*inverted triangles*). The *inset* shows the dependence of hole *g*-factor $g_{h||}$ on the GaAs thickness. *Open circles* show the data for the transition region of the sample with a composition gradient

periods from 2.2 to 7.5 nm are shown in Fig. 6.13a, b. The splitting between the two ODMR lines of excitons proportional to the isotropic exchange splitting Δ increases drastically with the decrease of the layer thickness. It was found that Δ in type II SLs depends rather on the SL period than on the composition [71, 72, 77]. The hole *g*-factor $g_{h||}$ depends mainly on GaAs quantum well width. Experimental data on the isotropic exchange splitting Δ in type II GaAs/AlAs SLs as a function of the SL period *P* and in type I systems (GaAs/AlAs SLs and GaAs/AlGaAs QWs) as a function the QW width are summarized in Fig. 5.13c. The inset shows the dependence of hole *g*-factor $g_{h||}$ on the GaAs thickness. These results can be used for a local diagnostics of SLs [74].

5.2.1.5 Type II-Type I Transitions

Since the type of the band alignment in GaAs/AlAs SLs depends on the layer thickness these systems provide a unique possibility to study type II-type I transition. Figure 5.14a shows the experimental dependences of circularly polarized luminescence on the magnetic field in the sample with the composition gradient. These dependences were measured at different positions of the excitation spot on the sample (x = 0–23 mm). 35 GHz microwaves were applied to the sample in continuous wave (cw) mode. Luminescence was excited non-resonantly far above the band gap and was detected within the zero-phonon luminescence line of SL in the direction of the static magnetic field. At the type II side (x = 0), LAC and

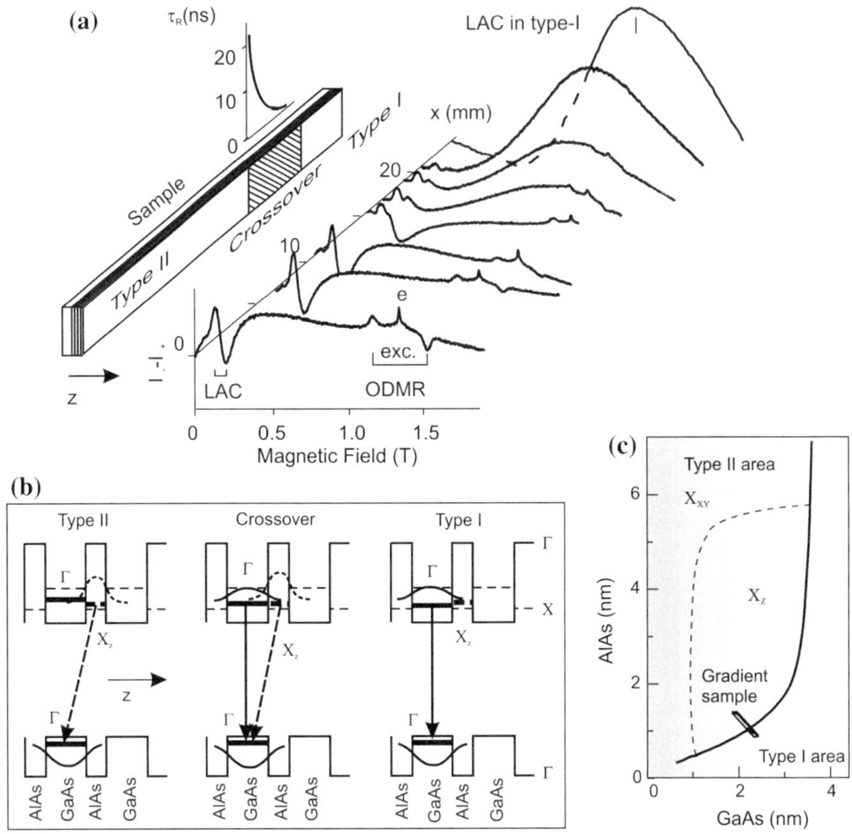

Fig. 5.14 **a** LAC and 35 GHz ODMR measured as variations of the circular polarization of luminescence at different positions of the excitation spot. T = 1.6 K, B || [001]. A fast decrease of the luminescence decay time τ_R in the transition region (x = 15–19 mm) is also shown. **b** Valence and conduction band structure together with the recombination transitions for the type II, type I and crossover regions of the investigated GaAs/AlAs superlattice with a composition gradient. *Narrow lines* show the energy of Γ (*solid lines*) and X (*dashed lines*) states in the bulk materials, *thick lines* correspond to the superlattice states. The calculated probability densities of the wave functions of electrons and holes in each quantum state are also shown. The axis z is the growth direction. **c** Map displaying the border between type I SLs and type II SLs with X_z and X_{xy} lowest conduction band states. The position of the sample with a composition gradient is also shown

ODMR of type II excitons is observed. In this record no ODMR of holes is detected because of faster hole spin relaxation. One can see that ODMR signals disappear with the shift of the excitation spot to the crossover region (x = 15–20 mm). τ_R decreases in this region from microseconds down to 0.3 ns [82].

A simplified diagram of the valence and conduction band structure for the sample with a composition gradient calculated within the envelope wave function approximation using the ECA4 program [83] is shown in Fig. 5.14b together with the recombination transitions. The idealized abrupt interfaces were used in these calculations. The energy positions of the X_z states in AlAs and the Γ states in GaAs layers are inverted for the both sides of the sample, i.e., the X_z-Γ crossover occurs with the change of the GaAs/AlAs composition along the sample.

The position of the gradient sample at the border between type II and type II SLs is shown in Fig. 5.14c. The change in PL spectra and variation of the emission lifetime measured for several positions of the excitation spots between $x = 0$ and $x = 23$ mm gave a clear evidence of a smooth transition from pseudodirect type-II to type-I SL. Since the electron and the hole are spatially separated in type II SL, the exchange splitting is believed to be much smaller than in type I systems, where the electron and the hole are in the same GaAs layer.

LAC and ODMR measured at the type II side ($x = 0$) and near the type II-type I transition ($x = 16$ mm) are different. LAC and ODMR for $x = 0$ can be described by the same set of parameters of the spin Hamiltonian (5.2): $g_{e\|} = 1.88$, $g_{h\|} = 2.5$, the exchange splittings are $\Delta = 20.7$ μeV; $\delta_1 = 6.1$ μeV; $\delta_2 < 1$ μeV. The spectra for $x = 16$ mm imply a coexistence of two exciton species with different exchange splittings and different lifetimes: (i) type II excitons with the parameters: $g_{e\|} = 1.88$, $g_{h\|} = 2.45$, $\Delta = 13.3$ μeV; $\delta_1 = 3.8$ μeV; $\delta_2 < 1$ μeV—these excitons can be observed both in ODMR and LAC; (ii) type-II—like excitons with $g_{e\|} = 1.88$, $g_{h\|} = 2.5$, $\Delta = 21$ μeV—no ODMR of these excitons is observed, although they are detected in LAC. Their lifetime can be estimated to be shorter than 100 ns.

The experimental spectra in a wider field range are shown in Fig. 5.15a. In the transition region wide microwave-independent resonances appear which shift to higher fields as the excitation spot approaches the type I end of the sample ($x = 20$–23 mm) where only LAC of type I excitons can be detected. The type I excitons have about an order of magnitude higher exchange splitting and their electron g-factor is strongly reduced: $g_{h\|} = 2.4$, $g_{e\|} = 0.9$, $\Delta = 170$ μeV, $\delta_1 \approx 30$ μeV. One can see from Figs. 6.14 and 6.15 that in the transition region type-I-like excitons appear. For "type-I-like" excitons (at $x = 18$ mm) the parameters ($g_{h\|} = 2.4$, $g_{e\|} = 1.1$, $\Delta = 150$ μeV) are close to those of type I excitons at $x = 23$ mm. The exchange splitting found for the type-II, type II-like and type I excitons in the gradient sample are shown in Fig. 5.13c by open circles.

Measurements of LAC in linearly polarized luminescence have shown that for the type II side of the gradient sample the lowest radiative exciton level is [110]-polarized. With an experimentally established direct link between the polarisation of the lowest optically allowed exciton level and the type of interface (normal, i.e., AlAs on GaAs or inverted, i.e., GaAs on AlAs) at which the exciton is localized [84, 85] one can conclude that type II excitons in our sample are mainly localized at the normal interface. The sign of LAC measured in linear polarization for type I side of the sample implies that the lowest optically allowed exciton levels are polarized along [1–10] similar to most of type I QWs and SLs studied [86]. This implies [81] a preferential orientation of the interface islands along [1–10].

It follows from the XRD data that in the gradient sample the mean period does not change significantly ($P = 3.3$ nm for $x = 0$ and $P = 3.2$ nm for $x = 23$ mm). As it was mentioned before, the isotropic exchange splitting of type-II excitons is mainly determined by the SL period and the hole g-factor depends in the first approximation on the GaAs width only. The close values of the exchange splittings for type II excitons at $x = 0$ ($\Delta = 20.2$ μeV) and type-II-like excitons ($\Delta = 21$ μeV) near the transition region ($x = 16$ mm) can be understood if to suppose that both excitons are localized in the regions with close local periods but different composition. Since the X-Γ mixing is proportional to $1/(E_X - E_\Gamma)^2$, where E_X and E_Γ are energies of the corresponding conduction band states [70], it strongly increases in the transition region. This results in much shorter lifetime of "type II-like" excitons but has little effect on the exchange splitting. It is to be noted that fine structure of heavy excitons was also studied by monitoring quantum beats in polarization of emission. This technique allows one to measure directly the splitting between the two exciton radiative levels. The measurements of quantum beats in the transition region of gradient sample [55] (at $x = 15.7$ mm) gave the splitting $\delta_1 = 6.3$ μeV and the decay time of luminescence about 28 ns [82]. The measurements of LAC and ODMR reveal a coexistence of type II excitons with the exchange splittings $\Delta = 13.3$ μeV, $\delta_1 = 3.8$ μeV, $\delta_2 \approx 0.1$ μeV and type II-like excitons which are not observed in ODMR because of their shorter lifetime (below some tenths of microseconds). The latter have the exchange splittings $\Delta = 21$ μeV, $\delta_1 \approx 7$ μeV in good agreement with the results of [82].

Type II excitons in the transition region have smaller exchange splitting than type II excitons at $x = 0$. With the dependences of exchange splitting on the SL period one can conclude that they are localized in the regions with ca. 1 monolayer larger local period and thicker AlAs layer. The open circles in Fig. 5.13c correspond to the transition region of the gradient sample.

At the crossover, recombination is determined by an interplay of the recombination rates and the probability for the electron to be at the X_z or Γ level. The HH-Γ recombination is faster, therefore coexistence of type II-like and type I-like excitons is possible only for X_z below Γ. With a decrease of $(E_X - E_\Gamma)$ the mixing of these states gives rise to the appearance of type I-like excitons for which an admixture of the Γ state is much larger. They have short lifetime and much stronger exchange splitting. LAC signals of type II-like and type I-like excitons are marked in Fig. 5.15a.

The study of the type II-type I transition allowed the first observation of LAC in type I QWs and SLs where short luminescence decay time prevents from using ODMR [75, 76]. The position of the LAC signal in type I systems are determined by the same exciton energy level scheme as ODMR and can be described by spin Hamiltonian (5.2). From the analysis of angular dependences of LAC one can obtain the parameters of the spin Hamiltonian. Figure 5.15b shows an example of angular dependence of the LAC signals in type I GaAs/AlAs QW of about 2.5 nm. The energy levels for two orientations of magnetic field are shown in the upper part of Fig. 5.15b. Cross-relaxation resonance shown in the energy level schemes was observed in addition to LAC, which provides additional information. The angular dependences allowed one to obtain the g-factors and exchange splitting parameters: $g_e = 0.4$, $g_{h||} = 2.45$, $g_{h\perp} \approx 0$, $\Delta = 170$ μeV, $\delta_1 = 65$ μeV; $\delta_2 < 10$ μeV.

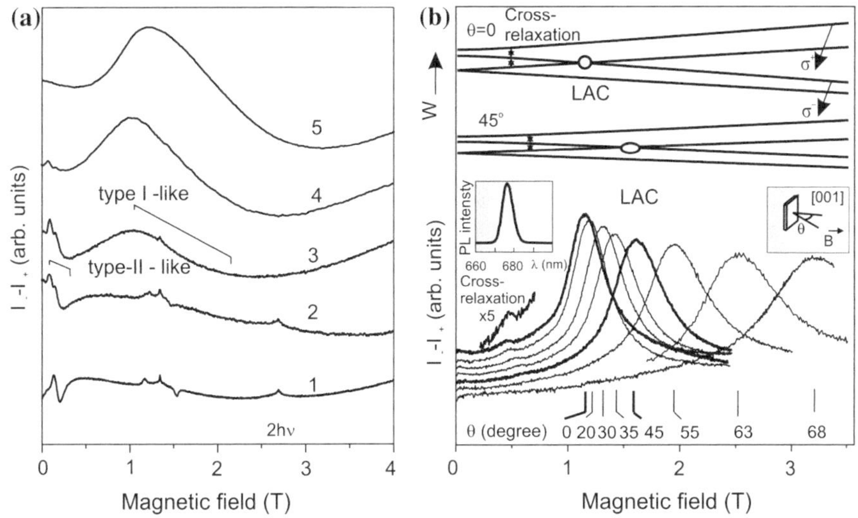

Fig. 5.15 **a** 35 GHz ODMR and LAC spectra measured at x = 0, 15, 17, 19, and 23 mm (*curves 1–5*, respectively) in the SL with the composition gradient. One can see a gradual disappearance of the characteristic ODMR and LAC of HH-X_z type II excitons and appearance of LAC of HH-Γ type I excitons at the X_z-Γ crossover region (x = 15–19 mm). The positions of the "type II-like" and "type I-like" excitons in the transition region are marked. T = 1.6 K, B || [001]. **b** Angular variation of level anticrossing in type I GaAs/GaAs QWs (ca. 2.5 nm) and calculated exciton energy levels for two orientations of magnetic field. Cross-relaxation resonance is also shown in the energy level schemes

5.2.1.6 Selective Study of the Opposite Interfaces in Type II SLs

The emission from the two optically allowed exciton levels is linearly polarized along [110] and [1-10] axes [70, 80]. The study of specially designed anisotropic SLs in which recombination appears only at one type of interface allowed one to establish a direct link between the order of the exciton radiative levels and the type of interface at which the exciton is localized [84, 85]: the lowest radiative level is [110]-polarized for the normal (AlAs on GaAs) interface and [1-10]-polarized for the inverted (GaAs on AlAs) interface. Due to state mixing at anticrossing of a more populated non-radiative exciton level with a less populated radiative level, a resonance increase of the emission with linear polarization of the corresponding radiative level is expected [87]. The low-field LAC resonance is detected in the polarization of the lowest exciton level. Since the order of the exciton radiative levels is inverted for the opposite interfaces, LAC detected in linear polarization should have opposite signs for the excitons localized at the opposite interfaces.

Figure 5.16a shows LAC signals in a usual "symmetrical" GaAs/AlAs SL detected in linearly polarized light with polarizations [110] and [1-10]. The obtained spectra imply the coexistence of two systems of exciton localized at the opposite interfaces. The fact that the LAC signals from the two classes of excitons

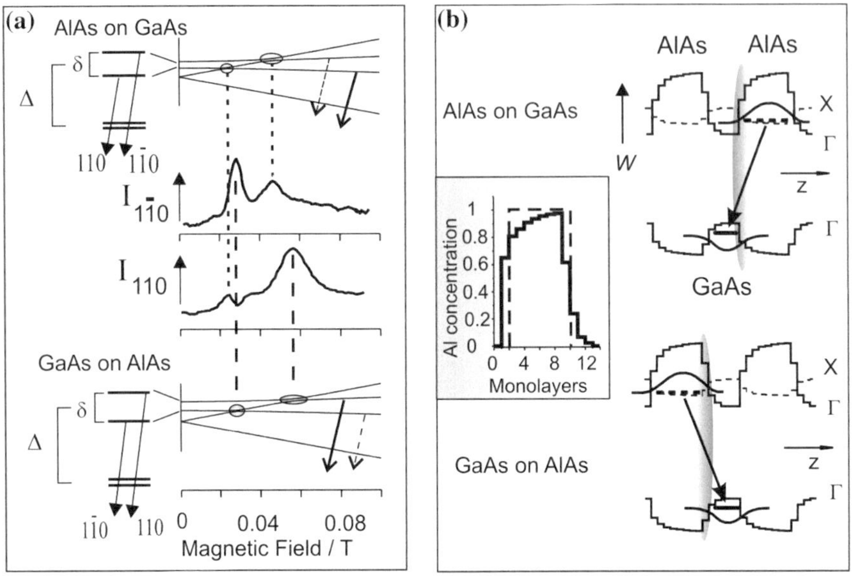

Fig. 5.16 **a** Energy levels and level anticrossing signals of excitons in type II GaAs/AlAs SL detected in [110] and [1–10] linearly polarized light. The observed LAC signals belong to two classes of excitons which are localized at the opposite interfaces and have different exchange splittings. **b** Composition profile of the interface (*inset*) and the conduction and valence band structure of a type II SL calculated with this composition profile. The calculated probability densities of the wave-functions of electrons and holes in each quantum state are also shown

appear at different magnetic fields implies that the excitons differ in the exchange splitting as shown in the energy level schemes in the Fig. 5.16a. The excitons at the inverted interface have ca. 20% larger exchange splitting.

This observation can be explained if we consider a real interface composition profile. Due to different surface segregation rates of gallium and aluminum, gallium penetrates into the AlAs layers forming the composition profile of the interface as shown in the inset in Fig. 5.16b [88, 89]. Calculations of the electron and hole wave functions with using ECA4 program [83] for a SL with this composition profile show that asymmetry in the GaAs/AlAs composition results in the asymmetry of the electron and hole envelope functions and a difference in the overlap integrals for recombination at the normal and inverted interface. The ratio of the squares of the overlap integrals, which was used as an estimate of the exchange interactions, is close to ratio of the experimentally determined exchange splittings. These observations confirm that in real SLs the interfaces are not symmetrical. On the other hand the fact that excitons at the opposite interfaces in type II SLs have different exchange splittings and a reversed order of the radiative levels allows separate studies of the normal and inverted interfaces.

5.2.1.7 Separately Localized Electron and Holes in Type II GaAs/AlAs SLs

Besides the ODMR and LAC of excitons with a definite value of the exchange splitting broad ODMR signals are observed (see Figs. 6.12b and 6.14). Their *g*-factor $g_{e\|} \approx 1.9$ corresponds to the X_z electrons in the AlAs layers. ODMR measured at $x = 0$ in the gradient sample is shown in the bottom of Fig. 5.17, in which ODMR recorded in several other type II SLs with different periods *P* and exchange splittings Δ are represented for comparison. It is seen that the splitting between the two ODMR lines of excitons strongly depends on the SL period, whereas the shape of the broad ODMR signal does not vary and is period-independent, in the first approximation. Similar broad ODMR of holes has been observed for some SLs.

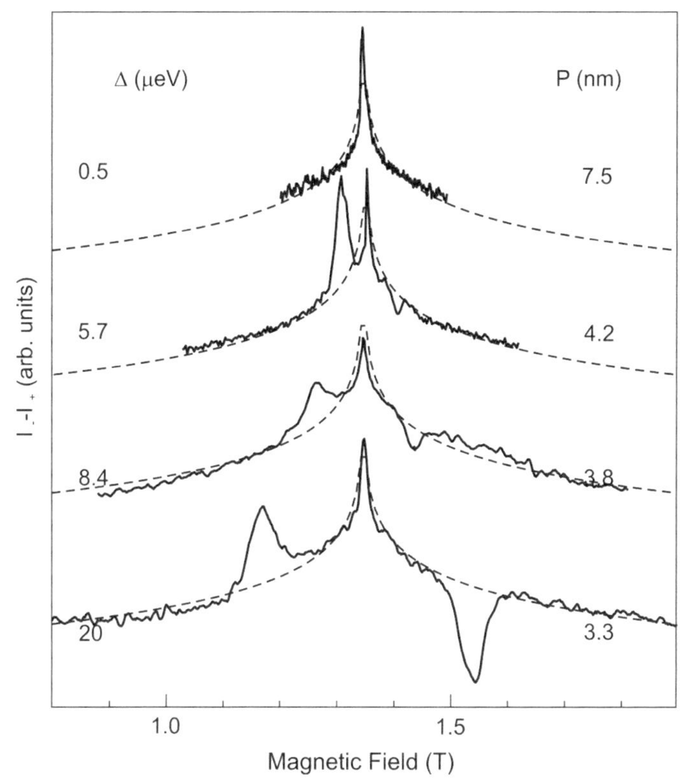

Fig. 5.17 ODMR spectra in the type II region of the sample E913 (*bottom*) and in a series of type II SLs with periods *P* showing a coexistence of ODMR lines originating from type II excitons with a definite exchange splitting Δ and broad ODMR of e-h pairs with a distribution of interpair separations. *Dashed lines* show electron ODMR signal calculated for a random 2-D distribution of interpair separations for electrons in AlAs and holes in adjacent GaAs layers. ν = 35 GHz, T = 1.7 K, B ∥ [001]

In Fig. 5.15a one can see additional ODMR signals corresponding to the double quantum transitions marked as "2 hν". They are detected at high enough microwave power (above 300 mW). Similar broad ODMR signals with the same shape and width have been observed both in double and triple magnetic field. These signals correspond to the EPR transitions with the absorption of two and three microwave quanta (effective frequencies, 70 and 105 GHz). The same shape of the broad electron signal for double and triple quantum ODMR [69, 90] proves that the line broadening is determined by a spreading of the zero-field (i.e., exchange) splittings but not *g*-factors. This implies the existence of a recombination of the X_z electrons in AlAs and heavy holes in GaAs that may be localized independently by the interface roughness potential and have a smooth distribution of the interpair separations. In contrast, for excitons the splitting of the ODMR lines corresponding to the electron spin flips in the exciton has a definite value for each SL. Interface roughness leads to a broadening of the exciton ODMR lines but since the excitons are localized as a whole the e-h separation is mainly determined by the SL period.

The dashed line in Fig. 5.17 shows the result of calculations of the lineshape of the electron ODMR for a random distribution of thermalized e-h pairs in the SL plane. The calculations were made similar to those for sallow donor-acceptor recombination in bulk GaAs [91] but for the case of two-dimensional (2-D) distribution of the e-h pairs. The values of the Bohr radius and exchange parameter used are 10 nm and 200 μeV, respectively; the ratio of the Bohr radius to the mean interpair separation was taken as 0.2. Broad signals with the shape similar to that observed in the ODMR spectra were also observed in LAC detected via linear polarization of luminescence in type II SLs [86] and are most probably due to LAC of e-h pairs. In the type II-type I transition region, ODMR of electrons is seen even after the disappearance of the type II excitons probably due to longer e-h recombination times. These signals are usually more pronounced in the SLs grown with a composition gradient which increases the probability of localization due to interface roughness.

To summarize, ODMR and LAC spectroscopy were applied to study *g*-factors and exchange splittings of localized excitons which can be considered as probes to study the interface microstructure in GaAs/AlAs and GaAs/GaAlAs QWs and SLs. Both type II and type I systems were investigated and also type II-type I transition which correspond to the X_z-Γ crossover of the conduction band states. *g*-factors, exchange splitings and the order of the exciton radiative levels were determined. Combination of ODMR and LAC spectroscopy made it possible to study shortly lived excited states in type I QWs and SLs. Selective spatially resolved investigations of the opposite interfaces in type II SLs were made and asymmetry in the interface composition profiles was revealed. The obtained results can be used for local diagnostics of QWs and SLs. Besides ODMR of excitons with a definite value of exchange splitting, ODMR ascribed to separately localized electrons and holes with a distribution of exchange splittings was detected. Multiquantum transitions in the ODMR spectra were used to analyze the electron-hole exchange interactions in GaAs/AlAs superlattices.

5.2.2 Self-organized Oriented Silver Halide Micro- and Nanocrystals Embedded in Crystalline Alkali Halide Matrix

5.2.2.1 Introduction

In the last decade, nanostructures have been successfully fabricated using self-organization effects common to strained heterosystems [92]. Nanocrystals can show a common orientation only in a crystalline matrix, as recently demonstrated for copper and silver halide nanocrystals embedded in an alkali halide matrix [93–97].

Silver halide microcrystals and nanocrystals, can be formed in growth of alkali halide single crystals heavily doped with silver halides [95–97]. Silver halides AgCl and AgBr have the same face-centerd cubic lattice as alkali halides KCl and KBr, slightly different lattice constants and the energy gaps larger by more than 5 eV. Thus, micrometer and nanometre size AgCl and AgBr crystals embedded in KCl and KBr matrices, respectively, can be considered as an array of self-organized microcrystals and nanocrystals (quantum dots) in a strained heterosystem.

Silver halides have some unique features and occupy a particular position in solid-state physics because their properties can be considered as of borderline nature between ionic and covalent bonding. Under UV light irradiation of AgCl and AgBr an electron is excited from the valence band into the conduction band and a hole is left in the valence band. Free electrons can be captured by some Coulombic core to form shallow electron centers (SECs), which are believed to play an important role in the latent image formation process. The pulsed ENDOR experiments have shown that the wave function of SEC in AgCl and AgBr single crystals is very diffuse with the Bohr radius of about 1.7 nm and have given evidence for a model of shallowly trapped electron center (shallow donor) which has the split-interstitial silver pair as a core [98]. This was confirmed by energy calculations [99]. In AgCl the hole is subjected to self-trap due to static Jahn-Teller (J-T) effect, to form the self-trapped hole (STH) [100]. The STH can be considered as the $(AgCl_6)^{4-}$ complex with approximately 84% of the hole wave function density located on the central silver ion and the four equatorial chlorine ions along the $\langle 100 \rangle$ directions in the plane perpendicular to the axis of the J-T distortion. STH in AgCl can capture an electron from the conduction band forming the self-trapped exciton (STE). STE, STH and SEC in bulk AgCl were successfully studied by ODMR, EPR, ENDOR and spin echo [100–102]. It was established by ENDOR that STE in AgCl consist of a very diffuse electron (with the Bohr radius of about 1.5 nm) attracted by a strongly localized STH that is virtually identical to the isolated STH [102]. In AgBr the nature of the localized holes is less known and it is believed that there is no hole self-trapping [103, 104]. So called "transition type excitons" were detected by ODMR in AgBr [104]. They are characterized by a singlet to triplet isotropic exchange splitting of $|J| = 1.9\ cm^{-1}$, which is much smaller than the value of $5.37\ cm^{-1}$ for STE in AgCl [69, 102].

The effects of confinement on shallow centers with Bohr radius comparable with the particle size are well known (see [92] for references). The influence of the nanoparticle size on deep level centers and local effects in solids is of fundamental importance, though to our knowledge it was much less studied. The J-T effect is one of the basic local effects in solids, rather sensitive to internal fields and variations of the electron-phonon interaction. STH and STE in bulk AgCl are classical J-T systems well studied in bulk AgCl by various radiospectroscopic techniques. Therefore, an investigation of AgCl and AgBr nano- and microcrystal systems embedded in a crystalline matrix seems to be very promising.

5.2.2.2 Experimental

The KCl:AgCl and KBr:AgBr single crystals were grown by the Stockbarger technique with 2–3 mol% silver in the melt. Transparent optical-quality KCl:AgCl and KBr:AgBr samples were cleaved from different parts of the grown crystal along the {100} planes and represented transparent single crystals without visible inclusions. Several KCl:AgCl and KBr:AgBr crystals were grown and studied, but the principal results of this chapter were obtained with two KCl:AgCl samples (No. 1 and No. 2) cleaved from the different parts of the same boule grown with 2 mol% silver in the melt and one KBr:AgBr crystal grown with 2 mol% silver in the melt.

35 GHz ODMR at a temperature of 1.6 K was detected from the luminescence which was excited by the UV light of a deuterium arc lamp with appropriate light filters and analysed with a grating monochromator. The microwave power in the cavity of an ODMR spectrometer was modulated at a sound frequency, and the microwave-induced changes in the luminescence intensity were detected using a lock-in detector and a grating monochromator. Reference measurements were made in AgCl and AgBr single crystals that were used for doping of KCl and KBr.

5.2.2.3 Self-organized AgCl Structures Embedded in KCl Crystalline Matrix

Figure 5.18a shows ODMR spectra and photoluminescence (inset) recorded at 1.6 K in two samples KCl:AgCl: no. 1 (spectra 1), no. 2 (spectra 2). The luminescence and ODMR spectra of bulk AgCl, taken under the same conditions, are shown for comparison in the lower part of Fig. 5.18a by a dashed line. Points in Fig. 5.18a (inset) represent the spectral dependences of the ODMR amplitude. Symbols || and ⊥ denote the centers with their axes parallel and perpendicular to the magnetic field, respectively. The spectra are recorded at magnetic field parallel to [001], $T = 1.6$ K, microwave power $P = 300$ mW and chopping frequency $f_{chop} = 80$ Hz.

The remarkable similarity of the ODMR spectra recorded for AgCl QDs in KCl sample no. 1 and for the bulk AgCl crystal was observed. The spectral dependence of the ODMR in AgCl QDs in KCl sample no. 1 (Fig. 5.18a, inset, spectrum 1) is

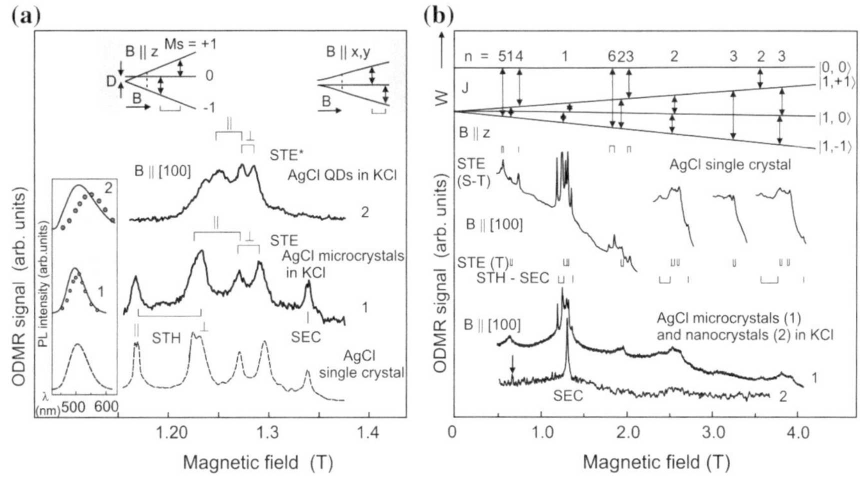

Fig. 5.18 **a** ODMR spectra and (*inset*) photoluminescence (*PL*) in two samples: no. 1 (the luminescence and ODMR (spectra 1)), no. 2 (spectra 2) of KCl:AgCl grown with 2 mol% of Ag in the melt. *Points* show the spectral dependences of the ODMR signals. For comparison the luminescence and ODMR spectra in bulk AgCl are shown by *dashed lines*. The position of the ODMR lines are marked for self-trapped excitons (STE), self-trapped holes (STH) and shallow electron centers (SEC) in AgCl microcrystals and bulk AgCl and for STE in AgCl nanocrystals (STE*). Symbols $\|$ and $\perp$ denote the centers with their axes parallel and perpendicular to the magnetic field. T = 1.6 K, ν = 35.2 GHz; P = 300 mW, f_{chop} = 80 Hz, **B** $\|$ [001]. **b** Energy level scheme for STE with the observed multiquantum transitions and ODMR measured at high microwave power (900 mW) in: **a** AgCl single crystal and **b** in AgCl microcrystals (*1*) and nanocrystals (2) embedded in KCl crystalline matrix), *n* denotes the number of microwave quanta. T = 1.6 K, ν = 35.2 GHz, f_{chop} = 80 Hz

also close to the PL spectrum of bulk AgCl. The ODMR lines positions are marked for STE, STH and SEC. The energy level diagrams for the triplet state of the STE and the observed EPR transitions are shown at the top in Fig. 5.18a. One can see that the ODMR lines of STH, STE and SEC in AgCl QDs in KCl sample no. 1 and bulk AgCl are well resolved due to strong anisotropy of *g*-factors and relatively large fine structure splitting *D*. Observation of bulklike ODMR in AgCl in KCl: AgCl can be considered as the direct confirmation of the formation of AgCl microcrystals (of micrometer size) in KCl matrix. These microcrystals have practically the same properties as bulk AgCl. The principal axes of STH and STE coincide with the ⟨100⟩ axes of KCl, which indicates that the AgCl crystals in KCl retain the symmetry of the matrix. The disorientation of the AgCl microcrystals as estimated from the ODMR spectra is below 5°.

The ODMR spectra of STE can be described by the general spin Hamiltonian

$$\hat{H} = \mu_B \vec{B} \cdot \overleftrightarrow{g}_e \cdot \hat{\vec{S}}_e + \mu_B \vec{B} \cdot \overleftrightarrow{g}_h \cdot \hat{\vec{S}}_h + \hat{\vec{S}}_e \cdot \overleftrightarrow{D} \cdot \hat{\vec{S}}_h + J \hat{\vec{S}}_e \cdot \hat{\vec{S}}_h, \qquad (5.6)$$

Table 5.1 Parameters of STH, STE and SEC in bulk AgCl crystals and AgCl micro- and nanocrystals embedded in KCl matrix

Crystal	STH		STE			SEC
	$g_{\|\|}$	$g_{\perp}$	$g_{\|\|}$	$g_{\perp}$	$\|D\|$ (MHz)	g
Bulk	2.147	2.040	2.0216	1.968	710	1.881
Microcrystal	2.147	2.040	2.020	1.966	710	1.88–1.90
Nanocrystals	2.016	1.974	1.992	1.965	335	~1.96

where all parameters have usual meaning [105], $\boldsymbol{S}_e$ and $\boldsymbol{S}_h$ are spins of electron and hole. Two spins 1/2 are coupled into a singlet ($S = 0$) and a triplet ($S = 1$) state. The splitting between the singlet and triplet states is equal to J. For STE in bulk AgCl the 35 GHz EPR spectra could be described by a spin Hamiltonian for the triplet state with axial symmetry and fine structure splitting D. For the principal axes along the $\langle 100 \rangle$ direction the g-factor of STE is

$$g_{\|} = (g_{\|h} + g_e)/2; \; g_{\perp} = (g_{\perp h} + g_e)/2. \tag{5.7}$$

This direct link between the g-factors of STE and those of STH and SEC strictly holds for bulk AgCl crystals and AgCl microcrystals. The experimental parameters for STH, STE and SEC in bulk AgCl and AgCl microcrystals embedded in KCl matrix are given in Table 5.1.

As one can see from Fig. 5.18a, the luminescence and ODMR spectra recorded for AgCl QDs in KCl sample no. 2 (spectra 2) are different compared with the spectra recorded for AgCl QDs in KCl sample no. 1 (spectra 1) and for the bulk AgCl crystal (dashed lines). Spectrum 2 contains ODMR lines which are marked in Fig. 5.18a as STE* and can be attributed to the triplet state. This conclusion was confirmed by an analysis of the angular dependences in (110) and (100) planes and by the observation of the forbidden transitions (not shown). The parameters of the triplet obtained by fitting the angular dependences and the spin Hamiltonian (5.6) are the same as those of the ODMR spectra ascribed in [95] to STE in AgCl nanocrystals embedded in a KCl matrix. Formation of silver halide nanocrystals with an average size of less than 10 nm in AgCl QDs in KCl and AgBr QDs in KBr crystals grown with 2 mol% silver in the melt was observed in [95] by atomic force microscopy. At higher chopping frequency the anisotropic signals with $S = 1/2$ were observed. For the ODMR of $S = 1/2$ and $S = 1$ centers in sample no. 2, both the anisotropy of the g-factors and the parameter D are considerably reduced.

The anisotropic centers with $S = 1/2$ and $S = 1$ were ascribed to STH and STE in AgCl nanocrystals embedded to KCl matrix, that is why they were labelled as STH* and STE* in Fig. 5.18a. The parameters of the ODMR spectra are listed in Table 5.1. As follows from (5.7), there is a direct link between the g-factors of STE and those of STH and SEC. It is seen from Table 5.1 that the g-factors for STH* and STE* in AgCl nanocrystals satisfy (5.7) if we suppose that the g-factor of SEC remains isotropic in the nanocrystals. In addition, this allows estimating the g-factor of SEC in AgCl nanocrystals, which exceeds the bulk AgCl value (Table 5.1). This correlation between the g-factors of the $S = 1/2$ and $S = 1$ centers strongly supports their identification as STH and STE in AgCl nanocrystals.

Although the arguments in favour of the assignment of the ODMR spectra in samples no. 2 to AgCl nanocrystals seem to be rather convincing, one can not totally exclude that these spectra may belong to point defects in KCl. Analysis of the EPR and ODMR spectra showed that the obtained ODMR spectra do not correspond to EPR of any known point defect or STE in KCl.

5.2.2.4 Multiquantum ODMR in AgCl Single Crystal and AgCl Nanocrystas (QDs) in KCl

Since STE in AgCl consists of a very diffuse electron attached by a strongly localized STH, the isotropic exchange splitting J between the singlet and the triplet states is too small to be observed in the optical spectra but it can be directly measured by EPR and ODMR. In [69], multiquantum singlet-to-triplet transitions corresponding to the absorption of up to seven microwave quanta (total energy $7 \times 35 = 245$ GHz) were found in ODMR of STE in bulk AgCl crystals. This finding allowed to measure J with extremely high accuracy: $J = -5.370 \pm 0.002\ \mathrm{cm}^{-1}$.

The energy level scheme of STE for B || [100] and the experimentally observed multiquantum ODMR transitions for STE, STH and SEC in bulk AgCl are shown in Fig. 5.18b. In addition to the transitions within the triplet state of STE, marked as "STE (T)", the lines which correspond to the singlet-to-triplet multiquantum transition, marked as "STE (S-T)" are observed. This surprising result was unambiguously proved by the measurements of ODMR at slightly different microwave frequencies, i.e., different energy of the microwave quanta. The directions and the values of the line shifts were different for different transitions, in complete agreement with the calculations [69]. Multiquantum ODMR in self-organized AgCl structures embedded in KCl crystalline matrix is shown in Fig. 5.18b. No singlet-to-triplet transitions were detected in this case.

Although multiquantum ODMR has been observed by different groups in several systems, the complete understanding of this effect is still missing. One of possible explanations of strong multiquantum ODMR [69] takes into account multiquantum transitions via real intermediate electronic states. According to this mechanism both the electric (E) and magnetic (B) components of the microwave field can be active in the multiquantum transitions including an interference effect of E and B. The intermediate states may appear as a result of vibronic interaction in a quasi-degenerate system of electronic states of the center. The possibility of successive transitions, including absorption of both E and B components, was shown previously. It is not excluded that the intermediate states are associated with the presence of the free carriers. It is probable that these states are really observed in the ODMR of AgCl as a broad background signal.

5.2.2.5 Electron-Hole Recombination Confinement in Self-organized AgBr Nanocrystals Embedded in Crystalline KBr Matrix

The phenomenon of spatial confinement of the electron–hole recombination in exchange-coupled pairs of recombining SEC and localized holes was revealed in AgBr nanocrystals embedded in the KBr matrix by ODMR [105–107]. It was found that the samples cleaved from different parts of KBr:Ag crystals grown from the melt (1–2 mol% AgBr) by the Bridgman technique contain self-organized AgBr microcrystals and/or nanocrystals [106, 107]. Similar to the AgCl microcrystals in KCl, the AgBr microcrystals in KBr retain the properties of bulk AgBr crystals. Formation of the AgBr nanocrystals in the KBr matrix was confirmed by the appearance of a characteristic peak of the exciton emission in the luminescence spectra and its blue shift relative to the exciton emission in bulk AgBr [95, 106, 107]. AgBr is an indirect band-gap material and the exciton emission is very weak. The exciton emission intensity increases more than 10^4 times in AgBr nanocrystals because of modification of the selection rules. A blue shift of the exciton emission appears because of spatial confinement effects. The value of the shift increases with a decrease in the nanocrystal size and can be used for characterization of the size. These effects were observed before for AgBr nanocrystals dispersed in gelatine, reverse micelles, etc. (see [108, 109] and references therein). Formation of AgBr micro and nanocrystals in glass was reported in [110] where a single ODMR line was observed for nanocrystals, which corresponds to a strong electron–hole exchange.

Figure 5.19a shows ODMR spectra in AgBr QDs in KBr crystals measured at different wavelengths within the emission band. For comparison the spectra measured in bulk AgBr that was used for doping are also shown. In spectral dependences of ODMR the broad outer lines in the region of electron and hole centers and central ODMR lines are shown. The spectra are recorded under the same conditions as in AgCl QDs in KCl (Fig. 6.18a). In contrast to AgCl QDs in KCl, only isotropic ODMR lines have been observed. The EPR spectra in bulk AgBr were identified as SEC ($g = 1.49$), a hole center ($g = 2.08$) [103] and a partly resolved doublet line between them as "intermediate-case exciton" with exchange splitting $J = -1.9\ \mathrm{cm}^{-1}$ [104]. The luminescence spectrum in bulk AgBr consist of a broad intrinsic band at 585 nm and a band 500 nm related to the residual iodine impurity in KBr [103]. The ODMR spectra in AgBr QDs in KBr crystals qualitatively have the same structure as in bulk AgBr: they are isotropic, there is a doublet central line and broad symmetrical distribution of lines in the region of electron and hole centers. With increasing wavelength the splitting of the central doublet increases and the maxima of broad ODMR lines become closer.

The shape of the ODMR spectrum of AgBr QDs significantly changes with an increase of the microwave chopping frequency (Fig. 5.19b) and microwave power: the separation of the maxima of the broad signals in the low and high fields increases, while the distance between the central ODMR peaks decreases. Such variations in the shape of ODMR spectra imply that they belong to

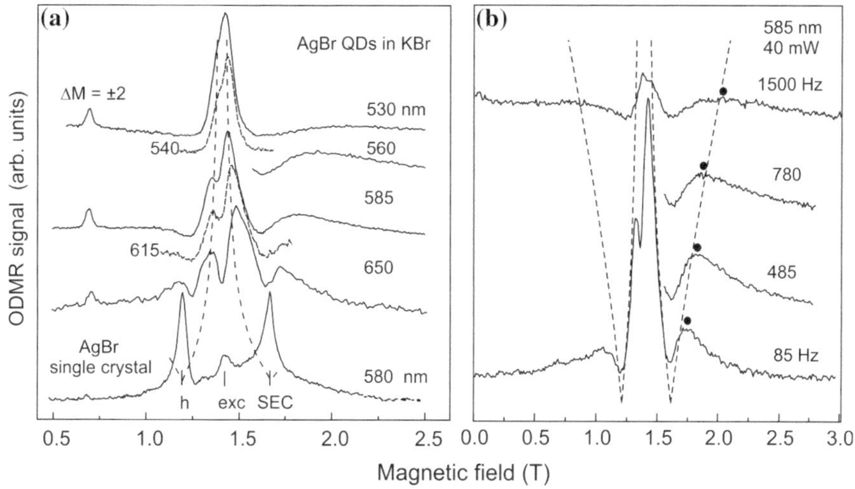

Fig. 5.19 **a** ODMR spectra for AgBr QDs embedded in a KBr crystalline matrix and in AgBr single crystal. ODMR spectra in AgBr QDs were measured at different wavelengths within the emission band. $\Delta M = \pm 2$ marks the forbidden transitions. T = 1.6 K; ν = 35.2 GHz; P = 300 mW; f_{chop} = 85 Hz; $B \parallel [001]$. **b** ODMR in the same sample recorded with wavelength of 585 nm and P = 40 mW for different chopping frequencies: f_{chop} = 85, 485, 780 and 1500 Hz. The spectra are vertically offset for clarity

exchange-coupled recombining electron–hole pairs with a distribution of the interpair separations.

In addition to the ODMR signals discussed above, the isotropic line at about 0.7 T attributed to the forbidden transitions $\Delta M_S = \pm 2$ of a triplet state have been observed. The forbidden transitions have also been observed in bulk AgBr single crystals, which confirmed the triplet state for the exciton.

The energy levels of a donor-acceptor (D–A) pair in a magnetic field B can be described using a spin Hamiltonian

$$\hat{H} = \mu_B g_e \vec{B} \cdot \hat{\vec{S}}_e + \mu_B g_h \vec{B} \cdot \hat{\vec{S}}_h + J \hat{\vec{S}}_e \cdot \hat{\vec{S}}_h, \tag{5.8}$$

where $S_e = S_h = 1/2$. Here, the first two terms describe the interaction of electron (on the donor) and hole (on the acceptor) with the magnetic field, while the third term describes the isotropic exchange interaction. In AgBr, the g values of donors and acceptors are isotropic. The positions (resonance fields) of the EPR transitions at a microwave frequency of 35 GHz, calculated by (5.8) for g_h = 2.07, g_e = 1.57 and various exchange interaction constants J, are presented in Fig. 5.20a in the same scale of fields as that used in Fig. 5.20b for the ODMR spectra. The calculations were performed using the "View EPR" program written by Grachev [10].

Variations of the ODMR spectra of AgBr QDs in KBr crystals can be explained as a change of the exchange interaction J with variation of the wavelength. Figure 5.20a presents calculated positions of 35 GHz EPR transitions for two

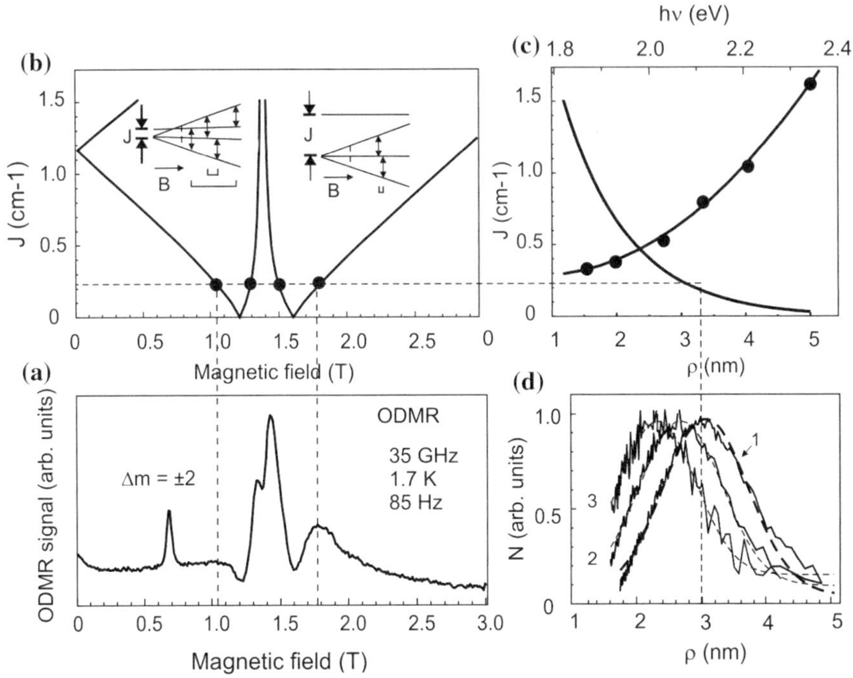

Fig. 5.20 Reconstruction of the distribution of recombining D–A pairs in AgBr nanocrystals from the shape of shallow donor ODMR. **a** ODMR spectrum recorded at $\lambda = 585$ nm; $\nu = 35.2$ GHz; $P = 300$ mW; $f_{chop} = 80$ Hz; T = 1.6 K. **b** Calculated positions of the 35 GHz EPR transitions for recombining pairs localized hole-SEC with different isotropic exchange splitting J; $S_h = S_e = 1/2$, $g_h = 2.08$ and $g_e = 1.49$; The *inset* shows the energy levels and EPR transitions for two values of the exchange splitting J. **c** Experimental dependence of exchange splitting J on the emission energy measured for the central doublet in ODMR of AgBr QDs in KBr (*filled circles*) and calculated dependence of J on the separation between the recombining partners ρ. **d** The obtained distribution of the number of recombining D–A pairs with respect to their spacing ρ determined from the ODMR spectra measured for AgBr QDs in KBr using three emission wavelengths (*1*) 585 nm, (*2*) 560 nm, and (*3*) 532 nm. *Dashed lines* show approximation by the Gaussian profiles

exchange coupled electron spins with isotropic exchange splitting J (these positions are also depicted in Figs. 6.20a, b). Inset shows the energy levels and EPR transitions for two values of the exchange splitting J. The filled circles correspond to the peak positions of the ODMR lines recorded in AgBr QDs in KBr at 587 nm as an example (Fig. 5.20b). It should be noted that for each wavelength there is some distribution of ODMR signals corresponding to different exchange J. This is confirmed by the dependences of the ODMR spectra on the microwave power and chopping frequency, which show that the response time decreases with an increase in exchange. The outer ODMR lines are very broad for electron and hole centers, for the central doublet they are narrower in agreement with a steeper slope of dependences in Fig. 5.20a. Figure 5.20c shows (filled circles) the experimental dependence of exchange splitting J on the emission energy. One can see a strong increase of the exchange energy J with luminescence energy.

For distant D–A pairs, in which the spacing ρ between donor and acceptor is large compared to the sum of their Bohr radii, the exchange interaction is weak. The ODMR spectrum of such a system must display two lines corresponding to the EPR of isolated donors and acceptors, as it is actually observed in the spectra of bulk AgBr and AgBr microcrystals in KBr. A decrease in the distance ρ gives rise to the exchange interaction J which leads to splitting of the energy levels of the D–A pair in a zero field and to splitting of the ODMR signals. In the ODMR spectrum, four lines correspond to each J value, whose positions vary with J as depicted in Fig. 5.20a. When the J value exceeds that for the Zeeman interaction, the D–A pair states split in a zero field can be described by the total spin $S = 0$ (singlet) and $S = 1$ (triplet). The corresponding ODMR spectrum must contain two lines, whose splitting, due to nonlinearity of the levels $S = 0$ and $S = 1$, $M_S = 0$, must tend to zero with increasing J. In contrast to the case of AgCl, the ODMR spectra of the triplet excitons in AgBr ($J = -1.9$ cm^{-1}) are isotropic and exhibit no splitting of the fine structure. Both the bulk crystals and the microcrystals of AgBr contain coexisting systems of the D–A pairs and the excitons with a fixed exchange splitting.

The exchange interaction depends on overlap of the wave functions of electrons and holes. When the Bohr radius of the donor is much greater than that of the acceptor ($a_D \gg a_A$), the exchange interaction constant exponentially depends on the D–A distance [111]: $J = J_0\exp(-2\rho/a_D)$, where J_0 is the limiting exchange interaction value. A similar exponential relation describes the rate of radiative recombination in the pair [112]: the emission due to the recombination of closer pairs is characterized by a higher recombination rate. This approach is applicable to AgBr crystals, since shallow electron centers are characterized by a hydrogenlike 1s wave function with a large Bohr radius $a_D = 1.7$ nm [113], while the wave function of a hole center is considered as localized.

In the presence of a Coulomb interaction, the emission wavelength decreases with increasing distance between the recombining centers [114]. It should be noted that a strong electron-phonon interaction in AgBr crystals leads to the appearance of broad bands in their PL spectra. The emission at a certain wavelength contains contributions from the D–A pairs with various distances between donors and acceptors, which is manifested as a distribution of exchange interactions in the ODMR spectra measured at a certain emission wavelength.

An analysis of the ODMR spectra measured at various emission wavelengths and chopping frequencies showed that the shape of the observed ODMR signals corresponds to a superposition of the signals from exchange-coupled D–A pairs (SEC and localized holes) with a certain distribution of exchange interactions related to the distribution of distances between donors and acceptors. Indeed, a decrease in the detection wavelength or an increase in the microwave power and/or the chopping frequency leads to a shift of the ODMR signal peak in accordance with the increase in magnitude of the exchange splitting.

In the bulk AgBr crystal, D–A pairs are predominantly encountered for which $J \approx 0$, while the fraction of pairs with nonzero exchange is rather insignificant and manifested only by the broadening and characteristic shape of the ODMR line of shallow electron centers and localized holes. The opposite situation is observed for AgBr nanocrystals, where no distant pairs are present because of small crystal size and only pairs featuring considerable exchange interactions are manifested in the ODMR spectra. Here, the region of *g* values corresponding to isolated donors and acceptors (ODMR of distant pairs) must display minimum signals. A shift of the ODMR minimum in the spectrum of AgBr QDs toward lower fields as compared to the line of shallow electron centers in the bulk AgBr is probably indicative of an increase in the *g* value of these centers in AgBr nanocrystals as a result of the spatial confinement.

It was established that the holes in AgCl crystals exhibit self-localization due to the Jahn–Teller effect [115]. In AgCl nanocrystals, the Jahn–Teller effect is partly suppressed which leads to a change in parameters of the spin Hamiltonian [96, 97]. It was previously accepted that no self-localization of holes takes place in the bulk AgBr. However, based on the results of this study, we believe that the holes in AgBr can be self-localized as well. However, in contrast to the situation in AgCl, the dynamic Jahn-Teller effect taking place in AgBr leads to isotropization of the *g* value, as observed in experiment. The *g* value of holes in AgBr is close to an average *g* value of the self-localized holes in AgCl. According to this approach, the exciton in AgBr possesses qualitatively the same structure as the self-localized exciton in AgCl in which the wave function of an electron trapped by a self-localized hole is close to the wave function of a shallow electron center. A smaller magnitude of the singlet–triplet splitting observed in AgBr can be explained by a more strongly delocalized wave function of the electron part of the exciton.

In the region of strong fields (B > 1.6 T), the ODMR spectrum exhibits only signals from the shallow electron centers. The ODMR signal amplitude is proportional to the number N of recombining pair with a given exchange *J* corresponding to the resonance magnetic field *B*. Using the results of calculations presented in Fig. 5.20a and the ODMR line shape, it is possible to reconstruct a distribution of the number N of recombining pairs with respect to the exchange interaction constant *J* and finally with using *J*(ρ) dependence in Fig. 5.20c to obtain N(ρ). The result of such reconstruction is presented by curve N(ρ) in Fig. 5.20d. With an allowance for the exponential dependence of the exchange interaction magnitude on the D–A distance ρ, one can also determine the distribution of D–A pairs with respect to their spacing.

Figure 5.20d shows the results of such calculations performed with $a_D = 2$ nm and $|J_0| = 5\ \text{cm}^{-1}$ for the ODMR spectra measured at the three emission wavelengths indicated above. The shape of the distribution profiles is close to Gaussian (dashed lines in Fig. 5.20d). As can be seen from these distributions, the emission from AgBr QDs in KBr contained no contribution due to the distant pairs, just as is expected for nanocrystals with dimensions on the order of several nanometers; it is also seen that the average ρ value decreases with the luminescence wavelength used to detect the ODMR signals.

It was natural to suggest that the most probable position of a shallow donor is at the center of an AgBr nanocrystal, while a hole most probably occurs at the surface of this crystal. Assuming that $L \approx 2\rho$, one can estimate the average size of nanocrystals from the resulting $N(J)$ distributions. As can be seen from Fig. 5.20d, the luminescence at a shorter wavelength is due to nanocrystals with a smaller average size. The total distribution of AgBr nanocrystals with respect to their dimensions in KBr samples (distribution of the AgBr nanocrystal size) can be determined from an ODMR spectrum measured using the total optical emission from the sample.

In contrast to the case of bulk AgBr and AgBr microcrystals, the ODMR spectrum of nanocrystals reveals no contribution due to localized excitons with a fixed exchange interaction magnitude. Thus, crystalline KBr boules grown from a KBr:AgBr melt with a large (1–2 mol%) concentration of AgBr impurity have been established to contain self-organized AgBr inclusions representing both microcrystals, retaining properties of the bulk material, and nanocrystals (QDs) in which significant role belongs to the spatial confinement effects. These effects are manifested by the maximum distance between recombining donor–acceptor pairs being restricted to the nanocrystal size and by a change in the g value of shallow electron donor centers. Based on an analysis of the exchange interactions in nanocrystals, a distribution of distances in the donor–acceptor pairs is determined and the dimensions of nanocrystals are estimated.

To summarize, self-organized microcrystals and nanocrystals (quantum dots) of AgCl and AgBr embedded in a KCl and KBr crystalline matrices and maintaining the orientation of the host lattice were studied by ODMR. It was unambiguously shown that self-organized microcrystalline silver halides can be grown inside alkali halide crystals with the properties of bulk crystals since the ODMR spectra of the embedded microcrystals were practically the same or close to those in bulk AgCl and AgBr and could be used as a “fingerprint” of AgCl and AgBr. For AgCl nanocrystals in KCl matrix the anisotropy of the g-factor both for isolated self-trapped holes and for self-trapped holes forming self-trapped excitons was found to be substantially reduced compared with those of bulk AgCl crystals. This implies a considerable suppression of the Jahn–Teller effect in nanoparticles. A rather general mechanism of the suppression of the Jahn–Teller effect in nanocrystals is proposed, taking into account the additional deformation field appearing because of the strong vibronic interaction at the interface.

Multiquantum transitions in the ODMR spectra were used to measure the singlet to triplet splitting in silver halide crystals.

It was concluded that the distribution of exchange interactions for electron-hole pairs and triplet excitons in AgBr QDs in KBr is due to the distribution of AgBr crystals sizes. The holes seem to be self-trapped in the AgBr because of the dynamical J-T effect. The exchange splitting increases for distant electron-hole pairs with the decrease of AgBr size. The spectra with the exchange splitting larger then that in bulk AgBr (1.9 cm^{-1}) seem to belong to AgBr nanocrystals. In contrast to AgCl, the wavelength of the luminescence in AgBr micro- and nanocrystals embedded in a KBr matrix decreases with decrease of AgBr crystal size.

5.3 Defects in Nanodiamonds: Application of High-Frequency cw and Pulse EPR, ODMR

Applications of EPR based techniques including high frequency electron spin echo (ESE), electron-nuclear double resonance (ENDOR) and optically detected magnetic resonance (ODMR) for a study of diamond nanostructures are considered. Spin is a purely quantum-mechanical object and spin phenomena begin to play a crucial role in the development of various instruments and devices based on nanostructures. Methods of magnetic resonance discovered by Zavoisky [116, 117] are the basic techniques for studying the spin phenomena in condensed matter and biological systems.

There are a variety of elegant techniques in which higher sensitivity and resolution are obtained by exciting two resonance transitions in the sample (double-resonance techniques). The great attractions of these techniques are that, in contrast to optical methods, they allow the identification of the dopants and provide information about the spatial distribution of the electronic wave function. This latter aspect is particularly attractive because it allows for a quantitative measurement of the effect of confinement on the shape and properties of the wave function.

In the case of an electron-nuclear double-resonance (ENDOR) a radio frequency is swept through the resonant frequency of the nuclei in the conditions of EPR and the nuclear-spin transitions are detected as changes in the EPR signal. ENDOR can be considered as a method for increasing the resolution of EPR spectra and as a technique for improving NMR sensitivity for a limited number of nuclei that are located near the paramagnetic defect.

In the case of optically detected magnetic resonance (ODMR) one of the resonances is not a paramagnetic resonance, but an optical resonance [118]. Optical absorption or emission is in some degree dependent on the population of the magnetic (spin) sublevels of the system under investigation. Magnetic resonance, by changing the populations of the spin sublevels, changes optical absorption or emission (intensity or polarization). In ODMR a microwave-induced repopulation of Zeeman sublevels is detected optically, i.e., there is a giant gain in sensitivity since energy of the optical quantum is by several orders of magnitude higher than the microwave energy and it becomes possible to detect a very small number of spins down to single spin. In addition, ODMR provides spatial selectivity of optical methods.

An understanding of the structure and constituents of defects in nanostructures is important since their presence can greatly affect the properties of the material. During the last two decades high-frequency (high-field) EPR and in particular pulsed EPR and pulsed ENDOR developed to a new fast advancing field in magnetic resonance spectroscopy (see [119] and references therein). Pulsed high-frequency EPR and ENDOR spectroscopy were shown to be excellent tools for the investigation of the electronic properties of semiconductor quantum dots (QDs) (see [5, 6] and references therein). Direct measurements of EPR and ENDOR in nanostructures are often difficult because of the small total number of spins, therefore ODMR is much better suited for a study of such systems. In combination with optical excitation, EPR (photo-EPR) can also yield information regarding optical transitions, energy-level positions, thus providing a bridge between optical and magnetic resonance spectroscopy.

This review covers the investigations of diamond nanoparticles by EPR, ENDOR and ODMR.

5.3.1 N and N_2 Centres in Nanodiamonds

The nanodiamond particles formed in the detonation of strong explosives, the so-called detonation nanodiamond (ND), are of particular interest. The detonation ND's are characterized by a narrow size distribution with a sharp diameter maximum at 4–5 nm, and each particle consists of a core with an ordered diamond lattice (sp^3 hybridized carbon atoms) and a shell. The surface and core shell structure of synthetic nanodiamond has been recently characterized by solid-state nuclear magnetic resonance (NMR) spectroscopy [120]. According to this NMR-based model, the nanodiamond particle has a diameter of 4.8 nm and contains close to 10,000 carbon and 200 nitrogen atoms. About 40% of carbons are in the 3.6-nm diameter ordered crystalline diamond core, and about 60% of carbons are in a seven-layer-thick, partially disordered shell. The nanodiamond surface carbons are bonded to H and OH groups.

A nonocrystalline core-shell model has also been proposed based on other experimental results that quantitatively are rather different from the NMR-based model (see references in [120]). Nevertheless, an overall conclusion is that detonation ND is a very complex system. Nanodiamond doping processes, formation and structure of intrinsic and impurity defects differ from those in bulk diamonds. In particular, the theoretical studies have shown that nitrogen impurities in ND seem to be metastable in contrast to bulk diamonds.

Nitrogen is the main impurity in diamonds and the form in which nitrogen is present in diamonds largely determines their properties and serves as the leading factor of the diamond classification. Nitrogen creates various paramagnetic centres in a diamond and exists as individual atoms and nitrogen clusters. Recently, a great interest has been inspired by the studies of nitrogen vacancy centres (NV defects) in a diamond, for which the magnetic resonance on single defects was successfully observed at room temperature [121] letting one even to speak of a "diamond era of spintronics" [122].

In this chapter, high-frequency electron-spin echo at W-band (94 GHz) have been used to study detonation ND with a size of 4.5 nm and detonation ND after high-temperature high-pressure sintering. The main goal of the study is to find EPR spectra of nitrogen related paramagnetic centres within of the diamond core of detonation ND and to solve the problem whether nitrogen donors are stable in detonation ND.

Figure 5.21a shows the electron spin echo (ESE) detected EPR signal observed in natural diamond nanocrystals with an average size of 150 nm. The high-frequency ESE studies allowed almost complete suppressing of the EPR signal of the broken bonds near the surface owing to a short relaxation time. These EPR spectra are ultimately informative and are explained by the presence of two types of nitrogen centres in the nanocrystals: the individual N^0 atoms (so called P1 centres) and nitrogen pairs N_2^+ [123]. The simulated ESR spectra are shown in the lower part of the Fig. 5.21a. The simulated EPR spectra were obtained with the parameters for the individual nitrogen atoms N^0 and nitrogen pairs N_2^+ in bulk diamond crystals. The agreement of the simulated and experimental spectra seen in figure definitely confirms the proposed interpretation of the EPR spectra and may serve as a reference to monitor the concentration of paramagnetic nitrogen in diamond nanocrystals. It also allows one to determine with a high accuracy the percentage of single atom and diatomic nitrogen centres in these crystals.

The results of EPR measurements in the source material of detonation ND are presented in Fig. 5.21b, curve (1) shows the ESE detected EPR signal at a frequency of 94 GHz in the detonation nanodiamond sample. There is only a slightly split very intensive central line with g factor of 2.0030 that belongs to unpaired electrons near the surface shell. To separate the ESE signals from the nanodiamond core and surface shell, different time sequences of microwave pulses were used because the relaxation characteristics of two objects are significantly different from each other.

ESE signal (1) has been measured with separation between the first and the second microwave pulse τ of 230 ns at 10 K. The high-gain spectrum ($\times 5$) for signal (1) shows that the central signals are accompanied by a set of weak lines, which become dominating when separation τ increases. Figure 5.21b, curves (2),

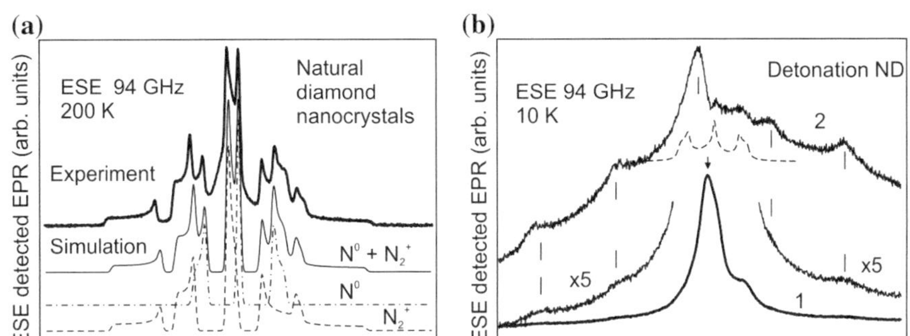

Fig. 5.21 The ESE-detected EPR signal at a frequency of 94 GHz in natural diamond nanocrystals (**a**) and in a detonation nanodiamond before sintering (**b**) recorded at a temperature of 10 K at the interval between the pulses τ = 230 ns (*1*) and 800 ns (2). The *dashed line* corresponds to the simulation of individual nitrogen donors N^0 in diamond nanocrystal powder

show ESE signal that have been measured with τ of 800 ns at 10 K. The intensive lines have disappeared and several new lines become visible. The five lines ESE spectrum (for the central line g = 2.0037) marked by bars dominates and additional lines are observed in interval of magnetic fields 3350–3360 mT. Increasing τ from 230 to 800 ns clearly leads to a considerable change in the ESE spectra that reveals a difference in spin-spin relaxation time T_2. These measurements demonstrate shorten of T_2 relaxation time for unpaired electrons in surface shell, T_2 is about 300 ns. For paramagnetic centres inside of diamond core T_2 relaxation time is much longer (longer then 1 ms). A possible candidate for five line spectrum is multivacancy complex with three unpaired electrons situated on the dangling bonds (S = 3/2), for example, three-vacancy chain or multi-oxygen-vacancy complex with three vacancies and three oxygen interstitials with zero field splitting about 250 MHz (R8) [123]. The most important result is observation of the EPR spectrum of N^0 centres inside of the diamond core of DND quantum dot.

5.3.2 *High-Density Nitrogen-Vacancy (NV) Ensembles Fabricated by Sintering Procedure of Detonation Nanodiamonds*

5.3.2.1 Electron Paramagnetic Resonance

Nitrogen-vacancy (NV) centre in diamond (consists of a nearest-neighbor pair of a nitrogen atom, which substitutes for a carbon atom, and a lattice vacancy) is one of the most prominent objects for applications in new generation of supersensitive magnetometers [123, 124], biosensors [125, 126], single photon sources [127], quantum computers [128]. Among others, magnetometry and biological applications demand very bright and photostable fluorescence from high-density nitrogen-vacancy ensembles located in diamond core of nanoparticles [129]. Fluorescence brightness strongly depends on the concentration of NV centres and the diamond quality. Although NV centres have effectively been generated in diamond by irradiation and following high temperature annealing [130, 131] they have not been observed in high concentrations in non-irradiated diamonds. Here, we show that enormously high concentrations of NV centres going as high as 0.1% (10^3 ppm) can be produced directly by high-pressure high-temperature sintering of detonation nanodiamonds. Intensive fluorescence spectra with quite well distinguished zero-phonon line related to negatively charged NV centres even at the room temperature were observed. Anisotropic optically detected magnetic resonance and electron paramagnetic resonance signals of NV centres and isolated nitrogen have been detected in single arrays showing that sintering leads to self-organization of nanodiamond particles in oriented micron-size arrays which include high concentrations of NV and isolated nitrogen centres. Moreover, significantly high coherence of the NV centre spin system was observed up to the room temperatures.

Production of detonation nanodiamonds (DND) through the detonation of strong explosives has been already for a long time commercially available technology [132, 133]. The DND nanocrystals consist of a core with an ordered diamond lattice (sp^3 hybridized carbon atoms) and mixed sp^2–sp^3 hybridized shell distorted by the large proportion of surface dangling bonds, impurities and dislocations are characterized by a narrow size distribution with a maximum at 4–5 nm [134]. The lattice parameter of the particle does not differ from in bulk diamond and relatively large surface area influences the properties of nanodiamond particles namely the location of impurities and native defects [135].

NV-containing diamond is the only known solid-state system where manipulation of the spin states of a single localized electron was realised [136–138]. Unique properties of this centre connect with its extreme photostablity and high sensitivity to magnetic fields. Usually NV centres are produced by irradiation of bulk diamond, micro and nanodiamonds with ions, neutrons or high-energy electrons [126, 130, 131, 139], aimed to the vacancies creation. Following annealing at the temperatures of about 800–900 °C optimal for the vacancies diffusion leads to trapping of the vacancies to the nitrogen impurity atoms and creation of NV centres. Another way is nitrogen ion implantation into the diamond lattice also followed by annealing [140]. These methods require highly sophisticated and costly equipment, which, therefore, hinders the easy availability of flourescent diamonds. In addition, high concentration of unwanted defects, which impair diamond quality, are created. Even though, there were several approaches to scale up the production of the fluorescent diamonds to make them more commercially available [126, 130], the irradiation methods are purely statistical and the effectiveness of creation of NV centres in nanodiamond with the size less than 20 nm is still under the question [141, 142]. In the most recent publication it was shown that photostable NV centres can be created by heavy electron irradiation (10^{20} cm^{-2}) and annealing in nanodiamond particles as small as 7 nm [131]. For free-space particles with the size of 5 nm the fluorescence becomes intermittent [143]. Chemical vapour deposition (CVD) technique also allow production of diamond nanocrystals containing NV centres, although the probability of creation of even one NV centre per particle of the size less than 40 nm is very poor [144].

Promising applications of NV-containing diamonds require spectrally and temporally stable emission from nanosized diamond crystals. The development of non-irradiative fabrication technique is essential for further development of technologies and applications based on the unique properties of the NV centres. In this paper we present the first results of observation of high-density NV ensembles created directly by high-pressure high-temperature (HPHT) sintering procedure. This technique does not require expensive equipment and can be done in two steps: purification of commercially available DND followed by the HPHT sintering procedure.

DND studied here were purified to reduce the metal inclusions. Before sintering DND were modified by C60 fullerenes with 99.5% purity and then sintered in a toroidal high-pressure chamber at temperature 800 °C and pressure 6 GPa for 11 s. The particle size distribution in arrays was analyzed by X-Ray diffraction technique, by which the average size of the particles was found to be of about 5.6 nm. Particle sizes have been found to grow with increasing of the sintering temperature by the scenario of oriented aggregation because of the absence of liquid-phase material transport [145].

To estimate the concentrations and spin-relaxation parameters room temperature high-frequency electron spin echo detected electron paramagnetic resonance (ESE detected EPR) measurements at 94 GHz were used. The ESE detected EPR spectra measured at room temperature on the single DND sintered array (left inset in Fig. 5.22a) consisted of two groups of lines (i) intensive slightly anisotropic central group of lines in the magnetic fields between 3345–3360 mT and (ii) highly anisotropic lines between 3240–3460 mT symmetrically disposed at the both sides of the central group of lines. The central group of the lines (see the right inset) belongs to the substitutional nitrogen centres N^0, which are incorporated in the DND crystalline core during detonation procedure, remaining stable and isolated [146]. The experimental angular dependencies for the group (ii) lines are shown for four different orientations of the sample with respect to the magnetic field, the positions of the lines for other orientations are shown with circles in Fig. 5.22a. The calculation of the angular dependencies for the group (ii) lines were performed using the following conventional spin Hamiltonian (see Chap. 1)

$$\hat{H} = \mu_B g \vec{B} \cdot \hat{\vec{S}} + D[\hat{S}_z^2 - \frac{1}{3} S(S+1)]$$

with g = 2.0028 and D = 958 cm^{-1} corresponding to those of the NV-centre [36]. Assuming that the spectrum marked by $\theta = 0°$ corresponds to the orientation of the magnetic field along the $\langle 111 \rangle$ axis of the crystalline core, the Euler angles α,β,γ for four $\langle 111 \rangle$ orientations of the diamond core were chosen: 0,0,0; 0,110,0; 120,110,0 and 240,110,0. The results of calculation are shown in Fig. 5.22a with a solid line. It can be noted that not all lines can be described by this calculation, except for B || $\langle 111 \rangle$ ($\theta = 0°$), thus one can suppose the twinning in the nanodiamond arrays. Chosen Euler angles for twins were 0,0,0; 0,250,0; 120,250,0 and 240,250,0, and the corresponding angular dependencies are shown with dotted line. Angular dependences simulation shows excellent agreement with experimental data. Thus, one can conclude that all EPR lines included in group (ii) of the spectra belongs to the NV^- centres, located within the diamond nano-crystalline core. It should be underlined that the concentration of the NV^- centres in sintered arrays was so extremely high that EPR spectra were observed even without photoexitation at room temperature.

The rough estimation of NV^- centres concentration was obtained from the EPR spectra. Sensitivity of the experimental setup was about 10^9 spins for the linewidth of 1 Gauss. Taking into consideration the width of each NV^- line and their signal/noise ratio, the detected number of NV^- spins amount to $\sim 10^{11}$–10^{12} for different fluorescent arrays. The average linear dimension of fluorescent arrays was ~ 10 μm. Thus, the concentration of spins should be $\sim 10^{19}$–10^{20} cm^{-3}. Similar calculations for the single nitrogen donors gave approximately the same

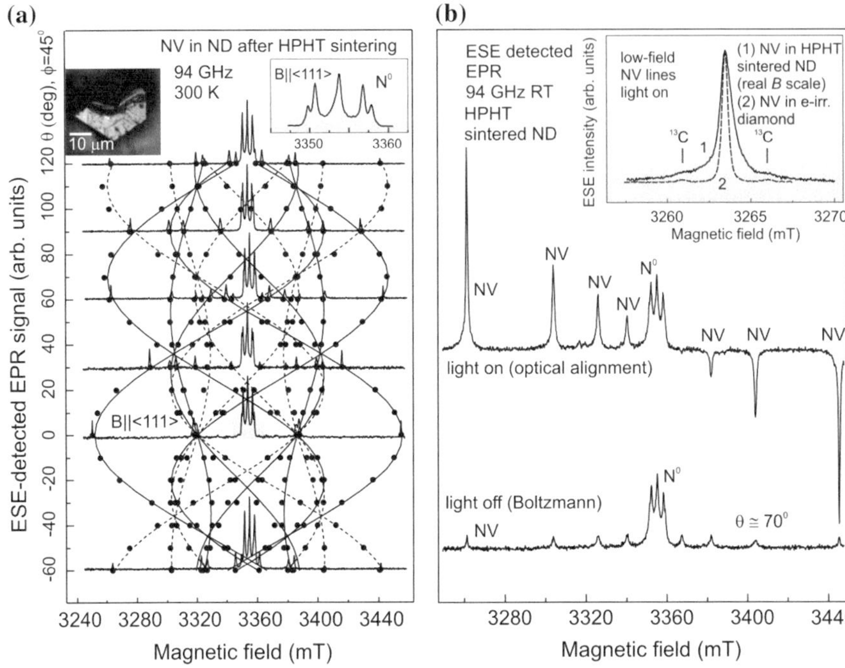

Fig. 5.22 **a** Angular dependence of the high-frequency ESE-detected EPR signal (94 GHz) at room temperature in a single cluster containing NV centres obtained by sintering a detonation nanodiamond at T = 800 °C and P = 6 GPa (*left inset* shows an image of the cluster). The *right inset* shows the central part of the EPR spectrum for B $\parallel \langle 111 \rangle$ corresponding to individual nitrogen centres N^0. The measured (*points*) and calculated (*lines*) dependences of the EPR spectra for the direction of the magnetic field B with respect to the $\langle 111 \rangle$ axis ($\varphi = 45°$) are shown. **b** The ESE-detected EPR signal at a frequency of 94 GHz in a detonation nanodiamond after sintering recorded in the dark (*bottom*) and with the light excitation (*top*). *Inset*: single lines of NV defects, recorded on an expanded scale. The *tick marks* indicate the peak positions of ^{13}C hyperfine lines. Magnetic field scale is shown for EPR spectrum of NV centres in HPHT sintered DND

concentrations of N^0 (10^{19}–10^{20} cm^{-3}). The concentration of the carbon atoms in bulk diamonds is $1.76 \cdot 10^{23}$ atoms/cm^3, which is three orders of magnitude higher than the observable concentration of the NV^- centres. Based on these rough estimations up to 0.1% of the carbon atoms should be replaced by NV^- centres. The sintered DND arrays are also including almost the same number of N^0 centres.

5.3.2.2 Optical Polarization and ODMR in High Magnetic Field

As discussed earlier, extremely high concentrations up to 0.1% of NV^- centres have been observed in sintered clusters of detonation nanodiamond (see [147, 148] and references therein). The results presented are opening new perspectives in the unique NV^--containing diamond fabrication. The 94 GHz ESE-detected EPR signal in a detonation nanodiamond after sintering recorded in the dark and with the light excitation are shown in Fig. 5.22b. An increase in amplitude and inversion of sign of the EPR lines are due to optical pumping effects which change the populations of the NV^- centre ground state triplet energy levels. Inset shows single lines of NV defects, recorded on an expanded scale. The tick marks indicate the peak positions of ^{13}C hyperfine lines. Magnetic field scale is shown for EPR spectrum of NV centres in HPHT sintered DND.

Figure 5.23 (insets) shows typical data from the sample, where (left) is the data obtained from the standard optical microscope under excitation with light of commercial solid-state 532-nm laser and (right) is a photoluminescence spectra at 2 and 300 K excited by the same laser. As shown some isolated arrays were fluorescent (~40%) and quite well distinguished zero-phonon line (ZPL) at 637 nm followed by prominent vibronic side bands was observable in the emission spectra. This optical resonance line is generally believed to correspond to the negatively charged NV^- centres [149]. Unambiguous confirmation of the existence of NV^- centres in the sintered arrays was obtained by performing low-temperature measurements and high-frequency optically detected magnetic resonance (ODMR). Low temperature experiments (Fig. 5.23) allowed narrowing of the ZPL at 637 nm as well as revealed the presence of the second type of NV centres inside material: neutral NV^0 with a ZPL at 575 nm [149]. This part of the emission spectrum is

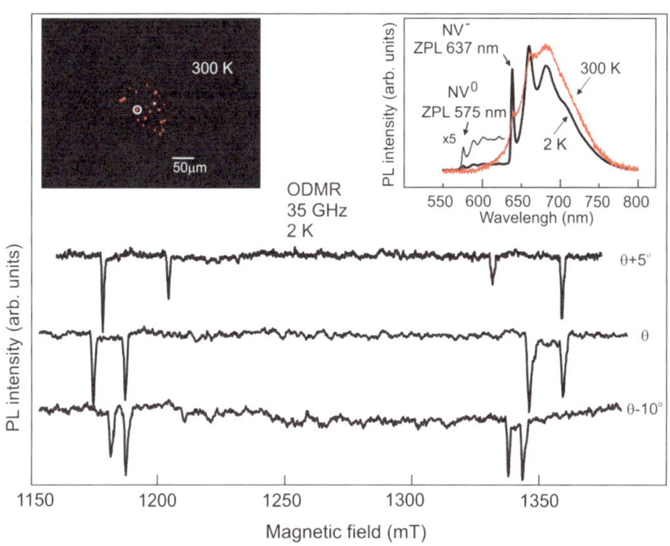

Fig. 5.23 Optically detected magnetic resonance spectra registered in single DND sintered array at Q-band for three different orientations of magnetic field with respect to NV^- axis, T = 2 K; *Insets*: Characterization of arrays produced by sintering of DND powder: optical microscope image under excitation with light of 532-nm laser at 300 K (*left*) and photoluminescence of single NV-containing array at 2 and 300 K (*right*)

shown in Fig. 5.23 (right inset) with the 5 times gain. The NV^-:NV^0 ZPL intensity ratio was estimated to be 25:1

ODMR recorded by monitoring of fluorescence allowed a selective study of only those defects, which are involved in the recombination process. The spectra were recorded at 35.2 GHz for three different orientations of the magnetic field with respect to the NV^- axis, T = 2 K (Fig. 5.23). The observation of strong ODMR effect proved existence of a spin-dependent mechanism in optical excitation cycle becoming apparent as fluorescence intensity reduction due to repopulation of ground sublevels at the moment of magnetic resonance. NV^- centre has a triplet ground state (S = 1) split by anisotropic dipolar interaction into doublet $M_S = \pm 1$ and singlet $M_S = 0$ [150]. In thermal equilibrium populations of the ground state sublevels are determined by Boltzmann statistics. Optical excitation scheme comprises the ^3A ground state, the ^3E excited state, and the metastable singlet state level which is generally believed to be ^1A level. Optical pumping with light resonant with the $^3A_2 \rightarrow {}^3E$, collects the population on the $M_S = 0$ sublevel of the ground state due to the nonradiative emission from $M_S = \pm 1$ excited state sublevels via singlet metastable level ^1A. As a result, the spin sublevel $M_S = 0$ of ^3A ground state is predominantly populated.

After switching on magnetic field oriented along the defect axis the ground state doublet sublevel splits into two separate $M_S = +1$ and $M_S = -1$ sublevels due to the Zeeman interaction. As the $M_S = 0$ level is predominantly populated upon optical pumping, one should observe the energy absorption for $M_S = 0 \rightarrow M_S = +1$ transition and energy emission for $M_S = 0 \rightarrow M_S = -1$ transition [151]. ODMR spectra shown in Fig. 5.23 reflect the fluorescence intensity reduction due to repopulation of $M_S = \pm 1$ sublevels at the moment of magnetic resonance. Anisotropy of the ODMR spectra recorded in the sintered array indicate the presence of preferential orientations of the magnetic field along the NV axis, which should not be observed in case of disoriented digit particles forming array. Thus, during the sintering procedure nanodiamonds particles are governed by the oriented attachment mechanism and as a result forming self-oriented arrays.

In ODMR experiments optical excitation influences all NV^- centres, whereas magnetic resonance conditions are applied only for one distinguished orientation of the defect, all other orientations (i.e., four possible orientations, corresponding to the N-V axes in diamond plus those of twins) appear to be in the nonresonance conditions. Thus the strength of the ODMR effect observed (Fig. 5.23) should be magnified for an order of magnitude.

For the NV centres fabrication the presence of both nitrogen and vacancy centres is essential. According to theoretical predictions [135] the nitrogen should be metastable within a DND core and its incorporation should strongly depend on the particle size, furthermore, the probability of creation of even one NV centre per particle by irradiation scales as the fifth power of the crystal size [142]. To date none of the theoretical models can explain the observed concentration of NV centres in the sintered DND arrays. In the purified DND material used as a source for sintering single nitrogen atoms together with multivacancy complexes were observed by high-frequency ESE-EPR technique [147]. Hence, both essential components (N and V) are present in the commercially available source DND material. We suggest the temperature used during the sintering procedure (800 °C) to be a reason of high-density NV ensembles fabrication, as this temperature is favourable for the vacancy migration towards single nitrogen. It should be also pointed out that NV-centres were not observed in the arrays sintered at the temperature of 1500 °C DND [147, 148], which is also natural as the vacancies are annealed at such a high temperatures.

Because of their extreme photostability and high sensitivity to magnetic fields, the NV^- centres in diamond are one of the most prominent objects for applications in a new generation of supersensitive magnetometers, biosensors, single photon sources, and quantum computers. Among others, magnetometry and biological applications demand very bright and photostable fluorescence from high-density NV^- ensembles located in the diamond core of nanoparticles (see Chap. 6).

Application of EPR, ENDOR and ODMR allowed revealing an enormously high concentration of NV^- ensembles created directly by a high-pressure high-temperature sintering procedure of detonation nanodiamond (DND) particles, which were produced through the detonation of strong explosives and have a size of 4–5 nm. In detonation nanodiamond used for sintering, both individual nitrogen atoms and vacancy complexes can be revealed by EPR. Thus, even the initial detonation nanodiamond has defects necessary for the formation of nitrogen vacancy centres.

5.3.2.3 Temperature Scanned Magnetic Resonance of NV Centres in Diamond and Nanodiamonds

New method for the detection of magnetic resonance signals versus temperature is developed on the basis of the temperature dependence of the spin Hamiltonian parameters of the paramagnetic system under investigation. The implementation of this technique is demonstrated on the nitrogen vacancy (NV) centres in diamonds.

Figure 5.24a shows the zero-field ODMR spectra of the NV defects measured at 25 °C by frequency scanning and by monitoring of PL intensity in diamond crystal. The PL was excited by a laser with a wavelength of 532 nm and detected in the ZFL of NV defects 637 nm and phonon side bands. The spectrum shows two intense central lines and these signals in accordance with the standard spin Hamiltonian describe the fine structure parameters $D = 2870$ MHz and $E = 3.5$ MHz for the 25 °C (for 260 °C: $D = 2842.5$ MHz and $E = 3.1$ MHz).

In addition two, relatively weak, side bands shifted to the low-frequency (LF) and high-frequency (HF) sides of the centre of the main double transition were observed. The side bands are attributed to the presence of interaction between triplet NV defect and substitutional nitrogen atom N_s which is deep donor in the neutral charge state and has a spin of $S = 1/2$ (coupled triplet-doublet molecular NV-N_s pair in diamond) and related to the strong hyperfine (HF) interaction with nuclei ^{14}N in N_s.

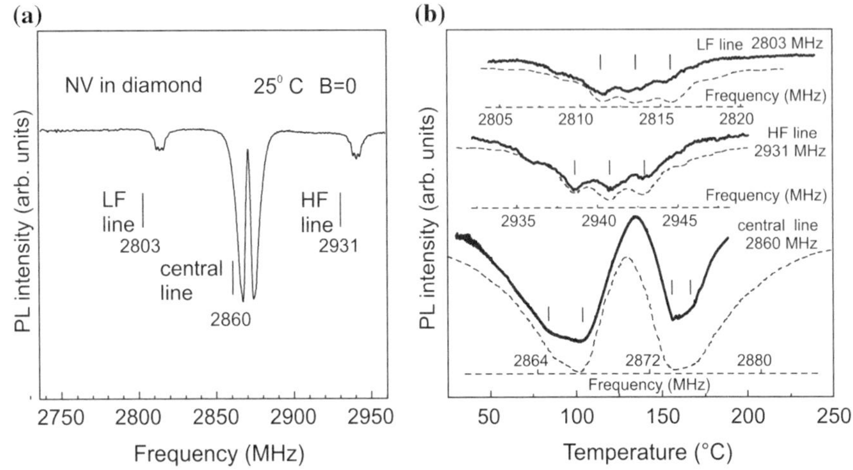

Fig. 5.24 **a** Zero-field ODMR spectra of the NV defects and NV–N_s pairs measured at 25 in diamond crystal. **b** Zero-field ODMR spectra of the NV defects and the NV–N_s pairs, recorded in zero magnetic field at fixed frequencies indicated in Fig. 5.24a for 25 °C by scanning temperature; the central part and side bands of the 25 °C measured ODMR spectrum of Fig. 5.24a are presented for comparison by *dashed line*

The temperature induced effects provide an opportunity to develop new methods of magnetic resonance. Figure 5.24b shows the ODMR spectra of NV defects and NV-N_s pairs recorded at zero magnetic field by scanning the temperature from 25 to 300 °C. The measurements were performed at three fixed frequencies which were chosen with low-frequency side of the ODMR lines recorded at 25 °C (marked by vertical bars in Fig. 5.24a), since the temperature will be raised from 25 to 300 °C, and the EPR line will move to lower frequencies. The central lines, LF and HF side bands were recorded at fixed frequencies 2860, 2803 and 2931 MHz, respectively. It is seen that the central signal, LF and HF side bands in the same way dependent on temperature, that is, this method of recording allows us to conclude that these spectra belong to the same paramagnetic molecular system that is not obvious from standard ODMR spectra. It also shows that there is additional structure in the ODMR spectra with the temperature scanning, this structure is indicated by vertical marks. This structure, on the one hand, are uniquely related to the strong HF interaction with nuclei ^{14}N in isolated centres of nitrogen N_s, on the other hand, shows a weak HF structure with ^{14}N nuclei in the NV defect. This seems to be direct evidence of two-way transfer of a nitrogen nuclear spin hyperfine interaction in coupled NV-N pairs in diamond.

The central part and sidebands of the spectrum of Fig. 5.24a is presented for comparison by dashed line in Fig. 5.24b, the distance between the two lines that are registered with the scan of frequency and temperature are chosen the same, which corresponds to 125 kHz/°C. It is evident that both registration methods give similar signals with an additional splitting due to ^{14}N HF structure of NV defect (see vertical bars), while the width of two lines, being the same in writing with the frequencies scan differ in width when recording the spectrum with a temperature scan. High-temperature line in the central part of the spectrum (Fig. 5.24b) is narrower, which indicates that with increasing sample temperature the temperature induced change of the fine splitting *D* increases. The temperature induced effects provide an opportunity to register locally with nanoscale resolution ultra fast small changes in temperature using as a probe nanodiamonds with NV defects.

5.3.3 Room-Temperature High-Field Spin Dynamics of NV Centres in Sintered Detonation Nanodiamonds

The sintering under high pressure and high temperature of the detonation nanodiamonds leads to the formation of the highly oriented arrays of the DND particles. The measurements of spin-lattice relaxation time (T_1) and spin-spin relaxation time (T_2) at the room temperature were performed for NV^- and N^0 centres to give the insight of characteristics of the spin interaction processes in fluorescent arrays incorporating giant concentrations of NV^- and N^0 centres.

Figure 5.25a shows Spin-lattice relaxation time T_1 curve (1) and spin-spin relaxation time T_2 curve (2) of NV defects in single DND sintered array without light excitation. The spin-lattice relaxation time T_1 was measured at 300 K in the strong magnetic field of 3447.5 mT (see Fig. 5.22b) using the pulse sequence inversion pulse $\pi - T - \pi/2 - \tau - \pi - \tau -$ echo, where the time T was incremented and τ was kept fixed. The T_2 was determined using the two-pulse sequence $\pi/2 - \tau - \pi - \tau -$ echo, where the echo amplitude was measured as a function of the time delay τ between the pulses. The first microwave $\pi/2$ pulse excites the coherences, i.e., the populations are converted into coherences between sublevels of the triplet state. In classical terms, this corresponds to changing the direction of the magnetization vectors, from longitudinal to transversal. For a fixed time period τ, the system is left unperturbed and the coherences evolve in the transverse plane. For the ensemble case, this will correspond to a spread in the transverse magnetizations due to different Larmor frequencies of different spin packages. Refocusing of the coherences is done by applying a microwave π-pulse. The values of $T_1 = 1.7$ ms and $T_2 = 1.6$ μs were obtained for nitrogen-vacancy centres single DND sintered array from the fits of the measured echo-decay curves. For nitrogen donors the corresponding relaxation times are $T_1 = 0.570$ ms и $T_2 = 0.55$ μs. Latter measurements were performed for the low field hyperfine component of the nitrogen line, for the central line the relaxation times are at about 15% shorter due to overlapping with the EPR signal of the surface centres. The relaxation time T_2 is long enough revealing the high coherence of the system, notwithstanding the fact that the concentrations of both NV^- and N^0 centres are very high.

Figure 5.25b (inset) shows a scheme of optical (532 nm) and microwave (3 mm) pulse sequence used to measure the spin dynamics in the ground state of NV

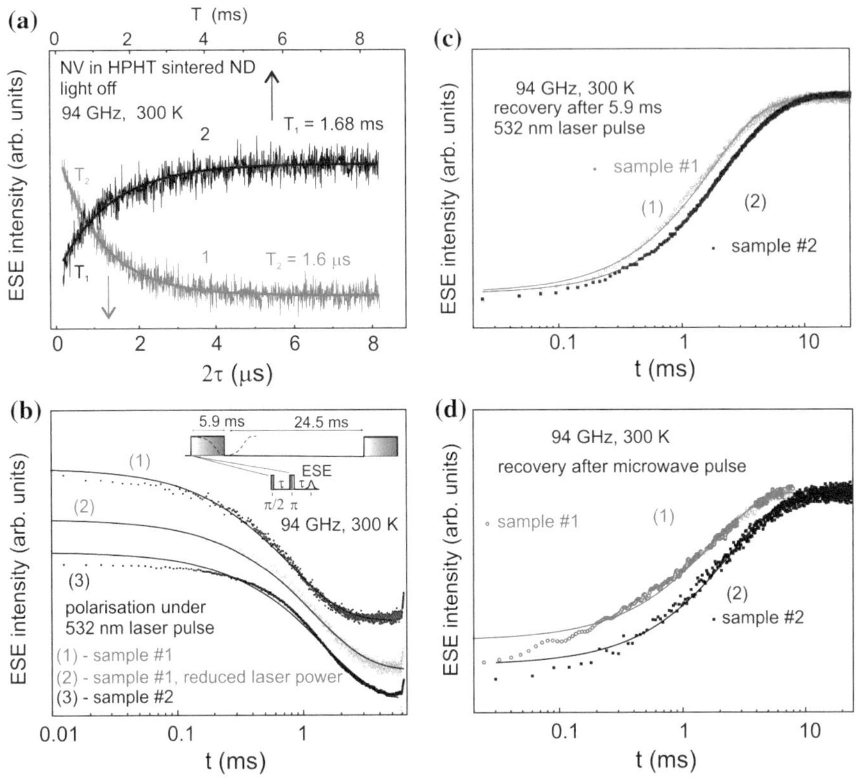

Fig. 5.25 **a** Spin-lattice relaxation time T_1 curve (*1*) and spin-spin relaxation time T_2 curve (*2*) of NV defects in single DND sintered array without light excitation. The spin-lattice relaxation time T_1 was measured at 300 K in the strong magnetic field of 3447.5 mT (see Fig. 5.22b) using the pulse sequence inversion pulse $\pi - T - \pi/2 - \tau - \pi - \tau -$ echo, where the time T was incremented and τ was kept fixed. The T_2 was determined using the two-pulse sequence $\pi/2 - \tau - \pi - \tau -$ echo, where the echo amplitude was measured as a function of the time delay τ between the pulses. The values of T_1 and T_2 times were obtained from the fits of the measured echo-decay curves. **b** Change the value of ESE signal under the influence of the 532 nm laser pulse with time: (1) single DND sintered array (sample #1), the maximum power; (2) sample #1, the reduced power; (3) natural diamond microcrystal (sample #2), maximum power. The ESE signal saturation for sample #1 is shown (curve 1) to be faster than for sample #2 (curve 3) with the same laser power. (inset) Time diagram of optical (532 nm) and microwave (3 mm) pulse sequences used to measure the spin dynamics in the ground state of NV defects. Pulse sequence repetition time 30.4 ms was chosen, which provided the full recovery after the laser pulse. Laser pulse duration 5.9 ms was chosen, which allowed to achieve the maximum output level of the polarization signal ESE. **c** The dependence of the ESE signal recovery to the equilibrium value after the 532 nm laser pulse: (1) sample #1; (2) sample #2. For convenience, all the equilibrium values were normalized to 1. **d** The dependence of the ESE signal recovery to the equilibrium value after the inverting microwave pulse π (standard sequence for the measurement of relaxation times—inversion pulse $\pi - T - \pi/2 - \tau - \pi - \tau -$ echo, where the time T was incremented and τ was kept fixed.): (1) sample #1, (2) sample #2. All the equilibrium values were normalized to 1. All the curves are measured in magnetic field of 3447.5 mT (see Fig. 5.22b) and roughly described by one-exponential functions, whose parameters are given in Table 5.2

defects. For probing the amplitude of the ESE signal $\pi/2 - 260$ ns $- \pi$ microwave pulse sequence was used, whose beginning synchronize with the start or the end of the laser pulse. Pulse sequence repetition time 30.4 ms was selected, which provided the full recovery after the laser pulse. Laser pulse with wavelength of 532 nm and 5.9 ms duration was used to polarize the NV defects into the $M_S = 0$ ground state and which allowed to achieve the maximum output level of the polarization. Change the value of ESE signal under the influence of the 532 nm laser pulse with time depicted in Fig. 5.25b for single DND sintered array (sample #1) with the maximum power (1); sample #1 with the reduced power (2) and natural diamond microcrystal (sample #2) with maximum power (3). From Fig. 5.25b, it can be seen that the ESE signal saturation for sample #1 (curve 1) to be faster than for sample #2 (curve 3) with the same laser power. Reduced power leads to the expected lengthening of dependence. Further dependences of the spin-echo signal recovery to the equilibrium value after the laser pulse, and after inverting microwave pulse π (standard sequence for the measurement of relaxation times—inversion pulse $\pi - T - \pi/2 - \tau - \pi - \tau -$ echo, where the time T was incremented and τ was kept fixed) were measured.

The normalized dependence of the ESE signal recovery to the equilibrium value after the 532 nm laser pulse in sample #1(1) and sample #2 (2) is observed in Fig. 5.25c. Figure 5.25d shows the normalized dependence of the ESE signal recovery to the equilibrium value after the inverting microwave pulse π in sample #1 (1) and sample #2 (2). All the curves can be roughly described by one-exponential functions (solid line), whose parameters are given in Table 5.2.

Nutation experiments were performed at room temperature to observe Rabi oscillations. This technique has been applied to controlling coherent manipulation of high-concentration NV defects spins in single sintered DND array of micron size (sample #1).

The upper part of Fig. 5.26 shows the pulse sequence used to measure Rabi oscillations: the first pulse $\Delta t - 100$ μs $- \pi/2 - 300$ ns $- \pi - \tau -$ echo. The lower part depicts experimental data for NV defects in sample #1. Curves (1) and (1′) show Rabi oscillations: ESE intensity versus nutation pulse length (Δt) measured at 3447.5 mT (see Fig. 5.22b) with microwave power of 40 mW under 532 nm laser

Table 5.2 Polarization time of NV spin sublevels in the ground triplet state by the 532 nm laser pulse

	Sample #1 full power	Sample #1 reduced power	Sample #2 full power
T^* (ms)	0.8	1.25	1.5
Recovery time of the equilibrium population of spin sublevels			
	Sample #1		Sample #2
Recovery time after laser pulse			
T_1 (ms)	1.8		2.45
Recovery time after microwave pulse			
T_1 (ms)	1.7		2.4
Decay of transversal magnetization			
T_2 (μs)	1.6		3.6

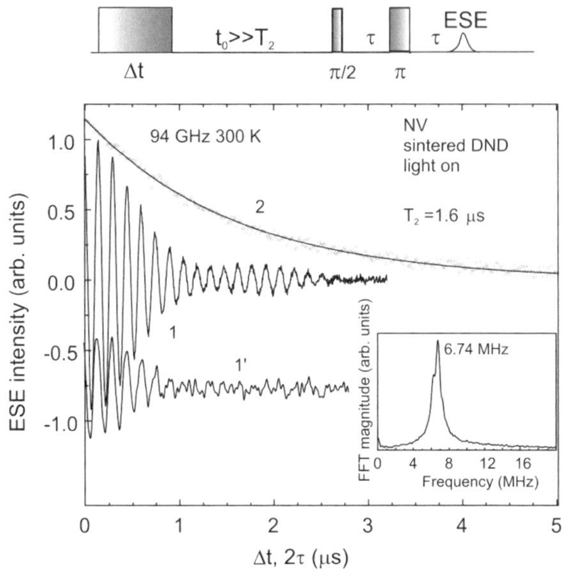

Fig. 5.26 The *upper part* of the figure shows the pulse sequence used to measure Rabi oscillations. The *lower part* depicts experimental data (1,1′) Rabi oscillations in sample #1: ESE intensity versus nutation pulse length (Δt) measured for sample #1 at 3447.5 mT (see Fig. 5.22b) with microwave power of 40 mW under 532 nm laser excitation (*1*) and without light (*1′*). *Inset* shows Fourier transform of (*1*). (*2*) The spin–spin T_2 relaxation time of NV defects in a single sintered DND cluster (sample #1) measured under 532 nm light excitation was determined using the two-pulse sequence $\pi/2 - \tau - \pi - \tau -$ echo, where the echo amplitude decay for NV defects was measured as a function of the time delay τ between the pulses in strong magnetic field of 3447.5 mT at 300 K. The *solid line* shows the approximation by one-exponential function with $T_2 = 1.6$ μs

excitation (1) and without light (1′) where decaying oscillations are clearly observed. Inset shows Fourier transform of (1). Rabi oscillations decay with a characteristic time constant τ_R, which depends on microwave power; τ_R was shown [16–18] is generally smaller than the coherence time (T_2) and inversely dependent on Rabi frequency Ω_R, which in turn depends on microwave power P as soon as oscillating component of the magnetic field B_1 is proportional to the square root of P. Here, α is a proportionality constant such that $1/\tau_R = 1/T_2 + 2\alpha\Omega_R$.

Closer inspection reveals more than one oscillation frequency, as confirmed by Fourier transformation (see inset). A distribution of frequencies is observed, with three close peaks at 6.27, 6.75 and 7.25 MHz. For EPR transitions, the Rabi frequency is proportional to the magnetic dipole transition matrix element $\Omega_R \propto B_1^2 \langle i|\boldsymbol{S}_+|j\rangle$. The presence of three nearby frequencies suggests excitation of multiple transitions simultaneously, probably on NV defects in different nanodiamonds with close orientations. Because of the axial symmetry of NV defect and the associated zero-field splitting of the NV defect ground state, the EPR signal depends strongly on the angle of the NV C_{3v} axis relative to the magnetic field. At least the eight main peaks due to fine splitting are observed (Fig. 5.22b). Thus, several transitions for NV defects with close orientation nanodiamonds overlap, corresponding to the various NV orientations in these ensemble measurements, and consequently multiple transitions are excited simultaneously. These different transitions have different transition matrix elements and therefore different Rabi frequencies. Consequently, a distribution of frequencies is observed, with three close maxima. The proximity of these Rabi frequencies indicates a high degree of DND ensemble orientation.

Inhomogeneity in exciting microwave fields have been suggested to be a microscopic basis for τ_R values smaller than T_2, [130–132], so the transient nutations experiments will not give a correct value for the spin dephasing time. Although accurate determination of τ_R is complicated by the some frequency distribution observed in Fig. 5.26 (inset), its upper limit (T_2) can be determined by measuring the decay of the echo intensity with increasing time delay between $\pi/2$ and π pulses (Hahn spin echo method) which eliminates the inhomogeneities related to the system. Figure 5.26 (2) shows a plot of the ESE decay at the strongest magnetic field of 3447.5 mT at 300 K for high-field fine-structure transition. The ESE intensity was shown to decay with approximately the same time constant at all fields. The spin–spin T_2 relaxation time of NV defects in a single sintered DND cluster (sample #1) measured under 532 nm light excitation was determined using the two-pulse sequence $\pi/2 - \tau - \pi - \tau -$ echo, where the echo amplitude decay for NV defects was measured as a function of the time delay τ between the pulses. The solid line shows the approximation by one-exponential function with $T_2 = 1.6$ μs.

5.3.4 Outlook

Isolated nitrogen centres N^0 and nitrogen pairs N^{2+} have been detected and identified, and their structure has been unambiguously determined by means of the high frequency EPR and ESE in natural diamond nanocrystals. In detonation ND and detonation ND after sintering, isolated nitrogen centres N^0 have been discovered in nanodiamond core. In addition EPR signals of multivacancy centres with spin 3/2 seem to be observed in nanodiamond core of detonation ND.

A giant concentration of nitrogen vacancy defects (up to 0.1%) has been revealed by the EPR, ESE and ODMR methods in a detonation nanodiamond sintered at high pressure and temperature. The results presented are opening new perspectives of NV-containing diamond fabrication, especially taking into account the high-density of single nitrogen atoms integrated in crystalline lattice and high coherence of the spin system.

In conclusion, the sintering procedure leads to the self-organization of the DND particles into the oriented arrays and the NV centres can be easily created without any post or prior irradiation. The concentrations of the NV centres that could be obtained by this technique are much higher then have been ever reported so far. It is shown that the spin-lattice T_1 and spin-spin relaxation T_2 for high-concentration NV defects in detonation nanodiamonds after HPHT sintering practically no different from the corresponding characteristics of NV defects introduced by the standard method of irradiation and subsequent annealing of commercial diamond single crystal containing a high concentration of nitrogen, illustrating the richness of this system for magnetometry and biological applications demand very bright and photostable fluorescence from high-density NV nanosized ensembles. The times of the polarization of the spin sublevels under the action of pulsed laser excitation and subsequent recovery of the equilibrium populations of the sublevels are close for both types of materials with NV defects.

References

1. de Mello Donega, C., Liljeroth, P., Vanmaekelbergh, D.: Physicochemical evaluation of the hot-injection method, a synthesis route for monodisperse nanocrystals. Small **1**, 1152–1162 (2005)
2. Baranov, P.G., Orlinskii, S.B., de Mello Donega, C., Schmidt, J.: High-frequency EPR and ENDOR spectroscopy on semiconductor quantum dots. Appl. Magn. Reson. **39**, 151–183 (2010) (and references therein)
3. Orlinskii, S.B., Schmidt, J., Baranov, P.G., Hofmann, D.M., de Mello Donega, C., Meijerink, A.: Probing the wave function of shallow Li and Na donors in ZnO nanoparticles. Phys. Rev. Lett. **92**, 047603 (2004)
4. Hofmann, D.M., Zhou, H., Pfisterer, D.R., Alves, H., Meyer, B.K., Baranov, P., Romanov, N., de Mello Donega, C., Meijering, A., Orlinskii, S., Blok, H., Schmidt, J.: Donors in ZnO nanocrystals. Phys. Stat. Sol. (c) **1**, 908–911 (2004)
5. Orlinskii, S.B., Schmidt, J., Groenen, E.J.J., Baranov, P.G., de Mello Donega, C., Meijerink, A.: Shallow donors in semiconductor nanoparticles: limit of the effective mass approximation. Phys. Rev. Lett. **94**, 097602 (2005)
6. Orlinskii, S.B., Blok, H., Groenen, E.J.J., Schmidt, J., Baranov, P.G., de Mello Donegá, C., Meijerink, A.: High-frequency EPR and ENDOR spectroscopy on semiconductor nanocrystals. Magn. Reson. Chem. **43**, S140–S144 (2005)
7. Orlinskii, S.B., Blok, H., Schmidt, J., Baranov, P.G., de Mello Donega, C., Meijerink, A.: Donor-acceptor pairs in the confined structure of ZnO nanocrystals. Phys. Rev. B **74**, 045204 (2006)
8. Orlinskii, S.B., Schmidt, J., Baranov, P.G., Rauh, D., Lorrmann, V., Dyakonov, V.: Identification of shallow Al donors in Al-doped ZnO nanocrystals: EPR and ENDOR spectroscopy. Phys. Rev. B **77**, 115334 (2008)
9. Orlinskii, S.B., Schmidt, J., Baranov, P.G., de Mello Donega, C., Meijerink, A.: Dynamic nuclear polarization of Zn^{67} and H^1 spins by means of shallow donors in ZnO nanoparticles. Phys. Rev. B **79**, 165316 (2009)
10. Meyer, B.K., Alves, H., Hofmann, D.M., Kriegseis, W., Forster, D., Bertram, F., Christen, J., Hoffmann, A., Straßburg, M., Dworzak, M., Haboeck, U., Rodina, A.V.: Feature article: bound exciton and donor–acceptor pair recombinations in ZnO. Phys. Stat. Sol. (b) **241**, 227 (2004)
11. Whitaker, K.M., Ochsenbein, S.T., Polinger, V.Z., Gamelin, D.R.: Electron confinement effects in the EPR spectra of colloidal n-Type ZnO quantum dots. J. Phys. Chem. C **112**, 14331–14335 (2008)

12. Ochsenbein, S.T., Feng, Y., Whitaker, K.M., Badaeva, E., Liu, W.K., Li, X., Gamelin, D.R.: Charge-controlled magnetism in colloidal doped semiconductor nanocrystals. Nat. Nanotechnol. **4**, 681–687 (2009)
13. Beek, W.J.E., Wienk, M.M., Janssen, R.A.J.: Hybrid solar cells from regioregular polythiophene and ZnO nanoparticles. Adv. Funct. Mater. **16**, 1112–1116 (2006)
14. Gonzales, C., Block, D., R.T. Cox, A. Herve, J.: Magnetic resonance studies of shallow donors in zinc oxide. Cryst. Growth **59**, 357–362 (1982)
15. Block, D., Herve, A., Cox, R.T.: Optically detected magnetic resonance and optically detected ENDOR of shallow indium donors in ZnO. Phys. Rev. B **25**, 6049R (1982)
16. Garces, N.Y., Giles, N.C., Halliburton, L.E., Cantwell, G., Eason, D.B., Reynolds, D.C., Look, D.C.: Production of nitrogen acceptors in ZnO by thermal annealing. Appl. Phys. Lett. **80**, 1334–1336 (2002)
17. Bennebroek, M.T., Poluektov, O.G., Zakrzewski, A.J., Baranov, P.G., Schmidt, J.: Structure of the intrinsic shallow electron center in AgCl studied by pulsed electron nuclear double resonance spectroscopy at 95 GHz. Phys. Rev. Lett. **74**, 442–445 (1995)
18. Hofmann, D.M., Hofstaetter, A., Leiter, F., Zhou, H., Henecker, F., Meyer, B.K., Orlinskii, S.B., Schmidt, J., Baranov, P.G.: Hydrogen: a relevant shallow donor in zinc oxide. Phys. Rev. Lett. **88**, 045504 (2002)
19. Orlinskii, S.B., Schmidt, J., Baranov, P.G., Bickermann, M., Epelbaum, B.M., Winnacker, A.: Observation of the triplet metastable state of shallow donor pairs in AlN crystals with a negative-*U* behavior: a high-frequency EPR and ENDOR study. Phys. Rev. Lett. **100**, 256404 (2008)
20. Blok, H., Orlinski, S.B., Schmidt, J., Baranov, P.G.: Overhauser effect of Zn^{67} nuclear spins in ZnO via cross relaxation induced by the zero-point fluctuations of the phonon field. Phys. Rev. Lett. **92**, 047602 (2004)
21. Meulenkamp, E.A.: Synthesis and growth of ZnO nanoparticles. J. Phys. Chem. B **102**, 5566–5572 (1998)
22. van Dijken, A., Meulenkamp, E.A., Vanmaekelbergh, D., Meijerink, A.: The kinetics of the radiative and nonradiative processes in nanocrystalline ZnO particles upon photoexcitation. Phys. Chem. B **104**, 1715–1723 (2000)
23. Hu, Z., Oskam, G., Searson, P.C.: Influence of solvent on the growth of ZnO nanoparticles. J. Coll. Interf. Sci. **263**, 454–460 (2003)
24. de Mello Donega, C.: The nanoscience paradigm: "Size matter!". In: de Mello Donega, C. (ed.) Nanoparticles: Workhorses of Nanoscience, pp. 1–12. Springer, Berlin, Heidelberg (2014)
25. de Mello Donega, C., Bol, A.A., Meijerink, A.: Time-resolved luminescence of $ZnS:Mn^{2+}$ nanocrystals. J. Lumin. **96**, 87–93 (2002)
26. Disselhorst, J.A.J.M., van der Meer, H.J., Poluektov, O.G., Schmidt, J.: A pulsed EPR and ENDOR spectrometer operating at 95 GHz. J. Magn. Reson. A **115**, 183–188 (1995)
27. Blok, H., Disselhorst, J.A.J.M., van der Meer, H., Orlinskii, S.B., Schmidt, J.: ENDOR spectroscopy at 275 GHz. J. Magn. Res. **173**, 49–53 (2005)
28. Mims, W.B.: Chapter 4. In: Geschwind, S. (ed.) Electron Paramagnetic Resonance, pp. 263–351. Plenum, New York (1972)
29. Park, C.H., Zhang, S.B., Wei, S.-H.: Origin of *p*-type doping difficulty in ZnO: the impurity perspective. Phys. Rev. B **66**, 073202 (2002)
30. Kutin, YuS, Mamin, G.V., Orlinskii, S.B., Bundakova, A.P., Baranov, P.G.: Effect of quantum confinement and influence of extra charge on the electric field gradient in ZnO. JETP Lett. **95**, 471–475 (2012)
31. Öğüt, S., Burdick, R., Saad, Y., Chelikowsky, J.R.: Ab initio calculations for large dielectric matrices of confined systems. Phys. Rev. Lett. **90**, 127401 (2003)
32. Melnikov, D.V., Chelikowsky, J.R.: Quantum confinement in phosphorus-doped silicon nanocrystals. Phys. Rev. Lett. **92**, 046802 (2004)
33. Rodina, A.V., Efros, A.L., Rosen, M., Meyer, B.K.: Theory of the Zeeman effect in semiconductor nanocrystals. Mater. Sci. Eng. C **19**(1–2), 435–438 (2002)

34. Overhauser, A.W.: Paramagnetic relaxation in metals. Phys. Rev. **89**, 689 (1953)
35. Denninger, G., Reiser, D.: Determination of electric-field gradients in semiconductors with high precision and high sensitivity. Phys. Rev. B **55**, 5073 (1997)
36. Dyakonov, V., Denninger, G.: Overhauser-shift measurements on Si: P near the metal-insulator transition. Phys. Rev. B **46**, 5008 (1992)
37. Morton, J.R., Preston, K.F.: Atomic parameters for paramagnetic resonance data. J. Magn. Reson. **30**, 577–582 (1978)
38. Schirmer, O.F., Zwingel, D.: The yellow luminescence of zinc oxide. Solid State Commun. **8**, 1559–1563 (1970)
39. Taylor, A.L., Filipovich, G., Lindeberg, G.K.: Electron paramagnetic resonance associated with Zn vacancies in neutron-irradiated ZnO. Solid State Commun. **8**, 1359–1361 (1970)
40. Rong, F.C., Barry, W.A., Donegan, J.F., Watkins, G.D.: Vacancies, interstitials, and close Frenkel pairs on the zinc sublattice of ZnSe. Phys. Rev. B **54**, 7779 (1996)
41. Barry, W.A., Watkins, G.D.: Exchange and radiative lifetimes for close Frenkel pairs on the zinc sublattice of ZnSe. Phys. Rev. B **54**, 7789 (1996). (and references therein)
42. Atherton, N.M.: Principles of Electron Spin Resonance. Ellis Horwood, PTR Prentice Hall, New York (1993)
43. Cox, R.T., Davies, J.J.: Electron-hole exchange interaction for donor-acceptor pairs in CdS determined as a function of separation distance by optically detected magnetic resonance. Phys. Rev. B **34**, 8591 (1986)
44. Poluektov, O.G., Donckers, M.C.J.M., Baranov, P.G., Schmidt, J.: Dynamical properties of the self-trapped exciton in AgCl as studied by time-resolved EPR at 95 GHz. Phys. Rev. B **47**, 10226 (1993)
45. Nirmal, M., Dabbousi, B.O., Bawendi, M.G., Macklin, J.J., Trautman, J.K., Harris, T.D., Brus, L.E.: Fluorescence intermittency in single cadmium selenide nanocrystals. Nature **383**, 802–804 (1996)
46. Volbers, N., Zhou, H., Knies, C., Pfisterer, D., Sann, J., Hofmann, D.M., Meyer, B.K.: Synthesis and characterization of ZnO:Co^{2+} nanoparticles. Appl. Phys. A **88**, 153–155 (2007)
47. Ochsenbein, S.T., Gamelin, D.R.: Quantum oscillations in magnetically doped colloidal nanocrystals. Nat. Nanotechnol. **6**, 112–115 (2011). (and references therein)
48. Morhain, C., Deparis, C., Laugt, M., Goiran, M., Golacki, Z.: Magnetic anisotropy of Co^{2+} as signature of intrinsic ferromagnetism in ZnO:Co. Phys. Rev. Lett. **96**, 017203 (2006)
49. D'Ambrosio, S., Pashchenko, V., Mignot, J.-M., Ignatchik, O., Kuzian, R.O., Savoyant, A., Golacki, Z., Grasza, K., Stepanov, A.: Competing exchange interactions in Co-doped ZnO: departure from the superexchange picture. Phys. Rev. B **86**, 035202 (2012)
50. Jedrecy, N., von Bardeleben, H.J., Zheng, Y., Cantin, J.-L.: Electron paramagnetic resonance study of $Zn_{1-x}Co_xO$: a predicted high-temperature ferromagnetic semiconductor. Phys. Rev. B **69**, 041308 (2004). (and references therein)
51. Pereira, A.S., Ankiewicz, A.O., Gehlhoff, W., Hoffmann, A., Pereira, S., Trindade, T., Grundmann, M., Carmo, M.C., Sobolev, N.A.: Surface modification of Co-doped ZnO nanocrystals and its effects on the magnetic properties. J. Appl. Phys. **103**, 07D140 (2008)
52. Ankiewicz, A.O., Gehlhoff, W., Martins, J.S., Pereira, B.S., Pereira, S., Hoffmann, A., Kaidashev, E.M., Rahm, A., Lorenz, M., Grundmann, M., Carmol, M.C., Trindade, T., Sobolev, N.A.: Magnetic and structural properties of transition metal doped zinc-oxide nanostructures. Phys. Status Solidi B **246**, 766–770 (2009)
53. Schneider, J., Sircar, S.R., Räuber, A.: Elektronen-spin-resonanz von Mn^{2+}-ionen im kubischen und trigonalen kristallfeld des ZnSZ. Natürforsch, **18a**, 980–993 (1963
54. Cavenett, B.C.: Optically detected magnetic resonance (O.D.M.R.) investigations of recombination processes in semiconductors. Adv. Phys. **4**, 475–538 (1981)
55. Baranov, P.G., Romanov, N.G.: ODMR study of recombination processes in ionic crystals and silicon carbide. Appl. Magn. Reson. **2**, 361–378 (1991); Baranov, P.G., Romanov, N.G.: Magnetic resonance in micro- and nanostructures. Appl. Magn. Reson. **21**, 165–193 (2001)

56. Baranov, P.G., Veshchunov, YuP, Zhitnikov, R.A., Romanov, N.G., Shreter, YuG: Optical detection of microwave resonance in germanium by means of luminescence of electron-hole drops. JETP Lett. **26**, 249–253 (1977)
57. Tomaru, T., Ohyama, T., Otsuka, E.: Optically detected cyclotron resonance in Ge and Si exchange interaction in impact dissociation. Appl. Magn. Reson. **2**, 379–395 (1991)
58. Godlewski, M., Chen, W.M., Monemar, B.: Optical detection of cyclotron resonance for characterization of recombination processes in semiconductors. Crit. Rev. Solid State Mater. Sci. **19**, 241–301 (1994)
59. Cavenett, B.C., Pakulis, E.J.: Optically detected cyclotron resonance in a GaAs/$Ga_{0.67}Al_{0.33}As$ superlattice. Phys. Rev. B **32**, 8449–8452 (1985)
60. Warburton, R.T., Michels, J.G., Nicholas, R.J., Harris, J.J., Foxon, C.T.: Optically detected cyclotron resonance of GaAs quantum wells: effective-mass measurements and offset effects. Phys. Rev. B **46**, 13394–13399 (1992)
61. Hofmann, D.M., Drechsler, M., Wetzel, C., Meyer, B.K., Hirler, F., Strenz, R., Abstreiter, G., Bohm, G., Weimann, G.: Optically detected cyclotron resonance on GaAs/$Al_xGa_{1-x}As$ quantum wells and quantum wires. Phys. Rev. B **52**, 11313–11318 (1995)
62. Chen, Y.F., Dai, Y.T., Fan, J.C., Lee, T.L., Lin, H.H.: Dependence of electron effective mass on alloy composition of InAlGaAs lattice matched to InP studied by optically detected cyclotron resonance. Appl. Phys. Lett. **67**, 1256–1258 (1995)
63. Baranov, P.G., Veshchunov, YuP, Romanov, N.G.: Sov. Phys. Solid State **22**, 2186–2188 (1980)
64. Romanov, N.G., Vetrov, V.A., Baranov, P.G.: Optical detection of EPR of self-trapped excitons using photostimulated luminescence of crystals. JETP Lett. **37**, 386–388 (1983)
65. Baranov, P.G., Bulanyi, M.F., Vetrov, V.A., Romanov, N.G.: Optically detected ESR in ZnS and ZnS: Mn crystals JETP Lett. **38**, 623–625 (1983)
66. Romanov, N.G., Dyakonov, V.V., Vetrov, V.A., Baranov, P.G.: Sov. Phys. Solid State **31**, 1899–1901 (1989)
67. Baranov, P.G., Vetrov, V.A., Romanov, N.G., Sokolov, V.I.: Sov. Phys. Solid State **27**, 2085–2086 (1985)
68. Baranov, P.G., Romanov, N.G., Vetrov, V.A., Oding, V.G.: In: Anastassakis, E.M., Joannopoulos, D. (eds.) Proceedings of the 20th International Conference on Physics Semiconductors (Thessaloniki, 1990), vol. 3, pp. 1855–1858. World Scientific (1991)
69. Romanov, N.G., Baranov, P.G.: Semicond. Sci. Technol. **9**, 1080–1085 (1994)
70. Ivchenko, E.L, Pikus, G.E.: Superlattices and other heterostructures. In: Cardona, M., Fulde, P., von Klitzig, K., Queisser, H.-J. (eds.) Springer Series in Solid State Sciences, vol. 110. Springer (1995)
71. van Kesteren, H.W., Cosman, E.C., van der Pool, W.A.J.A., Foxon, C.T.: Fine structure of excitons in type-II GaAs/AlAs quantum wells. Phys. Rev. B **41**, 5283–5292 (1990)
72. Baranov, P.G., Mashkov, I.V., Romanov, N.G., Lavallard, P., Planel, R.: Solid State Commun. **87**, 649–654 (1993)
73. van Kesteren, H.W., Cosman, E.C., Dawson, P., Moore, K.M., Foxon, C.T.: Order of the *X* conduction-band valleys in type-II GaAs/AlAs quantum wells. Phys. Rev. B **39**, 13426–13433 (1989)
74. Baranov, P.G., Romanov, N.G., Mashkov, I.V.: Phys. Solid State **37**, 1648–1654 (1995)
75. Romanov, N.G., Mashkov, I.V., Baranov, P.G., Lavallard, P., Planel, R.: Excitons in GaAs/AlAs superlattices near a type-II-type-I transition. JETP Lett. **57**, 802–807 (1993)
76. Romanov, N.G., Baranov, P.G., Mashkov, I.V., Lavallard, P., Planel, R.: Solid State Electron. **37**, 911–914 (1994)
77. Baranov P.G., Romanov, N.G.: In: Lockwood, D.J. (ed.) Proceedings of the 22nd International Conference on the Physics of Semiconductors, vol. 2, pp. 1400–1403. World Scientific (1994); Baranov, P.G., Romanov, N.G. Phys. Solid State **41**, 805–807 (1999)
78. Blackwood, E., Snelling, M.J., Harley, R.T., Andrews, S.R., Foxon, C.T.B.: Exchange interaction of excitons in GaAs heterostructures. Phys. Rev. B **50**, 14246–14254 (1994)

79. Harley, R.T., Snelling, M.J.: Magnetic-field dependence of exciton spin relaxation in GaAs/$Al_xGa_{1-x}As$ quantum wells. Phys. Rev. B **53**, 9561–9564 (1996)
80. Ivchenko, E.L., Kaminskii, AYu., Alleiner, I.L.: Exchange splitting of excitonic levels in types I and II superlattices. J. Exp. Theor. Phys. **104**, 609–616 (1993)
81. Gupalov, S.V., Ivchenko, E.L., Kavokin, A.V.: Fine structure of localized exciton levels in quantum wells. J. Exp. Theor. Phys. **86**, 388–394 (1998)
82. Gourdon, C., Lavallard, P.: Fine structure of heavy excitons in GaAs/AlAs superlattices. Phys. Rev B **46**, 4644–4650 (1992)
83. Teissier R.: Thèse de Doctorat, Université Paris VI, 231 pp. (1992)
84. Baranov, P.G., Mashkov, I.V., Romanov, N.G., Gourdon, C., Lavallard, P., Planel, R.: Magnetic resonance and anticrossing of levels of excitons trapped at opposite interfaces in type-II GaAs/AIAs superlattices. JETP Lett. **60**, 445–450 (1994)
85. Gourdon, C., Mashkov, I.V., Lavallard, P.: Planel R 1998. Phys. Rev. B **57**, 3955–3960 (1998)
86. Baranov, P.G., Romanov, N.G., Hofstaetter, A., Meyer, B.K., Scharmann, A., von Foerster, W., Ahlers, F.J., Pierz, K.: In: Alferov, Zh., Esaki, L. (eds.) Proceedings of the 7th International Symposium on Nanostructures: Physics and Technology, pp. 360–363. IOP Publ. Ltd., St. Petersburg (1999)
87. Ivchenko, E.L., Kaminskii, A.Yu.: Fiz. Tverd. Tela 37 1418–1428 (1995): Phys. Solid State **37**, 1417 (1995)
88. Bode, M.N., Ourmarzd, J.J.: Interfaces in GaAs/AlAs: perfection and applications. Vac. Sci. Technol. **10**, 1787–1794 (1992)
89. Jusserand, B., Mollot, F.: Long range gallium segregation in the AlAs layers of GaAs/AlAs superlattices. Appl. Phys. Lett. **61**, 423–428 (1992)
90. Hofstaetter, A., Scharmann, A., Schnorr, C., Baranov, P.G., Romanov, N.G., Ahlers, F.J., Pierz, K.: In: Alferov, Zh., Esaki, L. (ed.) Proceedings of the 4th International Symposium on Nanostructures: Physics and Technology, St. Petersburg, pp. 72–75 (1996)
91. Trombetta, J.M., Kennedy, T.A.: Optically detected magnetic resonance of shallow donors in GaAs. Phys. Rev. B **48**, 17031–17034 (1993)
92. Bimberg, D., Grundmann, M., Ledentsov, N.N.: Quantum Dot Heterostructures. Wiley, New York (1999)
93. Frohlich, D., Haselhoff, M., Reimann, K., Itoh, T.: Determination of the orientation of CuCl nanocrystals in a NaCl matrix. Solid State Commun. **94**, 189–194 (1995)
94. Haselhoff, M., Weber, H.-J.: Nanocrystal growth in alkali halides observed by exciton spectroscopy. Phys. Rev. B **58**, 5052–5061 (1998)
95. Vogelsang, H., Husberg, O., Kohler, U., von der Osten, W., Marchetti, A.: Exciton self-trapping in AgCl nanocrystals. Phys. Rev. B **61**, 1847–1852 (2000)
96. Baranov, P.G., Vikhnin, V.S., Romanov, N.G., Khramtsov, V.A.: Suppression of the local Jahn-Teller effect in nanostructures: self-trapped holes and excitons in AgCl nanocrystals. JETP Lett. **72**, 329–332 (2000)
97. Baranov, P.G., Vikhnin, V.S., Romanov, N.G., Khramtsov, V.A.: Oriented silver chloride microcrystals and nanocrystals embedded in a crystalline KCl matrix, as studied by means of electron paramagnetic resonance and optically detected magnetic resonance. J. Phys.: Condens. Matter **13**, 2651–2669 (2001)
98. Bennebroek, M.T., Arnold, A., Poluektov, O.G., Baranov, P.G., Schmidt, J.: Shallow electron centers in silver halides. Phys. Rev. B **54**, 11276–11289 (1996)
99. Baetzold, R.C., Eachus, R.S.: The possibility of a split interstitial silver ion in AgCl. J. Phys.: Condens. Matter **7**, 3991–3998 (1995)
100. Song, K.S., Williams, R.T.: Self-Trapped Excitons. Springer, Berlin, Heidelberg (1993)
101. Hayes, W., Owen, I.B.: J. Phys. C: Solid State Phys. **9**, L69-L73 (1976); Marchetti, A.P., Tinti, D.S.: Low-temperature photophysics of crystalline AgCl. Phys. Rev B **24**, 7361–7370 (1981)

102. Bennebroek, M.T., Arnold, A., Poluektov, O.G., Baranov, P.G., Schmidt, J.: Spatial distribution of the wave function of the self-trapped exciton in AgCl. Phys. Rev. B **53**, 15607–15616 (1996)
103. Marchertti, A.P.: The optically detected magnetic resonance of pure and doped silver bromide. J. Phys. C: Sol. Stat. Phys. **14**, 961–968 (1981); Burberry, M.S., Marchetti, A.P.: Low-temperature donor-acceptor recombination in silver halides. Phys. Rev. B **32**, 1192–1995 (1985)
104. Marchetti, A.P., Burberry, M.S., Spoonhower, J.P.: Characterization of an intermediate-case exciton in the 580-nm emission of Cd-doped and pure AgBr. Phys. Rev. B **43**, 2378–2383 (1991)
105. Baranov, P.G., Romanov, N.G.: Magnetic resonance in micro- and nanostructures. Appl. Magn. Reson. **21**, 165–193 (2001)
106. Baranov, P.G., Romanov, N.G., Preobrazhenskii, V.L., Khramtsov, V.A.: Electron-hole recombination confinement in self-organized AgBr nanocrystals in a crystalline KBr matrix. JETP Lett. **76**, 465–468 (2002)
107. Baranov, P.G., Romanov, N.G., Khramtsov, V.A., Vikhnin, V.S.: Silver halide micro- and nanocrystals embedded in an alkali halide matrix: suppression of the Jahn-Teller effect in nanoparticles. Nanotechnology **12**, 540–546 (2002)
108. Marchetti, A.P., Johanson, K.P., McLendon, G.L.: AgBr photophysics from optical studies of quantum confined crystals. Phys. Rev. B **47**, 4268–4276 (1993)
109. Rodney, P.J., Marchetti, A.P., Fauchet, P.M.: Effects of size restriction on donor-acceptor recombination in AgBr. Phys. Rev. B **62**, 4215–4217 (2000)
110. Marchetti, A.P., Rodney, P.J., von der Osten, W.: AgBr nanocrystals in glass: optical and ODMR investigations. Phys. Rev. B **64**, 132201 (2001)
111. Cox, R.T., Davies, J.J.: Electron-hole exchange interaction for donor-acceptor pairs in CdS determined as a function of separation distance by optically detected magnetic resonance. Phys. Rev. B **34**, 8591–8610 (1986)
112. Barry, W.A., Watkins, G.D.: Exchange and radiative lifetimes for close Frenkel pairs on the zinc sublattice of ZnSe. Phys. Rev. B **54**, 7789–7798 (1996)
113. Bennebroek, M.T., Arnold, A., Poluektov, O.G., Baranov, P.G., Schmidt, J.: Spatial distribution of the wave function of the self-trapped exciton in AgCl. Phys. Rev. B **53**, 15607–15616 (1996)
114. Dean, P.J.: Inter-impurity recombinations in semiconductors. In: McCaldin, J.D., Somorjaj, G. (eds.) Progress in Solid State Chemistry **8**, pp. 1–126. Pergamon Press, Oxford (1973)
115. Marchetti, P., Eachus, R.S.: The photochemistry and photophysics of the silver halides. In: Volman, D.H., Hammond, G.S., Neckers, D.C. (eds.) Adv. Photochem. **17**, pp. 145–216 (1992)
116. Zavoisky, E.: The paramagnetic absorption of a solution in parallel fields. J. Phys. **8**, 377–380 (1944)
117. Abragam, A., Bleaney, B.: Electron Paramagnetic Resonance of Transition Ions. Oxford University Press (Clarendon), London and New York (1970)
118. Baranov, P.G., Romanov, N.G., Poluektov, O.G., Schmidt, J.: Self-Trapped Excitons in Ionic-Covalent Silver Halide Crystals and Nanostructures: High-Frequency EPR, ESE, ENDOR and ODMR Studies. Appl. Magn. Reson. **39**, 453–486 (2010)
119. Schweiger, A., Jeschke, G.: Principle of Pulse Electron Paramagnetic Resonance. Oxford University Press (2001)
120. Fang, X.-W., Mao, J.-D., Levin, E.M., Schmidt-Rohr, K.: Nonaromatic core-shell structure of nanodiamond from solid-state NMR spectroscopy. J. Am. Chem. Soc. **131**, 1426–1435 (2009)
121. Gruber, A., Draebenstedt, A., Tietz, C., Fleury, L., Wrachtrup, J., von Borczyskowski, C.: Scanning confocal optical microscopy and magnetic resonance on single defect centres. Science **276**, 2012–2014 (1997)
122. Awschalom, D.D., Epstein, R., Hanson, R.: The diamond age of spintronics. Sci. Am. **84** (2007)

123. Ammerlaan, C.A.J.: Paramagnetic Centres in Diamond. Landolt-Börnstein, New Series III/41A2α, Springer
124. Maze, J.R., et al.: Nanoscale magnetic sensing with an individual electronic spin in diamond. Nature **455**, 644–647 (2008)
125. Chung, P.-H., et al.: Spectroscopic study of bio-functionalized nanodiamonds. Diamond Relat. Mater. **15**, 622–625 (2006)
126. Chang, Y.-R., et al.: Mass production and dynamic imaging of fluorescent nanodiamonds. Nat. Nanotech. **3**, 284–288 (2008)
127. Babinec, T.M., et al.: A diamond nanowire single-photon source. Nat. Nanotechn. **5**, 195–199 (2010)
128. Wrachtrup, J., Jelezko, F.: Processing quantum information in diamond. J. Phys.: Condens. Matter **18**, 807–824 (2006)
129. Taylor, J.M., et al.: High-sensitivity diamond magnetometer with nanoscale resolution. Nat. Phys. **4**, 810–816 (2008)
130. Boudou, J.-P., et al.: High yield fabrication of fluorescent nanodiamonds. Nanotechnology **20**, 235602 (2009)
131. Tisler, J., et al.: Fluorescence and spin properties of defects in single digit nanodiamonds. ACS Nano **3**, 1959–1965 (2009)
132. Greiner, N.R., Philips, D.S., Johnson, J.D., Volk, F.: Diamonds in detonation soot. Nature **333**, 440–442 (1988)
133. Shenderova, O., McGuire, G.: Ultra-nanocrystalline diamond: syntheses, properties and applications. In: Shenderova, O., Gruen, D. (eds.) Types of Nanocrystalline Diamond, pp. 79–114. Andrew, New York (2006)
134. Alekseenskiy, A., Baidakova, M., Osipov, V., Vul., A.: The fundamental properties and characteristics of nanodiamonds. In: Ho, D. (ed). Nanodiamonds: Applications in Biology and Nanoscale Medicine, pp. 55–77. Spinger (2009)
135. Barnard, A.S., Sternberg, M.: Substitutional nitrogen in nanodiamond and bucky-diamond particles. J. Phys. Chem. B **109**, 17107–17112 (2005)
136. Jelezko, F., Wrachtrup, J.: Single defect centres in diamond: a review. Phys. Stat. Sol. (a) **20**, 3207–3225 (2006)
137. Jelezko, F., Popa, I., Gruber, A., Tietz, C., Wrachtrup, J., Nizovtsev, A., Kilin, S.: Single spin states in a defect centre resolved by optical spectroscopy. Appl. Phys. Lett. **81**, 2160–2162 (2002)
138. Nizovtsev, A.P., Kilin, S.Y., Jelezko, F., Popa, I., Gruber, A., Wrachtrup, J.: NV centres in diamond: spin-selective photokinetics, optical ground-state spin alignment and hole burning. Physica B **340–342**, 106–110 (2003)
139. Mitra, Y.: Change of absorption spectra in type-Ib diamond with heavy neutron irradiation. Phys. Rev. B **53**, 11360 (1996)
140. Rabeau, J.R., Reichart, P., Tamanyan, G., Jamieson, D.N., Prawer, S., Jelezko, F., Gaebel, T., Popa, I., Domhan, M., Wrachtrup, J.: Implantation of labelled single nitrogen vacancy centres in diamond using N-15. Appl. Phys. Lett. **88**, 023113 (2006)
141. Vlasov, I., et al.: Nitrogen and luminescent nitrogen-vacancy defects in detonation nanodiamond. Small **6**, 687–694 (2010)
142. Smith, B.R., et al.: Five-nanometer diamond with luminescent nitrogen-vacancy defect centres. Small **5**, 1649–1653 (2009)
143. Bradac, C., Gaebel, T., Naidoo, N., Sellars, M.J., Twamley, J., Brown, L.J., Barnard, A.S., Plakhotnik, T., Zvyagin, A.V., Rabeau, J.R.: Observation and control of blinking nitrogen-vacancy centres in discrete nanodiamonds. Nat. Nanotechn. **5**, 345–349 (2010)
144. Rabeau, J.R., et al.: Single nitrogen vacancy centres in chemical vapor deposited diamond nanocrystals. Nano Lett. **7**, 3433–3437 (2007)
145. Kidalov, S.V., Shakhov, F.M., Vul, A.Ya., Ozerin, A.N.: Grain-boundary heat conductance in nanodiamond composites. Diamond Relat. Mater. **19**, 976–980 (2010)
146. Baranov, P.G., et al.: Electron spin resonance detection and identification of nitrogen centres in nanodiamonds. JETP Lett. **89**, 409–413 (2009)

147. Soltamova, A. et al.: Detection and identification of nitrogen defects in nanodiamond as studied by EPR. Phys. B: Cond. Matt. **404**, 4518 (2009)
148. Baranov, P.G., Soltamova, A.A., Tolmachev, D.O., Romanov, N.G., Babunts, R.A., Shakhov, F.M., Kidalov, S.V., Vul', A.Ya., Mamin, G.V., Orlinskii, S.B., Silkin, N.I.: Enormously high concentrations of fluorescent nitrogen-vacancy centres fabricated by sintering of detonation nanodiamonds. Small **7**, 1533–1537 (2011)
149. Davies, G., Hamer, M.F.: Optical studies of 1.945 eV vibronic band in diamond. Proc. R. Soc. Lond. A **348**, 285–298 (1976)
150. Rogers, L.G., Armstrong, S., Sellars, M.J., Manson, N.B.: Infrared emission of the NV centre in diamond: Zeeman and uniaxial stress studies. New J. Phys. **10**, 103024 (2008)
151. Harrison, J., Sellars, M.J., Manson, N.B.: Optical spin polarisation of the N-V centre in diamond. J. Lumin. **107**, 245–248 (2004)

Chapter 6
Perspectives of Applications of Magnetic Properties of Semiconductor Nanostructures and Single Defects

6.1 Manipulation of Single Spins by Optical and Microwave Spectroscopy

6.1.1 Introduction

The growing interest in manipulation and read-out of single-electron and nuclear spin states in semiconductors and semiconductor nanostructures is associated with possible applications in spintronics and solid state quantum information processing. The field of quantum computing has seen an explosive increase in experimental and theoretical work during the last decade. The advantage of quantum computing over classical computing lies in an exponential speed-up of certain calculations such as Fourier transformations and searching an unordered database [1]. On the other hand, as the size of modern computer chips approaches the atomic scale, it will become necessary, in the near future, to take into account the quantum properties of individual atoms [2]. The idea of an atomic scale quantum computer is not just building atomic logic elements, but also using quantum mechanical properties for computation. Because the quantum mechanical system can exist in a superposition of several states at once, this can be used for parallel data processing. On the other hand, quantum mechanical systems are difficult to handle, and several experimental obstacles must be taken into account.

Various specific requirements on hardware for quantum computation have been identified and summarized in the Di Vincenzo check list [3]. The most successful approach for testing quantum algorithms is via liquid state NMR [4–6], which is able to realize quantum algorithms with seven qubits [7]. The main problem related to bulk NMR quantum computing is the preparation of the initial state. The density matrix of liquid state NMR is based on the initial thermal distribution of spin states. Low polarization of the initial state results in a scaling problem, which is currently one of the main obstacles to building larger scale quantum computing devices [8]. Recently, using a single nuclear spin as a qubit was proposed [9]. Note that the

P.G. Baranov et al., *Magnetic Resonance of Semiconductors and Their Nanostructures*, Springer Series in Materials Science 253,
DOI 10.1007/978-3-7091-1157-4_6

thermalization problem can be efficiently solved in a quantum processor, which uses single spin states for computation because reading of the spin state is equivalent to the initialization of the system. If reading of the spin state occurs on a timescale faster than the spin relaxation time, then the state is pure even for a fully thermalized spin system. However, reading a single spin state is a difficult experimental challenge. Conventional EPR and NMR spectrometers are not suitable for single-spin experiments because of the low magnetic moment associated with single electron and nuclear spins.

The typical sensitivity of inductive read-out methods is limited to 10^{16}–10^{18} spins for NMR experiments and 10^{12}–10^{14} spins for EPR. It was realized in the late 1940s that the sensitivity of magnetic resonance can be enhanced by shifting the detection of the magnetic resonance effect into the optical domain. The first experiments of this type were reported in 1952 [10]. The polarization of the fluorescence of mercury vapours has been monitored upon excitation of the fine structure related transition. The optical excitation produces polarization of Zeeman sublevels of the ground state of the mercury atoms. Because the Δm selection rule holds for optical transitions, the emitted radiation has circular polarization. The application of the resonance RF field equalizes populations of the fine structure sublevels, resulting in change of the fluorescence polarization. At the end of the 1960s, optically detected magnetic resonance (ODMR) had been applied to solid state systems. The rate of decay of a photoexcited triplet state of organic molecules embedded in a solid host is specific to the particular spin sublevel. Hence, the phosphorescence intensity depends on the populations of the fine structure sublevels and the application of a resonant RF field results in a change of the phosphorescence intensity. The first solid state ODMR experiments on quinoxaline were reported in [11], and on phenanthrene in [12]. The important advantage of optical detection is the improvement of the sensitivity by seven orders of magnitude (the detection of 10^5 spins was reported in [13]). The ability to read out the state of a single nuclear spin is related to the recent achievements of single-molecule spectroscopy: an optical technique, which combines high resolution optical spectroscopy and fluorescence microscopy.

It was recognized in the early 1960s that the high absorption cross-section associated with the electronic transitions of impurities in low temperature solids can be used for ultrasensitive optical detection of aromatic compounds [14]. The detection limit for benzo[a]pyrene metabolite 50 amol was achieved for a 20 μl sample [15]. The combination of low temperature spectroscopy with high spatial resolution allowed pushing the detection limit to the ultimate frontier—single-molecule spectroscopy. The spectroscopy of single impurity molecules in low temperature solids has been carried out in pioneering works [16–19]. This was achieved by excitation of a small sample volume of pentacene-doped p-terphenyl crystal by a resonant narrow band laser. The laser-excited sample area contained thousands of dopant molecules. In order to detect individual chromophores, so-called spectral selection was applied. Individual molecules have been selected by tuning the laser frequency within the inhomogeneously broadened electronic transition of the dopant spectral site.

The transition to single-spin detection is rather obvious. Single-molecule spectroscopy can be combined with the ODMR technique in order to detect and manipulate spin states of single molecules. This spin state is a collective spin state of two unpaired electrons of the photoexcited triplet state of the organic molecule. Optical detection of such single molecular spins has been reported simultaneously by two working groups in 1993 [20–23]. In these experiments single dopant molecules were isolated by spectral selection, and standard ODMR techniques were applied to detect spin transitions among sublevels of the photoexcited pentacene triplet state. However, the photoexcited triplet states of organicmolecules are of limited interest for quantum information processing because of their short (microseconds) lifetime. The optical detection of a single paramagnetic defect in diamond opened a new perspective for single-spin based quantum computing in solids [9, 24].

6.1.2 Experimental Methods

The detection of single-molecule luminescence is based on two important experimental aims. First, the concentration should be kept low enough, and the excited spot small enough, that a single impurity can be isolated in the excitation spectrum. Second, the detection efficiency must be high enough for obtaining a signal that is higher than the dark count rate of the detector.

Figure 6.1 shows, as an example, the experimental setup used in the studies of single molecules [22, 23].

A laser, e.g., titanium sapphire (Ti-Saph) laser, pumped by an argon (Ar) laser, produces the excitation light with a precisely defined wavelength tunable from 700 to 800 nm. The laser light is then focused on the sample by means of a microscope objective, with a focal spot area of about 1 μm^2. The sample is kept inside a cryostat in liquid He. A scanning mirror allows one to move the focal point over an area of the sample of 200 μm^2. Changing the wavelength of the laser and the position of the focal point onto the sample makes it possible to select a molecule both spectrally and spatially.

The excited volume plays a crucial role in the experimental set-up. The signal-to-background ratio is inversely proportional to the excited volume, because all the illuminated host and guest molecules except the molecule whose fluorescence is assigned to the signal can be considered as background. A diffraction limited illuminated volume of about 1 μm^3 is achievable at room temperature, for which high numerical aperture objectives are commercially available (for a recent review see [25]). Room temperature experiments are usually based only on spatial selection of single impurities; therefore the concentration of guest molecules must be of the order of 10^{-11} M. These experiments require a simple set-up, but the low photostability of organic molecules under strong illumination is an important disadvantage for many organic systems. The photobleaching of organic molecules can be lessened by special treatment of the samples, allowing one to minimize the

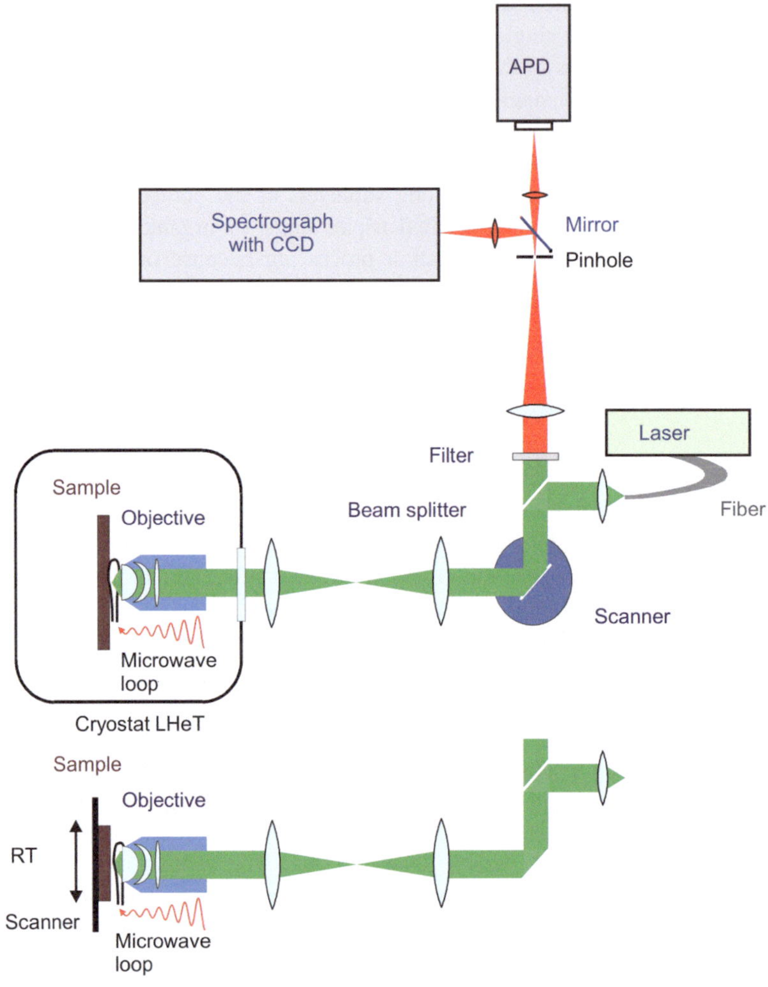

Fig. 6.1 Confocal microscopy and spectroscopy setup. *Lower panel* shows installation option for the room temperature. Microwave loop is used in magnetic resonance experiments

contact with atmospheric oxygen [26]. However, the typical lifetime of organic molecules remains below a minute. The only absolutely stable system reported so far is the vacancy–nitrogen defect centre in diamond, which will be discussed in the next section.

The light emitted from the sample is refocused into the pinhole because excitation volume and light source are in conjugate planes. For practical reasons the excitation and detection branch are separated by a dichroic mirror or by working in transmission. In front of the detector a second pinhole is placed symmetrically with respect to the entrance pinhole, i.e. in a conjugated plane with respect to the illumination pinhole and the sample volume. This way it is achieved that only light from the excitation volume is allowed to reach the detector, light from sample regions in front or behind the vocal volume is discriminated by the detection pinhole. As a result, a three dimensional selectivity is obtained with a spatial resolution which is limited by diffraction to about λ/2. A scanning the excitation volume across the sample allows to obtain three dimensional images. The reflected laser light is blocked by a long-pass filter, which is transparent for the red-shifted fluorescence. The strength of confocal microscopy resides in the use of an emission pinhole that eliminates all out of-focus light, increasing considerably the signal to noise ratio. In confocal microscopy, a pinhole acts as light source and is imagined into the sample.

The observation single impurity molecule was achieved by excitation of a small sample volume of pentacene-doped p-terphenyl crystal by a resonant narrow band laser (Fig. 6.2) [22, 23]. The laser-excited sample area contained thousands of dopant molecules. In order to detect individual molecules, a spectral selection was applied. Individual molecules have been selected by tuning the laser frequency within the inhomogeneously broadened electronic transition of the dopant spectral site.

The observation of narrow zero-phonon lines is restricted to a relatively limited number of host–guest systems, which shows the Spolskii effect. The origin of the Spolskii effect has been discussed in the literature (for recent work see [27–30]). To present a strong and stable ZPL, the impurity molecule must fit into an insertion site of the matrix, and its vibrational frequency, associated with local matrix vibration, must be as high as possible. This situation occurs when the size of the guest molecule corresponds to the size of the vacancy created by one or a few impurity molecules absent from the lattice [31].

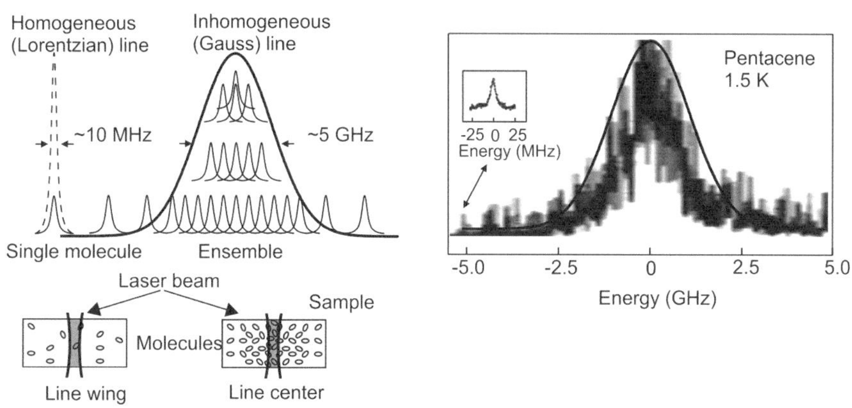

Fig. 6.2 Broad scan of the photoluminescence excitation spectra for pentacene in solid matrix at 1.5 K measured with a tunable laser, linewidth 3 MHz. The laser detuning frequency is referenced to the line centre. (*Inset*) Expansion low-power scan of a single molecule, the life-time limited width of about 10 MHz

The presence of spectrally stable and sharp zero-phonon lines is not sufficient for successful detection of a single molecule. The dopant molecules are detected via fluorescence emission. Hence, the optical transition must carry strong oscillator strength and only allowed singlet–singlet (or triplet–triplet) transitions can be considered. The study of single-molecule fluorescence requires consecutive detections of at least several thousand photons. Hence the host and guest molecules must be photochemically stable.

6.1.3 Photophysics of a Single Impurity in a Solid

The spectroscopy of single impurities isolated in a matrix has made important progress during the last decade and become the standard technique in a number of research laboratories. Although the first single-molecule experiment, reported in [16, 17], was based on absorption spectroscopy, all modern approaches are based on the detection of the fluorescence emission because of the superior signal-to-noise ratio (this technique was introduced in [18, 19]). The optical transition associated with the impurity molecule must be strong enough to produce a detectable fluorescence signal. Electronic transitions of an organic impurity can be approximated by a three-level system, including ground S_0 and excited S_1 singlet electronic states and the photoexcited triplet state T_1 (see Fig. 6.3). In a typical experiment, the molecule is illuminated with laser light in resonance with the $S_0 \rightarrow S_1$ transition and Stokes-shifted fluorescence emission is detected. After being excited in the S_1 state the molecule can either relax back to the ground state via fluorescence emission and internal conversion, or can be trapped in the triplet state via the intersystem crossing process. In the inset of the Fig. 6.3, the sublevels of the triplet state related to zero-field splitting are shown.

The dynamics of a molecule under coherent optical excitation can be described in terms of optical Bloch equations describing a pseudospin which performs Rabi oscillations between the ground and excited electronic states [32]. In the rotating wave approximation the evolution of the density vector for a three-level system is described by [32]

$$\begin{aligned} \dot{\sigma}_{11} &= k_{21}\sigma_{22} + k_{31}\sigma_{33} + \frac{i\Omega}{2}(\sigma_{21} - \sigma_{12}), \\ \dot{\sigma}_{22} &= -(k_{21} + k_{23})\sigma_{22} + \frac{i\Omega}{2}(\sigma_{12} - \sigma_{21}), \\ \dot{\sigma}_{33} &= k_{23}\sigma_{22} - k_{31}\sigma_{33}, \\ \dot{\sigma}_{12} &= -\frac{i\Omega}{2}\sigma_{11} + \frac{i\Omega}{2}\sigma_{22} + (i\Delta - \Gamma_2)\sigma_{12}, \end{aligned} \tag{6.1}$$

where a normalization condition for the populations holds: $\sigma_{11} + \sigma_{12} + \sigma_{33} = 1$. $\Gamma_2 = \frac{1}{T_2} = \frac{1}{2T_1} + \frac{1}{T_2^*}$ is the dephasing rate of the $S_0 \rightarrow S_1$ transition, T_1 and is the

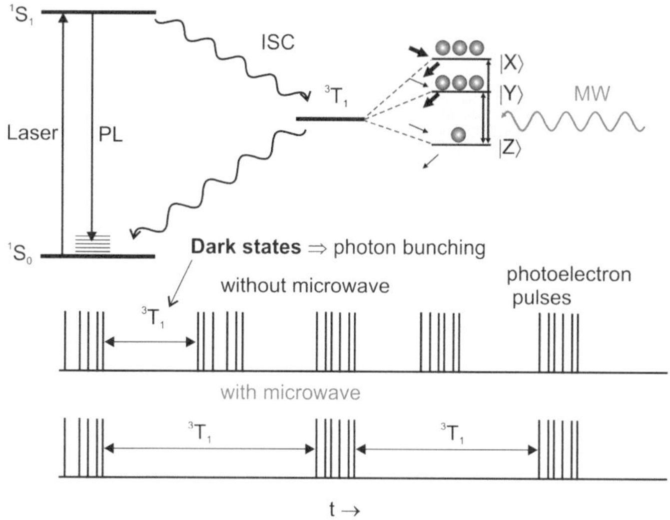

Fig. 6.3 The energy level scheme for a single organic molecule. The *inset* shows the fine structure sublevels of the triplet state

relaxation time, T_2^* is the pure dephasing time of the optical transition, Δ is the detuning of the excitation field from the frequency of the $S_0 \rightarrow S_1$ transition, k_{ij} are incoherent transition rates. Ω is the Rabi frequency of the resonant optical field: $\Omega = \frac{|\vec{\mu}\vec{E}|}{\hbar}$, where $\vec{\mu}$ is the transition dipole moment and $\vec{E}$ is the optical field. Here, level 3 is the metastable triplet level.

The Bloch equations can be solved analytically. The steady state solution $\dot{\sigma}_{11} = \dot{\sigma}_{22} = \dot{\sigma}_{33} = 0$ of the Bloch equations gives access to the steady state population of the excited state and the linewidth of the spectral line, corresponding to the $S_0 \rightarrow S_1$ transition. As was shown in [33], the linewidth and fluorescence emission rate of the molecule can be expressed as follows:

$$\Delta v(I) = \Delta v(0)\sqrt{1 + I/I_S}, \tag{6.2}$$

$$R(I) = \sigma_{22}\frac{1}{\tau_{Fl}}\phi_{Fl} = R_\infty \frac{I/I_S}{1+} I/I_S. \tag{6.3}$$

Here I is the excitation intensity, τ_{Fl} is the fluorescence lifetime, $\tau_{Fl} = 1/(k_{21} + k_{23})$, I_S is the saturation parameter, which can be expressed as

$$I_S = \frac{\varepsilon_0 c\hbar(k_{21} + \sum_i k_{23}^i)}{|\vec{\mu}|^2(2 + A)T_2}, \tag{6.4}$$

where $A = \sum_i k_{23}^i / k_{31}^i$, k_{23}^i and k_{31}^i are the intersystem crossing rates corresponding to the transitions to and from different fine structure sublevels of the photoexcited triplet state. Equations (6.3) and (6.4) can be combined to give the saturated emission rate:

$$R_\infty = \frac{\phi_{Fl}(k_{21} + \sum_i k_{23}^i)}{2 + A}. \tag{6.5}$$

The maximum emission rate of such a system is determined by the fluorescence quantum yield, but also by the rate of trapping to the metastable triplet state and by the rate of depopulation of this state. Hence the triplet state parameters play a crucial role in the choice of the system for single-molecule spectroscopy. For dibenzanthanthrene in a naphthalene matrix, $k_{23}^{x,y}$, k_{23}^z, $k_{31}^{x,y}$, k_{31}^z are 480, 5400, 30, 900 s^{-1}, respectively [34, 35]. The contribution of the triplet to saturation is small and the fully saturated signal is reduced by about 8% with respect to that of a pure two-level system. Note that the long living metastable state seriously affects the saturated signal even if the fluorescence quantum yield is high.

Figure 6.4 shows an experimental study of the linewidth of a single-molecule spectral line as a function of the excitation intensity. The results are in very good agreement with the expected power broadening law. The fit gave homogeneous widths of 25 MHz, with errors of a few megahertz. The inset to Fig. 6.4 shows an example of a single-molecule excitation line at weak and strong exciting intensity. The line is roughly Lorentzian and shows power broadening when the laser power is high.

The time distribution of photons emitted by a single molecule gives access to its internal photophysical processes. In order to describe the inhomogeneity of photons

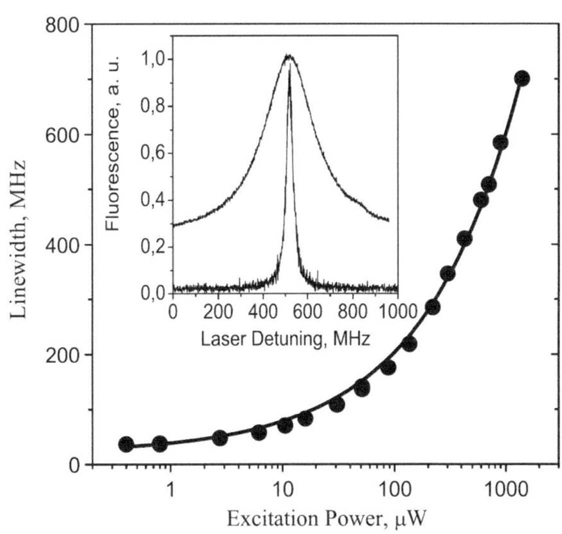

Fig. 6.4 The excitation power dependence of the excitation linewidth of a single dibenzoterrylene molecule in a naphthalene host at T = 1.6 K. The *inset* shows the fluorescence excitation lines recorded at high and low laser power

emitted by a single molecule, it is useful to introduce the second-order autocorrelation function $g^{(2)}(\tau)$, defined as follows:

$$g^{(2)}(\tau) = \langle I(t)I(t + \tau)\rangle / \langle I(t)\rangle^2, \tag{6.6}$$

where

$$\langle I(t)I(t+\tau)\rangle = \lim_{T\to\infty}\frac{1}{T}\int_0^T I(t)I(t+\tau)dt.$$

Here, $I(t)$ is the fluorescence intensity emitted by a single molecule at time t. As was shown in [36], the autocorrelation function can be deduced from the measurements of photocount pairs separated by a given time interval. The probability of detecting a pair of photons separated by an interval τ is proportional to the probability of finding the molecule in an excited state at time t and the probability that the molecule will be in the excited state at time $t + \tau$. There is also a connection between the autocorrelation function and the time dependent solution of the Bloch equations. The probability of finding the molecule in the excited state is proportional to the matrix element σ_{22}. The transient solution of the Bloch equation can be found by applying the Laplace transform technique. The solution for the correlation function with the triplet state contribution neglected can be expressed as follows:

$$g^{(2)}(\tau) = 1 - \exp(-\frac{(\Gamma_2 + k_{21})\tau}{2})(\frac{(\Gamma_2 + k_{21})}{2\Omega}\sin(\Omega\tau) + \cos(\Omega\tau)). \tag{6.7}$$

Here, Γ_2 is the dephasing rate of optical transition of the optical transition and k_{21} is the radiative decay rate of the excited state.

Figure 6.5 shows the results of a measurement of the correlation function for a single dibenzanthanthrene molecule isolated in a naphthalene host. There are

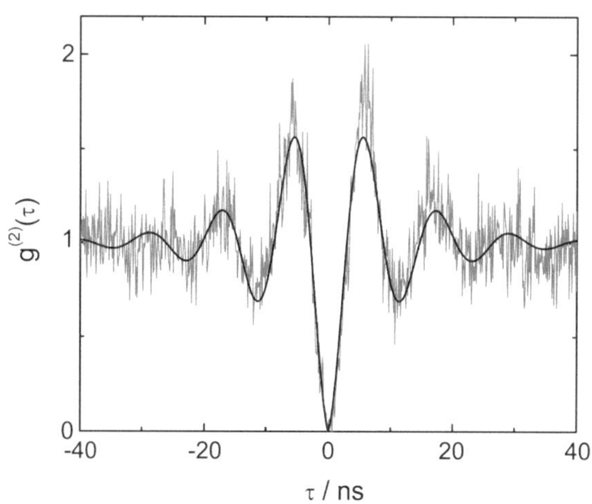

Fig. 6.5 The fluorescence intensity autocorrelation function for a single dibenzanthanthrene molecule isolated in a naphthalene host. The fit function follows (6.7)

several remarkable features visible from the correlation function. The zero-time value of the autocorrelation function tends to zero. Note that for the coherent light field, the zero-time value of $g^{(2)}$ is unity [37]. The value of the correlation function belowunity indicates the non-classical nature of the fluorescence emitted by a single molecule. The emission of photons by a single quantum system can be characterized by the so-called antibunching effect. The antibunching effect is related to projection type measurements, performed on a single quantum system. The observation of the first photon projects the system into the ground state. In order to emit the second photon the system must be excited again. The probability of emitting the second photon at time zero is zero because the system cannot emit a photon from the ground state. The situation will be different for a larger number of molecules. For an ensemble consisting of several molecules, there is a probability of obtaining a situation where more than two emitters are in the excited state. Therefore, there is a finite probability of simultaneous emission of photons and thus $g^{(2)}(0) > 0$. In general, the contrast is decreased by a factor of N, where N is the number of molecules.

It also can be seen from the Fig. 6.5 that the value of the correlation function increases to a higher value, showing damped oscillations. The Rabi oscillations correspond to the coherent evolution of resonantly driven two-level systems. The decay of the oscillations is related to a dephasing process, which occurs in the singlet excited state. The dephasing rate is mostly determined by the radiative decay in the temperature range between 1 and 10 K. At higher temperatures, the damping is stronger, because the pure dephasing processes related to electron–phonon interactions become active.

6.1.4 Magnetic Resonance of the Photoexcited Triplet States of Single Organic Molecules

Under continuous optical excitation, the average fluorescence emission is determined by the population and depopulation rates of the triplet state sublevels [see (6.5) and Fig. 6.3 (top panel)]. For the case of organic molecules the rates for the three sublevels differ significantly. This is the result of the high selectivity of the intersystem crossing process, which is related to the fact that the spin–orbit coupling can mix the singlet character only into specific triplet sublevels. Usually, the $|X\rangle$ and $|Y\rangle$ sublevels have a much higher probability of population than $|Z\rangle$ [Fig. 6.3 (inset)]. These two levels also have much shorter lifetimes. This creates a considerable population difference between the three triplet sublevels. Irradiation with microwaves resonant with either the $|X\rangle - |Z\rangle$ or $|Y\rangle - |Z\rangle$ transition leads to a redistribution of the population of the two levels involved in the resonance and hence to a change of the average lifetime of the triplet state. This in turn affects the population of the ground state and, since the system is excited continuously, leads to a change of the fluorescence intensity [Fig. 6.3 (lower panel)].

The Hamilton operator of the triplet system without a nuclear Zeeman terms is

$$\hat{H} = g_e \mu_B \hat{\vec{S}} \cdot \vec{B}_0 + \hat{\vec{S}} \cdot \overleftrightarrow{D} \cdot \hat{\vec{S}} + \sum_i \hat{\vec{S}} \cdot \overleftrightarrow{A}_i \cdot \hat{\vec{I}}_i, \tag{6.8}$$

where $\boldsymbol{S}$ is the electron spin operator (S = 1), $\boldsymbol{I}_i$ is the nuclear spin operator of the nucleus i, $\boldsymbol{D}$ is a fine structure tensor, μ_B is the Bohr magneton of the electron, g_e is the electron g-value, $\boldsymbol{B}_0$ is the external magnetic field, $\boldsymbol{A}_i$ is the hyperfine interaction tensor of nucleus i. The hyperfine interaction term includes all nuclei, intramolecular and intermolecular, coupled to the electron spin.

The first term, $g_e \mu_B \hat{\vec{S}} \cdot \vec{B}_0$, corresponds to the interaction of the electron spin with an externally applied magnetic field. The second term in the spin Hamiltonian, $\hat{\vec{S}} \cdot \overleftrightarrow{D} \cdot \hat{\vec{S}}$, leads to a zero-field splitting of the triplet state sublevels as a result of the (magnetic) dipole–dipole interaction of the two unpaired electron spins. The third part of the spin Hamiltonian describes the interaction of the electron spin with the surrounding nuclear spins.

The conventional ODMR technique can be applied in single-molecule studies when optical selection of single molecules is possible [21, 38–40]. Figure 6.6 shows the ODMR spectrum of a single pentacene molecule isolated in a para-terphenyl host at T = 1.6 K. The laser was tuned to the peak of the single-molecule fluorescence excitation line and the power was adjusted to saturate the optical $^1S_0 \rightarrow {}^1S_1$ transition. The fluorescence intensity was monitored as a function of the microwave frequency. The spectrum shows that, even in zero magnetic field, the triplet state of pentacene is split into the three zero-field eigenstates $|X\rangle$, $|Y\rangle$ and $|Z\rangle$. In Fig. 6.6 the $|Y\rangle - |Z\rangle$ and the $|X\rangle - |Z\rangle$ magnetic resonance transitions are observed as a decrease (up to 25%) of the fluorescence. This is caused by the increased population probability of the long lived $|Z\rangle$ level.

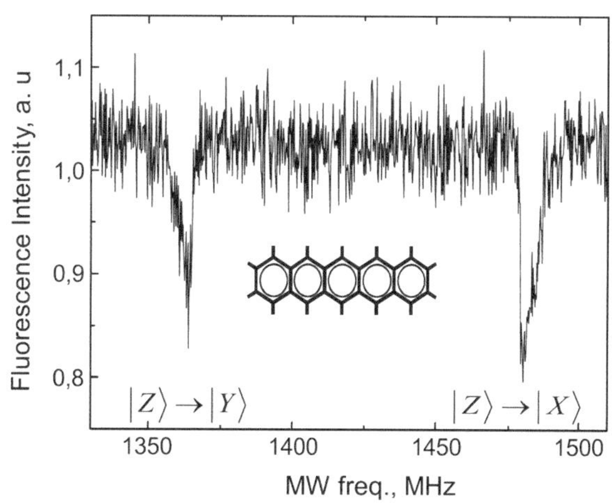

Fig. 6.6 The ODMR spectrum of a single pentacene molecule. The *inset* shows the chemical structure of pentacene

The third transition ($|X\rangle - |Y\rangle$) is much weaker due to unfavourable population and depopulation kinetics.

Note that the single-molecule ODMR lines show an asymmetric lineshape with a steep decrease towards higher microwave frequencies for both the single-molecule and the ensemble case. The lineshape results from the hyperfine interaction of the triplet electron spin with the pentacene proton spins ($I = 1/2$). Each proton can exist in one of its two nuclear spin states, which yields 2^{14} nuclear spin configurations.

The hyperfine interaction of each of these nuclear configurations causes a slight shift of the resonance. For a single molecule one would expect it to 'see' only one nuclear spin configuration and a very narrow magnetic resonance line to be observable. Apparently, the molecule experiences all of these configurations during the many optical pumping cycles which are needed to accumulate a sufficient signal-to-noise ratio. This is due to the dipolar coupling among the proton spins, which leads to a spin diffusion within the proton reservoir of the guest and the host. When the triplet magnetic moment is created, the 14 protons spins on the pentacene suddenly 'feel' the (second-order) hyperfine fields, which shift their resonance frequency away from the dipolar spectrum of the protons in the bulk of the crystal. Consequently, during the triplet lifetime, this configuration is frozen and the resonance frequency can only vary in a small interval $\Delta\nu$ determined by the flip-flop motions of the protons in the bulk. This interval can be estimated from the electron spin–spin relaxation time T_2 and amounts to $\Delta\nu = 1/(\pi T_2) \approx 150\,\text{kHz}$ [40, 41]. On return to the ground state, the hyperfine fields disappear and the pentacene protons are free to participate in the nuclear flip-flop motion. When the molecule is excited again into the triplet state, a new magnetic configuration is frozen, which corresponds to a different position of the zero-field resonance line. An estimate of the related timescales yields that the average time between two excitations into the triplet state is about 20 μs and that the mean residence time of the molecule in the triplet state is about 50 μs. For the inverse of the flip-flop rate one can estimate a value of about 30 μs which means that each time the molecule reappears in the triplet state it experiences a different nuclear configuration. Since some hundred thousand cycles are averaged for the spectrum in Fig. 6.4 the same linewidth is found as in ensemble experiments.

Single-molecule spectroscopy is performed by detection of fluorescence originating from strongly driven singlet–singlet transitions. This excitation–emission cycle is repeated millions of times per second in order to produce a high enough signal. Occasionally, if the single molecule is trapped in the metastable triplet state, the stream of emitted photons becomes interrupted. The fluorescence emission time trace of a single pentacene molecule is shown in Fig. 6.7. The length of the corresponding dark time interval of fluorescence emission is determined by the lifetime of the triplet state. Because the different triplet sublevels have different decay rates, the histogram of the dark time is determined by contributions of different triplet state sublevels. It was demonstrated that microwave-induced changes of the distribution of the dark intervals could be used to detect transitions between triplet sublevels. By synchronizing resonant microwave pulses with the quantum jumps of a terrylene molecule, the high contrast transient ODMR signal was detected [42].

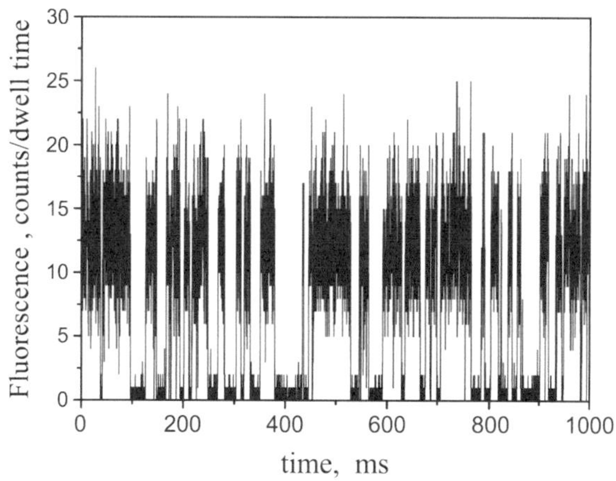

Fig. 6.7 The fluorescence signal from a single terrylene molecule

Although a number of classical experiments have been performed on the photoexcited triplet states of single molecules, including transient nutations, Hahn echo and electron–nuclear double-resonance (ENDOR) studies [20, 21, 40, 43, 44], these systems always require a time averaged read-out. When a photodetector records no counts, which corresponds to the beginning of the dark interval (see Figs. 6.3 and 6.7), it is impossible to determine in which sublevel of the triplet state the molecule is trapped. In order to obtain this information, it is necessary to know the duration of the dark interval and one needs to wait a while before the next photon burst is detected. But at this time, the molecule is no longer in the triplet state. Therefore, the measurement of the spin state always takes longer than the T_1 time of the spin state. On the other hand, quantum computing requires read-out schemes which are able to determine the spin state within the T_1 time. This was demonstrated for the nitrogen-vacancy (NV) defect centre in diamond, which is a system with a paramagnetic ground state. The NV centre in diamond will be discussed in the next section.

6.1.5 Conclusions and Outlook

The accurate measurement of a single spin state has two important aspects. First, single-spin magnetic resonance is a central point for any pure state based quantum computing scheme. Several promising techniques are currently under investigation. Recently, controlled electron spin injection and single-spin detection were demonstrated using electrical read-out in quantum dots [45, 46] and scanning tunnelling microscopy of organic molecules [47]. Important progress has been achieved in the field of magnetic resonance force microscopy [48, 49], which recently showed detection sensitivity of two electron spins [50]. Yet optical detection remains a unique technique, capable of demonstrating coherent EPR and

NMR in experiments on single quantum systems [51]. The next step will be to show coupling between several spins. This will allow achievement of two-qubit gates, which are basic elements for quantum computing.

The second important field is of more fundamental character. Experiments with single spins are suitable for experimental testing of quantum mechanics. Projective spin measurements on single quantum systems can be used in tests of the quantum Zeno effect and Bell's inequalities.

6.2 Single Spins in Diamond: Novel Quantum Devices and Atomic Sensors

6.2.1 Introduction

Coherent control and readout of single spin solids is attracting considerable attention owing to potential application of atomic scale technologies in novel information processing protocols and sensing at nanoscale [52]. Spins are particularly interesting candidates for achieving such control owing to the long coherence time. This isolation from the environment makes the readout of individual spins quite challenging. It was shown that magnetic moment associated with single quantum systems can be detected in transport measurements with single quantum dots [53] and defects in silicon [54]. Magnetic resonance force microscopy was also able to reach ultimate sensitivity regime [55]. Recently developed optical readout technique explore new avenue in single spin detection by combining high sensitivity of optical microscopy techniques and conventional magnetic resonance control methods. Following detection of single molecular spin [20, 21], quantum states associated with single defects in diamond were explored [24]. Experiments with single defects in diamond are particularly interesting owing to long coherence time associated with spins of colour centres. This chapter shows the basics of single spin detection and highlight importance of coherent control of spins in solid for novel quantum technologies.

6.2.2 NV Defects in Diamond

Diamond is exceptional material for technological applications including novel technologies that use quantum properties of matter. Apart from its extreme hardness and high thermal conductivity, diamond lattice is hosting a large number of optically active defects (colour centres). Owing to their importance for jeweler applications, many of these defects were studied using optical spectroscopy and magnetic resonance techniques since decades [56]. Several colour centres show strong optical transitions allowing to detect them at single site level using optical

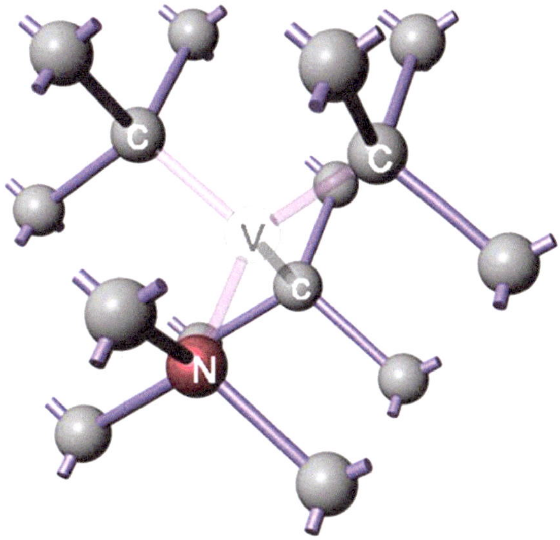

Fig. 6.8 Structure of NV centre in diamond

microscopy techniques [24]. The most studied defects in this context are nitrogen-vacancy (NV) defect. NV defect consists of nitrogen atom at the lattice site (see Fig. 6.8) and vacancy at the neighbor lattice position.

The structure of the defects shows C_{3v} symmetry with symmetry axe connecting nitrogen atom and vacancy. Electronic structure of NV centres is governed by dangling bonds electron belonging to vacancy and electrons of nitrogen atom. Neutral charge state of NV centre possesses 5 electrons (four form the dangling bonds and one additional electron from nitrogen). When electron donors are present in the lattice, NV centre can exist in negatively charged state having six electrons [57]. The negatively charged state possesses electronic spin 1 whereas neutral state is $S = 1/2$ system. The electronic states of negatively charged NV defect can be modeled using six electrons or two holes models [58]. The spin density of the ground state of NV defect shows that spin density is mostly concentrated at three carbon atoms surrounding vacancy [59]. In the excited state there is a significant shift of the electron spin density towards nitrogen atoms resulting in a stronger hyperfine coupling to ^{14}N nucleus.

NV defects can be found in natural nitrogen containing diamond and in synthetic diamonds grown by chemical vapor deposition (CVD) and high pressure high temperature (HPHT) techniques. It was also shown that single defects can be created artificially in synthetic diamonds using electron irradiation [60] and nitrogen implantation [61–63]. First approach relies on existence of substitutional nitrogen in the diamond lattice and diffusion of vacancies created by electron irradiation during annealing. The nitrogen vacancy centre in diamond is traditionally observed in radiation damaged nitrogen rich diamond. The centre is formed from the vacancy after annealing with temperatures larger than 600 °C.

The vacancy gets mobile and forms a stable NV complex under these conditions. Therefore position of implanted NV defect cannot be controlled with high precision. However this technique allow generation of dense ensemble of NV defects [64]. Well controlled generation of NV colour centres has been achieved by this approach [60]. Electron (400 keV) and Ga (30 keV) ion beams were used to generate localized areas of NV centres in Ib diamond. For 30 keV Ga ions the nominal penetration depth of ions inside the material is 15 nm. Patterns of NV centres have been generated with some ten thousand Ga ions used per irradiated dot. From the experiments the diffusion constant of vacancies in diamond has been determined to be D = 1.1 $(nm^2)/s$ The activation energy for vacancy diffusion is calculated to be 2.4 eV. Electrons with 400 keV penetrate some μm inside the diamond sample. At high irradiation dose an increased generation of NV^0 centres was observed. For a localized generation of NV defects these approaches do have the disadvantage of the relative high diffusion constant of the vacancy plus the large "natural" abundance of nitrogen. For generating NV defects with large spin dephasing times for example it would be preferential to implant defects into nitrogen free samples. This is possible by implanting nitrogen directly into type IIa diamonds. In a first attempt 2 meV nitrogen atoms have been implanted into type IIa diamond substrates [61] [Fig. 6.9 (left)]. STRIM calculations suggest that the ions should be found 1 μm below the surface. The lateral scattering in the end position of the nitrogen (straggling) should be 0.5 μm. With a displacement energy of 55 eV for carbon and a density of 3.5 g/cm^3 about 200 vacancies should be produced during a single nitrogen implantation. Indeed the optical spectra of the implantation areas do show mostly fluorescence emission from neutral vacancies prior to annealing. After annealing, fluorescence from mostly NV centres is found. In these studies the number of nitrogens implanted per spot has been decreased gradually. Single centre emission was observed when on average two ions are implanted in a single spot [see Fig. 6.9 (right)]. It is however difficult to ensure that the generated NV indeed is made from the nitrogen implanted and not from an abundant one. Even with a nitrogen concentration below 0.1 ppm there would be 30 native nitrogen atoms in a spherical volume of 150 nm surrounding the end range where the Bragg peak in the stopping power creates the maximum concentration of vacancies. Hence there is a considerable chance in Fig. 6.9 that the NV

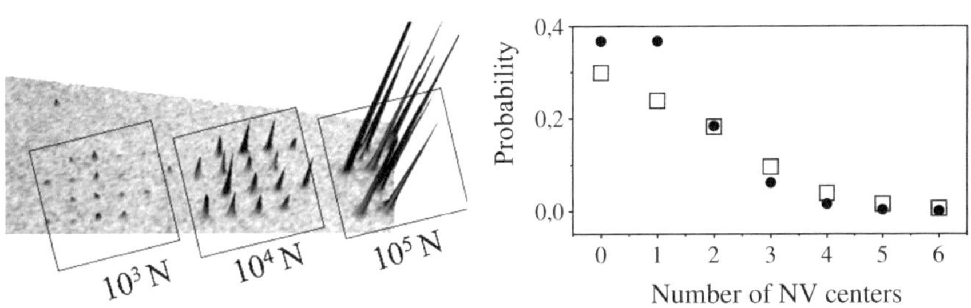

Fig. 6.9 (*Left*) Fluorescence image of a type IIa diamond irradiated with a N^+ ion micro beam at different nitrogen dosages. (*Right*) Probability to generate the given number of defects for a deposition of two nitrogen ions per spot. *Open squares* are measured data and *filled circles* are calculated values

fluorescence does not stem from an implanted nitrogen. To be able to separate implanted from impurity nitrogen, ^{15}N has been used as an implantation ion. ^{15}N has a nuclear spin angular momentum I = ½ with a characteristically different ODMR spectrum from ^{14}N [65]. Experimentally 14 keV $^{15}N_2^+$ ions have been implanted. From a comparison of the number ^{15}NV defect centres to the number of implanted ^{15}N the efficiency of NV defect generation has been calculated. Under the conditions employed in the experiment this efficiency was about 2.5%. It should be noted however, that a 14 keV implantation of N_2 results in a penetration depth of only a few nm. This close proximity to the surface might lead to a loss of vacancies due to diffusion to the surface. In any case the experiments demonstrate that single defects can be generated close to a diamond surface.

Implantation of single ions into nitrogen free crystals allows to reach positioning accuracy solely limited by straggling of nitrogen ions during implantation. Straggling can be controlled by appropriate choice of energy. Low energy is preferable for achieving high accuracy of implantation, but the yield of creation of NV defects is lower in this case owing to low number of created vacancies [65]. It was also shown that post irradiation of diamond with carbon ions allow to improve creation yield of nitrogen-vacancy centres [66]. Figure 6.10 shows pattern of implanted NV centres visualized by confocal microscopy.

6.2.3 *Optical Properties of NV Defects*

Excited state of NV centre is located 1.945 eV above ground state and is also spin triplet. In addition to the triplet ground and excited states negatively charged NV

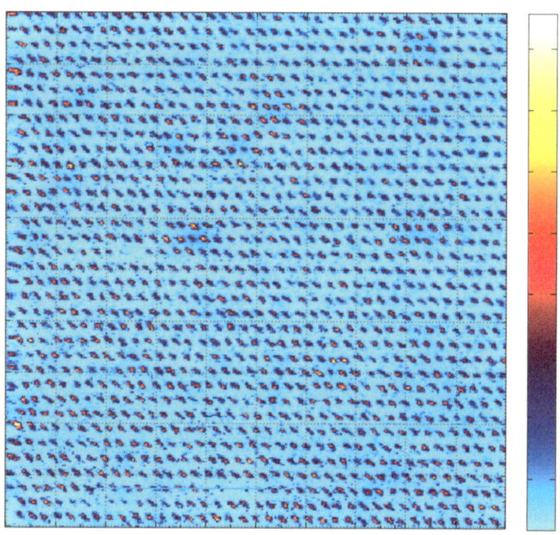

Fig. 6.10 Pattern of implanted NV centres visualized by confocal microscopy

centre has two metastable singlet states playing crucial role for optical spin polarization [67, 68]. The metastable singlet states can be detected in fluorescence as infrared emission [69]. The strongest optical transition for NV defect is related to the transition between ground and excited state triplets. The fluorescence lifetime of the excited triplet state is about 12 ns and fluorescence quantum yield is close to unity [70]. This allows detection of single colour centres by conventional confocal microscopy techniques [24].

The NV defect gives rise to a strong absorption at 1.945 eV (637 nm). At low temperature the absorption is marked by a narrow optical resonance line (zero phonon line) followed by prominent vibronic side bands.

Optical transition of NV centre is coupling to phonons resulting in the strong emission into phonon sidebands (intensity of the zero phonon line is only 4%) (see Fig. 6.11). Early ensemble studies show the strong inhomogeneous broadening of optical transition (the linewidth of zero phonon line of 1000 GHz) [71].

6.2.3.1 Single Defect Centre Experiments

Experiments on single quantum systems in solids have brought about a considerable improvement in the understanding of the dynamics and energetic structure of the respective materials. In addition a number of quantum optical phenomena, especially when light–matter coupling is concerned, have been investigated. As opposed to atomic systems on which first experiments on single quantum systems are well established, similar experiments with impurity atoms in solids remain challenging. Single quantum systems in solids usually strongly interact with their environment. This has technical as well as physical consequences. First of all single

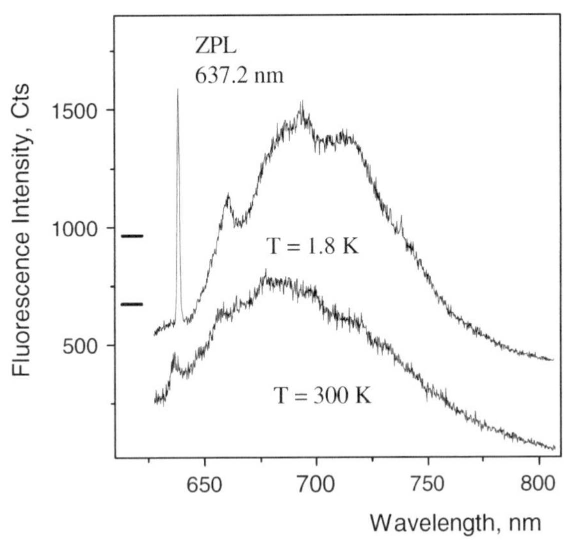

Fig. 6.11 Fluorescence emission spectra of single NV centres at room temperature and LHe temperatures. Excitation wavelength was 514 nm

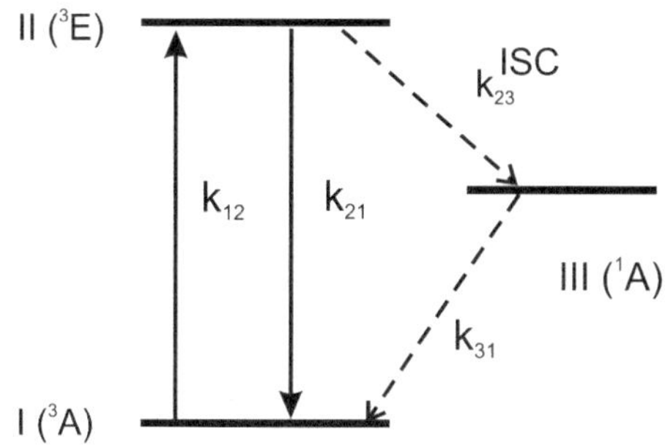

Fig. 6.12 Simplified three level scheme describing the optical excitation and emission cycle of colour centres (e.g. the NV centre where ^{3}A and ^{3}E are the triplet ground and excited state, ^{1}A is a metastable singlet state), *ISC* is intersystem crossing transitions. The *arrows* and k_{ij} denote the rates of transition among the various states. Allowed optical transitions between ground state and excited state are shown by *solid line*, nonradiative transitions are shown by *dashed lines*

solid state quantum systems are embedded in an environment which, for example, scatters excitation light. Given a diffraction limited focal volume usually the number of matrix atoms exceed those of the quantum systems by 10^6–10^8. This puts an upper limit on the impurity content of the matrix or on the efficiency of inelastic scattering processes like e.g. Raman scattering from the matrix. Various systems like single hydrocarbon molecules, proteins, quantum dots and defect centres have been analysed [72]. Except for some experiments on surface enhanced Raman scattering the technique usually relies on fluorescence emission. In this technique an excitation laser in resonance with a strongly allowed optical transition of the system is used to populate the optically excited state (e.g. the ^{3}E state for the NV centre), see Fig. 6.12.

Depending on the fluorescence emission quantum yield the system either decays via fluorescence emission or non-radiatively, e.g. via inter-system-crossing (ISC) to a metastable state (^{1}A in the case of the NV). The maximum numbers of photons emitted are given when the optical transition is saturated. In this case the maximum fluorescence intensity is given as

$$I_{max} = [k_{31}(k_{21} + k_{23})\Phi_F]/(2k_{31} + k_{23}). \tag{6.9}$$

Here k_{31} is the relaxation rate from the metastable to the ground state and k_{21} is the decay rate of the optically excited state, k_{23} is the decay rate to the metastable state and Φ_F marks the fluorescence quantum yield. For the NV centre I_{max} is about 10^7 photon/s. I_{max} critically depends on a number of parameters. First of all the fluorescence quantum yield limits the maximum emission. The maximum observable emission rate from the NV centre is around 10^5 photons/s which corresponds well to the value estimated above, if we assume a detection efficiency of 0.01. Single NV centres can be observed by standard confocal fluorescence microscopy in type Ib diamond. In confocal microscopy a laser beam is focussed onto a diffraction limited spot in the diamond sample and the fluorescence is collected

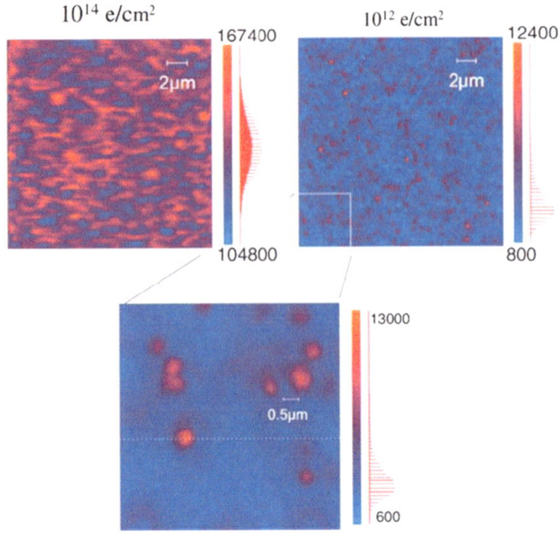

Fig. 6.13 Confocal fluorescence image of various diamond samples with different electron irradiation dosages

from that spot. Hence the focal probe volume is diffraction limited with a volume of roughly 1 μm^3. In order to be able to detect single centres it is thus important to control the density of defects. For the NV centre this is done by varying the number of vacancies created in the sample by e.g. choosing an appropriate dose of electron irradiation. Figure 6.13 shows confocal fluorescence image of various diamond samples with different electron irradiation dosages. Hence the number of NV centres depends on the number of vacancies created and the number of nitrogen atoms in the sample.

Figure 6.13 shows an image of a diamond sample for low electron irradiation dosage (10^{12} e/cm^2) where the number of defects in the sample is low enough to detect the fluorescence from single colour centres [24, 70]. As expected the image shows diffraction limited spots. From the image alone it cannot be concluded whether the fluorescence stems from a single quantum system or from aggregates of defects. To determine the number of independent emitters in the focal volume the emission statistics of the NV centre fluorescence can be used. The fluorescence photon number statistics of a single quantum mechanical two-level system deviates from a classical Poissonian distribution. If one records the fluorescence intensity autocorrelation function

$$g^2(\tau) = \langle I(t)I(t + \tau)\rangle / \langle I(t)\rangle^2, \tag{6.10}$$

for short time τ one finds $g^2(0) = 0$ if the emission stems from a single defect centre (see Fig. 6.14).

This is due to the fact that the defect has to be excited first before it can emit a single photon. Hence a single defect never emits two fluorescence photons simultaneously, in contrast to the case when a number of independent emitters are

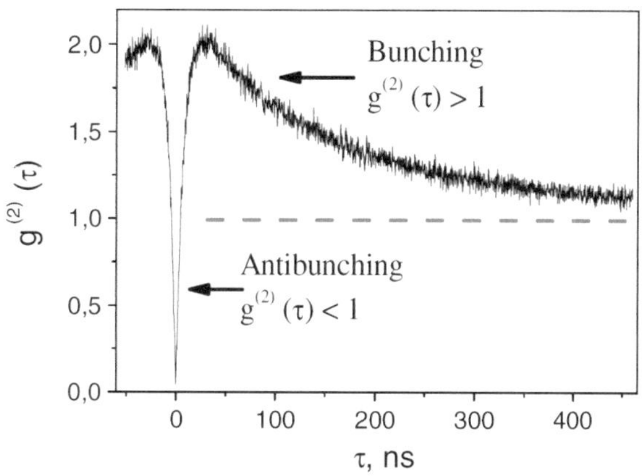

Fig. 6.14 Fluorescence intensity autocorrelation function of a single NV defect at room temperature

excited at random. If one adopts the three level scheme from Fig. 6.12, rate equations for temporal changes of populations in the three levels can be set up. The solution of the equations reproduce the dip in the correlation function $g^2(\tau)$ for $\tau \rightarrow 0$ shown in Fig. 6.14, which indicates that the light detected originates from a single NV [24, 70]. The slope of the curve around $\tau = 0$ is determinded by the pumping power of the laser k_{12} and the decay rate k_{21}. For larger times τ a decay of the correlation function becomes visible. This decay marks the ISC process from the excited triplet ^3E to the metastable singlet state ^1A. Besides the spin quantum jumps detected at low temperature the photon statistics measurements are the best indication for detection of single centres. It should be noted that the radiative decay time depends on the refractive index of the surrounding medium as $1/n_{medium}$. Because n_{medium} of diamond is 2.4 the decay time should increase significantly if the refractive index of the surrounding is reduced. This is indeed observed for NV centres in diamond nanocrystals [73]. It should benoted, that owing to their stability single defect centres in diamond are prime candidates for single photon sources under ambient conditions. Such sources are important for linear optics quantum computing and quantum cryptography. Indeed quantum key distribution has been successful with fluorescence emission from single defect centres [74]. A major figure of merit for single photon sources is the signal to background ratio, given (e.g.) by the amplitude of the correlation function at $\tau = 0$. This ratio should be as high as possible to ensure that a single bit of information is encoded in a single photon only. Unfortunately the NV centre has a broad emission range which does not allow efficient filtering of background signals. Besides application in single photon generation, photon statistical measurements also allow to derive conclusions on photoionization and photochromism of single defects. Most notably the NV centre is speculated to exist in two charge forms, the negatively charged NV with zero phonon absorption at 637 nm and the neutral from NV^0 with absorption around 575 nm [74, 75]. Although evidence existed that both absorption lines stem from the same defect no direct charge interconversion has been shown in bulk experiments. The best example for a spectroscopically resolved charge transfer in diamond is the vacancy, which exists in two stable charge states. In order to observe the charge transfer from NV to NV^0 photon statistical measurements similar to the ones described have been carried out, except for a splitting of photons depending on the emission wavelength [75, 76]. The two channel set up allows to detect the emission of NV^0 in one and NV in another detector arm. For delay time $\tau = 0$, $g^2(\tau)$ shows a dip, indicating the sub-Poissonian statistics of the light emitted. Hence it must be concluded that there is a continuous interconversion between the two spectral positions. Detailed time resolved experiments show that switching from NV^0 to NV^- is photoinduced whereas the reverse process NV^- to NV^0 occurs under dark conditions with a time constant between 0.3 and 3.6 μs.

Single centre experiments show that the line of individual color centres are narrow [77]. The linewidth of the transition of single centres strongly depends on the sample quality. In nitrogen-rich diamonds(type 1b) the linewidth is typically a few hundreds of MHz. The main source of spectral diffusion is most probably ionization process related to nitrogen donors (ionization energy 1.8 eV) [77]. For the case of ultrapure diamonds with concentration of nitrogen impurities on the order of ppb the linewidth approaches limit imposed solely by the lifetime of the excited state [78] (see Fig. 6.15).

The linewidth of the optical transition is significantly narrower than the splitting between spin state sublevels the NV centre. This opens the door towards manipulation and readout of the spin states using optical spectroscopy techniques. Spin orbit coupling lead to the mixing of the spin state in the excited states. As the results one of the sublevels of the excited state form lambda scheme with two ground state spins sublevels. The existence of this lambda type transition allows for observation of the electromagnetically induced transparency [79] non-destructive readout of the

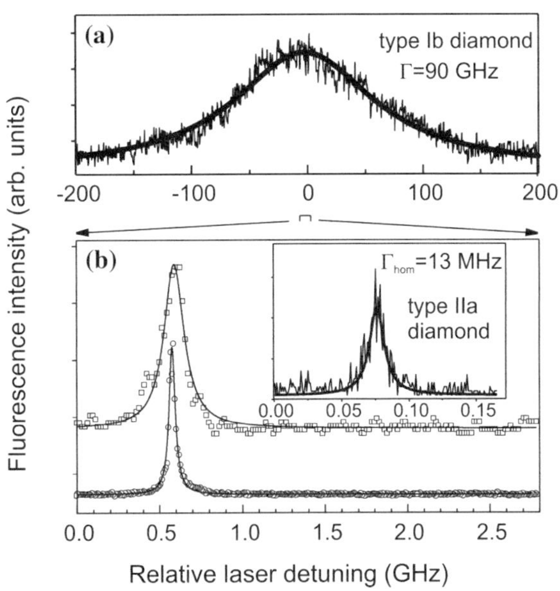

Fig. 6.15 Low temperature fluorescence excitation spectra of single NV defects. **a** Spectrum of single NV centre in diamond nanorystal (type 1b diamond) averaged over 5 min. **b** Spectra averaged during a few seconds. *Upper graph* shows spectrum of NV centre in type 1b diamond. *Lower graph* shows spectrum of NV defect in ultrapure type IIa diamond. The *inset* shows single scan spectrum obtained with low excitation power. The linewidth of NV spectrum approaches transform-limited value (12 MHz)

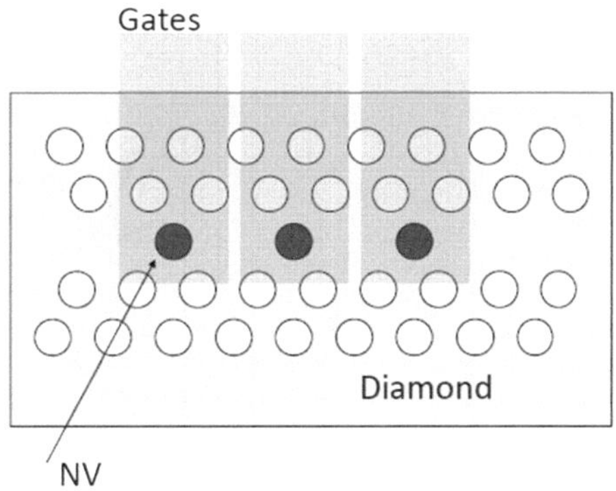

Fig. 6.16 Schematics of NV color centres coupled to photonic crystal cavity

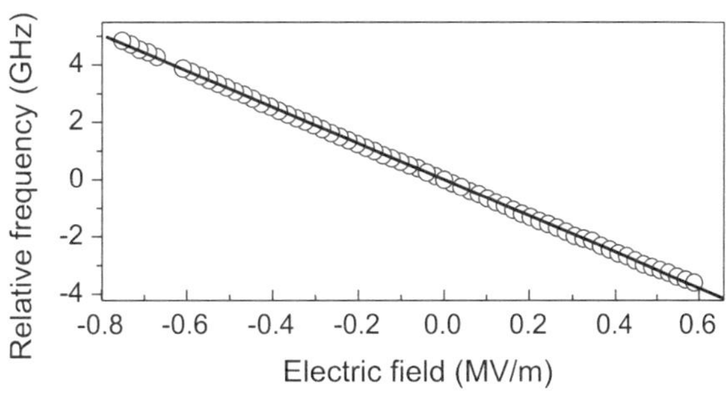

Fig. 6.17 Stark shift control of the NV centre spectral lines

spin state [80] and demonstration of entanglement between the photon and the spin states [81]. Intrinsic coupling between spins and photon are crucial for application of NV centres in the area of quantum communication.

Single defects placed in photonic structure can provide basis blocks for quantum repeaters and scalable quantum registers (Fig. 6.16). It is important to mention that intrinsic inhomogeneity present in diamond lattice might appear to be serious obstacle for above mentioned protocols. Important progress has been made towards control of optical transition using electric fields. The possibility of tuning was demonstrated experimentally within spectral range covering the whole inhomogeneous bandwidth (see Fig. 6.17) [78, 82]. Another important element of diamond nanophotonics platform is the ability to collect the photons efficiently. This can be achieved by building optical elements around single defects. Integrated diamond optics including solid immersion lenses and optical resonators is undergoing fast development period. It was shown that by using solid immersion lenses it is possible to increase collection efficiency of detection system by order of magnitude [83–85]. Furthermore, efficient extraction of photons made possible to show the effect of photon interference for two individual colour centres [86] (crucial blocks for quantum information processing protocols) and achieve single shot readout of single electron spins [87].

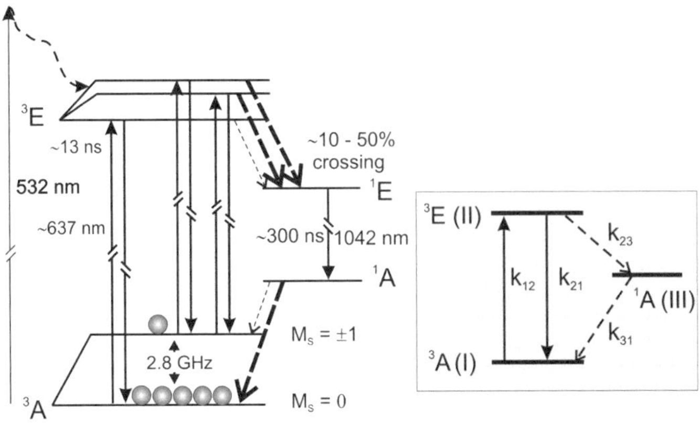

Fig. 6.18 Detailed energy level diagram of NV centre. Allowed optical transitions between triplet ground state (^{3}A) and triplet excited state (^{3}E) are shown by *solid line*. Nonradiative transitions are shown by *dashed lines*. The strength of the spin-selective intersystem crossing transitions is encoded by the thickness of the *arrows*. (*Inset*) Simplified three level scheme describing the optical excitation and emission cycle of single NV centres (see also Fig. 6.12). ^{3}A and ^{3}E are the triplet ground and excited state, ^{1}A is a metastable singlet state. The *arrows* and k_{ij} denote the rates of transition among the various states

6.2.4 Spin Properties and Spin Readout

Remarkable photophysical properties of NV defects offer the possibility for readout and manipulating of spin [88]. When combined with ultra sensitive optical detection technique, this provides the basis for single spin control [24]. Notably, optically assisted spin readout technique allow for efficient spin defection at room temperature. Optical transitions between the ground ^{3}A and excited ^{3}E states (see Fig. 6.18), under ambient conditions are strongly broadened and spectral selection of individual spin sublevels is not possible. However owing to spin orbit coupling, spin sublevels with magnetic quantum number $M_S = \pm 1$ have higher probabilities to undergo intersystem crossing to meta stable singlet states (^{1}E and ^{1}A in Fig. 6.13) [89, 90]. As the result fluorescence emission of NV centre is higher for $M_S = 0$ spin sublevel. In addition, intersystem crossing allows for efficient spin polarization owing to spin-selective decay of the metastable singlet state to $M_S = 0$ sublevel of the ground state triplet.

First experiments show that combination of optical readout and conventional magnetic resonance technique allows for reliable state control of individual spins [91]. Owing to spin free diamond lattice (the most abundant carbon isotope is nuclear spin free) long coherence times were detected in diamond with low concentration of paramagnetic impurities. It was also shown that rare ^{13}C nuclear spins can be explored as the resource for quantum information protocols. Hyperfine interaction between electron spins and nuclear spins located in so called "frozen core" make possible realization of two and three qubit gates [51, 92]. Nuclear spins located outside of the frozen core are source of decoherence. It was shown that growth of isotopically pure ^{12}C diamond allow to extend coherence time of single electron spins associated with NV centres to the values close to spin lattice relaxation time (2 ms) [93].

6.2.4.1 Spin Physics of Single Defects

The controlled generation of quantum states from individual quantum objects is a current research topic which receives considerable interest during the last decade. In part this is motivated by possible applications in quantum information processing. On the other hand, the simulation of quantum systems itself, for example to investigate the physics of quantum phase transitions, is of interest. Basically the control of a wave function of a collective quantum state requires control over the quantum state of interacting qubits: |

$$\psi = \sum_{i}^{N} a_i|\alpha_i\rangle. \tag{6.11}$$

The evolution of ψ is subject to unitary transformations:

$$\psi' = U\psi$$

In general it is necessary to be able to manipulate coherently each individual qubit and control the strength of interaction among them. This puts certain restrictions on the system parameters. To allow for nontrivial unitary operations U a certain phase coherence time together with interaction strength and speed of operation is required. While the interaction strength and control speed of individual qubits are limited by technical means, the dephasing times in solids are usually short. Spins are certainly among the most promising systems owing to long coherence times together with availability of fast control of individual qubits and relatively strong spin–spin coupling. Although such robust control of spin states plus adjustment of spin–spin interactions are common practice in electron and nuclear magnetic resonance, the measurement of single spin states is a fierce experimental challenge. Only a few solid state systems currently allow for single spin state detection. Most notably single spin state measurements have been successful in III–V quantum dots and in P centre defects in silicon single electron transistor (SET) structures. A system where single spin control and state measurement are well developed is the NV centre using optical technique. It is remarkable that for spins associated with defects in diamond, phase memory times can be long even under ambient conditions. As an example the electron spin lattice time is reported to be 1.8 ms at T = 300 K. The long dephasing times are attributed to the low phonon density of states in diamond even at room temperature. In the NV centre the spin state is detected via fluorescence. As discussed above the fluorescence intensity I_{max} depends on the spin state via the ISC rate k_{23}. Upon changing

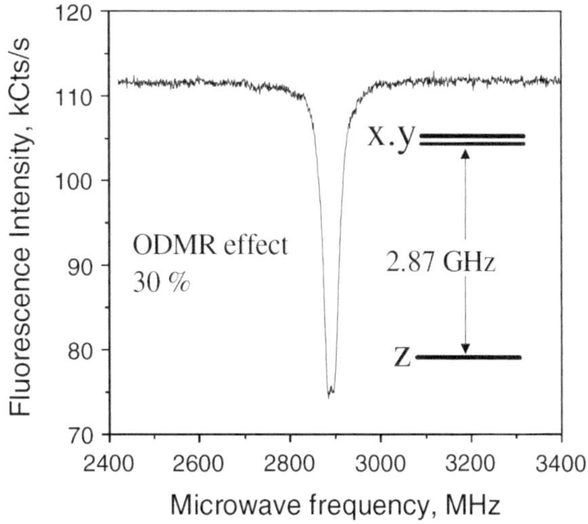

Fig. 6.19 Optically detected magnetic resonance (*ODMR*) spectrum of a single defect. The spectrum has been recorded at room temperature with 514 nm irradiation without an external B_0

the spin state this rate is changed from some kHz by more than three orders of magnitude towards some MHZ. Given the other parameters this results in a change of roughly 30% of I_{max}. Taking into account the photon shot noise and an average I_{max} of 10^5 photocounts per second, this change in fluorescence intensity can be detected with some ms averaging time.

Figure 6.19 shows an example of an optically detected magnetic resonance (ODMR) spectrum of a single NV defect. The general spin Hamiltonian describing the NV-defects spectrum is

$$\hat{H} = \mu_B \vec{B} \cdot \overleftrightarrow{g} \cdot \hat{\vec{S}} + \hat{\vec{S}} \cdot \overleftrightarrow{D} \cdot \hat{\vec{S}} + \hat{\vec{S}} \cdot \overleftrightarrow{A} \cdot \hat{\vec{I}} - g_I \mu_N \vec{B} \cdot \hat{\vec{I}} + \hat{\vec{I}} \cdot \overleftrightarrow{Q} \cdot \hat{\vec{I}}, \tag{6.12}$$

where all the terms have their usual meaning, S = 1, μ_B is the Bohr magneton, g_I is the nitrogen nuclear g-factor, μ_N is the nuclear magneton. The following terms are presented in the spin Hamiltonian: electron Zeeman interaction where $\overleftrightarrow{g}$ is the g-tensor, fine-structure interaction where $\overleftrightarrow{D}$ is the fine-structure interaction tensor, hyperfine interaction with N nucleus where $\overleftrightarrow{A}$ is the hyperfine structure coupling tensor, nuclear Zeeman interaction and the nuclear quadrupole interaction (only for ^{14}N, $I = 1$) where $\overleftrightarrow{Q}$ is the quadrupole-interaction tensor.

In general additional terms should be added to the spin Hamiltonian that describe hyperfine and nuclear Zeeman interaction with ^{13}C nuclei in the different shells near NV-center. From the spin Hamiltonian (6.12) one sees that some interactions are magnetic field dependent (the electron and nuclear Zeeman interactions), while others are not (the fine structure interaction, the hyperfine interaction, the nuclear quadrupole interaction). In order to separate these interactions from each other it will be often necessary to make EPR measurements with using various magnetic field/frequency settings including zero magnetic field. Since electronic g-factor of NV defect is almost isotropic the spin Hamiltonian can be simplified

$$\hat{H} = \mu_B g \vec{B} \cdot \hat{\vec{S}} + D[\hat{S}_z^2 - \frac{1}{3}S(S+1)] + E(\hat{S}_x^2 - \hat{S}_y^2) + \hat{\vec{S}} \cdot \overleftrightarrow{A} \cdot \hat{\vec{I}} - g_I \mu_N \vec{B} \cdot \hat{\vec{I}} + \hat{\vec{I}} \cdot \overleftrightarrow{Q} \cdot \hat{\vec{I}}, \tag{6.13}$$

where g is electronic g-factor of the NV centre (g = 2.003), D and E are the fine structure splitting parameters.

The spectrum in Fig. 6.19 has been taken without an external magnetic field. Hence only the fine structure and hyperfine term need to be considered. As mentioned above, due to the C_{3v} symmetry two of the three spin sublevels are degenerated (E = 0). Hence, only a single ODMR line is seen in the spectrum. Upon application of a magnetic field the two degenerated levels split and two lines become visible. The hyperfine coupling to the ^{14}N nucleus is not resolved in these spectra because of the large optical pumping rate used [94]. The spin density of the ground state electron spin wavefunction at the nitrogen nuclei is too low (2%) such that the hyperfine coupling to a ^{14}N nuclear is only around 2 MHz. Since the hyperfine and quadrupole coupling constants are about 2 and 5 MHz only, the corresponding splitting are easily masked by the homogeneous transition line width. For the NV centre this line width depends on the optical excitation intensity, since at least one of the levels is optically excited to the ^3E state. Since the optical Rabi frequency easily achieves some MHz, the line width correspondingly gets broad. This splitting and the concomitant nuclear quadrupole splitting is only resolved in a CW ODMR experiment when low laser and microwave excitation intensities are chosen. Such a well resolved spectrum is shown in Fig. 6.13. Three lines are visible, as expected for nuclei with I = 1 nuclear spin (^{14}N) angular momenta. The right panel in Fig. 6.20 shows the relevant energy level diagram together with the allowed transitions, marked by arrows.

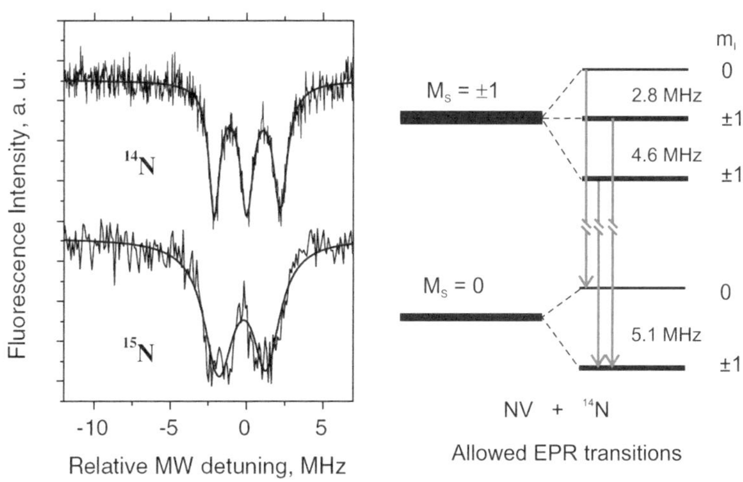

Fig. 6.20 (*Left*) ODMR spectra of NV centres with resolved ^{14}N and ^{15}N hyperfine structure. The spectra have been measured at room temperature without the application of an external magnetic field. (*Right*) Energy level schemes for the NV colour centre showing the hyperfine coupling (for ^{14}N) in the ground state spin substructure

The hyperfine structure in Fig. 6.20 has been analysed in detail [94] and corresponds to the known value of the hyperfine and quadruple coupling of the ^{14}N nucleus of the NV centre. It should be noted that these spectra provide an opportunity to verify the mechanism by which the defect has been generated. There are two mechanisms by which NV centres can be created in diamonds. First, vacancies are generated and the intrinsic nitrogen present in the material is used to create NV centres. Alternatively, the nitrogen atoms are implanted in nitrogen-free diamond and the vacancies which are generated during the implantation form NV defects. To ensure that a defect centre originates from an implanted nitrogen, ^{15}N isotope, which has a natural abundance of only 0.1% and is a I = 1/2 nucleus, can be used. A corresponding ODMR spectrum is shown in Fig. 6.20 and is clearly different from the ^{14}N case.

In [95] nitrogen g-factor, hyperfine, quadrupole (for $^{14}NV^-$) parameters for the NV^- defect were determined at room temperature with high precision. All interactions are axially symmetric about the C_{3v} ($\langle 111 \rangle$) NV symmetry axis. In this work the following values of the spin Hamiltonian parameters were measured.

A small anisotropic component of the electronic Zeeman interaction was found [95] to be $g_{\|} = 2.0029(2)$ and $g_{\perp} = 2.0021(2)$, where $g_{\|}$ lies along the $\langle 111 \rangle$ NV symmetry axis.

Nitrogen hyperfine and quadrupole (for $^{14}NV^-$) parameters are:

$$^{15}NV^- : A_{\|} = -2.14(7)\,\mathrm{MHz}, A_{\perp} = -2.70(7)\,\mathrm{MHz};\ a = -2.51(7)\,\mathrm{MHz}, b = 0.19(7); Q_{\|} = -5.01(6)\mathrm{MHz}.$$

$$^{15}NV^- : A_{\|} = +3.03(3)\,\mathrm{MHz}, A_{\perp} = +3.65(3)\,\mathrm{MHz};\ a = 3.44(3),\ b = -0.21(3).$$

Isotropic (a) and anisotropic (b) components of the hyperfine interaction for NV^- centres are also presented. The fact that the isotropic hyperfine component is small and of opposite sign to that expected for localization of the unpaired-electron probability density on the nitrogen nucleus indicates that this contribution arises through spin polarization. The unpaired-electron probability density is predominately localized on the three carbon neighbors (this $^3A_2$ state of the NV ground state does not involve orbitals related to nitrogen). This polarizes the core states of the nitrogen and, since the nuclear magneton for ^{14}N is positive, the Fermi contact term will be negative. It was also found a negative sign for the isotropic component of the ^{14}N hyperfine interaction from DFT calculations [59].

The ^{14}N might show a short coherence time because of its quadrupolar moment. This couples to lattice vibrations easily and hence causes phonon induced spin dephasing. Because of its fast decoherence and complex spin Hamiltonian the quantum state of the nitrogen nucleus is difficult to control. It is known however that the spin density at the three dangling bonds of the next nearest neighbour carbon atoms is largest. Roughly 70% of the electron spin density is expected here [96, 97]. The natural abundance of the ^{13}C (I = ½) nucleus is 1.1%. Hence in a not isotopically enriched diamond it is expected that roughly one out of thirty defects should show a hyperfine coupling to a ^{13}C nucleus in the first shell. In an external magnetic field the spin Hamiltonian describing this system is

$$\hat{H} = \mu_B g \vec{B} \cdot \hat{\vec{S}} + D[\hat{S}_z^2 - \frac{1}{3}S(S+1)] + E(\hat{S}_x^2 - \hat{S}_y^2) + \sum_i (\hat{\vec{S}} \cdot \overleftrightarrow{A}_i \cdot \hat{\vec{I}}_i - g_I \mu_N \vec{B} \cdot \hat{\vec{I}}_i), \tag{6.14}$$

where $\overleftrightarrow{A}_i$ and $\hat{\vec{I}}_i$ are the tensor of the hyperfine interaction of the NV centre unpaired electrons and nuclear moment with the surrounding ^{13}C nucleus *i*.

Such coupling to ^{13}C has been detected experimentally. A completely different set of hyperfine coupling parameters is measured when a ^{13}C nucleus is found in the shell of first nearest neighbours around the vacancy. Because of the much higher spin density of the electron at these carbon positions the measured coupling parameter is around 130 MHz (see Fig. 6.21). Unambiguous evidence of hyperfine interaction ("fingerprint" or "signature") with one ^{13}C nucleus (I = 1/2), with two ^{13}C nuclei (I_1 = 1/2, I_2 = 1/2), with three ^{13}C nuclei (I_1 = 1/2; I_2 = 1/2, I_3 = 1/2) and the absence of the ^{13}C nucleus (only ^{12}C, I = 0) in the first NV centre shell (the upper panel in Fig. 6.21).

Experimental ^{13}C hyperfine parameters for the NV^- defect which were assigned to the three carbon atoms neighboring the vacancy are [95]: $A_{\parallel}$ = 198.2(3) MHz and $A_{\perp}$ = 120.8(2) MHz. Experimental ^{13}C hyperfine parameters for the NV^- defect which were assigned to the hyperfine interaction with six equivalent carbon

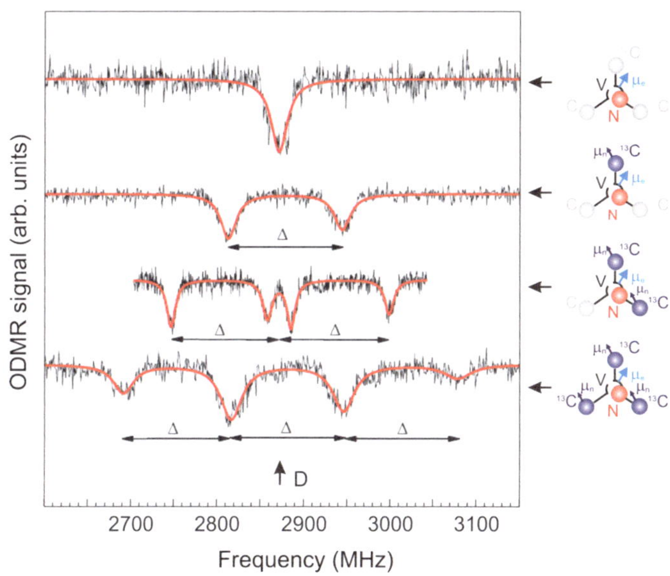

Fig. 6.21 ODMR spectra of single NV centres for $M_S = 0 \rightarrow M_S = -1$ transition which have been recorded at room temperature with 532 nm irradiation without the application of an external magnetic field. Unambiguous evidence of hyperfine interaction ("fingerprint" or "signature") with one ^{13}C nucleus (I = 1/2), with two ^{13}C nuclei (I_1 = 1/2, I_2 = 1/2), with three ^{13}C nuclei (I_1 = 1/2; I_2 = 1/2, I_3 = 1/2) and the absence of the ^{13}C nucleus (only ^{12}C, I = 0) in the first NV centre shell

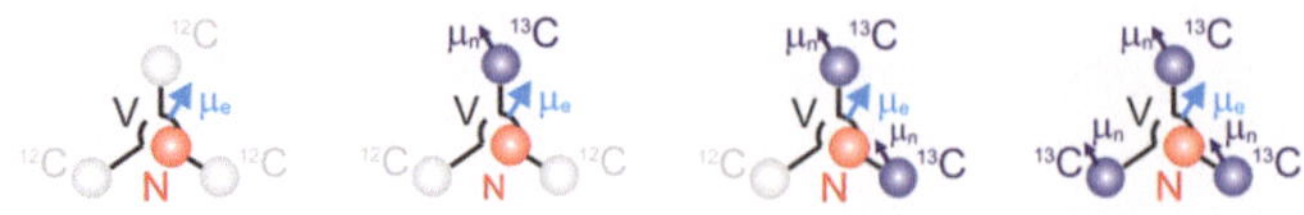

^{13}C-content	Probability to find zero ^{13}C in the first shell	Probability to find one ^{13}C in the first shell	Probability to find two ^{13}C in the first shell	Probability to find three ^{13}C in the first shell
1.1 % (natural abundance diamond)	96.7 %	3.2 %	0.04 %	10^{-4} %
8.4% (^{13}C enriched diamond)	76.9 %	21.1 %	1.9 %	0.06 %
measured	**81.3 %**	**17.2 %**	**1.1 %**	**0.4 %**

Fig. 6.22 Searching for ^{13}C in the 1st shell in diamond with natural abundance and ^{13}C enriched diamond

atoms at the third neighbor distance are: $A_{\parallel} = 18.49(5)$ MHz and $A_{\perp} = 13.26(5)$ MHz. Both the larger ^{13}C hyperfine interaction and the smaller ^{13}C hyperfine interaction are axially symmetric with $A_{\parallel}$ along a crystallographic $\langle 111 \rangle$ direction which is not the symmetry axis of the NV centre. These parameters are supported by recent DFT calculations [59].

The table in Fig. 6.22 presents searching for ^{13}C in the 1st shell in diamond with natural abundance and ^{13}C enriched diamond.

Figure 6.23 shows the ODMR spectrum of a single ^{13}C coupled defect in the magnetic field. Two EPR doublets with the separation of 126 MHz are visible due to hyperfine interaction with one ^{13}C. The spin system needs to be described by a six level system (instead of three levels). In first order (without taking into account hyperfine coupling to nitrogen) four EPR transitions between quantum states with identical nuclear spin quantum number are allowed. All transitions have identical transition strengths, and differences in ODMR contrast are related to frequency-selective transmission characteristics of the microwave line.

6.2.4.2 Excited Spin States

The NV^- centre are identified by a zero field magnetic resonance at 2.88 GHz [98] (see Fig. 6.19). This magnetic resonance occurs between the $M_S = 0$ and $M_S = \pm 1$ spin sub-levels of the spin triplet ground state ^3A and can be detected by either

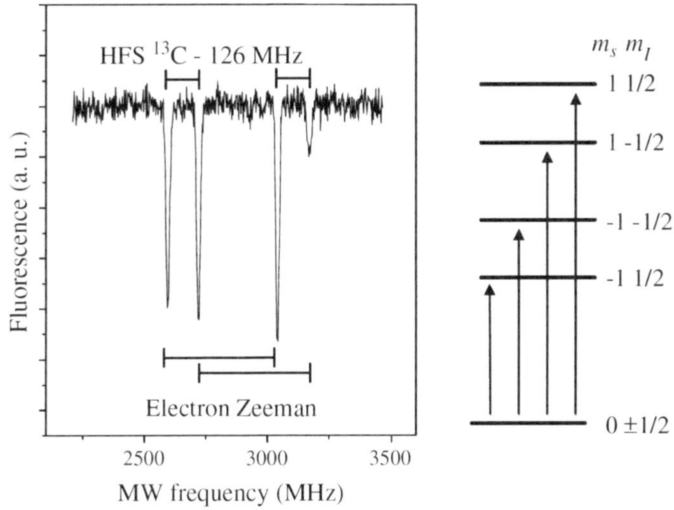

Fig. 6.23 (*Left*) ODMR spectrum of a single ^{13}C coupled defect in the magnetic field. (*Right*) The ground state energy level scheme for NV centres containing a single ^{13}C nucleus in the first coordination shell in the magnetic field

conventional electron paramagnetic resonance (EPR) [98] or optically detected magnetic resonance (ODMR) [99] techniques. A further ODMR at 1.42 GHz [70, 100, 101] is observed at room temperature (see Fig. 6.24) and is attributed to a zero field splitting of the spin triplet excited state ^{3}E that is analogous to that of ^{3}A. The magnetic resonances of ^{3}A and ^{3}E behave as would be expected of triplet spins in a trigonal crystal field and are characterised by the approximately isotropic electron g-factors 2.0028 [98] and 2.01 [100] respectively. In order to detect the spin sublevels in the excited state a resonance microwave field and an optical excitation were simultaneously applied [100]. Since the excited-state is short-lived (about 10 ns) the microwave field applied was strong enough to cause transitions within that lifetime.

The spin Hamiltonian (6.13) can describe the spin sublevels in the excited state with following parameters [100]: $D_{exc} = 1.425(3)$ GHz, $g_{exc} = 2.01(1)$, $E_{exc} = 70 \pm 30$ MHz, $A_{exc} = 61 \pm 4$ MHz (for ^{15}N). The g-factor isotropy indicates that the orbital angular momentum does not play a significant role in the excited state. The positions of EPR frequencies were measured by rotating a magnetic field around different crystal axes and the experimental results provide strong evidence that the ground and excited states exhibit the same orientations.

One can see that hyperfine interaction with N (^{15}N) in the excited-state is about twenty times larger than that of the ground-state. At magnetic fields near the excited-state spin level-crossing (about 50 mT) a dynamic nuclear polarization effect of hyperfine structure sublevels that changes the relative intensity of the two resonances (^{15}N, I = ½) by polarizing the nuclear spin of the nitrogen was observed [100]. This effect was suggested [100] can be used to simultaneously initialize the

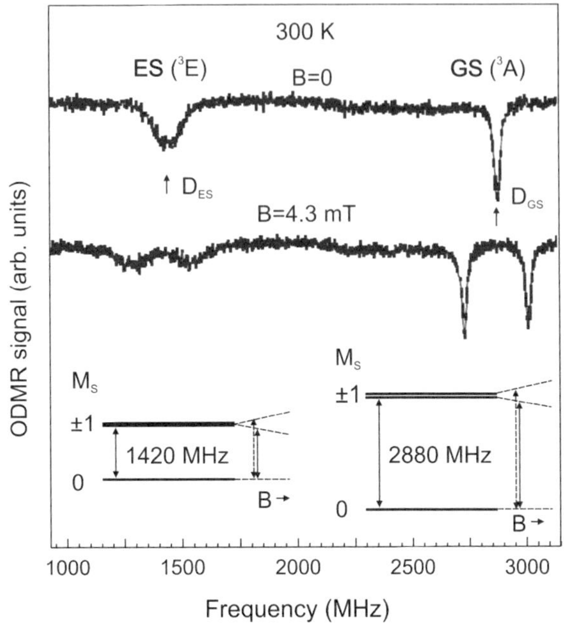

Fig. 6.24 Room temperature ODMR spectra [22, 54] of a single NV color centre at zero magnetic field (*upper trace*) and with a magnetic field of amplitude B = 4.3 mT applied along the NV symmetry axis which corresponds to a [111] crystal axis (*bottom trace*). EPRs are evidenced both in the ground state (*GS*) and in the excited state (*ES*) exhibiting the zero field splittings of the $M_S = 0$ and $M_S = \pm 1$ spin sub-levels of ^{3}A (2.88 GHz [98]) and ^{3}E (1.42 GHz [100])

NV-center electronic spin and the nitrogen nuclear spin for experiments with coupled spins in diamond.

The magnetic resonances of ^{3}A [103, 104] and ^{3}E [100, 101] also exhibit weak interactions with strain and electric fields that offers the ability to control the strain and electric fields within the material. It should be noted that the transverse anisotropy splitting E_{es} term is larger than an order of magnitude relative to that in the ground-state spin Hamiltonian. Thus this term is sensitive to the strain in the diamond. The local strain was shown in [100] can vary considerably within the same sample. A distinctly different value of E_{es} term, indicating a different value of local strain was observed by looking this term at different NV centres.

6.2.4.3 Dynamic Polarization of Single Nuclear Spins by Optical Pumping of NV Centres in Diamond at Room Temperature

An effective method to polarize single nuclear spins in diamond, based on optical pumping of a single nitrogen-vacancy (NV) defect and mediated by a level anti-crossing (LAC) in the excited state of the NV defect has been reported [102]. A nuclear spin polarization higher than 98% is achieved at room temperature for the ^{15}N nuclear spin associated with the NV centre, corresponding to μK effective nuclear-spin temperature. We then show simultaneous initialization of two nuclear spins (^{15}N and ^{13}C) in the vicinity of a NV defect. Such robust control of nuclear-spin states is a key ingredient for further scaling up of nuclear-spin based quantum registers in diamond.

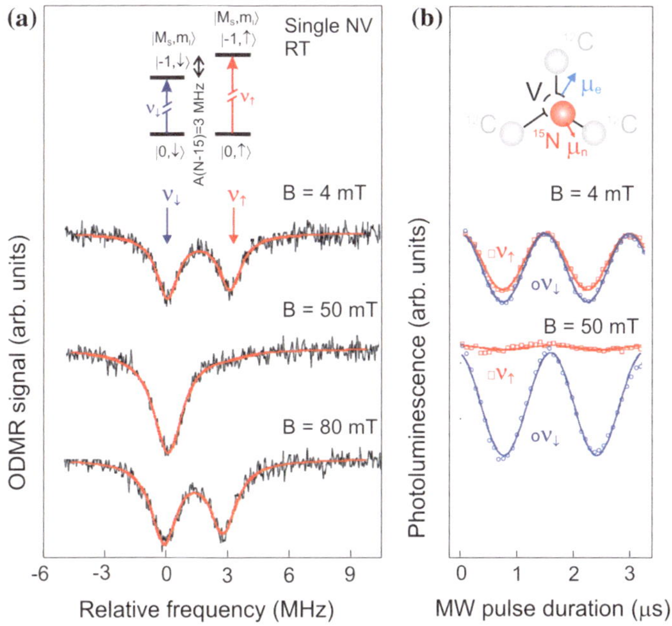

Fig. 6.25 **a** ODMR spectra recorded at different magnitudes of a magnetic field applied along the NV symmetry axis ([111] crystal axis) showing the ground state hyperfine structure associated with ^{15}N nuclear spin states $|\uparrow\rangle$ and $|\downarrow\rangle$. Nuclear-spin polarization is observed at the excited-state LAC (B = 50 mT). *Solid lines* are data fitting using Lorentzian functions. Identical results are obtained for transitions from $M_S = 0$ to the $M_S = +1$ manifold (data not shown). **b** Selective Rabi nutation measurements using microwave pulses at frequency $\nu_\downarrow$ [*blue* (*circle*) *points*] and $\nu_\uparrow$ [*red* (*square*) *points*]. The experiment is performed for B = 4 mT (*upper trace*) and B = 50 mT (*lower trace*). *Solid lines* are data fitting using cosine functions [102]

Ultrapure synthetic type IIa diamond crystals prepared using a chemical vapor deposition process were used. Single NV color centres were artificially created by implanting 7 meV ^{15}N ions and by annealing the sample for 2 h in vacuum at 800 °C. Those NV defects are associated to the ^{15}N isotope which is a I = 1/2 nucleus. The energy splitting resulting from hyperfine interaction between this nuclear spin and the electron spin is A_{GS} = 3.05 MHz in the ground state [65].

ODMR spectra of single NV color centres are recorded by applying microwaves and monitoring the photoluminescence intensity. In addition, a magnetic field is applied along the NV symmetry axis ([111] crystal axis). As shown in Fig. 6.25a, two electron paramagnetic resonances (EPR) are evidenced in ODMR spectra recorded at small magnetic field magnitude, each resonance being associated to a given orientation of the nuclear spin, $|\uparrow\rangle$ at frequency $\nu_\uparrow$ and $|\downarrow\rangle$ at frequency $\nu_\downarrow$. By keeping the magnetic field aligned but increasing its magnitude up to 500 G, which corresponds to the excited-state LAC, the EPR line at frequency $\nu_\uparrow$ disappears, indicating a strong polarization of the nuclear spin in state $|0, \downarrow\rangle$ (Fig. 6.25a) [100, 102].

The nuclear-spin polarization is measured as

$$P = \frac{I(v_\downarrow) - I(v_\uparrow)}{I(v_\downarrow) + I(v_\uparrow)}, \tag{6.15}$$

where $I(v_\uparrow)$ [respectively $I(v_\downarrow)$] is the integral of the EPR peak at frequency $v_\uparrow$ (resp. $v_\downarrow$). By fitting each ESR line with Lorentzian functions, we infer a polarization P = 0.98 ± 0:01. Owing the nuclear-Zeeman splitting between states $|0,\uparrow\rangle$ and $|0,\downarrow\rangle$ (≈200 kHz at B = 50 mT), such polarization corresponds to a μK effective nuclear-spin temperature.

In order to confirm this observation, selective Rabi nutations are performed for each EPR line using the standard pulse sequence described in [91]. At low magnetic field magnitude, the contrast of the Rabi nutation is almost identical for each resonance line, demonstrating that the two states $|0,\uparrow\rangle$ and $|0,\downarrow\rangle$ are populated with similar probabilities. Around the excited-state LAC, the contrast associated to state $|0,\uparrow\rangle$ vanishes whereas the one associated to state $|0,\downarrow\rangle$ becomes twice higher (Fig. 6.25b). This result constitutes another demonstration of the nuclear-spin polarization in state $|0,\downarrow\rangle$.

The nuclear spin polarization method was demonstrated can be extended to more than one nuclear spin. Figure 6.26 (upper trace) shows the ODMR spectrum of a single NV centre with a ^{13}C at the first coordination shell, leading to a hyperfine

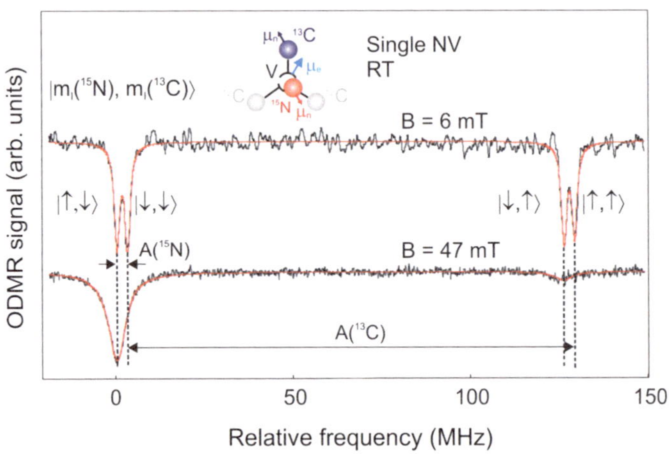

Fig. 6.26 ODMR spectra recorded for a single NV centre with a ^{13}C on the first coordination shell. At small magnetic field magnitude (B = 6 mT, *upper trace*), four EPR lines are observed, each line being associated to a given orientation $|^{15}N \downarrow_{or} \uparrow\rangle$ and $|^{13}N \downarrow_{or} \uparrow\rangle$ of the nuclear spins. Around the LAC (B = 47 mT, *lower trace*), ^{15}N and ^{13}C nuclear spins are both polarized. Note that the width of the EPR line is bigger for the measurement performed close to the LAC because of power broadening. However, such width would be small enough to resolve the 3 MHz hyperfine splitting related to ^{15}N (see *black dot lines*). *Red solid lines* are data fitting using Lorentzian functions [102]

splitting around 130 MHz in the ground state. In such spectrum, four EPR lines are observed, each being associated to a given orientation of ^{13}C and ^{15}N nuclear spins. By increasing the magnetic field magnitude up to the excited-state LAC, we observe an efficient polarization of both nuclear spins as only one ESR line remains visible [Fig. 6.26 (lower trace)], corresponding to a deterministic initialization of a three qubit quantum register by including the electron spin. From the data we infer a polarization P = 0.90 ± 0.01 for the double nuclear-spin system.

A mechanism to account for the observed nuclear-spin polarization was proposed in [102]. Assuming that the magnetic field B is perfectly aligned along the NV axis (z axis) and neglecting the nuclear-Zeeman splitting, using (6.14) the excited state spin Hamiltonian is given by

$$\hat{H} = \mu_B g_e B \hat{S}_z^2 + D_{ES}[\hat{S}_z^2 - \frac{1}{3}S(S+1)] + A_{ES}\hat{\vec{S}} \cdot \hat{\vec{I}},$$

where $\hat{\vec{S}}$ and $\hat{\vec{I}}$ are the electron and nuclear-spin operators, $D_{ES} = -1420$ MHz is the excited-state zero-field splitting, g_e the electron g factor, and $A_{ES} = +60$ MHz is the excited-state hyperfine coupling for ^{15}N. The eigenstates of the spin Hamiltonian (6.14) are $|0,\downarrow\rangle, |1,\uparrow\rangle, |+\rangle = \alpha|0,\uparrow\rangle + \beta|+1,\downarrow\rangle$ and $|-\rangle = \alpha|0,\uparrow\rangle + \beta|+1,\downarrow\rangle$, showing a LAC at $B_{LAC} \approx 50$ mT.

The position of the associated eigenenergies as well as the values of parameters α and β are represented as a function of the magnetic field magnitude can be calculated from spin Hamiltonian (6.14). At low magnetic field magnitudes, $\alpha \approx 1$ and $\beta \approx 0$. In such regime, optical transitions from the ground to the excited state ($^3A \rightarrow {}^3E$) are fully nuclear spin-conserving as no state mixing in the excited state is occurring. As a result, the nuclear spin is not polarized. Increasing the magnetic field magnitude close to the LAC, α and β begin to balance. The transition from $|0,\downarrow\rangle$ to the excited state remains nuclear spin conserving, whereas the transition from $|0,\uparrow\rangle$ results in $(\alpha|+\rangle + \beta|-\rangle)$ in the excited state. This superposition state then starts to precess between $\alpha|+\rangle + \beta|-\rangle = |0,\uparrow\rangle$ and $\alpha|+\rangle + \beta|-\rangle = (\alpha^2 - \beta^2)|0,\uparrow\rangle + 2\alpha\beta|+1,\downarrow\rangle$ at some frequency Ω. The maximum probability $p_{max}(B)$ to find the nuclear spin flipped from $|\uparrow\rangle$ to $|\downarrow\rangle$ within this precession is given by $p_{max}(B) = 4\alpha^2\beta^2$. The nuclear spin-flip has then a probability to be transferred to the ground state sublevel $|0,\downarrow\rangle$ by nonradiative intersystem crossing through the metastable singlet state responsible for electron spin polarization of the NV defect. Every subsequent excitation and decay cycles increase spin polarization in state $|0,\downarrow\rangle$.

All aforementioned arguments hold as well for the crossing of the levels $M_S = 0$ and $M_S = -1$. The evolution of polarization P as a function of the magnetic field magnitude shows that the dependence of P on the magnetic field magnitude is broad, showing that a precise adjustment of the magnetic field magnitude to the LAC is not required, e.g., P ≈ 95% for B ≈ 44 mT. Even small state mixing in the excited state leads to efficient nuclear-spin polarization through optical pumping. By saturating the optical transition, the speed of the polarization process is limited by the metastable state lifetime. The nuclear-spin polarization is, however, very sensitive to the magnetic field alignment along the NV axis. Inferring the evolution of the nuclear spin polarization as a function of the magnetic field angle would require more sophisticated numerical simulations of the NV color centre spin dynamics, as all spin states are partially mixed in that case. Note that ground state qubits keep high purity by working at the excited state LAC, since mixing of ground state sublevels occurs around B $\approx$ 100 mT. As a result local quantum operations in the ground state would not require switching off the magnetic field.

Summarizing, a new method was demonstrated to strongly polarize single nuclear spins in diamond at room temperature, which is experimentally simple to implement since it is only based on optical pumping. Such robust control of nuclear-spin states is one of the key ingredients for further scaling up of nuclear-spin based quantum register in diamond [92].

6.2.4.4 Coherence and Single Spin States

The generation of a coherent state superposition is achieved by a short microwave pulse in resonance with the transition in Fig. 6.19 or one of the transitions in Figs. 6.20 and 6.22. In order to generate a state superposition with arbitrary expansion coefficients of the two eigenstates e.g. $|\alpha\rangle$ and $|\beta\rangle$ one uses microwave pulses of variable length, such that $\psi(t) = \sin\Omega_{MW}t|M_s = 0\rangle + \cos\Omega_{MW}t|M_s = \pm 1\rangle$ (here Ω_{MW} is the microwave Rabi frequency). Depending on the magnitude of $\cos^2\Omega_{MW}t$ the fluorescence will change. Hence when plotting the fluorescence intensity as a function of pulse length a periodical variation of the fluorescence is seen [see Fig. 6.27 (left)].

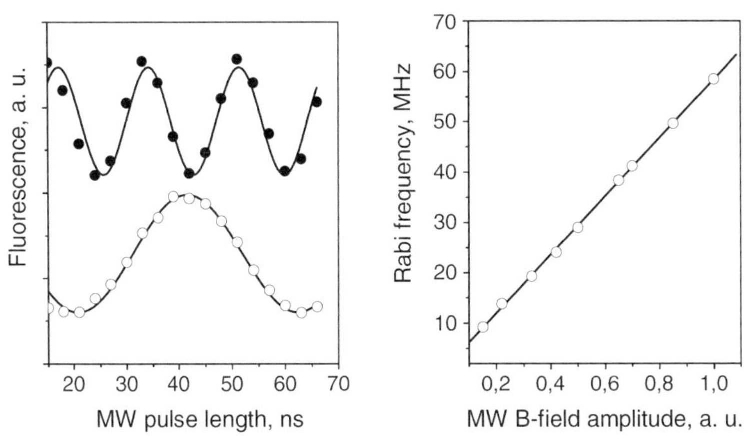

Fig. 6.27 Optically detected Rabi nutations of a single NV electron spin. The *points* represent experimental data and the line is a fit with a $\cos^2\Omega t$ function, where Ω is the spin Rabi frequency. Figure on *right*: Dependence of Rabi frequency on MW filed amplitude. The *solid line* represents a linear fit

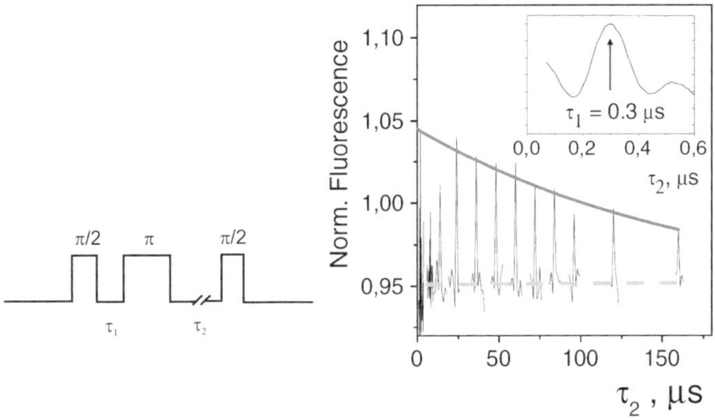

Fig. 6.28 Hahn echo trace of a single spin. The *left part* of the figure shows the microwave pulse sequence used. The *right part* depicts experimental data. The *inset* shows the Hahn echo itself. The main figure demonstrates the Hahn echo decay as a function of $\tau_1 = \tau_2$. Printed are the individual echoes at different delay times together with a fitted decay curve of the amplitude

The frequency of these oscillations (Rabi nutations) depends linearly on the MW field amplitude, as can be seen in Fig. 6.27 (right) [51]. Rabi frequencies of up to 140 MHz have been achieved with miniaturized coupling loops or wire structures. For the nutation curve in Fig. 6.27 (left) a decay of the amplitude is expected. The corresponding decay constant is related to the dephasing time T_2 but not equivalent to T_2. Rather T_2 has to be measured in the absence of any microwave field. This is achieved by the application of a Hahn echo sequence. In this pulse sequence all inhomogeneous distribution of resonance frequencies are refocused while fluctuation of transition frequency or random phase jumps cause an echo decay upon increasing the time between pulses. Figure 6.28 shows an example of a Hahn echo train with variable delay.

An echo decay is visible which can be fitted with a monoexponential decay time of 350 μs. The chief cause for dephasing in diamond are electron paramagnetic impurities in the lattice [105]. These impurities show dipolar coupling to the NV centre and hence may undergo energy conserving spin flip-flop processes with the NV spin. These processes result in a loss of phase memory of the NV spin. It has been demonstrated, that the NV centre T_2 time depends on the concentration of impurities in the lattice and the dephasing time was shown to decrease up to some hundred ns for nitrogen-rich diamond.

In defects which do show a hyperfine coupling to a ^{13}C nucleus in addition to electron spin also nuclear spin nutations can be detected [91]. Because nuclear spin wave functions do not couple to the optical transition outside of level anticrossing, all changes in nuclear spin wave function must be mapped into the electron spin states to be detectable. A single electron plus nuclear spin system is described by a four level system. To first order only electron spin resonance transitions with $\Delta M_S = 1$ and $\Delta m_I = 0$ are allowed, indicated by the two arrows in Fig. 6.29.

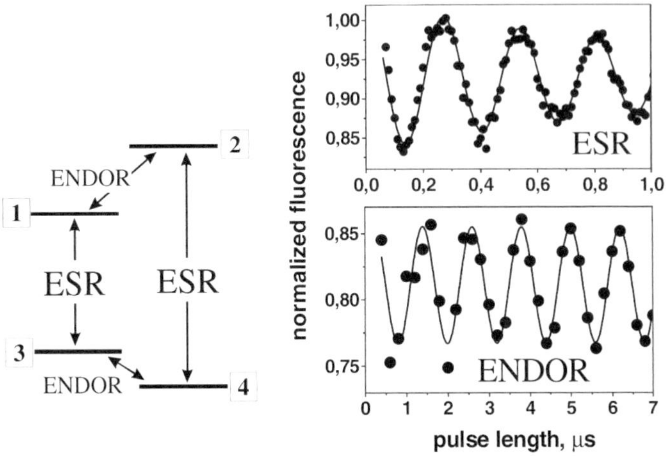

Fig. 6.29 Energy level scheme and Rabi nutations of a single NV electron spin coupled to a single ^{13}C spin. The *left part* of the figure shows the relevant spin levels for the coupled electron-nuclear spin system. The allowed electron spin resonance (*ESR*) and nuclear magnetic resonance (NMR) transitions are shown in the figure. The *right part* of the figure shows Rabi nutations of an electron spin (*upper trace*) and nuclear spin (*lower trace*)

In order to drive nuclear magnetic resonance transitions, radio frequency has to be irradiated at transition energy between level 1 and 2 (or level 3 and 4). For the 1–2 transition this corresponds to the hyperfine splitting observed in Fig. 6.21. ^{13}C nuclear relaxation times in diamond vary between 1.4 and 36 h. The T_2 time can be estimated from the width of ^{13}C NMR spectra to be on the order of ms for those nuclei that are not detuned from the dipolar nuclear spin bath. Hence, nuclear spin states should allow for coherent state preparation. To observe nuclear spin transients a microwave-radio frequency double resonance experiment has been carried out. The experiment comprises π pulses separated by time τ. During this time interval a radio frequency pulse of variable length in resonance with e.g. the 2–1 transition is applied. The strength of the EPR signal is measured on the 3–1 transition.

Figure 6.29 shows an example of a nuclear transient measured in this way. The amplitude of the oscillations corresponds to the amplitude of the ODMR signal itself, i.e. 30% of the fluorescence intensity. The approach corresponds to the well known electron-nuclear double resonance experiments.

With two spins at hand it is possible to carry out basic quantum computation experiments like, e.g. the conditional not gate (CNOT). It can be shown, that two gate operations are sufficient to perform all operations necessary for full quantum computation. These two gates are the single quantum bit NOT gate, which corresponds to an inversion of the bit value and the CNOT gate which is the inversion of one bit conditioned on the value of a second bit. In this nomenclature a single qubit corresponds to a single spin with either of the two eigenvalues $|0\rangle$ or $\langle 1|$ (spin-up or spin-down). A CNOT gate would flip, e.g. the electron spin depending on the state of the nuclear spin. Such a scenario can be easily realized in a situation shown in Fig. 6.22, i.e. in a 4 level system with coherent control over (at least) two transitions. For example the nuclear spin will be inverted only by RF irradiation in resonance with the 1–2 transition when the electron is in spin-up configuration, i.e. in state 1. A simplified version of the CNOT gate is the CROT transformation. The two operations are identical to each other except for a phase factor which can be achieved by a rotation around the z-axis. The CROT itself is only a π pulse. The action of the pulse only corresponds to an ideal gate in the limit of infinitely narrow spectral lines, i.e. long T_2 and rectangular microwave pulses. Under realistic conditions this is not the case. Performing more complex quantum information operations requires a certain precision of operations. Hence, it is useful to control the quality of gates.

Optical readout of single electron spin state leads to repolarization of electron spin. Owing to fast time scale of this process, so called "single shot" readout regime remains challenging under ambient conditions for the case of the electron spin. However nuclear spins can provide important resource for achieving high fidelity readout regime. It was shown that the state of the nuclear spins can be efficiently controlled by manipulation via electron spin of NV centres. The state of the electron spin can be efficiently mapped into nuclear spins and the readout repetitively [106]. During reach readout cycle the electron spin is repolarized, but nuclear spin state remains unchanged. Application of this readout cycle allows for projective readout of nitrogen nuclear spins of NV centre (Fig. 6.30) [107].

6.2.5 *Diamond Quantum Registers*

Long coherence time and possibly of single defect generating by ion implantation are crucial elements for building quantum logic based on individual impurity atoms. Two types of interactions can be employed for building scalable quantum registers. First is the optical coupling between qubits. The second type of coupling which can be explored is magnetic dipole coupling between qubits (see Fig. 6.31).

First proof-of-principle experiments towards implementation of scalable quantum register architectures were demonstrated recently. These experiments showing potential of diamond spins for quantum information were based on coupling of

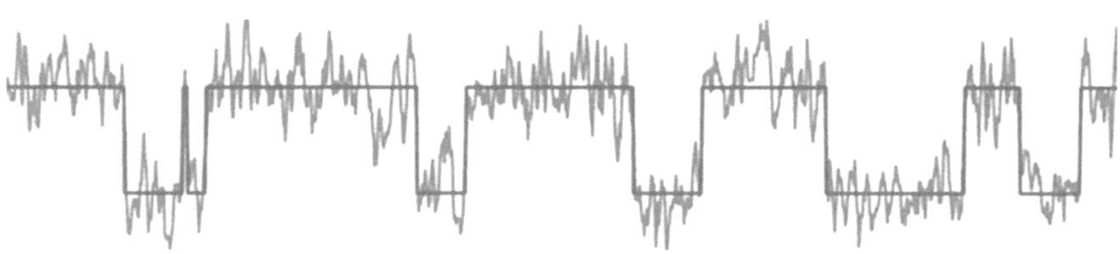

Fig. 6.30 Fluorescence time trace showing quantum jumps of single nitrogen nuclear spin

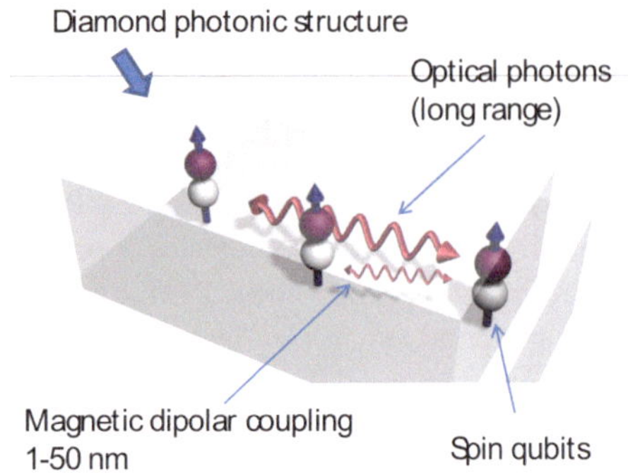

Fig. 6.31 Concept of the spin-diamond quantum register. Single optically active spins are embedded into a photonic structure allowing the use of long-range coupling via optical photons. In addition, short-range magnetic dipolar coupling between spins can be explored. The range at which such coupling prevails over decoherence depends on the coherence time of the electron spin

single electron spin to the nuclear spin associated with C13 nuclear spins [108, 109]. Although electron spin surrounded by a few nuclei might serve as model system for spin-based quantum register [110, 111], its scalability is limited owing to limited number of nuclear spins in so called "frozen core". Therefore longer range dipolar interactions between electron spins are more promising for use in quantum information processing protocols. Dipolar coupling between two implanted NV centres was employed for realization of quantum gate [112].

Recently schemes for probabilistic entanglement between color centres using optical channels were proposed [113] and important steps towards their realization using color centres in diamond were demonstrated. Two photon interference from distant NV defects was observed [86]. Further progress is critically dependent on the experimental ability to generate array of NV centres with high fidelity (for magnetic coupling) and ability to couple NV defects to optical cavities (for long range optical coupling). Several approaches were explored on the way to create integrated diamond photonics platform. First, NV centres in diamond were adjusted coupled to high Q resonators made from GaP [114]. Integrated diamond cavities were designed and fabricated [115, 116]. Coupling of colour centres to photons is crucial for building elements of quantum communications like quantum repeaters [117] and may improve readout fidelity for single spin measurements [118].

Besides conventional approaches for quantum computing (based on by quantum gates), there are other implementations called "one-way quantum computing" or measurement-based quantum computing. These techniques allow to create entangled states of large systems by measurement of qubits. Typically, the measurement is not performed on the qubits itself, but state information is transferred from the qubits to photons which interact in an interferometric scheme to erase the which-path information. As there is no need for direct interaction of qubits, and as photons can be transferred over large distances, measurement-based entanglement creation allows for entangling remote qubits in a distributed quantum computer. Technically, the requirements for realization of such entangling schemes are similar to the ones for spin detection: due to the combination of long spin coherence times and strong optical transitions, NV centres are promising candidates [119] high purity diamond reduces possible sources of dephasing, and coupling to optical cavities enhances emission into desired modes. Cavity coupling may be employed in two different regimes: in the weak coupling regime (coherent emission rate into cavity is faster than spontaneous decay of the dipole moment) one relies on the spontaneous emission enhancement by the Purcell effect [120] or dispersive shifts of the cavity resonance detected in reflection from the cavity [121]. Ultimate control over the emission process is gained in the strong coupling regime where the coherent coupling rate is larger than the incoherent decay channels; here stimulated Raman-type transitions allow for coherent emission into the cavity mode and for emission of transform-limited photons. In order to allow for two-photon interference as required for many entanglement schemes, the emitted photons have to be identical to a high degree. The technical preconditions are: long pure dephasing times of the excited state (need for ultrapure diamond), suppression of emission into phonon sidebands by enhancement of the zero phonon line (strong Purcell effect, i.e. cavities with small modal volume and high quality factor) and the ability to precisely tune the cavity mode to the emitter's optical resonance.

An increase in the number of qubits can be achieved due to magnetic dipolar coupling of several NV centres. The groups of NV defects can be created within the range of interaction of magnetic dipoles, thus one can control the interaction between defects. It can be achieved by performing nitrogen high precision ion implantation into diamond to generate NV centres. Within a pattern of defects, the strongest interaction is between adjacent centres, since the magnetic dipolar interaction depends on the distance between the NV centres and decreases rapidly with the distance.

Let us consider the NV diamond spin system after [122, 123]. The experimental system consists of two ^{15}NV centres separated by a distance of about 20–25 nm, with an effective mutual dipolar coupling of $E_{dd} \approx 5$ kHz after [124] (see Fig. 6.32 upper part). Each NV centre has an electron spin-S = 1 and a ^{15}N nuclear spin I = 1/2, hence the system exhibits $(3 \times 2)2 = 36$ energy levels in total. Individual addressing of both NVs' spin transitions is realized by different crystal field directions and proper magnetic field alignment resulting in a spectral separation of about 55 MHz between the individual NV transitions (see Fig. 6.33).

6.2.6 *Applications of Single Colour Centres for Novel Imaging Techniques*

There is much interest toward extending the principle of fluorescence microscopy down to the level of single molecules [125]. While fluorescent signals from single

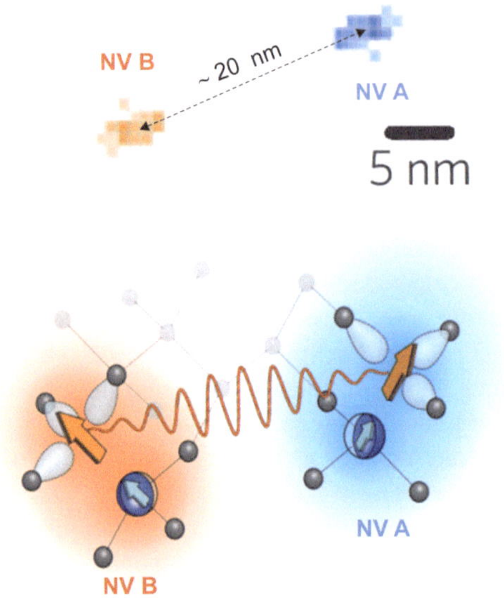

Fig. 6.32 Two qubit entanglement, (*top*) Optical super resolution—using of stimulated emission depletion (STED) microscopy and photoactivated localization microscopy (PALM). (*bottom*) Schematic of the NV-NV pair used in the experiment

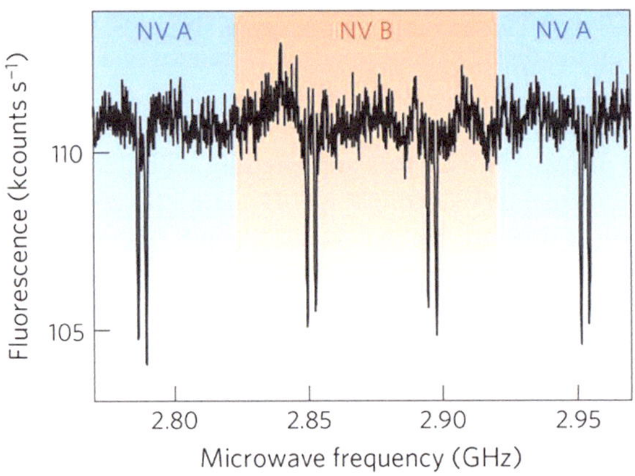

Fig. 6.33 Optically detected magnetic resonance (ODMR) spectrum of the NV pair, isotope ^{15}N (I = ½) was used in the experiment. The outer pairs of transitions correspond to NV A and the inner pairs to NV B. The splitting within one pair of 3 MHz is due to the hyperfine coupling with the ^{15}N nucleus. Spin transitions of different NV centres are separated by about 55 MHz due to different crystal field directions and proper magnetic field alignment

molecules can be detected using a confocal microscope and fluorescence microscopy, the resolution is not high enough to image the location of individual molecules. Conventionally, multi-colour dyes (or quantum dots) have been used to allow selective detection and localization of multiple dye molecules within the focal volume. However, the number of colours that can be cleanly resolved is limited to less than 10 because of the broad optical spectrum of fluorescence emitters at room temperature.

Recently much effort has been put into nonlinear resolution enhancement techniques, for example STED [126] which works by saturation of the fluorescence emitter and PALM [127] which works by activating only a portion of the dye molecules for each image and then post processing. While these techniques have achieved just over an order of magnitude improvement over the optical wavelength ($\sim$10's on nanometers), there is a barrier to reaching molecular scale resolution. This is due to the intensity near optical field nodes (STED) and anti-nodes (PALM) varies quadratically with position so that the extra factor of 10 needed to reach single molecule (or sub-macromolecule) resolution ($\sim$few nanometers) would be 100 times more difficult to achieve. Unique photostability of diamond defects can potentially allow to achieve this goal [128].

At the same time, nuclear magnetic resonance imaging (MRI) routinely produces images whose resolution is far below the wavelength of the microwave excitation. The apparatus and highly complex microwave pulse sequencing that is needed has already be engineered to the level where clinical application is routine. Therefore, to extend the resolution limit of fluorescence microscopy down to the single molecule scale, a miniature version of a MRI imaging system can be used where the dye molecule has an electron spin resonance (ESR) transition. NV colour centre in diamond are promising candidates for new MRI imaging because even single defects have detectable ESR signal at room temperature and its spin linewidth (which determines resolution) is very narrow [129]. The nanodiamond ESR tags are consumables and large quantities would eventually be needed. The key diamond technologies to be developed include functionalization and growth of ultra-pure diamonds with high NV yield. For functionalization, the surface of either crushed or CVD grown nanodiamonds must be modified by attaching an intermediate linker, for example silanes, that in turn can be used to attach functional groups such as biotin that can later be used to attach biologically specific tags. For growth all spins except the NV must be eliminated from the lattice.

Bioimaging application can profit from biocompatibility of nanodiamonds (recent studies show that diamond nanoparticles are not toxic [130]. The wavefunction of colour centres is localized within a few Angstroms allowing to produce fluorescent of small diamond nanoparticles without affecting stable strong fluorescence associated with NV centres. Fluorescence of nanodiamonds with typical size below 10 nm was reported [131, 132].

6.2.7 *Magnetometry with Single Diamond Spins*

One of the most promising applications of single colour centres is related to development of diamond-based electric and magnetic field sensors that are able to detect minute fields associated with single electron and nuclear spins and that can be positioned to achieve atomic scale spatial resolution. Atomic-scale spin sensors based on highly localized colour centres in diamonds can be placed in the close vicinity to external spins thus allowing distances to be reached at which their associated magnetic fields are sufficiently large to be measured.

The ultra sensitive detection of magnetic and electric fields with high sensitivity and nanometer spatial resolution is an outstanding challenge with strong impact on modern science and technology that cannot be underestimated. As a result, numerous attempts for sensitive detection of e.g. magnetic fields are known. Prominent examples are SQUIDS, atom vapour magnetometers, magnetic resonance force microscopy or Hall sensors. However, either the systems do not allow for miniaturization or they require special working conditions like ultrahigh vacuum or low temperature. This is why alternative magnetic field sensing devices based on diamond defects have gained considerable attention. Here, magnetic fields are sensed by localized electron spins which are then read-out optically (Fig. 6.34). The energy separation of spin states depends on an external magnetic field and the frequency is measured with optically detected magnetic resonance which allows detecting external fields. When used as a magnetic field measurement device, the sensitivity of the sensor is defined by the smallest shift in resonance frequency shift one can measure, which is limited by the coherence time of the spin transition. In addition, the NV sensor has the unique ability to switch between electric- or magnetic-field detection modes, making it a universal detector system for biology [133]. Although single diamond spins are known to have the longest coherence times among all solid state systems (reaching milliseconds for isotopically purified nuclear spin free ^{12}C diamond), coherence properties for defects close to the surface are less spectacular. Therefore recent demonstration of active decoupling techniques can be important for magnetometry applications [134–136].

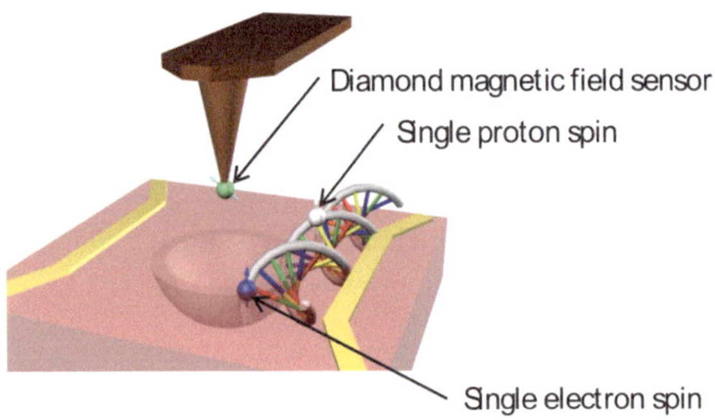

Fig. 6.34 Scanning probe single spin magnetometer

Quantum properties for single spins can be used to enhance sensitivity of diamond field sensor. It was shown that sensitivity and dynamic range of diamond magnetometers can be improved up to the limit solely imposed by Heisenberg uncertainty relations when quantum non-demolition technique s used for readout [137]. Entanglement between spins in diamond quantum registers can also improve sensitivity allowing Heisenberg scaling [138].

When combined with nano-positioning instrumentation the single spin NV defect can be used as an atomic size scanning probe vector magnetometer. The magnetic resonance imaging with single spin sensitivity will grant new information about the dynamics of a broad range of biological processes at the nanometer scale, e.g. unravel signal cascades responsible for deceases or accessing structure of single proteins under physiological conditions. Furthermore, it will potentially allow the resolution of the structure of single biomolecules under physiological conditions (using approaches similar techniques developed for liquid state NMR) [139]. First proof of principle demontrations were realized recently [140, 141].

6.2.8 Conclusions and Outlook

Quantum technology based on coherent control of diamond spins is rapidly developing during last two decades. Combination of ultra sensitive optical detection techniques, super resolution imaging and robust coherent control using NMR techniques are key ingredients towards building first quantum devices based on diamond. Ability to engineer defects with high accuracy and their long coherence time open new possibilities for so called hybrid quantum processors where NV centres are connected to different types of qubits. Interesting examples of such hybrid quantum processors are based on coupling of diamond spins to nanomechanical systems [142] and superconducting qubits [143]. Strong coupling between ensemble of spins in diamond and superconducting resonators [144, 145] and qubits [146] was demonstrated recently opening new avenues for solid state quantum information processing. There is no doubt that this field will develop rapidly in the nearest future resulting in developments of novel types of quantum technologies. Applications related to magnetometry and imaging techniques are expected to be developed further resulting in diamond based devices on the market in the next few years.

Most of experimental demonstrations were achieved using nitrogen-vacancy defects. It is worth to mention that diamond hosts more than 500 documented colour centres [147]. Recently nickel [148–150], chromium [151] and silicon-vacancy [152, 153] defects were identified as promising candidates for quantum technologies in diamond. The energy level structure for these systems is less studied and technology of engineering is less developed compared to NV defects. However attractive optical properties (strong zero photon line) are important for the integration of diamond quantum devices in photonics platform and imaging applications.

6.3 Quantum Effects in Carborundum: Application of Magnetic Resonance. Point Colour Centres in SiC as a Promising Basis for Nanostructure Single-Defect Resonance Spectroscopy with Room Temperature Controllable Spin Quantum States

Atomic-scale colour centres in bulk and nanocrystalline silicon carbide are promising for quantum information processing, photonics and sensing at ambient conditions. Their spin state can be initialized, manipulated and readout by means of optically detected magnetic resonance. It has been shown that there are at least two families of colour centres in the silicon carbide with S = 1 and S = 3/2, which have the property of optical alignment of the spin levels and allows a spin manipulation. For the S = 3/2 family, the ground state and the excited state were demonstrated to have spin S = 3/2 and a population inversion in the ground state can be generated using optical pumping, leading to stimulated microwave emission even at room temperature. By controlling the neutron irradiation fluence, the colour centres concentration can be varied over several orders of magnitude down to a single defect level. Furthermore, these atomic-scale spin centres can be also attractive for local or environment sensing. Several, separately addressable spin-3/2 centres have been identified in the same crystal for each polytype, which can be used either for magnetic field or temperature sensing. Some of these spin centres are characterized by nearly temperature independent zero-field splitting, making these centres very attractive for vector magnetometry. Contrarily, the zero-field splitting of the centres in the excited state exhibits a giant thermal shift, which can be used for thermometry applications. Finally coherent manipulation of spin states has been performed at room temperature and even at temperatures higher than room temperature by hundreds of degrees. Silicon carbide is taking on a new role as a flexible and practical platform for harnessing the new quantum technologies.

6.3.1 Introduction

Atomic-scale colour centres in bulk and nanocrystalline silicon carbide are promising for quantum information processing, photonics and sensing at ambient conditions. Their spin state can be initialized, manipulated and readout by means of optically detected magnetic resonance. It has been shown that there are at least two families of colour centres in the silicon carbide with S = 1 and S = 3/2, which have the property of optical alignment of the spin levels and allows a spin manipulation.

For the S = 3/2 family, the ground state and the excited state were demonstrated to have spin S = 3/2 and a population inversion in the ground state can be generated using optical pumping, leading to stimulated microwave emission even at room temperature. By controlling the neutron irradiation fluence, the colour centres concentration can be varied over several orders of magnitude down to a single defect level. Furthermore, these atomic-scale spin centres can be also attractive for local or environment sensing. Several, separately addressable spin-3/2 centres have been identified in the same crystal for each polytype, which can be used either for magnetic field or temperature sensing. Some of these spin centres are characterized by nearly temperature independent zero-field splitting, making these centres very attractive for vector magnetometry. Contrarily, the zero-field splitting of the centres in the excited state exhibits a giant thermal shift, which can be used for thermometry applications. Finally coherent manipulation of spin states has been performed at room temperature and even at temperatures higher than room temperature by hundreds of degrees. Silicon carbide is taking on a new role as a flexible and practical platform for harnessing the new quantum technologies.

Until recently, practical applications of semiconductors have been associated with using of defect ensembles. The unique quantum properties of the nitrogen–vacancy (NV) colour centre in diamond [24] have motivated efforts to find defects with similar properties in silicon carbide (SiC), which can extend the functionality of such systems not available to the diamond. NV colour centre—a nitrogen atom substituted for carbon with an adjacent lattice vacancy—is solid-state system where manipulation of the spin states of a single centre was realized at room temperature (RT) by means of optically detected magnetic resonance (ODMR). Spin-dependent optical excitation cycle is implemented to the NV centre that leads to the optical alignment of a triplet sublevels (S = 1) of the centre ground state. Such systems are the most prominent objects for applications in new generation of supersensitive magnetometers, biosensors, single photon sources [24, 77, 91, 154–157]. The diamond NV defect is in many ways the ideal qubit but it is currently quite difficult to fabricate devices from diamond. It remains difficult to gate these defects electrically. A search to find defects with even more potential ("better than excellent") has now been launched [158–163]. Silicon carbide is a compound wide-band-gap semiconductor with chemical, electrical, optical and mechanical properties that make this material very attractive for applications under extreme conditions and can open up a whole new world of scientific applications in spintronics.

A convincing point with SiC is that, similar to diamond, the stable spinless nuclear isotopes guarantee long dephasing times. Unusual polarization properties of various vacancy related centres in SiC (labelled as P3, P5, P6 and P7) were observed by means of electron paramagnetic resonance (EPR) under optical excitation and reported for the first time in the works of [164, 165], later in [158, 159, 166–171]. One of the main questions was to establish whether the observed EPR spectra belong to the ground or to excited state (similar problem existed before also for NV defects in diamond). EPR experiments performed at high frequency and at very low temperatures in darkness excluded the possibility of thermal or optically excited states and as a result it was proved that the EPR spectra of P3, P5, P6 and P7 defects belong to the ground state for all the defects [158, 159, 171]. It has been shown that there are at least two families of colour centres in the silicon carbide, which have the property of optical alignment of the spin levels and allows a spin manipulation at ambient conditions. (i) Family of silicon-carbon divacancy of the neighboring positions with covalent molecular bond and having a triplet ground state (S = 1). The symmetry of these centres is due to the direction of connection between the silicon and the carbon, zero-field splitting for these centres as in the case of NV-center in diamond is in the gigahertz range. (ii) Family of the centres which are formed by negatively charged silicon vacancies V_{si}^{-} in the paramagnetic state that is noncovalently bonded to the neutral carbon vacancy V_{C}^{0} in the non-paramagnetic state, located on the adjacent site along the SiC symmetry c axis having quadruplet ground and excited states (S = 3/2).

EPR, ODMR, electron spin echo (ESE) and electron nuclear double resonance (ENDOR) investigations presented here suggest that silicon vacancy (V_{Si}) related point centres in SiC possess properties the similar to those of the NV centre in diamond. Depending on the defect type, temperature, SiC polytype, and crystalline position, two opposite schemes have been observed for the optical alignment of the ground state spin sublevels population of the V_{Si}-related defects upon irradiation with unpolarized light. Spin ensemble of V_{Si}-related defects are shown to be prepared in a coherent superposition of the spin states even at room temperature. Zero-field (ZF) ODMR shows the possibility to manipulate of the ground state spin population by applying radiofrequency field and using the infrared optical pumping which is compatible with optical fibers. These altogether make V_{Si}-related defects in SiC very favorable candidate for spintronics, quantum information processing, magnetometry. In general, point colour centres in SiC can be considered as a promising basis for single-spin, single-photon spectroscopy with room temperature controllable quantum states.

The recent experiments demonstrated [158, 159, 162, 172–197] that several highly controllable defects exist in SiC, and some of them can be manipulated at room temperature or even higher. This paper presents a review of the recent studies of the defects with optically induced inverse population in three main SiC polytypes: hexagonal 4H-SiC and 6H-SiC and rhombic 15R-SiC. Silicon vacancy related colour centres in SiC are demonstrated to be a promising quantum system for single-spin and single-photon spectroscopy.

6.3.2 Experimental

Crystal of 4H-, 6H- and 15R-SiC polytypes were grown by the sublimation technique, they have been irradiated with fast neutrons to a dose up to 10^{18} cm^{-2} or electrons (1–2 meV). Typical concentration of V_{Si} related colour centres was $\sim 10^{15}$ cm^{-3}. The epitaxial SiC samples were also investigated. For experiments on single quantum object concentration varied in the range of several orders of magnitude by irradiation with electrons or neutrons. In addition, 6H-SiC samples of high crystalline quality have been grown by the modified Lely method. The high-temperature (2700 °C) seedless crystal growth is driven by the temperature gradients within the crucible, resulting in a pressure gradient and thus, in a mass transport. The process is followed by the subsequent fast cooling, which 'freezes' the defects within the lattice at low densities (for details see [198]). Taking into account the retrograde character of the nitrogen solubility with temperature, the doping level of the 6H-SiC crystals is below 10^{17} $cm^{-3}$and the compensation degree of nitrogen donors is ca. 20%. The V_{Si}-related defect concentration was estimated to be about 10^{12} cm^{-3}.

The EPR and ENDOR spectra were detected at X- (9.3 GHz) and W- (95 GHz) bands on a continuous wave (cw) and pulse spectrometers in the temperature range of 4–300 K. The ODMR experiments were performed in an X-band and Q-band (35 GHz) with direct optical access. In some experiments a tunable diode laser system (linewidth below 1 MHz) to resonantly excite into the zero-phonon line (ZPL) transition was used. The laser is focused onto the sample to a power density about 1 W/cm^2. The photoluminescence is passed through a 900 or 950 nm long-pass filter and detected by a fast Si photodiode or by photomultiplier. The microwave radiation was chopped, and the output signal at the photodiode was locked-in. The ODMR signal was obtained as a normalized change in photoluminescence (PL) of the phonon sideband. In some experiments confocal microscope and CCD camera were used.

The photo-induced microwave emission is measured in a commercial X-band spectrometer at nominal microwave power of 10 mW. A diode laser operating at 808 nm is used to excite all types of V_{Si} defect through phonon-assisted absorption at room temperature.

To link optical and EPR fingerprints of the colour centres a techniques known as flash-induced time-resolved EPR was used at the X-band which is also known as Directly-Detected EPR (DD-EPR) because the signal were taken directly from the microwave mixer, 100 kHz magnetic field modulation of the spectrometer was switched off. In field-swept experiments, after wide-band amplification the signal was sampled by a boxcar integrator (SR 250, Stanford Research Systems), triggered by the ashes. For optical excitation a parametric oscillator LP603 pumped by a Nd-YAG laser LQ 529B (Solar Laser Systems, Byelorussia) was used. The excitation flashes were 6 ns, ca. 1.5 mJ at the sample surface with 11 Hz repetition rate and 0.8 nm FWHM bandwidth. Following the flash, a boxcar sampled the changes in the microwave power reflected from the cavity. The signals were sampled for the time 0.15–1.5 μs after the flash.

6.3.3 Vacancy Related Atomic Scale Centres in SiC as a Promising Quantum System for Single-Spin and Single-Photon Spectroscopy

6.3.3.1 Two Families of Vacancy-Related Centres with S = 1 and S = 3/2 in SiC

Figure 6.35 shows two families of spin centres in 6H-SiC and 15R-SiC with unique mechanism of an optical alignment of the spin sublevels. The first family, labelled as a P6 and P7 after [165], is a silicon-carbon divacancy of the neighboring

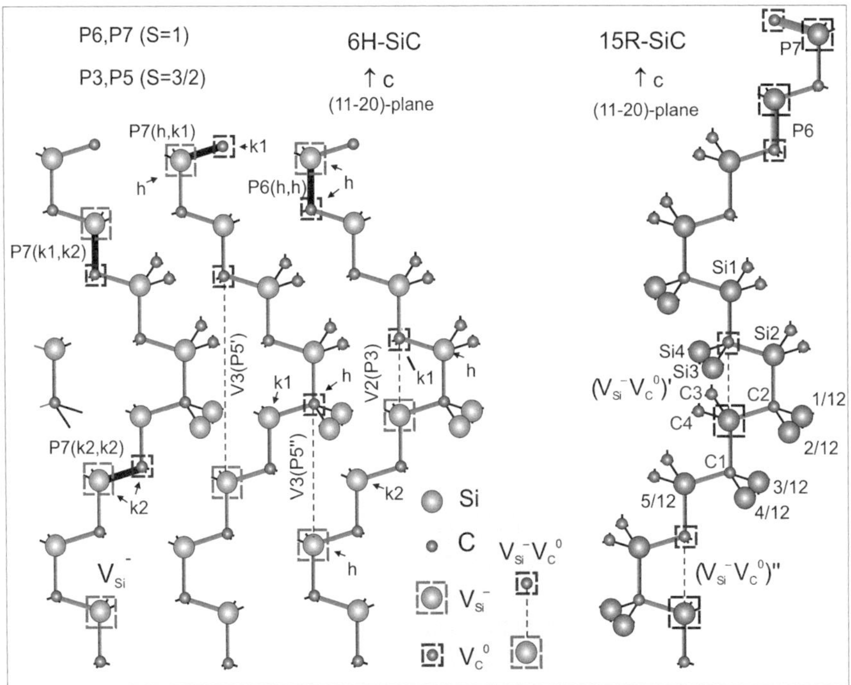

Fig. 6.35 Models showing possible configurations of two families of V_{Si}-related centres in 6H-SiC and 15R-SiC lattice in (11–20) plane: (i) S = 1, the nearest-neighbor (NN) divacancy with molecular bond (P6, P7); (ii) S = 3/2, negatively charged silicon vacancy that is noncovalently bonded to the neutral carbon vacancy, located on the adjacent site along the SiC symmetry c-axis (P3, P5)

positions with covalent molecular bond and having a triplet ground state (S = 1). The symmetry of this centre is due to the direction of connection between the silicon and the carbon, zero-field splitting for these centres as in the case of NV-center in diamond is in the gigahertz range. There are two types of the Si-C bonds in SiC: parallel to the c-axis and inclined at an angle of 70° giving rise for two possible modification of the defect—P6 and P7 centres. Each type of centre P6 or P7 is divided into several possible configurations, as there are different lattice positions in the crystal (k1, k2, h in 6H-SiC and k1, k2, k3, h1 and h2 in 15R-SiC). The second family of the centres, labelled as a V1, V3 (P5) and V2(P3) in 6H-SiC, which will be mainly discussed in this paper, is a negatively charged silicon vacancy V_{si}^{-} in the paramagnetic state that is noncovalently bonded to the neutral carbon vacancy V_{C}^{0} in the nonparamagnetic state, located on the adjacent site along the SiC symmetry c axis having quadruplet ground and excited states (S = 3/2).

Table 6.1 presents the available data on various parameters of the second family of colour centres with spin S = 3/2: zero-phonon line (ZPL) energy/wavelength at 10 K, labelled as V1, V2, V3 and V4 in the most common SiC polytypes; values of zero-field splitting (ZFS) Δ (Δ = 2D) and g-factor for each colour centre measured at room temperature for the ground state. Optically spin alignment level schemes at B = 0 are presented for RT. For V1, V3 centres in 6H-SiC, this scheme depends on the temperature and is inverted in the temperature of about 30 K.

Figure 6.36 shows EPR spectra of the S = 1 family (P6 and P7 centre) measured for two orientations θ = 0° (B || c) and θ = 70° in n-irradiated 6H-SiC crystal: (a) EPR spectra at 1.2 and 1.5 K in darkness recorded by the electron spin echo (ESE) technique at W-band; (b) X-band EPR spectra at 7 K recorded under optical illumination. The high-frequency EPR experiments were performed at low temperatures in total darkness, which excluded the possibility of thermal or optical population of the excited state and P6 and P7 centres and were concluded to have the triplet ground state. The intensities of the low and high-field fine-structure components measured in the EPR spectra by ESE at temperatures of 1.2 and 1.5 K sharply differ from each other because of a strong difference in the populations of triplet sublevels at low temperatures and large Zeeman splitting. Intensity ratio of these components gives direct information on the temperature of the sample and allows the positive sign of the fine-structure splitting D (for V2 centers) to be determined.

The EPR spectra can be fitted well by standard spin Hamiltonian (see Chap. 1)

$$H = g\mu_B \mathbf{B} \cdot \mathbf{S} + D[\mathbf{S}_Z^2 - 1/3S(S+1)],$$

where the first and the second terms correspond to the Zeeman interaction and fine structure splitting, respectively, μ_B is the Bohr magneton, g is electron g factor, and S_Z is the projection of the total spin on the symmetry axis of the centre. Without an external magnetic field (B = 0) the ground state is split due to the presence of the axial crystal field with fine-structure parameter D. For S = 1 spin state, the zero-field splitting Δ between $M_S = 0$ and $M_S = \pm 1$ sublevels is equal to D (Δ = D) and for S = 3/2 spin state, the ZFS between $M_S = \pm 1/2$ and $M_S = \pm 3/2$

Table 6.1 Zero-phonon line (ZFL) energy and wavelength (T = 10 K); values of zero-field splitting (ZFS) Δ (Δ = 2D) and g-factor of the each centre at room temperature for the ground-state spin centres of the second family with S = 3/2 in 4H-, 6H-, and 15R-polytypes of SiC

Polytype	4H-SiC		6H-SiC			15R-SiC		
ZFL	V1	V2	V1	V2	V3	V2	V3	V4
E (eV/nm)	1.438/862	1.352/917	1.433/865	1.397/887	1.368/906	139.9/886.5	1.372/904	1.352/917
Δ (MHz/ 10^{-4} cm^{-1}) D>0 (V2, V4) D<0 (V1, V3)	39/13	66/22	27/9	128/42.7	27/9	139.2/46.4	11.6/3.87	50.2/16.7
g-factor	2.0032	2.0032	2.0032	2.0032	2.0032	2.005(1)	2.005(3)	2.005(3)
Optically spin-alignment levels at RT, B = 0	–	±1/2	±1/2 ±3/2 at T < 30 K	±3/2	±1/2 ±3/2 at T < 30 K	±3/2	±1/2	±1/2

In addition optically spin alignment level schemes at B = 0 are presented

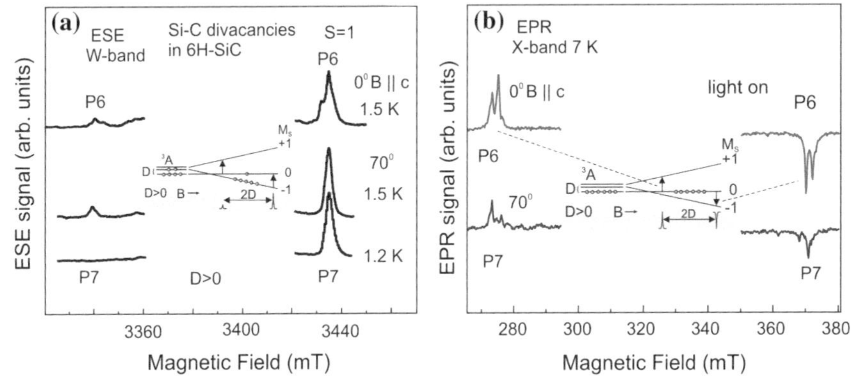

Fig. 6.36 **a** ESE-detected EPR spectra (W-band) of divacancy centres (P6 and P7) in n-irradiated 6H-SiC crystal measured for two orientations θ = 0° (B ‖ c) and θ = 70° at 1.2 and 1.5 K in darkness. **b** X-band cw EPR spectra measured at 7 K under optical illumination. *Insets* show the energy-levels schemes with S = 1 for darkness measurement with Boltzmann distribution (**a**) and with optical alignment of the spin sublevels where the spin sublevel M_S = 0 for the ground state is predominantly populated (**b**)

sublevels is equal to 2D (Δ = 2D, see parameters in Table 6.1). Spin state S = 3/2 is confirmed by ODMR, ESE and ENDOR measurements, observed angular dependencies show that all the centres are oriented along the c axis for each SiC polytype.

The EPR spectra of the S = 1 P6 centre in e-irradiated 15R-SiC crystal, which were first observed in the present study, are presented in Fig. 6.37 in darkness and under optical excitation with wavelength 789 nm. There are two types of centres (labelled as P6′ and P6″) with the following parameters: P6′ D = 46.85 mT and P6″ D = 45.60 mT, g factor of P6′ centre by a small amount more than g factor of P6″ centre. The optically induced inverse population of the P6′ and P6″ spin sublevels is clearly visible in the spectrum. In the figure also visible EPR signal of unidentified defect (UD) with S = 1 and the fine structure splitting D equal to 42.25 mT. An inverse population in this spectrum is not observed, however, one should take into account the fact that the divacancy centres have optical absorption and luminescence bands in the range of longer wavelengths. Low intensity of P6′ and P6″ signals may also be attributed by this reason. At least two types of P6 centre could be observed as there are different lattice positions of Si and C vacancies in the crystal (k1, k2, k3, h1 and h2 in 15R-SiC).

Figure 6.37 also shows EPR spectra of other centres, designated V2, V3 and V4, for which the population inversion were also observed. These spectra belong to the S = 3/2 family with axial symmetry along c-axis and smaller values of zero-field splitting (see Table 6.1). We will return to these centres in 15R-SiC later on, but first consider the S = 3/2 centres in 4H- and 6H-SiC polytypes. Possible configurations of these centres in 6H- and 15R-SiC are shown in Fig. 6.35.

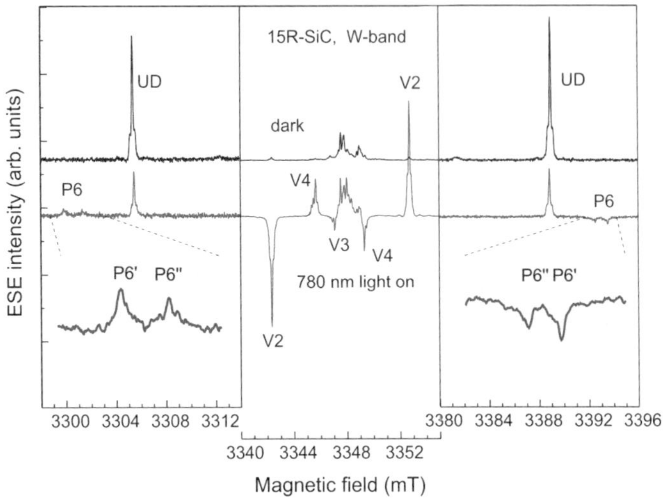

Fig. 6.37 ESE-detected EPR spectra (W-band) of divacancy P6 centres (P6′ and P6″) in e-irradiated 15R-SiC crystal measured at orientation θ = 0° (B ∥ c) at RT in darkness and under optical illumination with wavelength 789 nm

The cw X-band EPR spectra of the S = 3/2 family [colour centres V1, V3 (P3) and V2 (P5)] detected at 6H-SiC (a) and 4H-SiC (b) at different temperatures with the magnetic field parallel to the c-axis are shown in Fig. 6.38. The EPR spectra detected under continuous optical illumination (solid lines) and without optical illumination (dashed lines). Vertical bars indicate the positions of the lines for the

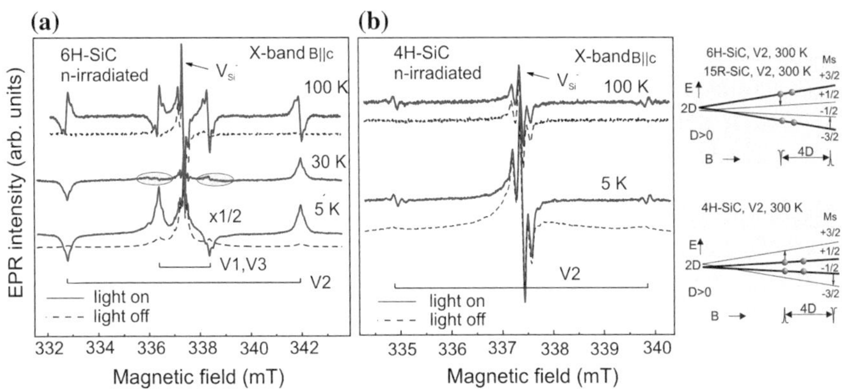

Fig. 6.38 cw X-band EPR spectra of the S = 3/2 family centres—V1, V3 (P3) and V2 (P5)—colour centres detected at 6H-SiC (**a**) and 4H-SiC (**b**) at different temperatures with the magnetic field parallel to the c-axis. Spectra detected under continuous optical illumination (*solid lines*) and without optical illumination (*dashed lines*). *Vertical bars* indicate the positions of the lines for the V_{Si}-related centres. (*right*) Two energy-levels schemes for opposite types of optical alignment of the spin sublevels for the ground state of the V_{Si}-related centres in magnetic field (S = 3/2)

V_{Si}-related centres. The central signal in Fig. 6.38a, b, marked by V_{Si}^{-}, accompanied by a set of hyperfine lines with a splitting of 0.29 mT and an intensity ratio to the central line of about 0.3. Such signal was attributed to the negatively charged silicon vacancy V_{Si}^{-} with S = 3/2 and a fine structure splitting D close to zero [171, 199]. From spectra demonstrated on Fig. 6.38a a saturation effect of the signal can be clearly seen in deviation from the first derivative of the EPR signal. Under optical pumping the intensity of the EPR spectra grows substantially and a phase reversal is observed for one of the two transitions in each pair of lines. For instance, V1, V3 centres in 6H-SiC at low temperatures (below 30 K) and V2 in 4H-SiC at all temperatures, demonstrate a phase for the high-field transition, while V1, V3 in 6H-SiC at high temperatures (above 30 K) and V2 in 6H-SiC at all temperatures demonstrate a phase reversal for the low-field transition. As a result of the optical pumping, the distribution of the populations of the spin sublevels in the ground state departs from a Boltzmann distribution. Even an inverse population is created between certain spin sublevels, and emission rather than absorption is detected for one of the transitions (high-field or low-field, depending on the polytype, crystal position, and temperature). Based on a study of the EPR spectra of the V_{Si}-related defects [158, 171] at 95 GHz and low temperatures (1.2–1.5 K) in full darkness, it was unambiguously shown that these spectra belonged to the high-spin ground state and that the sign of the fine-structure splitting D (at least, for V2 centres in 6H-SiC and 4H-SiC) is positive (D > 0). At a temperature of 30 K and with optical excitation the signal V1, V3 in 6H-SiC disappears indicating that at this temperature the equal spin sublevel populations are realized.

To explain the photokinetic processes leading to spin alignment under optical pumping, the presence of the excited metastable state is suggested and a spin-dependent intersystem crossing (ICS) between such state, the metastable state and the ground state can be proposed.

Figure 6.39a presents a typical low-temperature (T = 4 K) photoluminescence (PL) spectrum of 6H-SiC single crystal obtained with sub-bandgap excitation. The PL consists of sharp zero phonon lines and their sideband phonon replicas. Three of these ZPLs—labeled as V1, V2 and V3—originate from S = 3/2 colour centre ensemble with different configurations.

Light-enhanced EPR experiments are presented in Fig. 6.39b. The amplitudes of the V_{Si} EPR lines depend on the population difference between the particular spin sublevels involved. In the dark the difference is due to the Boltzmann factor and the amplitudes are negligibly small. The optical excitation of spin centre and following relaxation preferentially pump the system into certain spin sublevels of the S = 3/2 ground state. This results in light-enhanced EPR, as exactly observed in the experiment presented in Fig. 6.39b. The pair of outer lines at magnetic fields 331.9 mT and 341.0 mT appears in the EPR spectrum under optical excitation into the V2 ZPL with the energy 1.397 eV. The zero-field splitting in the V2 centre of 127 MHz (see Table 6.1), which is in agreement with the earlier reported value. Remarkably, when the excitation is not resonant with the V2 ZPL, no EPR lines are observable at these magnetic fields.

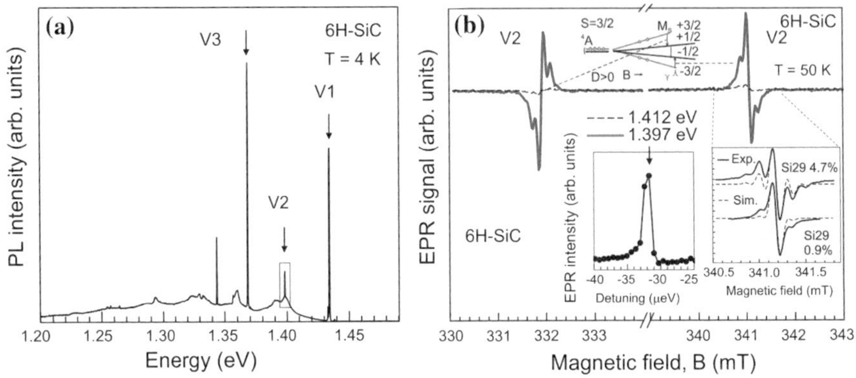

Fig. 6.39 **a** PL spectrum obtained under excitation with a He–Ne laser (E = 1.959 eV). The ZPLs V1, V2 and V3 of the corresponding V_{Si} related colour centres are labeled by *arrows*. ODMR was detected in the spectral range from 1.20 to 1.34 eV. **b** Light-enhanced X-band (9.4 GHz) EPR spectrum measured for orientation B || c under excitation into the V2 ZPL, E = 1.397 eV (*solid line*) and above-ZPL excitation, E = 1:412 eV (*dashed line*), recorded at temperatures T = 50 K. *Inset* (*top*): Energy-levels scheme for optical alignment of the spin sublevels for the ground state of the V2 colour centre in magnetic field, S = 3/2. *Inset* (*bottom*, *left*): The amplitude of the V2 EPR line at 341.0 mT as a function of the excitation energy with higher spectral resolution in the vicinity of the optical resonance. *Inset* (*bottom*, *right*): Comparison of the EPR signals recorded in a 6H-SiC crystal with natural ^{29}Si isotope content and in a 6H-SiC crystal with depleted ^{29}Si isotope content; *solid line* experiment, *dashed line* simulation

Inset (bottom, left) in Fig. 6.39b shows how the EPR amplitudes depend on the excitation energy. For energies E $\neq$ E(V2) the spin pumping is inefficient, and for E = E(V2) a very sharp resonance is detected. The spectral width of this line is about 2 μeV, which is comparable with the typical spectral linewidth of single defects. A similar behavior is observed for other colour centres in S = 3/2 family.

Based on the data of Fig. 6.39 an important conclusion can be drawn: the colour center spins are only addressed when the optical resonance and EPR conditions are simultaneously fulfilled. The contrast—i.e., the ratio of the EPR signal between on and off resonant optical excitation—is above 200. This is very similar to the double radio-optical resonance in atoms. The difference to atoms, is that due to the local environment, the optical resonance and EPR energies are individual for each defect. Additionally, they can be changed by local electric and magnetic fields in the range of about 50 μeV. This can eventually provide a spectroscopic tool to selectively address and manipulate coupled spin qubits by varying the excitation energy or alternatively by tuning the double radio-optical resonance conditions for a given spin qubit. EPR line can be substantially narrowed by changing the content of the ^{29}Si isotope with nuclear magnetic moment. This effect is shown in the inset (bottom, right), where comparison of the V2 EPR signals recorded in a 6H-SiC crystal with natural ^{29}Si isotope content and in a 6H-SiC crystal with depleted ^{29}Si isotope content; solid line—experiment, dashed line—simulation.

6.3.3.2 Room Temperature Coherent Spin Alignment of S = 3/2 Vacancy-Related Centres in 4H- and 6H-SiC

Further investigation of the V_{Si}-related centre properties were produced by means of DD-EPR at temperature T = 300 K. Signals of V_{Si}-related defect recorded at RT in the 4H and 6H-SiC for orientation of the magnetic field perpendicular to the c-axis is shown in Fig. 6.40 (top) after excitation of the samples by optical flash into the absorption band of the defect at 890 nm. Under optical pumping, a phase reversal is observed for the V2 centre in 4H-SiC even at RT. For 6H-SiC the a phase reversal is observed for V2 and for V1, V3 centres as well. The optical alignment scheme for 4H-SiC is the same as discussed before and shown on Fig. 6.38, though for 6H-SiC we obtain somewhat like RT maser effect even at zero magnetic field.

To determine time dependent spin properties of the optically aligned ground spin state of the vacancy-related defect, measurements of the light flash induced DD-EPR signal at the resonant field as a function of time delay after flash were performed. The results of such experiments for the V2 in 4H-SiC are presented in Fig. 6.40a. Figure 6.40b shows the transient nutations at RT for three values of microwave power P at resonant magnetic field B_0 = 321.5mT (low field transition). The transition nutation decays due to inhomogeneity of the B_1 microwave field over the sample. In addition, the resonance frequencies are also spread around some mean value of resonance Larmor frequency $\langle\omega_0\rangle$ leading to the inhomogeneous line broadening. Clearly, the observed oscillatory behavior demonstrates that the probed spin ensemble can be prepared in a coherent superposition of the spin states at

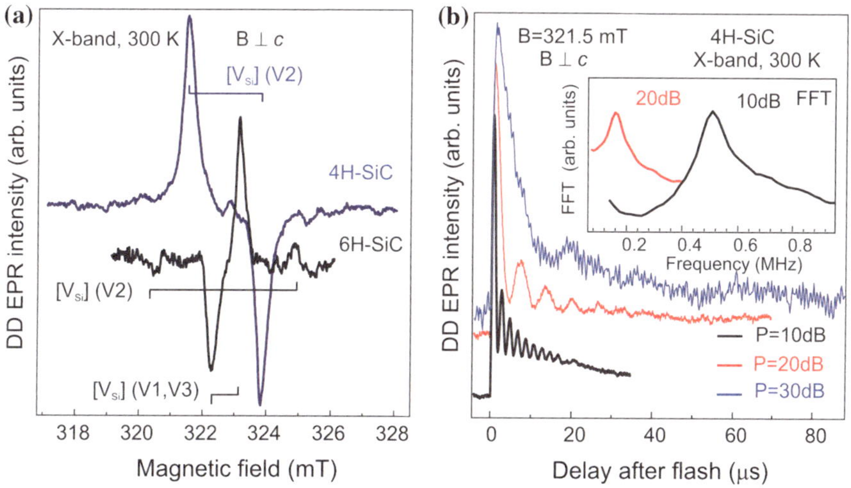

Fig. 6.40 **a** DD-EPR spectra of the V_{Si}-related centers detected in 4H-SiC and 6H-SiC at temperature T = 300 K. **b** Transient nutations for the defect in 4H-SiC at RT shown for three values of microwave power. (*Inset*) Corresponding fast Fourier transform (FFT)

resonant magnetic fields at RT. The population difference of spin states becomes modulated in time with the Rabi frequency given by $\omega_1 = \gamma B_1$, where γ is the gyromagnetic ratio for the electron. The Fourier transforms corresponding to observed oscillations are presented in the insets in Fig. 6.40 (bottom). The Rabi frequencies are 0.02, 0.16 and 0.5 MHz at P = 30, 20, and 10 dB, respectively. Rabi oscillations decay with a characteristic time constant τ_R that depends on the microwave power. Empirically, τ_R is generally smaller than the spin-spin relaxation time (T_2), thus, the lower limit of T_2 is about 80 μs at RT.

6.3.3.3 Zero-Field ODMR Experiments for S = 3/2 Spin Centres

Low temperature experiments. Low temperature zero-field ODMR experiments on quenched 6H-SiC sample were performed in order to show the possibility of manipulation of the ground state spin population by applying of the radiofrequency which corresponds for the ZFS of the silicon vacancy related defect. Here we demonstrate ZF ODMR effect for V2 and V3 ZFLs just as an example. Photoluminescence-excitation spectrum of V2 was recorded after single-mode Ti-sapphire laser excitation between 337.0 and 340.17 THz at 4 W/cm^2 and shown in Fig. 6.41a curve 1. Curve 2 in the same figure represents increasing of the PL intensity after resonant radio-frequency field 130 MHz have been applied. Spectrum in Fig. 6.41b shows the ODMR spectra obtained with excitation of the V2 ZPL and detection at 937 nm. The ODMR spectrum of V2 has its main feature at 130 MHz. As can be seen from the ODMR spectrum resonant radio-frequency field destroyed optically aligned ground state what causes the increasing of the corresponding photoluminescence. Figure 6.7c shows the high-resolution PL-excitation spectra of V3 colour centres in 6H-SiC with single-mode laser excitation between 330.72 and 330.85 THz at 1 W/cm^2 (1). Curve (2) is the PL-excitation spectrum obtained with a resonant RF field at 28 MHz and the same laser power. The ODMR spectra obtained with excitation of the V3 ZPL with the maximum ODMR effect detected at 958 nm is presented in Fig. 6.41d. The similar effects were observed for V1 line in 6H-SiC.

Strong ODMR effects on the intensity of the high-resolution fluorescence-excitation spectra obtained with a radio frequency of 28 MHz for the V1, V3 lines and 130 MHz for the V2 line were observed between 130 and 230% for different lines. The large ODMR effect indicates the presence of a bottleneck state, which can be emptied by the resonance RF field. Thus, RF quanta of 28 and 130 MHz can efficiently control optical quanta in the range of 330–350 THz. For the V2 line, saturation was not observed up to the available power of 4 W/cm^2, which makes this line the most promising for further detecting single defects.

The confocal arrangement for fluorescence-excitation spectroscopy and zero-field ODMR. To detect a small number of defects, even down to single defect, several conditions must be met. First, the excitation source must be at resonance with only one small group of defects (down to single defect) in the optical illuminated volume. Second, the PL from this small group of defects (single defect)

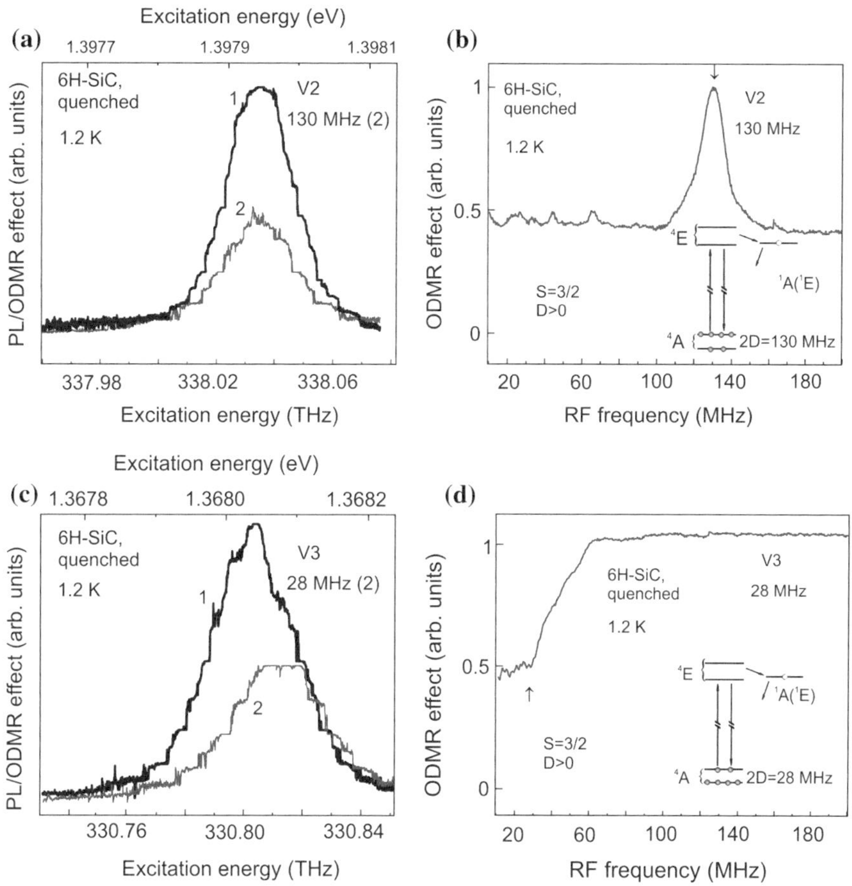

Fig. 6.41 **a** The high-resolution PL-excitation spectra of V2 colour centres in 6H-SiC with single-mode laser excitation between 337.95 and 341.07 THz at 4 W/cm^2 (*1*). Curve (*2*) is the PL-excitation spectrum obtained with a resonant RF field at 130 MHz and the same laser power. **b** The ODMR spectra obtained with excitation of the V2 ZPL with maximum ODMR effect detected at 937 nm. **c** The high-resolution PL-excitation spectra of V3 with single-mode laser excitation between 330.72 and 330.85 THz at 1 W/cm^2 (*1*). Curve (2) is the PL-excitation spectrum obtained with a resonant RF field at 28 MHz and the same laser power. **d** The ODMR spectra obtained with excitation of the V3 ZPL with the maximum ODMR effect detected at 958 nm

should be larger than the background signal. To meet the first requirement confocal optics is used. To meet the second requirement, the PL should be detected in the phonon-side part of the ZPL to suppress excitation light.

The V_{Si}-related defect has a broad phonon-side-band, which is well separated from the ZPLs, allowing this method of detection. Figure 6.8a shows the high-resolution fluorescence excitation spectra of the V3 ZPL in the thermally quenched 6H-SiC single crystal as detected at phonon side bands at 1.2 K using the confocal arrangement and excitation by the single-mode tunable laser with bandwidth of 0.5 MHz (1). Spectrum (2) shows the excitation spectrum obtained with the simultaneous presence of a resonant radiofrequency field at 28 MHz. One can see that the excitation with resonant radiofrequency of 28 MHz, drastically affects the intensity of the luminescence. It is evident from a comparison of Fig. 6.41a, b that the spectra in the sample measured with excitation of defects in a large spot of about 1 mm^3 (Fig. 6.41a) and in a small spot of about 1 μm^3 using the confocal arrangement (Fig. 6.42a), are similar.

Figure 6.42b shows, the high-resolution fluorescence-excitation spectra around the V1 region (top) and around the V3 region (bottom) in the thermally quenched and annealed at 750 °C 6H-SiC sample. The upper scales correspond to the V1 region (top) and the bottom scales to the V3 region (bottom). A remarkable result is the correlation between the groups of ZPLs of the inequivalent quasicubic sites. These lines can be described as arising from different positions of small groups of vacancy related centers near some extended nonparamagnetic defects thus indicating the possibility of the optical pumping of the qubits states.

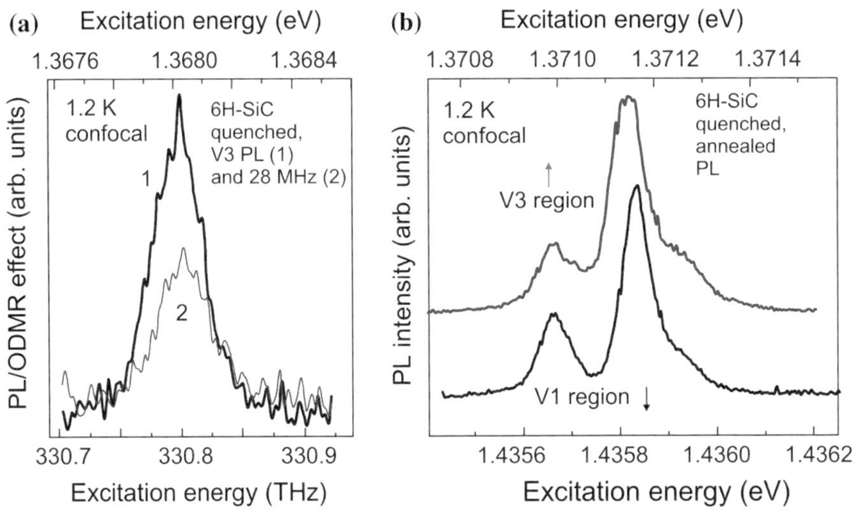

Fig. 6.42 The high-resolution, fluorescence excitation spectra of the V3 ZPL in the thermally quenched 6H-SiC sample no. 3. The single-mode laser excites the sample in the confocal arrangement and the detection is at the phonon band (*1*). The ODMR excitation spectrum obtained in the presence of a resonant rf field at 28 MHz at the same laser power (*2*). **b** High-resolution fluorescence excitation spectra are presented around the V1 region (*bottom*) and around the V3 region (*top*) in the thermally quenched and annealed 6H-SiC sample. The single-mode laser, with a bandwidth of 0.5 MHz, excites the sample in the confocal arrangement and the detection takes place at phonon replicas. The scales correspond to the V1 region (*bottom*) and the V3 region (*top*). There is a difference in the spectra recorded at different points in the crystal (excitation energy, the intensity ratio between the individual lines)

The high-resolution fluorescence experiments also indicate that the positions and the shape of the lines differ between the investigated samples apparently as a result of the annealing treatments. A short anneal at 750 °C changes the structure of ZPLs considerably. Several groups of lines in the V1 and V3 regions appear. The different components of the V1 and V3 lines show a shift and the V2 line disappears after heating up to 750 °C. This temperature of 750 °C is the annealing temperature of the silicon vacancy in irradiated SiC. Sequential annealing of the silicon vacancies was performed in order to reduce their amount and allowing the detection of the small groups of centers up to the single vacancy.

Thus, the annealing lead to the appearance of several groups of lines with different intensity and width. Surprisingly narrow ZPLs for Si vacancies with a width less then 0.05 meV are observed. To our knowledge, the line widths observed in these experiments are the narrowest of those detected so far in SiC. The narrowness of the lines observed evidences the small concentrations of the Si vacancies. It is known that annealing induces a migration of the silicon vacancies through the SiC lattice. As a result the vacancy can be captured by extended defect structures, like, e.g., stacking faults (SF) or dislocations. One of the problems that need to be investigated and understood in order to fully develop SiC-based technology is the occurrence of SF. Due to the small SF energy compared to other semiconductors such as Si or GaAs it is relatively easy to develop extended SF regions in SiC crystals, which, if electrically active, can seriously affect the device performance. SF, unlike point defects and surfaces, are not associated with broken or chemically perturbed bonds. Vacancies seem to tend to be attracted to a SF region due to strains.

The limit of the spatial resolution in the confocal experiments is about 1 μm3 and is a factor 100 larger than the defect spacing needed for entanglement. To address chosen pairs of qubits one can exploit the randomness of the interdefect distance in standard fabrication and doping. Light at different wavelengths excites different defect spins in the confocal volume, allowing manipulation of the entanglement of different qubits. Another attraction of the silicon vacancies in SiC is that the ZPLs occur in the range 850–920 nm, coinciding with the spectral window of silica glass optical fibers and biological systems.

Room-temperature experiments. The centre spins in SiC can be initialized and read out at ambient conditions (T = 300 K and B = 0), which is the basis for various sensing applications. Figure 6.43a shows a typical ODMR spectrum, i.e., relative change of the photoluminescence intensity ΔPL/PL as a function of applied RF at B = 0 in 6H-SiC. Two spin resonances at ν = 28 MHz and ν = 128 MHz agree well with zero-field spin splitting in the ground state (GS) of silicon vacancy related centres V3 (V1) and V2, respectively (see Table 6.1). Another spin resonances are observed at ν = 367 MHz (D = 183.5 MHz) and 1030 MHz (D = 515 MHz), RT data, we ascribe to the excited state (ES) of V3 (V1) and V2, respectively.

In order to examine the effect of temperature fluctuations, ODMR spectra have been measured in the temperature range from 10 to 320 K (Fig. 6.43b). As one can see, the zero-field splitting of the V2 and V3(V1) defects in the ground state is

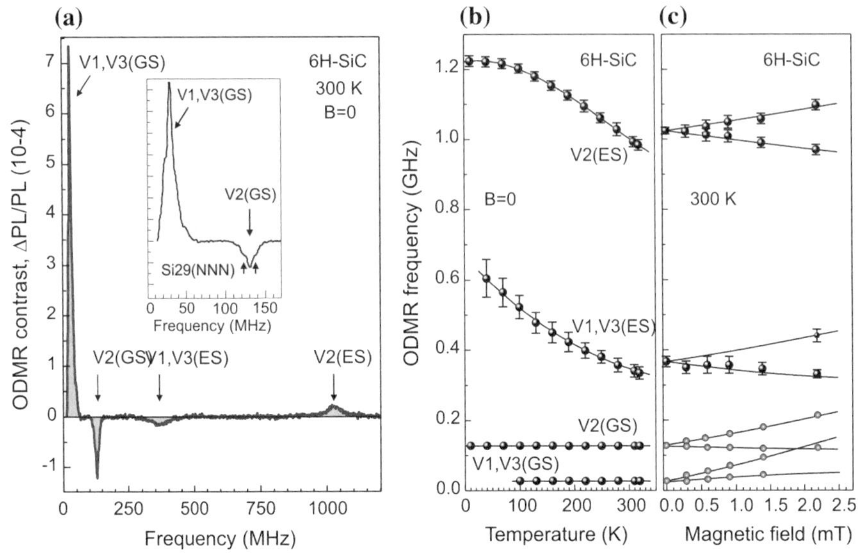

Fig. 6.43 **a** Room-temperature zero-field ODMR spectrum of V1, V2 and V3 colour centres in the ground state (GS) and in the excited state (ES) in 6H-SiC. *Inset* shows expanded-scale zero-field ODMR spectrum of V1, V2 and V3 colour centres in GS. **b** Zero-field splitting 2D for V1, V2, V3 spin centres in the ground state and in the excited states for 6H-SiC single crystal as a function of temperature. **c** ODMR frequencies as a function of magnetic field. Experimental data are shown by *symbols*. In all panels B⊥c

temperature independent within the accuracy of the experiment (a few kHz/K). In contrast to the GS, for the zero-field splitting of the V2 and V3(V1) defects in the excited state a reduction of its zero-field splitting of ca. 50%, from 0.6 down to 0.3 GHz has been observed when the temperature increases from 10 to 320 K. From the polynomial fit the thermal shift $\beta = -1.1$ MHz/K at T = 300 K was found for V2 excited state.

According to the magnetic field dependencies of Fig. 6.43c, all defects under consideration have the S = 3/2 ground and excited states.

We now consider the spin-3/2 colour centres in the 15R-SiC single crystal. First, we characterized 15R-SiC single crystal by measuring photoluminescence of spin centers. The PL spectrum was recorded under continuous illumination at wavelength of 532 nm and T = 10 K, and it exhibits four zero phonon lines that we labeled as V1, V2, V3 and V4 [Fig. 6.10 (right), Table 6.1].

Under optical excitation, EPR spectra which consist of three pairs of transitions labeled V2, V3, V4 (see Fig. 6.37) were observable up to the temperature T ~ 250 °C. Without optical excitation the signals were near the noise threshold. To determine the correspondence between PL and EPR spectra we used resonant optical excitation at the ZPL wavelength and measure the time resolved EPR response. All V1–V4 centres in 15R-SiC have their own optical fingerprint; however, resonant excitation into the V1 line (863.2 nm) did not give rise to the EPR signal; thus, the V1 centre is EPR silent and we do not discuss it in what follows. For each of the V2, V3, V4 signals we observed phase reversal for one of the two transitions in each pair of lines. Such behavior in the emission or absorption mode of the microwave power can be explained by induced spin alignment of the spin sublevels of the centres under optical excitation. The insets show two opposite the light-induced inverse population schemes of the spin sublevels of V2, V3 centres on the one hand and V4 centres on the other hand (there is evidence, that for V4 center $D < 0$).

The $S = 3/2$ colour centres in 15R-SiC were demonstrated can be optically addressed and read-out with high fidelity at room temperature by means of standard ODMR. The ODMR spectrum shows the relative change of the photoluminescence intensity ΔPL/PL as a function of applied radio frequency (Fig. 6.45). Resonances at $\nu = 13.1$ MHz, $\nu = 50.6$ MHz, and $\nu = 138$ MHz were observed and agree with ZFS of the V3, V4, and V2 centres determined from the EPR measurements (Fig. 6.44). Calculated frequency dependencies of the ODMR signals for $S = 3/2$ state at different strengths of magnetic fields ($B \parallel c$) coincided well with observed spectra (Fig. 6.45).

Changing the excitation energy and the optical registration window was believed can lead to the further increase of the ODMR contrast. Trial measurements of V4 ODMR signals in 15R-SiC under laser excitation at $\lambda = 808$ nm were made, and, as a result, about five times more intense signals than for the similar centres in other polytypes (1.2% for V4 centres) were observed.

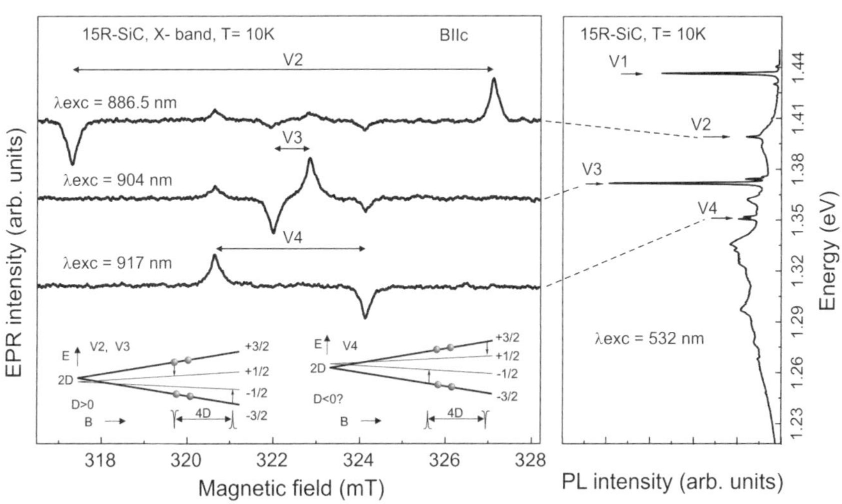

Fig. 6.44 (*Left*) X-band DD-EPR spectra measured in 15R-SiC at 10 K and $B \parallel c$ induced with ZFLs V2, V3 and V4. The *insets* show the light-induced inverse population of the spin sublevels of V2, V3, and V4 centres. (*Right*) PL spectrum of the spin centres in 15R-SiC with ZPL V1, V2, V3, V4

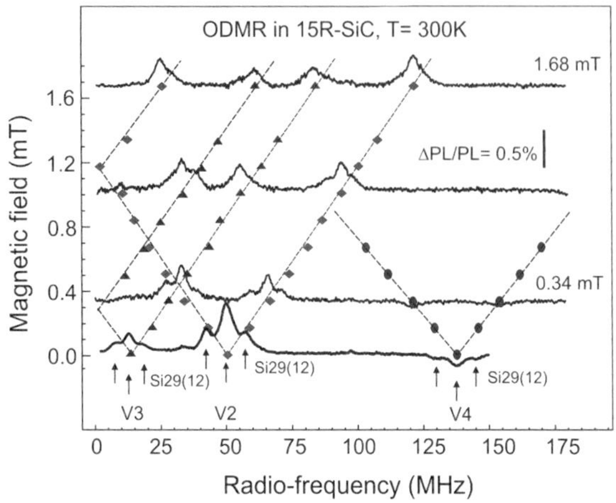

Fig. 6.45 The lower spectrum shows the ODMR signal of the V2, V3, V4 detected under laser excitation λ = 785 nm in the 0.05 mT external magnetic field applied to compensate the influence of Earth's magnetic field. The *vertical bar* indicates the ODMR contrast. Experimental radio-frequency dependence of the ODMR signals as a function of magnetic field is shown for each centre. *Dashed lines* are the calculated dependence using spin Hamiltonian for S = 3/2. Hyperfine interaction with ^{29}Si nuclear for twelve Si atoms in the next nearest-neighbour (NNN) shell of Si vacancy

We draw attention to the fact that the designations V2, V3 and V4 for 15R-SiC in the figures and the Table 6.1 do not coincide because of the inconsistency of the notation in different publications.

6.3.3.4 ESE and ENDOR Spectra Measured in 15R-SiC

As an example we consider ESE and ENDOR spectra measured in rhombic 15R-SiC for V2, V3 and V4 spin centres, but we note that many of the findings can also be used in other polytypes investigated. The observed hyperfine (HF) interactions were shown to be a strong evidence in favor of the model as a negatively charged silicon vacancy that is noncovalently bonded to the neutral carbon vacancy, located on the adjacent site along the SiC symmetry c-axis, i.e., the $V_{Si}^{-} - V_{C}^{0}$ model with S = 3/2. The microscopic structure of the centre is shown in Fig. 6.35. W-band ESE-detecte EPR spectra of the V2, V3, and V4 centres were measured for different orientations (θ) of the magnetic field B with respect to the c axis (Fig. 6.45a). Similarly to the case of 4H-SiC and 6H-SiC, two types of HF interactions were observed in the ESE spectra. The first type of the HF interactions occurs with the ^{13}C nucleus located in the nearest neighbor (NN) shell to the V_{Si}^{-} site. They are strongly anisotropic and reflect the tetrahedral symmetry of the nuclear spin locations. The HF lines arising from these interactions are shown in Fig. 6.46a: $^{13}C_1$ denotes the interaction with the carbon atom oriented along the c axis and $^{13}C_{2-4}$ denotes the interactions with atoms located in the basal plane with the bonds inclined by the angle $\theta = 71°$ relative to the c axis. The HF structure arising from such interactions is characterized by $A_{\|} = 3.02$ mT (84.6 MHz) and $A_{\perp} = 1.2$ mT (33.6 MHz), which match closely previously reported values for the V_{Si}^{-} centres in 4H-SiC and 6H-SiC [16, 19, 20, 49]. The second type of the HF interactions occurs with the ^{29}Si nucleus located in the next nearest neighbor (NNN) shell to the V_{Si}^{-}. These interactions with A = 0.297 mT are shown in the inset in Fig. 6.46a. Changes in the orientation of the magnetic field did not induce the line splitting; only the strong anisotropy of the linewidth was observed.

Our discussion is based on the ENDOR spectra of the V2 centres. Figures 6.46b and 6.47a, b show the ENDOR spectra recorded by monitoring the intensity of the ESE, following microwave $\pi/2$ pulses, as a frequency function of the pulse, applied between the second and third microwave pulses [200]. In Fig. 6.46 allowed dipole magnetic transitions with $\Delta M_S = \pm 1$: $3/2 \leftrightarrow 1/2$ are indicated by lf (low field) and $-3/2 \leftrightarrow -1/2$ are indicated by hf (high field).

Observed HF interactions with silicon and carbon nuclei can be described by $\Sigma \mathbf{S}\cdot\mathbf{A}_i\cdot\mathbf{I}_i$ term in the spin Hamiltonian. Here, $\mathbf{A}_i$ are tensors, which describe the HF interaction with the ith Si or C atoms located at different neighbor shells of the Si sites and C sites. ENDOR transition frequencies determined by the selection rules $\Delta M_S = 0$ and $\Delta m_I = \pm 1$ are given by [201]:

$$\nu_{ENDORi} = h^{-1}\left|M_s[a_i + b_i(3\cos^2\theta - 1)] - g_{ni}\mu_n B\right|, \quad (6.16)$$

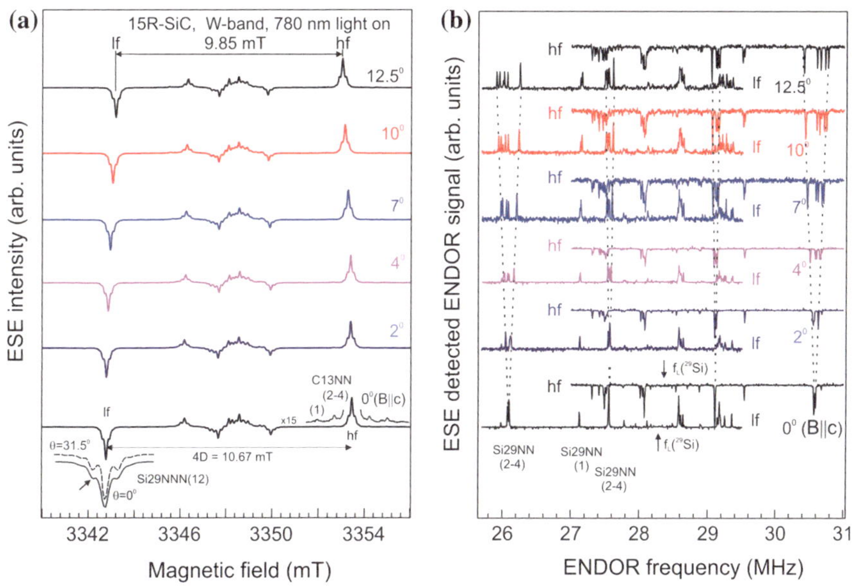

Fig. 6.46 W-band angular dependence ESE detected EPR (**a**) and ENDOR (**b**) spectra of the V2 colour centres in single 15R-SiC crystal

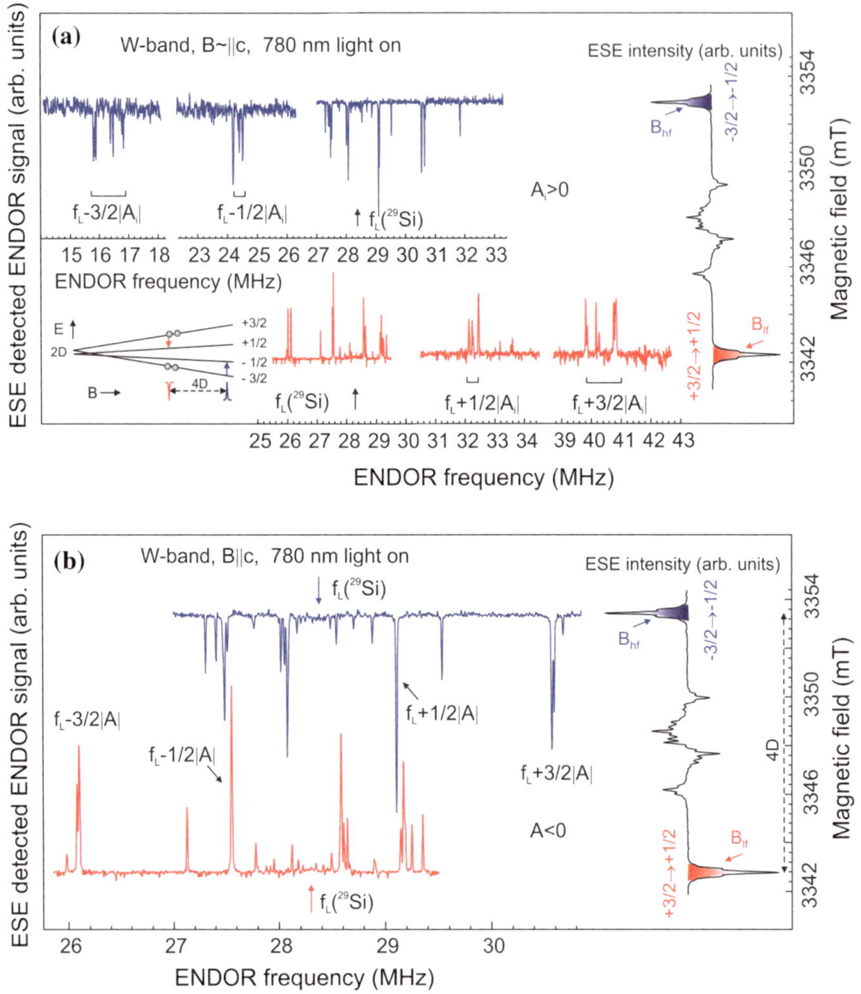

Fig. 6.47 ESE detected ENDOR spectra of the V2 colour centres in 15R-SiC measured at two angles θ between the B and c axis: **a** θ ≅ 2–3° (B ∼ //c) and **b** θ = 0° (B //c) for the low-field (lf) and high-field (hf) transitions indicated in optically induced ESE spectra shown at the *right*. **a** Transitions $f_L + 1/2|A_j|, f_L + 3/2|A_j|$ and $f_L - 1/2|A_j|, f_L - 3/2|A_j|$ correspond to HF interactions with the NNN Si atoms with respect to V_{Si}^-. **b** Transitions $f_L + 1/2|a_j|, f_L + 3/2|a_j|$ and $f_L - 1/2|a_j|, f_L - 3/2|a_j|$ correspond to HF interactions with the NN Si atoms with respect to V_C^0 $^{29}Si_1$ and $^{29}Si_{2\text{-}4}$ (see Fig. 6.35) indicate ENDOR lines corresponding to the presence of ^{29}Si atoms in the NN shell of the V_C^0. For the NNN Si atoms with respect to V_{Si}^- all the lines are grouped together because of the almost isotropic hyperfine interaction. The *inset* shows the light-induced inverse population of the spin sublevels of V2 centres

where a_i and b_i are isotropic and anisotropic parts of the HF interaction with the ith nucleus, θ is the angle between the external magnetic field B and the HF interaction tensor, $g_{ni}\mu_{ni}B/h$ is the Larmor frequency f_L, g_{ni}, and μ_{ni} are the g factor of nucleus i and its nuclear magneton (g_n is negative for ^{29}Si and positive for ^{13}C). For axial symmetry the HF interaction in terms of the principal values is given by $A_{||} = a + 2b, A_{\perp} = a - b$.

All lines observed in the spectra in Figs. 6.46b and 6.47a, b can be attributed to the HF interactions between colour centre electron spin and ^{29}Si nuclear spin. Equation (6.16) predicts that interaction with each ith nucleus induces two sets of ENDOR transitions located at the distances of $3/2A_i$ and $1/2A_i$ for the lf EPR line, $-3/2A_i$ and $-1/2A_i$ for the hf EPR line, from the Larmor frequency f_L. Thus, the ENDOR lines observed are due to the HF interactions with ^{29}Si, because the f_L of the ^{29}Si (I = 1/2, abundance 4.7%) is about 28.2 MHz (marked by arrows in Figs. 6.46b and 6.47a, b). HF interactions with ^{29}Si nuclear magnetic moments in the NNN shell observed in the ESE spectra can also be seen in the ENDOR spectra (Fig. 6.47a), labeled ^{29}Si (NNN). The position of these lines corresponds to the negative spin density of the electronic wave function on the silicon nucleus.

ENDOR signals that arose due to the positive spin density on the Si nucleus were also observed. These strongly anisotropic signals labeled as $^{29}Si_1$ and $^{29}Si_{2-4}$ are shown in Figs. 6.46b and 6.47a, b. HF interactions with ^{29}Si located in NNN and more distant shells around the silicon vacancy of the S = 3/2 centre are almost isotropic, meanwhile, interactions with ^{13}C located in the NN shell around the silicon vacancy are anisotropic. In the ENDOR spectra the HF interactions on ^{29}Si exhibit anisotropic dependence typical of the HF interactions between ^{13}C located in the NN shell around V_{Si}^- (see EPR spectrum in Fig. 6.46a). To explain such anisotropy, we need to identify Si atoms that have the same symmetry as the C atoms. Such a configuration can be found only at the tetrahedron vertices around the carbon site. The position of the ENDOR lines labeled as $^{29}Si_1$ and $^{29}Si_{2-4}$ in Fig. 6.46b agrees well with the proposed configuration and reflects the HF interactions with axial (Si_1) and basal (Si_{2-4}) nuclear spins. The constants of the HF interactions are relatively large $|A_{||}| = 2.2$ MHz ($\approx$0.08 mT) and $|A_{\perp}| = 1.3$ MHz ($\approx$0.05 mT) and describe well the anisotropy of the linewidth observed in the ESE.

These HF interactions characterized by the positive spin density on the ^{29}Si nuclei can be explained if the spin density is located on four Si nuclei placed around nonparamagnetic neutral V_C^0. This implies that the V2 centre is formed by both nonparamagnetic V_C^0 and paramagnetic S = 3/2 V_{Si}^-. Spin density is caused by the spin polarization (similar to the core polarization for transition metals [202, 203]), and arises from an exchange interaction with S = 3/2 that leads to the partial decoupling of coupled covalent bonds of the V_C^0 site. The presence of the V_C^0 distorts the crystal lattice, which in turn lowers the symmetry of the V_{Si}^-. Because the V2 centre is characterized by the largest ZFS we can conclude that V_{Si}^- and V_C^0 are located closer to each other than in the case of V3, V4 centres. The duplication of lines observed in the ENDOR spectra can be explained by the presence of two similar centres with slightly different parameters of the HF interactions. Based on our experiments we suggested the model of the V2 centres shown in Fig. 6.35. The model represents two possible configurations $(V_{Si}^- - V_C^0)'$ and $(V_{Si}^- - V_C^0)''$, depending on the positions of the vacancies in the lattice.

6.3.3.5 Spin-Echo Measurements of a Spin-Lattice (T_1) and Spin-Spin (T_2) Relaxation Times of the S = 3/2 Colour Centres

Both T_1 and T_2 relaxation curves of the V2, V3, V4 centres were measured at the room temperature using standard pulsed X-band Electron Spin Echo (ESE) detected EPR technique. The measurements were provided on the low-field EPR transition ($M_S = 3/2 \leftrightarrow M_S = 1/2$) of the each centre under light illumination of the sample with wavelength of 785 nm and magnetic field applied parallel to the *c* axis. Here we presented experimental results obtained on the V2 centres.

To measure T_1 the ESE signal intensity was monitored by applying inversion-recovery pulse sequence $\pi - \Delta T - \pi/2 - \tau - \pi$, where ΔT was varied from 900 ns up to desired value (5–10 T_1 value, i.e. ca. 1 ms) and τ was kept at 200 ns. Measured data are shown on Fig. 6.48a. T_2 measurements was done using Hahn-echo decay sequence $\pi/2 - \tau - \pi$, τ was varied from 200 ns up to 40 μs. Measured data are shown on Fig. 6.48b. The values of T_1 = 80 μs and T_2 = 10 μs were obtained from the fits of the measured ESE decay curves by $y = B^*(1 - \exp(-\Delta T/T_1))$ and $y = A^*\exp(-2\tau/T_2)$, respectively. The fits are shown on Fig. 6.48 by solid curves.

6.3.3.6 Level-Anticrossing ODMR Spectroscopy of S = 3/2 Spin Centres in Silicon Carbide Single Crystals and Based Nanostructures

A sharp variation of the IR photoluminescence intensity in the vicinity of a level anticrossing (LAC) in an external magnetic field was observed for S = 3/2 colour centres in all the polytypes under investigations [204–206]. This LAC can be used

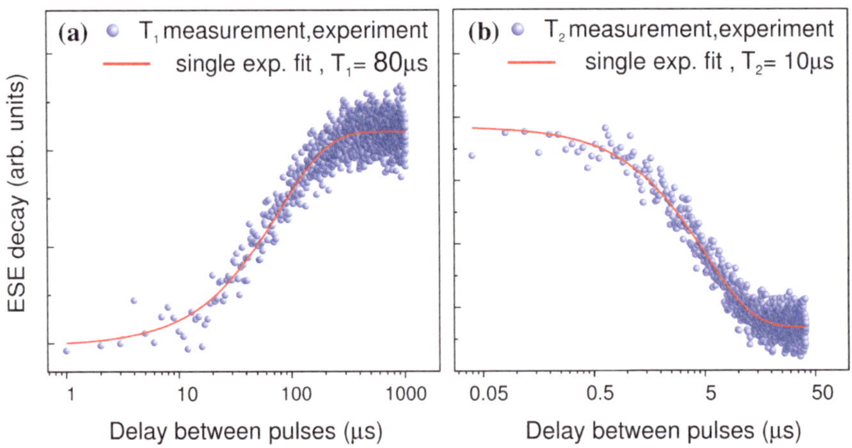

Fig. 6.48 Experimentally obtained spin echo decay curves (*balls*) at the room temperature. **a** Spin-lattice relaxation T_1 curve of the V2 centres. **b** Spin-spin relaxation T_2 curve of the V2 centres. The fits of the measured echo decay curves represent by *solid lines*

for a purely all-optical sensing of the magnetic field. A distinctive feature of the LAC signal is weak dependence on the direction of the magnetic field that allows one to monitor the LAC signals in the non-oriented systems, such as powder materials. In addition there is no need to determine the orientation of the crystal or nanocrystal in the sensing measurements.

These findings are directly translated to a working application, namely an all-optical magnetometry with nT sensitivity. This is a general concept of all-optical sensing without RF fields as it can be used to measure various physical quantities, such as temperature and axial stress, through their effect on the zero-field splitting and hence on the magnetic fields corresponding to the LACs. The results may potentially be applied for biomedical imaging and geophysical surveying, especially when RF fields cannot be applied. Furthermore, the proposed method is not restricted to magnetic sensing and can potentially be extended for radiofrequency-free sensing of other physical quantities, particularly temperature and axial stress. An intriguing possibility is to image the PL from a SiC wafer onto a CCD camera to visualize magnetic and temperature fields with temporal and spatial resolution.

Figure 6.49 shows LAC spectra for the ground state (GS) and the excited state (ES) of colour centres in 6H-SiC single crystal. The inset shows two LAC magnetic

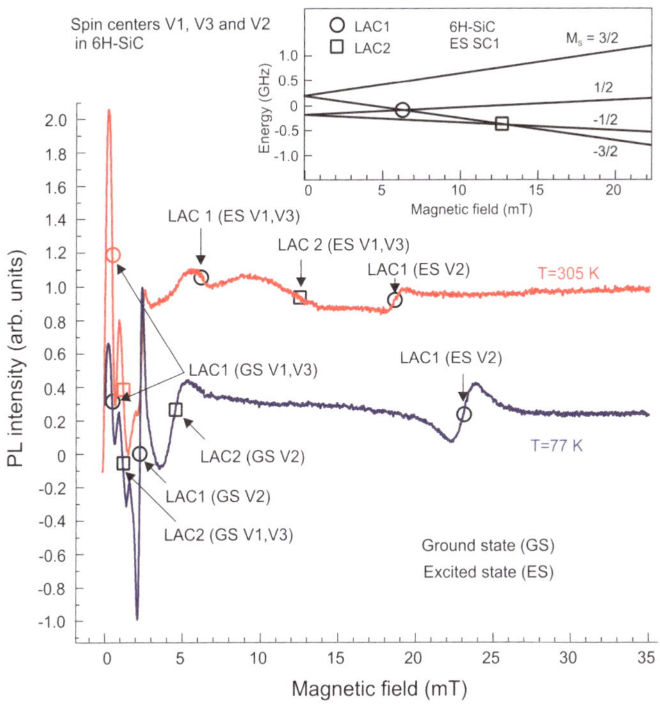

Fig. 6.49 Experimentally obtained LAC spectra for the ground state (GS) and the excited state (ES) of V1, V3 and V2 colour centres in 6H-SiC single crystal. (*Inset*) Two LAC magnetic fields at positions of $B = D/g\mu_B$ (LAC1) and $B = 2D/g\mu_B$ (LAC2) for the spin sublevels of V1(V3) centre in the excited state

fields at positions of $B = D/g\mu_B$ (LAC1) and $B = 2D/g\mu_B$ (LAC2) for the spin sublevels of V1(V3) centre in the excited state. The fine structure parameters D for of the V1 and V3 centers in the ground state are the same (Table 6.1), also the fine structure parameters seem to coincide for the excited state. A surprising result was obtained for the quadruplet ground and excited states: the zero-field splitting of the ground state is independent on temperature making this state very attractive for vector magnetometry, whereas for the zero-field splitting in the excited state a strong temperature dependence was observed, which can be used for thermometry applications.

A sharp variation of the photoluminescence intensity in the vicinity of the level anticrossing, has been recently used [206] for a purely all-optical sensing of the magnetic field; dc magnetic field sensitivity better than 100 nT $Hz^{-1/2}$ within a volume of 3×10^{-7} mm^3 has been achieved at room temperature. It was demonstrated that this contact less method is robust at high temperatures up to at least 500 K. As this approach does not require application of radiofrequency fields, it is scalable to much larger volumes. For an optimized light-trapping waveguide of 3 mm^3 the projection noise limit is below 100 fT $Hz^{-1/2}$.

An all-optical thermometry technique based on the energy level anticrosings in silicon vacancy centers in silicon carbide has been recently proposed [207]. This technique exploits a giant thermal shift of the excited state zero-field splitting, which is equal 2.1 MHz/K, and does not require radiofrequency fields. A temperature sensitivity of 100 $mK/Hz^{1/2}$ within a detection volume of approximately 10^{-6} mm^3 has been estimated. Using level anticrossings in the ground and excited states, an integrated magnetic field and temperature sensor can be implemented using the same center.

6.3.3.7 Coherent Control of Single Spins in Silicon Carbide at Room Temperature

It is reported [195] the characterization of photoluminescence and optical spin polarization from single silicon vacancy related defects (S = 3/2 family) in SiC, and demonstrate that single spins can be addressed at room temperature. Coherent control of a single defect spin was realized and long spin coherence times under ambient conditions was demonstrated. The study provides evidence that SiC is a promising system for atomic-scale spintronics and quantum technology. At the same time it has been demonstrated magnetic resonance imaging on a single defect with spin S = 1 family in SiC [194].

The negatively charged silicon vacancy (V_{Si}) in SiC is known to have the quartet manifold of S = 3/2 in both ground and excited states [169, 182, 191, 199, 208]. While optically excited, alteration of these spin states by ESR can result in a change in PL intensity, thus allowing optically detected magnetic resonance (ODMR) [162, 209]. In order to create single silicon vacancies (V_{Si}) in a negatively charged state, a commercially available high-purity semi-insulating 4H-SiC, which is lightly

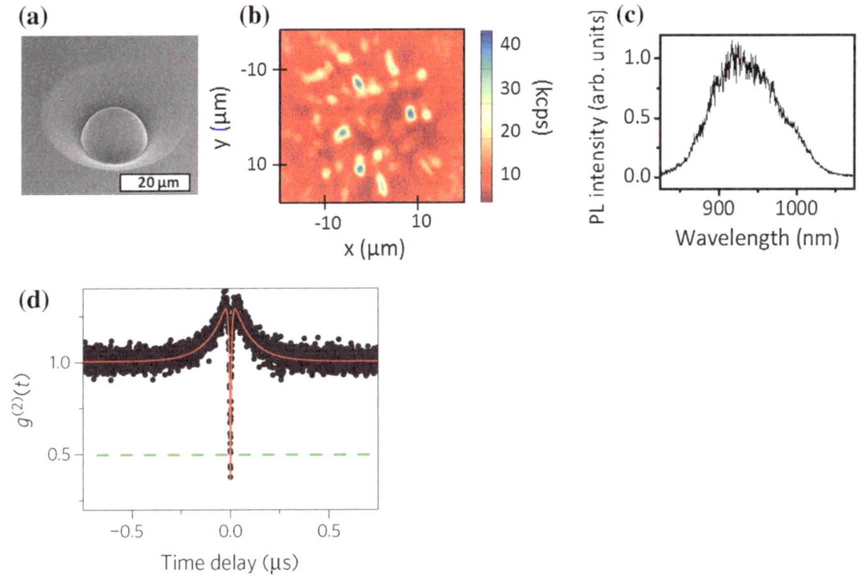

Fig. 6.50 **a** Optical detection of fluorescing single V_{Si} defects created in 4H-SiC at room temperature [210]. **a** Scanning electron microscope image of the fabricated SIL on the SiC surface. **b** Confocal fluorescence image scanned around the fabricated hemispherical SIL with 12 mW laser excitation at 730 nm. The *color scale* indicates the measured fluorescence intensity in the unit of kilo counts per second (kcps). The bright spots correspond to single V_{Si}. **c** PL spectrum of a single V_{Si} collected by 1 mW 730 nm laser excitation. **d** Autocorrelation measurement of a single V_{Si} defects measured at 0.1 mW optical power. Experimental data are plotted after background correction and deconvoluting timing jitter of the APDs (≈1.39 ns). The *red curve* is a fit based on the three-states model and the green line indicates $g^{(2)} = 0$

n-doped, was used as a substrate. Because a low density of created defects is necessary for single defect detection, 2 meV electron irradiation at a low electron fluence of 10^{13} cm^{-2} was used. This high energy but low dose electron irradiation successfully created V_{Si} homogeneously through the whole substrate (~0.5 mm thick) at a concentration of ~10^{11} cm^{-3}. This results in an average separation of a few micrometers between V_{Si} [195]. Since the PL intensity of a single defect emitter is a key to determine the measurement speed, it is helpful to create special structures, which can enhance the photon collection efficiency. The fabricated structure was a hemispherical solid immersion lens (SIL) at the surface by focused ion beam milling [210] as shown in Fig. 6.50a. The photon collection efficiency, which is limited by the total internal reflection at the flat crystal surface, could be four-fold enhanced [195]. A few micrometer separation among V_{Si} defects together with the enhanced photon detection efficiency allowed optical addressing of single V_{Si} defects as shown in Fig. 6.50b. The observed single V_{Si} PL spectrum in 4H-SiC (Fig. 6.50c) was similar to that of V2 centers, one of V_{Si} defects in two inequivalent lattice cites, observed at room temperature [162]. Autocorrelation measurement of a single V_{Si} defects measured at 0.1mW optical power is shown in Fig. 6.50d.

Experimental data are plotted after background correction and deconvoluting timing jitter of the APDs (≈1.39 ns). The red curve is a fit based on the three-states model and the green line indicates $g^{(2)} = 0$. Note that this defect center sometimes appears as a TV2a center in literature [209] or V2 center in this Section.

The optical single spin detection was first tested by a simple method. As previously reported for an ensemble [162, 182, 205], ODMR spectra could be obtained by scanning frequency of applied oscillating magnetic fields under continuous optical excitation as in Fig. 6.51a. Two distinct spin resonance transitions corresponding to $|M_s = +3/2\rangle \leftrightarrow M_s = +1/2$ and $|M_s = -1/2\rangle \leftrightarrow M_s = -3/2$, respectively, can be observed when a 5 mT static magnetic field was applied along the spin quantization axis, known to be parallel to the crystal c-axis. Another transition between $|M_s = \pm 1/2\rangle$ was not observable due to equal population induced by optical excitation, as reported previously [162, 182, 205]. Coherent spin signals were also measured by detecting spin Rabi oscillations. To this end, a few microsecond long laser pulse was first applied for optical spin polarization, followed by roughly one microsecond long idle time to ensure ground state occupation. A RF pulse resonant to, for example, $|M_s = +3/2\rangle \leftrightarrow M_s = +1/2$ was applied, and its length was being changed to induce Rabi oscillation at various RF power. Spin state was then measured by integrating a few hundred nanosecond long PL response to a readout laser pulse as in Fig. 6.51b [210]. In order to also test how long spin coherence can be observed, a spin Hahn-echo decay was also measured at a 27 mT axial magnetic field. A standard two pulse sequence, $\pi/2 - \tau - \pi - \tau$ was applied together with a $\pi/2$ pulse for projective readout. The total free precession

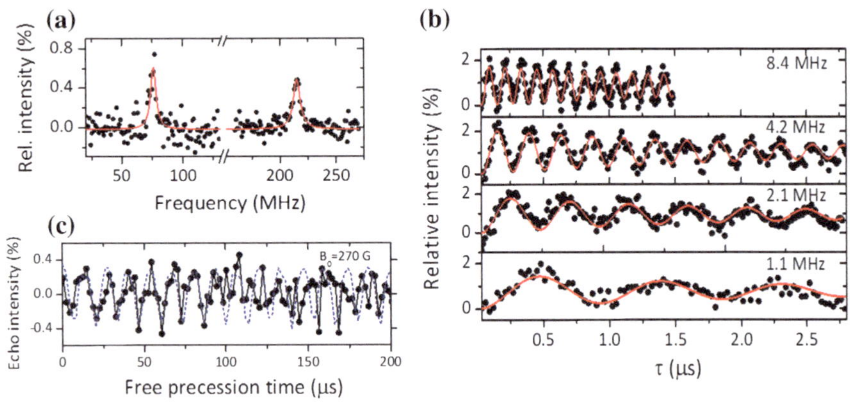

Fig. 6.51 Optically detected spin state of a single V_{Si} defect at ambient condition. **a** ODMR sepctrum from a single V_{Si} defect at 5 mT external magnetic field parallel to the c-axis. *Black dots* Measured data expressed as a relative fluorescent intensity $\Delta PL/PL_{off}$, where PL_{off} is the PL intensity off-resonance condition. *Red curves* Lorentzian fits. **b** Optically detected spin Rabi oscillation. Solid red curves are fitted by damping sinusoidal functions. Rabi frequencies obtained by fitting at various RF power are indicated. **c** Optically detected spin Hahn-echo decay showing ESEEM at 27 mT external magnetic field parallel to the c-axis. The *blue curve* shows results of a simulation [187]

time, 2τ, was increased to measure a Hahn-echo decay as shown in Fig. 6.51c. By comparing the amplitude of the electron spin echo envelope modulation (ESEEM) arising from a ^{29}Si nucleus [187], the lower limit of the coherence time, T_2, was determined to be 160 μs [210]. This long coherence time was expected theoretically because spin flip-flop processes between two most abundant nuclei, ^{29}Si and ^{13}C are suppressed at several hundred gauss external magnetic fields. In addition a longer mutual distance results in a decoherence rate similar to or even slower than that in natural diamond [187].

The effective nuclear spin bath concentration of SiC is similar to that of diamond. Consequently, the SiC nuclear spin bath should give a similar electron spin decoherence rate to that of ^{13}C in diamond. The estimation predicts a very long coherence time, of as much as milliseconds [195].

6.3.4 Conclusions and Outlook

Families of homotypic colour centers in silicon carbide exhibiting attractive spin properties were revealed. Optically induced alignment (polarization) of the ground-state and excited-state spin sublevels of the colour centers in 4H-, 6H- and 15R-SiC was observed at RT. In distinction from the known NV defect in diamond, two opposite schemes for the optical spin alignment of $S = 3/2$ centres were realized upon illumination with unpolarized light. The alignment schemes depending on the crystal polytype, temperature and structure of the spin centre. Observed Rabi nutations persist for 80 μs at RT and evidence that the probed spin ensemble can be prepared in a coherent superposition of the spin states at resonant magnetic fields at RT. In addition the electron spin of the colour centers can be manipulated by low-energy radio field 30–130 MHz which is compatible with the NMR imaging. The accent is made on the colour centres, which are optically active in the near infrared spectral region, which is preferential for potential in vivo biological applications due to the deepest tissue penetration and which is compatible with fiber optics. The concept of sensing is based on variants of the ODMR technique with sensitivity down to a single-spin. Demonstrated spin properties of the colour centres open up new avenues for quantum computing and quantum sensing. The optically induced population inversion of spin states leads to stimulated microwave emission, which can be used to implement solid-state masers and extraordinarily sensitive radiofrequency amplifiers.

References

1. Shor, P.W.: Why haven't more quantum algorithms been found? J. ACM **50**, 87–90 (2003)
2. Datta, S.: Electronic Transport in Mesoscopic Systems. Cambridge University Press, New York (1997)
3. Di Vincenzo, D.P.: Quantum computation. Science **270**, 255–261 (1995)
4. Gershenfeld, N.A., Chuang, I.L.: Bulk spin-resonance quantum computation. Science **275**, 350–356 (1997)
5. Gershenfeld, N., Chuang, I.L.: Quantum computing with molecules. Sci. Am. **278**, 66–71 (1998)
6. Brassard, G., Chuang, I., Lloyd, S., Monroe, C.: Quantum computing. Proc. Natl. Acad. Sci. U.S.A. **95**, 11032–11033 (1998)
7. Vandersypen, L.M.K., Steffen, M., Breyta, G., Yannoni, C.S., Sherwood, M.H., Chuang, I. L.: Experimental realization of Shor's quantum factoring algorithm using nuclear magnetic resonance. Nature **414**, 883–887 (2001)
8. Warren, W.S.: The usefulness of NMR quantum computing. Science **277**, 1688–1689 (1997)
9. Wrachtrup, J., Kilin, S.Y., Nizovtsev, A.P.: Quantum computation using the C-13 nuclear spins near the single NV defect centre in diamond. Opt. Spectrosc. **91**, 429–437 (2001)
10. Brossel, J., Bitter, F.: A new "double resonance" method for investigating atomic energy levels. Application to Hg $^3P_1$. Phys. Rev. **86**, 308–316 (1952)
11. Schmidt, J., van der Waals, J.H.: Optical detection of zero-field transitions in phosphorescent triplet states. Chem. Phys. Lett. **2**, 640–642 (1968)
12. Sharnoff, R.J.: Collisional transfer of rotational energy. Chem. Phys. **46**, 3262–3263 (1967)
13. Clarke, R.H.: Triplet State ODMR Spectroscopy, p. 185. Wiley, New York (1982)
14. Personov, R.I.: Quantitative determination of 3,4-benzopyrene via the line fluorescence spectra at 77 K. Zh. Anal. Khim. **17**, 506 (2003)
15. Weeks, S.J., Gilles, S.M., D'Silva, A.P.: Detection of benzo[a]pyrene metabolites by laser-excited Shpol'skii spectrometry. Appl. Spectrosc. **45**, 1093–1100 (1991)
16. Moerner, W.E., Kador, L.: Optical detection and spectroscopy of single molecules in a solid. Phys. Rev. Lett. **62**, 2535–2538 (1989)
17. Moerner, W.E., Orrit, M.: Illuminating single molecules in condensed matter. Science **283**, 1670–1676 (1999)
18. Orrit, M., Bernard, J.: Single pentacene molecules detected by fluorescence excitation in a *p*-terphenyl crystal. Phys. Rev. Lett. **65**, 2716–2719 (1990)
19. Basche, T., Moerner, W.E., Orrit, M., Wild, U.P.: Single Molecule Optical Detection, Imaging and Spectroscopy. Verlag-Chemie, Munich (1997)
20. Köhler, J., Dosselhorst, J.A.J.M., Donckers, M.C.J.M., Groenen, E.J.J., Schmidt, J., Moerner, W.E.: Magnetic resonance of a single molecular spin. Nature **363**, 242–244 (1993)
21. Wrachtrup, J., von Borczyskowski, C., Bernard, J., Orrit, M., Brown, R.: Optical detection of magnetic resonance in a single molecule. Nature **363**, 244–245 (1993)
22. Jelezko, F., Wrachtrup, J.: Read-out of single spins by optical spectroscopy. J. Phys.: Condens. Matter **16**, R1089–R1104 (2004)
23. Köehler, J.: Magnetic resonance of a single molecular spin. Phys. Rep. **310**, 261–339 (1999)
24. Gruber, A., Drabenstedt, A., Tietz, C., Fleury, L., Wrachtrup, J., von Borczyskowski, C.: Scanning confocal optical microscopy and magnetic resonance on single defect centres. Science **276**, 2012–2014 (1997)
25. Moerner, W.E., Fromm, D.P.: Methods of single-molecule fluorescence spectroscopy and microscopy. Rev. Sci. Instrum. **74**, 3597–3619 (2003)
26. English, D.S., Furube, A., Barbara, P.F.: Single-molecule spectroscopy in oxygen-depleted polymer films. Chem. Phys. Lett. **324**, 15–19 (2000)
27. Rebane, K.K., Rebane, I.: Peak value of the cross-section of zero-phonon line's absorption. J. Lumin. **56**, 39–45 (1993); Rebane, K.K.: Purely electronic zero-phonon line as the

foundation stone for high resolution matrix spectroscopy, single impurity molecule spectroscopy, persistent spectral hole burning. J. Lumin. **100**, 219–232 (2002)
28. Matsushita, M., Bloess, A., Durand, Y., Butter, J.Y.P., Schmidt, J., Groenen, E.J.J.: Single molecules as nanoprobes. A study of the Shpol'skii effect. J. Chem. Phys. **117**, 3383–3390 (2002)
29. Bloess, A., Durand, Y., Matsushita, M., Schmidt, J., Groenen, E.J.J.: A single-molecule study of the relation between the resonance frequency and the orientation of a guest molecule in a Shpol'skii system. Chem. Phys. Lett. **344**, 55–60 (2001)
30. Bloess, A., Durand, Y., Matsushita, M., Verberk, R., Groenen, E.J.J., Schmidt, J.: Microscopic structure in a Shpol'skii system: a single-molecule study of dibenzanthanthrene in n-tetradecane. J. Phys. Chem. A **105**, 3016–3021 (2001)
31. Shpolski, E.V.: Problem of the origin and structure of the quasilinear spectra of organic compounds at low temperatures. Usp. Fiz. Nauk **77**, 321–326 (1962)
32. Bernard, J., Fleury, L., Talon, H., Orrit, M.: Photon bunching in the fluorescence from single molecules: a probe for intersystem crossing. J. Chem. Phys. **98**, 850–859 (1993)
33. Ambrose, W.P., Bashe, T., Morner, W.E.: Detection and spectroscopy of single pentacene molecules in a p-terphenyl crystal by means of fluorescence excitation. J. Chem. Phys. **95**, 7150–7163 (1991)
34. Jelezko, F., Lounis, B., Orrit, M.: Pump–probe spectroscopy and photophysical properties of single di-benzanthanthrene molecules in a naphthalene crystal. J. Chem. Phys. **107**, 1692–1702 (1997)
35. Lounis, B., Jelezko, F., Orrit, M.: Single molecules driven by strong resonant fields: Hyper-Raman and subharmonic resonances. Phys. Rev. Lett. **78**, 3673–3676 (1997)
36. Brown, R., Wrachtrup, J., Orrit, M., Bernard, J., von Borczyskowski, C.: Kinetics of optically detected magnetic resonance of single molecules. J. Chem. Phys. **100**, 7182–7191 (1994)
37. Loudon, R.: The Quantum Theory of Light. Oxford University Press, Oxford (1983)
38. Vogel, A., Gruber, A., Wrachtrup, J., von Borczyskowski, C.: Determination of intersystem crossing parameters via observation of quantum jumps on single molecules. J. Phys. Chem. **99**, 14915–14917 (1995)
39. Silbey, R.J.: ESR on a single molecule. Nature **363**, 214 (1993)
40. Wrachtrup, J., von Borczyskowski, C., Bernard, J., Orrit, M., Brown, R.: Optically detected spin coherence of single molecules. Phys. Rev. Lett. **71**, 3565–3568 (1993)
41. van Strien, A.J., Schmidt, J.: An EPR study of the triplet state of pentacene by electron spin-echo techniques and laser flash excitation. Chem. Phys. Lett. **70**, 513–517 (1980)
42. Brouwer, A.C.J., Groenen, E.J.J., Schmidt, J.: Detecting magnetic resonance through quantum jumps of single molecules. Phys. Rev. Lett. **80**, 3944–3947 (1998)
43. Köhler, J., Brouwer, A.C.J., Groenen, E.J.J., Schmidt, J.: Fluorescence detection of single molecule magnetic resonance for pentacene in *p*-terphenyl. The hyperfine interaction of a single triplet spin with a single ^{13}C nuclear spin. Chem. Phys. Lett. **228**, 47–52 (1994)
44. Köhler, J., Brouwer, A.C.J., Groenen, E.J.J., Schmidt, J.: Single molecule electron paramagnetic resonance spectroscopy: hyperfine splitting owing to a single nucleus. Science **268**, 1457–1460 (1995)
45. Hanson, R., Witkamp, B., Vandersypen, L.M.K., van Beveren, L.H.W., Elzerman, J.M., Kouwenhoven, L.P.: Zeeman energy and spin relaxation in a one-electron quantum dot. Phys. Rev. Lett. **91**, 196802 (2003)
46. van der Wiel, W.G., De Franceschi, S., Elzerman, J.M., Fujisawa, T., Tarucha, S., Kouwenhoven, L.P.: Electron transport through double quantum dots. Rev. Mod. Phys. **75**, 1–22 (2002)
47. Durkan, C., Welland, M.E.: Electronic spin detection in molecules using scanning-tunneling-microscopy-assisted electron-spin resonance. Appl. Phys. Lett. **80**, 458–460 (2002)
48. Wago, K., Botkin, D., Yannoni, C.S., Rugar, D.: Force-detected electron-spin resonance: adiabatic inversion, nutation, and spin echo. Phys. Rev. B **57**, 1108–1114 (1998)

49. Berman, G.P., Tsifrinovich, V.I.: Modified approach to single-spin detection using magnetic resonance force microscopy. Phys. Rev. B **61**, 3524–3527 (2000)
50. Mamin, H.J., Budakian, R., Chui, B.W., Rugar, D.: Detection and manipulation of statistical polarization in small spin ensembles. Phys. Rev. Lett. **91**, 207604 (2003)
51. Jelezko, F., Gaebel, T., Popa, I., Domhan, M., Gruber, A., Wrachtrup, J.: Observation of coherent oscillation of a single nuclear spin and realization of a two-qubit conditional quantum gate. Phys. Rev. Lett. **93**, 130501 (2004)
52. Ladd, T.D., Jelezko, F., Laflamme, R., Nakamura, Y., Monroe, C., O'Brien, J.L.: Quantum computers. Nature **464**, 45–53 (2010)
53. Elzerman, J.M., Hanson, R., van Beveren, L.H.W., Witkamp, B., Vandersypen, L.M.K., Kouwenhoven, L.P.: Single-shot read-out of an individual electron spin in a quantum dot. Nature **430**, 431–435 (2004)
54. Xiao, M., Martin, I., Yablonovitch, E., Jiang, H.W.: Electrical detection of the spin resonance of a single electron in a silicon field-effect transistor. Nature **430**, 435–439 (2004)
55. Rugar, D., Budakian, R., Mamin, H.J., Chui, B.W.: Single spin detection by magnetic resonance force microscopy. Nature **430**, 329–332 (2004)
56. Collins, A.T.: The detection of colour-enhanced and synthetic gem diamonds by optical spectroscopy. Diamond Relat. Mater. **12**, 1976–1983 (2003)
57. Goss, J.P., Jones, R., Briddon, P.R., Davies, G., Collins, A.T., Mainwood, A., van Wyk, J. A., Baker, J.M., Newton, M.E., Stoneham, A.M., Lawson, S.C.: Comment on Electronic structure of the N-V centre in diamond: theory. Phys. Rev. B **56**, 16031–16032 (1997)
58. Maze, J.R., Gali, A., Togan, E., Chu, Y., Trifonov, A., Kaxiras, E., Lukin, M.D.: Properties of nitrogen-vacancy centres in diamond: the group theoretic approach. New J. Phys. **13**, 025025 (2011)
59. Gali, A., Fyta, M., Kaxiras, E.: Ab initio supercell calculations on nitrogen-vacancy centre in diamond: electronic structure and hyperfine tensors. Phys. Rev. B **77**, 155206 (2008)
60. Martin, J., Wannemacher, R., Teichert, J., Bischoff, L., Kohler, B.: Generation and detection of fluorescent color centres in diamond with submicron resolution. Appl. Phys. Lett. **75**, 3096–3098 (1999)
61. Meijer, J., Burchard, B., Domhan, M., Wittmann, C., Gaebel, T., Popa, I., Jelezko, F., Wrachtrup, J.: Generation of single color centres by focused nitrogen implantation. Appl. Phys. Lett. **87**, 261909 (2005)
62. Pezzagna, S., Rogalla, D., Becker, H.W., Jakobi, I., Dolde, F., Naydenov, B., Wrachtrup, J., Jelezko, F., Trautmann, C., Meijer, J.: Creation of colour centres in diamond by collimated ion-implantation through nano-channels in mica. Phys. Stat. Sol. (a), 2017–2022 (2011)
63. Rabeau, J.R., Chin, Y.L., Prawer, S., Jelezko, F., Gaebel, T., Wrachtrup, J.: Fabrication of single nickel-nitrogen defects in diamond by chemical vapor deposition. Appl. Phys. Lett. **86**, 131926 (2005)
64. Boudou, J.P., Curmi, P.A., Jelezko, F., Wrachtrup, J., Aubert, P., Sennour, M., Balasubramanian, G., Reuter, R., Thorel, A., Gaffet, E.: High yield fabrication of fluorescent nanodiamonds. Nanotechnology **20**, 359801 (2009)
65. Rabeau, J.R., Reichart, P., Tamanyan, G., Jamieson, D.N., Prawer, S., Jelezko, F., Gaebel, T., Popa, I., Domhan, M., Wrachtrup, J.: Implantation of labelled single nitrogen vacancy centres in diamond using N-15. Appl. Phys. Lett. **88**, 023113 (2006)
66. Naydenov, B., Richter, V., Beck, J., Steiner, M., Neumann, P., Balasubramanian, G., Achard, J., Jelezko, F., Wrachtrup, J., Kalish, R.: Enhanced generation of single optically active spins in diamond by ion implantation. Appl. Phys. Lett. **96**, 163108 (2010)
67. Nizovtsev, A.P., Kilin, S.Y., Jelezko, F., Popa, I., Gruber, A., Tietz, C., Wrachtrup, J.: Spin-selective low temperature spectroscopy on single molecules with a triplet-triplet optical transition: application to the NV defect centre in diamond. Opt. Spectrosc. **94**, 848–858 (2003)
68. Nizovtsev, A.P., Kilin, S.Y., Jelezko, F., Popa, I., Gruber, A., Wrachtrup, J.: NV centres in diamond: spin-selective photokinetics, optical ground-state spin alignment and hole burning. Physica B **340**, 106–110 (2003)

69. Rogers, L.J., Armstrong, S., Sellars, M.J., Manson, N.B.: Infrared emission of the NV centre in diamond: Zeeman and uniaxial stress studies. New J. Phys. **10**, 103024 (2008)
70. Jelezko, F., Wrachtrup, J.: Single defect centres in diamond: a review. Phys. Stat. Sol. (a) **20**, 3207–3225 (2006)
71. Redman, D., Brown, S., Rand, S.C.: Origin of persistent hole burning of N-V centres in diamond. J. Opt. Soc. Am. B-Opt. Phys. **9**, 768–774 (1992)
72. Orrit, M.: Single-molecule spectroscopy: the road ahead. J. Chem. Phys. **117**, 10938–10946 (2002)
73. Beveratos, A., Brouri, R., Gacoin, T., Poizat, J.-P., Grangier, P.: Nonclassical radiation from diamond nanocrystals. Phys. Rev. A **64**, 061802 (2001)
74. Alleaume, R., Treussart, F., Messin, G., Dumeige, Y., Roch, J.-F., Beveratos, A., Brouri-Tualle, R., Poizat, J.-P., Grangier, P.: Experimental open-air quantum key distribution with a single-photon source. New J. Phys. **6**, 92 (2004)
75. Mainwood, A., Stoneham, A.M.: The vacancy (V − 0, V+, V−) in diamond: the challenge of the excited states and the GR2–GR8 lines. In: Proceedings of the 13th International Conference on Defects in Insulating Materials—ICDIM 96, pp. 99–102 (1997)
76. Gaebel, T., Domhan, M., Wittmann, C., Popa, I., Jelezko, F., Rabeau, J., Greentree, A., Prawer, S., Trajkov, E., Hemmer, P.R., Wrachtrup, J.: Photochromism in single nitrogen-vacancy defect in diamond. Appl. Phys. B: Laser Optics **82**, 243–246 (2006)
77. Jelezko, F., Popa, I., Gruber, A., Tietz, C., Wrachtrup, J., Nizovtsev, A., Kilin, S.: Single spin states in a defect centre resolved by optical spectroscopy. Appl. Phys. Lett. **81**, 2160–2162 (2002)
78. Tamarat, P., Gaebel, T., Rabeau, J.R., Khan, M., Greentree, A.D., Wilson, H., Hollenberg, L.C.L., Prawer, S., Hemmer, P., Jelezko, F., Wrachtrup, J.: Stark shift control of single optical centres in diamond. Phys. Rev. Lett. **97**, 083002 (2006)
79. Santori, C., Tamarat, P., Neumann, P., Wrachtrup, J., Fattal, D., Beausoleil, R.G., Rabeau, J., Olivero, P., Greentree, A.D., Prawer, S., Jelezko, F., Hemmer, P.: Coherent population trapping of single spins in diamond under optical excitation. Phys. Rev. Lett. **97**, 247401 (2006)
80. Buckley, B.B., Fuchs, G.D., Bassett, L.C., Awschalom, D.D.: Spin-light coherence for single-spin measurement and control in diamond. Science **330**, 1212–1215 (2010)
81. Togan, E., Chu, Y., Trifonov, A.S., Jiang, L., Maze, J., Childress, L., Dutt, M.V.G., Sorensen, A.S., Hemmer, P.R., Zibrov, A.S., Lukin, M.D.: Quantum entanglement between an optical photon and a solid-state spin qubit. Nature **466**, 730–734 (2010)
82. Bassett, L.C., Heremans, F.J., Yale, C.G., Buckley, B.B., Awschalom, D.D.: Electrical tuning of single nitrogen-vacancy centre optical transitions enhanced by photoinduced fields. Phys. Rev. Lett. **107**, 266403 (2011)
83. Castelletto, S., Harrison, J.P., Marseglia, L., Stanley-Clarke, A.C., Gibson, B.C., Fairchild, B.A., Hadden, J.P., Ho, Y.L.D., Hiscocks, M.P., Ganesan, K., Huntington, S.T., Ladouceur, F., Greentree, A.D., Prawer, S., O'Brien, J.L., Rarity, J.G.: Diamond-based structures to collect and guide light. New J. Phys. **13**, 025020 (2011)
84. Marseglia, L., Hadden, J.P., Stanley-Clarke, A.C., Harrison, J.P., Patton, B.R., Ho, Y.L.D., Naydenov, B., Jelezko, F., Meijer, J., Dolan, P.R., Smith, J.M., Rarity, J.G., O'Brien, J.L.: Nanofabricated solid immersion lenses registered to single emitters in diamond. Appl. Phys. Lett. **98**, 189902 (2011)
85. Siyushev, P., Kaiser, F., Jacques, V., Gerhardt, I., Bischof, S., Fedder, H., Dodson, J., Markham, M., Twitchen, D., Jelezko, F., Wrachtrup, J.: Monolithic diamond optics for single photon detection. Appl. Phys. Lett. **97**, 241902 (2010)
86. Bernien, H., Childress, L., Robledo, L., Markham, M., Twitchen, D., Hanson, R.: Two-photon quantum interference from separate nitrogen vacancy centres in diamond. Phys. Rev. Lett. **108**, 043604 (2012)
87. Robledo, L., Childress, L., Bernien, H., Hensen, B., Alkemade, P.F.A., Hanson, R.: High-fidelity projective read-out of a solid-state spin quantum register. Nature **477**, 574–578 (2011)

88. Vanoort, E., Manson, N.B., Glasbeek, M.: Optically detected spin coherence of the diamond N-V centre in its triplet ground-state. J. Phys. C: Solid State Phys. **21**, 4385–4391 (1988)
89. Doherty, M.W., Manson, N.B., Delaney, P., Hollenberg, L.C.L.: The negatively charged nitrogen-vacancy centre in diamond: the electronic solution. New J. Phys. **13**, 025019 (2011)
90. Harrison, J., Sellars, M.J., Manson, N.B.: Optical spin polarisation of the N-V centre in diamond. J. Lumin. **107**, 245–248 (2004)
91. Jelezko, F., Gaebel, T., Popa, I., Gruber, A., Wrachtrup, J.: Observation of coherent oscillations in a single electron spin. Phys. Rev. Lett. **92**, 076401 (2004)
92. Neumann, P., Mizuochi, N., Rempp, F., Hemmer, P., Watanabe, H., Yamasaki, S., Jacques, V., Gaebel, T., Jelezko, F., Wrachtrup, J.: Multipartite entanglement among single spins in diamond. Science **320**, 1326–1329 (2008)
93. Balasubramanian, G., Neumann, P., Twitchen, D., Markham, M., Kolesov, R., Mizuochi, N., Isoya, J., Achard, J., Beck, J., Tissler, J., Jacques, V., Hemmer, P.R., Jelezko, F., Wrachtrup, J.: Ultralong spin coherence time in isotopically engineered diamond. Nat. Mater. **8**, 383–387 (2009)
94. Popa, I., Gaebel, T., Domhan, M., Wittmann, C., Jelezko, F., Wrachtrup, J.: Energy levels and decoherence properties of single electron and nuclear spins in a defect centre in diamond. Phys. Rev. B **70**, 201203 (2004)
95. Felton, S., Edmonds, A.M., Newton, M.E., Martineau, P.M., Fisher, D., Twitchen, D.J., Baker, J.M.: Hyperfine interaction in the ground state of the negatively charged nitrogen vacancy centre in diamond. Phys. Rev. B **79**, 075203 (2009)
96. Luszczek, M., Laskowski, R., Horodecki, P.: The ab initio calculations of single nitrogen–vacancy defect centre in diamond. Phys. B **348**, 292–298 (2004)
97. Pushkarchuk, V.A., Kilin, S.Y., Nizovtsev, A.P., Pushkarchuk, A.L., Borisenko, V.E., von Borczyskowski, C., Filonov, A.B.: Ab initio modeling of the electronic and spin properties of the [NV](−) centres in diamond nanocrystals. Opt. Spectrosc. **99**, 245–256 (2005)
98. Loubser, J.H.N., van Wyk, J.A.: Electron spin resonance in annealed type 1b diamond. Diamond Res. **11**, 4–7 (1977); Loubser, J.H.N., van Wyk, J.A.: Electron spin resonance in the study of diamond. Rep. Progress Phys. **41**, 1201 (1978)
99. van Oort, E., Manson, N.B., Glasbeek, M.: Optically detected spin coherence of the diamond N-V centre in its triplet ground state. J. Phys. C: Solid State Phys. **21**, 4385–4391 (1988)
100. Fuchs, G.D., Dobrovitski, V.V., Hanson, R., Batra, A., Weis, C.D., Schenkel, T., Awschalom, D.D.: Excited-state spectroscopy using single spin manipulation in diamond. Phys. Rev. Lett. **101**, 117601 (2008)
101. Neumann, P., Kolesov, R., Jacques, V., Beck, J., Tisler, J., Batalov, A., Rogers, L., Manson, N.B., Balasubramanian, G., Jelezko, F., Wrachtrup, J.: Excited-state spectroscopy of single NV defects in diamond using optically detected magnetic resonance. New J. Phys. **11**, 013017 (2009)
102. Jacques, V., Neumann, P., Beck, J., Markham, M., Twitchen, D., Meijer, J., Kaiser, F., Balasubramanian, G., Jelezko, F., Wrachtrup, J.: Dynamic polarization of single nuclear spins by optical pumping of nitrogen-vacancy color centres in diamond at room temperature. Phys. Rev. Lett. **102**, 057403 (2009)
103. van Oort, E., Glasbeek, M.: Electric-field-induced modulation of spin echoes of N-V centres in diamond. Chem. Phys. Lett. **168**, 529–532 (1990)
104. van Oort, E., van der Kamp, B., Sitters, R., Glasbeek, M.: Microwave-induced line-narrowing of the N-V defect absorption in diamond. J. Lumin. **48–49**, 803–806 (1991)
105. Kennedy, T.A., Colton, J.S., Butler, J.E., Linares, R.C., Doering, P.J.: Long coherence times at 300 K for nitrogen-vacancy centre spins in diamond grown by chemical vapor deposition. Appl. Phys. Lett. **83**, 4190–4192 (2003)
106. Jiang, L., Hodges, J.S., Maze, J.R., Maurer, P., Taylor, J.M., Cory, D.G., Hemmer, P.R., Walsworth, R.L., Yacoby, A., Zibrov, A.S., Lukin, M.D.: Repetitive readout of a single electronic spin via quantum logic with nuclear spin ancillae. Science **326**, 267–272 (2009)
107. Neumann, P., Beck, J., Steiner, M., Rempp, F., Fedder, H., Hemmer, P.R., Wrachtrup, J., Jelezko, F.: Single-shot readout of a single nuclear spin. Science **329**, 542–544 (2010)

108. Childress, L., Dutt, M.V.G., Taylor, J.M., Zibrov, A.S., Jelezko, F., Wrachtrup, J., Hemmer, P.R., Lukin, M.D.: Coherent dynamics of coupled electron and nuclear spin qubits in diamond. Science **314**, 281–285 (2006)
109. Dutt, M.V.G., Childress, L., Jiang, L., Togan, E., Maze, J., Jelezko, F., Zibrov, A.S., Hemmer, P.R., Lukin, M.D.: Quantum register based on individual electronic and nuclear spin qubits in diamond. Science **316**, 1312–1316 (2007)
110. Nizovtsev, A.P., Kilin, S.Y., Jelezko, F., Gaebal, T., Popa, I., Gruber, A., Wrachtrup, J.: A quantum computer based on NV centres in diamond: optically detected nutations of single electron and nuclear spins. Opt. Spectrosc. **99**, 233–244 (2005)
111. Wrachtrup, J., Kilin, S.Y., Nizovtsev, A.P.: Quantum computation using the C-13 nuclear spins near the single NV defect centre in diamond. Opt. Spectrosc. **91**, 429–437 (2001)
112. Neumann, P., Kolesov, R., Naydenov, B., Beck, J., Rempp, F., Steiner, M., Jacques, V., Balasubramanian, G., Markham, M.L., Twitchen, D.J., Pezzagna, S., Meijer, J., Twamley, J., Jelezko, F., Wrachtrup, J.: Quantum register based on coupled electron spins in a room-temperature solid. Nat. Phys. **6**, 249–253 (2010)
113. Cabrillo, C., Cirac, J.I., Garcia-Fernandez, P., Zoller, P.: Creation of entangled states of distant atoms by interference. Phys. Rev. A **59**, 1025–1033 (1999)
114. Barclay, P.E., Fu, K.M., Santori, C., Beausoleil, R.G.: Hybrid photonic crystal cavity and waveguide for coupling to diamond NV-centers. Opt. Express **17**, 9588–9601 (2009)
115. Bayn, I., Meyler, B., Lahav, A., Salzman, J., Kalish, R., Fairchild, B.A., Prawer, S., Barth, M., Benson, O., Wolf, T., Siyushev, P., Jelezko, F., Wrachtrup, J.: Processing of photonic crystal nanocavity for quantum information in diamond. Diamond Relat. Mater. **20**, 937–943 (2011)
116. Bayn, I., Meyler, B., Salzman, J., Kalish, R.: Triangular nanobeam photonic cavities in single-crystal diamond. New J. Phys. **13**, 025018 (2011)
117. Childress, L., Taylor, J.M., Sorensen, A.S., Lukin, M.D.: Fault-tolerant quantum repeaters with minimal physical resources and implementations based on single-photon emitters. Phys. Rev. A **72**, 052330 (2005)
118. Young, A., Hu, C.Y., Marseglia, L., Harrison, J.P., O'Brien, J.L., Rarity, J.G.: Cavity enhanced spin measurement of the ground state spin of an NV centre in diamond. New J. Phys. **11**, 013007 (2009)
119. Benjamin, S.C., Lovett, B.W., Smith, J.M.: Prospects for measurement-based quantum computing with solid state spins. Las. Phot. Rev. **3**, 556–574 (2009)
120. Su, C.H., Greentree, A.D., Hollenberg, L.C.L.: Towards a picosecond transform-limited nitrogen-vacancy based single photon source. Opt. Express **16**, 6240–6250 (2008)
121. Young, A., Hu, C.Y., Marseglia, L., Harrison, J.P., O'Brien, J.L., Rarity, J.G.: Cavity enhanced spin measurement of the ground state spin of an NV centre in diamond. New J. Phys. **11**, 013007 (2009)
122. Dolde, F, Bergholm, V, Wang, Y, Jakobi, I, Naydenov, B, Pezzagna, S, Meijer, J, Jelezko, F, Neumann, P, Schulte-Herbrüggen, T, Biamonte, J, Wrachtrup, J: High-fidelity spin entanglement using optimal control. Nat. Commun. **5** (2014)
123. Yamamoto, T., Müller, C., McGuinness, L.M., Teraji, T., Naydenov, B., Onoda, S., Ohshima, T., Wrachtrup, J., Jelezko, F., Isoya, J.: Strongly coupled diamond spin qubits by molecular nitrogen implantation Phys. Rev. B **88**, 201201(R) (2013)
124. Dolde, F., Jakobi, I., Naydenov, B., Zhao, N., Pezzagna, S., Trautmann, C., Meijer, J., Neumann, P., Jelezko, F., Wrachtrup, J.: Room-temperature entanglement between single defect spins in diamond. Nat. Phys. **9**, 139–143 (2013)
125. Moerner, W.E., Orrit, M.: Illuminating single molecules in condensed matter. Science **283**, 1670–1676 (1999)
126. Willig, K.I., Rizzoli, S.O., Westphal, V., Jahn, R., Hell, S.W.: STED microscopy reveals that synaptotagmin remains clustered after synaptic vesicle exocytosis. Nature **440**, 935–939 (2006)

127. Brown, T.A., Fetter, R.D., Tkachuk, A.N., Clayton, D.A.: Approaches toward super-resolution fluorescence imaging of mitochondrial proteins using PALM. Methods **51**, 458–463 (2010)
128. Rittweger, E., Han, K.Y., Irvine, S.E., Eggeling, C., Hell, S.W.: STED microscopy reveals crystal colour centres with nanometric resolution. Nat. Photonics **3**, 144–147 (2009)
129. Shin, C., Kim, C., Kolesov, R., Balasubramanian, G., Jelezko, F., Wrachtrup, J., Hemmer, P. R.: Sub-optical resolution of single spins using magnetic resonance imaging at room temperature in diamond. J. Lumin. **130**, 1635–1645 (2010)
130. Liu, K.K., Cheng, C.L., Chang, C.C., Chao, J.I.: Biocompatible and detectable carboxylated nanodiamond on human cell. Nanotechnology **18**, Artn 325102 (2007)
131. Bradac, C., Gaebel, T., Naidoo, N., Sellars, M.J., Twamley, J., Brown, L.J., Barnard, A.S., Plakhotnik, T., Zvyagin, A.V., Rabeau, J.R.: Observation and control of blinking nitrogen-vacancy centres in discrete nanodiamonds. Nat. Nanotechnol. **5**, 345–349 (2010)
132. Tisler, J., Balasubramanian, G., Naydenov, B., Kolesov, R., Grotz, B., Reuter, R., Boudou, J.P., Curmi, P.A., Sennour, M., Thorel, A., Borsch, M., Aulenbacher, K., Erdmann, R., Hemmer, P.R., Jelezko, F., Wrachtrup, J.: Fluorescence and spin properties of defects in single digit nanodiamonds. ACS Nano **3**, 1959–1965 (2009)
133. Dolde, F., Fedder, H., Doherty, M.W., Nobauer, T., Rempp, F., Balasubramanian, G., Wolf, T., Reinhard, F., Hollenberg, L.C.L., Jelezko, F., Wrachtrup, J.: Electric-field sensing using single diamond spins. Nat. Phys. **7**, 459–463 (2011)
134. De Lange, G., Wang, Z.H., Riste, D., Dobrovitski, V.V., Hanson, R.: Universal dynamical decoupling of a single solid-state spin from a spin bath. Science **330**, 60–63 (2010)
135. Hall, L.T., Hill, C.D., Cole, J.H., Hollenberg, L.C.L.: Ultrasensitive diamond magnetometry using optimal dynamic decoupling. Phys. Rev. B **82**, 045208 (2010)
136. Laraoui, A., Meriles, C.A.: Rotating frame spin dynamics of a nitrogen-vacancy centre in a diamond nanocrystal. Phys. Rev. B **84**, 161403 (2011)
137. Waldherr, G., Beck, J., Neumann, P., Said, R.S., Nitsche, M., Markham, M.L., Twitchen, D. J., Twamley, J., Jelezko, F., Wrachtrup, J.: High-dynamic-range magnetometry with a single nuclear spin in diamond. Nat. Nanotechnol. **7**, 105–108 (2012)
138. Cappellaro, P., Lukin, M.D.: Quantum correlation in disordered spin systems: applications to magnetic sensing. Phys. Rev. A **80**, 032311 (2009)
139. Degen, C.L.: Scanning magnetic field microscope with a diamond single-spin sensor. Appl. Phys. Lett. **92**, 243111 (2008)
140. Balasubramanian, G., Chan, I.Y., Kolesov, R., Al-Hmoud, M., Tisler, J., Shin, C., Kim, C., Wojcik, A., Hemmer, P.R., Krueger, A., Hanke, T., Leitenstorfer, A., Bratschitsch, R., Jelezko, F., Wrachtrup, J.: Nanoscale imaging magnetometry with diamond spins under ambient conditions. Nature **455**, 645–648 (2008)
141. McGuinness, L.P., Yan, Y., Stacey, A., Simpson, D.A., Hall, L.T., Maclaurin, D., Prawer, S., Mulvaney, P., Wrachtrup, J., Caruso, F., Scholten, R.E., Hollenberg, L.C.L.: Quantum measurement and orientation tracking of fluorescent nanodiamonds inside living cells. Nat. Nanotechnol. **6**, 358–363 (2011)
142. Rabl, P., Cappellaro, P., Dutt, M.V.G., Jiang, L., Maze, J.R., Lukin, M.D.: Strong magnetic coupling between an electronic spin qubit and a mechanical resonator. Phys. Rev. B **79**, 041302 (2009)
143. Twamley, J., Barrett, S.D.: Superconducting cavity bus for single nitrogen-vacancy defect centres in diamond. Phys. Rev. B **81**, 241202 (2010)
144. Amsuss, R., Koller, C., Nobauer, T., Putz, S., Rotter, S., Sandner, K., Schneider, S., Schrambock, M., Steinhauser, G., Ritsch, H., Schmiedmayer, J., Majer, J.: Cavity QED with magnetically coupled collective spin states. Phys. Rev. Lett. **107**, 060502 (2011)
145. Kubo, Y., Ong, F.R., Bertet, P., Vion, D., Jacques, V., Zheng, D., Dreau, A., Roch, J.F., Auffeves, A., Jelezko, F., Wrachtrup, J., Barthe, M.F., Bergonzo, P., Esteve, D.: Strong coupling of a spin ensemble to a superconducting resonator. Phys. Rev. Lett. **105**, 140502 (2010)

146. Zhu, X.B., Saito, S., Kemp, A., Kakuyanagi, K., Karimoto, S., Nakano, H., Munro, W.J., Tokura, Y., Everitt, M.S., Nemoto, K., Kasu, M., Mizuochi, N., Semba, K.: Coherent coupling of a superconducting flux qubit to an electron spin ensemble in diamond. Nature **478**, 221–224 (2011)
147. Zaitsev, A.M.: Optical Properties of Diamond: A Data Handbook. Springer, Berlin, New York (2001)
148. Gaebel, T., Popa, I., Gruber, A., Domhan, M., Jelezko, F., Wrachtrup, J.: Stable single-photon source in the near infrared. New J. Phys. **6**, 98 (2004)
149. Orwa, J.O., Aharonovich, I., Jelezko, F., Balasubramanian, G., Balog, P., Markham, M., Twitchen, D.J., Greentree, A.D., Prawer, S.: Nickel related optical centres in diamond created by ion implantation. J. Appl. Phys. **107**, 093512 (2010)
150. Siyushev, P., Jacques, V., Aharonovich, I., Kaiser, F., Muller, T., Lombez, L., Atature, M., Castelletto, S., Prawer, S., Jelezko, F., Wrachtrup, J.: Low-temperature optical characterization of a near-infrared single-photon emitter in nanodiamonds. New J. Phys. **11**, 113029 (2009)
151. Muller, T., Aharonovich, I., Lombez, L., Alaverdyan, Y., Vamivakas, A.N., Castelletto, S., Jelezko, F., Wrachtrup, J., Prawer, S., Atature, M.: Wide-range electrical tunability of single-photon emission from chromium-based colour centres in diamond. New J. Phys. **13**, 075001 (2011)
152. Neu, E., Arend, C., Gross, E., Guldner, F., Hepp, C., Steinmetz, D., Zscherpel, E., Ghodbane, S., Sternschulte, H., Steinmuller-Nethl, D., Liang, Y., Krueger, A., Becher, C.: Narrowband fluorescent nanodiamonds produced from chemical vapor deposition films. Appl. Phys. Lett. **98**, 243107 (2011)
153. Wang, C.L., Kurtsiefer, C., Weinfurter, H., Burchard, B.: Single photon emission from SiV centres in diamond produced by ion implantation. J. Phys. B-At. Mol. Opt. Phys. **39**, 37–41 (2006)
154. Jelezko, F., Wrachtrup, J.: Single defect centres in diamond: a review. Phys. Status Solidi A **203**, 3207–3225 (2006)
155. Awschalom, D.D., Flatté, M.E.: Challenges for semiconductor spintronics. Nat. Phys. **3**, 153–159 (2007)
156. Hanson, R., Awschalom, D.D.: Coherent manipulation of single spins in semiconductors. Nature **453**, 1043–1049 (2008)
157. Koenraad, M., Flatté, M.E.: Single dopants in semiconductors. Nat. Mater. **10**, 91–100 (2011)
158. Baranov, P.G., Il'in, I.V., Mokhov, E.N., Muzafarova, M.V., Orlinskii, S.B., Schmidt, J.: EPR identification of the triplet ground state and photoinduced population inversion for a Si-C divacancy in silicon carbide. JETP Lett. **82**, 441–443 (2005)
159. Baranov, P.G., Bundakova, A.P., Borovykh, I.V., Orlinskii, S.B., Zondervan, R., Schmidt, J.: Spin polarization induced by optical and microwave resonance radiation in a Si vacancy in SiC: a promising subject for the spectroscopy of single defects. JETP Lett. **86**, 202–206 (2007)
160. Weber, J.R., Koehl, W.F., Varley, J.B., Janotti, A., Buckley, B.B., Van de Walle, C.G., Awschalom, D.D.: Quantum computing with defects. Proc. Natl. Acad. Sci. U.S.A. **107**, 8513–8518 (2010)
161. DiVincenzo, D.: Quantum bits: better than excellent. Nat. Mater. **9**, 468–469 (2010)
162. Baranov, P.G., Bundakova, A.P., Soltamova, A.A., Orlinskii, S.B., Borovykh, I.V., Zondervan, R., Verberk, R., Schmidt, J.: Silicon vacancy in SiC as a promising quantum system for single defect and single-photon spectroscopy. Phys. Rev. B **83**, 125203 (2011)
163. Smart, A.G.: Silicon carbide defects hold promise for device-friendly qubits. Phys. Today **65**, 10–11 (2012)
164. Veinger, A.I., Il'in, V.A., Tairov, Yu.M., Tsvetkov, V.F.: Investigation of thermal defects in silicon carbide by the ESR method. Sov. Phys. Semicond. **13**, 1385 (1979)
165. Vainer, V.S., Il'in, V.A.: Electron spin resonance of exchange-coupled vacancy pairs in hexagonal silicon carbide. Sov. Phys. Solid State **23**, 2126–2133 (1981)

166. von Bardeleben, H.J., Cantin, J.L., Vickridge, I., Battistig, G.: Proton-implantation-induced defects in n-type 6H- and 4H-SiC: an electron paramagnetic resonance study. Phys. Rev. B **62**, 10126–10134 (2000)
167. von Bardeleben, H.J., Cantin, J., Henry, L., Barthe, M.: Vacancy defects in *p*-type 6H-SiC created by low-energy electron irradiation. Phys. Rev. B **62**, 10841–10846 (2000)
168. Wagner, M., Magnusson, B., Chen, W.M., Janzen, E., Sörman, E., Hallin, C., Lindström, J. L.: Electronic structure of the neutral silicon vacancy in 4H and 6H SiC. Phys. Rev. B **62**, 16555–16560 (2000)
169. Mizuochi, N., Yamasaki, S., Takizawa, H., Morishita, N., Ohshima, T., Itoh, H., Isoya, J.: Continuous-wave and pulsed EPR study of the negatively charged silicon vacancy with S = $^3/_2$ and C_{3v} symmetry in *n*-type 4H-SiC. Phys. Rev. B **66**, 235202 (2002)
170. Carlos, W.E., Garces, N.Y., Glaser, E.R., Fanton, M.A.: Annealing of multivacancy defects in 4H-SiC. Phys. Rev. B **74**, 235201 (2006)
171. Orlinski, S.B., Schmidt, J., Mokhov, E.N., Baranov, P.G.: Silicon and carbon vacancies in neutron-irradiated SiC: a high-field electron paramagnetic resonance study. Phys. Rev. B **67**, 125207 (2003)
172. Koehl, W.F., Buckley, B.B., Heremans, F.J., Calusine, G., Awschalom, D.D.: Room temperature coherent control of defect spin qubits in silicon carbide. Nature **479**, 84–87 (2011)
173. Soltamov, V.A., Soltamova, A.A., Baranov, P.G., Proskuryakov, I.I.: Room temperature coherent spin alignment of silicon vacancies in 4H- and 6H-SiC. Phys. Rev. Lett. **108**, 226402 (2012)
174. Riedel, D., Fuchs, F., Kraus, H., Vath, S., Sperlich, A., Dyakonov, V., Soltamova, A.A., Baranov, P.G., Ilyin, V.A., Astakhov, G.V.: Resonant addressing and manipulation of silicon vacancy qubits in silicon carbide. Phys. Rev. Lett. **109**, 226402 (2012)
175. Fuchs, F., Soltamov, V.A., Vath, S., Baranov, P.G., Mokhov, E.N., Astakhov, G.V., Dyakonov, V.: Silicon carbide light-emitting diode as a prospective room temperature source for single photons. Sci. Rep. **3**, 1637 (2013)
176. Castelletto, S., Johnson, B.C., Boretti, A.: Quantum effects in silicon carbide hold promise for novel integrated devices and sensors. Adv. Opt. Mater. **1**, 609–625 (2013)
177. Falk, A.L., Buckley, B.B., Calusine, G., Koehl, W.F., Dobrovitski, V.V., Politi, A., Zorman, C.A., Feng, P.X.L., Awschalom, D.D.: Polytype control of spin qubits in silicon carbide. Nat. Commun. **4**, 1819 (2013)
178. Castelletto, S., Johnson, B.C., Ivady, V., Stavrias, N., Umeda, T., Gali, A., Ohshima, T.: A silicon carbide room-temperature single-photon source. Nat. Mater. **13**, 151–156 (2013)
179. Hain, T.C., Fuchs, F., Soltamov, V.A., Baranov, P.G., Astakhov, G.V., Hertel, T., Dyakonov, V.: Excitation and recombination dynamics of vacancy-related spin centers in silicon carbide. J. Appl. Phys. **115**, 133508 (2014)
180. Castelletto, S., Johnson, B.C., Zachreson, C., Beke, D., Balogh, I., Ohshima, T., Aharonovich, I., Gali, A.: Room temperature quantum emission from cubic silicon carbide nanoparticles. ACS Nano **8**, 7938–7947 (2014)
181. Muzha, A., Fuchs, F., Tarakina, N.V., Simin, D., Trupke, M., Soltamov, V.A., Mokhov, E. N., Baranov, P.G., Dyakonov, V., Krueger, A., Astakhov, G.V.: Room-temperature near-infrared silicon carbide nanocrystalline emitters based on optically aligned spin defects. Appl. Phys. Lett. **105**, 243112 (2014)
182. Kraus, H., Soltamov, V.A., Riedel, D., Vath, S., Fuchs, F., Sperlich, A., Baranov, P.G., Dyakonov, V., Astakhov, G.V.: Room-temperature quantum microwave emitters based on spin defects in silicon carbide. Nat. Phys. **10**, 157–162 (2014)
183. Calusine, G., Politi, A., Awschalom, D.D.: Silicon carbide photonic crystal cavities with integrated color centers. Appl. Phys. Lett. **105**, 011123 (2014)
184. Klimov, P.V., Falk, A.L., Buckley, B.B., Awschalom, D.D.: Electrically driven spin resonance in silicon carbide color centers. Phys. Rev. Lett. **112**, 087601 (2014)

185. Falk, A.L., Klimov, P.V., Buckley, B.B., Ivady, V., Abrikosov, I.A., Calusine, G., Koehl, W.F., Gali, A., Awschalom, D.D.: Electrically and mechanically tunable electron spins in silicon carbide color centers. Phys. Rev. Lett. **112**, 187601 (2014)
186. Kraus, H., Soltamov, V.A., Fuchs, F., Simin, D., Sperlich, A., Baranov, P.G., Astakhov, G. V., Dyakonov, V.: Magnetic field and temperature sensing with atomic-scale spin defects in silicon carbide. Sci. Rep. **4**, 5303 (2014)
187. Yang, L.-P., Burk, C., Widmann, M., Lee, S.-Y., Wrachtrup, J., Zhao, N.: Electron spin decoherence in silicon carbide nuclear spin bath. Phys. Rev. B **90**, 241203 (2014)
188. Soltamov, V.A., Yavkin, B.V., Tolmachev, D.O., Babunts, R.A., Badalyan, A.G., Davydov, VYu., Mokhov, E.N., Proskuryakov, I.I., Orlinskii, S.B., Baranov, P.G.: Optically addressable silicon vacancy-related spin centers in rhombic silicon carbide with high breakdown characteristics and ENDOR evidence of their structure. Phys. Rev. Lett. **115**, 247602 (2015)
189. Zwier, O.V., O'Shea, D., Onur, A.R., van der Wal, C.H.: All-optical coherent population trapping with defect spin ensembles in silicon carbide. Sci. Rep. **5**, 10931 (2015)
190. Falk, A.L., Klimov, P.V., Ivady, V., Szasz, K., Christle, D.J., Koehl, W.F., Gali, A., Awschalom, D.D.: Optical polarization of nuclear spins in silicon carbide. Phys. Rev. Lett. **114**, 247603 (2015)
191. Carter, S.G., Soykal, Ö.O., Dev, P., Economou, S.E., Glaser, E.R.: Spin coherence and echo modulation of the silicon vacancy in 4H-SiC at room temperature. Phys. Rev. B **92**, 161202 (2015)
192. Simin, D., Fuchs, F., Kraus, H., Sperlich, A., Baranov, P.G., Astakhov, G.V., Dyakonov, V.: High-precision angle-resolved magnetometry with uniaxial quantum centers in silicon carbide. Phys. Rev. Appl. **4**, 014009 (2015)
193. Lee, S.-Y., Niethammer, M., Wrachtrup, J.: Vector magnetometry based on $S = {}^3/_2$ electronic spins. Phys. Rev. B **92**, 115201 (2015)
194. Christle, D.J., Falk, A.L., Andrich, P., Klimov, P.V., Ul Hassan, J., Son, N.T., Janzen, E., Ohshima, T., Awschalom, D.D.: Isolated electron spins in silicon carbide with millisecond coherence times. Nat. Mater.**14**, 160–163 (2015)
195. Widmann, M., Lee, S.-Y., Rendler, T., Son, N.T., Fedder, H., Paik, S., Yang, L.-P., Zhao, N., Yang, S., Booker, I., Denisenko, A., Jamali, M., Momenzadeh, S.A., Gerhardt, I., Ohshima, T., Gali, A., Janzen, E., Wrachtrup, J.: Coherent control of single spins in silicon carbide at room temperature. Nat. Mater. **14**, 164–168 (2015)
196. Fuchs, F., Stender, B., Trupke, M., Simin, D., Paum, J., Dyakonov, V., Astakhov, G.V.: Engineering near infrared single-photon emitters with optically active spins in ultrapure silicon carbide. Nat. Commun. **6**, 7578 (2015)
197. Lohrmann, A., Iwamoto, N., Bodrog, Z., Castelletto, S., Ohshima, T., Karle, T.J., Gali, A., Prawer, S., McCallum, J.C., Johnson, B.C.: Single-photon emitting diode in silicon carbide. Nat. Commun. **6**, 7783 (2015)
198. Tairov, Y.M., Tsvetkov, V.F.: Investigation of growth processes of ingots of silicon carbide single crystals. J. Cryst. Growth **43**, 209–212 (1978)
199. Wimbauer, T., Meyer, B.K., Hofstaetter, A., Scharmann, A., Overhof, H.: Negatively charged Si vacancy in 4H SiC: a comparison between theory and experiment. Phys. Rev. B **56**, 7384–7388 (1997)
200. Mims, W.B.: In: Geschwind, S. (ed.) Electron Paramagnetic Resonance, pp. 227–376. Plenum, New York (1972)
201. Spaeth, J.-M., Niklas, J.R., Bartram, R.H.: Electron nuclear double resonance. In: Spaeth, J.-M., Niklas, J.R., Bartram, R.H. (eds.) Structural Analysis of Point Defects in Solids, pp. 139–168. Springer, Berlin, Heidelberg (1992)
202. Freeman, A.J., Frankel, R.B. (eds.): Hyperfine Interactions. Academic Press, New York (1967)
203. Abraham, A., Bleaney, B.: Electron Paramagnetic Resonance of Transition Ions. Clarendon Press, Oxford (1970)

204. Astakhov, G.V., Simin, D., Dyakonov, V., Yavkin, B.V., Orlinskii, S.B., Proskuryakov, I.I., Anisimov, A.N., Soltamov, V.A., Baranov, P.G.: Spin centres in SiC for quantum technologies. Appl. Magn. Reson. **47**, 793–812 (2016)
205. Anisimov, A.N., Babunts, R.A., Kidalov, S.V., Mokhov, E.N., Soltamov, V.A., Baranov, P. G.: Spin centres in SiC for all-optical nanoscale quantum sensing under ambient conditions. JETP Lett. **104**, 83–88 (2016)
206. Simin, D., Soltamov, V.A., Anisimov, A.N., Babunts, R.A., Tolmachev, D.O., Mokhov, E. N., Trupke, M., Tarasenko, S.A., Sperlich, A., Baranov, P.G., Dyakonov, V., Astakhov, G. V.: All-Optical dc nanotesla magnetometry using silicon vacancy fine structure in isotopically purified silicon carbide. Phys. Rev. X **6**, 031014 (1–12) (2016)
207. Anisimov, A.N., Simin, D., Soltamov, V.A., Lebedev, S.P., Baranov, P.G., Astakhov, G.V., Dyakonov, V.: Optical thermometry based on level anticrossing in silicon carbide. Nat. Sci. Rep. (2016)
208. Isoya, J., Umeda, T., Mizuochi, N., Son, N.T., Janzén, E., Ohshima, T.: EPR identification of intrinsic defects in SiC. Phys. Status Solidi B **245**, 1298–1314 (2008)
209. Sörman, E., Son, N.T., Chen, W.M., Kordina, O., Hallin, C., Janzén, E.: Silicon vacancy related defect in 4H and 6H SiC. Phys. Rev. B **61**, 2613–2620 (2000)
210. Jamali, M., Gerhardt, I., Rezai, M., Frenner, K., Fedder, H., Wrachtrup, J.: Microscopic diamond solid-immersion-lenses fabricated around single defect centers by focused ion beam milling. Rev. Sci. Instrum. **85**, 123703 (2014)

Correction to: Retrospectives: Magnetic Resonance Studies of Intrinsic Defects in Semiconductors

Correction to:
Chapter 3 in: P.G. Baranov et al., *Magnetic Resonance of Semiconductors and Their Nanostructures*, Springer Series in Materials Science 253, https://doi.org/10.1007/978-3-7091-1157-4_3

The original version of chapter 3 was inadvertently published with incorrect cross citations of the references. The chapter has been modified now with correct references cross citations corresponding to the references list.

The updated version of this chapter can be found at
https://doi.org/10.1007/978-3-7091-1157-4_3

P.G. Baranov et al., *Magnetic Resonance of Semiconductors and Their Nanostructures*, Springer Series in Materials Science 253,
https://doi.org/10.1007/978-3-7091-1157-4_7

Index

P.G. Baranov et al., *Magnetic Resonance of Semiconductors and Their Nanostructures*, Springer Series in Materials Science 253, DOI 10.1007/978-3-7091-1157-4

D

E

MIX
Papier aus verantwortungsvollen Quellen
Paper from responsible sources
FSC® C105338

If you have any concerns about our products,
you can contact us on
ProductSafety@springernature.com

In case Publisher is established outside the EU,
the EU authorized representative is:
Springer Nature Customer Service Center GmbH
Europaplatz 3, 69115 Heidelberg, Germany

Printed by Libri Plureos GmbH
in Hamburg, Germany